GENETICS

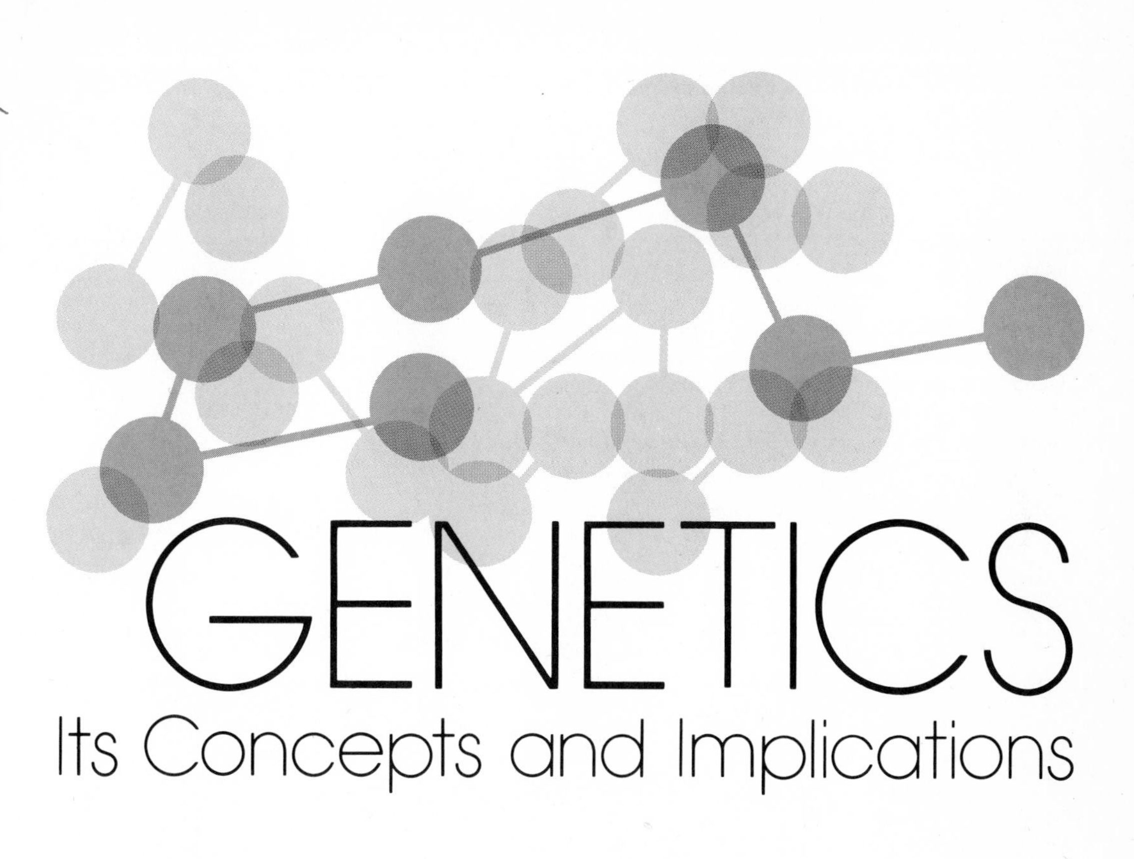

GENETICS

Its Concepts and Implications

ANNA C. PAI

Department of Biology

HELEN MARCUS-ROBERTS

Department of Mathematics and Computer Science

Montclair State College

PRENTICE-HALL, INC., Englewood Cliffs, New Jersey 07632

Library of Congress Cataloging in Publication Data

Pai, Anna C (date)
Genetics, its concepts and implications.

Includes bibliographies and index.
1. Genetics. 2. Genetics—Social aspects.
I. Marcus-Roberts, Helen, joint author. II. Title.
QH430.P35 575.1 80-39616
ISBN 0-13-351007-7

Cover photograph: A color solarization of a scanning electron micrograph of an isolated metaphase chromosome. (Courtesy Dr. Wayne Wray, Baylor College of Medicine.)

Editorial/production supervision by Zita de Schauensee

Cover and chapter-opening design by Jayne Conte

Art production by Diane Doran Sturm

Manufacturing buyer: John Hall

Printed in the United States of America

10 9 8 7 6 5 4 3 2 1

Prentice-Hall International, Inc., *London*
Prentice-Hall of Australia Pty. Limited, *Sydney*
Prentice-Hall of Canada, Ltd., *Toronto*
Prentice-Hall of India Private Limited, *New Delhi*
Prentice-Hall of Japan, Inc., *Tokyo*
Prentice-Hall of Southeast Asia Pte. Ltd., *Singapore*
Whitehall Books Limited, *Wellington, New Zealand*

TO DAVID, BEN, AND MIKE
AND
TO THE MEMORY OF WALTER MARCUS

Summary Contents

Complete Contents

TWO The Transmission of Traits: Simple Mendelian Inheritance 32

THREE Probability and Statistics in Genetics 59

FOUR The Interaction of Genes 89

FIVE Sex Chromosomes and Sex Determination 120

SIX Cytogenetics 153

SEVEN Linkage and Chromosome Mapping 186

TEN Molecular Genetics II. Protein Synthesis and the Genetic Code 303

ELEVEN Regulation of Gene Action in Prokaryotes 342

TWELVE Gene Regulation in Eukaryotes 369

FIFTEEN Mutation, Mutagenesis, and Repair 473

SIXTEEN Population Genetics 499

Preface

An organism is judged by biologists to be living if it is capable of (1) self-reproduction, (2) metabolism, and (3) mutation. All these phenomena which distinguish living from nonliving matter are now known to involve genes and gene action. Therefore, central as the gene is to life processes, so the science of genetics, the study of the gene and gene action, has become central to the training of a biologist. Furthermore, important as genetic concepts are to biologists, these concepts have also achieved a unique degree of interest in the eyes of the general public as a result of growing public awareness of the relevance of genetic studies to every living organism including humans.

Historically, there has existed a sometimes considerable gap of communication and/or interest between the laboratory sciences and a society which more often than not felt only remotely affected by the activity within the laboratories. More than any other science in modern times, genetics has bridged this gap. Geneticists recognize that their work has bearing on many aspects of the natural world and human society. By informing the nonscientific community about the potential of genetic research, geneticists themselves have encouraged a broadening of communication between scientists and the public which is both necessary and desirable. A geneticist today must not only be knowledgeable but also able to relate his knowledge to the community at large.

In this textbook, the authors aim to present the concepts of genetics as a focal topic in biology to students who may advance to specialize in one of its many subdivisions. The first seven chapters deal with various aspects of heredity, such as the roles of mitosis and meiosis, patterns of transmission, chromosome structure and mapping, and gene interaction. A thorough but simply presented discussion of the application of probability and statistics to the analysis of heredity (as in genetic counseling) is also included. Students are introduced to current techniques being used in these areas, such as somatic cell fusion in mapping human chromosomes.

With a firm understanding of the basic concepts of heredity after study of these first chapters, we turn our attention to the physical structure of genes, and how they function. Historically, the Age of Molecular Genetics began with the discovery that microorganisms such as bacteria and viruses can be used in genetic studies, and accordingly, we begin our chapters on molecular genetics with a discussion of the genetics of bacteria and viruses. From there we proceed to the ingenious experiments which gave us the structure of the double helix of DNA, and the relationship between protein synthesis and gene function.

More recently, much attention has been focused on the manner in which gene activity is regulated, because we are now aware that many biological phenomena, both normal and abnormal, are due to aspects of gene regulation. For example, immune phenomena and the transformation of normal cells into malignant cells in the development of cancer all reflect enormous complexity of regulative processes, and the results when regulation becomes abnormal. Therefore, we include these and other topics in our text, even though they are topics heretofore seldom found in genetics texts.

The controversy which has recently arisen in both the scientific and lay communities over the implications of certain experiments in genetic engineering will be considered at length. With our ever-increasing understanding of life processes, the debate on genetic engineering clearly illustrates the concerns of many as we approach the point where we may be able to control those processes.

We center our attention in the last few chapters on various aspects of mutation: its molecular basis, natural repair processes, agents that cause mutations, and tests for mutagens. Fluctuation in the frequencies of mutations and genes in populations leads from a consideration of population genetics to the genetic basis of evolutionary change.

It is our feeling that a discussion of our evolutionary origins is important for a better understanding of the human species and its part in the natural world. We hope that including a discussion of the genetic basis to race formation and IQ in this context will bring to light misconceptions which have caused so much unnecessary grief to our society. Our concern for the societal implications of genetics carries into the final chapter which explores the promises and perils of genetic research and methodology in the near future.

To present genetic concepts and their implications and yet have a textbook which can be handled within the limits of a one-semester course, we have stressed principles and ideas in the text. Students who desire more details (for example, of experiments) should consult the extensive bibliography listed at the end of each chapter. Central ideas are emphasized by special print within the text. In addition, a number of appendixes have been provided which deal in greater depth with some aspects of material covered in the text. Answers to some of the problems and an extensive glossary are included in the appendixes.

We hope that students who use this textbook will share both in our excitement over developments in a fascinating and rapidly moving science, and in our concern that the new knowledge will be used with wisdom.

We wish to acknowledge the persuasive encouragement of Mary Ann Richter, who was directly responsible for the initiation of this project. Her untimely death cut short a promising career in publishing and removed a bright spot from the lives of all who knew her.

We further wish to thank the staff of Prentice-Hall and reviewers for their helpful comments. We would like to express our appreciation to the following reviewers: Audrey Barnett, C. William Birky, Jr., Allyn A. Bregman, Loy V. Crowder, Robert G. Fowler, Stephen L. Goldman, David Knauft, Paul A. Roberts, Howard Rosen, and R. C. Vrijenhoek. Thanks also to Marie Hromin for verifying the calculations in Chapter 16; to Sue Rowley, Victoria Berutti, Jane Freund, and Roger Korey for their help in preparing the manuscript; and not least, to my (ACP) students who used the manuscript and contributed constructive suggestions.

We are grateful to the Literary Executor of the late Sir Ronald A. Fisher, F.R.S., to Dr. Frank Yates, F.R.S., and to Longman Group Ltd., London, for permission to reprint portions of Table IV from their book *Statistical Tables for Biological, Agricultural and Medical Research.* (Sixth edition, 1974.)

In addition, Anna Pai acknowledges the moral support of her family and Dina Campos, which was invaluable. Helen Marcus-Roberts wishes to acknowledge the encouragement of her parents and friends. In particular she wishes to acknowledge the love, encouragement, and help of her husband, Fred, and the love of her daughter, Sarah.

Livingston, New Jersey
Westfield, New Jersey

Anna C. Pai
Helen Marcus-Roberts

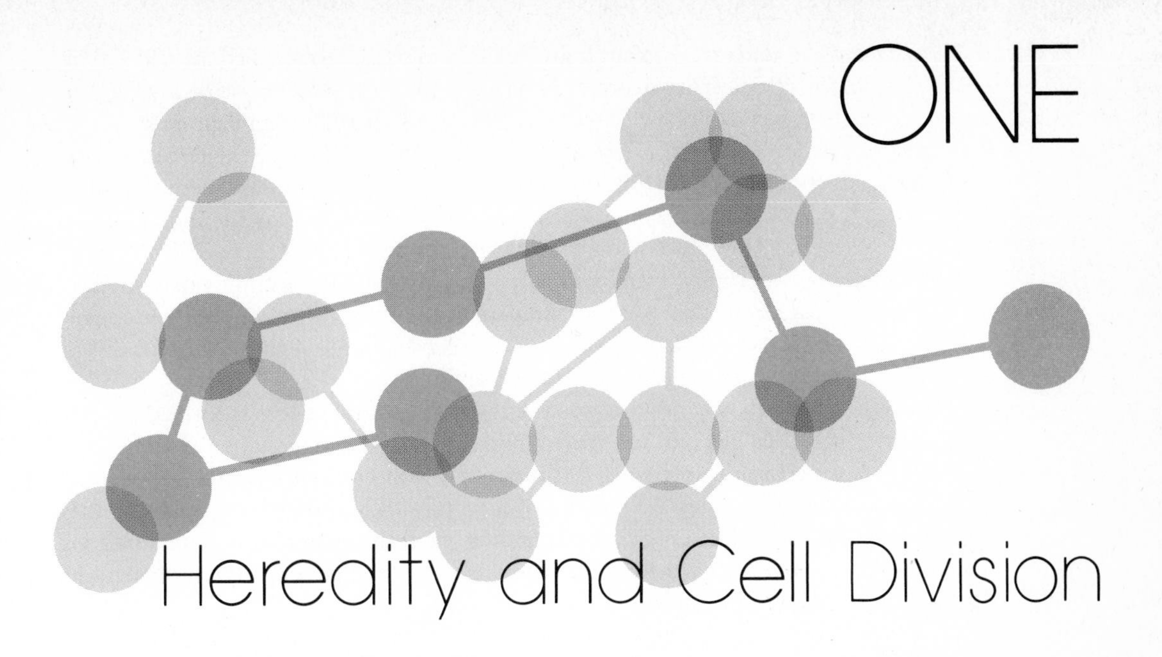

ONE

Heredity and Cell Division

INTRODUCTION

There is no field more basic to the understanding of living organisms than the science of genetics. It is a science as complex and diverse, as fascinating and beautiful, as the life forms that exist on our planet. And because we students of the gene, the unit that determines the structure and function of all living organisms, are no less the products of its actions, one of the most appealing aspects of our studies is their potential to help us better understand ourselves.

Our discussion of the many subdivisions of genetics will begin with heredity and its biological basis, for this is the aspect of gene action that is the most familiar to all. It is the area of genetics which deals with the ability of an organism to transmit its species characteristics and some of its own unique traits to its offspring.

History

Developments in biology in the latter half of the nineteenth century laid the foundations for our understanding of the physical basis for the transmission of inherited traits. That "like begets like" in the natural world has been apparent since humans evolved the ability to think rationally, but not until improvements in microscopes and their use in the 1800s could scientists begin to gather clues to the nature of the mechanism that ensures this kind of biological continuity.

Cytological studies. Research by many biologists established by 1855 that every cell indeed comes from a pre-existing cell. In 1883, in the first of a series of essays, August Weissmann conjectured that all cells of an adult organism must therefore have descended from the original cell of that individual's life, the fertilized egg cell. The production of that fertilized egg cell, or *zygote*, and its subsequent development, must therefore contain some mechanism which ensures that the individual is similar to its parents. Biologists assumed that there must exist factors (determinants) which in some way are transmitted from parents to offspring, determining its inherited traits. Studies by a number of great researchers of the time, including Walther Flemming and Oscar Hertwig, led further to the correct assumption that the mechanism of transmission had to be associated with cell division.

The basis of this assumption was the discovery that a zygote is formed by the union of two cells, the egg and the sperm, and that the first stage of development of any zygote is a period of unceasing cell division which continues (in the process of growth and repair) even after the individual is mature. What, then, occurs during the events of cell division (called *mitosis*) that allows the transmission of determinants so that even though a cell physically splits, each daughter cell contains the information necessary for its development into a particular kind of cell?

Cytological observations in the 1870s pointed to chromosomes as the logical vehicles for the transmission of determinants. Every species of plant and animal appeared to have a specific number of chromosomes in the cells of every member of that species. Here was the likely mechanism for genetic continuity: no matter how many times a cell and its daughter cells divide, the number of chromosomes distributed to the new cells remains identical to the number found in the original cell, or zygote.

On the other hand, the germ cells, or gametes, usually contained half the number of chromosomes present in the adult cells of the parents. The number of chromosomes in gametes is reduced to one-half the parental number through a special kind of cell division, *meiosis*, found only in maturing germ cells. In the 1880s, studies by Edouard van Beneden on the fertilization of nematode eggs indicated that the zygote receives half its chromosomes from each parent.

Furthermore, embryologist Theodor Boveri conducted a series of experiments on sea urchin fertilization in which he altered the number and combinations of chromosomes in the zygotes, and discovered that not only was the correct *number* of chromosomes essential for normal development, but also the correct *combination*. These results led Boveri to conjecture that since there are qualitative differences between chromosomes, the chromosomes must contain different genetic determinants.

Genetic evidence. By the turn of the century, another significant development occurred, which when related to the cytological phenomena we have mentioned, firmly established cell division as the physical basis for the transmission of inherited traits. At this time, almost simultaneously in three different countries, researchers rediscovered the work of an obscure Austrian monk named Gregor Mendel (Fig. 1-1). DeVries of Holland, Tschermak of Austria, and Correns of Germany reintroduced to the scientific community the long-ignored concepts first proposed by Mendel in 1865.

Figure 1-1. Gregor Mendel. (American Museum of Natural History, New York.)

Mendel had labored for many years over the inheritance of traits in the garden pea plant. His extensive experiments had led him to conclude that specific traits such as stem length are determined by pairs of genetic determinants, which he referred to simply as units. Today, of course, we refer to them as genes, a term first introduced by W. L. Johanssen in the early 1900s. Pairs of genes that determine a particular trait are known as alleles. Mendel's studies on crosses of the pea plants also revealed that allelic genes need not be identical.

He found, for example, that there are two different alleles determining stem lengths, one which results in very tall plants, and one which determines that the plants be very short. Further, if an individual plant contained a pair of alleles determining stem length that were indeed different, the plant grew to be tall. (This indicated that not only could "partner" genes be different, but that some, such as the gene for tallness, determine traits which are expressed over others.) Traits that are expressed over others are referred to as *dominant*. Traits determined by alleles whose expression is suppressed are termed *recessive*. We will be discussing this concept of dominance and recessiveness in more detail later; for now, the significant point is that the alleles which determine a particular trait can be qualitatively different.

Mendel was not only a biologist, but also a trained mathematician, and he therefore analyzed his data mathematically. His analysis led him to conjecture that each of the partner genes has an equal chance to be transmitted to the offspring. For example, in the alleles determining stem length, if a parent plant possesses one allele for tallness and one for shortness, half its offspring will inherit the allele for tallness and half will inherit the allele for shortness.

Further, in studies of combinations of traits, such as stem length and pod color in peas, Mendel found that the transmission of one pair of alleles determining one trait is totally independent of the transmission of another pair of genes that determine another trait. The independent inheritance results from a *random assort-*

ment of chromosomes in all the germ cells of an individual to be transmitted to the next generation.

The Chromosomal Basis of Heredity

The relationship between Mendel's concepts and the cytological discoveries that we have mentioned was evident to biologists in the early 1900s, and in 1903, Boveri and a pioneer in the field of genetics, Walter S. Sutton, independently published papers defining the relationship. Their proposals are now referred to as the Sutton–Boveri Hypothesis. The following major points of the relationship have now been firmly established as fact:

1. Genes are carried on chromosomes, and both genes and chromosomes exist in pairs. Members of a pair of chromosomes carry genes determining the same traits.
2. The members of a pair of both genes and chromosomes are of maternal and paternal origin, and are transmitted separately to the next generation of offspring.
3. The process of meiosis results in the precise halving of the chromosome content of the cells such that one member of each pair is transmitted to the daughter cells.
4. A germ cell may contain any combination of maternal and paternal genes and chromosomes, since the transmission of the members of any one pair of chromosomes is independent of the transmission of the members of any other pairs.
5. Since chromosomes are qualitatively different, germ cells are produced containing random combinations of genetically different chromosomes.
6. Since there are more traits in complex organisms than there are chromosomes, every chromosome contains more than one gene, and genes on a particular chromosome are usually inherited together.

MEIOSIS

Let us turn now to the details of the dramatic events of the two kinds of cell divisions, which in many ways are still incompletely understood. We shall begin with a discussion of meiosis. During our description of the events of this process, we shall point out their relationship to basic concepts of heredity; however, discussion of some of the cellular changes that occur during the entire cycle will be omitted, so that we may concentrate on the behavior and distribution of the chromosomes, our main concern as students of genetics.

It bears emphasizing that the main function of meiosis is to provide genetic variability and to reduce the number of chromosomes in half, but in a specific combination, namely, with a representative of each pair of chromosomes. We refer to the basic species number of chromosomes as the *diploid*, or $2n$, number. Immature germ cells are diploid, as are the body (or somatic) cells of the mature organisms. At the completion of meiosis, all gametes contain half the diploid number, which we call the *haploid*, or *n*, number of chromosomes.

For convenience and ease of reference, the events of meiosis are discussed in stages although the process is a continuous one. There are actually two divisional events, enabling one immature germ cell to produce four mature gametes (Fig. 1-2).

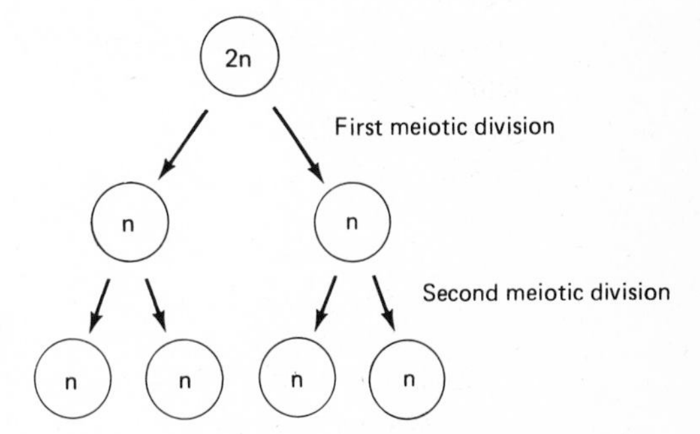

Figure 1-2. One diploid (2n) immature germ cell can give rise to four haploid (n) mature germ cells as a result of meiosis in males. In females only one of the four haploid cells is functional.

The stages of meiosis prior to the first division are called *interphase I*, *prophase I*, *metaphase I*, *anaphase I* and *telophase I*. *Cytokinesis*, or the actual division of the cell into two daughter cells, follows telophase I. The daughter cells then synchronously enter into *interphase II*, *prophase II*, *metaphase II*, *anaphase II* and *telophase II*, followed by the second and final division. Although the general sequence of events during meiosis is similar, male gametes usually give rise to four mature germ cells, whereas immature ova, give rise to four cells of which only one is functional. We shall discuss this phenomenon in more detail later in the chapter.

Interphase I

Interphase I is very important in cell division although not much can be detected visually under the light microscope at this stage. It is at this time that the cell is actively synthesizing an exact duplicate of every chromosome it possesses. Figure 1-3 shows an electron micrograph of chromatin in an interphase nucleus. The chromosomes are not visible because they are very much attenuated at the time of duplication, and form a granular network of chromatin.

Prophase I

The first visualization of chromosomes marks the beginning of prophase I. Their appearance as more discrete structures is due to a progressive coiling of the strands. Because prophase I is a relatively complicated stage in meiosis, it has been further divided into a number of substages.

Leptotene. The first appearance of chromosomes in prophase I constitutes the leptotene stage. At this time the strands are very thin, and although the chromosomes are duplicated, one can discern them only as single-stranded (Fig. 1-4).

Zygotene. There next occurs a movement of the chromosomes through the nucleoplasm to pair with partners, beginning the zygotene stage. The members of a

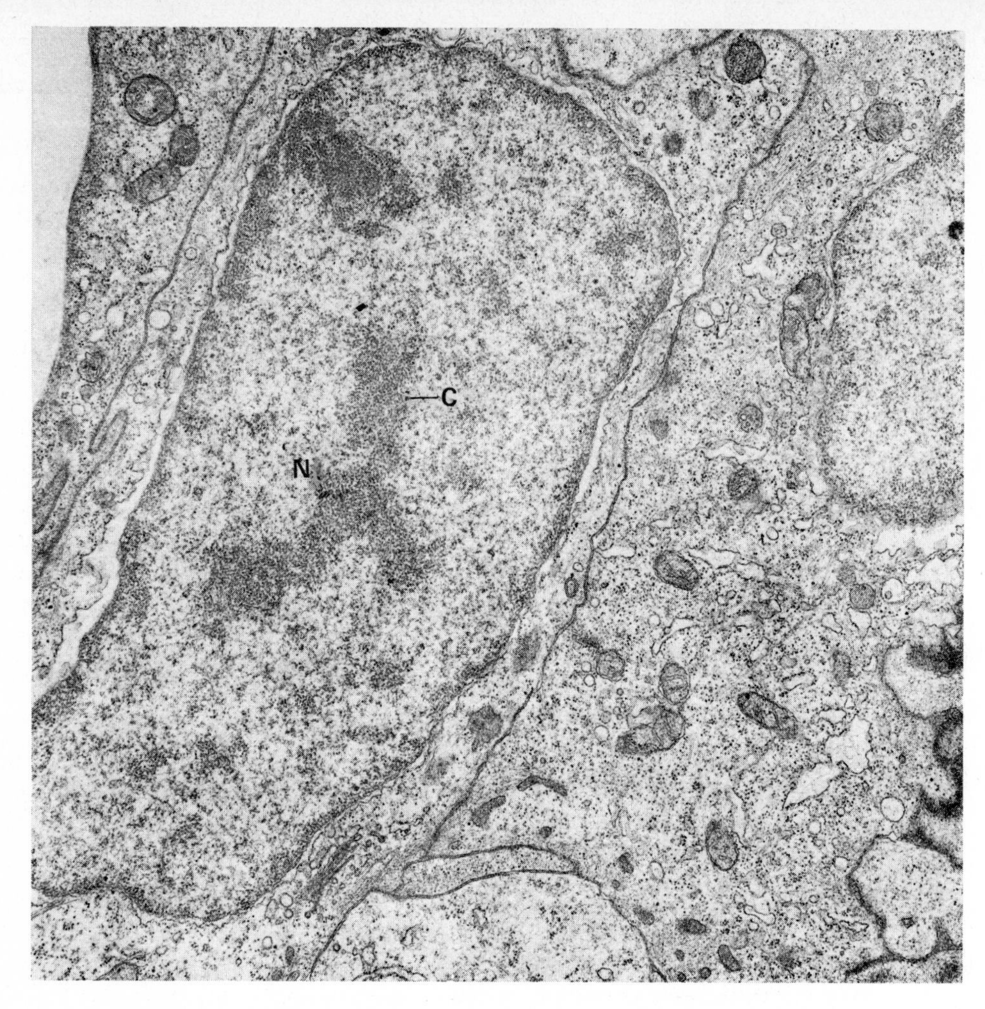

Figure 1-3. Electron micrograph of an interphase nucleus of the fish *Fundulus heteroclitus* showing the nucleus (N) and chromatin (C) 5000×. (Photograph courtesy Stephen Koepp, Montclair State College, N. J.)

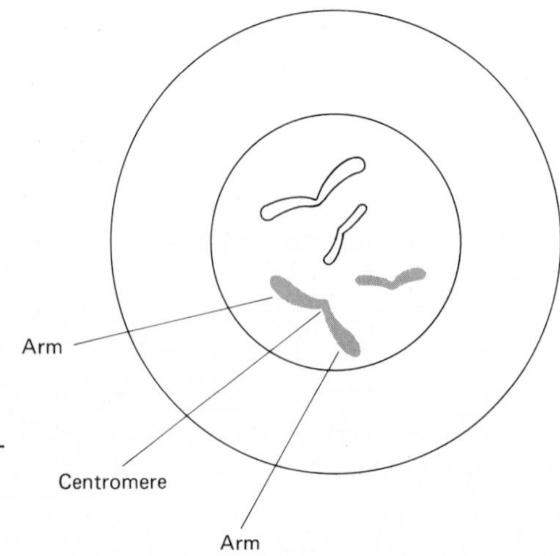

Figure 1-4. Chromosomes in leptotene of prophase I in a cell with a diploid number of 4.

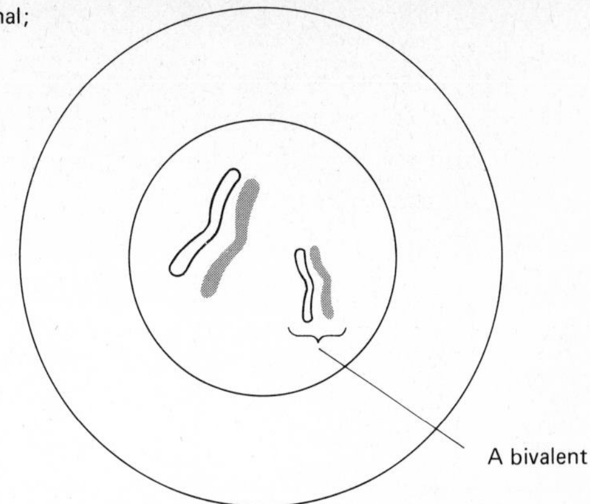

Figure 1-5. Synapsis of homologous chromosomes in zygotene of prophase I.

pair of partner chromosomes are visibly similar to each other in size and in the position of centromeres. The process of pairing is called *synapsis*. With synapsis, it is obvious that chromosomes in cells of sexually reproducing organisms do exist in pairs, called *homologs* (Fig. 1-5). The existence of chromosomes in pairs is what one would expect in view of the concept first formalized by van Beneden that every individual inherits an equal number of chromosomes from both parents. In other words, every pair of homologous chromosomes is constituted of a maternal and a paternal chromosome.

During zygotene, the chromosomes continue to coil and become thicker and shorter. With the completion of synapsis, the two homologs are very closely associated, and the paired chromosomes are termed *bivalents*. In human cells at this stage there would be 23 bivalents, or 46 chromosomes.

CONCEPT OF HOMOLOGS AND ALLELES: IT HAS BEEN ESTABLISHED THAT ALLELES ARE FOUND AT THE SAME POSITION, OR LOCUS, ON EACH OF A PAIR OF HOMOLOGOUS CHROMOSOMES. IN PEA PLANTS, THE PAIR OF CHROMOSOMES ON WHICH THE ALLELES DETERMINING STEM LENGTH ARE LOCATED WOULD CARRY THE GENES AT EXACTLY THE SAME POSITION (FIG. 1-6). SINCE EVERY CHROMOSOME CONTAINS MORE THAN ONE GENE, EVERY GENE DOWN THE LENGTH OF A CHROMOSOME HAS ITS COUNTERPART IN THE SAME POSITION ON THE HOMOLOG. In fact, this linear arrangement of genes is believed to determine the success of synapsis, which is not a haphazard pairing of homologs, but a *gene-to-gene pairing*. Alteration of the order of genes on one of the homologous pairs may affect the process of synapsis, and thus lead to unsuccessful meiosis.

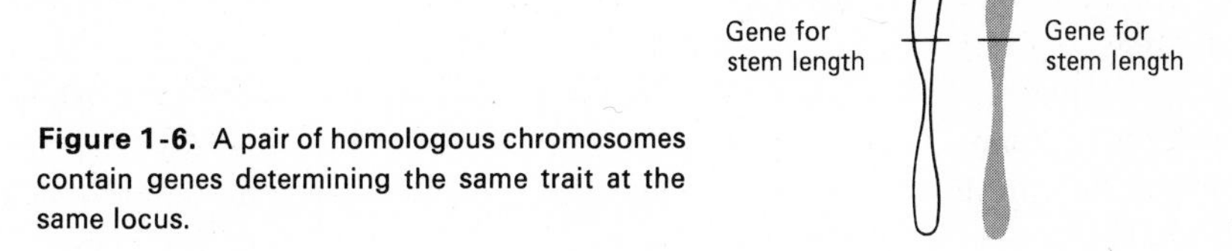

Figure 1-6. A pair of homologous chromosomes contain genes determining the same trait at the same locus.

As we mentioned earlier, alleles are involved with the determination of the same trait. Remember, however, that alleles need not be the same gene, as Mendel pointed out. The gene for tallness in peas is allelic to the gene for shortness. In mice, for example, there is a dominant gene determining the production of coat pigmentation (C) and recessive alleles to this gene that result in colors, such as chinchilla (c^{ch}) and absence of color, albino (c). Such existence within members of a species of many different forms of a gene is termed *multiple allelism.* In diploid species, however, a given individual carries only two of these alleles.

Pachytene. Following the completion of synapsis, the chromosomes continue to shorten and thicken. At this point in prophase I, *pachytene*, an organized structure known as the *synaptinemal complex* can be seen between the closely associated homologs. This complex appears to consist of a central element and two lateral elements, composed of nucleic acids and proteins synthesized after synapsis of the chromosomes (Fig. 1-7).

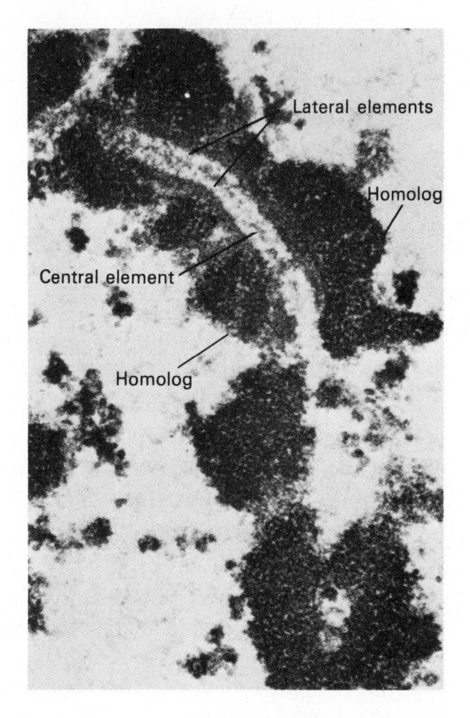

Figure 1-7. Electron micrograph of a synaptinemal complex in pachytene. (Photo: T. F. Roth in E. D. Garber, 1972. *Cytogenetics*, cover photo. McGraw-Hill, New York.)

The role of the synaptinemal complex has yet to be determined, but it is thought that there may be some relationship between its formation, which begins in zygotene and ends in pachytene, and the process of synapsis, which occurs at the same time.

Diplotene. In the next substage of prophase I, *diplotene*, the synaptinemal complex disappears. Concomitantly, the two strands of the duplicated chromosomes, which until now had been visually indistinguishable, seem to repulse each other, making the double nature of the chromosomes apparent for the first time. The two identical strands of each duplicated chromosome remain attached to each other at the centromere and are called *sister chromatids* (Fig. 1-8).

Actually, the only time most chromosomes can ever be visualized under the light microscope is in prophase and later stages of cell divisions. Most light photo-

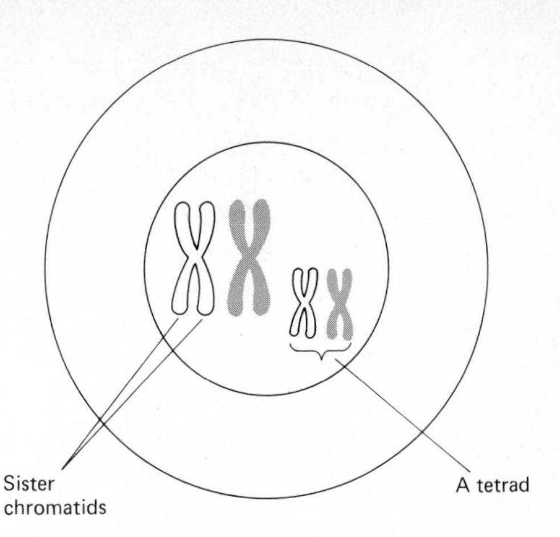

Figure 1-8. Duplicated chromosomes in diplotene.

micrographs are therefore of duplicated, double-stranded, chromosomes. Bear in mind, then, that in cells that are not dividing but are otherwise biologically active, the chromosomes are single structures.

With the appearance of the sister chromatids, there are now four strands in close association due to synapsis of homologs. Bivalents with four strands together are sometimes referred to as *tetrads*. Although the sister chromatids of each duplicated chromosome are quite apparent, we still consider each chromosome to be a single unit, because they remain attached by the centromere and are not individually functional. Therefore, in diplotene, the diploid number of chromosomes are still present. *In fact, an easy way of determining the number of chromosomes in a cell is to count the number of centromeres*, rather than the number of chromosome threads. In diplotene of meiosis in human cells, there are still 46 chromosomes and 23 tetrads, although there are 92 chromatids.

During diplotene, another phenomenon is frequently observed, which is illustrated in Fig. 1-9. At certain regions along the length of some nonsister chromatids—that is, a chromatid strand of one homolog and a chromatid of the other homolog—contact can be seen. As the strands in contact sometimes form crosslike

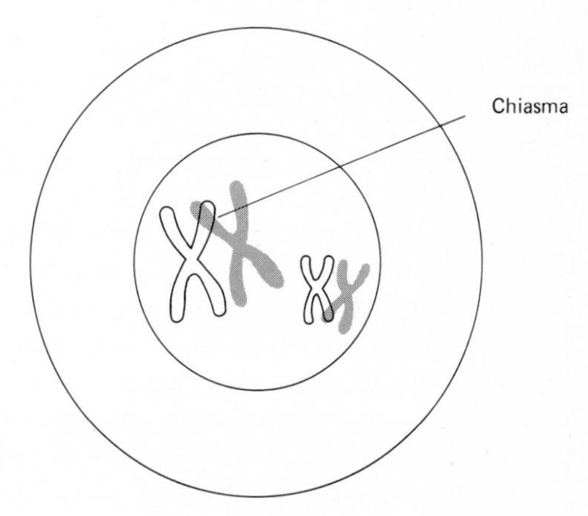

Figure 1-9. Presence of chiasmata in tetrads during diplotene.

configurations, they have been termed *chiasmata* (sing., *chiasma*). There is evidence, which we will discuss later, of exchange of genetic material between the two chromatids at chiasmata. This exchange is known as *crossing-over* in eukaryotes.

CONCEPT OF CROSSING-OVER AND GENETIC VARIABILITY: IN DIPLOID ORGANISMS, MANY GERM CELLS UNDERGO MATURATION AT THE SAME TIME, ESPECIALLY IN MALES. CROSSING-OVER IN THE DIFFERENT CELLS IS APPARENTLY A RANDOM PROCESS. IN OTHER WORDS, IF WE ASSUME 100 GERM CELLS ARE UNDERGOING MEIOSIS, THE CHIASMATA FOUND IN ANY ONE OF THEM WILL LIKELY DIFFER FROM THOSE FOUND IN ANY OTHER. THIS RESULTS IN AS MANY DIFFERENT COMBINATIONS OF MATERNAL AND PATERNAL GENES ON THE CHROMATIDS INVOLVED AS THERE ARE DIFFERENT CHIASMATA. THE COMPLETION OF MEIOSIS WILL FIND ALL FOUR THREADS OF A TETRAD IN DIFFERENT GAMETES; THUS IF THE CHROMATIDS ARE GENETICALLY DIFFERENT AFTER A CROSSOVER EVENT, ALL FOUR GAMETES WILL ALSO BE DIFFERENT. THE PROCESS OF CROSSING-OVER IS THEREFORE ONE OF THE SOURCES OF GENETIC VARIABILITY WHICH IS DERIVED FROM MEIOSIS. FIG. 1-10 ILLUSTRATES THIS IMPORTANT PRINCIPLE.

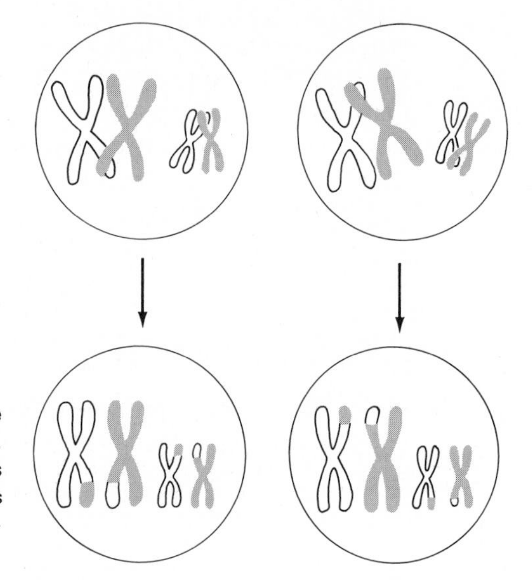

Figure 1-10. Different germ cells from the same individual show random formation of chiasmata. The functional germ cells derived from these cells will be genetically different, as the chromatids involved in crossing-over will have different combinations of genes.

Why chiasmata form is not well understood. What is known is that they appear to be random, and that all four chromatids in a tetrad may be involved in a number of cross-over events. Interestingly, in some species crossing-over has been found to occur more frequently in cells of females than of males. An extreme example of this phenomenon can be seen in males of the fruitfly *Drosophila melanogaster*, in which no crossing-over at all occurs. There is some evidence that the synaptinemal complex may have some influence on chiasma formation. Electron microscopy has shown that there is no synaptinemal complex between the homologs of *Drosophila* males in prophase I.

Diakinesis. The process of coiling continues throughout prophase I and reaches its maximum in diplotene and the final substage of prophase I, diakinesis. As chiasmata are formed, and exchanges completed, the contact of chromatids ends in most regions. This decrease in the number of chiasmata is progressive, beginning in diplotene and ending in diakinesis. However, the homologs are usually connected by one last chiasma which seems to traverse the chromatids until it reaches one end. This phenomenon is known as *terminalization*, and its function or regulation is not fully understood. At the end of prophase I, then, the tetrads are composed of non-identical sister chromatids (if crossing-over has occurred; otherwise the sister chromatids are identical) and in most cases the bivalents remain together by a chiasma at one end (Fig. 1-11).

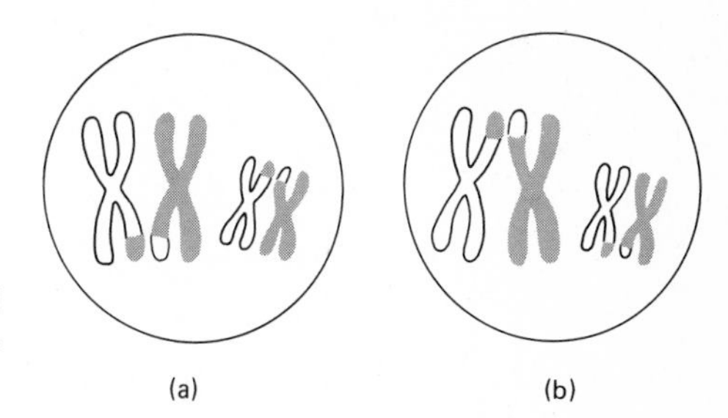

Figure 1-11. Diakinesis in prophase I. We shall continue to follow the cell in (a) to illustrate subsequent events in meiosis.

Metaphase I

Many cytological changes are occurring at the same time as the events described above. Important to our discussion is the formation of the meiotic spindle. In animal cells the poles of the spindle are formed by centrioles that have migrated to opposite sides of the cell. In plant cells there are no centrioles. From the poles radiate microtubules forming a football-shaped structure which we call the *spindle*. The microtubules are called *spindle fibers*. Some fibers seem to span the length of the spindle. Those that attach to chromosomes appear to be shorter and connect only to the chromosome (Fig. 1-12).

Chromosomes are no longer confined to a nucleus, as the nuclear membrane

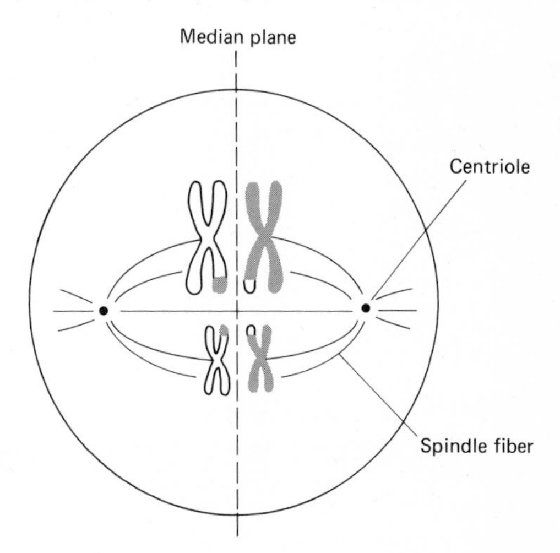

Figure 1-12. Metaphase I. Tetrads are attached to spindle fibers at the centromere region. Note: centrioles are not seen in plant cells, but are present in animal cells. Plant cells tend to be more rectangular than animal cells.

has disintegrated. The tetrads become attached by the centromeres of the chromosomes to spindle fibers in the median, or equatorial, plane (Fig. 1-13). Note that the tetrads are so aligned that the homologs are on different sides of the median plane, one closer to one pole of the spindle, the other closer to the other pole. Terminal chiasmata are lost during this stage in meiosis.

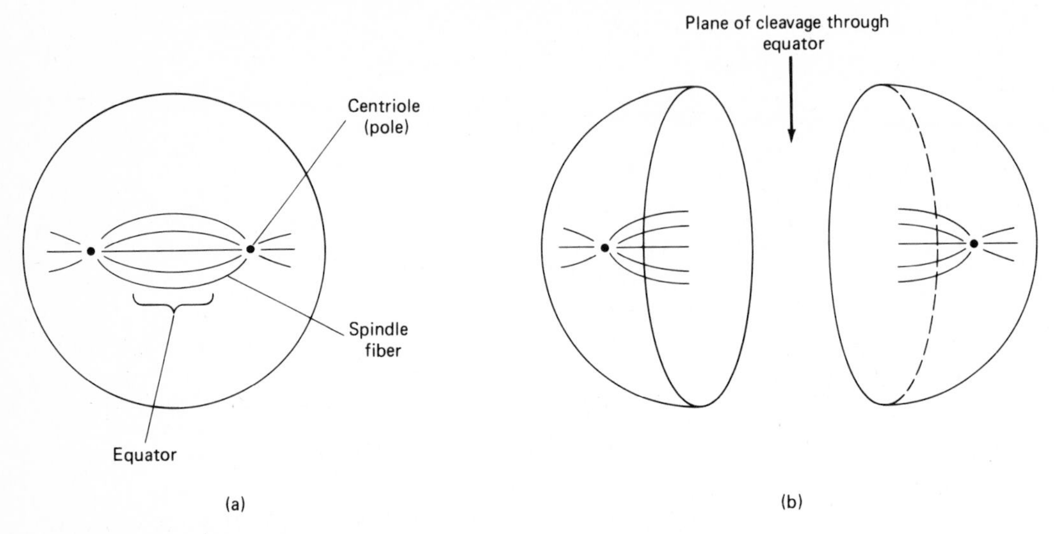

Figure 1-13. (a) Spindle formation. (b) The plane of cleavage of a dividing cell. (From A. Pai, 1974. *Foundations of Genetics, A Science for Society,* Fig. 2.4. McGraw-Hill, New York.)

Concept of random assortment and genetic variability: The manner in which tetrads are aligned on the spindle is precise with regard to the position of the homologs in relation to the poles. They are always positioned so that the partner chromosomes are on different sides of the median plane of the spindle, which is eventually the plane of division (Fig. 1-14). However, the side of the median plane upon which the maternal and the paternal chromosome of each tetrad is located is entirely random. Consequently, the combinations of maternal and paternal chromosomes on each side of the equator of the spindle differ in different gametes.

If we look at two pairs of chromosomes from a cell undergoing meiosis, it is evident that they can be aligned in two different ways with regard to the maternal and paternal chromosomes. The maternal chromosome of one pair can be on the same side of the median plane as the paternal chromosome of the other pair; or, both maternal chromosomes can be on the same side. As this is a completely random process, we can expect that of 100 gametes undergoing meiosis, half will have one alignment, and half the other alignment.

This holds true for all tetrads in a cell. In humans, germ cells would have 23 tetrads at this stage in meiosis. All would align randomly (with regard to their paternal or maternal derivation) on either side of the median plane. The total number of

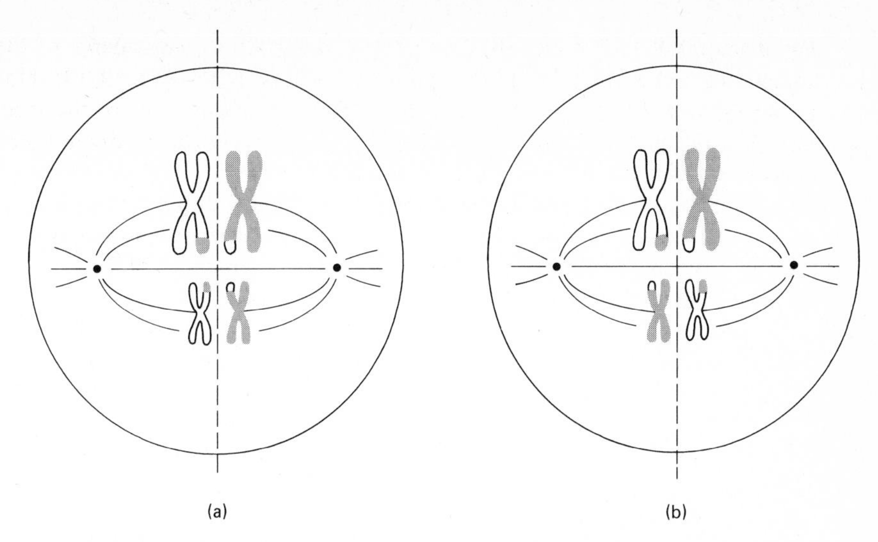

Figure 1-14. Two possible alignments of two pairs of chromosomes during metaphase I. Division of the cell in diagram (a) would result in daughter cells containing mainly maternal or paternal chromosomes. The cell in diagram (b) would give rise to a mixture of maternal and paternal chromosomes. This is the physical basis for random assortment.

possible combinations would be 2^{23}, or 8,388,608! (See p. 47 for a more detailed explanation of this point.)

The genetic significance of this phenomenon is profound, for it means that the assortment of chromosomes will be different from one germ cell to another, when division is completed. *Random assortment* of chromosomes is therefore a major source of genetic variability provided by the events of meiosis. Since it was first recognized by Mendel, random assortment is frequently referred to as one of the Mendelian Laws of Heredity. We shall discuss specific examples of random assortment in heredity studies in Chapter 2.

Anaphase I

The loss of all terminal chiasmata marks the beginning of anaphase I. During this stage, homologous chromosomes are separated from one another. The separation is due to movement of the spindle fibers (to which the chromosomes are attached by the centromeres) pulling the chromosomes toward opposite poles.

The actual mechanism of movement of the chromosomes is still a controversial subject among scientists. Some feel that the chromosomes may be pulled to one or the other pole by a sliding mechanism similar to the movement of myofibrils in skeletal muscle cells; others postulate that the microtubules are constantly undergoing assembly and dissembly at the ends, and that this molecular rearrangement causes the movement of the chromosomes (Salmon, 1975).

Photomicrography techniques reveal that the centromere regions of the chromosomes precede the arms toward the poles, with the arms trailing behind. As with all the movements seen in cell division, this is a dynamic process; time-lapse photography shows a bubbling of the cytoplasm which gives the appearance of great expenditure of energy by the cell. As a result of this movement, each of the two groups of chromosomes being pulled to opposite ends of the cell can be seen to have exactly one-half the total number of chromosomes, with one representative of each original pair of chromosomes in each group. A single duplicated chromosome, now separated from its homolog, is called a *dyad* (dy—referring to the two chromatids making up each chromosome). The direction of movement of chromosomes is always perpendicular to the plane of division (Fig. 1-15).

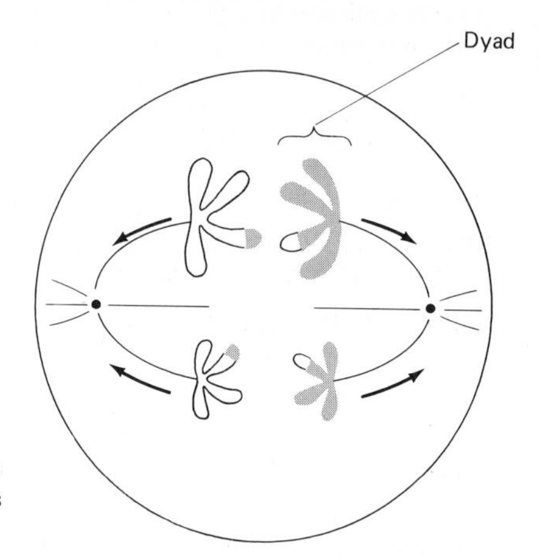

Figure 1-15. Anaphase I. Tetrads are split, with homologs moving toward opposite poles (arrows).

CONCEPT OF SEGREGATION: THE MOVEMENT OF DYADS TO DIFFERENT PARTS OF THE CELL SERVES AS THE PHYSICAL BASIS FOR A SECOND MENDELIAN LAW OF HEREDITY, NAMELY, SEGREGATION. AS A RESULT OF THE SEPARATION OF HOMOLOGS, PAIRS OF ALLELES ARE ALSO SEPARATED, SINCE THEY ARE LOCATED ON HOMOLOGS. THE SEPARATION OF ALLELES IS WHAT IS MEANT BY SEGREGATION, THAT NORMALLY EACH GERM CELL WILL CONTAIN ONLY ONE OF EVERY PAIR OF ALLELES IN THE CELL. THIS PROCESS IS OF COURSE NECESSARY TO MAINTAIN THE DIPLOID NUMBER FOLLOWING UNION OF TWO HAPLOID GAMETES IN FERTILIZATION. NOTE, TOO, THAT THE ALIGNMENT OF HOMOLOGS DURING METAPHASE I AND THEIR SUBSEQUENT SEPARATION IN ANAPHASE I GUARANTEES THAT THE REDUCTION IN THE NUMBER OF CHROMOSOMES FOLLOWING DIVISION IS NOT HAPHAZARD, BUT THAT ONE REPRESENTATIVE OF EACH PAIR OF HOMOLOGS WILL BE PRESENT IN EACH GAMETE. SINCE CHROMOSOMES DO CARRY DIFFERENT GENES, THIS IS ESSENTIAL TO ENSURE THAT THE ZYGOTE RECEIVES THE FULL COMPLEMENT OF GENES FOR NORMAL DEVELOPMENT.

Telophase I

Telophase I is marked by the approach of the two clusters of chromosomes to the poles. At the plane of division, the cleavage furrow appears in animal cells; in plant cells the cell plate begins to form (Fig. 1-16).

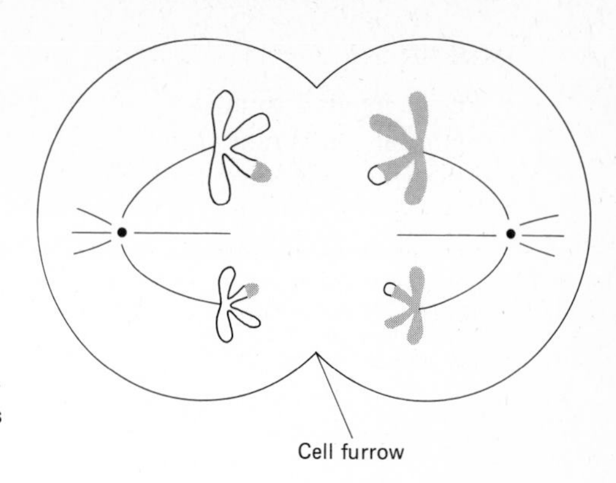

Figure 1-16. Telophase I. Spindle fibers continue to draw chromosomes toward poles. Cells begin to constrict, leading to division.

Division I

With the completion of division (*cytokinesis*), the original gamete has been transformed into two daughter cells, each containing the haploid number of chromosomes. Because of this reduction in chromosome number, the first division of meiosis is sometimes referred to as the *reductional division* (Fig. 1-17).

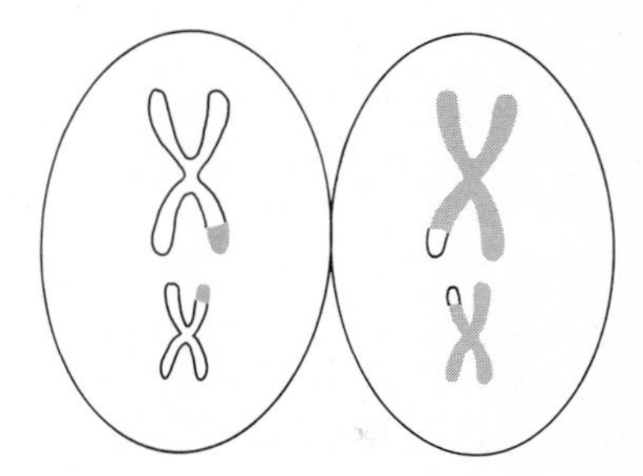

Figure 1-17. Cytokinesis. Completion of division resulting in two haploid daughter cells.

Interphase II

The events following the first division vary from organism to organism. In some, the chromosomes once again become attenuated and nuclear membranes form. In others, no significant changes occur and the cells enter into prophase II. In either case there is no further duplication of genetic materials. The chromosomes are still duplicated; they are still in the form of dyads. The events that take place following reductional division occur synchronously in both daughter cells.

Prophase II

In cells that do enter a second interphase, the chromosomes reappear because of coiling in prophase II; they appear as in Fig. 1-17. None of the other events of prophase I (such as synapsis or crossing-over) occur, because there are no longer homologous pairs of chromosomes in the now haploid cells.

Metaphase II

Once again a spindle is formed in the cytoplasm, and with the disappearance of the nuclear membrane, the chromosomes migrate to the median plane. Remember that there are now no tetrads, only dyads. As the chromosomes are not homologs, they line up at random on the metaphase plate, again being attached to spindle fibers by the centromere. Human cells at this stage of meiosis would show 23 dyads (Fig. 1-18).

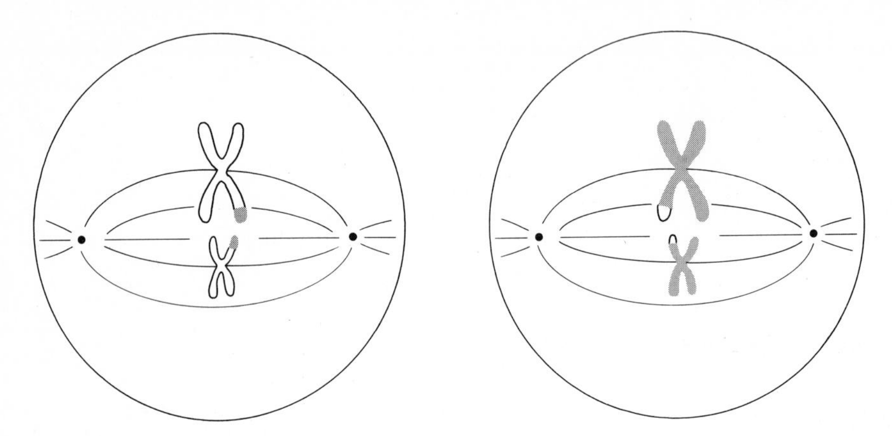

Figure 1-18. Metaphase II. Dyads are attached to spindle fibers in the median plane.

Anaphase II

At this stage, movement of spindle fibers causes a separation of sister chromatids at the centromeric region. Each strand now has its own functional centromere and is considered a chromosome (Fig. 1-19).

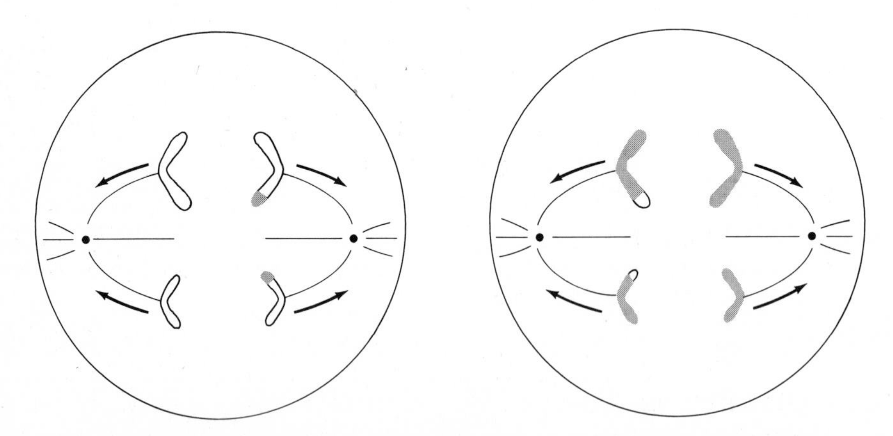

Figure 1-19. Anaphase II. The centromeres divide, and sister chromatids are pulled to opposite poles (arrows).

Telophase II

Telophase II merely continues the process of separation of the two groups of chromosomes. Again, cytological changes such as a furrow of the cell membrane in animal cells, and initiation of cell wall formation in plant cells, signal the onset of cytokinesis. Note that in each cluster the number of chromosomes is haploid (Fig. 1-20).

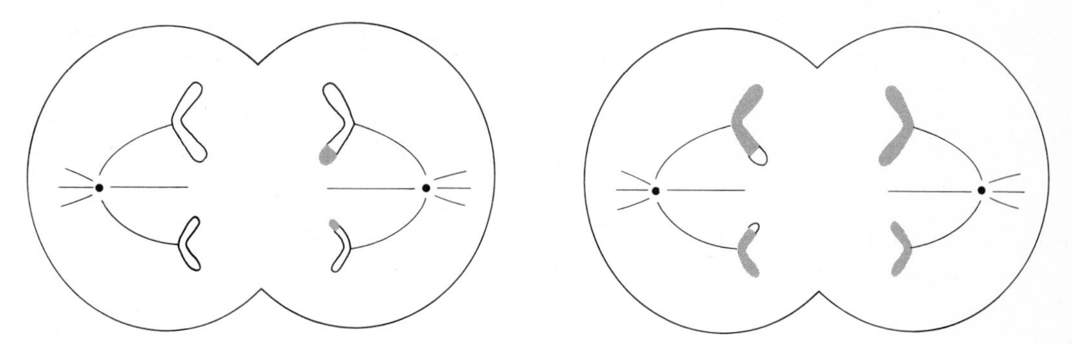

Figure 1-20. Telophase II.

Division II

The second division of the daughter cells occurs, producing four haploid cells. Note that this division has not changed the number of chromosomes in each cell from the number in prophase II cells. Following the second division the daughter cells now have single-stranded chromosomes and still have the haploid number. For this reason the second division is considered an *equational division* (Fig. 1-21).

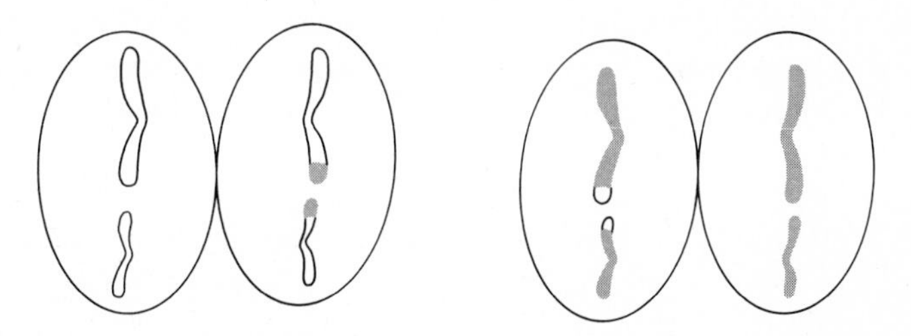

Figure 1-21. Division II. The completion of meiosis results in four haploid cells. Note that because of crossing-over and segregation all four cells possess different combinations of genes and chromosomes.

GENETIC SIGNIFICANCE OF MEIOSIS: AT THE COMPLETION OF MEIOSIS, FOUR HAPLOID CELLS HAVE BEEN DERIVED FROM ONE DIPLOID CELL. TO REITERATE THE SIGNIFICANT POINTS OF MEIOSIS FOR OUR STUDY OF HEREDITY, A GERM CELL CONTAINS ONE REPRESENTATIVE OF EACH PAIR OF CHROMOSOMES IN THE PARENT CELL. EVERY ONE OF THE

FOUR DAUGHTER CELLS FROM THE SAME PARENT CELL IS MOST LIKELY TO CONTAIN DIFFERENT COMBINATIONS OF CHROMOSOMES AND GENES, BECAUSE OF THE RANDOM ALIGNMENT OF MATERNAL AND PATERNAL CHROMOSOMES AT METAPHASE I. FURTHER, THE GENETIC DIFFERENCES OF DIFFERENT DAUGHTER CELLS ARE INCREASED BY THE PHENOMENON OF CROSSING-OVER (SEE AGAIN FIG. 1-21).

GAMETOGENESIS

In sexually reproducing organisms, meiosis occurs as we have described during the maturation of germ cells, or gametogenesis. In some species, such as humans, however, some cytological differences occur in the maturation of sperm and egg; these will be discussed briefly here, as there are consequences of a genetic nature.

Spermatogenesis

The events of meiosis during the maturation of sperm are very similar to those that we have described. Each division gives rise to equal-sized daughter cells (it is a quirk of our culture that we never speak of son cells!) which then must undergo a series of fascinating cytological changes before they actually become functional sperm. We refer to immature sperm cells as *spermatogonia.* Those that have gone through the stages prior to the first division are called *primary spermatocytes*; those that have already become haploid through reductional division are *secondary spermatocytes.* The completion of meiosis produces cells called *spermatids*; not until the cells have taken on the physical shape of mature sperm do we consider them *spermatozoa.* A summary of both spermatogenesis and oogenesis is given in Fig. 1-22.

The cellular changes which convert spermatids to functional spermatozoa are illustrated in Fig. 1-23. Essentially, the cells slough off almost all their cytoplasm. Cell organelles take on new functions and new positions in the mature sperm. For example, one of the centrioles elongates to form the flagellum. Mitochondria become clustered in the neck region and serve as the energy source for the highly active mature cell. The Golgi body is believed to develop into the acrosome, a saclike structure on the tip of the sperm's head containing enzymes that aid in penetration of the egg. Although we have known of these changes since the nineteenth century, the mechanism which regulates them is still a mystery. As is the case with any biological phenomenon, however, the answer will eventually be traced back to gene activity in the maturing cells.

Oogenesis

In certain species, including humans, the maturation of eggs is cytologically quite distinct from spermatogenesis though the process of meiosis is the same (see again Fig. 1-22). As with sperm, we classify eggs by the stages of meiosis that they have completed. Diploid immature cells are called *oogonia*; those that have completed

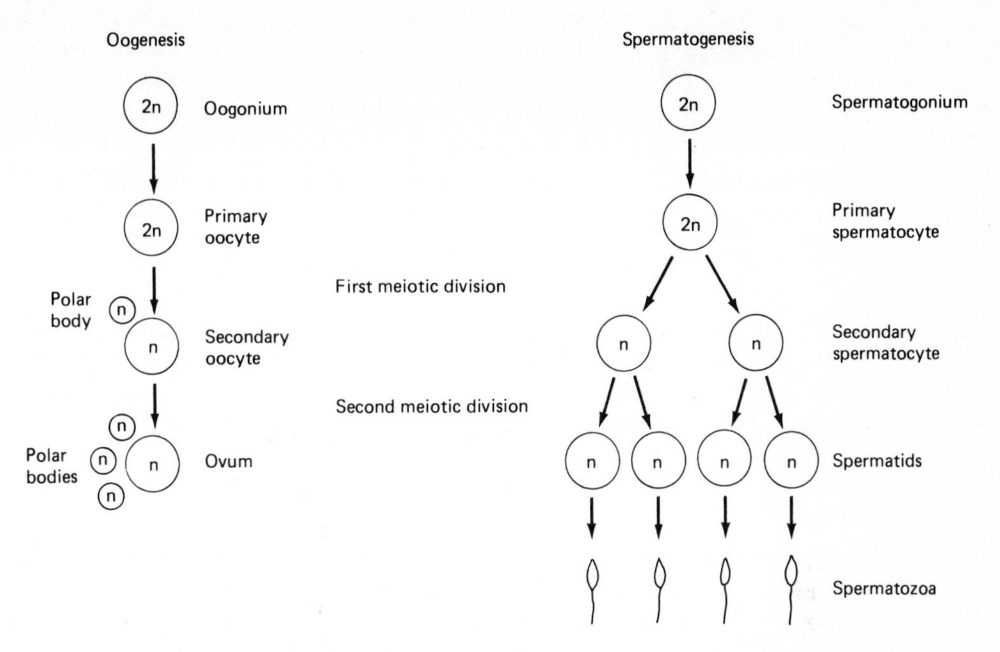

Figure 1-22. Summary of genetic and cytological changes during gametogenesis.

the stages prior to reductional division are *primary oocytes*. Following reductional division they are called *secondary oocytes*, and completion of the equational division results in the mature *ova*. One obvious cytological difference between spermatogenesis and oogenesis is that division is unequal in oogenesis. The meiotic spindle is formed near the periphery of the cell, so that at cytokinesis there is formed one very small cell called a *polar body* and one large cell that will eventually become the functional ovum. The second division of the large cell is again acentric, resulting in another polar body and the mature haploid ovum. The first polar body may or may not undergo a second division. Biologically it matters little, because the polar bodies are usually incapable of normal development if they are fertilized.

It is important to realize that even though only one of the four cells resulting from meiosis is functional, no exception occurs to either the law of segregation or that of random assortment. The homologs still are separated at anaphase I to cause segregation of alleles; and the combination of chromosomes eventually found in the ovum is apparently entirely random.

The conservation of most of the cytoplasm in the ovum is in direct contrast to the discarding of almost all the cytoplasm in the maturation of sperm. Embryologists have discovered that the conservation of cytoplasm in the ovum is essential for normal development of the zygote. It appears now that the regulation of the first stage of development of the zygote, which as we mentioned before is cell division, is determined by information stored in the cytoplasm of the egg. This information is genetic in nature, and we shall discuss it in more detail later. It is probably the lack of this information in the polar bodies that renders them nonfunctional as germ cells.

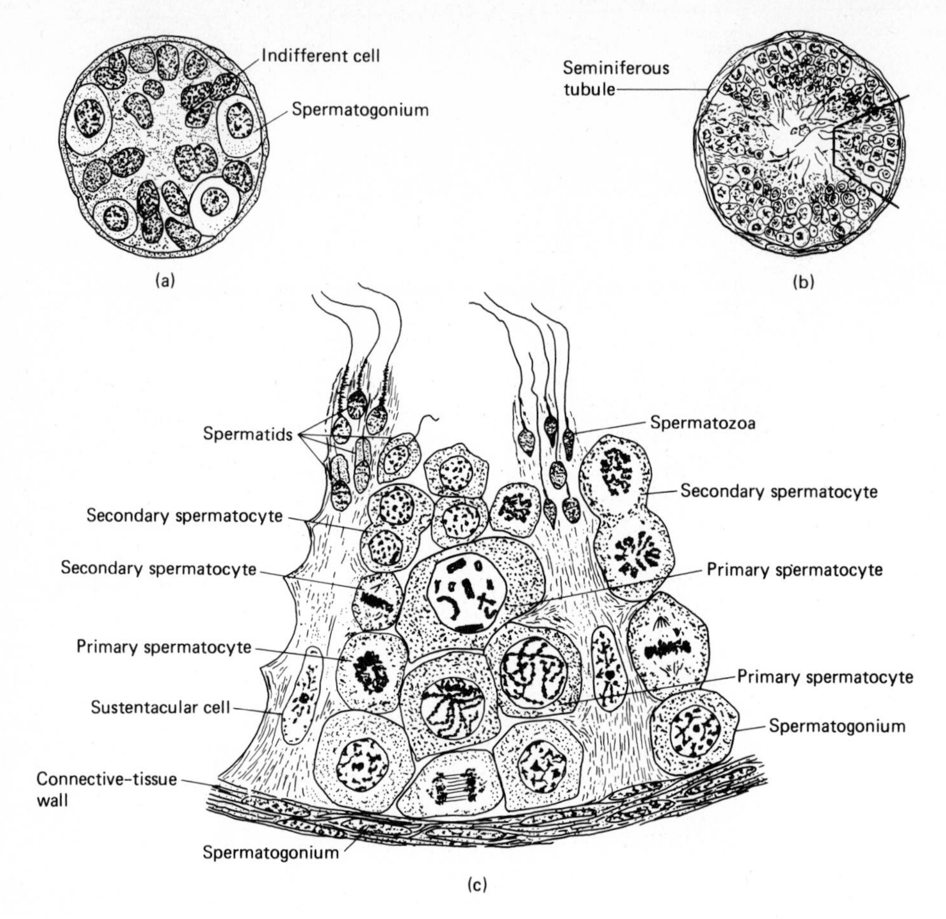

Figure 1-23. Human testis tubules, in transverse section: (a) newborn (×400); (b) adult (×115); (c) detail of the area outlined in (b) (×900), showing stages of spermatogenesis. (From L. B. Arey, 1966. *Developmental Anatomy*, p. 41. Saunders, Philadelphia.)

Genetic Consequences of Aging in Ova

In complex organisms such as humans, there is another biological difference between oogenesis and spermatogenesis which is of concern to geneticists. This is the fact that human females are born with all the germ cells that they will ever have, some 400,000 in the ovaries of the newborn. In the adult, the process of maturation occurs under hormonal control (Fig. 1-24). Studies have shown that as cells age, chromosomes are affected in some way that causes their movement in cell division to be abnormal. An aberration of the meiotic process known as *nondisjunction* is more frequently found in older cells. This abnormality results in homologs not separating from one another in anaphase I, or in chromatids not separating properly in anaphase II. At the completion of meiosis, cells have too many or too few of that pair of chromosomes, as illustrated in Fig. 1-25. Should such cells be fertilized the zygote will be

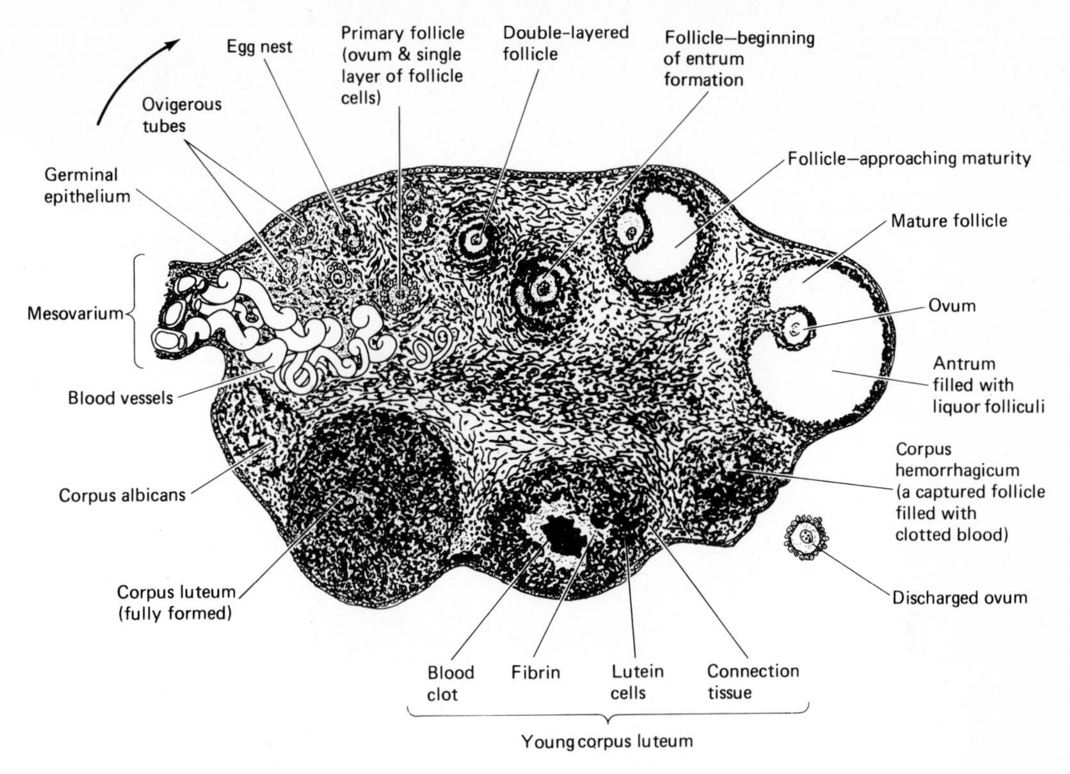

Figure 1-24. Life cycle of an egg and its follicle, shown in a diagram of the mammalian ovary (Patten). Start at the arrow and follow the stages clockwise around the ovary. (From L. B. Arey, 1966. *Developmental Anatomy,* Fig. 17, p. 32. Saunders, Philadelphia.)

abnormal, and in many cases the abnormal chromosome content is a lethal condition. Thus the older a woman is, the older her cells, and the greater likelihood of nondisjunction.

One of the most familiar conditions in humans that results from nondisjunction is Down's syndrome, formerly referred to as mongoloid idiocy (Fig. 1-26). Many individuals with this condition are found to have 47 chromosomes, presumably because abnormal meiosis resulted in two homologs in one ovum. The relation between the incidence of Down's syndrome and maternal age is statistically significant. In women younger than 25, the incidence of Down's syndrome is 1 per 2000 live births; in women 45 and older, the incidence of Down's syndrome is 1 in 50 live births. A progressive increase in frequency occurs in mothers between the ages of 25 and 45 (Hamerton, 1971).

Although there is evidence that nondisjunction can occur in spermatogenesis (which we will discuss later under the topic of chromosomal abnormalities), it occurs with far less frequency than in females, and no statistical correlation has been found between the incidence of Down's syndrome and paternal age. This lower incidence of nondisjunction is probably due to the constant production of new spermatogonia

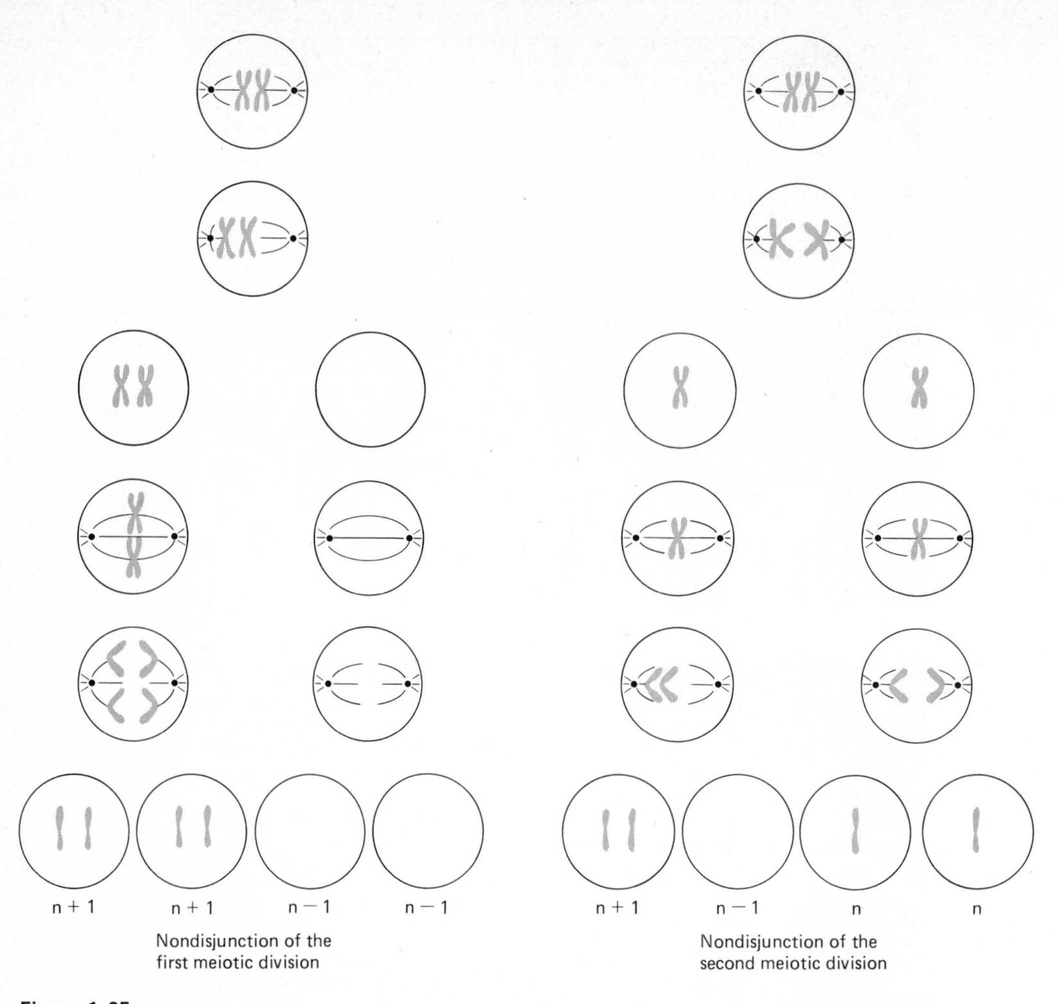

Figure 1-25

from stem cells that continue to divide, and never mature into spermatozoa. This type of cell division is mitosis; it occurs in all cells of the body and provides genetic continuity. We shall discuss this type of cell division next.

MITOSIS

Union of a sperm and egg in fertilization produces a zygote which contains the diploid, species number of chromosomes. One member of each pair of homologs is contributed by each parent. Zygotes of multicellular species must then undergo the process of mitosis many times before the organism reaches the adult stage, and even then local cell divisions must occur for normal replacement and function, as in the case of spermatogonia being constantly supplied in human males. The magnitude of

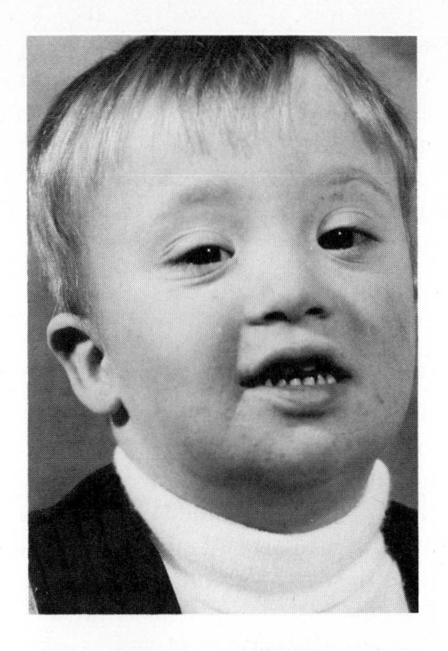

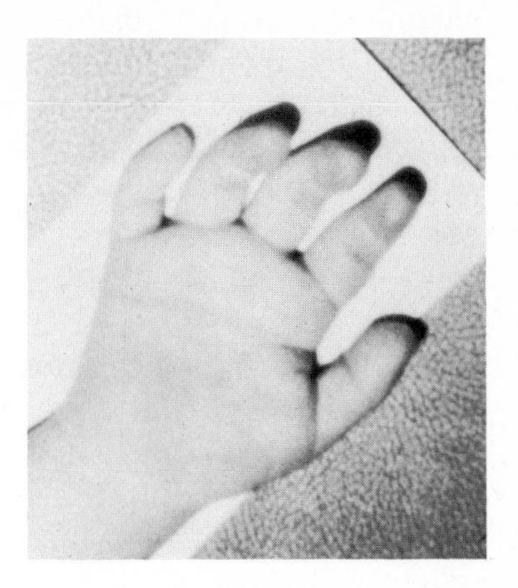

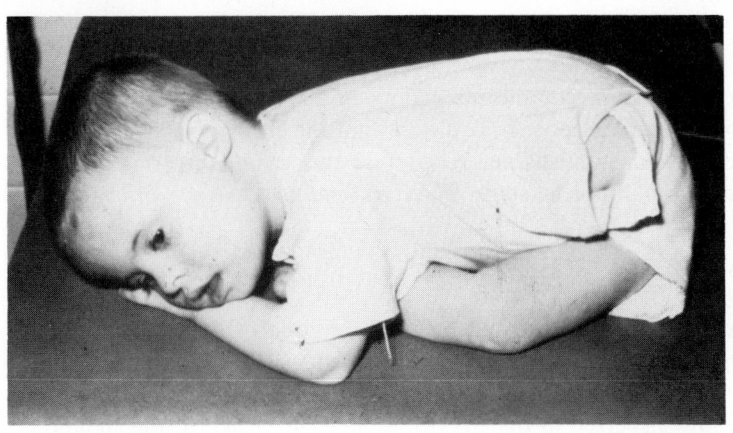

Figure 1-26. Down's Syndrome patients. (From A. Pai. 1974. *Foundations of Genetics,* Fig. 11.1. McGraw-Hill, New York.)

the numbers of mitotic divisions and the exactness with which they occur are truly awesome.

There are about 10^{14} cells in adult humans. Even without considering the replacement of cells, this suggests that more than 46 mitotic divisions must occur to produce such a large number of cells. But in addition, replacement of cells occurs constantly in adult humans. Consider, for example, that red blood cells alone in adult humans must be replaced at the rate of about 2,500,000 every second of the day and night (Villee and Dethier, 1971).

As with meiosis, the process of mitosis has for convenience been divided into stages, and these are named the same as those in meiosis: interphase, prophase, metaphase, anaphase, telophase, and division. Since one of the differences between mitosis and meiosis is that only one divisional event occurs in mitosis, it is not necessary to give the stages Roman numerals as in meiosis. Figure 1-27(a) illustrates diagrammatically the entire process of mitosis; Figure 1-27(b) presents photomicrographs of actual cells in the different stages of mitosis.

Interphase

Interphase of mitosis is very much like that of meiosis. The chromosomes cannot be seen as discrete structures, but form a chromatin network. At this time there is synthesis of new genetic material, and the chromosomes are duplicated. J. H. Taylor and coworkers (Taylor *et al.* 1957) performed autoradiography experiments which gave definitive evidence of duplication during interphase of mitosis. In studying cultures of dividing plant cells of the broad bean, *Vicia faba*, they used radioactive precursors of chromosomal material, such as tritiated thymidine. They then exposed photographic film to these cultures at different stages of mitosis, the radioactive material being visualized as dark grains on the developed film. Their results showed the incorporation of the precursors into new chromosomal material during interphase (Fig. 1-28).

Prophase

With the duplication completed, coiling of the chromosomes makes them visible under the light microscope. In addition, that the chromosomes have been duplicated is immediately distinguishable; recall that this is not discernible in prophase I of meiosis until the diplotene stage. *There is usually no synapsis of homologs in prophase of mitosis.*

Metaphase

Metaphase of mitosis is marked by the formation of a mitotic spindle, a structure very similar to the spindle we described in meiosis. Since synapsis does not occur, the homologs are not in close association with each other; rather, all chromosomes are randomly dispersed in the equatorial plane, attached to the spindle fibers by the centromeres as in meiosis.

Anaphase

In anaphase of mitosis, as in anaphase II of meiosis, there is a splitting of the centromeres and the sister chromatids are pulled to opposite poles. Since they are now single-stranded, the chromatids are considered chromosomes. Again, the movement of chromosomes is in a direction perpendicular to the plane of division.

Telophase

The two clusters of chromosomes continue to separate as they draw toward their respective poles. Again, cytological changes signaling the onset of cytokinesis, such as the construction of the cell membrane, become apparent.

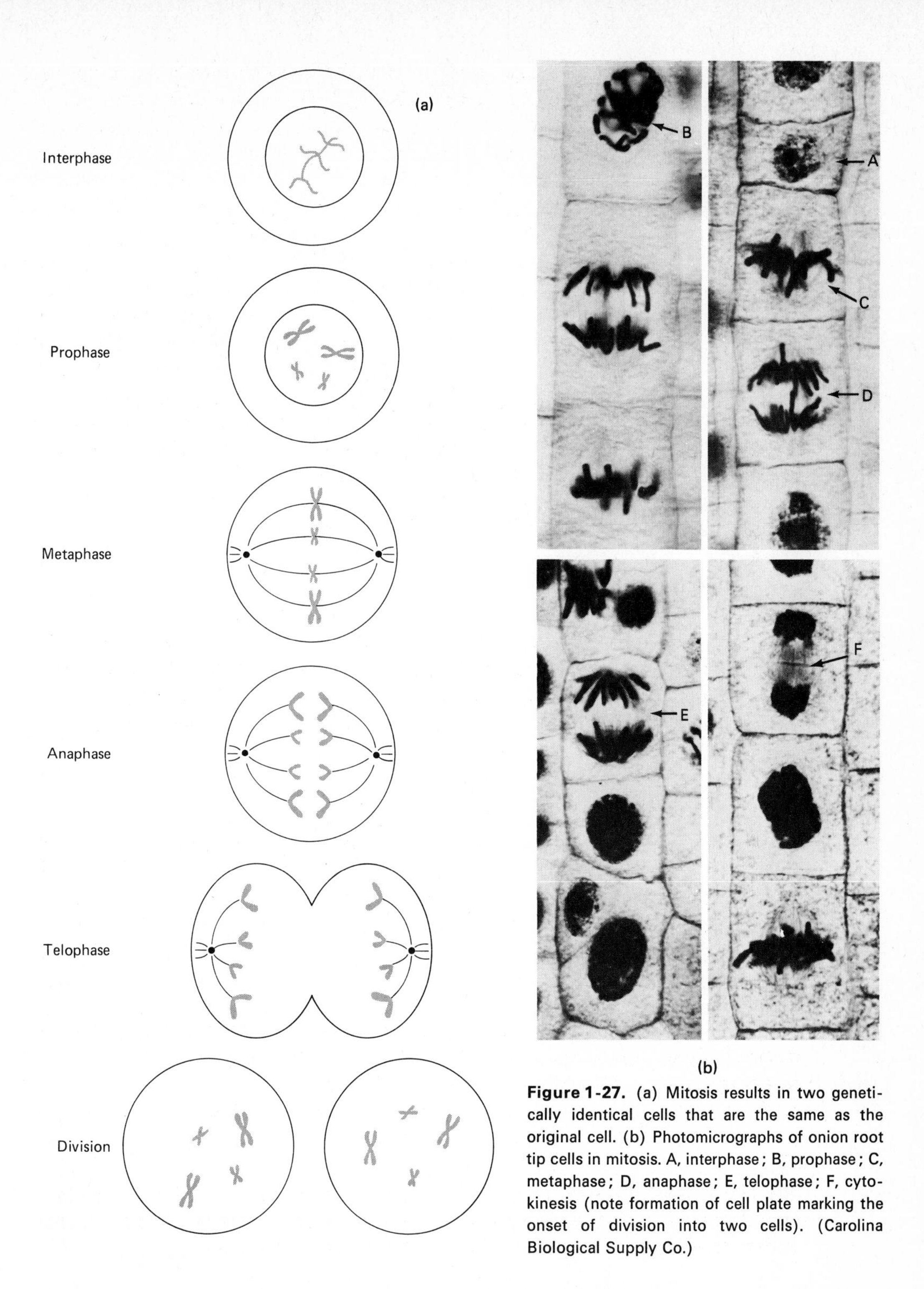

Figure 1-27. (a) Mitosis results in two genetically identical cells that are the same as the original cell. (b) Photomicrographs of onion root tip cells in mitosis. A, interphase; B, prophase; C, metaphase; D, anaphase; E, telophase; F, cytokinesis (note formation of cell plate marking the onset of division into two cells). (Carolina Biological Supply Co.)

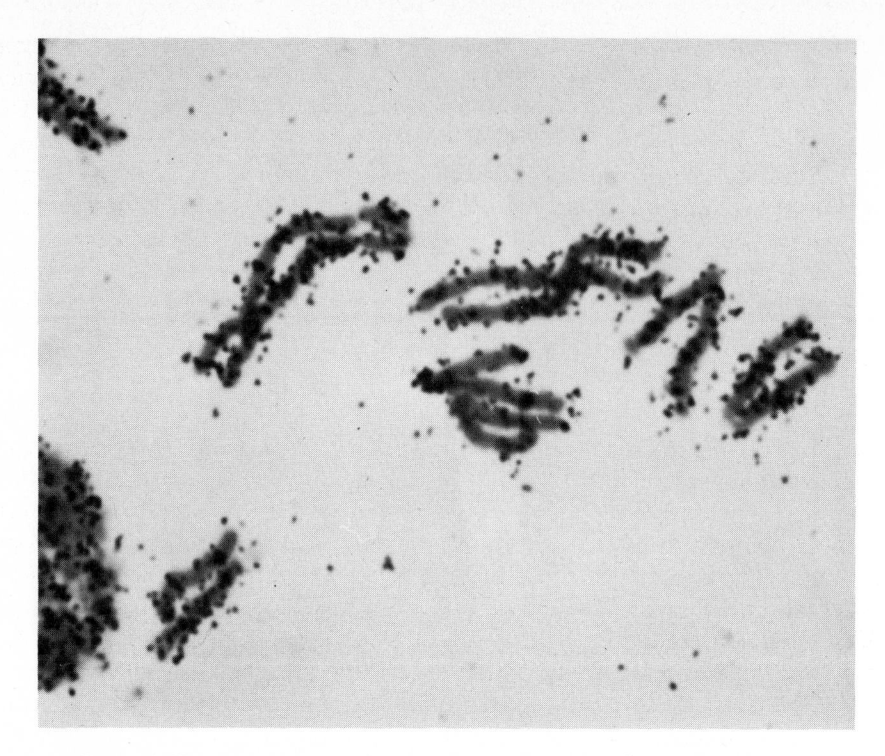

Figure 1-28. Autoradiograms of chromosomes from *Vicia faba* cells which had duplicated in medium containing tritiated thymidine. Presence of radioactive material is seen in the dark spots, and interpreted as representing duplication during interphase. (From J. H. Taylor, 1963. *Molecular Genetics* part I, J. H. Taylor, ed., pp. 74, 75. Academic Press, New York; photograph courtesy J.H. Taylor, Florida State University.)

Division

The separation of the two halves of the cell into independent daughter cells completes the process of mitosis. Each daughter cell has the same number of chromosomes as the parent cell and the same combination of chromosomes.

Genetic significance of mitosis: The production of genetically identical daughter cells following the completion of mitosis stems directly from the fact that no synapsis of homologs occurs in prophase. In the absence of synapsis there is usually no crossing-over between homologous chromatids or genetic exchange resulting in nonidentical sister chromatids. Later we shall discuss some exceptions to this statement, but generally we can consider sister chromatids in mitotically dividing cells to be identical.

Further, without synapsis there is no formation of tetrads, and consequently in metaphase no alignment of maternal and paternal chromosomes on different sides of the plane of division leading to a random assortment of chromosomes in daughter cells. What does occur is a positioning of each chromosome so that at anaphase the centromeres split, and sister chromatids are pulled into different daughter cells. Thus mitosis serves as the mechanism for genetic continuity.

As an aid toward clearly comprehending differences in the distribution of chromosomes during meiosis as contrasted with that during the less complex series of events in mitosis, we again suggest that you focus on the *centromeres*. During anaphase I of meiosis, the centromeres do not split; the sister chromatids remain together. This causes *homologs* to be pulled into different daughter cells in anaphase I. In anaphase of mitosis, the centromeres do split, allowing *sister chromatids* to be pulled into different cells. Figure 1-29 summarizes and contrasts some of the stages of meiosis and mitosis. The second division of meiosis is similar to mitosis in that centromeres of dyads in anaphase II are split apart, and the sister chromatids pulled into different cells.

The Cell Cycle

In studying the behavior of cells undergoing mitosis, biologists discovered that most cells show a cyclic pattern of synthesis and division which is now known as the *cell cycle*. It has been found that following each division, cells enter interphase, which can be said to have three substages or phases itself, with regard to the synthesis of new genetic material. There is first a period in which no synthesis of genetic material can be detected. This period is called G_1, the G referring to a "gap" in synthesis of chromosomes, although the cell is metabolically active otherwise. G_1 is followed by the duplication of chromosomes, which is the S (standing for synthesis) period of interphase. After another gap, G_2, the cell then enters prophase and completes the stages of mitosis to divide into two cells. This last part of the cell cycle is called the M (for mitosis) period. The amount of time that cells spend in the various periods of the cell cycle varies among species, and even among tissues in the same species.

Twinning

Knowledge of the genetic consequences of meiosis and mitosis has led to the understanding of the phenomenon of multiple births in species such as humans, which normally produce one young at each birth event. Single births occur because during gametogenesis in the female, usually only one egg cell reaches maturity per monthly cycle.

Following fertilization, the zygote commences a period of unceasing mitotic activity. Occasionally, for reasons still poorly understood, there occurs a separation of the cells that result from the first divisions. If the separation occurs fairly early in development, the separated clusters of cells can continue development in a normal manner, resulting in two normal individuals. Since mitosis results in genetically identical cells, the two individuals are genetically identical also, and at birth we refer to them as *identical twins*. Geneticists also refer to identical twins as *monozygotic* twins. In cases of identical triplets and more, one must assume that more than one separation of embryonic cells occurred.

There also can occur the phenomenon of multiple births of *nonidentical siblings* (*sibling* refers to all the children of one generation in a family). Nonidentical twins are referred to as *fraternal* or *dizygotic*. This we believe to result from the ovulation of more than one egg cell, and their subsequent fertilization by separate sperms. In recent times, such multiple births have become more common because of

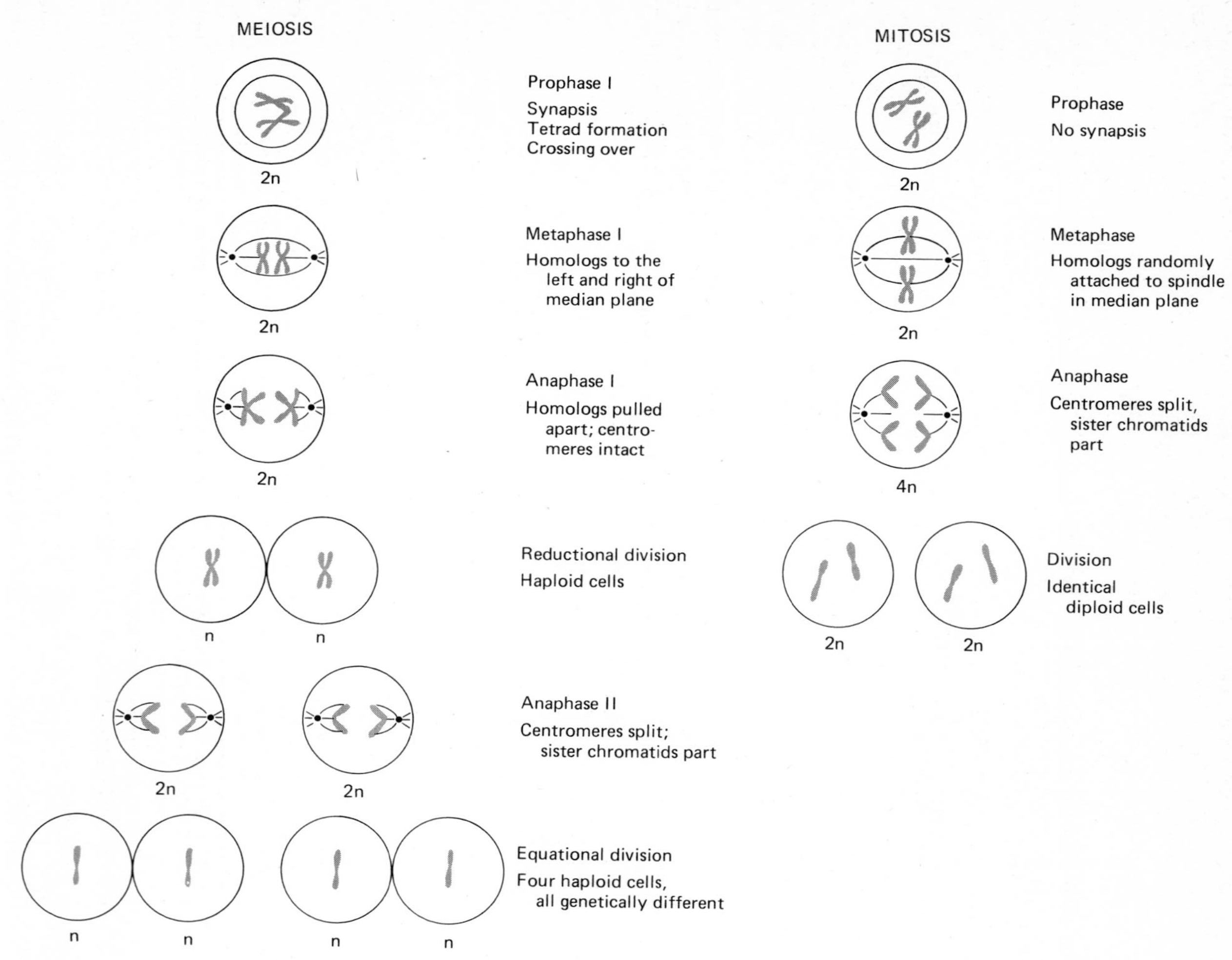

Figure 1-29. Contrasting events in meiosis and mitosis of a cell with one pair of chromosomes.

modern medical treatment of some kinds of infertility. Some women suffer from hormonal irregularities which do not allow the normal process of oogenesis to be completed. Hormonal treatment of such patients sometimes results in a physiological imbalance to the other extreme, so that more than one egg cell is ovulated and fertilized. Since different gametes are involved here, the zygotes are genetically different also, just as are siblings conceived and born in different years.

PROBLEMS

1. Using different colors to distinguish maternal from paternal chromosomes, draw cells from an individual whose diploid number is 6 in all stages of mitosis and meiosis. How many different combinations of chromosomes can you expect to find in the mature gametes when meiosis is completed, assuming no crossing-over? How many different combinations of chromosomes can you expect to find in the daughter cells following the completion of mitosis?

2. The diploid number of chromosomes in a human cell is 46. How many chromosomes and chromatids would you expect to find in cells at the following stages: (a) primary spermatocyte (b) spermatid, (c) secondary oocyte, (d) first polar body, (e) anaphase I, (f) telophase II, (g) prophase II, (h) prophase of mitosis, (i) prophase I, (j) telophase of mitosis?

3. Assume that the following diagrams are of cells of sexually reproducing organisms seen under the light microscope at different stages of division. Give for each the stage you think the diagram depicts, whether it is mitosis and/or meiosis, and the diploid number of the cell:

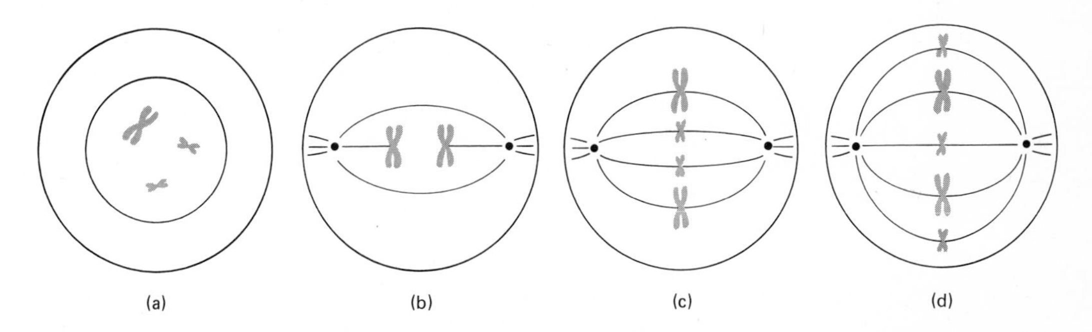

4. Down's syndrome is frequently the result of an extra chromosome #21, giving the patient a chromosome number of 47. Diagram what you think would be the result of meiosis in the germ cells of such an individual. If the germ cells were fertilized by normal gametes, what ratio of normal and abnormal offspring would you expect to find?

5. Define the following terms: tetrad, bivalent, homologs, dyad, alleles.

6. What is the genetic significance of synapsis?

7. In view of the mechanics of meiosis, why are hybrid offspring of some interspecies crosses, such as the mule (from donkey × horse), sterile? Such

interspecific hybrids are rare in nature. What reasons can you think of to explain why crosses of individuals from different species are unsuccessful, even though fertilization may be achieved?

8. Theodor Boveri's experiments mentioned on page 2 involved the fertilization of sea urchin ova by two instead of the usual single sperm. This can be accomplished by adding a large excess of sperm to the eggs in seawater. A large percentage of the dispermic zygotes showed abnormal development. How might dispermic fertilization interfere with normal cell division? With normal development?

9. What advantages can you think of for the survival of the species that sexually reproducing organisms hold over asexually reproducing organisms (which simply divide mitotically)?

10. Why are identical twins truly genetically identical, and nonidentical twins genetically different? The Dionne quintuplets were found to be identical. How would you explain this phenomenon?

11. *Parthenogenesis* is the process of development of an unfertilized egg. Yet parthenogenetic rabbits are found to be diploid. How might this occur? What can you say about the genetic constitutions of germs cells from such a rabbit?

REFERENCES

BOVERI, T., 1902. "On Multipolar Mitosis as a Means of Analysis of the Cell Nucleus." Translated and reprinted in B. H. Willier, and J. M. Oppenheimer, eds., 1964. *Foundations of Experimental Embryology.* Prentice-Hall, Inc., Englewood Cliffs, N.J.

HAMERTON, J., 1971. *Human Cytogenetics.* Vol. II, p. 201. Academic Press, New York.

MENDEL, G., 1865. "Experiments in Plant Hybridization." Translated in J. Peters, ed., 1959. *Classic Papers in Genetics.* Prentice-Hall, Inc., Englewood Cliffs, N.J.

SALMON, E. D., 1975. "Spindle Microtubules: Thermodynamics of in vivo Assembly and Role in Chromosome Movement." *Ann. N.Y. Acad. Sci.* 253.

SUTTON, W. S., 1903. "The Chromosomes in Heredity." Biol. Bulletin 4. Reprinted in M. L. Gabriel, and S. Fogel, eds., 1955. *Great Experiments in Biology.* Prentice-Hall, Inc., Englewood Cliffs, N.J.

TAYLOR, J. H., P. S. WOODS, AND W. L. HUGHES, 1957. "The Organization and Duplication of Chromosomes as Revealed by Autoradiography Studies Using Tritium-labeled Thymidine." Proc. Nat. Acad. Sci. U.S. 43: 122–128.

VILLEE, C. A., and V. C. DETHIER, 1971. *Biological Principles and Processes,* p. 571. Saunders, Philadelphia.

For a thorough presentation of scientific thought of the late nineteenth- and early twentieth-century decades in biology, turn to the following classic work which includes references to original papers, many of which, however, are in German and French.

WILSON, E. B., 1925. *The Cell in Development and Heredity.* Macmillan., New York.

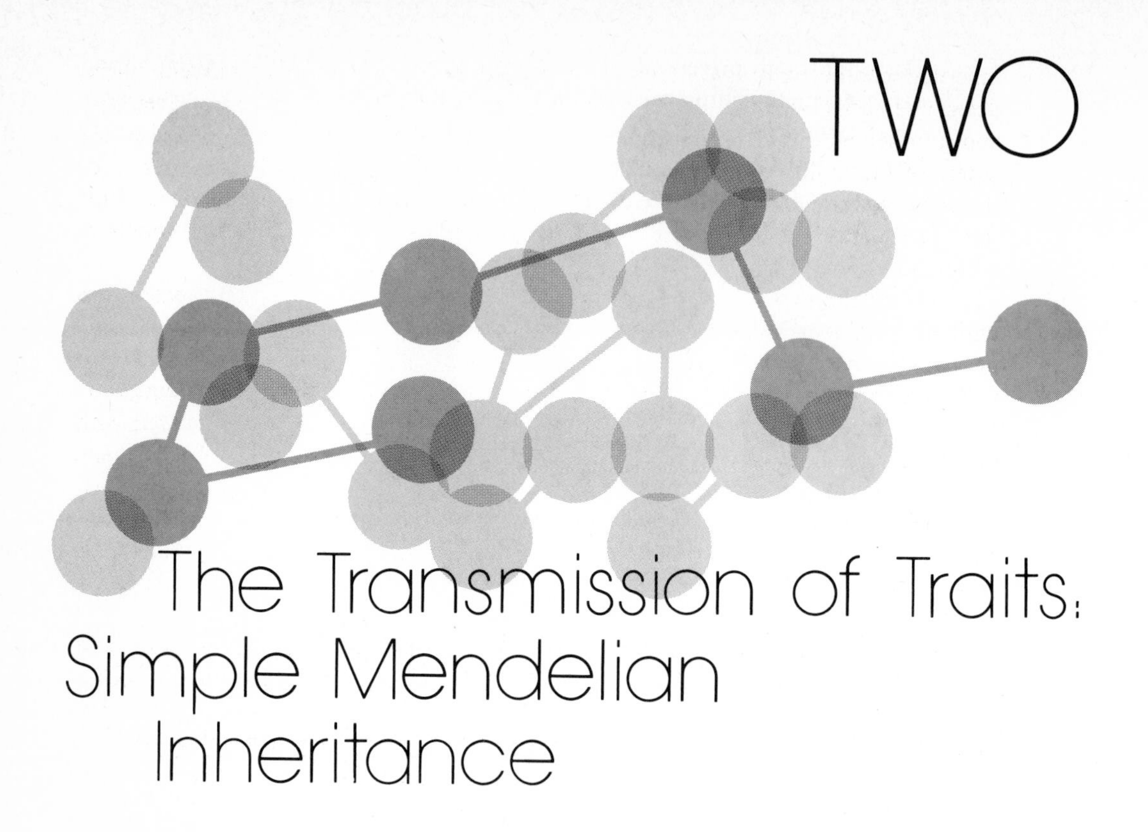

TWO

The Transmission of Traits: Simple Mendelian Inheritance

MENDELIAN GENETICS

Prior to the mid-nineteenth century, even great scientists were unable to grasp the basic concepts of heredity because they believed that the transmission of traits was a vague process at best, involving the haphazard mixing of undefined elements in the blood during sexual reproduction. Indeed we still retain in our speech idioms which refer to blood heredity, such as "blood brother," "bad blood," and so forth. The establishment of cell division as the physical basis for the transmission of traits, together with the rediscovery of the Mendelian laws, signaled the beginning of the science of genetics.

The great contribution to genetics by Mendel and by the cytologists and embryologists that we have mentioned, lay in their hypothesizing, and later verifying scientifically, that in fact the transmission of traits is based on discrete, definitive units, and that the process is consistent to the point of being predictable. Actually, modern geneticists analyze the transmission of inherited traits by essentially the same method as that detailed by Mendel in his papers published in 1865.

Homozygosity and Heterozygosity

The inheritance of stem length in pea plants, a characteristic which Mendel studied, serves as a good illustration of the transmission of a simple trait. By considering various crosses involving this trait we can gain some insight into the transmission of inherited characteristics. In Chapter One we indicated that two different alleles determine variations of this trait, one resulting in tall plants (about 6 to 7 feet tall), and another in short plants (about $1\frac{1}{2}$ to 2 feet tall). Traditionally, geneticists use letters of the alphabet to designate genes. For this trait we can use the letter *T* for the allele determining tallness, and *t* for the allele determining shortness. Use of the same letter specifically connotes allelism, the capital letter signifying the dominant allele and the lower-case letter the recessive allele.

Mendel bred the plants which he used in his experiments for many generations until he obtained "true-breeding" plants, meaning plants that would produce only progeny uniform for a given characteristic (such as tallness), whether they were self-pollinated or cross-pollinated with other plants true-breeding for that characteristic. Similarly, true-breeding short plants, when self-pollinated or cross-pollinated with other true-breeding short plants, would produce only short offspring. We can now explain this phenomenon of true-breeding by assuming that these plants possessed two identical alleles for stem length. The true-breeding tall plants can thus be designated *TT*; the true-breeding short plants, *tt*. In modern terminology, such individuals are referred to as *homozygotes*; they are said to be *homozygous* for the allele under consideration.

Germ cells from homozygotes would, upon completion of meiosis, yield gametes carrying the same allele for that trait; for example, all gametes of *TT* homozygotes would carry the allele *T*. Crosses of such plants would produce progeny that are also homozygous, that is, all *TT*, since each parent would contribute an identical allele, *T*. Hence, they are true-breeders (Fig. 2-1). The same explanation, of course, holds for the true-breeding short plants.

What would be the result of a cross between a tall homozygote and a short homozygote? We designate this cross in the following manner: $TT \times tt$. Since all the gametes of the *TT* parent would carry the dominant allele, and all the gametes of the *tt* parent would carry the recessive allele, the offspring of this cross must inherit a combination of the two genes and therefore would be *Tt*. (Fig. 2-2). (By tradition, geneticists write the dominant allele first.) Individuals forming a pair of different alleles determining a particular trait are said to be *heterozygotes*, or *heterozygous* for that pair of genes. (Another frequently used term which is synonymous with heterozygote is *hybrid*. An individual heterozygous for one particular pair of genes is called a *monohybrid*; an individual heterozygous for 2 pairs of genes, a *dihybrid*, etc.)

Law of Segregation

The fact that an inherited trait is determined by a *pair* of genes in every generation led Mendel to the logical deduction that each parent can transmit only *one* of each pair to its offspring. If each parent transmitted both members of each pair of genes to the next generation, each offspring would have four genes determining the trait, and in each succeeding generation the number of genes would increase geometrically.

The only way to explain how the number of genes determining traits remains constant from one generation to another is to assume that the members of a pair of alleles are transmitted *separately*. This is known as Mendel's *Law of Segregation*. The

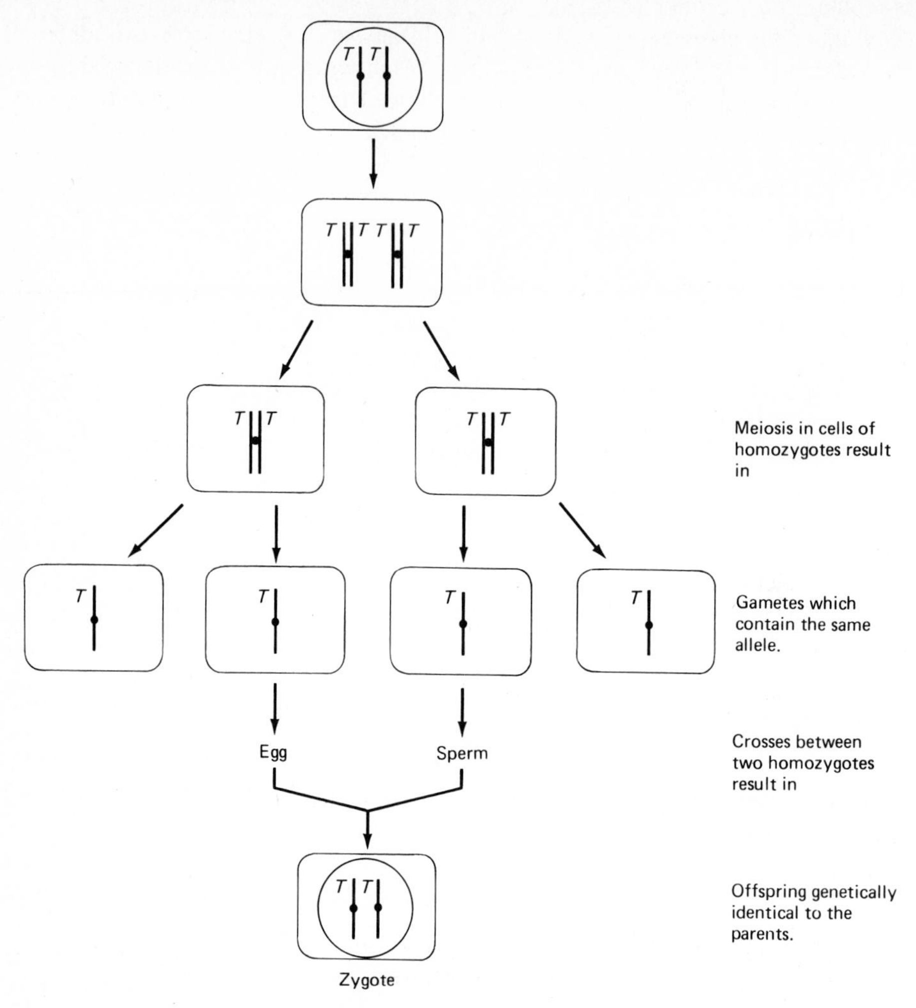

Figure 2-1. The genetic basis for true-breeding in Mendel's pea plants: homozygosity. Note that not all stages of meiosis have been included.

physical basis for this law, as we have discussed previously, is the movement of homologs into different daughter cells beginning at Anaphase I of meiosis (p. 14).

Dominance and Recessiveness

In his extensive studies, Mendel found that plants heterozygous for the genes determining stem length were as tall as the plants homozygous for *T*. He therefore concluded, as we mentioned in Chapter 1, that some genes are dominant and others recessive. THE CONCEPT OF DOMINANCE AND RECESSIVENESS UNDERLIES A VERY IMPORTANT PRINCIPLE IN THE ANALYSIS OF INHERITED TRAITS, NAMELY, THAT TWO INDIVIDUALS IDENTICAL IN APPEARANCE WITH REGARD TO A PARTICULAR TRAIT, MAY BE QUITE DIFFERENT IN THEIR GENETIC CONSTITUTIONS. GENETICISTS USE TWO TERMS TO MAKE THIS

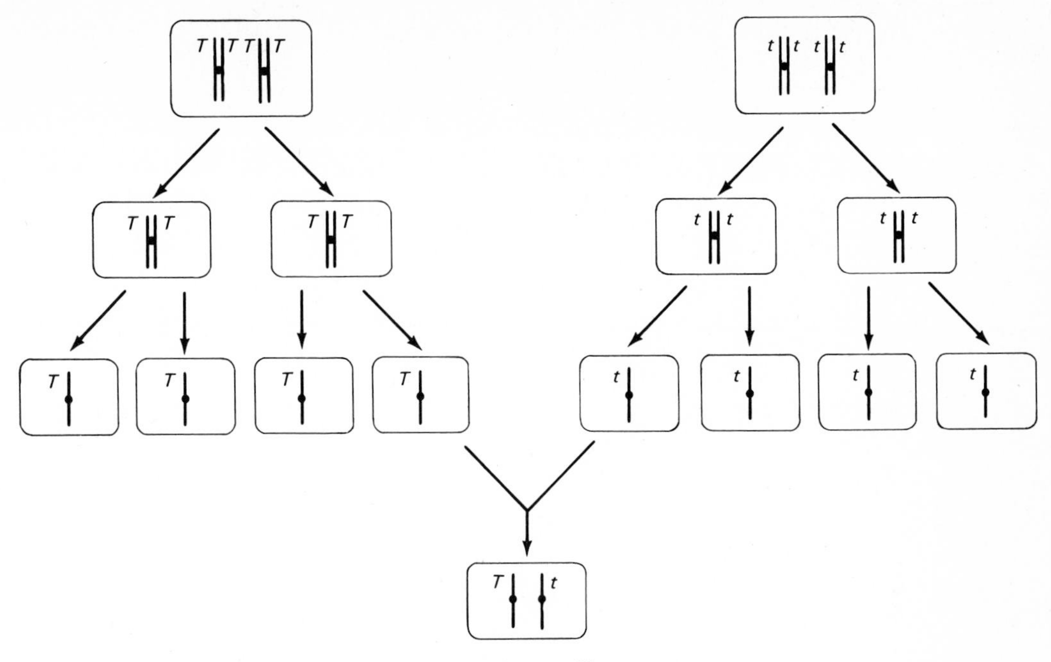

Figure 2-2. A cross between homozygous dominant plants and homozygous recessive plants results in heterozygous progeny.

ESSENTIAL DISTINCTION BETWEEN APPEARANCE AND GENETIC CONSTITUTION: PHENOTYPE REFERS TO APPEARANCE; GENOTYPE, TO GENETIC CONSTITUTION.

The Testcross

It is essential to distinguish phenotype and genotype in any genetic discussion because the genotype determines not only the individual's appearance, but also what genes can be transmitted to the progeny of that individual. For example, a *TT* homozygote will produce all tall offspring regardless of the genotype of the plant to which it is crossed since all the progeny will inherit at least one *T* allele. This is of course not the case for heterozygotes.

Consider the following cross: *Tt* × *tt*. The Law of Segregation predicts that half the gametes produced by the heterozygote carry the *T* allele and half the *t* allele. All the gametes of the short homozygote carry the *t* gene. Assuming random fertilization, we can expect that the progeny will have two genotypes, *Tt* and *tt*, in equal numbers (Fig. 2-3); the genotypic ratio is then 1 *Tt*: 1 *tt*, and the phenotypic ratio would be the same as the genotypic ratio in this cross. This type of cross is used frequently by geneticists to determine whether an individual with a dominant phenotype is homozygous or heterozygous, for appearance of the recessive phenotype in the progeny would prove the individual to be heterozygous. Such a cross, between an individual of undetermined genotype and a homozygous recessive individual, is termed a *testcross*. Multiple testcrosses are also used to determine the genotypes of

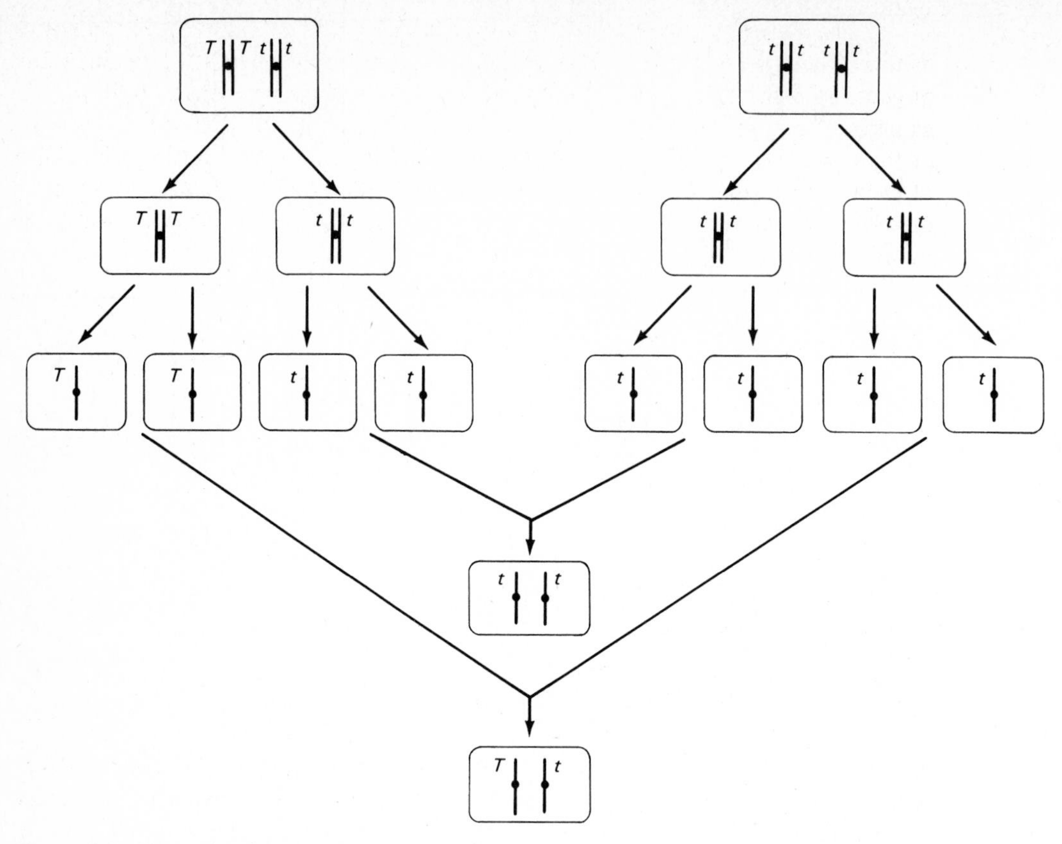

Figure 2-3. The results of a testcross.

organisms expressing two or three or more dominant traits; they are crossed to organisms homozygous recessive for each trait.

The Monohybrid Cross

Let us now consider a cross between two monohybrids, symbolized by $Tt \times Tt$. Each parent will produce two types of gametes. Half will contain the dominant alleles and half the recessive allele. The randomness of fertilization dictates that a *T* gamete (a gamete carrying the *T* gene) from one parent has an equal chance of fertilizing a *T* gamete or *t* gamete from the other parent; the same holds for the *t* gamete. This results in four possible combinations of genes in the offspring, which can be illustrated as follows:

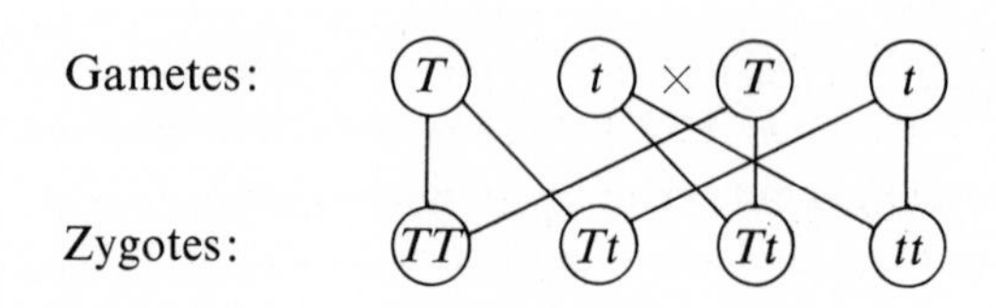

The Punnett Square. The classical geneticist R. C. Punnett devised another method, which we now call the *Punnett Square*, of analyzing the possible combinations of genes in zygotes. Using the Punnett Square, a cross is represented as follows: The gametes of one parent are written along the top line of the square, and the gametes of the other parent are written usually down the left side of the square. The alleles of each parental gamete are written in the appropriate boxes, which then represent the offspring genotypes. The crosses we have previously discussed may be written in Punnett Squares as follows:

$TT \times tt$

	T	*T*
t	*Tt*	*Tt*
t	*Tt*	*Tt*

Cross between two different homozygotes

$Tt \times Tt$

	T	*t*
T	*TT*	*Tt*
t	*Tt*	*tt*

Monohybrid cross

Genetic Ratios. From the Punnett Square for the monohybrid cross, we see that the phenotypic ratio found in the offspring of a monohybrid cross will approximate a ratio of 3 tall : 1 short, whereas the genotypic ratio should be 1 *TT* : 2*Tt* : 1 *tt*. This was the ratio obtained by Mendel in his experiments—not just for stem length, but for a total of seven different traits. For example, he found that the production of round seeds is a dominant trait, and that the recessive plants produced wrinkled seeds. A cross of monohybrids for this pair of alleles yielded a 3 round : 1 wrinkled phenotypic ratio after two generations, the same as that observed in the monohybrid cross for stem length. It was the continual appearance of this 3 : 1 ratio in his breeding experiments that led Mendel to assume correctly that he was dealing with pairs of units that segregate and unite at random through the processes of sexual reproduction (Table 2-1.)

Mendel also proposed some terminology which we still use. The first cross made in an experiment is referred to as the *parental*, or *P*, *generation.* The offspring of this cross are designated as the F_1 generation ($F =$ filial). Offspring of the F_1 are the F_2 generation; the next generation is the F_3; and so on.

The Dihybrid Cross

As we discussed the crosses involving one pair of alleles, it probably occurred to you, as it did to Mendel, that a multicellular organism certainly possesses more than one inherited trait. Can the transmission of *combinations* of traits be studied in a manner similar to that of individual traits? Mendel pursued the answer to this question by breeding plants that were homozygous for genes determining different traits.

As mentioned before, tall stem length is dominant over short stem length, and round seed is dominant over wrinkled seed. We can use a cross between plants homozygous dominant for tallness and round seed and plants homozygous recessive

Table 2-1 Experiments by Mendel on seven traits in the garden pea plant. Note the approximate 3 : 1 ratio found in the F_2 generation in each experiment.

Trait	*F_1 phenotype*	*F_2*	*Ratio*
Axial/Terminal flowers	All axial	651 axial 207 terminal	3.14 : 1
Inflated/Constricted pods	All inflated	882 inflated 299 constricted	2.95 : 1
Yellow/Green cotyledons	All yellow	6022 yellow 2001 green	3.01 : 1
Round/Wrinkled seeds	All round	5474 round 1850 wrinkled	2.96 : 1
Colored/White flowers	All colored	705 colored 224 white	3.15 : 1
Green/Yellow pods	All green	428 green 152 yellow	2.82 : 1
Long/Short stems	All long	787 long 277 short	2.84 : 1

for shortness and wrinkled seed to illustrate the analysis of the inheritance of two different traits. The cross is symbolized as follows: $TTWW \times ttww$. Figure 2-4 depicts the results of meiosis in germ cells of these two parental types.

Two pairs of chromosomes are involved, one pair carrying the genes for stem length, the other the genes for seed structure. As you can see, all germ cells of the dominant parental type contain the genes *TW*; all gametes of the other parent *tw*. Combination of these genes by fertilization yields zygotes with the genotype *TtWw*; individuals thus heterozygous for two pairs of genes are called *dihybrids*. Mendel found that all such F_1 generation dihybrids resemble the dominant parent, indicating that traits which are dominant when studied individually remain dominant when studied in combination with other traits. This observation is generally true, although other genetic phenomena do exist that can mask the expression of dominant genes (see Chapter 4). Mendel did not encounter any of these phenomena, however, and it was fortunate for genetics that he did not, as such factors would have rendered his analysis difficult, if not impossible.

Law of Random Assortment

A cross of dihybrids produces a somewhat more complicated array of genotypes and phenotypes of offspring. Figure 2-5 depicts the results of meiosis in gametes of cells heterozygous for two different pairs of genes located on different pairs of chromosomes. As a result of the random alignment of tetrads during metaphase I (p. 11) and the segregation of homologs at anaphase I, reductional division results in every possible combination of chromosomes (and therefore genes) in the gametes. We would expect the four possible germ cells of the dihybrid to occur in equal numbers. The gene combinations in the gametes are: *TW*, *tw*, *Tw*, *tW*. Since fertilization is random, the combinations of genes and traits in the offspring will also be random.

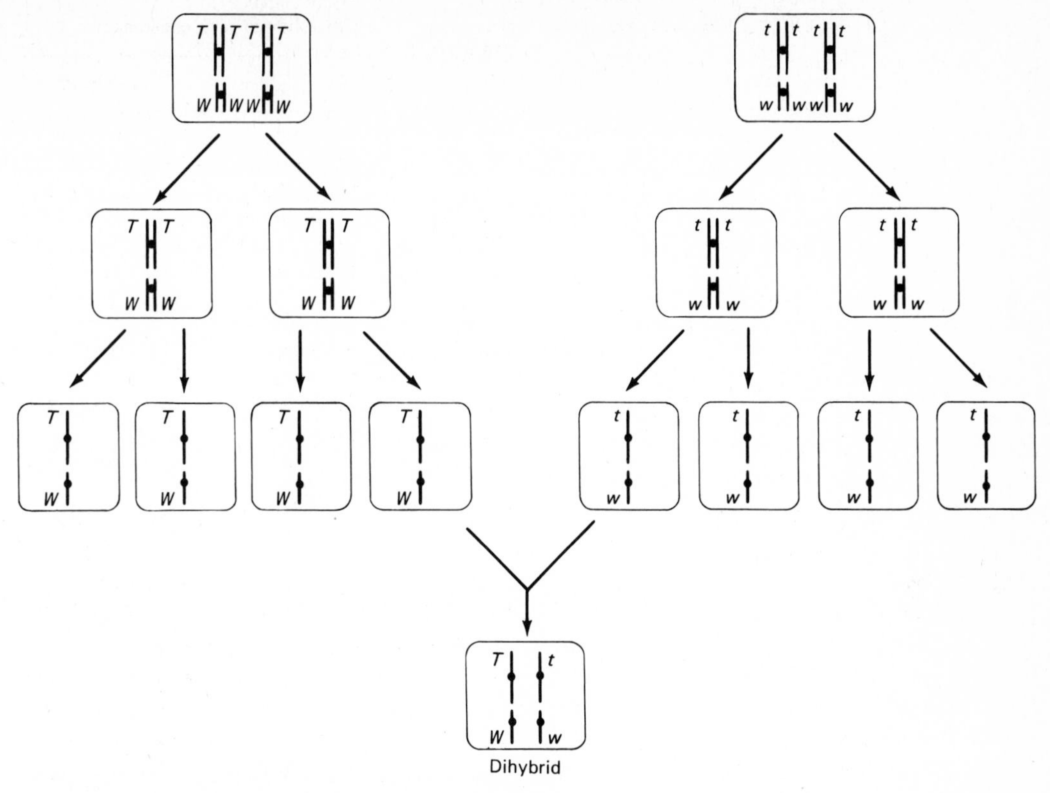

Figure 2-4. A cross between an individual homozygous dominant for two pairs of genes and an individual homozygous recessive for two pairs of genes produces a dihybrid, an individual heterozygous for both pairs of genes.

We refer to this as *random,* or *independent, assortment.* A Punnett Square of this cross is

	TW	*tw*	*Tw*	*tW*
TW	*TTWW*	*TtWw*	*TTWw*	*TtWW*
tw	*TtWw*	*ttww*	*Ttww*	*ttWw*
Tw	*TTWw*	*Ttww*	*TTww*	*TtWw*
tW	*TtWW*	*ttWw*	*TtWw*	*ttWW*

The phenotypic ratio of the offspring is 9 tall/smooth : 3 tall/wrinkled : 3 short/smooth : 1 short/wrinkled. This 9 : 3 : 3 : 1 phenotypic ratio is classically associated with the offspring of dihybrid crosses. Again, Mendel found the same ratio in

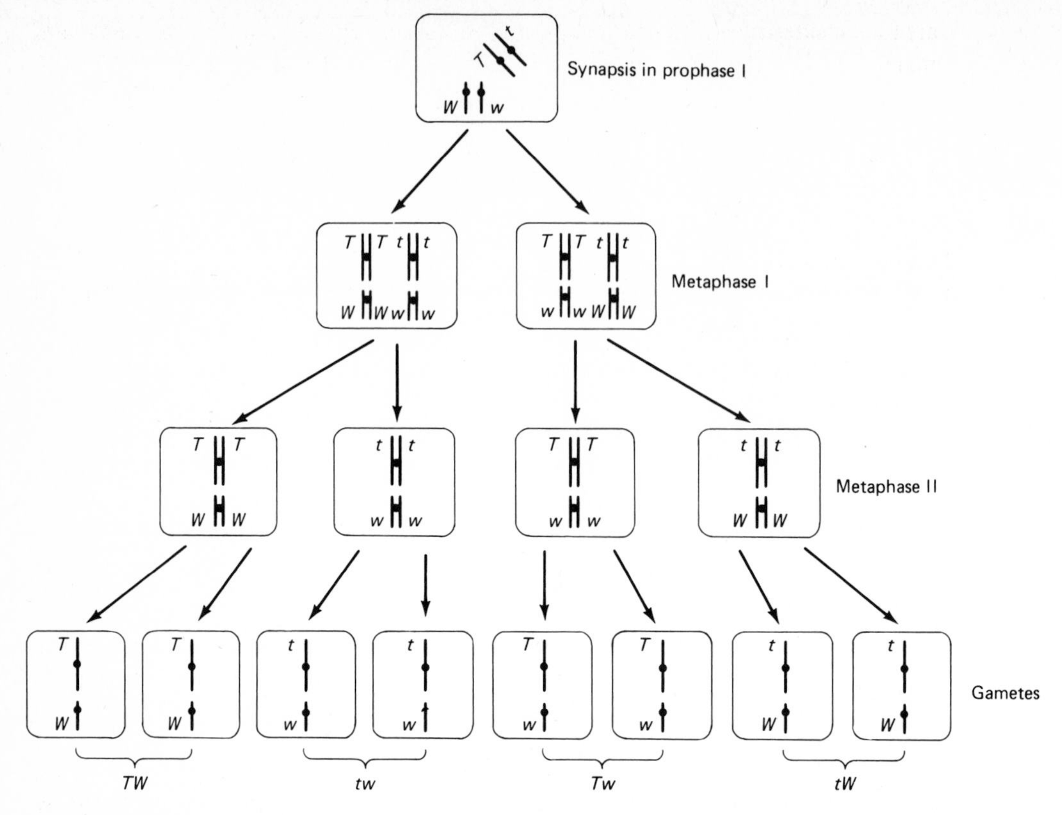

Figure 2-5. Meiosis in cells of a dihybrid results in four different combinations of genes in the gametes.

his data for dihybrid crosses regardless of which two traits he combined in his experimental plants.

Note that although the combined phenotypic ratio is 9: 3: 3: 1, each trait is still observed in a 3: 1 ratio expected of monohybrid crosses. Again, this is evidence that the transmission of the alleles of one pair of genes is independent of the transmission of the other pair.

Forked-line Method (Method of Tree Diagrams)

The Punnett Square has already been presented as a method for determining the possible combinations of genes in zygotes of certain genetic crosses. Another simpler, less tedious method which allows us to determine both genotypic and phenotypic proportions in zygotes is the *forked-line method.* This method uses a simple diagram that has been likened to the shape of a tree (albeit a tree on its side!), hence its designation by statisticians as a *tree diagram.*

Suppose we wish to determine how many possible classifications of people must be considered if we are interested in studying people by sex (male, female), eye color (blue, brown, green, hazel), and hair color (blond, red, brown, black). We start by considering all human beings, and we first categorize them by sex (male and female):

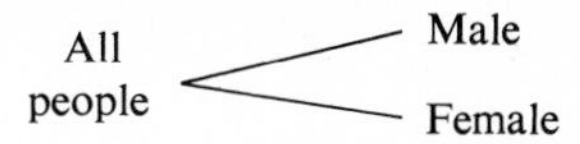

Then we subdivide each of the sexes by eye color into blue, brown, green, and hazel, producing the following "tree," or forked-line, diagram given in Fig. 2-6. We

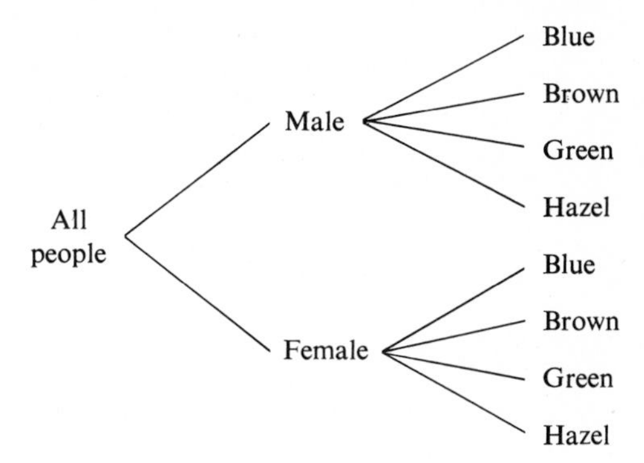

Figure 2-6

see that if we follow branches in the tree, the first branch is males with blue eyes, the second branch is males with brown eyes, and so on. Furthermore, using the same approach we can subdivide the blue-eyed males by hair color and the blue-eyed females by hair color and similarly subdivide all the other classifications, finally obtaining the tree diagram in Fig. 2-7. It now becomes clear that we have subdivided people into 32 groups and that each of the branches corresponds to one of the 32 classifications. The top branch corresponds to blond-haired, blue-eyed males, the last branch corresponds to black-haired, hazel-eyed females, and so on. This method is a general one, and can very easily be applied to many different genetic crosses.

In monohybrid crosses. Consider the cross of two monohybrid tall plants, that is, $Tt \times Tt$. Using the forked-line m thod, we represent all possible alleles from one hybrid tall plant, T and t, at the first stage of the tree, and those from the second plant, T and t, at the second stage of the tree. Since it is possible for the T allele from plant #1 to be transmitted to an offspring with either the T or the t allele from plant #2, and similarly for the t allele from plant #1 to be transmitted with either the T or the t allele from plant #2, we obtain the diagram given in Fig. 2-8. If we follow the first branch, we see that the zygote receives the T alleles from both parents and is therefore a homozygous (TT) tall plant. The second and third branches result in heterozygous tall plants (Tt), and lastly the bottom branch yields homozygous short plants (tt). Since we assume that the T alleles and the t alleles are transmitted in equal numbers as a result of meiosis, we can actually determine the genotypic ratios directly from the tree diagram. The genotypic results of a monohybrid cross $Tt \times Tt$ are thus seen to be $1\,TT : 2\,Tt : 1\,tt$. From this tree diagram we can also

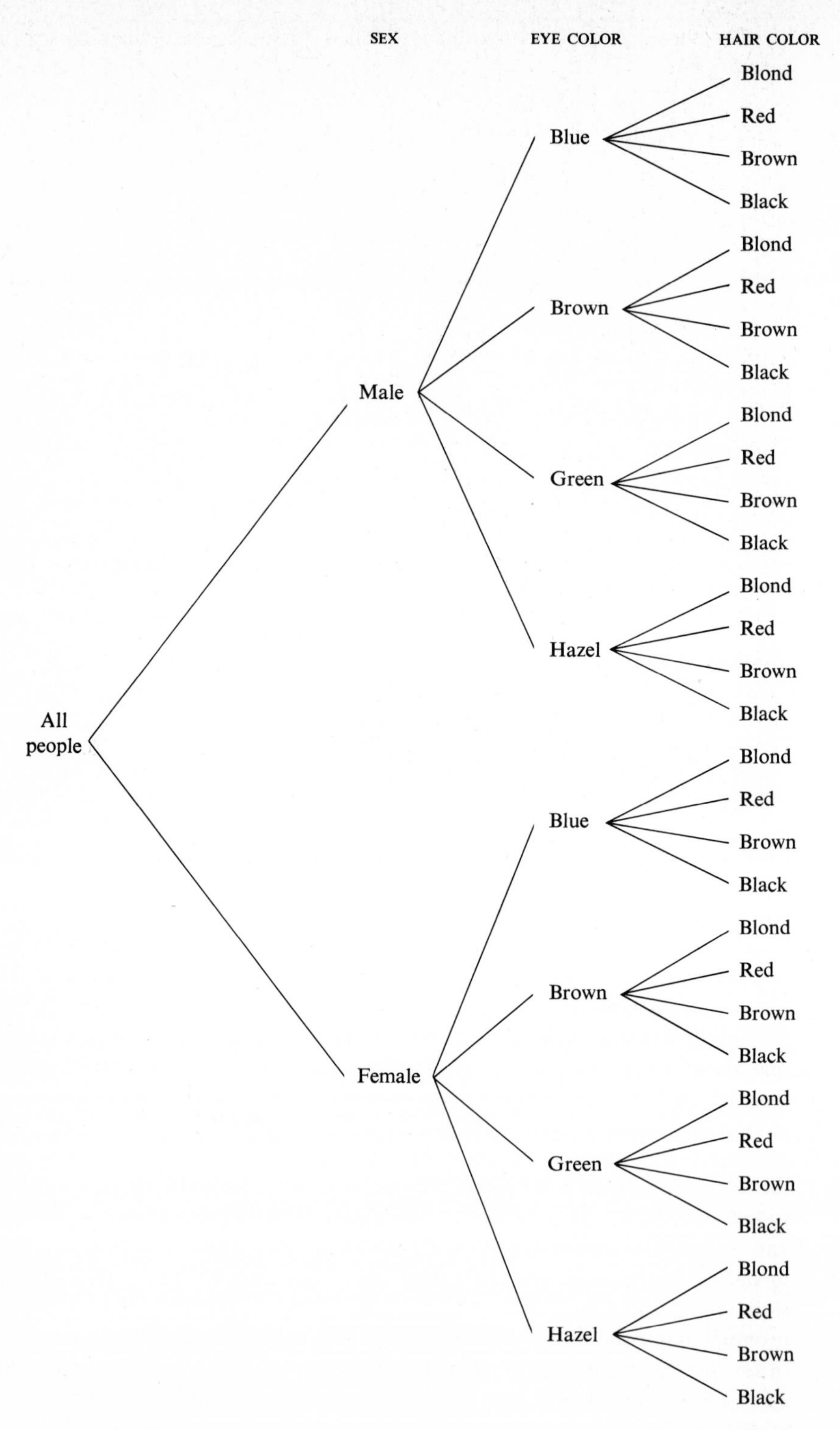

Figure 2-7. Forked-lined (tree diagram) for classification by sex, eye color, and hair color.

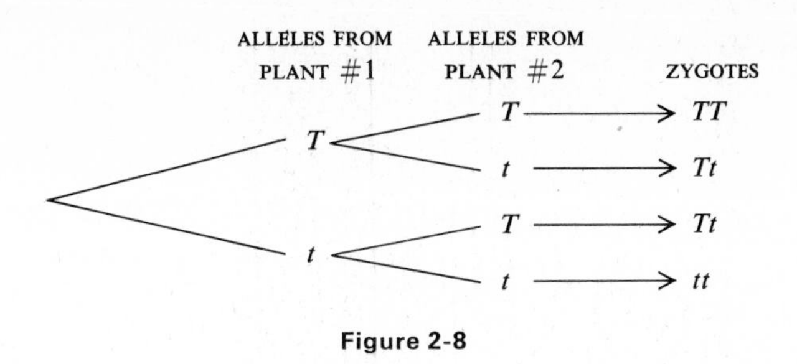

Figure 2-8

obtain the phenotypic outcomes directly. Since it is assumed that T is completely dominant, phenotypically we obtain 3 tall : 1 short, or equivalently we can say that the probability of a tall plant is $\frac{3}{4}$ and the probability of a short plant is $\frac{1}{4}$.

In dihybrid crosses. This same method can be used to determine phenotypic and genotypic ratios for dihybrid, trihybrid, or even more complex crosses. Using the Punnett Square for a dihybrid cross $TtWw \times TtWw$, we are required to enumerate all possible gametes (see page 39), that is, TW, Tw, tW, tw for both plants (where T is dominant tall, t recessive short, W dominant round, and w recessive wrinkled). However, using the forked-line method and the Law of Random Assortment (assuming these alleles for height and seed coat are on different chromosomes), we can break this cross down into two simpler problems and then combine them. The Law of Random Assortment tells us that we can consider the results of the $Tt \times Tt$ and the $Ww \times Ww$ matings separately and then combine the two results. That is, we can consider each of the outcomes of the $Tt \times Tt$ cross, and then, using the forked-line method, consider that they can combine with each of the possible offspring of the $Ww \times Ww$ mating. Schematically, we get:

Each distinct offspring of $Tt \times Tt$	combines with	Each distinct offspring of $Ww \times Ww$

On page 37 we derived the genotypic results of the mating $Tt \times Tt$; the genotypic results of $Ww \times Ww$ would be similar (1 WW : 2 Ww : 1 ww). Combining these two results, we get the genotypic results illustrated in Fig. 2-9. The relative numbers of the different genotypes of the zygotes can be calculated quite easily by considering each trial separately. As we have established, each monohybrid cross yields a genotypic ratio of 1 : 2 : 1. If we incorporate these ratios into Fig. 2-9, we get Fig. 2-10. In this figure we have determined the ratios in the progeny of the dihybrid cross, by multiplying the numbers along the branches to get the numbers for the combinations. That is, if there are 2 Tt and they combine with 2 Ww, they will result in 4 $TtWw$; similarly if the 2 Tt combine with 1 WW, the result is 2 $TtWW$.

We of course obtain the same relative numbers of each genotype as those given by the Punnett Square (see page 39), but with much less effort. If we are interested only in the phenotype, the power of the forked-line method becomes even more obvious. We can look at the phenotypic proportions of each trait separately and then combine them into a single tree diagram, as in Fig. 2-11, yielding the typical 9 : 3 : 3 : 1 ratio of a dihybrid cross.

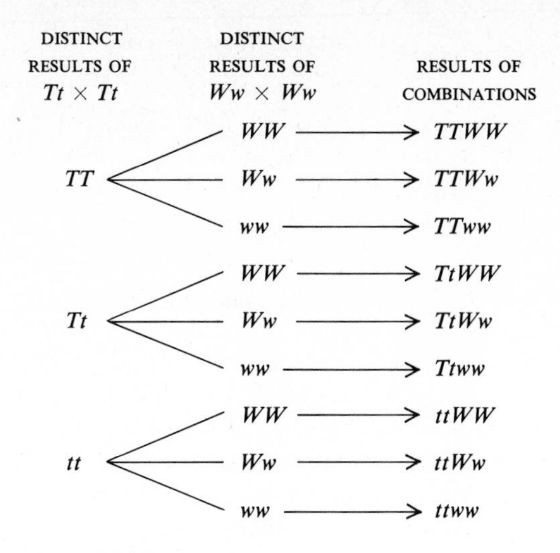

Figure 2-9. Possible genotypes for dihybrid cross *TtWw* × *TtWw*.

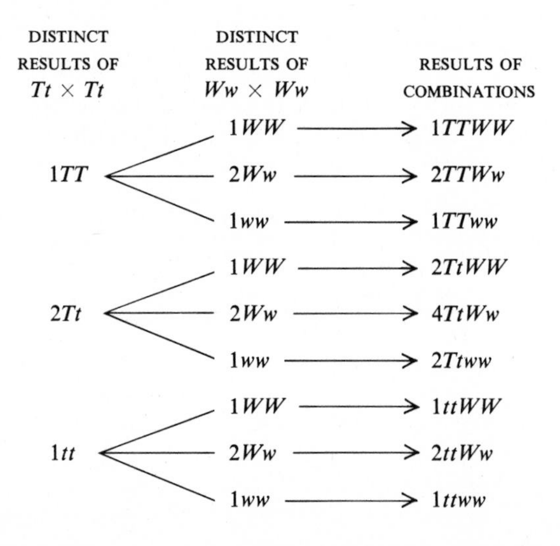

Figure 2-10. Relative number of possible genotypes for dihybrid cross *TtWw* × *TtWw*.

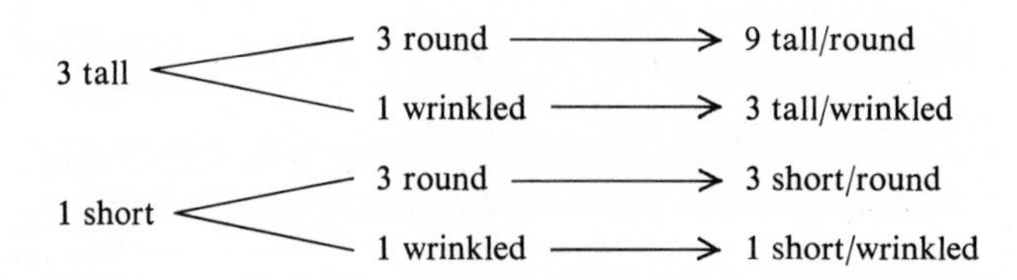

Figure 2-11. Relative number of each phenotype for dihybrid cross *TtWw* × *TtWw*.

In trihybrid crosses. The reader might wish to apply the same method to determine the genotypic and phenotypic ratios in a trihybrid cross (see Exercise 12).

Value of the forked-line method. Using the forked-line method for a trihybrid cross is somewhat cumbersome, but it is still much simpler than the Punnett Square, which requires the enumeration and then the mating of all possible gametes. For n hybrid traits (with random assortment), this means 2^n traits and therefore a square of size

$$\underset{\text{Possible gametes from male}}{2^n} \times \underset{\text{Possible gametes from female}}{2^n} = 2^{2n} = 4^n.$$

For example, in a trihybrid, Fig. 2-12 shows the different alignments of tetrads at metaphase I, resulting in eight ($2^3 = 8$) different gametes; a Punnett Square for a

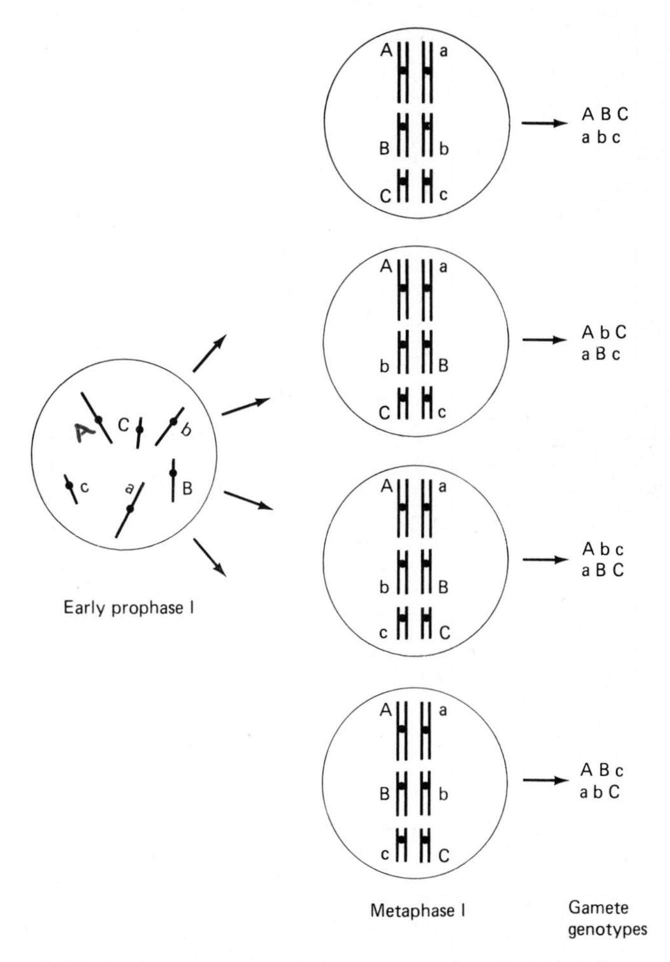

Figure 2-12. Random assortment of chromosomes of a trihybrid during gametogenesis.

trihybrid cross would therefore contain 8×8 ($2^3 \times 2^3 = 2^{2(3)} = 4^3$), or 64, progeny squares. To obtain the phenotypic proportions using the Punnett Square would require studying all 64 squares, an exorbitant amount of work.

The forked-line method uses the Law of Random Assortment and therefore requires knowledge only of the offspring of matings of individuals bearing a single trait. Genotypically, if there are n traits in an n-hybrid cross, this means

$$3 \times 3 \times \ldots \times 3 = 3^n$$

possible genotypes (branches). The number 3 here represents the three possible genotypes of the zygotes for each trait, that is, homozygous dominant, heterozygous, and homozygous recessive. Phenotypically, if there is complete dominance, using the forked-line method we have to consider only

$$\underset{\substack{\text{1st}\\\text{trait}}}{2} \times \underset{\substack{\text{2nd}\\\text{trait}}}{2} \ldots \times \underset{\substack{n^{\text{th}}\\\text{trait}}}{2} = 2^n$$

possible lines. The number 2 here represents the two possible phenotypes for each trait, that is, the dominant or the recessive. Obviously, since 2^n and 3^n are much smaller numbers than 4^n, the amount of work for the geneticist is considerably reduced by the forked-line method. [These results are summarized in Table 2-2, part (a).]

Table 2-2

(a)			(b)		
Number of hybrid gene pairs	*Number of distinct geno-types*	*Number of distinct pheno-types*	*Number of pairs of chromo-somes*	*Number of possible gametes*	*Number of possible zygotes*
1	$3^1 = 3$	$2^1 = 2$	1	2^1	$2^1 \times 2^1 = 4$
2	$3^2 = 9$	$2^2 = 4$	2	2^2	$2^2 \times 2^2 = 2^4$
3	$3^3 = 27$	$2^3 = 8$	3	2^3	$2^3 \times 2^3 = 2^6$
⋮	⋮	⋮	⋮	⋮	⋮
n	3^n	2^n	n	2^n	$2^n \times 2^n = 2^{2n}$

Multiplication Rule

The general method of counting genotypic classes ($3 \times 3 \times \ldots \times 3 = 3^n$) and phenotypic classes ($2 \times 2 \times \ldots \times 2 = 2^n$) involves the so-called *multiplication rule*: If an event occurs in n_1 ways, and a totally independent event can occur in n_2 ways, the number of ways both events can occur at the same time is

$$n_1 \times n_2$$

We are able to apply the multiplication rule in studying the results of dihybrid and trihybrid crosses. If the transmission of the alleles for one trait is considered one event, and the transmission of the alleles for another trait is considered a second event, then because of the Law of Random Assortment the two events can be considered independent. In the trihybrid cross, the transmission of the third pair of alleles would be a third independent event.

We shall consider here one more application of the multiplication rule which is of interest to the geneticist. One observation, well-known to scientists and laymen alike, which historically has caused confusion regarding the transmission of inherited traits is that great variability exists among the offspring of complex organisms such as human beings. Every individual possesses the characteristics of its species, of course, but differences among brothers and sisters in human families, for example, seemed to imply a lack of precision in the transmission of traits.

By employing our knowledge of the Mendelian Laws and of the events of meiosis, as well as the multiplication rule, we can satisfactorily explain this variability. Excluding environmental effects, basically the explanation resides in the numbers of genetically different gametes produced by the parents. By using the multiplication rule, we shall now see how the numbers of different gametes can be calculated.

Consider first a cell from a genetically simple organism, one with only 2 different pairs of chromosomes. If the chromosomes of pair 1 are designated A_1 and A_2, and those of pair 2 are designated B_1 and B_2, two possible gametes can be formed with each pair. Altogether, four possible gametes can result from the two pairs: A_1B_1, A_1B_2, A_2B_1, A_2B_2. We see this from the forked-line diagram given in Fig. 2-13.

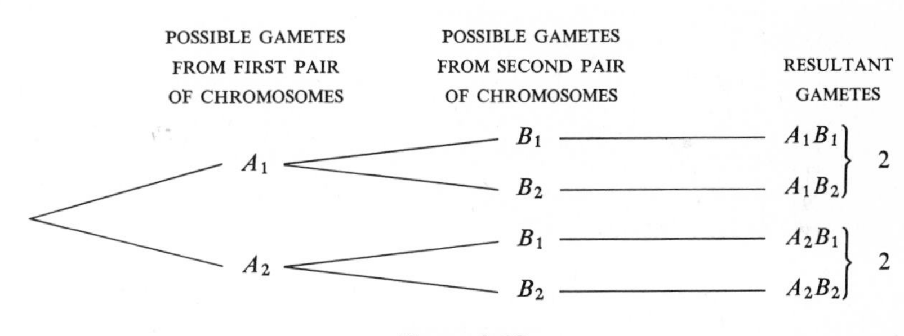

Figure 2-13

Thus, the number of possible gametes from the pair is given by $2 + 2 = 2(1 + 1) = 2 \times 2 = 4$.

This result can also be obtained directly by using the multiplication rule. The first event consists of choice of an A gamete, the second event of choice of a B gamete. There are 2 choices for the first event, 2 for the second, so $2 \times 2 = 4$ choices in all. If we continue this analysis, assuming there are n pairs of chromosomes, we see that there are $2 \times 2 \ldots \times 2 = 2^n$ possible gametes; these results are summarized in Table 2-1(b). Human beings have 23 pairs of chromosomes; therefore $2^{23} = 8{,}388{,}608$ possible genetically distinct gametes can be produced by each individual, assuming heterozygosity at one locus per chromosome!

Furthermore, since fertilization is random, any one of the 2^{23} possible eggs produced by a female can be fertilized by any of the 2^{23} possible sperms produced by a male. Therefore $2^{23} \times 2^{23}$ possible fertilized egg cells can result, or 70,368,744,177,664 possible combinations of sperm and egg to produce a zygote! This is an incredible number of possibilities, and clearly explains why so much variability exists even within one family.

Mendel's Contributions to Genetics

The preceding discussion essentially summarizes the basic concepts of Mendelian genetics, based on experiments performed by Mendel in the mid-nineteenth century. Perhaps the most significant concept proposed by Mendel was that heredity is determined by discrete units which are transmitted in a predictable pattern. But Mendel also realized that certain criteria must be met in the planning of any experimentation dealing with the analysis of inherited traits. It was perhaps his awareness of the need to satisfy these criteria that allowed Mendel to succeed in understanding heredity where so many others before him had failed.

The first criterion is to choose only those traits which can be clearly discerned in the experimental organism. Mendel called them "constant differentiating characters." For example, the gene for tallness resulted in plants that were uniformly $6\frac{1}{2}$ to 7 feet tall. If the trait could be modified from one generation to another so that plants of the same genotype would vary in height, then there would be no way to gather meaningful data. A 3- or 4-foot plant could not be easily classified as tall or short.

The experimental organisms must not be exposed to situations where crosses might be made accidentally, therefore involving foreign genes. For example, Mendel realized that the plants he studied had to be protected from "the influence of all foreign pollen," for if foreign genes were introduced the progeny would not reflect the transmission of genes from the intended parents only.

Mendel also recognized that to analyze the inheritance of traits, one must follow the transmission pattern through many generations. Therefore, it is necessary to plan crosses that will not diminish the fertility of hybrids and their offspring. It was only through the ratios found in the F_2 generations, the progeny of monohybrid and dihybrid crosses, that Mendel discerned the classical ratios we have discussed.

To this day geneticists must satisfy these criteria to produce valid data in their experiments. We have achieved our greatest progress using organisms, such as microorganisms, that best satisfy these guidelines. Conversely, organisms that do not, such as humans, are those about which we know the least.

PEDIGREE ANALYSIS

The one aspect of transmission genetics that is unique to humans is that experimental crosses cannot be made to test the genotypes of individuals; other problems include the small numbers of offspring obtained from a mating, and the relatively long generation span. Thus classical methods are inadequate for the analysis of inherited traits in our species.

To establish the genetic basis of an inherited trait in humans, geneticists have devised a means of analysis known as the *pedigree*. This technique uses symbols to represent as many members of a family as one can study, so that a pattern of transmission is traced beginning with the present generation and going backward in time.

Pedigree Symbols

In the traditional use of symbols, circles represent women and squares represent men. A marriage or mating is shown by a horizontal line drawn between a circle and a square:

Offspring from a mating are suspended from the horizontal line of an inverted *T* which is perpendicular to the marriage bar:

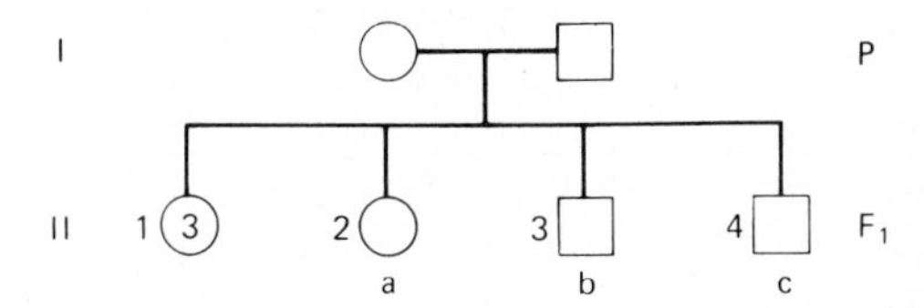

Numbers next to the symbols of a generation of children, called a *sibship*, indicate order of birth. To conserve space, a number inside a circle or square indicates the number of children of that sex. The above family thus has 4 girls and 2 boys. Generations are often designated by roman numerals, and individuals by letters or numbers beneath their symbols.

Individuals affected by the trait of interest are usually represented by colored-in symbols, or in some way distinguished from others in the family pedigree:

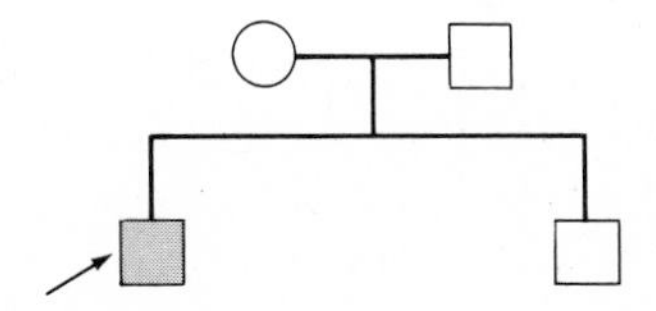

An arrow pointing to a symbol usually indicates the individual who brought the condition to the attention of the geneticist. Such a person is called either the *proband* or the *propositus* (*m.*; *proposita*, f.). Some other pedigree symbols are shown in Fig. 2-14. An exceptional trait which is subsequently inherited usually arises from alterations of genes in germ cells which are known as *mutations*.

Pattern of Transmission of a Dominant Trait

To establish the genetic basis of a trait, a geneticist gathers as much information about as many family members as possible, for frequently the pattern of transmis-

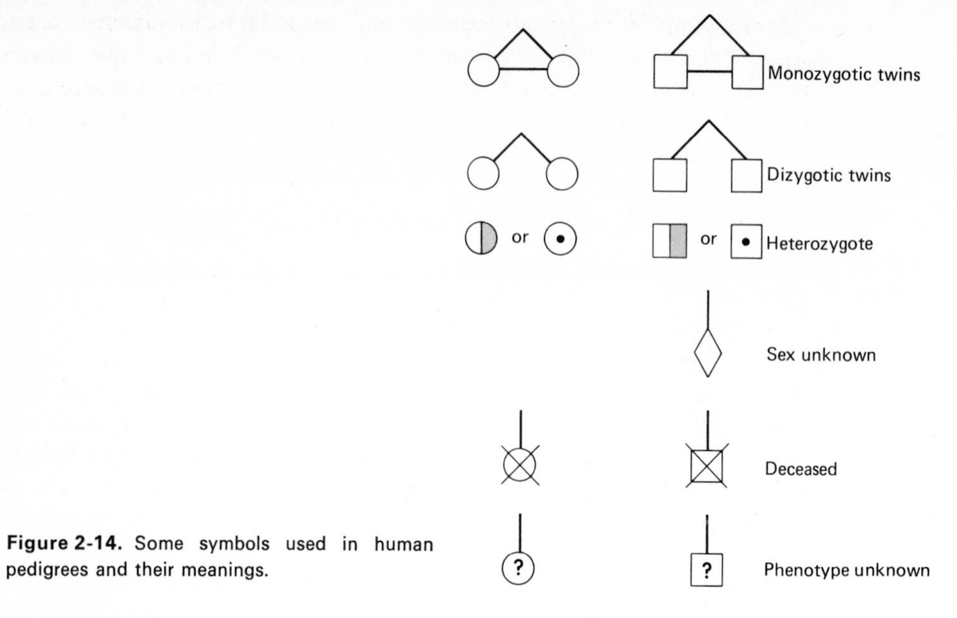

Figure 2-14. Some symbols used in human pedigrees and their meanings.

sion will indicate the nature of the gene involved. Figure 2-15, a pedigree typical of a dominant trait, has the following characteristics:

1. Every affected individual has an affected parent.
2. If a branch of the family manifests transmission of the trait, there is no skipping of generations.
3. Unaffected individuals do not transmit the trait to the next or subsequent generations.

A brief consideration of these characteristics reveals that they are to be expected, based on our discussion of dominance and the chromosomal basis of heredity. If a dominant trait is being transmitted, any parent who passes the gene to

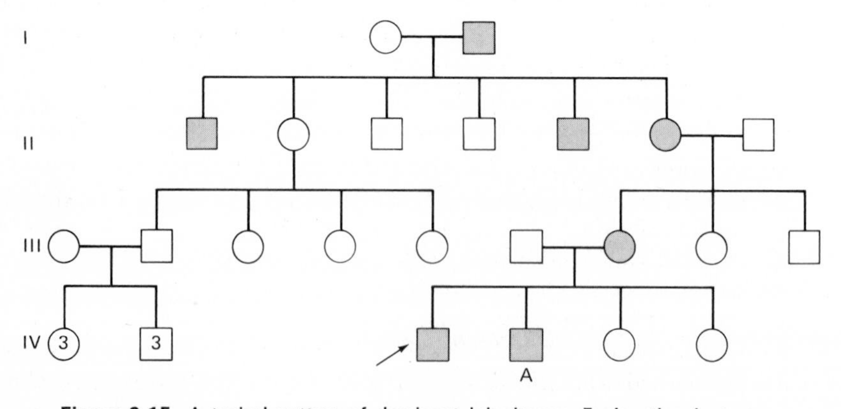

Figure 2-15. A typical pattern of dominant inheritance. Explanation in text.

an offspring must necessarily express the trait also. This explains the first and second characteristics of a pedigree of a dominant trait. Conversely, if an individual does not manifest the trait, he or she must be homozygous recessive, and therefore cannot transmit the gene to subsequent generations; hence, characteristic number 3.

Note that the unaffected F_1 female in Fig. 2-15 is shown to have children, although no husband is indicated. This does not imply parthenogenesis—only that the spouse's phenotype is unknown, or does not affect the analysis of the pedigree.

Pattern of Transmission of a Recessive Trait

Figure 2-16 illustrates a pattern of transmission typical of a recessive gene. Note that the recessive trait, especially if it is a rare one, appears far less frequently than does a dominant trait. This is of course because heterozygotes are phenotypically normal, or dominant, in appearance. There is skipping of generations owing to the masked presence of the recessive gene. Note that *A* and *B* are first cousins; marriage between related people increases the frequency of expression of recessive traits, a phenomenon which will be discussed in Chapter 16.

These are two examples of Mendelian traits in humans, that is, traits which are transmitted in a pattern analyzable by Mendelian laws of heredity. There is an additional pattern of transmission which indicates the presence of an inherited trait that is determined by genes known as *sex-linked*, but we shall defer our discussion of this pattern for the chapter on the genetic basis of sex determination.

The Pedigree as a Tool for Prediction

Besides their importance in diagnosing a condition as an inherited one, pedigrees are also useful in predicting phenotypes of subsequent generations. For example, suppose individual *A* in Fig. 2-15 wishes to know whether he could transmit his condition to his children. Since he is affected, he obviously must have the dominant gene; however, he can only be heterozygous for the gene since his father was normal, or homozygous recessive. Therefore individual *A* has a 50% chance of transmitting the dominant gene to any of his offspring. If we assume that he will marry a normal woman, we can say that the likelihood of his having affected children would be $\frac{1}{2}$, or 50%.

What of the woman marked *C* in Fig. 2-16? First we have to establish the likelihood that she is carrying the recessive gene, since we cannot phenotypically distinguish heterozygotes from homozygous normals. The fact that she has an affected sister establishes the genotypes of her parents as heterozygous. In a monohybrid cross, the expected offspring are in the ratio 1 homozygous dominant : 2 heterozygous : 1 homozygous recessive. The last category does not apply to *C* as the pedigree shows *C* to be normal. Therefore the chance that she is heterozygous, as opposed to homozygous normal, is 2 : 1, or $\frac{2}{3}$. If she marries a homozygous normal man, she has of course no chance of producing an affected offspring, but what are the chance that she will produce an affected offspring if she marries a heterozygote? The answer is $\frac{2}{3} \times \frac{1}{4} = \frac{1}{6}$ ($\frac{2}{3}$ = chance that she is heterozygous; $\frac{1}{4}$ = chance that two heterozygotes produce a homozygous recessive child). The multiplication rule is used since the chance that she is heterozygous, and that she will produce an affected child, are two independent "events." We shall discuss this in more detail in Chapter 3.

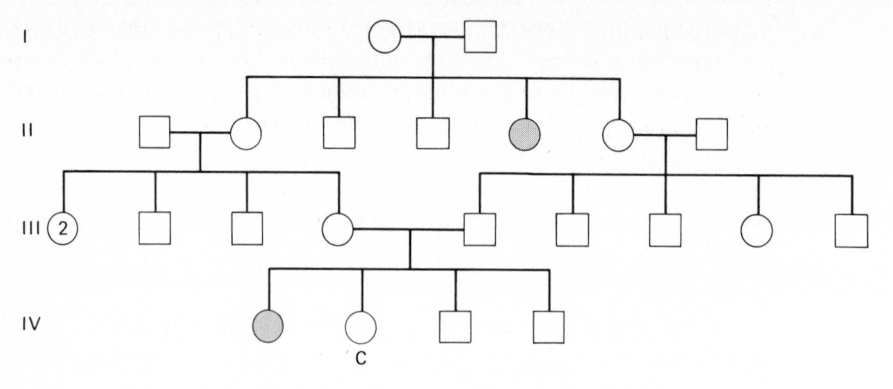

Figure 2-16. A typical pattern of recessive inheritance. Explanation in text.

Such analysis of pedigrees enables geneticists to counsel people who know that they may have an unfavorable gene in the family, and are concerned about themselves and their children. If enough information can be obtained about relatives to draw up a pedigree, the probability of individuals inheriting the gene or transmitting it to their children can be estimated. This type of analysis has given rise to a new field known as *genetic counseling*. The genetic counselor, however, can only *analyze the situation*; he or she *cannot make the decision*, if one is to be made, as to whether a person *at risk*—that is, with a high probability of transmitting an unfavorable gene—should have children. We shall discuss further aspects of genetic counseling in Chapter 19.

Limitations on the Usefulness of Pedigree Analysis

Although of great help in the study of human genetics in the past, pedigree analysis does have limitations. For one thing, it must rely on the memory and honesty of the people involved, some of whom are not always rational regarding the existence of an undesirable inherited trait in the family. A common experience of genetic counselors advising families with a heritable disease is encountering feelings of guilt in parents who have produced a child with a genetic defect. Such guilt often leads to difficulties in obtaining a true picture of the family situation. The analysis of inherited conditions, especially the more serious ones, frequently requires not only genetic counseling, but also the help of psychologists and sociologists.

Another problem resides in the fact that we cannot use the pedigree as a tool unless a number of individuals in several generations in a family have been affected, providing us with a pattern. But what of those situations in which only one person in a family is affected—can we assume that this is an inherited disease? If so, we can say very little as to whether it is a new (or *spontaneous*) dominant mutation, or whether recessive genes are involved and the propositus was the first in the family to receive two copies of the gene. In addition, as we shall discuss in Chapter 4, certain gene interactions may alter the expression of genes (phenomena which Mendel never encountered), possibly resulting in a misleading pattern of transmission in a pedigree.

Furthermore, geneticists must never forget that a condition which looks for all the world like a known inherited disease may in fact *not* be due to a mutant gene. Certain environmental factors, operating during critical periods of gestation, can lead to symptoms indistinguishable from conditions known to be genetically determined. This phenomenon is known as *phenocopy*.

A well-known illustration of phenocopy is the story of the thalidomide babies. There exists a rare recessive gene in humans called *phocomelia*, which results in babies born with stunted arms and legs. In the 1960's, a large number of children were born with stunted arms and legs very similar to those produced by *phocomelia*. The frequency of this occurrence (in the thousands), and the fact that it was restricted to European countries, led scientists to search for a cause deriving from the environment rather than from mutations.

The factor was eventually found to be a newly developed sleeping medicine, thalidomide. When administered to women in early pregnancy, the drug resulted in an interference with development of the fetal extremities and in other deformities. Figure 2-17 illustrates the great similarity in appearance between phocomelia and its phenocopy, the thalidomide syndrome.

If phenocopies can be so similar to actual genetic conditions, how can we distinguish one from the other? With experimental animals, crosses can be repeated and test crosses made, in the hope that the ratios of offspring will give some indication of the genetic basis for the trait.

For example, assume that a blood disorder has been found in a strain of rabbits. When affected individuals are mated, the offspring ratio is approximately 2 affected : 1 normal. What is the most likely hypothesis to explain this? The ratio appears to be a classical 1 : 2 : 1 ratio from a monohybrid cross with one class of genotype missing. The most likely hypothesis to put forth, then, would be that of a single pair of alleles, one of which has a dominant effect on white blood cells, and which is perhaps lethal in the homozygote.

The geneticist must then perform experimental crosses to confirm this hypothesis. One would be to repeat the above cross with many affected animals and study litters at progressively earlier stages to find abnormal embryos. A second experiment would be to cross affected rabbits with normal ones. On the basis of our hypothesis, what ratio of progeny can be expected? Since we assume dominance of the gene causing abnormal white blood cells, the cross would essentially be a testcross, and the expected ratio would be 1 affected : 1 normal. Incidentally, there does exist a dominant gene in rabbits causing abnormal white blood cells in heterozygotes and lethality in homozygotes. It is known as the Pelger anomaly, and a similar condition has also been found in humans (see reference to Srb, Owen, and Edgar).

In humans, of course, this kind of experimentation is not possible. We must work with available information and make an analysis as best we can. For affected individuals we can provide only symptomatic treatment of their condition.

Because of the serious limitations to pedigree analysis, geneticists in this and related fields have turned to such aids as mathematical tools to estimate the risks which affected families face in producing another affected child. This will be our topic for the next chapter.

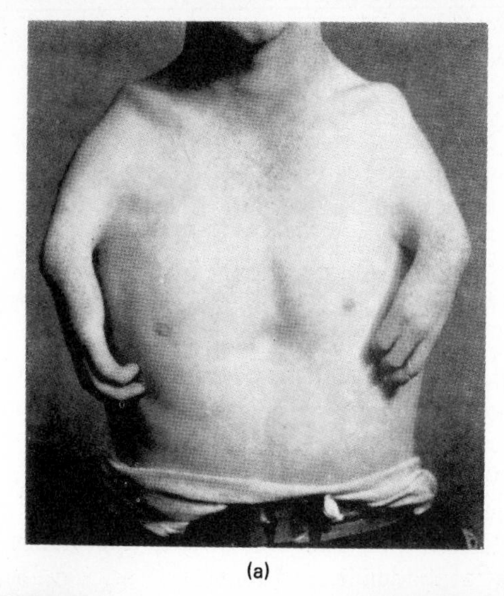

(a)

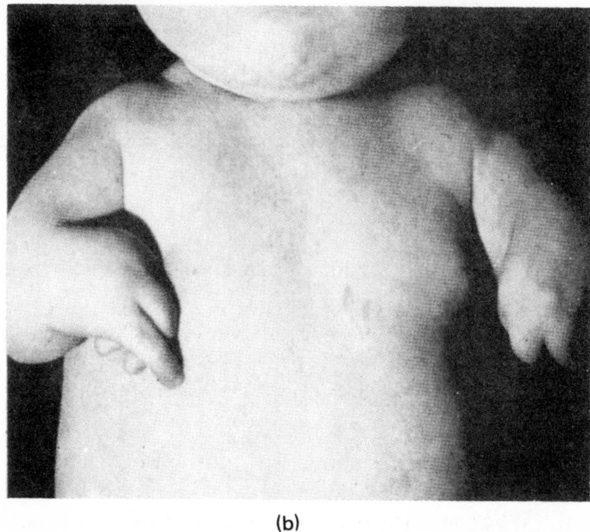

(b)

Figure 2-17. A case of phenocopy. (a) Phocomelia. (b) Thalidomide syndrome. (From R. E. Cooke, ed., 1968. *The Biologic Basis of Pediatric Practice.* McGraw-Hill, New York, in A. C. Pai, 1974. *Foundations of Genetics,* p. 84. McGraw-Hill, New York.)

PROBLEMS

1. Mendel established the dominance and recessiveness of a number of inherited traits in pea plants. Specifically, tallness is dominant over shortness, and round seed coat is dominant over wrinkled. These traits are determined by the following alleles, respectively: *T*, *t*, *W*, *w*.

Assume that the following crosses were performed, using plants of unknown genotypes. From the results obtained, give the most probable genotypes of the parental plants:

Parental phenotypes	Offspring phenotypes Tall/ round	Tall/ wrinkled	Short/ round	Short/ wrinkled
Tall/Round × Short/Wrinkled	66	70	73	69
Short/Round × Tall/Wrinkled	123	0	0	0
Short/Round × Tall/Wrinkled	84	80	0	0
Tall/Wrinkled × Short/Round	99	104	104	96

2. In humans, freckles is dominant to no freckles; the ability to taste PTC is dominant to nontasting. A couple, both of whom have freckles and are tasters, marry. The woman's mother had no freckles and was a nontaster. The man's father had no freckles, and his mother was a nontaster. Using the following letters to designate the alleles: $F=$ freckling, $f=$ nonfreckling, $T=$ tasting, $t=$ nontasting, show the expected phenotypes and genotypes of the offspring of this cross using (a) the Punnett Square and (b) the forked-line method.

3. In *Drosophila melanogaster*, the fruitfly, normal eye color (B) is dominant over brown eye color (b); normal wing shape (V) is dominant over vestigial wing (v).

(a) Assume that a testcross is made between dihybrid and homozygous recessive flies. If 1600 offspring are obtained, what are the expected phenotypes and in what numbers would you expect to find offspring of the different phenotype classes? What would your answers be for the genotypes?

(b) Assume that dihybrid crosses are made, and that again 1600 offspring are obtained. What are the expected phenotypes and in what numbers would you expect to find offspring of the different phenotype classes? What would your answers be for the genotypes?

4. What explanation can you give for the large number of different alleles determining variations in certain traits such as eye color in man? How do you explain the lack of variation in other of our species' traits, such as one head and two arms?

5. Diagram testcrosses for a dihybrid; for a trihybrid. What genotypic and phenotypic ratios can you expect to obtain?

6. (a) Analyze the following pedigrees to determine whether the most likely mode of inheritance is dominant or recessive.

(b) Briefly give reasons for your answer to (a).

(c) What is the most likely genotype of the individuals designated x, y, z?

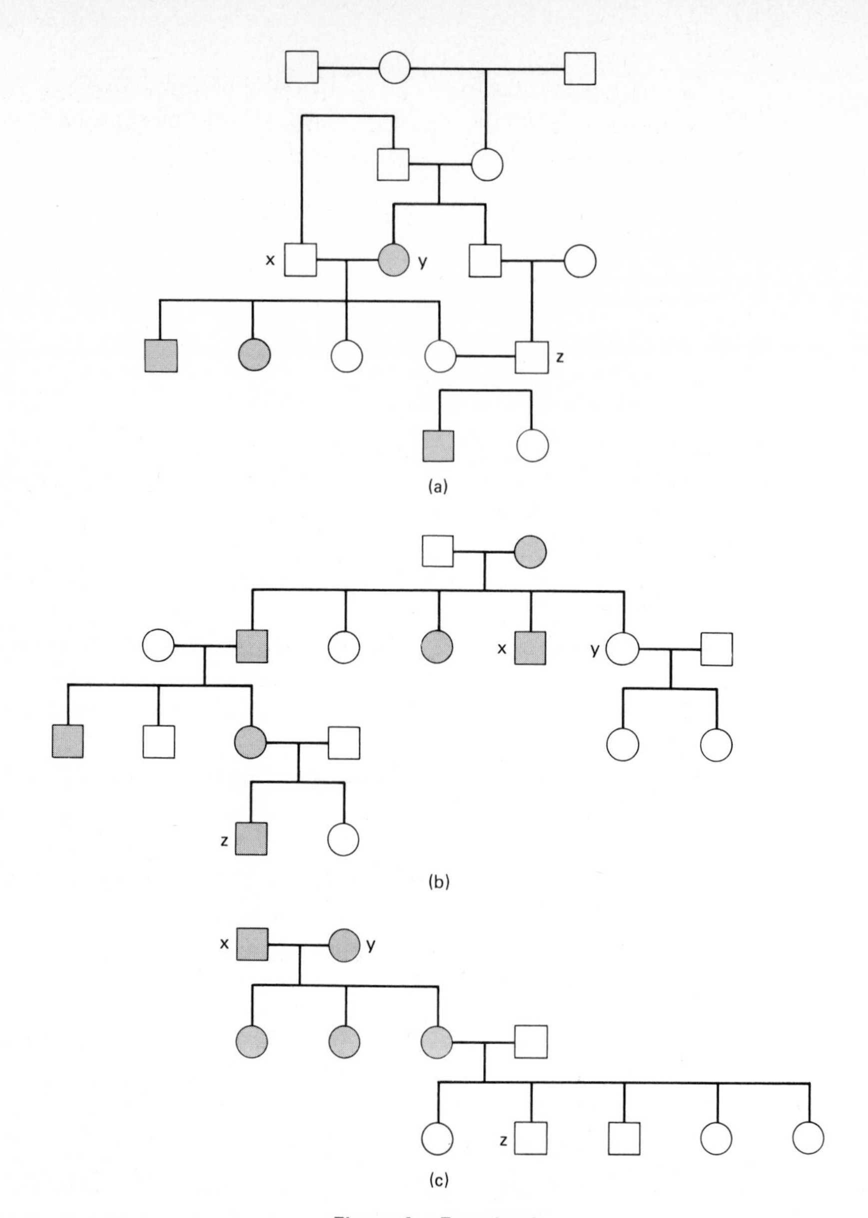

Figure for Exercise 6.

7. Assume that a newborn mouse is found to have skeletal abnormalities, whereas all its littermates are normal. Assume further that the parents are mice from the same *inbred* strain (a strain that has been bred to be homozygous at all loci so that scientists have genetically identical animals with which to work). Detail the crosses you would make to determine whether the condition is an inherited one, and what you would expect from these crosses if the condition is determined (a) by a dominant gene, (b) by a recessive gene, or (c) by environmental factors.

8. If friends came to you for genetic advice because they had a child with congenital abnormalities, what would you tell them regarding all the possible interpretations on the basis of this condition?

9. Cows have 30 pairs of chromosomes, spider monkeys have 17 pairs, and oppossums have 11 pairs.

(a) How many distinct germ cells (sperm or egg cells) are possible for cows, spider monkeys, and oppossums?
(b) If the male and female of each species can be assumed to have completely different chromosomes, what is the maximum number of distinct zygotes for each species?

10. In the mating of two hybrid tall plants, $Tt \times Tt$ (where T is dominant tall, and t is recessive short), the possible genotypic results are TT (tall), Tt (tall), and tt (short). Thus in a mating of plants hybrid for 1 trait there are 3 distinct genotypes.

(a) If plants hybrid for 2 traits (such as height and color of seed) are mated, how many distinct genotypes are possible?
(b) If plants hybrid for 3 traits are mated, how many distinct genotypes are possible?
(c) If plants hybrid for 5 traits are mated, how many distinct genotypes are possible?
(d) If plants hybrid for n traits are mated, how many distinct genotypes are possible?

11. In a mating of two hybrid tall plants, $Tt \times Tt$ (where T is dominant tall and t is recessive short), the phenotypes are tall (TT and Tt) and short (tt) plants. Thus in a mating of plants hybrid for 1 trait there are 2 distinct phenotypes.

(a) If plants hybrid for 2 traits, both of which exhibit complete dominance, are mated, how many distinct phenotypes are possible?
(b) If plants hybrid for 3 traits, all of which exhibit complete dominance, are mated, how many distinct phenotypes are possible?
(c) If plants hybrid for 5 traits, all of which exhibit complete dominance, are mated, how many distinct phenotypes are possible?
(d) If plants hybrid for n traits, all of which exhibit complete dominance, are mated, how many distinct phenotypes are possible?

12. Use the forked-line (tree-diagram) method to determine the genotypic and phenotypic ratios for the trihybrid cross $TtWwGg \times TtWwGg$. The two plants are heterozygous for stem length (tall is dominant; short is recessive), seed structure (round is dominant; wrinkled is recessive), and pod color (green pod color is dominant; yellow is recessive).

13. Draw the gametes of a $TtWw$ dihybrid plant during the following stages of meiosis:

(a) Zygotene of Prophase I (b) Metaphase I (c) Prophase II

Show the chromosomes and label them with the genes.

14. Draw cells from a plant with the genotype *TtWw* during: (a) Anaphase I; (b) Telophase II; (c) Anaphase of mitosis. Show the chromosomes and label them with the genes.

REFERENCES

MENDEL, GREGOR, 1865. "Experiments in Plant Hybridization." Translated in J. A. Peter, ed., 1959, *Classic Papers in Genetics*, pp. 1–20, Prentice-Hall, Inc., Englewood Cliffs, N.J.

SRB, A. M., R. D. OWEN, and R. S. EDGAR, 1965. *General Genetics*, 2nd ed., W. H. Freeman and Co., San Francisco.

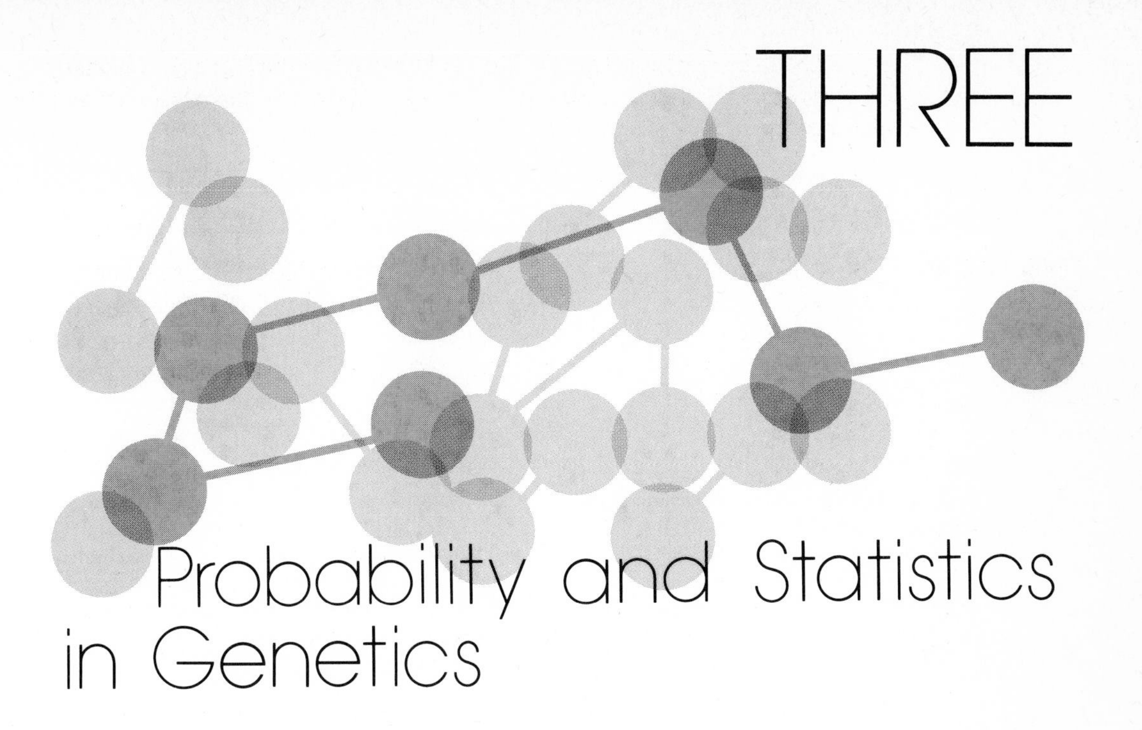

THREE

Probability and Statistics in Genetics

INTRODUCTION

Honored today as the "Father of Genetics," Mendel contributed not only concepts which formed a foundation for modern genetics, but also a method of analysis which still remains the primary method in transmission genetics. The analysis of inherited traits is essentially an inductive one in which data are obtained and then a statistical analysis made to discern some pattern or model that will clarify the genetic basis of the trait under study. Later, statistical analysis is again used to test the model developed.

STEPS IN MODELING

Mendel's methodology provides classic examples of the development of models and theories. If we consider the pattern of transmission of a trait to be a model, in that the pattern enables us to make predictions, we can schematically summarize the steps in modeling taken by Mendel and many other scientists, as in Fig. 3-1.

From his large numbers of observations (see Table 2-1), Mendel was able to formulate a model and propose the Law of Segregation and the Law of Random Assortment. Although these proposed laws were to become extremely important

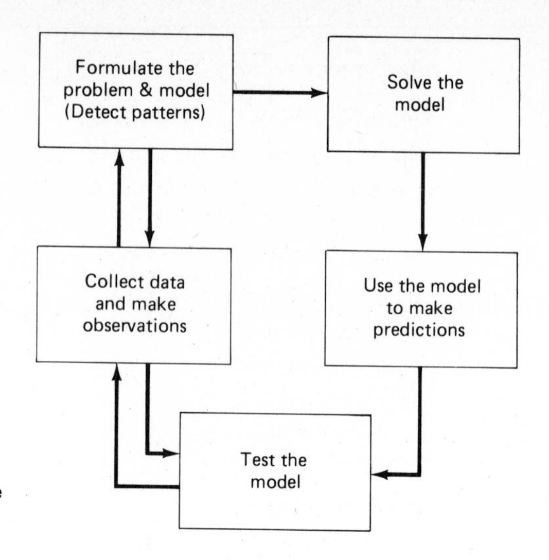

Figure 3-1. Schematic representation of the modeling process.

contributions, they were at first simply hypotheses or models that had to be tested and verified.

In the following sections we will consider simple mathematical and statistical tools that will enable us to develop models of (propose hypotheses for) observable phenomena, use these models to predict the genetic bases of the phenomena, and ultimately test whether our predictions concur with observations.

PROBABILITY IN GENETIC ANALYSIS

In genetics, questions are posed about the probability that particular crosses will produce a tall plant, a child with albinism or sickle-cell anemia, and so forth. Segregation and the random assortment of genes create the potential for many possible genotypes and phenotypes to be represented in the offspring of crosses of heterozygotes. Genetic events, however, cannot be predicted with absolute certainty; rather, we can know only that a certain phenomenon has a certain likelihood of occurring. The measure of chance or likelihood that a particular event will occur is referred to as a *probability*.

To measure probability we have to determine all the possible occurrences that could occur in a given situation and then relate the chance of a *particular* occurrence happening to all possible occurrences. To formalize the concept of probability, we develop the notion of an experiment and its outcomes. An *experiment* (*statistically speaking*) is a process which produces some outcome or result that we can observe or "imagine." In fact we can describe all possible outcomes, but on any given trial of the experiment we do not know in advance what the outcome will be; hence, we cannot predict with absolute certainty. Broadly defined in this way, an experiment can encompass a variety of situations and phenomena. For example, consider the following experiments:

Experiment	*Possible outcomes*
1. Tossing a coin	Head, tail
2. A monohybrid cross, $Tt \times Tt$	The genotypes TT, Tt, tT, tt[1]
3. A dihybrid cross, $TtWw \times TtWw$ (e.g., $T =$ tall, $t =$ short, $W =$ round, $w =$ wrinkled, T and W are dominant)	The genotypes $TTWW$, $TTWw$, $TTwW$, $TtWW$, $tTWW$, $TtWw$, $tTWw$, $TtwW$, $tTwW$, $TTww$, $Ttww$, $tTww$, $ttWW$, $ttWw$, $ttwW$, $ttww$[1]
4. Cross of two people who are carriers for albinism (i.e., a monohybrid cross $Aa \times Aa$, as albinism is recessive and normal pigmentation is dominant)	3 normally pigmented children, 1 albino child; The genotypes AA, Aa, aA, aa[1]
5. Determining the sex and birth orders of two children within a family	MM, MF, FM, FF

These particular experiments all have one thing in common: All the outcomes are equally likely; on any given trial of the experiment, one and only one of the outcomes will occur. In Experiment 1, if we have a fair coin we expect that the outcomes head and tail will occur an equal number of times, and hence we consider each of these outcomes to be equally likely. The genetic crosses of Experiment 2 result in four equally likely outcomes, the genotypes TT, Tt, tT, and tt, three of which are phenotypically indistinguishable. Experiment 3 has 16 equally likely outcomes, which fall into 4 different groups corresponding to 9 tall round, 3 tall wrinkled, 3 short round, and 1 short wrinkled.

We must introduce one more idea in order to define the notion of probability more formally. An *event* is one or a combination of the possible outcomes in an experiment. For example, in Experiment 5 the combination of the outcomes MM, MF, and FM represents the event of having at least one male child. We may now define probability as follows. In an experiment which has n possible outcomes, each of which is assumed equally likely, the probability of an event E occurring is defined as

$$\text{Probability of an event } E = \frac{n(E)}{n},$$

where $n(E)$ is the number of outcomes that make up the event E.

In the tossing of a fair coin, for example, there are two possible equally likely outcomes, namely, a head or a tail. The probability of obtaining a head when the coin is tossed would be one chance out of two possible outcomes, that is, $n(E) = 1$, and $n = 2$, so that the probability of a head $= \frac{1}{2}$. In the monohybrid cross,

[1]We have here and in the following examples listed genes in order giving that obtained from the first parent first, and that obtained from the second parent second. Later, we shall not distinguish outcomes Tt or tT, reading both as Tt, in which case TT and Tt may or may not be equally likely.

$Tt \times tT$, one outcome out of the four equally likely possible outcomes of the cross, (tt), corresponds to a homozygous recessive offspring. Thus the probability of such an offspring is given by $\frac{n(E)}{n} = \frac{1}{4}$. Similarly, the probability of the event of obtaining a tall round plant from the dihybrid cross would be nine out of the total of sixteen equally likely possible offspring, or $\frac{9}{16}$.

In exactly the same manner, if we know the genotypes of the parents, we can calculate the probability of obtaining an offspring of given genotype in a human family. In our example, suppose that two people are heterozygous for albinism. We can then calculate the probability of these two people producing an albino child. The probability would be 1 out of 4, or $\frac{1}{4}$, as in any monohybrid cross, since of the four equally likely outcomes (AA, Aa, aA, aa), only one outcome yields an albino child.

Such probability calculations represent an extremely important aspect of genetic studies. As we mentioned in Chapter 2, in human genetics they serve as the foundation for genetic counseling, a medically related service through which people from families with known genetic disorders can seek information on the probability of their having inherited an abnormal gene or of their transmitting the condition to their children.

These probabilities can best be interpreted in the following manner. If we repeat the experiment of interest many times, the probability of a particular event (or phenomenon) occurring is simply a measure of the overall percentage of the time that that event will occur. That is, if we flip a coin 100 times we expect about 50 heads, but on a single trial we may get a head or we may not. If we consider families in which both parents are carriers for albinism and in which they have 8 children, we can expect that the *average* number of albino children per family is 2, that is, $\frac{1}{4}$ of 8. In any *particular* family of 8 children there may be more or fewer than 2 albino children, but the average will be 2. Similarly the probability of an event occurring is a relative measure of how often the event will occur. But on any one trial of the experiment we cannot be sure of the occurrence or nonoccurrence of the event. This lack of absolute certainty explains why genetic counseling can be difficult.

Note that the probabilities of all the possible outcomes of any experiment add up to 1. For example, the sum of the probabilities for the possible outcomes of the monohybrid cross would be $\frac{1}{4}$ (the probability of obtaining TT) + $\frac{1}{4}$ (the probability of obtaining Tt) + $\frac{1}{4}$ (the probability of obtaining tT) + $\frac{1}{4}$ (the probability of obtaining tt) = 1. *From our definition of probability it follows that the probability of an outcome of an experiment cannot be either negative or greater than* 1. These rules of probability are constantly used by geneticists in predicting the results of genetic crosses.

The Probability of Combinations of Events

Frequently it becomes very cumbersome to enumerate all possible outcomes of an experiment in order to determine the probabilities of various events occurring. This is especially true if we are interested in the probability of a combination of events occurring. Let us consider the simple example of a couple planning to have a family of two children who wish to know the probability that they will have two females. (We can calculate this probability by considering the equally likely outcomes, but we shall simplify the effort involved.) We are dealing in this case with two events which are *independent*; that is, the occurrence of one event does not affect in any way the occurrence of the other. The sex of the first born in no way affects the sex of the second child. (Similarly, if a coin is flipped twice, the outcomes in terms of heads and tails are totally independent from trial to trial.)

To determine the probability of the joint occurrence of independent events, we employ the following simple probability rule: *The probability of the occurrence of two independent events is the product of their individual probabilities.* With regard to the probability of having two female children, since the probability of having a female child is $\frac{1}{2}$, the probability of having two female children is $\frac{1}{2}$ (the probability that the first born is a female) $\times$ $\frac{1}{2}$ (the probability that the second born is a female) $= \frac{1}{4}$. We can also arrive at this result by using a *tree diagram* to enumerate the outcomes and then incorporating the probabilities. The possible outcomes for the sexes of the two children in a family are illustrated in Fig. 3-2. We know that the probability of a

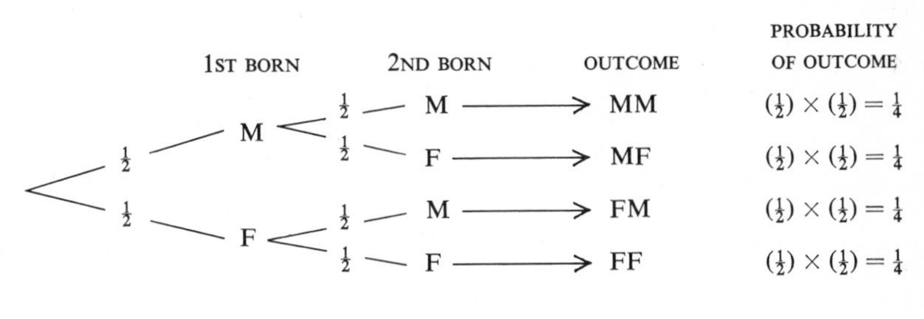

Figure 3-2.

male is $\frac{1}{2}$ and of a female is $\frac{1}{2}$, and we incorporate this into Fig. 3-2 by putting $\frac{1}{2}$ on the appropriate branches of the tree. We can determine the probabilities of each of the outcomes by noting that since the constituent events are independent, we can multiply the probabilities along each branch to obtain the probabilities of the outcomes in the last column of Fig. 3-2. (When using the tree diagrams to determine probabilities of combinations of independent events, it is important to make sure that the sum of the probabilities emanating from a point is 1. The probabilities for all the outcomes should also sum to 1.)

Probabilities of combinations of genetic events can be calculated in the same manner. In what follows, we will no longer distinguish genotypes such as *Tt* and *tT*, writing them both as *Tt*. As a result, *TT* and *Tt* may no longer be equally likely. In the dihybrid plant *TtWw*, the transmission of the alleles determining stem length (*T*, *t*) is independent of the transmission of the alleles determining roundness of the seeds (*W*, *w*) because of Mendel's Law of Random Assortment. Suppose that we wish to know the probability of a dihybrid cross producing a plant of the specific genotype *Ttww*. By using the probability rule for independent events, we can calculate this probability easily: The probability of obtaining *Tt* from a cross of *Tt* $\times$ *Tt* is $\frac{1}{2}$; the probability of obtaining *ww* from a cross of *Ww* $\times$ *Ww* is $\frac{1}{4}$; since these traits occur independently, the probability of obtaining both gene combinations in the same individual (*Ttww*) is $\frac{1}{2} \times \frac{1}{4} = \frac{1}{8}$.

The rules of probability represent another labor-saving method for geneticists, as the rules eliminate the need to construct Punnett Squares to determine the possible outcomes of crosses before calculating the probability in question. The preceding example of a specific genotype in the offspring of a dihybrid cross would have necessitated constructing the Punnett Square and counting how many of the 16 boxes of the square contained individuals of that genotype.

Applying the rules of probability to a typical situation in a human family, let us assume that a couple is known to be heterozygous both for albinism (*Aa*) and for a dominant trait for the ability to taste the chemical PTC (phenylthiocarbamide) (*Tt*). Phenotypically, both have normal pigmentation and both are *tasters*. (People homozygous for the recessive allele for tasting are unable to detect PTC and are called *nontasters*.) Determine all the possible phenotypes that could occur in the children of this couple and their corresponding probabilities. The results of the tree diagram for the children of this couple are as shown in Fig. 3-3.

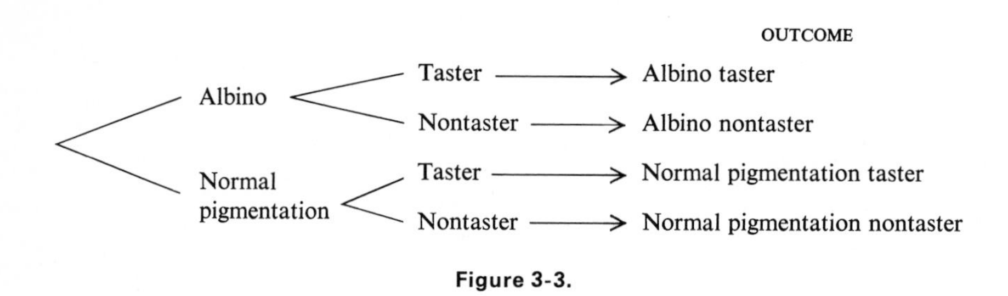

Figure 3-3.

Since albinism is a recessive condition, we would expect the probability of producing an albino child from a cross of hybrids to be $\frac{1}{4}$. Since tasting is a dominant trait and we are here interested in phenotype and not genotype, the ability to taste would appear in $\frac{3}{4}$ of the offspring and nontasting in $\frac{1}{4}$. Completing the tree diagram by incorporating the probabilities for the possible phenotypes to be expected among their children and multiplying the probabilities along each branch we get Fig. 3-4.

We can extend this example even further and calculate the probability of having a male child who is albino and a taster. Since these three events are independent, we simply enlarge our tree diagram to include the sex of the offspring, and incorporate

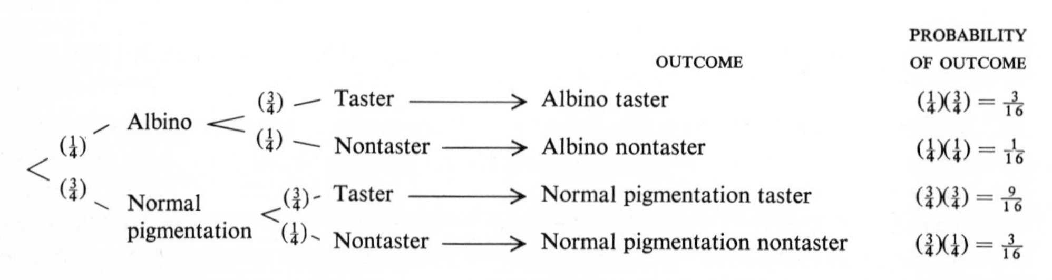

Figure 3-4.

the appropriate probabilities into Fig. 3-5. The probability of a male albino taster is determined simply by multiplying along the correct branch of the tree diagram. That is, $(\frac{1}{4}) \times (\frac{3}{4}) \times (\frac{1}{2}) = \frac{3}{32}$.

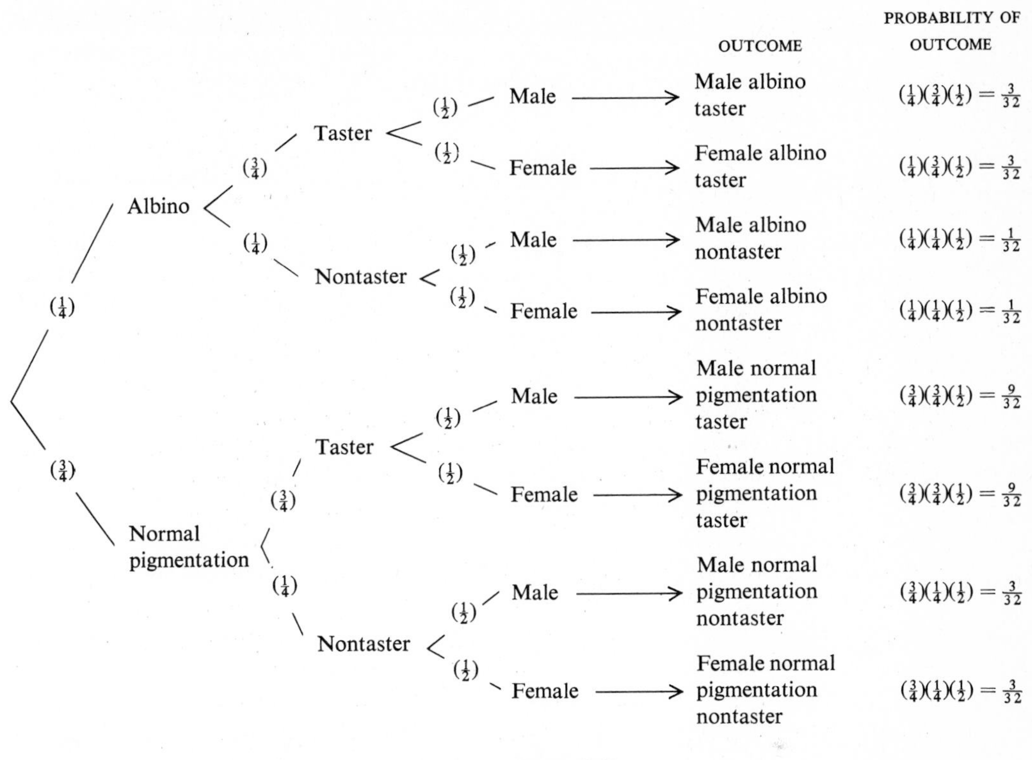

Figure 3-5.

These examples illustrate how tree diagrams and the rule for the probability of the occurrence of two independent events can be extremely useful to geneticists in determining possible outcomes of genetic crosses and their relative chance of occurring.

The Probability of Mutually Exclusive Events

Suppose that the couple in the preceding example were interested in the probability of their having a child who is normally pigmented and a taster *or* an albino child who is not a taster. Here we are asking a different question: What is the probability of one event *or* the other occurring? In this case, we first note that it would be impossible to have a child who is both normally pigmented and an albino. Such events, which cannot possibly both occur, are known as *mutually exclusive events.* (Note that the branches of tree diagrams correspond to mutually exclusive events by virtue of the structure of the diagrams.) If events are mutually exclusive, it is very easy to find the probability of one event or the other occurring. *If two events are mutually exclu-*

sive, the probability of one or the other event occurring is the sum of the probabilities of the two separate events.

To answer the question of the probability of having a child who is normally pigmented and a taster or a child who is albino and not a taster we consider that:

Probability of an albino nontaster $= (\frac{1}{4}) \times (\frac{1}{4}) = \frac{1}{16}$

Probability of a normally pigmented taster $= (\frac{3}{4}) \times (\frac{3}{4}) = \frac{9}{16}$

and therefore

Probability of an albino nontaster *or* a normally pigmented taster $=$
$(\frac{1}{16}) + (\frac{9}{16}) = \frac{10}{16} = \frac{5}{8}$

The same answer can be obtained by studying the tree diagram of Fig. 3-4. We simply pick out the branches of the tree that correspond to the events of interest, and if we confirm that the events are mutually exclusive, we add up the probabilities: $\frac{1}{16}$ for albino nontaster and $\frac{9}{16}$ for normally pigmented taster and get $\frac{10}{16} = \frac{5}{8}$.

Suppose instead that the couple wanted to know the probability of having a normally pigmented nontaster female or an albino taster male. These events are clearly mutually exclusive; therefore it is simply a matter of determining the probabilities of the individual events and then adding them. Figure 3-5 indicates that the probability of a normally pigmented nontaster female is $\frac{3}{4} \times \frac{1}{4} \times \frac{1}{2} = \frac{3}{32}$ and the probability of an albino taster male is $\frac{1}{4} \times \frac{3}{4} \times \frac{1}{2} = \frac{3}{32}$. Therefore the probability of having a normally pigmented nontaster female or an albino taster male is $\frac{3}{32} + \frac{3}{32} = \frac{6}{32} = \frac{3}{16}$.

It is important to note that if the events of interest are not mutually exclusive, one cannot simply add probabilities. For example, the probability of having an albino child or a child who is a nontaster is *not* equal to the probability of an albino plus the probability of a nontaster since the events albino and nontaster are not mutually exclusive. Calculation of such probabilities is considerably more complex and will not be considered here.[2]

USE OF THE BINOMIAL DISTRIBUTION IN GENETICS

The methods we have used so far have dealt mostly with the possible outcomes of a single mating. Often, however, the genetic counselor will be interested in determining the probabilities of the various outcomes possible in *repeated* matings, since families generally are comprised of several children. In this section we shall discuss techniques for analyzing how a single trait may be transmitted if there are several children or offspring in a particular family or experiment.

To start with a simple example, suppose that a couple is planning to have two children, and are interested in determining the probability that they will have two boys, a boy and a girl, or two girls. We can use a tree diagram to enumerate all possible birth orders and to calculate the probabilities, as illustrated in Fig. 3-6.

[2]For a discussion, see for example, Daniel (1974).

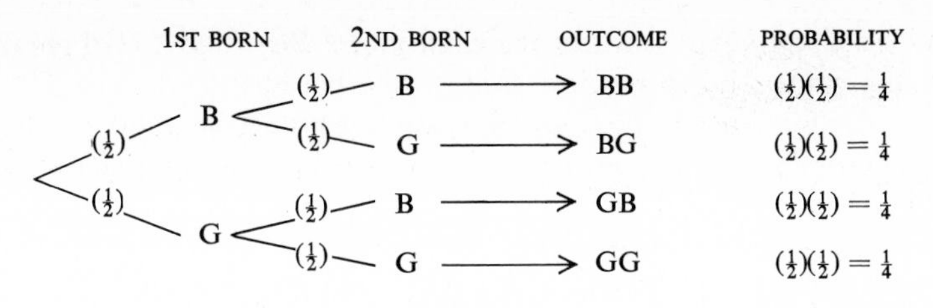

Figure 3-6.

Note that since the sex of each child born is independent of that of any previous children, we can calculate the probability of events corresponding to a branch of the tree diagram simply by multiplying the probabilities along the appropriate branch. Therefore the probability of having two boys is $(\frac{1}{2}) \times (\frac{1}{2}) = \frac{1}{4}$. The event of a boy and a girl is actually composed of two events, Boy Girl or Girl Boy (which are mutually exclusive events), so that the probability of a boy and a girl is just $(\frac{1}{2}) \times (\frac{1}{2}) + (\frac{1}{2}) \times (\frac{1}{2}) = \frac{1}{4} + \frac{1}{4} = \frac{1}{2}$. Lastly, the probability of two girls is $(\frac{1}{2}) \times (\frac{1}{2}) = \frac{1}{4}$.

Let us extend this discussion to a couple planning to have three children and see if we can find a way to calculate these probabilities without having to draw a tree diagram each time. Considering all possible outcomes for a family with 3 children, we get the following tree diagram and probabilities given in Fig. 3-7. Again, note that

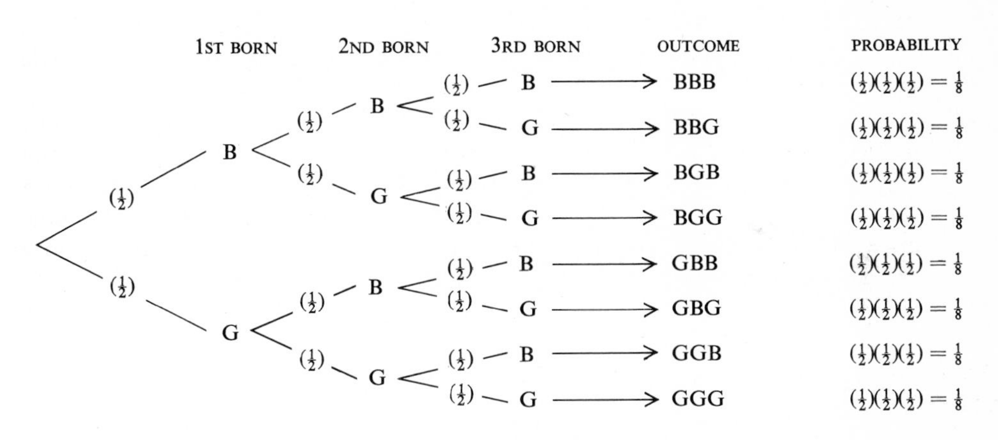

Figure 3-7.

since the events are independent, we multiply probabilities along each branch. We can summarize these results in the following way:

Probability of 3 Boys = Probability of BBB $= \frac{1}{8}$

Probability of 2 Boys and 1 Girl = Probability of BBG + Probability of BGB + Probability of GBB $= \frac{1}{8} + \frac{1}{8} + \frac{1}{8} = \frac{3}{8}$

Probability of 1 Boy and 2 Girls = Probability of BGG + Probability of GBG + Probability of GGB $= \frac{1}{8} + \frac{1}{8} + \frac{1}{8} = \frac{3}{8}$

Probability of 3 Girls = Probability of GGG $= \frac{1}{8}$

Careful study of the probability of 2 Boys and 1 Girl reveals some very interesting points. First, we note that 3 possible birth orders—BBG, BGB, and GBB—can result in 2 boys and a girl, and next, that the probability of each event is the same: $(\frac{1}{2})(\frac{1}{2})(\frac{1}{2})$. So we are actually interested in the sum of the probabilities of realizing each birth order, that is, $(\frac{1}{2})(\frac{1}{2})(\frac{1}{2}) + (\frac{1}{2})(\frac{1}{2})(\frac{1}{2}) + (\frac{1}{2})(\frac{1}{2})(\frac{1}{2}) = 3 \times (\frac{1}{2})(\frac{1}{2})(\frac{1}{2}) = 3 \times (\frac{1}{2})^3$.

Similarly, the probability of 2 girls and 1 boy is $(\frac{1}{2})(\frac{1}{2})(\frac{1}{2}) + (\frac{1}{2})(\frac{1}{2})(\frac{1}{2}) + (\frac{1}{2})(\frac{1}{2})(\frac{1}{2}) = 3 \times (\frac{1}{2})^3$. We see, therefore, that $(\frac{1}{2})^3$ is just the probability of the realization of one of the possible birth orders, and since there are three birth orders with the same probability that yield 2 girls and 1 boy, the total probability is the number of possible birth orders times the probability of obtaining one of the birth orders. (Note that the same holds for 2 boys and 1 girl.) In fact, we can apply this same approach to determine the probability of having 3 girls. There is only one birth order that results in 3 girls; therefore the probability of having 3 girls is one times the probability of having 3 girls, which is $1 \times (\frac{1}{2})^3 = \frac{1}{8}$.

Let us now consider a couple who plan to have a family of three children, and who know that they are both carriers for albinism. The probability that a single child will be an albino is $\frac{1}{4}$, but what are the probabilities in the case of three children? We start by enumerating the possibilities with a tree diagram and then obtaining the probabilities. Using A to denote albino and N, normal pigmentation, the possible outcomes in terms of pigmentation of the children are given in Fig. 3-8. Recall that

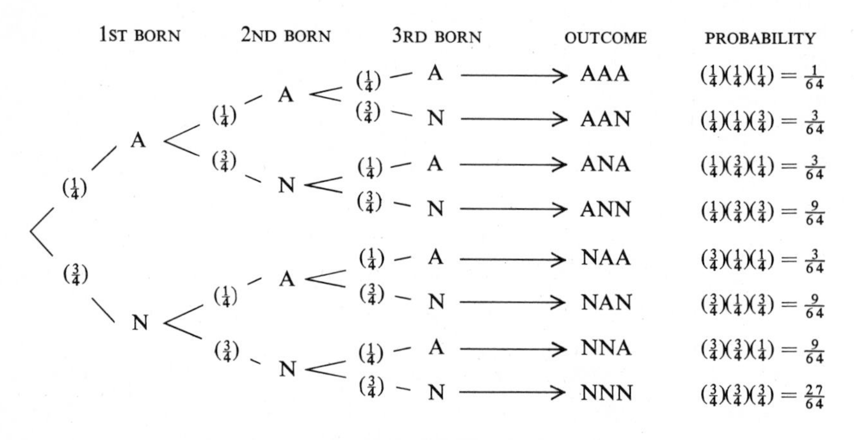

Figure 3-8.

in the previous example we discovered, after determining the probability of one of several possible outcomes comprising an event, that the other outcomes also comprising the same event would have the same probability. Therefore the total probability of an event equals the number of outcomes that comprise the event times the probability of the realization of one of the outcomes. We can verify this generalization by summarizing the events that could result from the mating of a couple who are both carriers for albinism and who plan to have three children. The probabilities are as follows:

Probability of no albino children = Probability of NNN = $(\frac{3}{4})(\frac{3}{4})(\frac{3}{4}) = 1 \times (\frac{3}{4})^3 = .421875$.

Probability of 1 albino child and 2 normally pigmented children = Probability of ANN or NAN or NNA = $(\frac{1}{4})(\frac{3}{4})(\frac{3}{4}) + (\frac{3}{4})(\frac{1}{4})(\frac{3}{4}) + (\frac{3}{4})(\frac{3}{4})(\frac{1}{4}) = (\frac{1}{4})(\frac{3}{4})^2 + (\frac{1}{4})(\frac{3}{4})^2 + (\frac{1}{4})(\frac{3}{4})^2 = 3 \times (\frac{1}{4})(\frac{3}{4})^2 = .421875$ (which corresponds to the number of outcomes which result in 1 albino and 2 normally pigmented children times the probability that one of these outcomes will occur).

Probability of 2 albino children and 1 normally pigmented child = Probability of AAN or ANA or NAA = $(\frac{1}{4})(\frac{1}{4})(\frac{3}{4}) + (\frac{1}{4})(\frac{3}{4})(\frac{1}{4}) + (\frac{3}{4})(\frac{1}{4})(\frac{1}{4}) = (\frac{1}{4})^2(\frac{3}{4}) + (\frac{1}{4})^2(\frac{3}{4}) + (\frac{1}{4})^2(\frac{3}{4}) = 3 \times (\frac{1}{4})^2(\frac{3}{4}) = .140625$ (which corresponds to the number of outcomes which result in 2 albino and 1 normally pigmented child times the probability that one of those outcomes will occur).

Probability of 3 albino children = Probability of AAA = $(\frac{1}{4})(\frac{1}{4})(\frac{1}{4}) = 1 \times (\frac{1}{4})^3 = .015625$ (which corresponds to 1 times the probability of having 3 albino children, which follows the pattern we have observed).

We can therefore consider each of the preceding probabilities in the following way:

$$\begin{pmatrix}\text{Probability of}\\ \text{having } k \text{ albino}\\ \text{children and}\\ n-k \text{ normally}\\ \text{pigmented}\\ \text{children}\end{pmatrix} = \begin{pmatrix}\text{Number of}\\ \text{orders with}\\ k \text{ albinos}\\ \text{and } n-k\\ \text{normally}\\ \text{pigmented}\\ \text{children}\end{pmatrix} \times \begin{pmatrix}\text{Probability}\\ \text{of having}\\ \text{an albino}\\ \text{child}\end{pmatrix}^{\text{\# of albino children}} \times \begin{pmatrix}\text{Probability}\\ \text{of having a}\\ \text{normally}\\ \text{pigmented}\\ \text{child}\end{pmatrix}^{\text{\# of normally pigmented children}}$$

The power to which each of the probabilities above is raised is equal to the number of offspring of that type in the event of interest. Our only remaining task is to find an easy way to determine the number of orders that result in the event of interest, since using tree diagrams to enumerate all possibilities can become very tedious and cumbersome. There is a mathematical formula that can be used to count the number of possible orders of occurrence of two different objects or events. Suppose we have 4 people, 2 men and 2 women. We can calculate the number of possible arrangements of these 2 men and 2 women simply by doing the following multiplication:

$$\frac{4 \times 3 \times 2 \times 1}{2 \times 1 \times 2 \times 1} = \frac{4 \times 3}{2 \times 1} = \frac{12}{2} = 6$$

We can verify this calculation simply by enumerating all the possibilities:

MMFF	FFMM
MFMF	FMFM
MFFM	FMMF

Clearly, using the preceding formula is much simpler than trying to enumerate all the possibilities.

The following notation, known as *factorial* notation, simplifies this method even further. It is a "shorthand" way to denote the multiplication of consecutive integers. If we want to multiply $3 \times 2 \times 1$, we write $3!$ (which we read as *three factorial*). To indicate the multiplication of $5 \times 4 \times 3 \times 2 \times 1$ we simply write $5!$. In general, if we want to multiply $n \times (n-1) \times (n-2) \times (n-3) \times \ldots \times 2 \times 1$, we write $n!$. (The notation ... simply indicates to proceed in the same pattern.) Using this notation the number of possible arrangements of 2 men and 2 women is given by $\frac{4!}{2!\,2!}$. In factorial notation the number of orders of 2 albino children and 1 normally pigmented child is given by $\frac{3!}{2!\,1!} = \frac{3 \times 2 \times 1}{2 \times 1 \times 1} = 3$. This is precisely the number we derived by using the tree diagram and enumerating (see again Fig. 3-3).

In general, if we have n objects, of which k are of one type (male, albino, etc.) and the remaining n — k are of another type (women, normally pigmented, etc.), the number of arrangements (orders) of these objects is given by

$$\frac{n!}{k!\,(n-k)!}$$

It is important to note two facts about the use of factorial notation. First, $0! = 1$, by definition. Second, $(n-k)!$ does not equal $n! - k!$. (This is easy to verify if we consider the following example. $(5-2)! = (3)! = 3 \times 2 \times 1 = 6$; but $5! = 5 \times 4 \times 3 \times 2 \times 1 = 120$ and $2! = 2 \times 1 = 2$ and $120 - 2 = 118$.

We can now construct a general formula for calculating probabilities of the type we have been discussing. If the couple who are carriers for albinism were planning to have n children, for example, we could determine the probability of their having k albino children and $n-k$ normally pigmented children by using the following formula:

$$\frac{n!}{k!\,(n-k)!}\left(\frac{1}{4}\right)^k\left(\frac{3}{4}\right)^{n-k},$$

where $\frac{n!}{k!\,(n-k)!}$ is the number of possible orders of k albino and $n-k$ normally pigmented children;

$\frac{1}{4}$ is the probability of an albino child resulting from a single mating;

$\frac{3}{4}$ is the probability of a normally pigmented child resulting from a single mating;

k is the number of albino children;

$n-k$ is the number of normally pigmented children.

We can verify that this formula works by checking the case for three children which we have already discussed.

In families that transmit genes causing serious illness and even death, as in the case of sickle-cell anemia, it is obviously of importance to calculate the probability of producing affected children so that potential parents may better consider the risks involved. Suppose that a couple heterozygous for sickle-cell anemia want four children. What would be the probability that exactly one of them might be affected? If the couple is heterozygous for sickle-cell anemia, from a single mating the probability is $\frac{1}{4}$ that they will have a child with sickle-cell anemia, and $\frac{3}{4}$ that they will have a child without sickle-cell anemia. Using our general formula we calculate the probability that exactly 1 of 4 children will be affected as follows:

$$\frac{4!}{1!\,3!}\left(\frac{1}{4}\right)^1\left(\frac{3}{4}\right)^3 = 4\left(\frac{1}{4}\right)^1\left(\frac{3}{4}\right)^3 = \frac{108}{256} = \frac{27}{64} = .421875$$

Similarly if we want to know the probability of having 2 children with sickle-cell anemia and 2 children without, we calculate:

$$\frac{4!}{2!\,2!}\left(\frac{1}{4}\right)^2\left(\frac{3}{4}\right)^2 = 6\left(\frac{1}{4}\right)^2\left(\frac{3}{4}\right)^2 = \frac{54}{256} = \frac{27}{128} = .2109$$

The formula thus makes it very easy to calculate probabilities, and eliminates the necessity of writing various symbols and lines to calculate the probability of a particular event occurring, as in the tree-diagram method (Fig. 3-9).

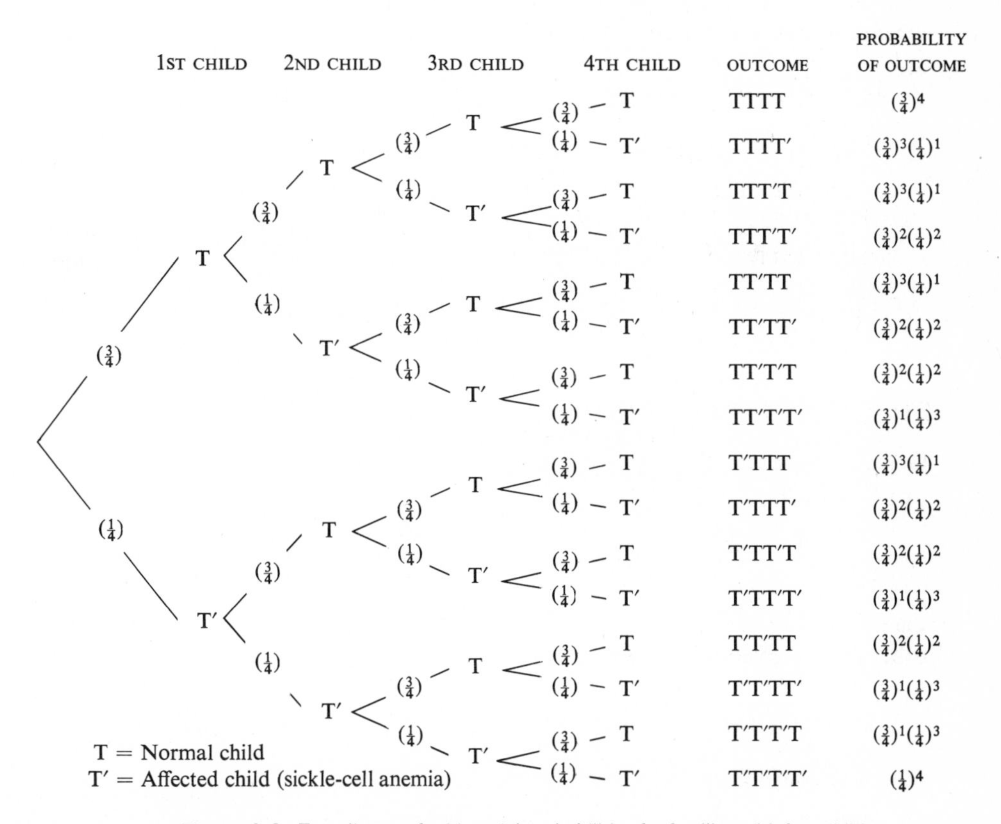

Figure 3-9. Tree diagram for binomial probabilities for families with four children.

When we deal with inherited diseases, the basic question of interest generally involves the probability of having *at least* one child afflicted with the disease—for example, what is the probability of having 1, 2, 3, or 4 children with sickle-cell anemia? Applying our formula, we calculate the probability of 1 sickle-cell child and 3 normal children plus the probability of 2 sickle-cell children and 2 normal children plus the probability of 3 sickle-cell children and 1 normal child plus the probability of all 4 children having sickle-cell and we get

$$\frac{4!}{1!\,3!}\left(\frac{1}{4}\right)^1\left(\frac{3}{4}\right)^3 + \frac{4!}{2!\,2!}\left(\frac{1}{4}\right)^2\left(\frac{3}{4}\right)^2 +$$

$$\frac{4!}{3!\,1!}\left(\frac{1}{4}\right)^3\left(\frac{3}{4}\right)^1 + \frac{4!}{4!\,0!}\left(\frac{1}{4}\right)^4\left(\frac{3}{4}\right)^0 =$$

$$\frac{108}{256} + \frac{54}{256} + \frac{12}{256} + \frac{1}{256} = \frac{175}{256} = .6836$$

Therefore, in a family of 4 children born to heterozýgous parents the probability of their having at least one child with sickle-cell anemia is 0.6836.[3] This is a rather high probability, and might cause the parents to seriously consider the consequences of their plans.

All the examples we have discussed so far have had four similar properties, which can be summarized as follows:

1. An experiment has two outcomes; for example, male or female, albino or normally pigmented, having sickle-cell anemia or not having sickle-cell anemia. We generally refer to such cases as having *dichotomous* outcomes.
2. The experiment is repeated for a number of trials under identical conditions; for example, several children born to the same couple.
3. The probability of one outcome is complementary to the other; that is, the sum of their probabilities is always 1; e.g., $\frac{1}{2} + \frac{1}{2} = 1$; $\frac{1}{4} + \frac{3}{4} = 1$. In general we say that the probability of one outcome is p, and the probability of the other outcome is $1 - p = q$. Furthermore these probabilities are constant and do not vary from trial to trial.
4. The trials are independent.

Genetic situations or experiments that satisfy these conditions are known as *binomial experiments*. In setting up such experiments, we would like to predict in advance the probabilities associated with obtaining k occurrences of one of the dichotomous outcomes and $n - k$ occurrences of the other outcome. Such probabilities are referred to as *binomial probabilities*. For a general binomial experiment with n identical trials, the probability of k occurrences of one type (outcome) and $n - k$ occurrences of the other type is given by:

$$\frac{n!}{k!\,(n-k)!}\,p^k q^{n-k},$$

[3]This probability can also be obtained by calculating

$$1 - \text{Prob [0 affected children]} = 1 - \frac{4!}{0!\,4!}\left(\frac{1}{4}\right)^0\left(\frac{3}{4}\right)^4 = 1 - \left(\frac{3}{4}\right)^4 = 1 - \frac{81}{256} = \frac{175}{256} = .6836$$

where $\frac{n!}{k!(n-k)!}$ is the number of possible arrangements of k of type 1 and $n - k$ of type 2;

p is the probability of type 1 in a single trial;

k is the number of type 1;

q is the probability of type 2 in a single trial;

$n - k$ is the number of type 2.

When one wishes to calculate the probability of a particular event occurring when the number of trials involved is large, the benefits of our formula will become abundantly clear. For example, it was reported in one New York City hospital that of 20 live births (in a two-day period), all 20 were girls. What is the probability of this event occurring? (Even though we are not dealing with one family, the conditions for a binomial experiment are satisfied.) In this case we find that $n = 20$ (20 identical trials), $k = 20$ (since we are interested in 20 girls), $n - k = 20 - 20 = 0$ (20 girls and 0 boys), that $p = \frac{1}{2}$ (since for a single fertilization of an egg the probability of a girl is $\frac{1}{2}$) and that $q = 1 - p = 1 - \frac{1}{2} = \frac{1}{2}$. The probability of 20 girls in 20 live births is obtained by substituting, yielding

$$\frac{20!}{20!(20-20)!}\left(\frac{1}{2}\right)^{20}\left(\frac{1}{2}\right)^{0} = 1 \times \left(\frac{1}{2}\right)^{20} \times 1 = \frac{1}{1{,}048{,}576} = .0000009536$$

This is a rare occurrence, needless to say! It is interesting to note that even though the probability is incredibly small it does not mean that the event cannot occur.

The binomial formula can be very useful for calculating probabilities and gaining some insight into the likelihood that particular events will occur. Scientists and genetic counselors alike find it a valuable tool for understanding various phenomena.

THE CHI SQUARE GOODNESS OF FIT TEST

In the preceding sections we have discussed a number of techniques that can be used to predict outcomes of genetic crosses. We have also pointed out that in working with probabilities, we cannot predict with absolute certainty the results of crosses, nor do we expect that ideal ratios will necessarily be attained in experiments with living organisms because of segregation and random assortment.

For example, as we have mentioned, Mendel made extensive observations on different traits in plants. From the data in Table 2-1 and the summary of some of the patterns he observed in Table 3-1, it is clear that Mendel dealt with large numbers of plants, which allowed him to discern patterns of inheritance. Had he studied organisms such as humans, which produce very few offspring per generation, it is doubtful that Mendel would have been able to discern specific ratios. (Also, the generation time in humans is too long for a single researcher to follow more than a few generations, making it impossible to discover patterns in the same manner as Mendel did.)

Table 3-1 Summary of Some of the Data in Table 2-1

Trait	*Number of* F_2 *offspring*	*Ratio*
Round/Wrinkled seed	5474 Round/1850 Wrinkled	2.96 to 1
Colored/White flowers	705 Colored/224 White	3.15 to 1
Yellow/Green cotyledons	6022 Yellow/2001 Green	3.01 to 1
Green/Yellow pods	428 Green/152 Yellow	2.82 to 1
Inflated/Constricted pods	882 Inflated/299 Constricted	2.95 to 1

We can illustrate the difficulty in formulating models when dealing with small numbers of observations by considering the family data presented in Table 3-2. Even though we expect a 1 to 1 ratio of females to males (that is, an equal number of males and females) in the families, it is extremely difficult to detect such a pattern in the ratios given in Table 3-2. Actually, in humans the sex ratio of birth is not exactly 1 to 1, but for all practical purposes, this ratio is a good approximation. In fact, for the data in Table 3-2, it is difficult to detect any pattern whatsoever.

Table 3-2 Number of Male and Female Offspring

Family	*Number of male children*	*Number of female children*	*Male/female ratio*
1	2	3	1 to 1.5
2	1	2	1 to 2
3	1	0	1 to 0
4	2	1	2 to 1
5	1	1	1 to 1
6	3	2	1.5 to 1
7	2	1	2 to 1
8	2	2	1 to 1
9	3	1	3 to 1

Consider families 8 and 9. Both have four children, but whereas for family 8 the ratio is 1 to 1, for family 9 it is 3 to 1. The difference between 3 males and 1 female and 2 males and 2 females has a considerable effect when the total number of observations is small. But if we consider families with 100 children, a difference of one offspring is hardly noticeable. For example, in a family with 53 males and 47 females, the ratio is 1.13 to 1; for a family with 52 males and 48 females the ratio is 1.08 to 1.

In the plants Mendel studied, if the results for round and wrinkled seeds had been 5473 round and 1851 wrinkled rather than the 5474 round and 1850 wrinkled which he actually obtained, the ratio would still have been 2.96 to 1, still extremely close to 3 to 1. Therefore, if organisms with large numbers of progeny are studied in transmission genetics, slight variations and deviations will not play as significant a role as in the case of organisms that produce small numbers of progeny.

Looking again at Table 3-1, we note that none of the ratios is *exactly* 3 to 1. Some are very close to 3 to 1, such as the ratio in the offspring of crosses between yellow/green cotyledons, round/wrinkled seeds, and inflated/constricted pods, whereas the offspring of crosses involving colored/white flowers and green/yellow pods differ somewhat more from the ratio 3 to 1. *How different from* 3 *to* 1 *could the ratios be and still be considered a reasonable representation of a* 3 *to* 1 *pattern*? The term used to denote the degree to which *observed* results agree with *expected* ratios is *goodness of fit*.

Consider Mendel's experiment for green and yellow pods, in which he studied 580 plants. In his results Mendel observed 428 green pods and 152 yellow pods, a ratio of 2.82 to 1. But suppose instead that Mendel had observed 387 green pods and 193 yellow pods, a 2.05 to 1 ratio—he probably would have had serious doubts that his hypothesis of a 3 to 1 ratio was valid. What if Mendel had observed something such as 417 green pods and 163 yellow pods, a ratio of 2.56 to 1? How large a deviation can we accept and still consider that a hypothesized ratio reflects the actual situation? Whenever a pattern is observed or a model proposed, it is necessary to consider how to validate and confirm it. In genetics we must be able to test whether a proposed model is valid and whether it explains our observations. In other words, we must be able to test its goodness of fit.

Let us consider a very simple assumption and discuss how we go about making predictions from the proposed assumption, collecting data, and then testing whether our observations can be explained reasonably by the predictions. Assume that we have a die, and want to test whether it is a fair die. The term *fair* suggests that it is a balanced die and that the probability of getting any one of the six faces uppermost when the die is tossed is $\frac{1}{6}$. That is, the probabilities of "rolling" a 1, 2, 3, 4, 5, or 6 are all the same, $\frac{1}{6}$. In order to test this assumption, we can perform an experiment and collect data. We have discussed previously the need to make a large number of observations, since a few tosses of the die would not be likely to suggest any clear-cut patterns. We therefore decide to toss the die 300 times. The number of 1's, 2's, 3's, 4's, 5's, and 6's from a sample "experiment" are recorded in Table 3-3. How do we test whether these observations were likely to have been the result of tossing a fair die?

Table 3-3 Results of Tossing a Die 300 Times

Uppermost face of the die	1	2	3	4	5	6	*Total*
Observed number of tosses with face uppermost	45	52	58	37	48	60	300

Let us predict, based on the assumption of a fair die, what we would *expect* to occur if we tossed our die 300 times. More specifically, if we have a fair die, how many of the 300 tosses would we expect to result in a 1, how many in a 2, and so on? If we toss a die 300 times and we expect equally likely outcomes for each of the six faces, it follows that $\frac{1}{6}$ of 300 would be the number of 1's expected, $\frac{1}{6}$ of 300 would be the number of 2's expected, and so on. Therefore our predictions from the assumptions are that we would expect each face to appear uppermost 50 times. We combine the observed results and the expected results (derived from the assumptions) in Table 3-4.

Table 3-4 Observed and Expected Results When Tossing a Die 300 Times

Uppermost face of the die	1	2	3	4	5	6	*Total*
Observed number of tosses with face uppermost	45	52	58	37	48	60	300
Expected number	50	50	50	50	50	50	300

How do we now compare the observed and expected results? One possible criterion for deciding how well the observations (based on the experiment) agree with the expected values (based on the assumption) is to look at the difference between the observed and expected results. We let O_i represent the observed number for the ith category, and E_i the expected number for the ith category. For example, when $i = 1, O_1 = 45$, and $E_1 = 50$; for $i = 2, O_2 = 52$, and $E_2 = 50$, and so on. Since we are interested in comparing our observations and the expected values, it makes sense to consider the difference between the observed values and the expected values, that is, $O_i - E_i$ for each value of i. These results are given in Table 3-5. To get some measure of how large or small these deviations are, we might total the difference $O_i - E_i$ for all values of i. That is, for the six values of i, we get $(-5) + (+2) + (+8) + (-13) + (-2) + (+10) = 0$. Does this mean that there is no difference between the observed and expected values? There *is* a difference, since the observed and expected values are not equal, so there must be an explanation of why we obtained the value 0. If we write the difference in terms of the original values, we get $(45 - 50) + (52 - 50) + (58 - 50) + (37 - 50) + (48 - 50) + (60 - 50)$. We can rewrite these values as $(45 + 52 + 58 + 37 + 48 + 60) - (50 + 50 + 50 + 50 + 50 + 50) = 300 - 300$.

Table 3-5 Deviations Between Observed and Expected Values in Die Experiment

Uppermost face	1	2	3	4	5	6
O_i	45	52	58	37	48	60
E_i	50	50	50	50	50	50
$O_i - E_i$	−5	+2	+8	−13	−2	+10

The first set of numbers adds up to 300, the number of trials of the experiment. So does the second set, since our expected values were based on 300 trials. It is not surprising then that the difference of the two sums of numbers is 0—in fact, this will *always* happen.

It is convenient at this point to introduce some notation. Mathematicians and statisticians often add a series of numbers, so they use $\sum$ (the Greek capital sigma) to give instructions to add numbers. In our example we would write $\sum (O_i - E_i)$. We can also give further instructions, for example to sum the quantities $(O_i - E_i)$ when $i = 1, i = 2, \ldots, i = n$, and we denote this by writing $\sum_{i=1}^{n} (O_i - E_i)$. In our problems we will be summing over all values, so that we write in general $\sum_i (O_i - E_i)$, or just $\sum (O_i - E_i)$.

We have seen that no matter what the observed values, then $\sum (O_i - E_i)$, the sum of all the deviations $O_i - E_i$, the sum will always be zero. Nevertheless, considering the deviations between observed and expected values is intuitively appealing. Is there a way to consider this deviation without obtaining values that will sum to zero? What we really need to consider is *how large* each deviation $O_i - E_i$ is, while disregarding whether it is positive or negative. There are two common ways of doing this: considering the absolute value[4] $|O_i - E_i|$ or considering the square $(O_i - E_i)^2$. Then we can consider either the sum $\sum |O_i - E_i|$ or $\sum (O_i - E_i)^2$. Both approaches add nonnegative numbers and would yield zero only if all deviations $O_i - E_i$ are zero. The first approach is intuitively very appealing but theoretical considerations make it a difficult one to work with.

What about the second approach? The square of any number is a positive value. What would small values of $\sum (O_i - E_i)^2$ suggest as opposed to large values of $\sum (O_i - E_i)^2$? If the agreement between observed and expected values is good (that is, the model is a good predictor of what is observed), the deviations between O_i and E_i should be small, and hence $\sum (O_i - E_i)^2$ should be small. If the model is a poor predictor of what is observed, the deviations between observed and expected values will be large, and hence $\sum (O_i - E_i)^2$ will also be large. Therefore, we will need to determine criteria for deciding what values of $\sum (O_i - E_i)^2$ we consider large and what values we consider small. Before discussing this point, we have to examine one further point that is extremely important in determining whether observed and expected values agree. In Table 3-4 we considered the results of tossing a die 300 times. Let us now suppose that the die has been tossed 3000 times. Both the old and the new sets of data are included in Table 3-6. We know that if the die is fair, making the six faces equally likely to appear, in 300 tosses of the die we expect each face to appear 50 times. In 3000 tosses of a fair die we expect each face to appear 500 times. Considering the deviations between observed and expected values in both cases, we obtain the data in Table 3-7.

If we study both cases, that is, tossing the die 300 times and 3000 times, we find that the deviations $O_i - E_i$ for both cases are the same, and therefore the values

[4]Absolute value makes all negative numbers positive, that is

$$|a| = \begin{cases} a & \text{if } a \geq 0 \\ -a & \text{if } a < 0 \end{cases}$$

Table 3-6 Results of Tossing Die

Face uppermost	*Results of 300 tosses*	*Results of 3000 tosses*
1	45	495
2	52	502
3	58	508
4	37	487
5	48	498
6	60	510

Table 3-7 Calculation of $\sum (O_i - E_i)^2$

Face uppermost	*300 tosses*		$O_i - E_i$	$(O_i - E_i)^2$
	Observed	*Expected*		
1	45	50	−5	25
2	52	50	2	4
3	58	50	8	64
4	37	50	−13	169
5	48	50	−2	4
6	60	50	10	100
			$\sum (O_i - E_i)^2 =$	366

Face uppermost	*3000 tosses*		$O_i - E_i$	$(O_i - E_i)^2$
	Observed	*Expected*		
1	495	500	−5	25
2	502	500	2	4
3	508	500	8	64
4	487	500	−13	169
5	498	500	−2	4
6	510	500	10	100
			$\sum (O_i - E_i)^2 =$	366

for $\sum (O_i - E_i)^2$ are the same. But if we look at the discrepancy between the observed and expected values, we see that there is a difference even though our measure seems to suggest that they are the same. Does it make sense to consider a deviation of 5, when the expected value is 50, to be comparable to a deviation of 5 when the expected value is 500? Or is a deviation of 2, relative to an expected value of 50, comparable to a deviation of 2 when the expected value is 500? It is very important to note that even though the *values* of the numbers are the same, *relatively* they are not comparable. (Similarly, when a newborn baby gains five pounds, the baby is almost doubling its weight, but for an adult a gain of 5 pounds represents a very small fraction of overall body weight.) Therefore, how do we compensate for such differences?

If we consider each deviation squared *relative to its own expected value*, that is, $\frac{(O_i - E_i)^2}{E_i}$, we can then scale the deviation appropriately. For example, in the 300-toss experiment the first deviation would be $\frac{(45 - 50)^2}{50} = \frac{(-5)^2}{50} = \frac{25}{50} = .5$; whereas in the 3000-toss experiment we would get $\frac{(495 - 500)^2}{500} = \frac{(-5)^2}{500} = \frac{25}{500} = .05$—a more realistic measure of the magnitude of the deviation between observed and expected.

The measure used for testing the agreement between predictions from a model and the experimental observations is given by the quantity X^2 (chi square), where $X^2 = \sum \frac{(O_i - E_i)^2}{E_i}$. For the case of 300 tosses, $X^2 = \frac{(45 - 50)^2}{50} + \frac{(52 - 50)^2}{50} + \frac{(58 - 50)^2}{50} + \frac{(37 - 50)^2}{50} + \frac{(48 - 50)^2}{50} + \frac{(60 - 50)^2}{50} = .50 + .08 + 1.28 + 3.38 + .08 + 2.00 = 7.32$. For the 3000 tosses, $X^2 = \frac{(495 - 500)^2}{500} + \frac{(502 - 500)^2}{500} + \frac{(508 - 500)^2}{500} + \frac{(487 - 500)^2}{500} + \frac{(498 - 500)^2}{500} + \frac{(510 - 500)^2}{500} = .050 + .008 + .128 + .338 + .008 + .200 = .732$. We can see that $X^2 = \sum \frac{(O_i - E_i)^2}{E_i}$ more accurately reflects the agreement between prediction and observations. But can we say that the value for the 300-toss experiment does *not* reflect agreement? What about another similar experiment, for which the values are given in Table 3-8? For these values $X^2 = \frac{(57 - 50)^2}{50} + \frac{(47 - 50)^2}{50} + \frac{(59 - 50)^2}{50} + \frac{(39 - 50)^2}{50} + \frac{(43 - 50)^2}{50} + \frac{(55 - 50)^2}{50} = .98 + .18 + 1.62 + 2.42 + .98 + .50 = 6.68$. What do we say about these values?

Table 3-8 Results of 300 Additional Tosses of a Die

Uppermost face	1	2	3	4	5	6
O_i	57	47	59	39	43	55
E_i	50	50	50	50	50	50
$O_i - E_i$	+7	−3	+9	−11	−7	+5

How do we compare $X^2 = 7.32$ and $X^2 = 6.68$? We can say that the agreement in the second case ($X^2 = 6.68$) is better than in the first case ($X^2 = 7.32$), but does that mean that in the first case we can conclude that we have an unfair die?

We need a method for deciding what to do with the calculated values of X^2. When is a value of X^2 so large that we cannot consider there to be reasonable agreement between the prediction of the model and the observations? What is this cutoff value for X^2? Can we each pick our own cutoff value?

Imagine having to take medicine whose efficacy was determined purely by an experimenter choosing his own cutoff value! It is obviously important to have some universal procedures for making such decisions, for determining a cutoff value.

These procedures should result in our selecting a value that is universally agreed upon (given prior agreement about the level of certainty we want to attain in our conclusion). Exceeding the cutoff should cause us to doubt that there is agreement between our expected values and observed values, or in other words, to doubt that our model is valid for the observations we have made.

It is important to note that we can *never be absolutely certain* in our decision. Our observations may appear to be so extreme that we doubt the model could possibly be appropriate for what we have observed. We must realize that fluctuation and variation are always present in observable phenomena. (For example, if a coin is flipped 10 times and the number of heads is recorded, and then the process is repeated by 100 people, it is possible—though not likely—that one person will get 0 heads in 10 flips. We cannot conclude with absolute certainty that cheating occurred, but we might have serious suspicions.) As a further example, we saw that the probability of 20 girls being born in 20 births was .00000095367, nevertheless it did occur in a New York City hospital. Therefore, in deciding on the validity of a model we may be faced with such large values of X^2 that we doubt agrèement could be possible, but in fact our model may reflect reality in spite of our observations. A comment on this point is illustrated in Fig. 3-10. Therefore when we pick a cutoff value, it is entirely possible that we could erroneously reject a model that is actually appropriate (type I error); or, we may fail to reject an inappropriate model (type II error). Hence there are two possible kinds of error.

The methodology we are about to develop will enable us to determine the risk we take of incorrectly rejecting an appropriate model. When testing a model a researcher may decide to accept a 5% risk of rejecting a model when in fact it is really valid. The researcher determines the acceptable level of risk depending on the seriousness of the situation. In some cases a risk as low as 1% or even 0.1% might be appropriate. This distinction can be understood by considering the following two situations. (1) Suppose you go to a physician to determine whether you have a cold or not. What risk are you willing to take of an incorrect diagnosis? How serious would it be if the physician told you that you do not have a cold when you actually do have a cold? How serious would it be if the physician told you that you have a cold when you actually do not? (2) Suppose you go to a physician to determine whether you have

Figure 3-10. A cartoon comment on absolute certainty. (Cartoon by Johnny Hart in *Scientific American,* March 1976; reprinted by permission of Field Newspaper Syndicate.)

lung cancer or not. What risk are you now willing to take of an incorrect diagnosis? How serious would it be if the physician told you that you do not have lung cancer when you actually do have lung cancer? How serious would it be if the physician told you that you have lung cancer when you actually do not? Ideally, one wants to minimize both kinds of errors; but this is very difficult since decreasing one type of error increases the other. We shall concentrate on the first kind of error, that of rejecting our model when it is actually true. The risk of error that one is willing to take depends entirely on the situation. Historically the values most commonly used are 10%, 5%, and 1%.

In summary, we must determine a value of X^2 that we will deem to be so large that we doubt there could be agreement between the observations and the expected values derived from our model. But when we choose the value of X^2, we do so knowing that we are taking a risk of making an error. The beauty of the approach is that it enables us to measure the risk we are taking, and to state the risk clearly. The risk is known as the *level of significance*. The procedure for testing models is known as the χ^2 *Goodness of Fit Test*, where χ is the Greek letter "Chi."

How do we select the cutoff values for a predetermined risk, that is, for a specified level of significance? The probability that the calculated value of X^2 will exceed a given number depends on the number of categories being studied. These probabilities are incorporated in χ^2 tables, with a different set of values for each number of *degrees of freedom* (df)—the number of df is in general one less than the number of categories. With 2 categories, there is 1 df. Consider the following:

	O_i	E_i
Tall:	80	75
Short:	20	can *only* be 25
	100	100

There is one degree of freedom, since once one of the expected values is determined the other derives automatically from our knowledge of what the expected values must total. In our die-tossing example there are 6 categories (the faces of the die: 1, 2, 3, 4, 5, 6) and hence 5 df. In general we write $X^2 \sim \chi^2_{k-1}$ where k is the number of categories.[5] In this example, $X^2 \sim \chi^2_5$. What good does this do us? If we can establish beforehand the ideal range and frequency of X^2 values, we can collect data and calculate their X^2 value, then compare our value of X^2 to all possibilities. In doing this we can decide whether our value of X^2 is a commonly occurring value, or an extreme value. In fact we can determine the probability of obtaining an X^2 value as large or larger than the one we have calculated from our data. It can be shown that with 5 df, the probability that $X^2 \geq 1.15$ is .95, the probability that $X^2 \geq 1.61$ is .90, the probability that $X^2 \geq 9.24$ is .10 and the probability that $X^2 \geq 11.07$ is .05.

What can we do with such information? Suppose we choose 11.07 to be our cutoff value. That is, any time we perform this experiment, if we calculate from the data a value of X^2 that is larger than 11.07, we will consider that the disagreement

[5]Please note the distinction between the notation X^2 and the notation χ^2. The former stands for a random sum of differences, the latter for an ideal distribution.

between observed and expected values is so large that we doubt the model was appropriate. If our X^2 value is less than 11.07, we will not have sufficient evidence to doubt the validity of the model. If we were to do this repeatedly and in fact if we have the correct model, about 5% of our X^2 values would tend to be greater than 11.07 purely by chance. Very often when our data yield a value of X^2 greater than 11.07 we will legitimately have come upon cases where the model is not valid and then be making the correct decision. The point is, if we do have the correct model (which we never know for sure) and we use 11.07 as our cutoff point, the probability of rejecting our model when it is actually true is .05. This is the way we quantify our risk of error.

Similarly, if we want to take a 1% risk of rejecting our model when the model is actually true, we want to find the value of X^2 at which the probability that X^2 is greater than some number c is equal to .01. By consulting an appropriate table we will find that with 5 df, the cutoff value we want is 15.09—that is, the probability that X^2 is greater than or equal to 15.09 is .01. Therefore, if we take 15.09 to be our cutoff value, and if our model is valid, the probability of rejecting our model when the model is true is .01. By using X^2, we can establish a criterion for rejecting our model, and we can also determine the probability of rejecting the model when it is actually true—that is, the probability (at the stated level of significance) of committing an error (the type I error).

How do we determine the cutoff values and the corresponding probabilities of error? There are available tables of χ^2 distribution incorporating different degrees of freedom, and for various probabilities a χ^2 distribution is given in Table 3-9.

Table 3-9 The Chi Square Table

	Probability								
df	0.95	0.90	0.70	0.50	0.30	0.20	0.10	0.05	0.01
1	0.004	0.016	0.15	0.46	1.07	1.64	2.71	3.84	6.64
2	0.10	0.21	0.71	1.39	2.41	3.22	4.61	5.99	9.21
3	0.35	0.58	1.42	2.37	3.67	4.64	6.21	7.82	11.35
4	0.71	1.06	2.20	3.36	4.88	5.99	7.78	9.49	13.28
5	1.15	1.61	3.00	4.35	6.06	7.29	9.24	11.07	15.09

Abridged from Table IV of Ronald A. Fisher and Frank Yates: *Statistical Tables for Biological, Agricultural and Medical Research*, published by Longman Group Ltd., London (previously published by Oliver and Boyd, Edinburgh), and by permission of the authors and publishers. (Sixth edition, 1974.)

The values under the heading df give the degrees of freedom, the values across the top give the probabilities, and the body of the table gives the cutoff values. If we look at the row corresponding to 5 degrees of freedom and follow the column under .05, the value 11.07 lies at the intersection of the row and column. The probability that X^2 is greater than or equal to 11.07 is .05. If df = 2, and we want to determine the cutoff value for a 10% risk of rejecting the model when it is actually true (10% level of significance), the intersection of the column with the heading .10 and the row with df 2 gives 4.61. Thus, the probability that X^2 is greater than or equal to 4.61 is .10. If there are 3 degrees of freedom we find that the cutoff value when the level of significance is .05 is 7.82—that is, the probability that X^2 is greater than or equal to 7.82 is .05.

We return now to our examples of tossing the die to illustrate how the whole procedure works. We are interested in ascertaining whether we have a fair die. Prior to performing the experiment we determine the risk (level of significance) we are willing to take—that is, how big a risk are we willing to take in saying that the die is unfair when it is actually fair? (If we were dealing with a professional gambler, we might want our risk to be about .0001, otherwise .05 might be reasonable.) Suppose we decide that we are willing to take a 5% risk. From the χ^2 table, with 5 degrees of freedom (since there are 6 categories), we determine that the cutoff value is 11.07. We are now ready to perform the experiment, calculate X^2, and make a decision regarding the die. For the experiment of 300 tosses of the die the data yielded an X^2 value of 7.32. Comparing 7.32 to 11.07, we can conclude that at the 5% risk of error we do not have sufficient evidence to doubt that we have a fair die.

Our similar analyses of the experiment with 3000 tosses of the die, yielding $X^2 = .732$[6] and of the second experiment with 300 tosses, yielding $X^2 = 6.68$, also lead us to the conclusion that at the 5% level of significance there is not sufficient evidence to conclude that we have an unfair die. (Note that if we want to take a smaller risk, that is, less than 5%, the cutoff value must be larger.)

It is always important to indicate the level of significance, since the conclusion may be altered for a different level of significance. For example, in the 300-toss experiment, if the level of significance is .30, the cutoff value is 6.06, making 7.32 too high and giving us reason to doubt that the die is fair.

USE OF THE CHI SQUARE TEST IN GENETICS

Let us assume again that we are studying inherited traits in pea plants, specifically the trait of seed color, in which Yellow is dominant and Green is recessive, and that of Round Seed is dominant and Wrinkled Seed is recessive. Homozygous Yellow/Round plants are crossed with homozygous Green/Wrinkled plants. The F_1 generation are all dominant in phenotype, that is, producing yellow, round seeds. We are interested in testing the hypothesis[7] of normal dihybrid segregation, that is, 9 : 3 : 3 : 1, for the F_2 generation. We choose the .05 level of significance. An F_2 generation is produced with the following results:

Yellow/Round	Yellow/Wrinkled	Green/Round	Green/Wrinkled	Total
2,504	853	881	292	4,530

In order to determine the cutoff value for X^2, we note that there are 3 degrees of freedom (since there are four categories), and upon consulting the χ^2 table we find that the cutoff value is 7.82. That is, for any $X^2 > 7.82$, we will, with a 5% risk, claim that the model is incorrect.

[6]In fact this value of X^2 is so small we might wonder if the data has been fudged.
[7]Statisticians would formally refer to this as the null hypothesis.

In order to calculate X^2 we must determine the expected number of each type of seed. From the ratio 9 : 3 : 3 : 1 we can determine what proportion of the total seeds will be of each type. We can translate the ratio into probabilities by summing the numbers and then dividing each number by the sum, that is, $9 + 3 + 3 + 1 = 16$ and therefore in terms of probabilities we can express the same ratio as $\frac{9}{16} : \frac{3}{16} : \frac{3}{16} : \frac{1}{16}$. We can then use these probabilities to determine the expected values. If there are 4530 seeds and if the predictions from the model are valid, we expect $\frac{9}{16}$ of the 4530 seeds or 2548.1 to be round and yellow; $\frac{3}{16}$ of 4530, or 849.4, to be yellow and wrinkled; $\frac{3}{16}$ of 4530, or 849.4, to be green and wrinkled; and $\frac{1}{16}$ of 4530, or 283.1, to be green and wrinkled. We calculate X^2 from the computations of Table 3-10. Therefore $X^2 = .763 + .015 + 1.176 + .280 = 2.234$. Comparing this value to 7.82, we find that we do not have sufficient evidence to doubt our hypothesis of normal dihybrid segregation in a 9 : 3 : 3 : 1 ratio.[8,9]

Table 3-10 Calculation of X^2

Trait	O_i	E_i	$O_i - E_i$	$(O_i - E_i)^2$	$\frac{(O_i - E_i)^2}{E_i}$
Yellow/Round	2504	2548.1	−44.1	1944.81	$\frac{1944.81}{2548.1} = .763$
Yellow/Wrinkled	853	849.4	3.6	12.96	$\frac{12.96}{849.4} = .015$
Green/Round	881	849.4	31.6	998.56	$\frac{998.56}{849.4} = 1.176$
Green/Wrinkled	292	283.1	8.9	79.21	$\frac{79.21}{283.1} = .280$

Robertson (1937) gives data which include the F_2 progeny of a barley cross from F_1 normal green plants. The observed characteristics are Non-2-row versus 2-Row, and normal Green versus Chlorina plant color. We can use Robertson's data (Table 3-11) to test the model of normal dihybrid segregation, that is, a 9 : 3 : 3 : 1 ratio.

We use our model to determine the expected number of barley plants with each trait. For a total of 1898 plants, we expect that $\frac{9}{16}$, or 1067.6, will be Green/Non-2-row. We summarize in Table 3-12. We now determine the level of significance (risk) and the corresponding cutoff value for X^2. Let us assume that we choose .01 as our level of significance. Since there are four categories, there are 3 degrees of freedom and hence from the χ^2 table we determine that the cutoff value for X^2 is 11.35. In Table 3-13 we include the figures necessary to calculate X^2. Therefore $X^2 = 11.416 + 11.835 + 19.310 + 11.794 = 54.355$.

[8]Our description of how to calculate X^2 is not completely accurate if there are just two distinct groups. In the latter case, it is in principle necessary to use the so-called Yates Correction Factor. (See Daniel, 1974) Often in practice this correction factor is disregarded and we shall disregard it.

[9]The reader should be aware that the χ^2 Goodness of Fit Test cannot be used when the observations are in terms of percentages or there are too few observations.

Table 3-11 F_2 Progeny of a Barley Cross from F_1 Normal Green Plants

Trait	*Observed number of plants*
Green/Non-2-row	1178
Green/2-Row	291
Chlorina/Non-2-row	273
Chlorina/2-Row	156
Total	1898

Table 3-12 Expected Values in Barley Cross

Trait	O_i	E_i
Green/Non-2-row	1178	$\frac{9}{16}(1898) = 1067.5$
Green/2-Row	291	$\frac{3}{16}(1898) = 355.9$
Chlorina/Non-2-row	273	$\frac{3}{16}(1898) = 355.9$
Chlorina/2-Row	156	$\frac{1}{16}(1898) = 118.6$

Table 3-13 Calculation of X^2 for Barley Experiment

Trait	O_i	E_i	$\frac{(O_i - E_i)^2}{E_i}$
Green/Non-2-row	1178	1067.6	$(1178 - 1067.6)^2/1067.6 = 11.416$
Green/2-Row	291	355.9	$(291 - 355.9)^2/355.9 = 11.835$
Chlorina/Non-2-row	273	355.9	$(273 - 355.9)^2/355.9 = 19.310$
Chlorina/2-Row	156	118.6	$(156 - 118.6)^2/118.6 = 11.794$

Comparing the calculated value of X^2 (54.355) to the cutoff value (11.35), we would reject the claim that we have dihybrid segregation as evidenced by the 9 : 3 : 3 : 1 ratio. Formally, we state that at the .01 level of significance, we have sufficient evidence to reject the hypothesis of 9 : 3 : 3 : 1. Such a conclusion would suggest to the researcher that further work is necessary to develop a new model that will explain the observations; predictions can then be made from the new model, and ultimately it can be tested. In short, the modeling process must be started all over again. Through this process, an explanation was found to explain why the 9 : 3 : 3 : 1 ratio did not adequately describe the observations in Robertson's experiments; the type of inheritance involved will be discussed in Chapter 6. In many respects, science is a continuing process of developing new models based on creative insights and painstakingly gathered experimental information.

PROBLEMS

1. A normal woman whose father was albino marries a man who is albino. What phenotypic proportions can be expected among their offspring? What is the probability that they will have (a) an albino child, (b) a normally pigmented child?

2. What proportions may be expected in the offspring of two heterozygotes for brachydactyly (short fingers, which is caused by a dominant gene)? What is the probability of their having (a) a child with short fingers, (b) a child with normal fingers?

3. What is the probability that the offspring of two heterozygotes for brachydactyly will be male with brachydactyly?

4. A woman who is heterozygous for brachydactyly and albinism marries a man who is heterozygous for albinism and has normal-sized fingers. List (a) the genotypes and (b) the phenotypes of the offspring of such a mating and the corresponding probabilities. If you are interested in the sex of the offspring of such a mating, how would the probabilities in each case be modified?

5. If the probability of having trait A is $\frac{2}{3}$ and the probability of having trait B is $\frac{3}{4}$, and the traits are independent, what is the probability that

(a) a person will have both traits;
(b) a person will have neither trait;
(c) a person will have exactly one trait?

6. List the corresponding probabilities of (a) the genotypes and (b) the phenotypes of the offspring of the mating $TtGg \times TtGg$.

7. If a couple plan to have five children, what is the probability that they will have three boys and two girls?

8. Dark hair is dominant over blond hair. If parents both heterozygous for hair color have four children:

(a) What is the probability that all the children will be blond?
(b) What is the probability that four children will have dark hair?
(c) What is the probability they will have two dark-haired children and two blond children?

9. The inability to taste PTC (phenylthiocarbamide) is inherited as a recessive trait and so is red hair. If parents both heterozygous for these traits have three children:

(a) What is the probability they will have three children who are red-haired and who cannot taste PTC?
(b) What is the probability that these children will be girls?

10. In humans, polydactyly (extra fingers) is due to a dominant gene. When one parent is polydactylous, but heterozygous, and the other parent is normal, what is the probability that if they have three children (a) all will have polydactyly; (b) none will have polydactyly; (c) one will have polydactyly and two will be normal?

11. If the parents in problem 10 were also heterozygous for short fingers (brachydactyly) list the phenotypes of the offspring of such a mating and the corresponding probabilities.

12. Intestinal polyposis, an abnormality of the human large intestine, is dependent on a dominant gene *P*, and Huntington's chorea, a neurological disorder, is determined by a dominant gene *H*. A man carrying the gene *P* (genotype *Pphh*) marries a woman carrying the gene *H* (genotype *ppHh*). Assume that *P* and *H* are on different chromosomes. If they have three children, what is the probability that all three children will be normal?

13. Assume that parents both heterozygous for brachydactyly (caused by a dominant gene) are planning a family of 10 children. Determine all possible outcomes for the number of children out of 10 with or without brachydactyly and for each outcome determine the corresponding probability.

14. In a cross between ivory and red snapdragons the following counts are observed in the F_2 generation:

Phenotype	*Number of plants*
Red	30
Pink	55
Ivory	25

On the basis of these data, can segregation be assumed to occur in the simple Mendelian ratio 1 : 2 : 1?

15. The crossing of two hybrids of a flower species yields the results shown below. Are these results consistent with the expected proportion 9 : 3 : 3 : 1?

Magenta Flower/ Green Stigma	Magenta Flower/ Red Stigma	Red Flower/ Green Stigma	Red Flower/ Red Stigma
120	49	36	12

16. In a certain cross of two varieties of pea, genetic theory leads us to expect half the seeds to be wrinkled and half to be round. In a sample of 800 seeds we find 440 wrinkled. Is there sufficient evidence to doubt the genetic theory?

17. Of 64 offspring of a certain cross between guinea pigs, 34 are red, 10 are black, and 20 are white. According to the genetic model, these numbers should be in the ratio 9 : 3 : 4. Are the data consistent with the model?

18. In an experiment with the shepherd's purse (*Capsella bursapastoris*), the model predicted a ratio of 15 plants with triangular seed capsules to 1 with ovoid capsules. The observed numbers were 141 to 19. What would you say about the model?

19. In an experiment with coat color in mice, a ratio of 9 agouti : 3 black : 4 albino was expected. The observed numbers were 95 agouti, 35 black, and 30 albino. Does the model seem appropriate?

REFERENCES

Mendelian Genetics

MENDEL, GREGOR, 1865. "Experiments in Plant Hybridization." Translated in J. A. Peters, ed., 1959. *Classic Papers in Genetics*, pp. 1–20. Prentice-Hall, Inc., Englewood Cliffs, N.J.

Probability and Statistics in Genetics

ALDER, H. L., and E. B. ROESSLER, 1972. *Introduction to Probability and Statistics*, 5th ed. W. H. Freeman & Company Publishers, San Francisco.

BISHOP, O. N., 1967. *Statistics for Biology*. Houghton Mifflin, Boston.

DANIEL, W. W., 1974. *Biostatistics: A Foundation for Analysis in the Health Sciences*. Wiley & Sons, Inc., New York.

HILL, A. B., 1961. *Principles of Medical Statistics*. Oxford University Press, New York.

MOSIMANN, J. E., 1968. *Elementary Probability for the Biological Sciences*. Prentice-Hall, Inc., Englewood Cliffs, N.J.

ROBERTSON, D. W., 1937. "Maternal inheritance in barley." *Genetics* 22: 104–113.

SCHEFLER, W. C., 1969. *Statistics for the Biological Sciences*. Addison-Wesley, Reading, Mass.

FOUR

The Interaction of Genes

INTRODUCTION

In the preceding chapters we have referred to the manner and pattern of transmission of a number of genetically determined traits in different organisms. All the traits discussed thus far show complete dominance or recessiveness and follow the expected patterns established by Mendel in his studies on the garden pea plant. As we mentioned before, traits determined by a single pair of alleles transmitted according to expected patterns are referred to as *simple Mendelian traits*.

If all inherited conditions were as clearcut as those we have already described, analysis of the genetic determination of all heritable characteristics would be simple. Yet we know that most inherited traits in complex plants and animals are still beyond the complete understanding of modern geneticists. In humans, the inheritance of traits such as intelligence, height, and weight, for example, are still not completely understood. The reason is that they are not simple Mendelian traits.

This fact became apparent in the years following the rediscovery of Mendel's work at the turn of the twentieth century. Classical experiments to discern the genetic determination of various inherited traits in a number of organisms showed significant variation from the ratios obtained in garden peas. In fact, it appeared that simple Mendelian traits are in the minority. As more such data accumulated, researchers realized the enormity of Mendel's good fortune in having chosen to study a particular plant and particular traits which involved mostly the simplest genetic patterns. This in no way demeans the brilliant monk's accomplishments, for few would have been able to equal his understanding of the significance of his data.

IN THE FOLLOWING SECTIONS WE SHALL BE DISCUSSING A NUMBER OF GENETIC PHENOMENA WHICH RESULT IN OFFSPRING RATIOS THAT DIFFER FROM THE PREDICTIONS OF MENDELIAN THEORIES. WE SHALL SEE HOW THE INTERACTION OF ALLELES CAN CAUSE HETEROZYGOTES TO VARY IN PHENOTYPE FROM INDIVIDUALS HOMOZYGOUS FOR THE DIFFERENT ALLELES, AS IN INCOMPLETE DOMINANCE AND CODOMINANCE.

ANOTHER OF THE IMPORTANT DIFFERENCES BETWEEN THESE PHENOMENA AND THE ONES THAT MENDEL OBSERVED IS THAT WITH FEW EXCEPTIONS, EVERY PAIR OF GENES HE STUDIED DETERMINED A SINGLE TRAIT. GENETICISTS HAVE SINCE FOUND THAT ONE PAIR OF GENES CAN AFFECT A WIDE VARIETY OF TRAITS. ALSO, WHERE MENDEL STUDIED TRAITS THAT APPEARED TO BE DETERMINED BY ONLY ONE PAIR OF ALLELES, WE SHALL DISCUSS IN THIS CHAPTER A NUMBER OF TRAITS EACH DETERMINED BY MORE THAN ONE PAIR OF GENES IN INTERACTION.

INTERACTION BETWEEN ALLELES

Incomplete Dominance

All the characteristics described in the preceding chapters were determined by alleles that were either completely dominant or completely recessive. In some cases, however, alleles can interact in such a way that the phenotype of the heterozygote can be distinguished from either homozygote. *Incomplete*, or *partial*, *dominance* is the term used to describe one type of interaction in which neither allele shows dominance over the other, and the heterozygote expresses a phenotype intermediate to those of the homozygotes.

Four o'clock flowers. In the early 1900s, C. Correns, one of the rediscoverers of Mendelian genetics, described the results of a series of crosses that he had made with four o'clock plants (*Mirabilis jalapa*). When homozygous red-flowered plants were crossed to homozygous white-flowered plants, the F_1 were uniformly pink.

A cross of the F_1 plants produced a generation of F_2 plants which included all three phenotypes—red, pink, and white—in a 1 : 2 : 1 ratio (Fig. 4-1). It appeared that the two alleles in the heterozygote were interacting in such a way as to produce a pigment different from that in either homozygote. Note that in situations where incomplete dominance applies, the phenotypic ratio is also the genotypic ratio, and the genotype can be ascertained directly by observation of the phenotype.

Andalusian fowl. Many other traits determined by incomplete dominance have since been observed. In Andalusian fowl, a species of chicken, a cross of homozygous black with homozygous white individuals produces only blue-gray offspring. The F_2 generation shows a phenotypic ratio of 1 black : 2 blue-gray : 1 white, indicating that this is again a case of incomplete dominance.

Incomplete dominance in humans. A number of traits in humans are believed to be determined by incomplete dominance. Curly-haired individuals homozygous for a particular allele when mated to straight-haired individuals produce progeny

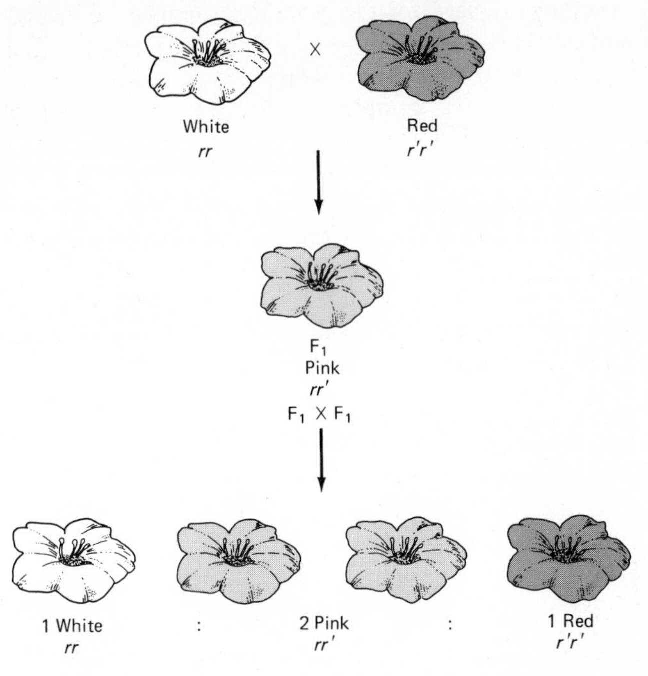

Figure 4-1. Incomplete dominance in four o'clock flowers.

with wavy hair. Similarly, voice tone is determined by homozygosity for one allele (y^1) in the bass or alto, and by homozygosity for a second allele (y^2) in the tenor or soprano; the heterozygotes are baritones and mezzo-sopranos (y^1y^2). Note that symbols for incompletely dominant alleles are usually written in the same lower-case letters, with superscripts to distinguish one from the other.

Codominance

A different kind of interaction between alleles can be found in traits determined by *codominant* genes. As the term implies, an individual heterozygous for two codominant alleles manifests expression of both traits determined by the genes in an undiluted or unmodified manner.

ABO blood group in humans. The classic example of codominant inheritance is the genetic determination of human blood groups. Immunological substances on the surface of our red blood cells determine blood type. We shall be discussing the immune aspects of blood types in a later chapter, but the genetic determination of different blood types is of interest here because it is of a codominant nature.

Shortly after the turn of the century, Nobel laureate K. Landsteiner and his coworkers discovered the best known of the blood groups found in human red blood cells, the ABO blood group. They analyzed this group to be determined by a multiple allelic series, of which two alleles were codominant. Codominant genes are usually designated by the same capital letter, with superscripts for different alleles. The symbol commonly used for the genes determining ABO blood types is the letter I. Codominant alleles of this series are designated I^A and I^B, signifying the A and B blood types respectively.

There is also a completely recessive allele, i, which in the homozygous condition results in the absence of either A or B substance in the red blood cells, producing type O individuals. Type A people can thus be one of two genotypes, either I^AI^A or I^Ai; similarly, the type B genotype can be either I^BI^B or I^Bi. As you have probably discerned, codominance determines that heterozygotes for the I^A and I^B alleles are people classified as having type AB blood. In I^AI^B individuals, the A and B blood substances are both present on the red blood cells, with no modification of their chemical properties. Table 4-1 summarizes the phenotype and genotype combinations of the ABO blood group.

Table 4-1 Genetic Determination of the ABO Blood Types

Phenotype	*Genotype*
A	I^AI^A, I^Ai
B	I^BI^B, I^Bi
AB	I^AI^B
O	ii

MN *blood group.* Another blood group found on our red blood cells is the MN group. Individuals are of the blood types M, N, or MN. There is no recessive allele in this case, and so the corresponding genotypes are *MM*, *NN*, and *MN*, as the *M* and *N* alleles are codominant. The symbolism here is rather unusual, in that different letters are used to denote alleles. This is really an exception to the rule of using the same letter for alleles, but it simply reflects a traditional practice of symbolizing the MN blood group with different letters, which is followed in all the literature.

Actually, a large number of blood groups have been discovered (see reference to Race and Sanger 1975), many of which involve codominant alleles. As Table 4-2 indicates, some of the blood groups are multiple allelic series with a large number of alleles; note that the ABO and MN blood groups are determined by 5 and 20+ alleles, respectively. Recent tests have differentiated slight variations in immunological properties in the blood substances, for example among type A people, indicating that there may exist even more than one I^A allele.

Rh *blood group.* The two most commonly known blood groups of those listed in Table 4-2 are the ABO and Rh groups. The reason is that the substances of these groups produce the strongest immunological reactions, as we shall discuss in Chapter 12. The Rh blood group, however, is genetically quite different from the ABO or MN groups. As you may know, an immunological substance found on the surface of red blood cells in Rhesus monkeys was also found to be present on red blood cells of 85% of the white people of New York City (Landsteiner and Weiner 1940). People whose cells have the Rhesus factor are referred to as *Rh positive* (Rh+); those whose cells do not are *Rh negative* (Rh−).

Table 4-2 Some of the Human Blood Groups

Blood group system	*Date discovered*	*Probable number alleles*
ABO	1900	5+
MN	1927	20+
P	1927	4
Rh	1940	30+
Lutheran	1945	3
Kell	1946	6
Lewis	1946	2
Duffy	1950	3
Kidd	1951	3
Diego	1955	2
Yt	1956	2
I	1956	2
Xg	1962	2
Dombrock	1965	2

From M. Strickberger, 1976. *Genetics*, 2nd ed., Table 9-4, p. 172. Macmillan, New York.

It was originally thought that the presence of the Rh substance is determined by a dominant gene, and the lack of it, by homozygosity for the recessive allele; it has since been discovered that the situation is much more complex. Two lines of interpretation of the genetic determination of the Rh blood groups have been proposed by A. S. Weiner and by R. A. Fisher. Weiner and his coworkers believe that there are a series of multiple alleles (at least 9) at one locus, which interact to determine Rh+ or Rh− conditions. This system is sometimes referred to as the Rh_0 system. Approximately half the alleles are recessive and half codominant. Heterozygosity or homozygosity for any of the dominant alleles lead to the Rh+ trait, whereas any combination of recessive alleles appears to result in Rh−.

Fisher and his coworkers, on the other hand, feel that the data better fit a model involving the interaction of genes at three different loci, designated *C*, *D*, and *E* (Race 1948). The presence of a dominant allele at any of the loci usually results in Rh+ trait.

It should be pointed out here that although crosses involving traits determined by incomplete dominance or codominance result in phenotypic ratios of offspring that differ from those obtained by Mendel in dealing with completely dominant traits, the alleles nonetheless follow the laws of segregation and pairs of alleles on different chromosomes show random assortment.

For example, suppose a man with type AB blood and curly hair marries a woman who has type A blood and wavy hair. What phenotypes can be expected among their offspring and in what proportions? If we designate the curly-hair allele by the symbol *c*, and its straight-hair allele by c', the genotype of the male would be I^AI^Bcc. Let us assume that the mother of the woman in the problem was known to have type O blood. The genotype for the blood group of the woman would have to be I^Ai, as she would have received the *i* allele from her mother. Her complete genotype for the two traits would then be $I^Aic'c$. The results of the cross $I^AI^Bcc \times I^Aic'c$ are shown in the following Punnett Square:

	I^Ac'	I^Ac	ic'	ic
I^Ac	$I^AI^Ac'c$	I^AI^Acc	$I^Aic'c$	I^Aicc
I^Bc	$I^BI^Ac'c$	I^BI^Acc	$I^Bic'c$	I^Bicc

2 Type A/Wavy hair
2 Type A/Curly hair
1 Type AB/Wavy hair
1 Type AB/Curly hair
1 Type B/Wavy hair
1 Type B/Curly hair

Lethal Genes

Genetic conditions also exist in which the heterozygote manifests a phenotype different from those of either homozygote for two alleles, which are not due either to complete or to incomplete dominance. These include conditions determined by genes which cause death to the organism of a particular genotype and are therefore known as *lethal genes*.

Curly wings in Drosophila. In the fruitfly *Drosophila melanogaster*, for example, a certain gene produces curly wings. In the homozygote for this gene, death occurs at an early stage in development. The individual heterozygous for the *Curly* gene and a normal allele is viable, but has abnormal curly wing tips, which distinguish it from the normal homozygote.

In situations in which an allele determines a phenotype different from the normal, or wild-type, trait, geneticists frequently designate the normal allele by +, and the abnormal allele by a letter or letters. In this case, since the effect on the wing is dominant, the gene for curly wings is symbolized by *Cy*, and its normal allele as +. The results of a cross between two heterozygotes would be as diagrammed in Fig. 4-2.

There are two points of interest here. One is that since *Cy*/*Cy* homozygotes die, the expected ratio of the offspring of two heterozygotes is different from that in the classical Mendelian monohybrid cross, being 2 curly : 1 normal rather than 3 curly : 1 normal, since the class of *Cy* homozygous offspring dies in development and is never hatched. The second is that whereas the effect on the wingtip is dominant, lethality must be considered recessive as the heterozygotes are viable. This is an example of two separate effects, one dominant and one recessive, determined by one pair of alleles.

The T locus in mice. A fascinating series of alleles, which include lethal genes, exist in mice. In 1927, a geneticist with the lyrical name N. Dobrovolskaia-Zavadskaia found a wild mouse which had a noticeably short tail. A cross of the short-tailed mouse to a normal one produced short-tailed and normal-tailed mice in equal numbers. Crosses of the short-tailed mice to each other resulted in a ratio of 2 short-tailed : 1 normal-tailed among the progeny.

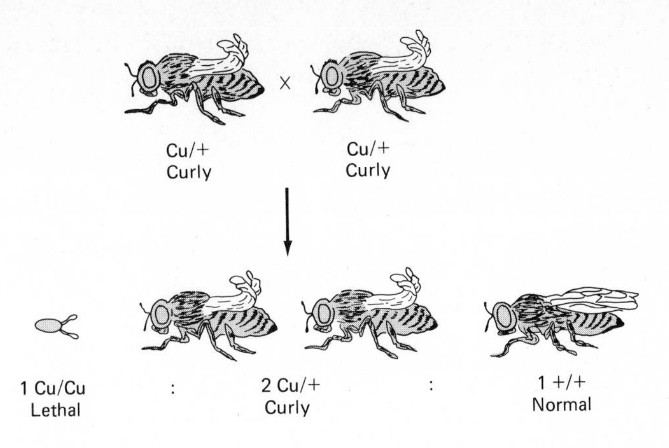

Figure 4-2. Results of a cross between flies heterozygous for the curly gene which is lethal in homozygotes.

A study of litters of crosses between short-tailed mice at embryonic stages showed a significant number of embryos dying at early stages in gestation and being resorbed. These data thus established that the gene determining the short-tailed condition is dominant in its effects on the vertebral column, and is lethal in the homozygous condition. The gene was named Brachyury and given the symbol *T*. (As you read further in genetic literature you will find that geneticists often take poetic license in the naming and symbolic designation of genetic conditions!)

The *T* locus was later found to be a multiple allelic series. In addition to the *T* and + alleles, a large number of *t* alleles were discovered by Dobrovolskaia-Zavadskaia and later by L. C. Dunn, S. Gluecksohn-Waelsch, and coworkers. The *t* alleles are referred to as recessive because in combination with the normal allele, or $+/t$, the mice were normal-tailed and viable. Many of the *t* alleles were found to be lethal, affecting homozygotes at different stages in development and in different ways. For example, t^0t^0 homozygotes die at the fifth day of development due to abnormalities of gastrulation; t^9t^9 embryos die at around the ninth day of gestation due to abnormalities of axial structures.

Other *t* alleles, however, such as t^3, were found to be viable. Homozygotes for t^3 are not only viable, but also normal-tailed. Since t^3t^3 mice cannot be distinguished from $+/t$ heterozygotes or $+/+$ animals, you may wonder how such viable *t* alleles were recognized. The presence of any *t* allele can be detected by its interaction with *T*: mice heterozygous for *T* and any *t* gene, lethal or viable, have no tail at all! In some way as yet not understood, *t* alleles enhance the effect of the *T* gene in interfering with normal tail development.

This effect produces a very interesting progeny ratio when mice heterozygous for *T* and the same lethal *t* gene are mated to each other, for example, if we made the following cross: $T/t^0 \times T/t^0$. Since T/T and t^0/t^0 offspring would die, only tailless heterozygotes would be obtained from such crosses. This situation, in which both classes of homozygotes are never obtained because of lethality, is known as a *balanced lethal system*; Fig. 4-3 illustrates this phenomenon.

The description of the interaction between various alleles of the *T* locus again

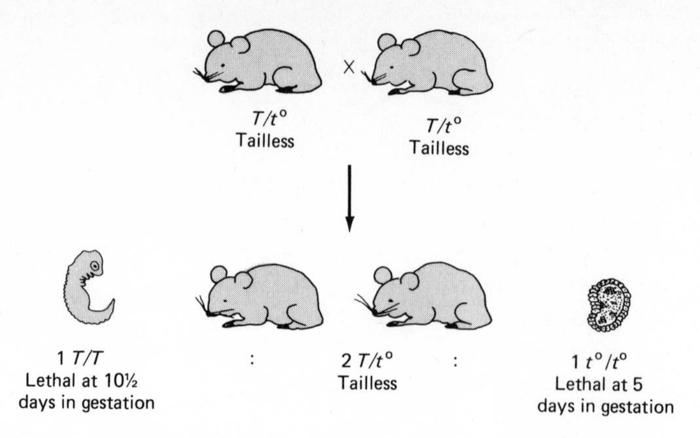

Figure 4-3. A balanced lethal system in mice.

points up the complexity of categorizing genes as being dominant or recessive. The effect of T on the vertebral column appears to be dominant and t recessive; yet t enhances the effect of T to produce taillessness in T/t mice. In addition, heterozygotes for different t alleles, such as t^0/t^9, are normal-tailed and viable. This type of interaction, in which two mutations interact to produce a "normal phenotype," is called *complementation*. However, complementation between different recessive t alleles is not complete, as heterozygous ($t^x t^y$) males are frequently sterile. Crosses between heterozygotes for different t alleles have revealed that the numerous recessive mutations at this locus fall into six complementation groups (Bennett 1975). Figure 4-4 summarizes the phenotypes associated with different combinations of the alleles at the T locus.

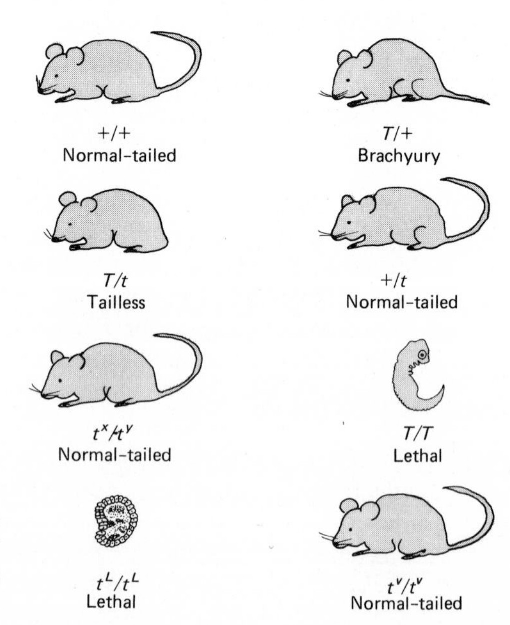

Figure 4-4. A summary of genotypes and associated phenotypes of allele combination at the T locus. (t^x, t^y symbolizes any two different t alleles; t^L = lethal t alleles; t^v = viable t alleles.)

Segregation Distortion

The T locus genes are also an illustration of a genetic phenomenon which, though rare, exists and must be pointed out as an apparent exception to the law of segregation. One curious fact discovered by geneticists as they pursued the study of this series of alleles was the high incidence of t alleles in wild populations of mice. Normally if a detrimental gene such as a lethal gene exists in a population, it tends to decrease in frequency over the years simply because individuals that inherit them from both parents die and do not transmit them to subsequent generations. In practically every wild mouse population studied, however, a significant number of mice carried t alleles.

One explanation for this phenomenon was found to be an exceptionally high rate of mutation at the T locus, leading to the existence of different alleles. In laboratory mice, mutation from t^x to t^y alleles occurred at much greater frequencies than did mutation at other loci.

Another explanation was also discovered for the high frequency of t genes by D. Bennett, W. Braden, and others. Male heterozygotes of the genotypes $+/t$ or T/t transmitted the t allele to their offspring far more frequently than the wild-type, or T, allele. The abnormal ratio of t to $+$ or T alleles has been found to be as distorted as 90 : 1! Such deviation from the expected equal transmission of a pair of alleles has been called *segregation distortion.* Note, however, that this distortion is not necessarily due to abnormal meiosis (implied by the word *segregation*), but to an effect on the transmission of sperm carrying a particular allele. M. I. Sherman has proposed, therefore, that a more appropriate term would be *transmission ratio distortion.* For a review of all the effects of genes of the T locus, some of which we have not mentioned here, see reference to Sherman and Wudl (1977).

Segregation distortion is also known to exist in *Drosophila melanogaster.* L. Sandler and coworkers discovered a natural population in which male heterozygotes for two different alleles at a particular locus seemed to transmit one allele much more frequently, some 5 to 10 times more often, than the other. They designated the unusual gene *SD* for *Segregation Distorter.* It was later discovered that the *SD* gene causes a dysfunction of sperm carrying the allele; thus this would again be better designated transmission ratio distortion. Some evidence (see reference to Kettaneh and Hartl, 1976) indicates that the dysfunction may be related to an absence of a particular group of histones (basic proteins) which are necessary for sperm head condensation. Accordingly, *SD*/*SD* males are sterile.

Peacock et al. (1975) have reported segregation distortion involving a special pair of *Drosophila* chromosomes. This is a *heteromorphic* (physically unlike) pair of chromosomes, which is called the *sex chromosomes,* found in male *Drosophila*; one is known as the X and the other the Y chromosome. Female *Drosophila* have two chromosomes in their cells. Sex chromosomes shall be the topic of discussion for Chapter Five. For now, of interest is the fact that one strain of flies studied by Peacock and coworkers show an excess transmission of X-bearing sperm compared to Y-bearing sperm. Furthermore, this phenomenon is affected by temperature changes; when the flies are raised at 16°C rather than the usual 25°C, the distortion disappears.

As we pointed out above, segregation and transmission ratio distortion are not common phenomena and most pairs of alleles follow the law of segregation. Nonetheless, their existence indicates that geneticists must be aware of this possible cause of unexpected ratios when dealing with inherited traits.

INTERACTION BETWEEN NONALLELIC PAIRS OF GENES

The preceding sections of this chapter have dealt with genetic phenomena involving the interaction of alleles. There exist many examples of inherited traits which are determined by the interaction of pairs of nonallelic genes. In the following sections we shall be discussing a number of these interesting phenomena.

Pleiotropy

In several of the inherited conditions discussed in the preceding sections, one pair of alleles appear to determine more than one trait. In the *Curly Wing* mutation, for example, the gene not only affects wing structure, but in the homozygote, interferes with development. In mice, alleles of the *T* locus affect a large number of functions such as tail formation, transmission ratios, mutation rate, sterility, and developmental processes. In complex organisms this ability of single pairs of alleles to affect a large number of characteristics is frequently observed and is known as *pleiotropy.*

Pleiotropy can be explained by the integration of development and function of the different tissues and organ systems of eukaryotes. If a basic cell type such as red blood cells is affected, then other tissues and organs depending on this cell type may also become abnormal. Pleiotropic effects of a particular pair of alleles are thus actually a reflection of abnormalities of many cells and functions determined by different pairs of genes. These abnormalities are secondary to the original one caused by the gene pair in question.

Sickle-cell anemia. A good example of pleiotropy in humans is in the effects of a well-known inherited disease of the blood, sickle-cell anemia. The mutant gene that determines this disease is usually considered recessive because the lethal form of anemia is found only in sickle-cell homozygotes. Sickle-cell anemia does not cause death at embryonic stages, but rather after the individual is born, in contrast to the lethal genes we discussed above. Although a severe anemia is the cause of death, a large number of other physical infirmities are also associated with the disease, as shown in Fig. 4-5.

The *primary action* of the *sickle-cell anemia* gene is to cause the production of abnormal hemoglobin. We define the term primary action as the direct effect of the gene on cell function. The blood abnormality in turn causes a wide syndrome of secondary ailments in sickle-cell anemia patients, such as heart failure, paralysis, etc. Pleiotropy is again a phenomenon which Mendel did not happen to encounter in his studies; as you may recall, his experiments dealt mainly with pairs of alleles which seemed to affect only one particular trait, such as stem length, but had no apparent effect on any other characteristics.

Because of the pleiotropic effects of the *sickle-cell anemia* gene, it is again difficult to refer to the gene as dominant or recessive. Certainly from the point of

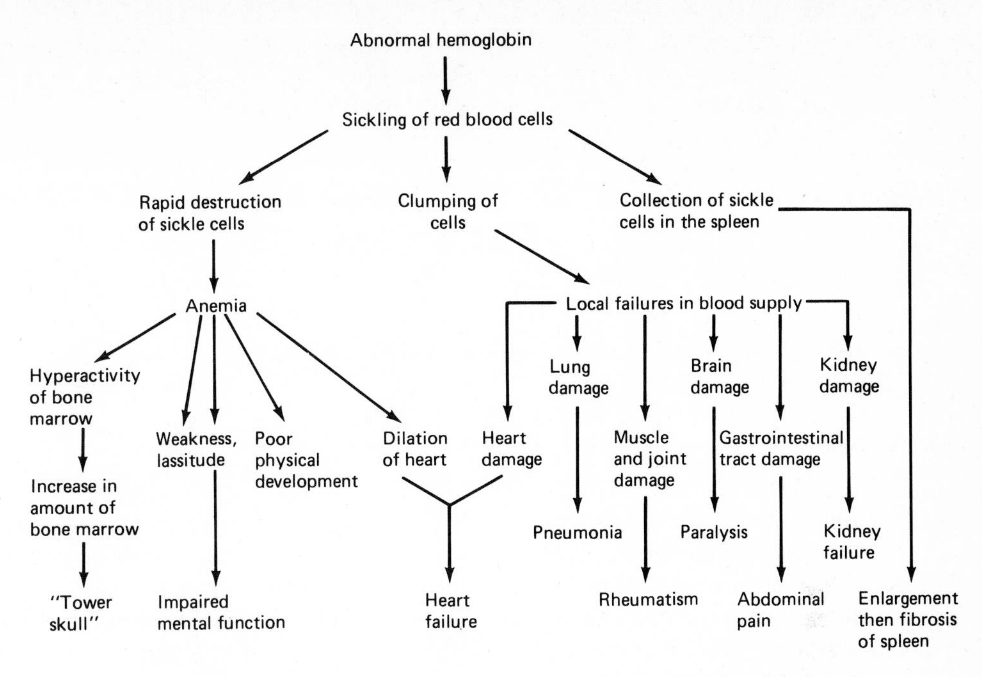

Figure 4-5. A summary of the wide syndrome of abnormalities found in homozygotes for the sickle-cell anemia gene. This serves as a clear example of pleiotropy. From M. W. Strickberger, 1976. *Genetics* 2nd ed., p. 262. Macmillan, New York. Copyright © 1976 by Monroe Strickberger.

view of lethality, the gene is recessive, as heterozygotes survive and function normally. On the other hand, abnormal hemoglobin caused by the mutant gene as well as normal hemoglobin can be detected in the red blood cells of heterozygotes, leading to the term *Sickle-Cell trait* to describe the effect on carriers of the mutant gene (see reference to Ingram, Chapter 15). The effect on hemoglobin must therefore be considered codominant.

Alteration of Phenotype

Comb shape in chickens. One of the first examples of the interaction of different gene pairs in the determination of one particular trait was reported by W. Bateson and R. C. Punnett in the early 1900s. They studied comb shape in different strains of chickens. Some strains, such as the Wyandotte, have what are called *rose* combs; Brahma chickens have *pea* combs.

When Bateson and Punnett crossed true-breeding Wyandottes with true-breeding Brahmas, they found that the offspring had yet a third type of comb, called *walnut.* Chickens with walnut combs were then crossed to produce an F_2 generation, and the phenotypic ratio obtained was the following: $\frac{9}{16}$ Walnut, $\frac{3}{16}$ Rose, $\frac{3}{16}$ Pea, $\frac{1}{16}$ Single. All four types of comb are illustrated in Fig. 4-6.

The phenotypic ratio of the F_2 progeny clearly indicated that a dihybrid cross was in effect. The inheritance of comb shape in chickens, then, was one of the first examples discovered of more than one pair of genes interacting to determine a single

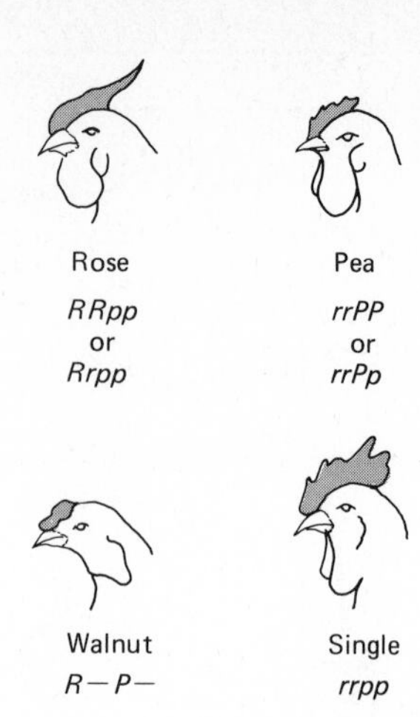

Figure 4-6. Comb types found in chickens studied by Bateson and Punnett.

trait. The Walnut phenotype must be determined by the presence of one dominant allele in each gene pair. Rose and pea combs would be the results, respectively, of the presence of a dominant gene of one of the two pairs of genes involved, and of the homozygous recessive condition for the other gene pair. The fourth comb type, which appeared in the F_2 generation, can be assumed to be manifested in animals homozygous recessive for both gene pairs.

Coat color in mice. One of the best-known examples of a single trait which is determined by the interaction of many gene pairs is that of coat color in mice. A large number of genes are involved here. For example, one pair of genes is known to determine whether the animal will have black coat color or brown. The gene for black color is completely dominant to the gene for brown, and we accordingly symbolize the respective alleles as *B* and *b*.

In addition, a separate locus determines whether the coat color will be pure black or brown, or whether the color will be speckled with yellow. This is the *Agouti* locus, and again there are two alleles, *A* and *a*, which are completely dominant and recessive, respectively. Mice which are heterozygous or homozygous dominant for the *Agouti* gene possess hairs which have a yellow band; homozygous recessive mice have hairs that lack this band and are pure black or brown, a trait that geneticists refer to as *nonagouti*, or *self*. What we consider the wild-type phenotype in coat color in mice is actually black agouti. It is the same color found in common field mice, an overall grayish color. Brown agouti mice, whose genotype would be *bbA-*, are called *cinnamon* because of the alteration of the coat color from pure brown or brown self to a yellowish brown hue.

There are in fact many genes other than those of the *Black* and *Agouti* loci which are active in the production of coat color in mice. Table 4-3 summarizes a few of these genes, including the *Black and Agouti* alleles, and their effects on phenotype.

A mouse with the genotype *A-bbddss* would have a coat which is dilute cinnamon in color, with white spots. Considering all the above loci, then, the wild-type coat color would be determined by *A-B-D-S-*, and this is only a partial list of genes known to determine coat color in mice!

Table 4-3 Coat Color Genes in Mice

Genotypes	*A-*	*aa*	*B-*	*bb*	*D-*	*dd*	*S-*	*ss*
Phenotypes	Agouti	Self	Black	Brown	Nondilute	Dilute	Nonspotted	Spotted

Eye pigmentation in Drosophila. A similar situation exists in *Drosophila melanogaster*, in which a cross of dihybrids for two different eye color mutations results in four different phenotypes among the offspring. The genes in question are the *scarlet* and *brown* genes, each of which is completely recessive (in its visible expression) to its normal allele. In the homozygous condition, scarlet mutants have eyes which are a brighter red color than the normal wild-type red eye; brown mutants manifest a darker color than the normal wild-type color.

A cross of homozygous scarlet flies with homozygous brown flies produces an F_1 generation which is uniformly wild-type. When F_1 flies are mated, the F_2 generation shows the classical dihybrid ratio of phenotypes: $\frac{9}{16}$ wild-type, $\frac{3}{16}$ scarlet, $\frac{3}{16}$ brown, and $\frac{1}{16}$ white eyes. The ratio of course confirms that *brown* and *scarlet* are mutations at entirely different loci and segregate independently. Here again we see the interaction of two pairs of nonallelic genes interacting in the determination of a single trait, namely, eye pigment formation.

It has been established by I. Ziegler and others that normal wild-type eye color in *Drosophila* is actually due to the presence of several chemicals which can be grouped into two classes: the pteridines, which are red, and ommochromes, which are brown. The *scarlet* mutation apparently causes a blockage in the production of ommochromes, resulting in the presence of only red pigment in eyes of *scarlet* homozygotes and accounting for the bright red phenotype. Conversely, the mutation at the *brown* locus evidently interferes in some way with the production of pteridines. The eye which has only the brown pigment thus appears darker than normal.

The white-eyed flies found among the F_2 progeny can be assumed by their proportion to be the doubly homozygous recessive class. In view of our understanding of the metabolism of eye pigmentation in *Drosophila*, it becomes clear that in this group, production of both pteridines and ommochromes is blocked, resulting in the absence of pigment formation and the white-eye phenotype (Fig. 4-7).

STUDIES SUCH AS THESE ON EYE PIGMENT MUTATIONS IN *Drosophila* WERE IMPORTANT NOT ONLY FOR INCREASING OUR UNDERSTANDING OF THE GENETIC DETERMINATION OF THIS PARTICULAR CHARACTERISTIC, BUT ALSO BECAUSE THEY ESTABLISHED THAT ALTERATIONS OF GENES RESULT IN INTERFERENCE WITH NORMAL METABOLIC REACTIONS. THE RELATIONSHIP BETWEEN GENES AND METABOLISM ESTABLISHED BY TRANSMISSION GENETICS SERVED AS THE FOUNDATION FOR LATER RESEARCH THAT UNCOVERED THE TRUE NATURE OF GENES AND OF GENE ACTION.

Modification of Phenotype

The preceding examples of gene interaction to produce a particular phenotype involved discrete pairs of genes. That is, it was possible to analyze the ratios of

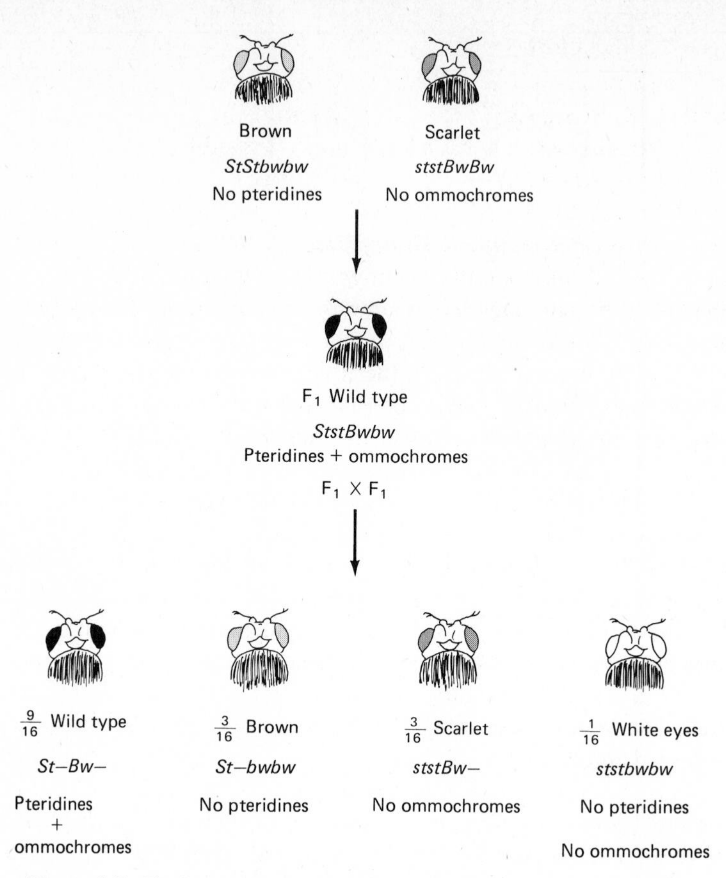

Figure 4-7. The interaction of nonallelic eye pigment genes in *Drosophila*.

crosses involving comb shape, for example, and to discern that two pairs of randomly assorting genes were involved. There are a number of cases of inherited traits, however, in which different pairs of genes are obviously involved, but in which the exact number and nature of these genes cannot be easily ascertained.

Penetrance. One such phenomenon is known as *penetrance*, a term used to refer to the actual manifestation of a particular trait when the genes determining that trait are known to be present in an individual. For example, if an individual has inherited a dominant gene, we expect the trait to be expressed; if it is, we refer to the trait as one with *full penetrance*. All the dominant traits which Mendel studied in pea plants, for example, were fully penetrant. It has been discovered, however, that not all traits are expressed as expected. In some cases expression is only partial, and we refer to this as *incomplete penetrance*; in others the trait is not seen at all, a situation referred to as *nonpenetrance*, or *lack of penetrance*.

C. Stern has described an example of lack of penetrance in humans. A dominant gene (*D*) causes camptodactyly (a stiff pinky) due to abnormal musculature controlling the joints of the little finger. Although dominant, the trait is not always seen in individuals who are believed to have inherited that gene; in these people, the gene is nonpenetrant. In some people it is expressed on only one hand, and so is considered incompletely penetrant. The difference in penetrance of a trait in different individuals is assumed due to interaction of the gene with other genes.

Expressivity. A somewhat different situation exists where *expressivity* is affected. Where penetrance refers to whether or not a trait is expressed, expressivity refers to variations in the expression of a particular trait. A well-known example of expressivity involves the inherited ability of humans to taste a chemical, phenylthiocarbamide, commonly referred to as PTC. The ability to taste PTC is determined by a single gene pair, and is a dominant trait; homozygous recessive individuals cannot taste the substance at all.

Variations exist among tasters, however, in their sensitivity to the substance, as shown in Table 4-4. The data in the table were obtained in a study in which tasters sampled varying strengths of solutions of PTC.

Table 4-4 Taste Thresholds for PTC

Solution mgm/l	*Number of persons*	*Solution mgm/l*	*Number of persons*
1300	31	10.2	51
1300	37	5.1	73
650	21	2.5	73
225	26	1.3	29
162	23	0.6	18
81.2	10	0.3	6
40.6	13	0.16	1
20.3	29		

From E. W. Sinnott, L. C. Dunn, and T. Dobzhansky, 1958. *Principles of Genetics*, p. 131. McGraw-Hill, New York.

As indicated by the data, there is a wide degree of expressivity of the taster trait. Some people were very insensitive and needed as much as 1300 mgm/l or more of the substance in solution before they could detect its presence. One person was obviously highly sensitive, responding to a solution with only 0.16 mgm/l. Yet all these tasters by genotype are presumed to have inherited the same dominant gene for tasting.

Why there are different degrees of expressivity (or lack of penetrance) of traits among individuals assumed to be of the same or similar genotype for certain genes is not clearly understood. One would have to say that at least in some of the instances of penetrance and expressivity the explanation surely lies in the interaction of the genes in question with other genes that modify their expression.

Modifier genes. Genes which are believed to interact in such a way as to modify the expression of other genes are referred to by a comfortably ambiguous term, *modifier genes*. These genes, which can so drastically alter the expression of a trait so that it is not expressed at all, cannot be analyzed with classical methods of transmission genetics for the simple reason that their expression does not follow Mendelian patterns of transmission. There is no way, then, for the geneticist to determine which genes or even how many pairs of genes are involved in varying the expression of the taster gene, for example, or how many are involved in suppressing penetrance of camptodactyly.

Epistasis. In organisms possessing lethal genotypes, the expression of the lethal gene of course prevents the expression of all other genes by terminating the life of the victim. There also exist instances in inherited traits in which the expression of one pair of genes blocks the expression of other genes without causing death, a phenomenon known as *epistasis*. Note that epistasis is the interaction between non-allelic pairs of genes, and should therefore be clearly distinguished from complete dominance, in which a dominant trait is expressed instead of a trait determined by a recessive allele.

One of the effects of epistasis is to modify expected Mendelian ratios. Let us assume a cross of dihybrids, *AaBb* × *AaBb*. Further assume that the genotype *aa* is epistatic to the expression of either the *B* or *b* alleles. The phenotypic ratio one would expect from this cross would be 9 (*A-B-*): 3 (*A-bb*): 4 (*aaB-*, *aabb*).

Albinism, or the total lack of pigmentation, is a good example of epistasis. It is determined by a completely recessive gene. Recall that many pairs of genes determine coat color in mice; however, if a mouse is homozygous for the *albino* gene, the mouse is an albino regardless of its genotype for coat color. For example, a mouse known to be *BBAA* for the *Black* and *Agouti* genes would be wild-type or black agouti if it is heterozygous or homozygous dominant at the *albino* locus. If it is homozygous for the *albino* gene (*c*), and has the genotype *BBAAcc*, it would be albino. In this situation, we say the *albino* gene is epistatic over the *Black* and *Agouti* genes.

The explanation for this effect is again metabolic. The *albino* mutation causes

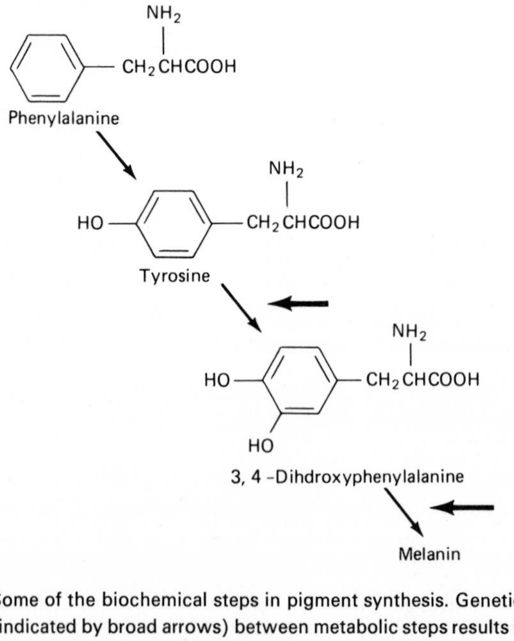

Figure 4-8. Some of the biochemical steps in pigment synthesis. Genetically determined blocks (indicated by broad arrows) between metabolic steps results in albinism.

a block in the ability of pigment cells to complete the synthesis of the pigment melanin (Fig. 4-8). Therefore, although the genes for variations in coat color exist, if no color can be produced, these genes will not be expressed. The same situation has been found in humans and in many other species of animals and plants. The total lack of pigment is reflected not only in white hair, but also in pink eyes, since the absence of pigment in the iris allows the pinkness of retinal blood vessels to be seen (Fig. 4-9).

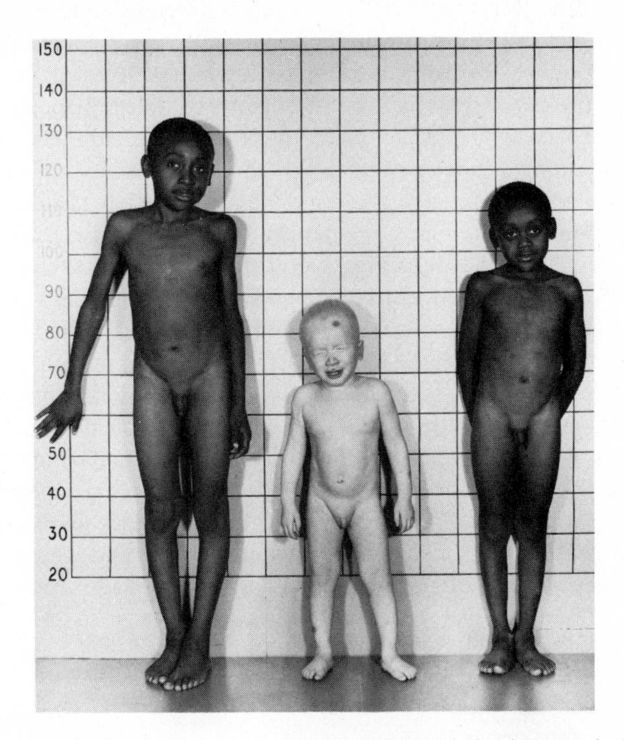

Figure 4-9. Albinism in humans. The albino child is squinting because the lack of pigmentation in his eyes causes him to be especially sensitive to light. From A. C. Pai, 1974. *Foundations of Genetics,* Fig. 5.2, p. 95. McGraw-Hill, New York.

POLYGENIC TRAITS AND QUANTITATIVE INHERITANCE

The traits we have discussed so far have been readily separable into distinct groups. Pea plants were clearly either tall (6 to 7 feet) or short (9 to 18 inches); or the seeds of pea plants were clearly either green or yellow. Traits such as these are very easy to classify and therefore easy to study. As we mentioned before, his choice of such traits to study was one of the reasons for Mendel's success. The traits discussed so far in this chapter have also been clearly classifiable into groups; that is, blood type MM, MN, or NN; Rh factor Rh+ or Rh−; A, B, AB, or O blood type; wavy or curly hair; and hence these traits have been studied extensively and with success. Traits easily classified into distinct groups are known as *qualitative traits*.

If we look at people around us, however, we find that many characteristics are not so easily classified into distinct groups. As an example, consider height, weight, or intelligence. These traits vary *continuously*. We can classify people into two weight groups (heavy and light), but this classification does not adequately describe the possibilities. Similarly, classification into three groups (light, medium, or heavy) does not really serve to describe the possibilities much better.

The difference between weight and blood type is that weight is really a continuously varying variable. Adults might weigh from 80 to 400 pounds or more, and in fact one can always measure an individual's weight more precisely by getting a more accurate measuring instrument (consider scales used in a chemistry lab vs. those used in the produce department of the supermarket). Such continuously varying, or *quantitative characteristics* are much more complex than are qualitative characteristics. Measuring instruments are often used to distinguish quantitative characteristics, revealing differences that are often imperceptible to the observer.

If one empirically studies these quantitative traits by obtaining frequency distributions, a very interesting pattern starts to emerge. Examples of such frequency distributions are shown in Figs. 4-10 through 4-13; in each case, we observe a "bell-shaped" curve. The frequencies of people in the categories are such that we start with a few people at the low extreme, build up toward the middle, then start decreasing toward the high extreme. A very interesting phenomenon is that this basic pattern occurs regardless of the particular intervals chosen. This basic "normal" ("bell-shaped") curve seems to model adequately many phenomena (traits) in nature.

A good illustration of this is one of the first studies of a system showing quantitative inheritance, that of wheat kernel color, reported by Nilsson-Ehle in 1909. Again, he was repeating experimental crosses similar to those of Mendel's, using wheat plants which produced white kernels and wheat plants which produced red kernels. The results of some of the crosses he made are as follows:

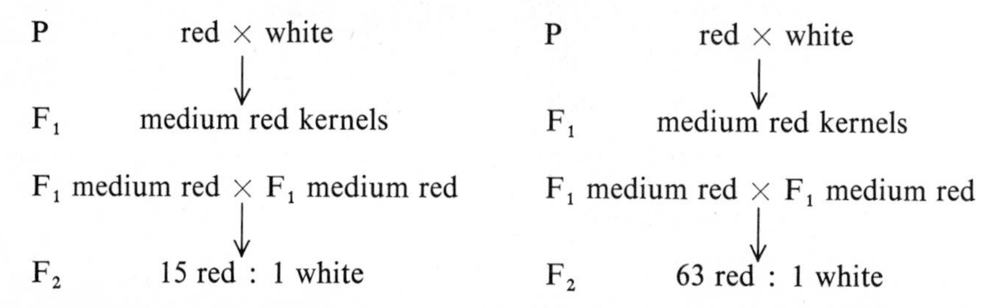

Although the phenotypic ratios of the F_2 generations were quite different from any obtained by Mendel, Nilsson-Ehle recognized two significant points. One was that F_2 plants produced varying shades of redness in their kernels, from light red to the same dark intensity of color as the parental generation red plants. The second point was that the ratios of reds and whites in one experiment related directly to the ratio expected from a dihybrid cross [(9 + 3 + 3) : 1], and in the second experiment related directly to the ratio expected from a trihybrid cross (see Problem 2-13).

Let us follow his analysis of the simpler cross, dealing with two pairs of genes, since the principle is the same for the trihybrid cross. We can assume, as Nilsson-Ehle did, that the parental cross was $R^1R^1R^2R^2 \times r^1r^1r^2r^2$. The F_1 generation, then, were

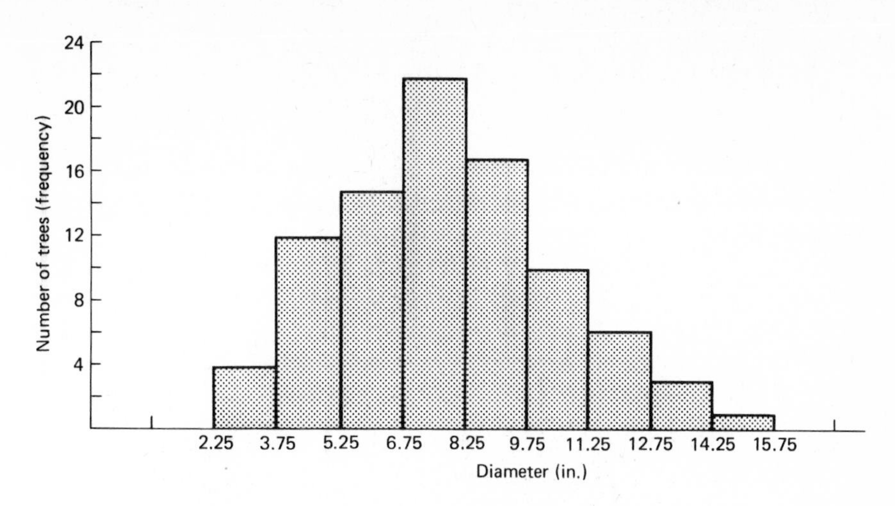

Figure 4-10. Diameters of oak trees 65 in. off the ground.

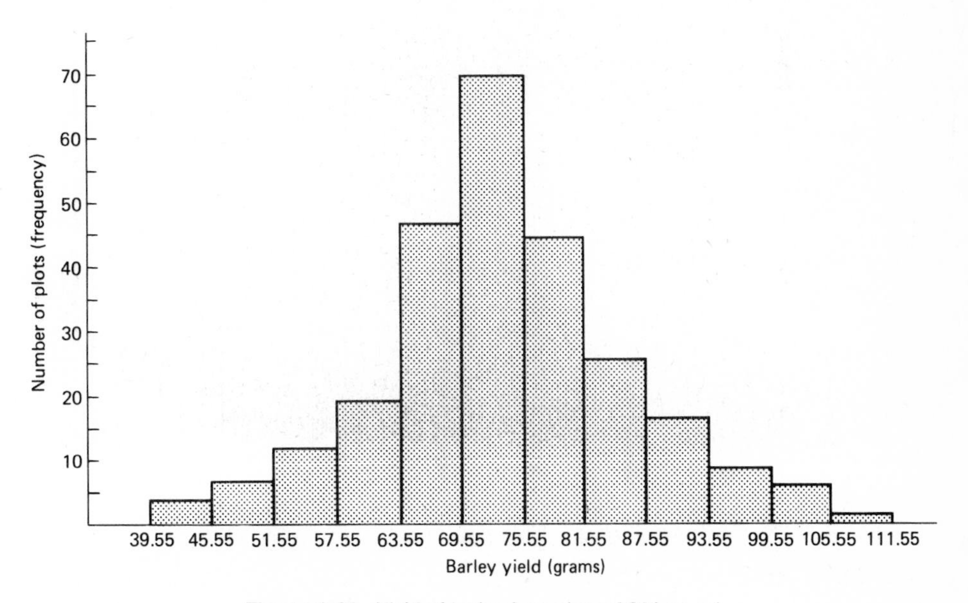

Figure 4-11. Yield of barley from plots of 200 sq. yd.

genotypically $R^1r^1R^2r^2$. Assuming further that the number of dominant genes determines the intensity of redness in kernels in an additive manner, this would explain the medium red color in kernels of all F_1 plants, as they possess half as many dominant genes as the parental red plants.

Results for a Punnett Square of the cross of F_1 plants are given in Table 4-5. We see that the total number of F_2 red plants can be classified so that they are in a ratio of 1 dark red : 4 medium dark red : 6 medium red : 4 light red : 1 white. The same analysis showed that the results of the second experiment could be attributed

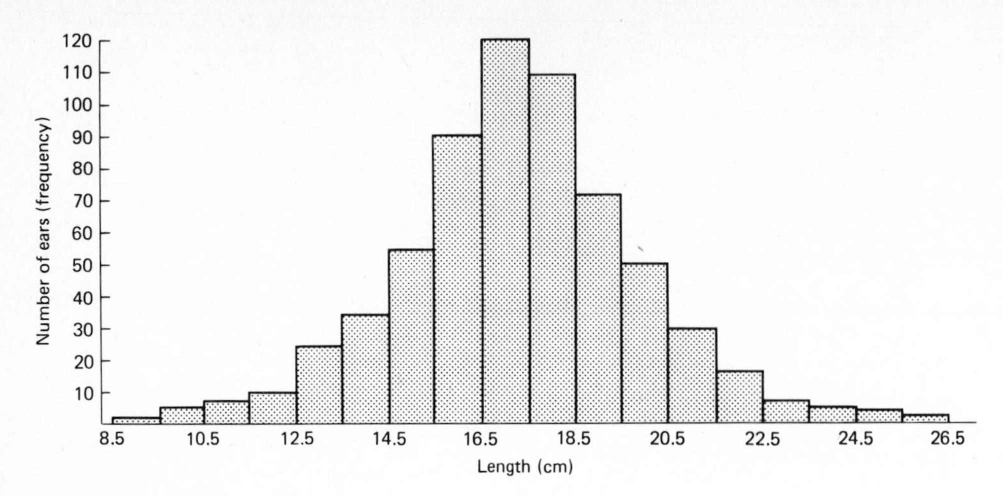

Figure 4-12. Lengths of ears of corn.

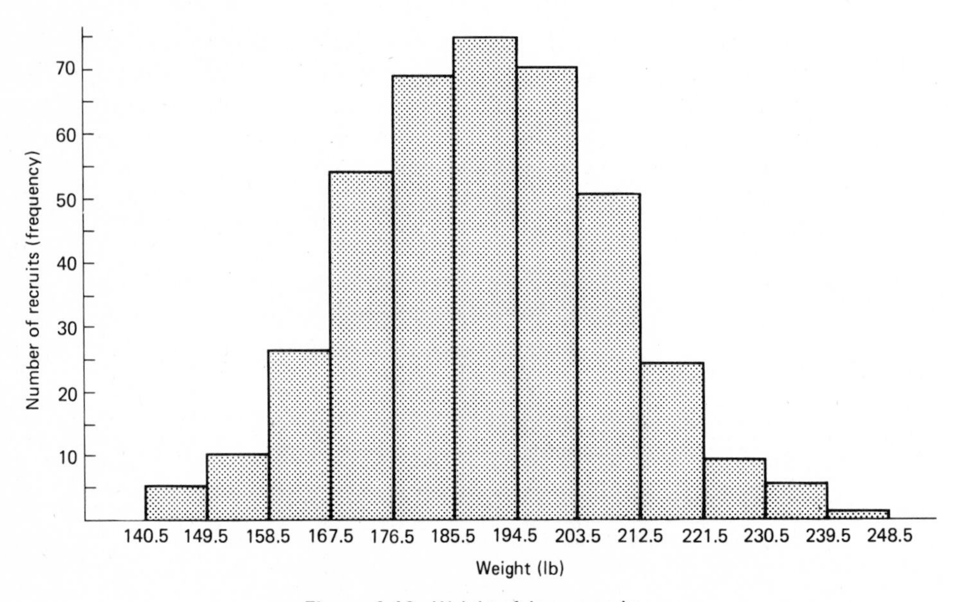

Figure 4-13. Weight of Army recruits.

to a parental cross of $R^1R^1R^2R^2R^3R^3 \times r^1r^1r^2r^2r^3r^3$, and a trihybrid cross of F_1 plants to produce the F_2 generation. In each case a plot of the relative number of each phenotypic class forms a bell-shaped curve.

Such empirical regularities suggest some common basis of inheritance. Geneticists, after careful study, have suggested that quantitative inheritance might in fact be explained by a combination of a polygene theory (as in wheat kernel color) and environmental factors. Quantitative traits might in fact be controlled by two, three, four, or even 40 or more allele pairs. (It is very interesting to note that if this theory

Table 4-5 Punnett square for $R^1r^1R^2r^2 \times R^1r^1R^2r^2$

	R^1R^2	R^1r^2	r^1R^2	r^1r^2
R^1R^2	$R^1R^1R^2R^2$ Dark Red	$R^1R^1R^2r^2$ Medium Dark Red	$R^1r^1R^2R^2$ Medium Dark Red	$R^1r^1R^2r^2$ Medium Red
R^1r^2	$R^1R^1R^2r^2$ Medium Dark Red	$R^1R^1r^2r^2$ Medium Red	$R^1r^1R^2r^2$ Medium Red	$R^1r^1r^2r^2$ Light Red
r^1R^2	$R^1r^1R^2R^2$ Medium Dark Red	$R^1r^1R^2r^2$ Medium Red	$r^1r^1R^2R^2$ Medium Red	$r^1r^1R^2r^2$ Light Red
r^1r^2	$R^1r^1R^2r^2$ Medium Red	$R^1r^1r^2r^2$ Light Red	$r^1r^1R^2r^2$ Light Red	$r^1r^1r^2r^2$ White

is interpreted statistically, one arrives at a theoretical justification of the empirical observations that are described by the normal curve.)

How does the polygene theory describe what is observed? We will use height as another example. Assume that height is governed by one allele and that in the parental generation, P, one parent is considered tall (homozygous) and the other parent short (homozygous). The offspring of such a mating (the F_1 generation) should be in between the two extremes of height (Fig. 4-14). If we now consider the results of mating the F_1 offspring and we assume no dominance, we get the result of Fig. 4-15 for the relative frequency of offspring in the F_2 generation.

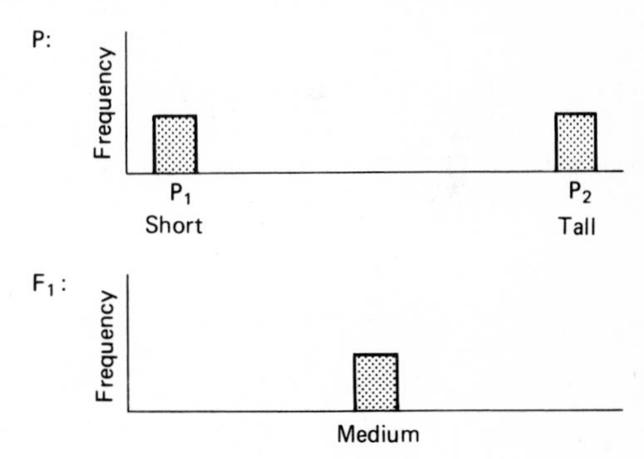

Figure 4-14. Relative frequency of heights in F_1 generation if one parent is *TT* and the other is *tt*.

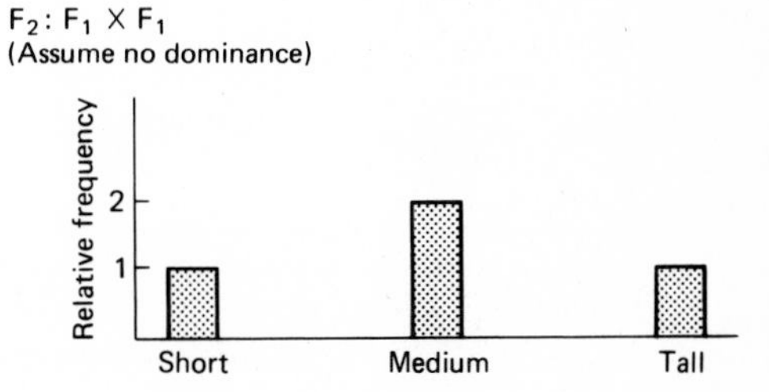

Figure 4-15. Relative F_2 frequency with one pair of alleles.

Next, suppose that two pairs of alleles govern height, and that one parent in the P generation is doubly homozygous tall and the other parent is doubly homozygous short. Then in the first generation, all offspring have medium height. However, in the second generation, we get a distribution as in Fig. 4-16. Note that there are five phenotypically distinct groups.

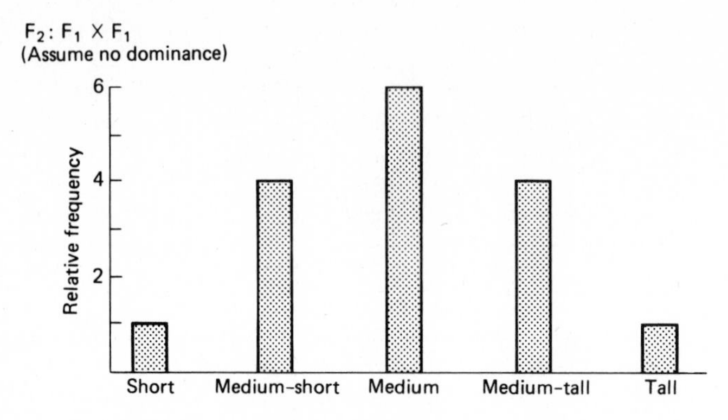

Figure 4-16. Relative F_2 frequency with two pairs of alleles.

If we have three pairs of alleles, we get seven phenotypically distinct groups in the second generation (see Fig. 4-17). With four pairs of alleles the number of phenotypic classes is shown in Fig. 4-18. Notice that as the number of pairs of alleles increases, it is increasingly difficult to distinguish the phenotypic classes. It can be proven mathematically that as the number of gene pairs increases, and we plot relative

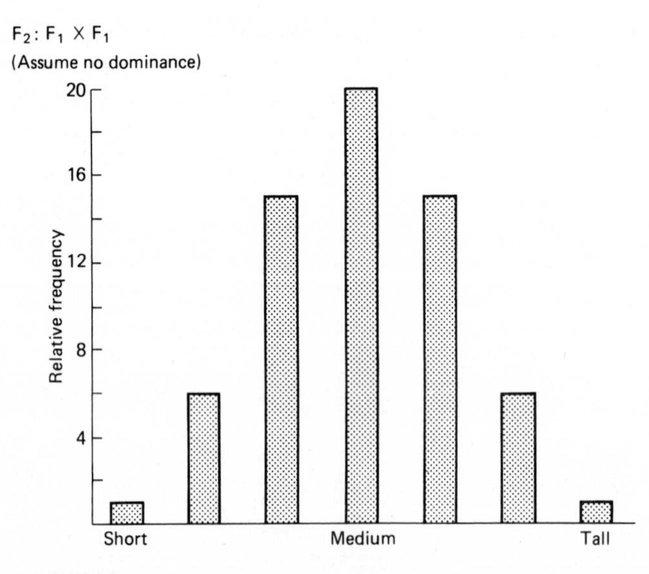

Figure 4-17. Relative F_2 frequency with three pairs of alleles.

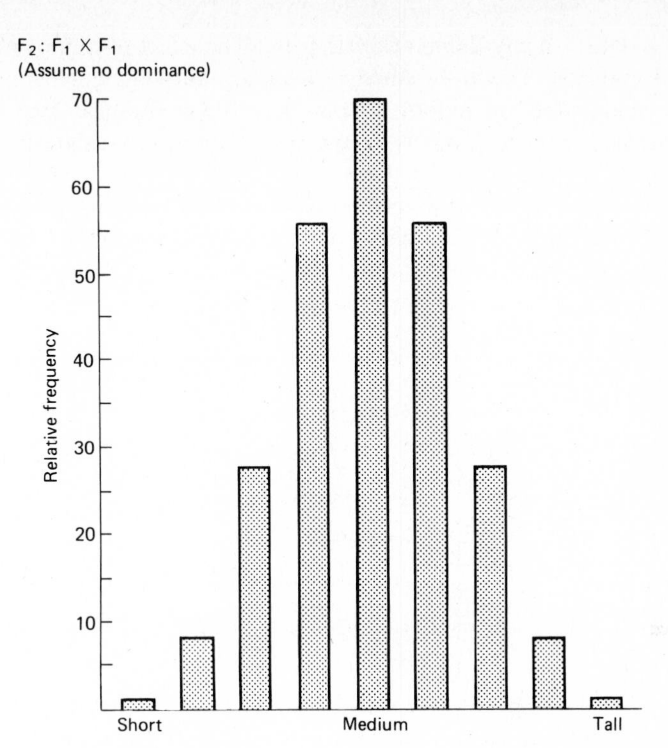

Figure 4-18. Relative F_2 frequency with four pairs of alleles.

frequency of phenotypic classes against size, we get a "bell-shaped" curve as in Fig. 4-19.[1]

If we use this explanation for the bell-shaped curve, we find a large number of phenotypic categories. However, we still have a classification into categories rather than a continuous variation. Why does a real-world variable such as height seem to vary continuously? The large number of phenotypic classes are made possible by

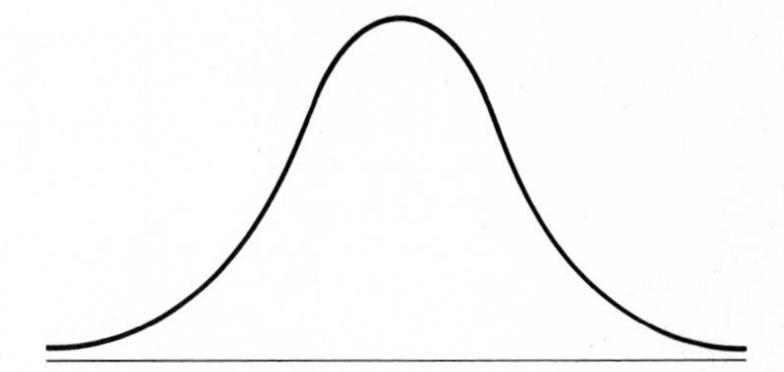

Figure 4-19. Ideal normal (bell-shaped) curve.

[1]Mathematically we determine the relative numbers in each phenotypic class by using the binomial distribution. If n is the number of pairs of alleles, we determine the relative frequencies by expanding $(p + q)^{2n}$. If n gets large it can be proven that the mathematical equation for the bell-shaped curve (normal curve) is the correct ideal model to describe the relative frequencies in each phenotypic class.

the action of many different allele pairs. The effect of any one allele would be hard to distinguish. Therefore since each allele can contribute only slightly toward the expression of the quantitative (continuously varying) trait, the effect of the environment seems to be relatively larger than that of any single gene or gene pair. For example, we know that fertilizers are used to increase the corn ear size, or that diet can affect the height or weight of human beings.[2] Phenotypically the corn or human being falls into a certain class, but the environment (food or fertilizer) introduces additional variability. As a result the corn or human being may fall in between two phenotypes or be shifted into a higher or lower phenotype. Thus, we get characteristics that vary continuously.

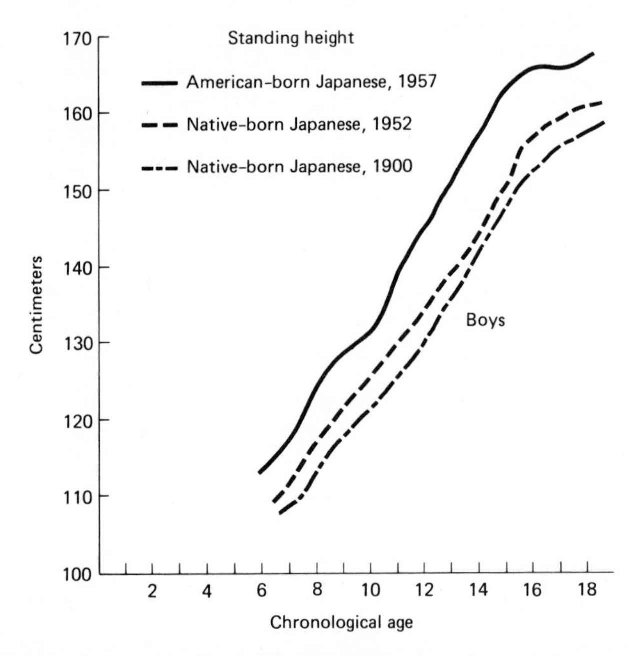

Figure 4-20. Average standing height of American-born Japanese boys compared with height of boys in Japan in 1900 and 1952. (Redrawn from W. W. Greulich, 1958. "Growth of children of the same race under different environmental conditions." *Science* 127 : 515–516 ; copyright 1958 by the American Association for the Advancement of Science.)

[2] As for evidence that environment influences height in human beings, there are many statistics which show that while people are becoming taller as the generations pass, when people migrate to the United States their children on the average are taller than the comparable generation in the native homeland. These two points are well illustrated by Fig. 4-20. There are many possible explanations, but since genetic factors are not varying, it appears that the environment influences height. Specifically, it would appear that diet may have an effect on height. We shall have more to say on environmental effects later.

Let us consider a specific example and see how the polygenic theory can be used to explain continuously varying or quantitative traits. Consider the length of corn ears. We can consider the distribution of the F_2 generation if we make various assumptions about the number of pairs of alleles that determine height. For the parental generation consider the mating of pure long (26 cm) and pure short (8 cm). The result of such a mating will result in F_1 hybrids of about 17 cm. In the F_2 generation we can expect the following if we make various assumptions. Suppose first that there is one pair of alleles. For discussion we assume that a recessive gene contributes 4 cm to length and a dominant gene 13 cm; the distribution of genotypes by generation is given in Table 4-6. If we plot the relative frequency distribution of the F_2, we get Fig. 4-21.

Table 4-6 Lengths of Ears of Corn in the F_1 and F_2 Generation (One Pair of Alleles)

P:			*TT* 13 + 13 26 cm	×	*tt* 4 + 4 8 cm
F_1:			*Tt* 13 + 4 17 cm		
F_2:	*TT* 13 + 13 26 cm Tall	:	2*Tt* 13 + 4 17 cm Medium	:	*tt* 4 + 4 8 cm Short

Suppose next that there are two pairs of alleles. Suppose a recessive gene contributes 2 cm to length and a dominant gene $6\frac{1}{2}$ cm. We then have the distribution of genotypes given in Table 4-7. The relative frequency distribution of phenotypes (lengths) is given by Fig. 4-22.

Suppose next that we have three pairs of alleles and that a recessive gene contributes $1\frac{1}{3}$ cm to length and a dominant gene $4\frac{1}{3}$ cm. We then have the distribution of genotypes given in Table 4-8. The relative frequency distribution of phenotypes is plotted in Fig. 4-23.

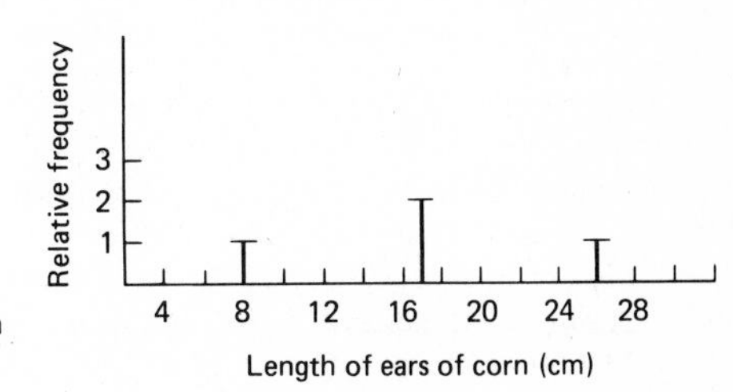

Figure 4-21. Relative frequency plot of F_2 in Table 4-6.

Table 4-7 Lengths of Ears of Corn in the F_1 and F_2 Generation (Two Pairs of Alleles)

P: $\underbrace{6\frac{1}{2} + 6\frac{1}{2} + 6\frac{1}{2} + 6\frac{1}{2}}_{26\text{ cm}}$ ($T_1T_1T_2T_2$) $\times$ $\underbrace{2 + 2 + 2 + 2}_{8\text{ cm}}$ ($t_1t_1t_2t_2$)

F_1: $\underbrace{6\frac{1}{2} + 2 + 6\frac{1}{2} + 2}_{17\text{ cm}}$ ($T_1t_1T_2t_2$)

F_2:

	T_1T_2 (13)	T_1t_2 ($8\frac{1}{2}$)	t_1T_2 ($8\frac{1}{2}$)	t_1t_2 (4)
T_1T_2 (13)	(26) $T_1T_1T_2T_2$	($21\frac{1}{2}$) $T_1T_1T_2t_2$	($21\frac{1}{2}$) $T_1t_1T_2T_2$	(17) $T_1t_1T_2t_2$
T_1t_2 ($8\frac{1}{2}$)	($21\frac{1}{2}$) $T_1T_1T_2t_2$	(17) $T_1T_1t_2t_2$	(17) $T_1t_1T_2t_2$	($12\frac{1}{2}$) $T_1t_1t_2t_2$
t_1T_2 ($8\frac{1}{2}$)	($21\frac{1}{2}$) $T_1t_1T_2T_2$	(17) $T_1t_1T_2t_2$	(17) $t_1t_1T_2T_2$	($12\frac{1}{2}$) $t_1t_1T_2t_2$
t_1t_2 (4)	(17) $T_1t_1T_2t_2$	($12\frac{1}{2}$) $T_1t_1t_2t_2$	($12\frac{1}{2}$) $t_1t_1T_2t_2$	(8) $t_1t_1t_2t_2$

The relative frequency distribution of phenotypes determined by 5 pairs of alleles is plotted in Fig. 4-24 [see Problem 11(d)]. If we refer back to our example, we note that in all cases the lengths in the F_2 generation lie between the lengths of the original P generation, 8 cm and 26 cm. Thus, as we increase the number of pairs of alleles, the number of phenotypic classes is increased, and we approach a continuous curve.

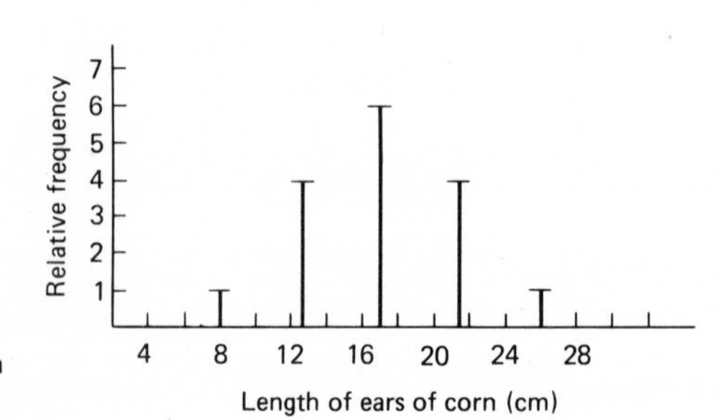

Figure 4-22. Relative frequency plot of F_2 in Table 4-7.

Table 4-8 Lengths of Ears of Corn in the F_1 and F_2 Generations (Three Pairs of Alleles)

P: $\underbrace{\overset{T_1T_1T_2T_2T_3T_3}{4\tfrac{1}{3} + 4\tfrac{1}{3} + 4\tfrac{1}{3} + 4\tfrac{1}{3} + 4\tfrac{1}{3} + 4\tfrac{1}{3}}}_{26} \times \underbrace{\overset{t_1t_1t_2t_2t_3t_3}{1\tfrac{1}{3} + 1\tfrac{1}{3} + 1\tfrac{1}{3} + 1\tfrac{1}{3} + 1\tfrac{1}{3} + 1\tfrac{1}{3}}}_{8}$

F_1: $\underbrace{\overset{T_1t_1T_2t_2T_3t_3}{4\tfrac{1}{3} + 1\tfrac{1}{3} + 4\tfrac{1}{3} + 1\tfrac{1}{3} + 4\tfrac{1}{3} + 1\tfrac{1}{3}}}_{17\text{ cm}}$

F_2:

	$T_1T_2T_3$ (13)	$T_1T_2t_3$ (10)	$T_1t_2T_3$ (10)	$t_1T_2T_3$ (10)	$T_1t_2t_3$ (7)	$t_1T_2t_3$ (7)	$t_1t_2T_3$ (7)	$t_1t_2t_3$ (4)
$T_1T_2T_3$ (13)	(26)	(23)	(23)	(23)	(20)	(20)	(20)	(17)
$T_1T_2t_3$ (10)	(23)	(20)	(20)	(20)	(17)	(17)	(17)	(14)
$T_1t_2T_3$ (10)	(23)	(20)	(20)	(20)	(17)	(17)	(17)	(14)
$t_1T_2T_3$ (10)	(23)	(20)	(20)	(20)	(17)	(17)	(17)	(14)
$T_1t_2t_3$ (7)	(20)	(17)	(17)	(17)	(14)	(14)	(14)	(11)
$t_1T_2t_3$ (7)	(20)	(17)	(17)	(17)	(14)	(14)	(14)	(11)
$t_1t_2T_3$ (7)	(20)	(17)	(17)	(17)	(14)	(14)	(14)	(11)
$t_1t_2t_3$ (4)	(17)	(14)	(14)	(14)	(11)	(11)	(11)	(8)

In summary, quantitative traits are determined by varying numbers of gene pairs interacting with each other. The larger the number of genes involved, the more difficult it is to discern differences between phenotypes. Patterns of transmission are then obscured, and such traits cannot be analyzed with classical Mendelian methods.

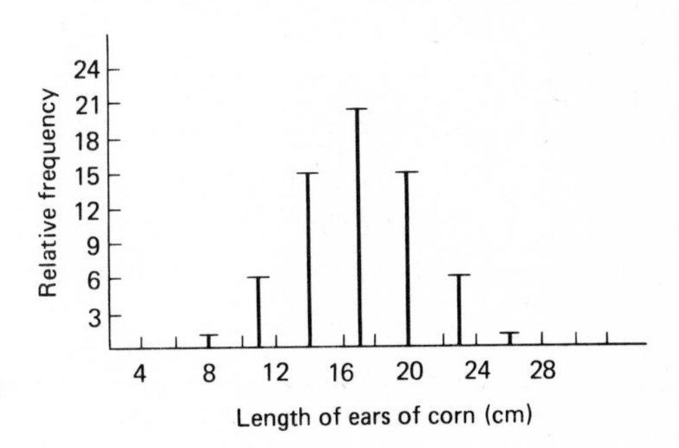

Figure 4-23. Relative frequency plot of F_2 in Table 4-8.

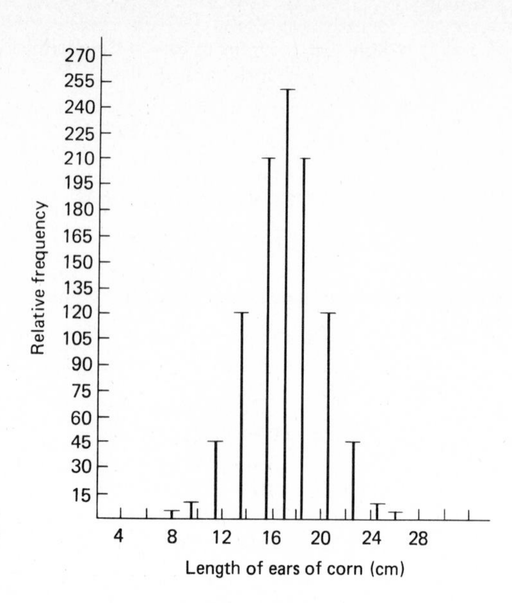

Figure 4-24. Relative frequency plot of lengths of ears of corn in the F_2 generation (five pairs of alleles).

PROBLEMS

1. In four o'clock plants, what progeny ratio can be expected from a cross between plants with pink flowers and plants with white flowers?

2. An abnormality in chickens results in curling of feathers, leading to a condition with the descriptive name of frizzle fowl. When two such frizzled fowls are mated, all the offspring are fine frizzled fowl.

(a) Detail the first experimental cross that you would make to establish that the frizzled condition is genetic.

(b) Assume that you have read a report that the condition is determined by an allele which is incompletely dominant to the normal allele determining flat feathers. What crosses would you make to confirm this, and what progeny ratios would you expect from your crosses?

3. What phenotypes would you expect in the children of a couple whose genotypes for the ABO group are the following: $I^Ai \times I^AI^B$? What phenotypic ratios would you expect among their children?

4. Suppose that the couple in problem 3 also have the following genotypes for the MN blood group: $MMI^Ai \times NMI^AI^B$? What phenotypic ratios would you expect for both blood groups among their children?

5. Mice from an albino strain are mated to mice from a strain with brown coat color. Their offspring are black. The F_1 are mated, and F_2 offspring are found to be in the following ratio: 9 black : 3 brown : 4 albino. Give the genotypes of the albino and brown parents for the *Black* and *albino* loci.

6. Assume albino mice from a homozygous strain are mated to homozygous wild-type mice (*AABBCC* for the *Agouti*, *Black*, and *albino* loci). The progeny are all wild-type (black agouti). The F_1 are mated and produce an F_2 generation which was recorded to be the following:

50 black agouti
16 black self
21 albino

What is the genotype of the homozygous strain of albino mice? Diagram all crosses, giving the genotypes of all animals.

7. In a colony of Brachyury mice ($T/+$), a tailless mouse was found in one litter of newborns. When matured, the tailless mouse was tested by mating to a tailless mouse from the T/t^9 strain.

(a) What phenotypes and in what ratios would you expect among the offspring if the new tailless mouse was T/t^x (t^x being a different t allele)?

(b) On the other hand, suppose the new tailless mouse happened to be also T/t^9? What would you expect to find among the offspring?

8. A white-eyed fly from the F_2 generation of a cross between scarlet and brown flies is mated to an unknown fly with wild-type eye color. The progeny are recorded as follows: 163 scarlet eye : 159 wild-type eye. Diagram the crosses, showing the genotypes of all flies involved.

9. Without specifying phenotypes, assume a number of crosses of different homozygous organisms are made and an F_2 generation produced with the following results: (a) 13 : 3, (b) 9 : 3 : 4, (c) 12 : 3 : 1, (d) 15 : 1, (e) 9 : 6 : 1, (f) 1 : 2 : 2 : 4 : 1 : 2 : 1 : 2 : 1, (g) 3 : 6 : 3 : 1 : 2 : 1, (h) 9 : 3 : 3 : 1. Suppose that the F_1 individuals which produced the above F_2 were testcrossed to homozygous recessive individuals. What phenotypic ratios would you expect in the testcross progeny? (From Srb, Owen, Edgar, Problem 1-25.)

10. An albino man and woman were married and had six children. All the children were normally pigmented. Without invoking infidelity, what genetic explanation would you give for this, since genes determining albinism are known to be recessive?

11. Verify the values of the length of the F_2 ears of corn if

(a) There is one pair of alleles and the dominant gene contributes 13 cm and the recessive gene contributes 4 cm.

(b) There are two pairs of alleles and the dominant gene contributes $6\frac{1}{2}$ cm and the recessive gene 2 cm. (Why $6\frac{1}{2}$ and 2?)

(c) There are three pairs of alleles and the dominant gene contributes $4\frac{1}{3}$ cm and the recessive gene $1\frac{1}{3}$ cm.

(d) There are 5 pairs of alleles and the dominant gene contributes 2.6 cm and the recessive gene .8 cm.

12. Suppose there are 8 pairs of alleles in the previous example of corn length.

(a) Determine the contribution of the dominant gene.

(b) Determine the contribution of the recessive gene.

(c) Determine the frequency distribution of lengths of ears of corn in the F_2.

REFERENCES

BATESON, W., and R. C. PUNNETT, 1905–1908. "Experimental Studies in the Physiology of Heredity." Reports to the Evolution Committee of the Royal Society. Reprinted in J. A. Peters, ed., 1959. *Classic Papers in Genetics*. Prentice-Hall, Inc., Englewood Cliffs, N.J.

BENNETT, D., 1975. "The *T* Locus of the Mouse." *Cell* 6: 441–454.

BRADEN, A. W. H., 1958. "Influence of Time of Mating on the Segregation Ratio of Alleles at the *T* Locus in the House Mouse." *Nature* 181: 786–787.

CORRENS, C., 1912. *Die Neuen Vererbungsgesetze*. Born-traeger, Berlin.

DOBROVOLSKAIA-ZAVADSKAIA, N., 1927. "Sur la Mortification spontanée de la queue chez la souris nouveau-née et sur l'existence d'un charactère (facteur) hereditaire non-viable." *CR Soc. Biol.* 97: 114–116.

DUNN, L. C., and S. GLUECKSOHN-SCHOENHEIMER, 1950. "Repeated Mutations in One Area of a Mouse Chromosome." *Proc. Nat. Acad. Sci.* 36: 233–237.

FISHER, R. A., cited by R. R. Race, 1944. "An Incomplete Antibody in Human Serum." *Nature* 153: 771–772.

GLUECKSOHN-WAELSCH, S., 1963. "Lethal Genes and Analysis of Differentiation." *Science* 142: 1269–1276.

KETTANEH, N. P., AND D. L. HARTL, 1976. "Histone Transition During Spermiogenesis Is Absent in *Segregation Distorter* Males of *Drosophila Melanogaster*." *Science* 193: 1020–1021.

LANDSTEINER, K., AND P. LEVINE, 1927. "Further Observations on Individual Differences of Human Blood." *Proc. Soc. Exp. Biol. Med.* 24: 941–942.

LANDSTEINER, K., AND A. S. WEINER, 1940. "An Agglutinable Factor in Human Blood Recognized by Immune Sera for Rhesus Blood." *Proc. Soc. Exp. Biol. N.Y.* 43: 223.

PEACOCK, W. J., G. L. G. MIKLOS, AND D. J. GOODCHILD, 1975. "Sex Chromosome Meiotic Drive Systems in *Drosophila Melanogaster*. Abnormal Spermatid Development in Males with a Heterochromatin-deficient X Chromosome." *Genetics* 79: 613–634.

RACE, R. R., 1948. "The Rh Genotypes and Fisher's Theory." *Blood* 3 *Suppl.* 2: 27–42.

______, AND R. SANGER, 1975. *Blood Groups in Man*, 6th ed. Lippincott, Philadelphia.

SANDLER, L., Y. HIRAIZUMI, AND I. SANDLER, 1959. "Meiotic Drive in Natural Populations of *Drosophila Melanogaster*. I, The Cytogenetic Basis of *Segregation Distortion*." *Gen.* 44: 231–249.

SHERMAN, M. I., AND L. R. WUDL, 1977. "T Complex Mutations and Their Effects" in M. I., Sherman, ed., *Concepts in Mammalian Embryogenesis*. MIT Press, Boston.

SRB, A. M., R. D. OWEN, R. S. EDGAR, 1965. *General Genetics*, 2nd ed. W. H. Freeman & Company Publishers, San Francisco.

STERN, C., 1973. *Principles of Human Genetics*, 3rd ed. W. H. Freeman & Company Publishers, San Francisco.

STURTEVANT, A. H., 1913. "The Himalayan Rabbit Case, with Some Considerations on Multiple Allelomorphs." Reprinted in L. Levine, ed., 1971. *Papers on Genetics*. C. V. Mosby, St. Louis, Mo.

WAGNER, R. P., AND H. K. MITCHELL, 1964. *Genetics and Metabolism*. John Wiley, New York.

WEINER, A. S., E. B. GORDON, AND L. HANDMAN, 1949. "Heredity of the Rh Blood Types." *Am. J. Hum. Gen.* 1: 127–140.

ZEIGLER, I., 1961. "Genetic Aspects of Ommochrome and Pterin Pigments." *Adv. in Genetics* 10: 349–403.

FIVE

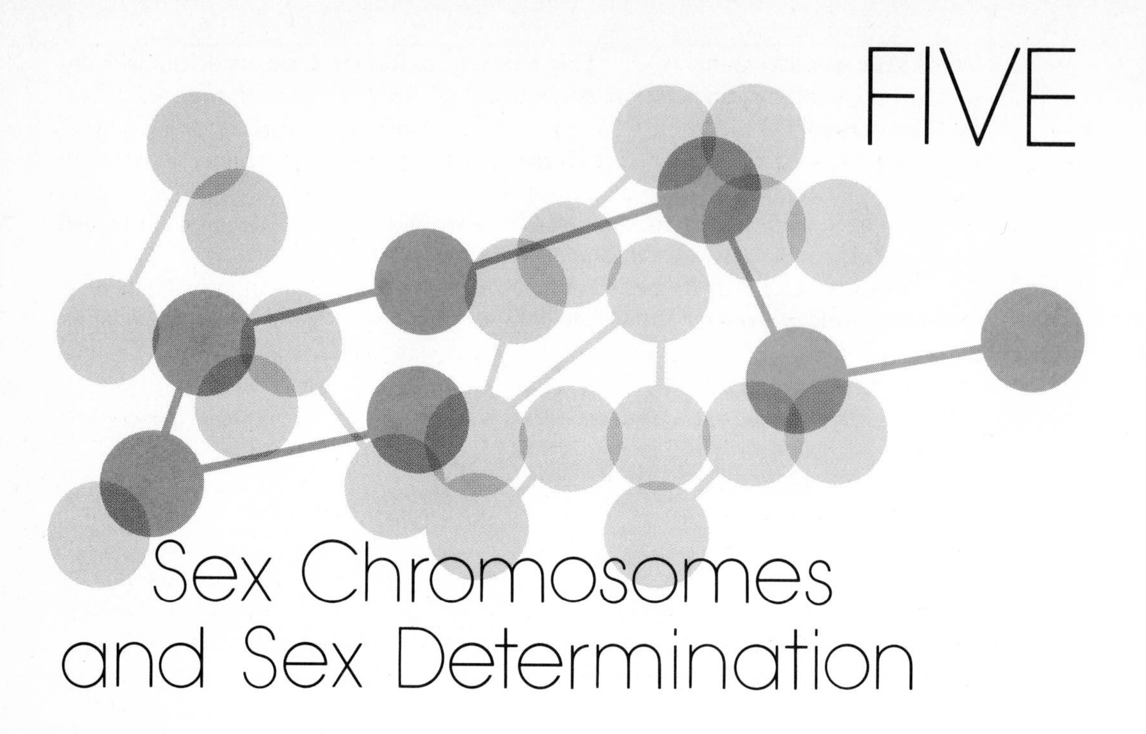

Sex Chromosomes and Sex Determination

INTRODUCTION

In the preceding chapter we discussed a number of genetic phenomena which are very difficult to analyze in the classical Mendelian manner. In this chapter we shall explore yet another phenomenon which also remains largely a mystery: the genetic mechanism for the determination and development of different sexes in sexually reproducing organisms.

Implicit in the theories established by classical geneticists that chromosomes serve as the vehicles for transmission of genes from one generation to another is that chromosomes exist in homologous pairs, one partner of each pair contributed by the mother, and one by the father. Obviously, then, diploid organisms should all have even numbers of chromosomes. Yet in 1891, H. Henking reported that studies of cells of males in the Hemipteran *Pyrrhocoris* showed there to be an odd number of "chromatin elements." Cells of females were found to have an even number of chromosomes, one more than males.

SEX CHROMOSOMES

The biological significance of this phenomenon was not discovered until 1902, when Clarence McClung and others studied various insects (see Wilson 1925). McClung postulated that the extra "element" which he called the *accessory chromosome* and which later came to be known as the *X element*, was in some way involved with the determination of sex. This hypothesis was supported by the studies of E. B. Wilson in 1905 on Hemipteran genera such as *Protenor* and *Anasa*. Wilson found that females of these insects possessed one more chromosome in their cells than did males. For example, in *Protenor belfragi*, the female number of chromosomes was found to be 14; the male number, 13. Wilson then proposed that the X element be called the *sex chromosome*.

In such species, the male is referred to as the *heterogametic sex*, because of the difference in chromosome numbers in sperm cells after meiosis. One-half the sperm would carry 6 chromosomes, and these would give rise to male offspring; the other half would have 7 chromosomes and produce female offspring. Females, since they would have 7 chromosomes in all mature ova, are accordingly referred to as the *homogametic sex*.

Studies on other insect orders such as Coleoptera and Diptera, however, showed that sex is not necessarily determined by a quantitative difference in chromosomes. Wilson reported that both females and males of species in these orders possess the same number of chromosomes. For example, there are four pairs in the fruitfly, *Drosophila melanogaster*. But it was noticed that in the males of these organisms, the homologs comprising one of the four pairs showed distinct morphological differences. (Recall that our definition of homologs included the aspect of physical similarity.) In *Drosophila* males, this unlike pair were considered the sex chromosomes, and referred to as the *X* and *Y* chromosomes, the Y chromosome being the one with the hook. Although sperm of *Drosophila* all have the same haploid number of four chromosomes, we still consider the male as the heterogametic sex, because half of the sperm contain the X chromosome and half the Y.

TRANSMISSION OF SEX-LINKED TRAITS

If sex chromosomes are nonmatching, how does the difference affect the transmission of traits determined by genes located on these chromosomes? If the chromosomes are morphologically different, can one assume that they are also genetically different? If so, they would not carry alleles at all loci. Then how might the expression of dominance and recessiveness at these loci be affected, since we have seen that this and other phenomena depend on the interaction between alleles?

X-Linkage in Drosophila

T. H. Morgan and the white-eye mutation. The first definitive answer to these questions came from the laboratory of Thomas Hunt Morgan in 1909. Morgan (Fig. 5-1) was a giant in transmission genetics, and was the first of a long line of geneticists who have been awarded the Nobel Prize for their contributions to the understanding of basic life processes. Since the early work was primarily centered on traits in *Drosophila*, our first understanding of many basic concepts was of those that apply to the fruitfly. As you shall see later in the chapter, these phenomena vary from species to species. For example, sex determination in humans is different from that in *Drosophila*.

Figure 5-1. Thomas Hunt Morgan. (The Bettmann Archive.)

Morgan had found a male fruitfly in his colony which had white eyes in striking contrast to the normal wild-type red color. To determine the genetic basis for this trait, Morgan mated the fly to normal wild-type females. The result of the crosses included 1237 wild-type males and females and 3 white-eyed males. The white-eyed males he correctly assessed to be due to spontaneous new mutations to White-eye, and were ignored in his analysis. Since the F_1 were all wild-type, it was assumed that the White-eyed trait was completely recessive to wild-type.

When the F_1 were mated to produce an F_2 generation, the following data were recorded:

2459 red-eyed females
1011 red-eyed males
782 white-eyed males
0 white-eyed females

Although it appeared that the White-eyed trait was "limited" to males, Morgan performed experiments that proved otherwise. He crossed the original white-eyed male to some of the F_1 females. (Such a cross, between a parent and an offspring, is referred to as a *backcross*.) The results of this backcross were as follows:

129 red-eyed females
132 red-eyed males
88 white-eyed females
86 white-eyed males

In view of the cytological evidence for structural differences between the X and Y chromosomes in *Drosophila* males, Morgan realized that the data could be interpreted along Mendelian lines if one assumption is made: that the male possesses only one copy of genes carried on sex chromosomes, whereas the female has the diploid number. Since females have two X chromosomes and males one, it was logical to infer that the gene determining white eyes was a recessive mutation of a gene on the X chromosome, which is masked by a dominant allele in the females. THUS ALL GENES ON THE X CHROMOSOME, WHETHER DOMINANT OR RECESSIVE, ARE EXPRESSED IN MALES.

With this assumption we can satisfactorily explain the progeny results obtained by Morgan. If we symbolize the wild-type gene for eye color as X^W (the X indicates we are speaking of X-linked genes and the superscript denotes the specific allele), the *White-eye* mutation as X^w, and the Y chromosome simply as Y, the crosses can be diagrammed as shown in Fig. 5-2. It should be pointed out here that X-linked genes can also be symbolized by a letter as we have done all along. For example, a white-eyed female would be designated X^wX^w or *ww*; a wild-type male would be either X^WY or *W*Y. Whichever system of symbols you prefer to use, it is essential that you designate the Y to obtain the correct expected results of a cross.

Morgan referred to this inheritance as "sex-limited" because of the absence of white-eyed females in his original cross. Later, as evidence accumulated to confirm his theory that such genes are on the X chromosome, the term widely used was *sex-linked* inheritance, or *sex-linkage*. The term *sex-limited* is now used in a different

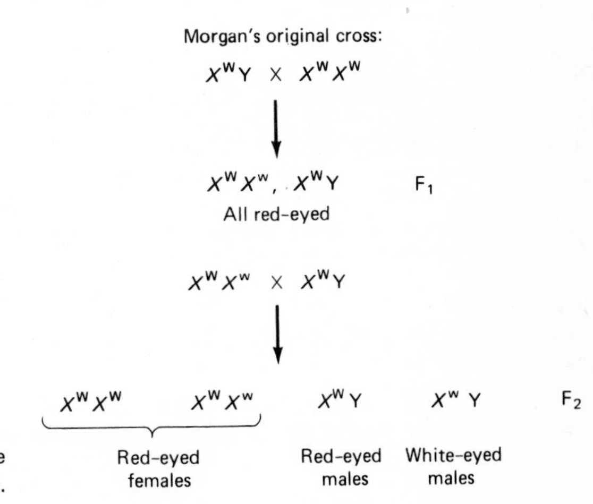

Figure 5-2. Morgan's analysis of the inheritance of the white-eye mutation, an X-linked recessive.

context (p. 141). Recently, geneticists have preferred to use the more explicit term *X-linkage*, for as we shall discuss later in the chapter, there is evidence for active genes on the Y chromosome in some organisms.

All the inherited traits we have discussed prior to this chapter were determined by genes present in pairs, such as in Mendel's experiments and in albinism. If genes are located on true pairs of homologs, they are referred to as *autosomal*, as we distinguish chromosomes that exist in pairs by calling them *autosomes*. It is important to specify traits as either autosomal or X-linked because of the expression of recessive X-linked genes in males, whereas females must be homozygous for recessive X-linked traits if they are to be expressed, just as in autosomal traits. Males are thus said to be *hemizygous* for X-linked traits. They cannot be either homozygous or heterozygous, since by definition these terms imply the presence of two alleles.

Because of these male-female differences in the expression of recessive X-linked traits, the "crisscross" pattern of transmission, that is, from grandfather to grandson, is very typical of X-linked traits.

C. B. Bridges and nondisjunction. At the time that Morgan presented his analysis of the *white-eye* mutation, geneticists still disagreed over whether genes are actually carried on chromosomes. Circumstantial evidence that they are, including the transmission of X-linked traits, was abundant. Shortly after the publication of Morgan's studies in 1916, one of Morgan's students (who himself became a renowned geneticist), C. B. Bridges, published further studies on the *white-eye* mutation. Bridges' work not only confirmed Morgan's analysis but also provided unequivocal evidence that chromosomes are indeed the vehicles for the inheritance of traits.

Bridges noticed that in crosses of white-eyed females with normal males ($X^wX^w \times X^W$Y), where one might expect only white-eyed males, an occasional white-eyed female and red-eyed male appeared. These were very rare, on the order of one in every 2000–3000 offspring. Bridges reasoned that the white-eyed females must have received both X chromosomes from the mother in order to be homozygous, and the red-eyed males the normal allele from the X chromosome of the father.

This unusual transmission of sex chromosomes could be explained if an abnormality of meiosis occurred that resulted in two types of eggs, some containing both X chromosomes and some with no X chromosome. Let us digress for a moment to discuss again this process, known as *nondisjunction* (mentioned previously in Chapter One). Only with an understanding of nondisjunction can we appreciate the significance of Bridges' experiments.

The stage of meiosis during which homologs normally segregate is anaphase I, and the stage for separation of chromatids is anaphase II. For reasons as yet not clearly understood, the process of separation sometimes does not occur, and this failure is called *nondisjunction*. If nondisjunction of the X chromosomes occurs at anaphase I, the results would be as illustrated in Fig. 5-3: All gametes would contain an abnormal number of chromosomes. If we assume that we are dealing with a pair of X^w chromosomes, $\frac{1}{2}$ the gametes would contain X^wX^w, and $\frac{1}{2}$ 0, or no X chromosomes at all.

Suppose that nondisjunction occurred at anaphase II—what then would be the effect on the gametes? Again we can diagram this, as shown in Fig. 5-4. We see that at the end of meiosis, $\frac{1}{4}$ of the eggs would be X^wX^w, $\frac{1}{4}$ 0, and $\frac{1}{2}$ X^w, or normal.

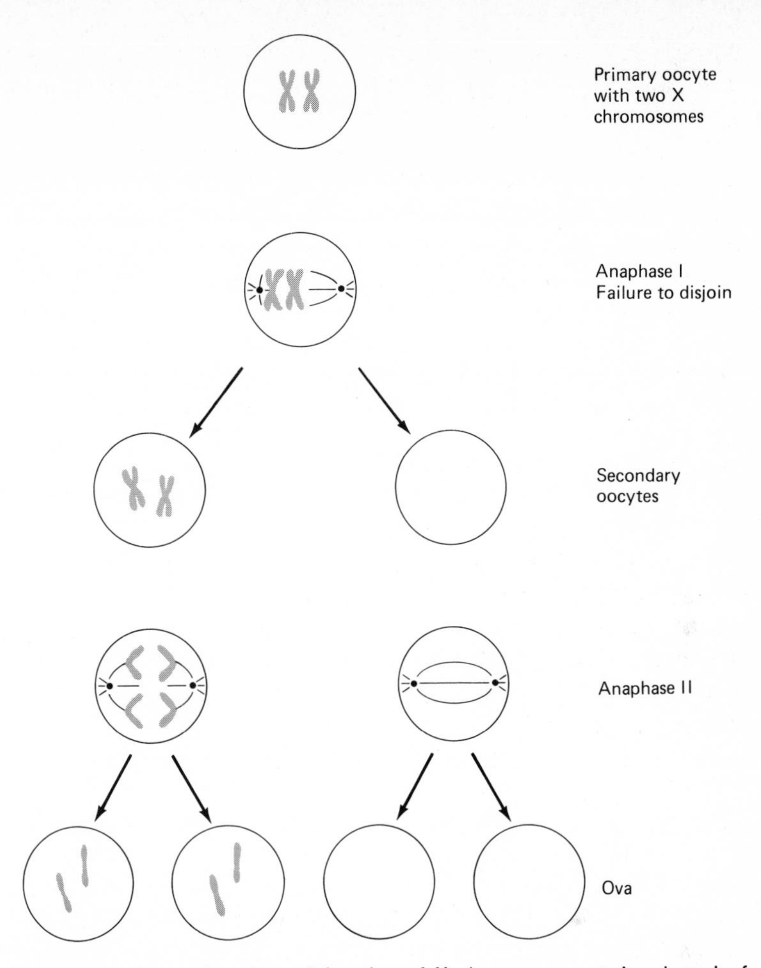

Figure 5-3. The results of nondisjunction of X chromosomes at Anaphase I of meiosis. All four ova are abnormal in sex chromosome number. Two have none, and two are XX.

When nondisjunction occurs in cells that are normally diploid at the onset of division, we refer to this as *primary nondisjunction.* This abnormality of division is not restricted to meiosis, nor to sex chromosomes. It can involve autosomes, and it can occur during mitosis. In Chapter 6 we shall discuss the effects of abnormal numbers of autosomes in cells.

To return to Bridges' experiment, you can see that the results of fertilization of eggs with abnormal sex chromosome numbers by normal sperm would result in the following offspring:

	X^W	Y
X^wX^w	$X^WX^wX^w$	X^wX^wY
0	X^W0	Y0

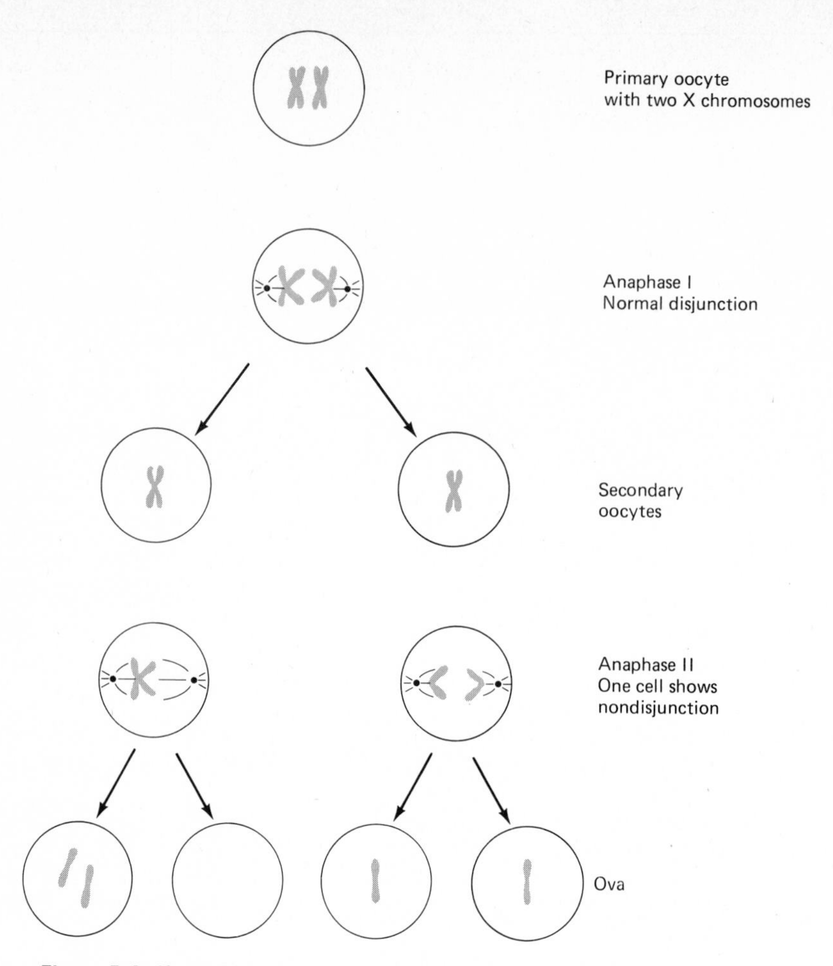

Figure 5-4. If nondisjunction occurs at Anaphase II, $\frac{1}{4}$ of the ova have 2 X chromosomes, $\frac{1}{4}$ have no X chromosomes, and $\frac{1}{2}$ are normal haploid ova.

If flies with two or more X chromosomes are female, the Punnett Square shows that white-eyed females would actually possess two X chromosomes and one Y chromosome; the red-eyed males would have only one sex chromosome, the X from the father. An examination of the cells of the exceptional females and males provided cytological support for Bridges' interpretation. White-eyed females had nine chromosomes, including two X's, and red-eyed males had seven chromosomes, with only one X. Both the XXX and the Y0 conditions were found to be lethal.

Bridges further confirmed his analysis by crossing the unusual white-eyed females with normal males. Because of the odd number of sex chromosomes, such females could be expected to produce four different combinations of sex chromosomes in their ova: X^w, X^wY, X^wX^w, and Y (Fig. 5-5). The presence of extra sex chromosomes in the ova is referred to as *secondary nondisjunction* because it is due to an abnormal number of chromosomes in the cells of white-eyed females, not to an original failure of chromosomes to disjoin. The expected results of the cross, X^wX^wY $\times$ X^WY, are

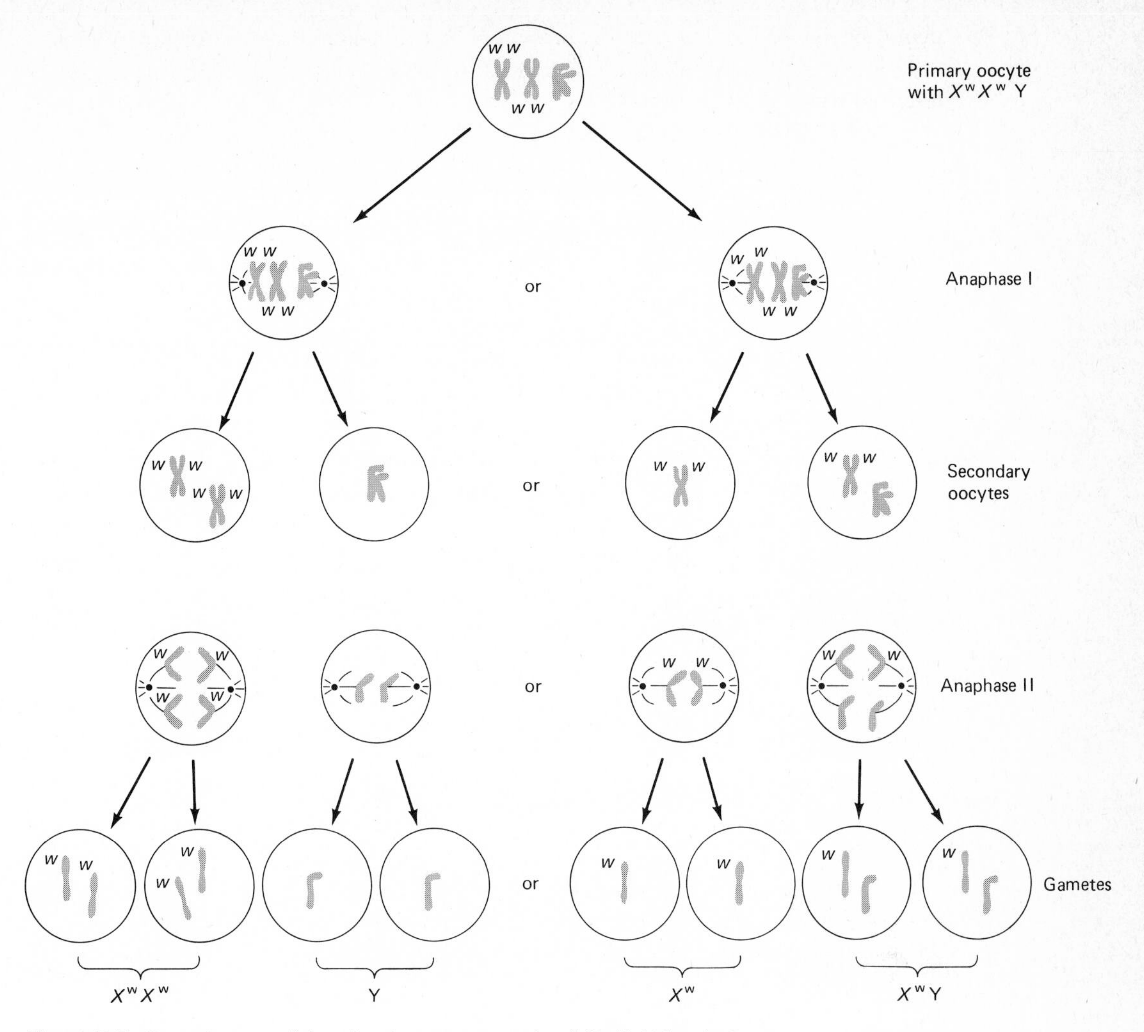

Figure 5-5. Secondary nondisjunction in gametogenesis of X^wX^wY female *Drosophila.*

diagrammed in the following Punnett Square:

	X^w	X^wY	X^wX^w	Y
X^W	X^WX^w	X^WX^wY	$X^WX^wX^w$	X^WY
Y	X^wY	X^wYY	X^wX^wY	YY

As he predicted, Bridges obtained in the offspring more red-eyed females (X^WX^w, X^WX^wY) than white-eyed (X^wX^wY), and more white-eyed males (X^wY, X^wYY) than red-eyed (X^WY). Again, cytological studies showed that the chromosome numbers were as predicted according to the phenotypes of the flies. Through these experiments, Bridges offered conclusive evidence for the X-linked nature of the White-eye mutation and the role of chromosomes as the physical basis for the transmission of inherited traits.

L. V. Morgan and the attached X. Another classical *Drosophila* experiment involving X-linkage and abnormal numbers of sex chromosomes was that done by T. H. Morgan's wife, L. V. Morgan (L. V. Morgan 1922). In studying the recessive trait *yellow body color*, which had shown a pattern of transmission typical of X-linkage, L. V. Morgan discovered an exceptional yellow female which, when mated to normal males, produced only yellow females and normal (gray) males. One would have expected exactly the opposite phenotypes in the two sexes of the offspring. (Why? Diagram the cross and obtain the expected ratio of offspring.) Furthermore, the yellow daughters, when mated to normal males, produced only yellow daughters and normal males.

L. V. Morgan hypothesized that in this particular case, there must be a permanent nondisjunction of the X chromosomes (also referred to as *attached X* chromosomes) in the exceptional yellow females, a hypothesis confirmed by later cytological investigation. The genotypes of the flies can be designated as follows: $\widehat{X^yX^y}\mathrm{Y} \times X^Y\mathrm{Y}$. We denote the attached X's by the mark ^. Results of the cross are given in the following Punnett Square:

	$\widehat{X^yX^y}$	Y
X^Y	$\widehat{X^yX^y}X^y$	$X^Y\mathrm{Y}$
Y	$\widehat{X^yX^y}\mathrm{Y}$	YY

Note that in this strain normal male offspring are actually inheriting their Y chromosome from the mother and not the father. The father donates his Y chromosome to the $\widehat{XX}$Y females. This is of course in contrast to the normal situation, in which the gametes of the father determine the sex of the offspring.

X-Linkage in Humans

Humans also have a pair of X and Y sex chromosomes. The pattern of transmission of X-linked traits is very similar to the pattern described above for X-linked traits in *Drosophila*. We can thus predict the transmission of X-linked conditions along the same lines of analysis. For example, a well-known X-linked trait is color-blindness, which studies have shown to be recessive to normal color vision.

Color vision in humans is dependent upon the presence of three primary color vision pigments—red, green, and blue—in the retina of the eye. Persons with deficiencies in these pigments have problems distinguishing various colors. There is a rare autosomal form of total color-blindness, but various types of partial color-blindness, frequently referred to as *red-green* color blindness, are the most common. For the most part, these are X-linked recessive conditions. It is unclear at present whether one or two different loci on the X chromosome are involved (see reference to Stern

1973), but for our discussion, we shall assume that color-blindness is determined by alleles at a single locus.

What can we expect to find among the offspring of a woman of normal vision whose father was color-blind, if she marries a normal man? Since she is of normal vision, we can assume that she carries the normal allele on her maternal X chromosome. Since all X-bearing sperm from her father must carry the gene for color-blindness, she would obviously have to be a heterozygote, or $X^C X^c$. Her husband's genotype must be X^CY since he is normal, and the following Punnett Square diagrams the expected genotypes of the offspring:

	X^C	Y
X^C	$X^C X^C$	X^CY
X^c	$X^C X^c$	X^cY

The phenotypic ratio expected from this cross would be 3 normal : 1 color-blind child. Breaking it down into sexes, there would be no color-blind daughters and half the sons would be color-blind and half normal. The absence of alleles on the Y chromosome in humans again results in the expression of all genes on the X chromosome in males, a situation identical to that which we have already discussed in *Drosophila.*

Hemophilia is another X-linked condition in humans, and its expression is found predominantly in men. Hemophilia, color-blindness, and the great majority of X-linked traits are in fact determined by recessive genes, so that females would have to be homozygous to manifest the condition. This is the explanation for the well-known fact that X-linked traits are much more frequently expressed in men than in women.

X-Linkage in Pedigrees

Whether an inherited trait is autosomal or X-linked can frequently be detected in the analysis of pedigrees. Turn back once again to Figs. 2-8 and 2-9. Both those pedigrees were of autosomal genes, and the transmission pattern clearly indicates this. In Fig. 2-8 an affected male in the P generation transmitted a dominant gene to two sons and a daughter; therefore it is not X-linked as the sons would not have inherited the trait having received only the Y from their father. In Fig. 2-9 the affected daughters both had unaffected fathers. This would not be X-linkage as, to be homozygous, the girls would have had to inherit one X from their fathers, in which case the fathers should manifest the trait also.

Figure 5-6 is a real pedigree involving an X-linked trait. This is the well-known transmission of hemophilia in the royal families of Europe in the nineteenth and twentieth centuries. Queen Victoria herself was the first detectable carrier of the gene, which she then transmitted to a number of her offspring and in turn to her descendants. Note that only males are hemophiliacs. This is a characteristic of X-linked recessive traits that is the most obvious in such pedigrees.

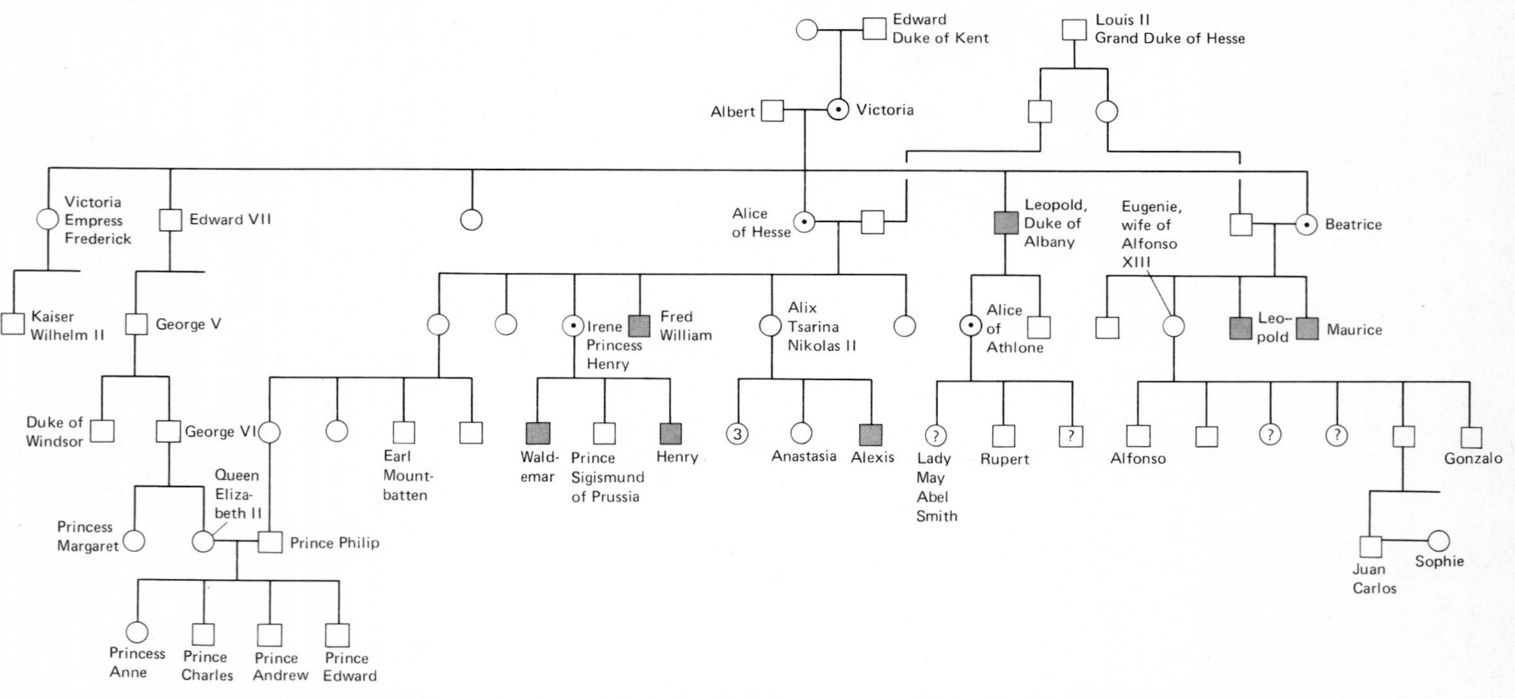

Figure 5-6. Pedigree of royal families of Europe showing transmission of hemophilia, an x-linked recessive trait. The gene was apparently transmitted from Queen Victoria to several descendants.

(A couple of points of historical rather than scientific interest brought out by the pedigree: Queen Elizabeth II and Prince Philip are cousins. Historians also believe that the hemophiliac condition of Tsar Nicholas' only son, Alexis, led the Tsar and his wife to turn to the monk Rasputin for help, and this contributed directly to the development of the Russian Revolution. Now back to genetics!)

Y-Linkage

In *Drosophila* and humans, much research has revealed a large number of X-linked genes in both species. Note that the X-linked genes we have discussed so far do not determine traits related to sex, or to the development of reproductive organs or secondary sexual characteristics. Further, as we have mentioned above, none of these genes appear to have alleles on the Y chromosome. Since its discovery, geneticists have searched to little avail for the presence of genes on the Y chromosome which can be traced with the tools of transmission genetics.

Y-linked genes would of course have to be transmitted only from father to son in species such as humans and normal *Drosophila*. One would expect not only a father-to-son transmission, but also that the trait could not be transmitted from father to daughter, nor would any of the daughter's sons inherit the gene, unless the daughter happened to marry a man with the same Y-linked gene.

There have been a few rare cases in *Drosophila*, humans, and other organisms which show a pattern of transmission that suggests *holandric* inheritance, a term meaning Y-linkage. In *Drosophila* the *Bobbed* gene, which causes a shortening of bristles and alterations in abdominal markings, is believed to have a locus on both sex chromosomes which is found near the centromere of both the X and Y chromosomes. In fish there are Y-linked traits such as a black pigment spot in the guppy *Lebistes reticulatus* (see reference to Donamraju 1965).

In India, studies of a family in which men have unusually long hairs on the pinna of the ears seem to show such a pattern of inheritance; however, the genetic data are not unequivocal, and some question still remains. An antigen (a substance which provokes an immune response, see p. 386) found only in males of various species of mammals such as mice and men, is presumed to be determined by a Y-linked gene. For the most part, however, there appears to be little homology between the X and Y chromosomes in most organisms.

THE X CHROMOSOME AND SEX DETERMINATION

If the X-linked genes and most holandric genes studied so far do not appear to be involved in sex determination, what *is* the actual role of the X and Y chromosomes in relation to maleness and femaleness? In *Drosophila*, part of the answer can be found if we return to the classical experiments by Bridges with XXY females. The very fact that flies of this genotype were female although they contained a Y chromosome indicates that the Y is not in itself a determinant of maleness. Furthermore, recall that the X0 flies are also male. They are, however, also sterile. Therefore in *Drosophila*, the Y chromosome does not appear to determine maleness, although its presence is necessary for normal fertility of male flies.

The Balance System in Drosophila

If the Y chromosome does not determine maleness in *Drosophila*, what does? Bridges again provided the answer to this in a series of ingenious experiments using *Drosophila* with abnormal chromosome constitutions (Bridges 1925). He isolated and used female flies that possessed three entire sets of chromosomes, a condition known as *triploidy*. (The suffix *-ploidy* refers always to *sets* of chromosomes. Thus by definition, triploidy implies that there are three copies of every chromosome; it is not a random collection of chromosomes that total to the $3n$ number.)

Bridges designated the three sex chromosomes by 3X, and the three sets of autosomes by 3A. Thus the triploid female flies were 3X3A in genotype. Presence of the odd set of chromosomes leads to different combinations of autosomes and sex chromosomes in the gametes following meiosis. There would be four combinations of chromosomes in the gametes, as shown in Fig. 5-7. When mated to a normal male,

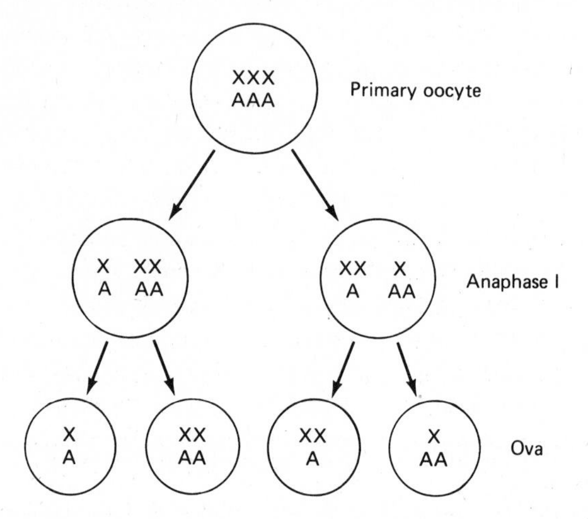

Figure 5-7. Abbreviated meiosis of cells of triploid flies. Each letter X refers to one X chromosome. Each letter A refers to one set of autosomes.

these four gametes, AX, AAX, AXX, AAXX would be fertilized by sperm carrying either the X or the Y chromosome, and the offspring are diagrammed in the following Punnett Square:

	AX	AAX	AXX	AAXX
AX	AAXX	AAAXX	AAXXX	AAAXXX
AY	AAXY	AAAXY	AAXXY	AAAXXY

The phenotypes of the progeny were recorded and correlated with cytological studies of their cells to determine the numbers and combinations of autosomes and sex chromosomes. The results are given in Table 5-1.

Table 5-1 Some Results of Bridges' Experiments with Triploid Flies

Chromosome combinations	*Sex*
AAXX	Female
AAXY	Male
AAAXX	Intersex (female)
AAAXY	Supermale
AAXXX	Superfemale
AAXXY	Female
AAAXXX	Female
AAAXXY	Intersex (male)

As Bridges pointed out, the Y chromosome does not appear to determine maleness. RATHER, SEX IS DETERMINED IN *Drosophila* BY A BALANCE BETWEEN THE NUMBER OF SETS OF AUTOSOMES AND THE NUMBER OF X CHROMOSOMES. There must be genes on the X chromosome which cause a tendency to femaleness, and genes on the autosomes which lead to the development of maleness. The net male tendency of a set of autosomes is less than the net female tendency of an X chromosome, since flies with an equal number of X chromosomes and sets of autosomes, for example, AAXX and AAAXXX, are female.

The flies that were classified as intersexes showed the development of both male and female parts, with some showing more maleness and others more femaleness depending on their genotypes. They did not, however, possess normally functioning male and female organs, and were sterile. The "super" males and females were so named only because the presence of extra autosomes or sex chromosomes threw the balance of development farther toward one or the other sex. The term is really a misnomer—instead of being able to leap tall buildings in a single flap of the wings which the term implies, the superfemale condition is lethal for the most part, and supermales are sterile!

Drosophila Gynandromorphs

More evidence for the nature of the X-linked inheritance and the chromosomal balance theory of sex determination in *Drosophila* can be found in flies with an interesting aberration known as *gynandromorphism.* This is a condition in which the flies are combinations of genetically different cells, some male and some female. First studied in 1919 by the two giants of *Drosophila* genetics, T. H. Morgan and C. B. Bridges, gynandromorphs appear because of an unequal distribution of X chromosomes during the first mitotic division following fertilization of an egg. If one of the X chromosomes in the zygote which is a normal female XX lags behind on the spindle during division, one of the daughter cells would be XX and female, but the other would be X0 and male.

It is a peculiarity of *Drosophila* embryogenesis that the two cells resulting from the first mitotic division are the progenitors of all cells to form the left and the right sides of the body, respectively. In the fly with unequal distribution of sex chromosomes this leads to the development of cells on one half of its body that are XX and female, and the other half, X0 and male. We refer to individuals whose cells show a mixture of different combinations of chromosomes as *genetic mosaics*. If such flies are from fertilized eggs heterozygous for X-linked recessive genes such as *white eyes* and *miniature wings*, the female half would appear as wild type and the male half would express the mutant phenotypes as illustrated in Fig. 5-8.

Figure 5-8. A *Drosophila* gynandromorph. Cells and structures on the left are XX and female; on the right X0 and male. The recessive X-linked genes for white eyes and miniature wings are expressed only in the right half, as female cells are heterozygous and wild type. (See Gardner, *Principles of Genetics* 5th ed., Fig. 4.22, p. 126.)

THE Y CHROMOSOME AND SEX DETERMINATION

Although it had been known for some time that humans and other mammals have the XY system of sex chromosomes, the role of the mammalian Y chromosome remained unclear until the 1950s, and then was found to be different from its role in the fruitfly. Because of Bridges' work on *Drosophila*, one could have interpreted the situation as men being male because of the presence of one X in balance with one set of autosomes. It was not until improvements in techniques allowed more accurate observations of mammalian chromosomes that the differences in sex determination between *Drosophila* and mammals became apparent.

Role of the Y Chromosome in Mammals

It was necessary to develop special procedures for staining human chromosomes in order for cytologists to clearly visualize the number and gross structure of the chromosomes. We shall discuss the structure of chromosomes in detail in Chapter 6, but for now we shall restrict our discussion to the contributions that such studies made to our understanding of sex chromosomes and their function in humans.

In the 1950s, techniques became available to scientists for the study of chromosomes of normal and abnormal human cells. In 1959, papers were published by C. E. Ford and others and by P. A. Jacobs and J. A. Strong, in which they reported on the effects of X0 and XXY sex chromosome combinations, respectively. The X0 patients suffered from abnormalities known as Turner's syndrome or gonadal dysgenesis, and were phenotypically female (Fig. 5-9). The XXY patients had abnormalities called Klinefelter's syndrome, and were phenotypically male (Fig. 5-10). Note that this situation is directly contrary to the maleness of X0 and femaleness of XXY *Drosophila*.

The phenotypes of these patients indicated that the Y chromosome determines

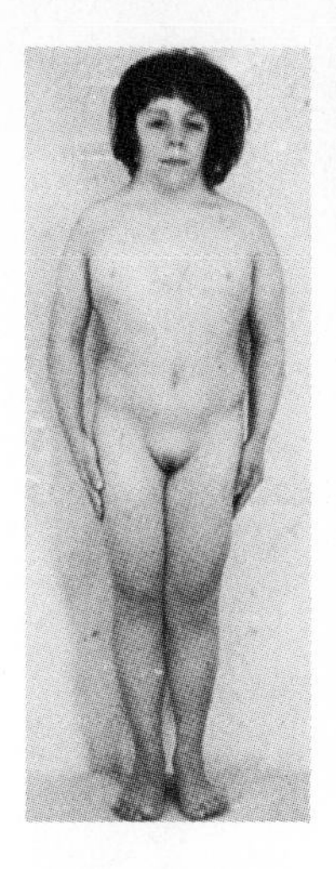
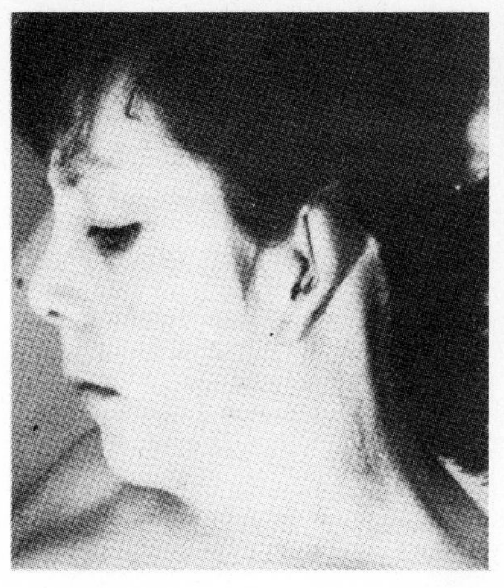
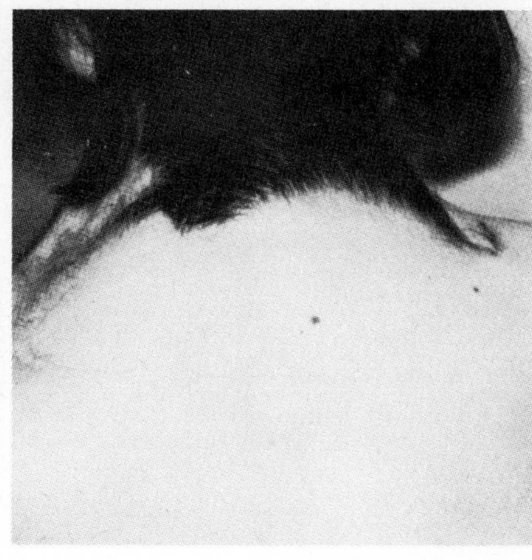

Figure 5-9. (a) A patient with Turner's syndrome (X0); (b) typical "webbing" of neck region; (c) low hairline on neck. (From J. de Grouchy and C. Turleau, 1977. *Clinical Atlas of Human Chromosomes*, p. 233. John Wiley, New York.)

maleness. IN HUMANS, PRESENCE OF THE Y CHROMOSOME CAUSES MALENESS, AND ABSENCE OF THE Y CHROMOSOME RESULTS IN FEMALENESS. It is not a balanced system of sex determination as in *Drosophila*. Since the 1950s a wide variety of abnormal human sex chromosome combinations have been found, and regardless of the number of X chromosomes present in the cells, the individual is usually phenotypically male if there is at least one Y, as shown in Table 5-2. For example, XXXY and XXXXY are male, although abnormalities (including mental retardation) increase in severity with each extra copy of the sex chromosome.

The role of the Y chromosome deduced for humans is now believed to apply to most mammals. In the same year that the human chromosomal abnormalities

Table 5-2 Abnormal Sex Chromosome Constitutions and Phenotypic Effects in Humans

	Sex phenotype	*Fertility*	*Sex chromosome constitutions*
Normal male	Male	+	XY
Normal female	Female	+	XX
Turner's syndrome	Female	−	X0
Klinefelter's syndrome	Male	−	XXY
XYY syndrome	Male	+	XYY
Triple X syndrome	Female	±	XXX
Triple X-Y syndrome	Male	−	XXXY
Tetra X syndrome	Female	?	XXXX
Tetra X-Y syndrome	Male	−	XXXXY
Penta X syndrome	Female	?	XXXXX

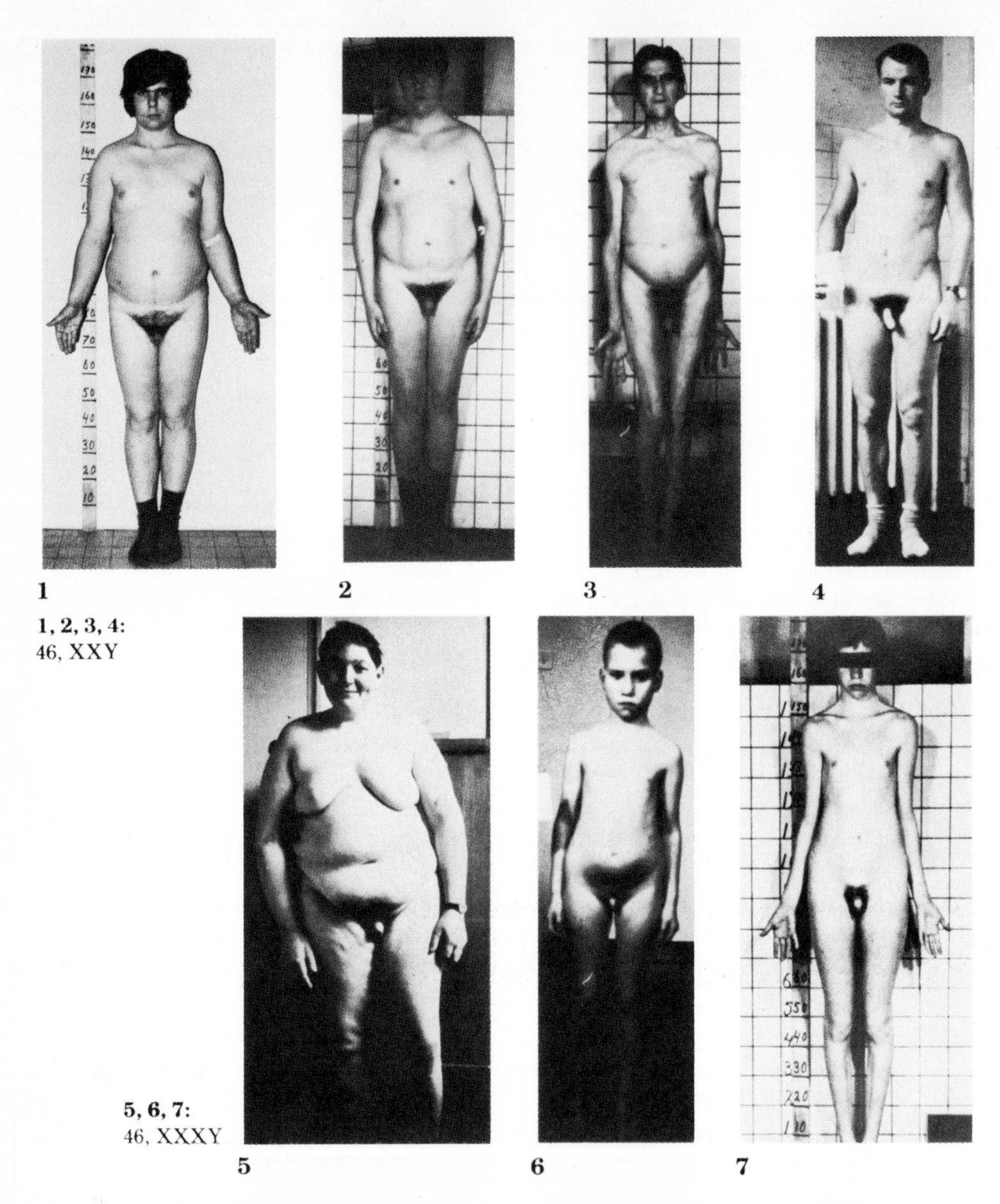

Figure 5-10. Patients with Klinefelter's syndrome. All have extra X chromosomes. (From J. de Grouchy and C. Turleau, 1977. *Clinical Atlas of Human Chromosomes*, p. 245. John Wiley, New York.)

were reported, W. J. Welshons and L. B. Russell presented evidence that the Y chromosome also determines maleness in mice, and that its absence results in femaleness.

They studied an X-linked dominant trait in mice designated Tabby (*Ta*), which causes a darkness of the coat color. Mice of the genotype $X^{Ta}X^{Ta}$ were as dark as X^{Ta}Y male mice, but female heterozygotes $X^{Ta}X^{+}$ have a mottled effect so that their coat color shows areas of dark color mixed with areas of light color. Mice with the X^+X^+ genotype have normally light pigment.

If $X^{Ta}X^+$ females are crossed to X^+Y males, we should expect only $X^{Ta}X^+$ or X^+X^+ female offspring; however, Welshons and Russell found female mice among the progeny with the same dark pigment as females homozygous for the gene. The actual results of the cross gave the following phenotypic ratio : 1 light female : 1 mottled female : 1 dark male. These data are compatible with the assumption that the unusual dark females were actually $X^{Ta}0$. A cross of $X^{Ta}0 \times X^+$Y would give the following results:

	X^{Ta}	0
X^+	X^+X^{Ta}	X^+0
Y	X^{Ta}Y	Y0

If we assume that the Y0 class of offspring is lethal, the expected ratio of this cross would be exactly the phenotypic ratio that was obtained. These results are strongly supportive of the determination of maleness by the Y chromosome and of femaleness by the absence of the Y chromosome in mice.

The Y Chromosome in Development

The exact manner in which the Y chromosome determines maleness in humans and other mammals is not yet known. However, a brief description of the development of the urogenital system and secondary sex characteristics in human embryos may give us some idea as to the role of the genes which must exist on the Y chromosome for normal sex development.

Figure 5-11 shows an illustration of the human embryonic reproductive system at around six weeks of gestation. At this stage the gonads are sexually undifferentiated; they are the same in both male and female embryos. There are two sets of ducts, the Mullerian and the Wolffian (or mesonephric) ducts.

If the embryo is genetically male, the internal portion of the gonad, the medulla, will develop, and the gonad becomes a testis. The Mullerian ducts degenerate, and the Wolffian ducts become part of the sexual duct system of the male, namely, the vas deferens. On the other hand, if the embryo is female, the more peripheral portion of the gonad, known as the cortex, develops and the gonad becomes an ovary. The Mullerian ducts continue to develop into the oviducts and the uterus of the female reproductive tract, and the Wolffian ducts degenerate (Fig. 5-12).

The one or more genes of the Y chromosome which determine maleness are most likely active at the stage of development when the undifferentiated gonad develops into a testis. Some scientists theorize that the maleness gene and the Y antigen gene are one and the same (Silvers and Wachtel 1977). Once the testis has formed, it will produce male hormones that then cause male secondary sex characteristics to develop. There is no evidence at hand, however, to indicate that all the genes involved in sex determination are located on only the Y chromosome. That there are probably genes on the X chromosome that cause a tendency toward femaleness is indicated by the

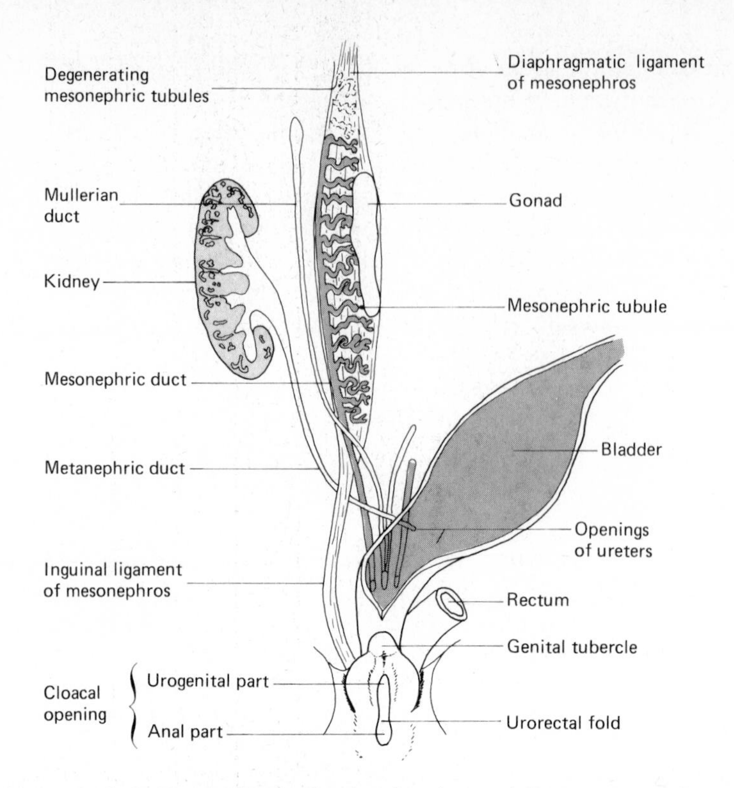

Figure 5-11. Schematic diagram showing plan of urogenital system at an early stage when it is still sexually undifferentiated. (From B. M. Patten, 1953. *Human Embryology* 2nd ed. Fig. 358, p. 575. McGraw-Hill, New York.)

fact that Klinefelter males frequently exhibit gynecomastia, or a tendency to breast development. Further, some geneticists believe that a set of genes, possibly on the autosomes, also are involved in the development of secondary sex characteristics (Ohno *et al.* 1971).

To summarize, in humans it is believed that one or more Y-linked genes are responsible for the differentiation of the embryonic gonad into a testis. Testicular hormones, possibly in conjunction with the activity of other (autosomal?) genes then cause development of male secondary sex characteristics. Absence of male hormones leads to the development of female characteristics.

Testicular Feminization

There is one condition in both humans and mice which involves normal secondary sex development that appears to be determined by a single locus. Known as *testicular feminization*, this condition has been found to be X-linked in mice, but whether it is X-linked or autosomal in humans is still in question. Individuals with the syndrome have female secondary sex characteristics; however, they possess only rudimentary vaginas, and testes rather than ovaries. Examination of their cells reveals that they are XY in sex chromosome constitution.

Studies of such persons have further shown that the testes are functional, as

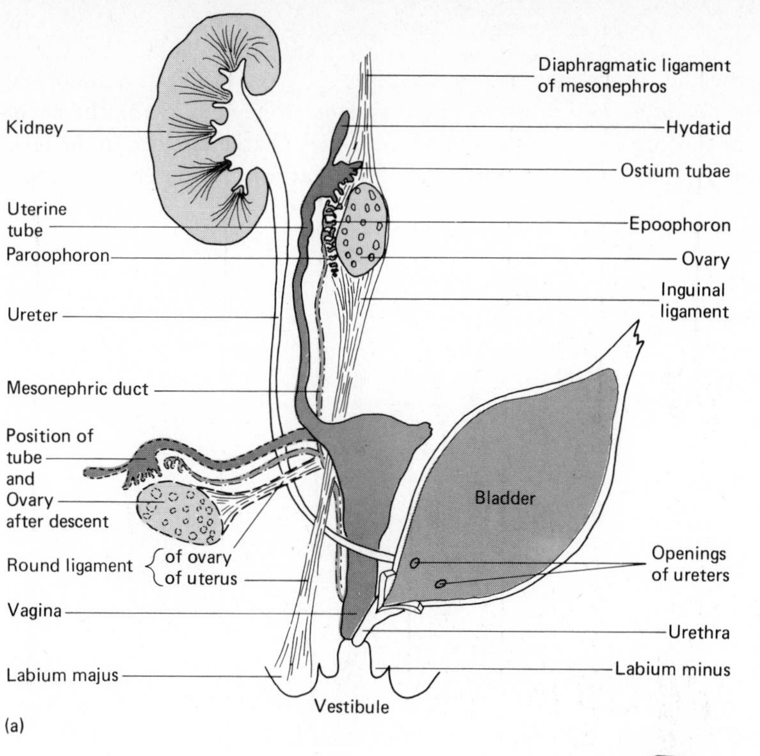

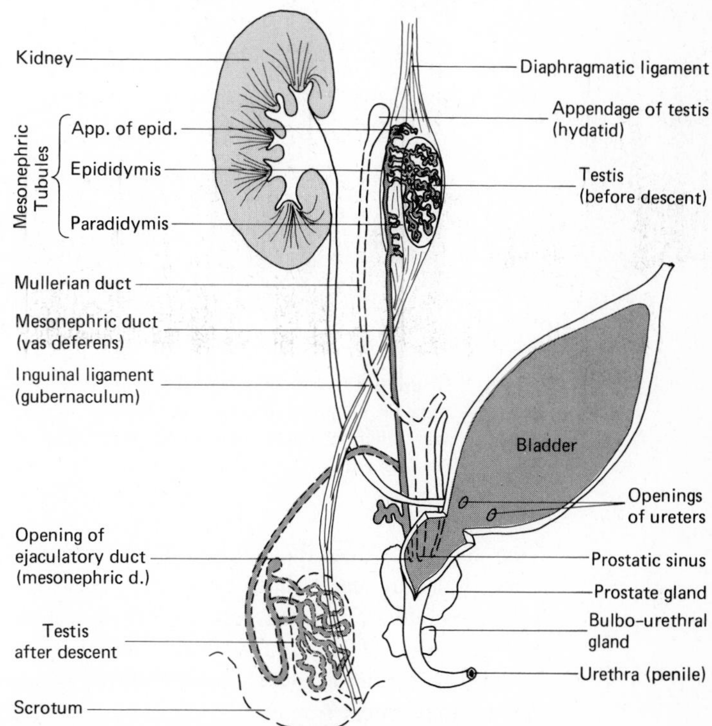

Figure 5-12. Schematic diagram showing plan of (a) developing female reproductive system, (b) developing male reproductive system. (From B. M. Patten, *Human Embryology* 2nd ed., Figs. 362, 361 respectively, pp. 579, 578. McGraw-Hill, New York.)

they do produce the male hormone testosterone, although gametogenesis is abnormal. Normally testosterone is broken down to dihydrotestosterone (Fig. 5-13), the chemical that causes various organs to develop male secondary characteristics. In the testicular feminization syndrome, dihydrotestosterone utilization is apparently defective (Fig. 5-14).

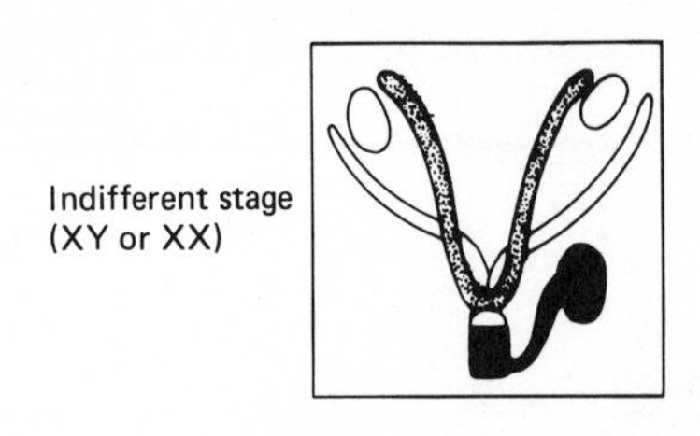

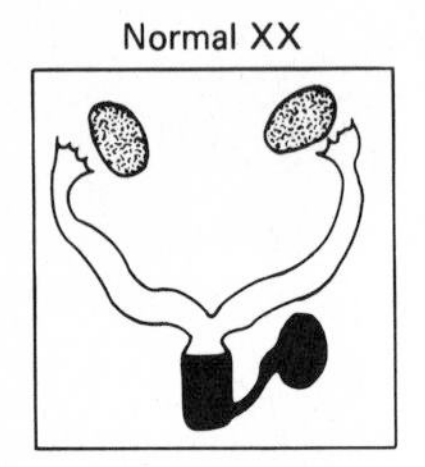

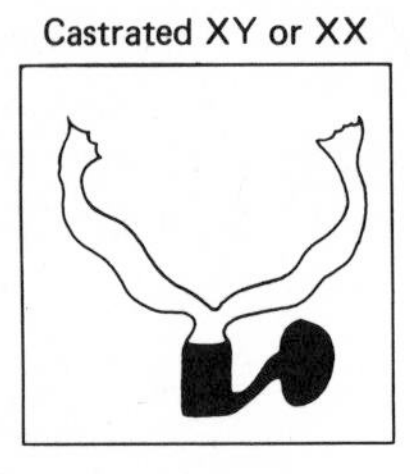

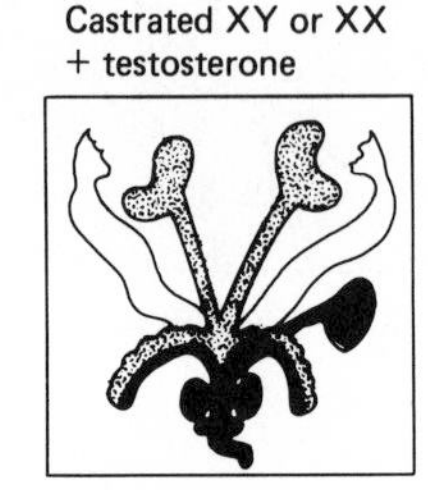

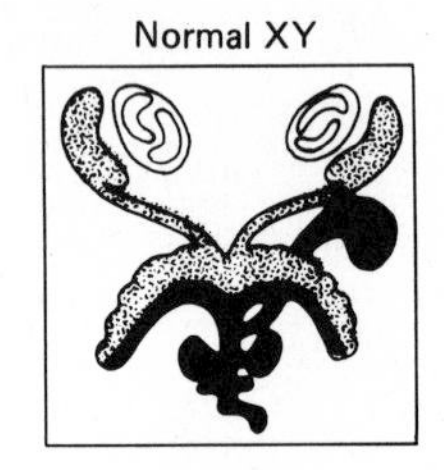

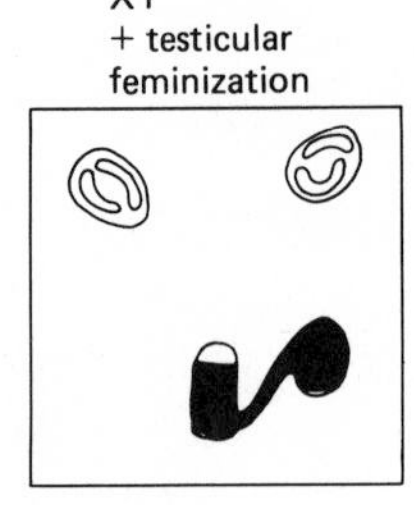

Figure 5-13. A scheme showing the role of testosterone and the effect of the *Tfm* mutation on the embryonic development of the Wolffian duct (shaded), Mullerian duct (outlined), and the urogenital sinus (black). (From S. Ohno et al., 1971. *Hereditas* 69: 108, Fig. 1.)

Not only does this condition illustrate the role of a single locus in sex development, it also indicates that in mammals there is apparently an inherent tendency for female secondary sex characteristics to develop. Along this line, Ohno has shown that castration of XX and XY rabbit embryos at early stages results in the development of the Mullerian ducts (female) into oviducts and uterus, and in degeneration of the Wolffian ducts (male). Perhaps the function of male hormones is therefore to suppress the tendency to femaleness in mammalian development.

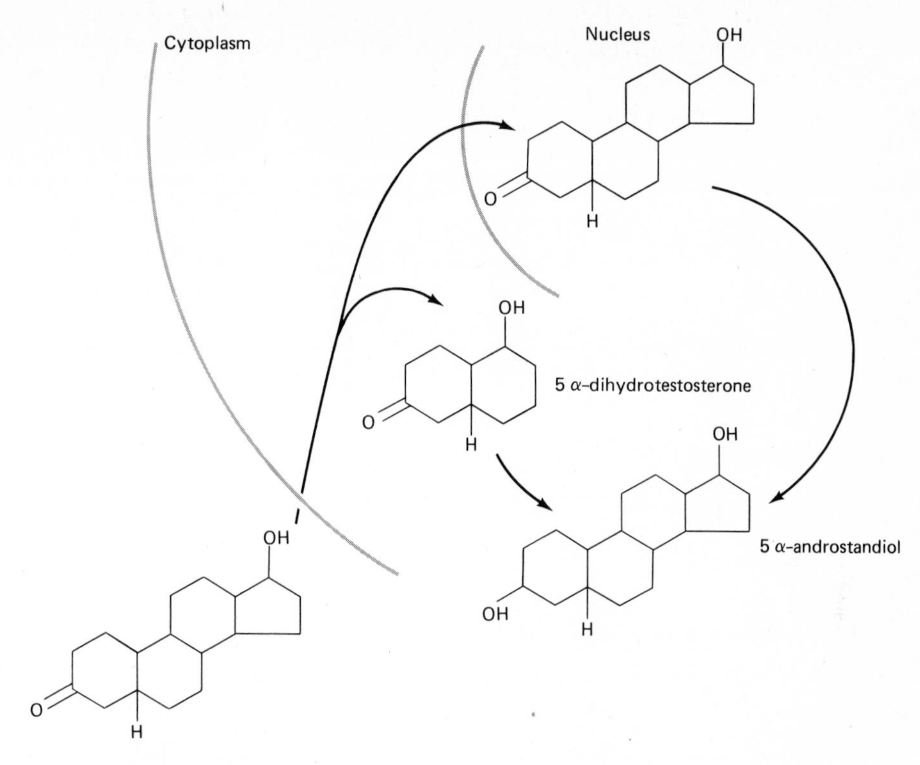

Figure 5-14. The metabolism of testosterone. (From S. Ohno et al., 1971. *Hereditas* 69:107–124; on p. 111, Fig. 2.)

Sex-influenced and Sex-limited Traits

Because of the differences in sex hormones, and their powerful role in normal development, there are a few rare traits which seem to be expressed predominantly in one or the other sex, but are due to autosomal genes. They are *not* sex-linked. Perhaps the best known of these traits, which have been referred to as *sex-influenced*, is that of balding. The loss of hair is believed to be determined by an autosomal gene which behaves as a dominant gene in men and as a recessive gene in women; consequently, the expression of baldness is far more frequent in men than in women. Presumably the difference in effect of the balding gene is due to hormonal differences in males and females. There are also traits whose expression is found exclusively in one or the other sex, for example, milk production in females. These traits are considered *sex-limited*, although some ambiguity exists in the literature. Some use the terms sex-influenced and sex-limited interchangeably.

Genetic Mosaicism in Humans

Another line of evidence indicating the possible effects of the X chromosome in enhancing female development has arisen in the study of human genetic mosaics. We had previously described genetic mosaicism in *Drosophila* gynandromorphs. Genetic mosaics found in humans are similarly called *gynanders* (adjective, *gynandric*); however, the cells with different sex chromosome constitutions are usually not positioned bilaterally. When speaking of humans, another term commonly used as a synonym to gynander is *hermaphrodite*, an interesting combination of the name of the Greek god Hermes and the goddess Aphrodite. Table 5-3 shows the combinations of genetically different cells found to date in human genetic mosaics. Note that all individuals whose cells have no Y chromosome appear to be female; those with cells containing a Y chromosome, regardless of the number of X chromosomes, appear male.

Table 5-3 Human Sex Chromosomal Mosaics

Female	*Male*	*Gynandric*
X0/XX	XY/XXY	X0/XY
X0/XXX	XY/XXXY	X0/XYY
XX/XXX	XXXY/XXXXY	X0/XXY
XXX/XXX	XY/XXY/?XXYY	XX/XY
X0/XX/XXX	XXXY/XXXXY/XXXXXY	XX/XXY
XX/XXX/XXXX		XX/XXYY
		X0/XX/XY
		X0/XY/XXY
		XX/XXY/XXYYY

From C. Stern, 1973. *Human Genetics*, 3rd ed., p. 519. W. H. Freeman & Company Publishers, San Francisco.

In view of the developmental history of the reproductive system of the mammalian embryo, it is understandable that the presence of cells with chromosome combinations determining femaleness with cells that normally determine maleness would interfere with developmental processes.

The control of the genes of the X and Y chromosomes over normal sex development must rely on a fine balance of factors which is obviously disrupted in genetic mosaics. The developmental potential of the mammalian embryo to form both female and male structures is activated, with the consequent presence of, for example, both testicular and ovarian tissue in the gonads of gynanders. In some cases, testicular tissue is found in one gonad and ovarian tissue in the other. The gonads, however, are usually not fertile, and there is no record of human hermaphrodites that have the capability of "self fertilization" as is found in some lower phyla of animals such as Annelida, and in the plants (Money 1973).

Researchers have also found that if the cells of different genetic constitutions are not present in equal numbers, the genetic mosaic may resemble the sex determined by the most abundant type of cell. Detection of the condition at an early age can aid in the correction of abnormalities through surgery and hormone treatment.

How do genetic mosaics arise in humans? In the case of genetic mosaics such as the X0/XX or X0/XY persons, one can conjecture that a loss of one of the sex chromosomes during mitosis in early development occurred, in a manner similar to that in the *Drosophila* gynandromorphs. Mitotic nondisjunction may account for cells in which extra sex chromosomes are present. The X0/XXX females, for example, may have resulted from nondisjunction of the first mitotic division following fertilization.

There is also evidence that a genetic mosaic may result from double fertilization, that is, the fertilization of two cells by two different sperms. Although polar bodies are normally not involved in reproduction, it has been postulated that in rare instances an ovum may be fertilized by an X-bearing sperm and a polar body by a Y-bearing sperm, or vice-versa. The embryo would then be a composite of XX/XY cells.

THE BARR BODY AND X INACTIVATION

In the early 1950s, M. L. Barr and his coworkers discovered that comparisons of stained interphase nuclei from human somatic cells revealed differences in the cells of normal men and of normal women. The difference was that female cells contained a darkly staining chromatin body which was absent in normal male cells (Fig. 5-15).

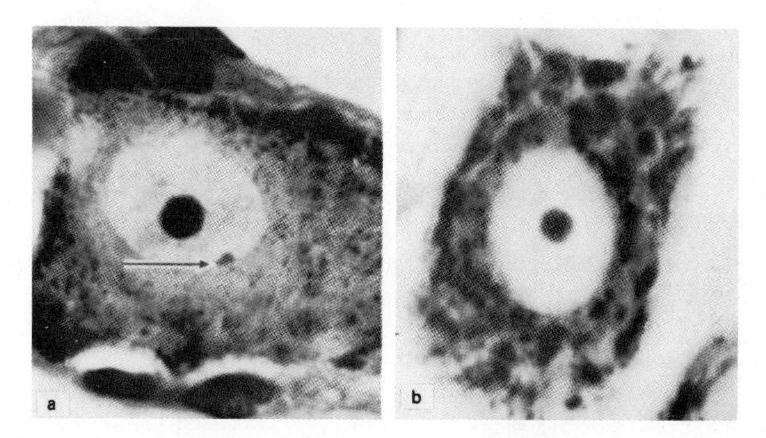

Figure 5-15. Sex chromatin in neurons of the cat. (a) Sex chromatin at the nuclear membrane. (b) Chromatin negative. (From J. L. Hamerton, 1971. *Human Cytogenetics*, Fig. 6-1a, 6-1d, p. 133. Academic Press, New York.)

Cytologists have used the term *heterochromatin* for chromatin material which remains dense and darkly staining during interphase. We shall discuss heterochromatin further in Chapter 12 in our treatment of gene regulation.

The heterochromatin which Barr found has since been named for him, and is called the *Barr body*. Autoradiography techniques revealed that the Barr body is an X chromosome which is highly coiled during interphase. It is known to be present in cells of placental mammals and marsupials, but has not been found in any other group of animals or in plants.

The Barr Body in Human Cells

Further studies revealed that the number of Barr bodies in human cells was a good indicator of the number of X chromosomes in the cells. Turner's syndrome patients (X0) show no Barr bodies in their cells, whereas Klinefelter's syndrome patients (XXY) do have a Barr body. It is apparent, then, that human female cells contain one fewer Barr body than the number of X chromosomes. Table 5-4 further illustrates

Table 5-4 Relationship of Barr Bodies to Number of X Chromosomes

Sex chromosome constitution	*Number of Barr bodies*
XX	1
XY	0
X0	0
XXY	1
XXX	2
XYY	0
XXXY	2
XXXXY	3

this point. It must be noted, however, that the staining of Barr bodies is a tricky technique, as only a certain percentage of cells in even normal females will show the Barr body. If the sex chromosome constitution of a person is in question, it is necessary to study the chromosomes of cells in division to be sure of a diagnosis.

The Lyon Hypothesis

The genetic significance of the Barr body was first postulated by British geneticist Mary F. Lyon, and shortly after by Lliane B. Russell and her coworkers. In 1962 Lyon proposed that the Barr body, being highly coiled, is genetically inactive. This would explain a situation which has puzzled geneticists for some time, namely, that although female mammals have twice the number of X-linked genes than males, there is no evidence of any quantitative differences between X-linked traits in females and males. For example, it is known that an X-linked gene in humans determines the production of the enzyme glucose-6-phosphate dehydrogenase (g-6-pd). The enzyme is found in approximately the same amount in normal female and male cells.

When we discuss autosomal abnormalities in Chapter 6, you will see that the presence of an extra autosome always leads to severe abnormalities and in most cases death, and the same is true when an autosome is missing. The inactivation of one X chromosome may be a method of *dosage compensation* in human cells to maintain equal gene activity of X-linked genes in both males and females. The regulation of gene activity for dosage compensation in organisms such as *Drosophila*, in which there is no heterochromatic X chromosome, is believed to be at the chromosomal level; the mechanism, however, is still unknown (Stewart and Merriam 1975).

THE LYON HYPOTHESIS, AS THIS THEORY IS NOW CALLED, PROPOSES THAT A

RANDOM INACTIVATION OF ONE OF THE X CHROMOSOMES IN ALL CELLS OF A FEMALE EMBRYO OCCURS VERY EARLY IN GESTATION, AFTER A FEW MITOTIC CYCLES HAVE BEEN COMPLETED. Whether the inactivation is random is currently under investigation. B. R. Migeon and J. F. Kennedy (1975) reported that cells from 4- to 6-week-old embryos already have one X inactivated. The mechanism of this inactivation is still a matter of conjecture (Lyon 1971). However, studies have shown that at stages prior to the inactivation, including during gametogenesis, both X chromosomes must be active to ensure the survival of oogonia. On the other hand, if spermatogonia contain more than one X chromosome, they die. This is circumstantial but strong evidence that both X chromosomes are indeed active during gametogenesis (Lyon 1974).

Many lines of evidence have supported the theory of X inactivation. For example, it is now known that female calico cats, which have mottled dark and yellow coats, owe their coloration to the fact they are actually heterozygous for two different X-linked alleles determining the two colors. The different patches of dark or yellow hair represent areas in which one or the other of the alleles is inactive due to the coiling of the X chromosome on which it is located. Incidentally, there are also male calico cats, but they have been found to be XXY.

Studies on cells from a woman known to be heterozygous for two different X-linked alleles that determine the production of two different forms of the enzyme g-6-pd, showed that approximately half the woman's cells produced one enzyme, and the other half produced the other form (Davidson *et al.* 1963). This is interpreted as another indication that only one of every pair of X-linked alleles is active in all cells.

Although the inactivation of one X chromosome is now widely accepted, the mechanism, as we noted above, is not yet understood. It is curious, for example, that Turner's syndrome females who have only one X chromosome are abnormal if inactivation is normal and essential. Some geneticists believe that the inactivation may be a progressive process, and that alleles on both chromosomes must be active at some of the later stages in gestation for normal development to occur. Work is in progress at many laboratories on this very interesting genetic phenomenon.

The Barr Body, Transsexuals, and Athletics

It may be of interest to digress from this discussion of the genetics of sex determination and the function of sex chromosomes to mention a recent controversy regarding the use of the Barr body test for sex chromosomes. The situation described is a good example of how developments in genetics can affect our society directly. Women in sports have until recently suffered derision for being "manlike" if they were athletic or muscular beyond the criteria of femininity which tradition has established in our society. The true sex of female athletes has been a subject of debate (serious and not-so-serious) and suspicion. In 1955, a German named Dora Ratjen, who set a women's world record in high-jumping in 1938, admitted to being a man and being forced by the Nazi state to compete as a woman (see Larned 1976).

This disclosure and the attitudes mentioned above led to the utilization of the Barr body test in international athletic competitions such as the Olympic Games. To some athletes, the test represents an intrusion on their civil rights; to others it is necessary to prevent unfair competition with persons who are not biologically female.

Because of the effects of the different sex hormones, muscle and bone structure is quite different in males and females. In fact, a gold and bronze medalist of the 1964 Olympics, a Polish sprinter named Eva Klobukowska, had to return her medals and was declared ineligible for further competition because she was later found to be a genetic mosaic.

Another much-publicized controversy of a similar nature took place in women's athletics in 1976, when a transsexual, Dr. Renee Richards, was allowed to compete in a women's tennis tournament. Transsexuals are in some cases quite different from genetic mosaics. The term *transsexual* refers to a person who has had an operation to remove gonads and alter external organs in order to take on the appearance of the opposite sex. Such individuals also undergo hormone therapy which contributes to the development of the desired secondary sex characteristics.

In some cases, transsexuals are genetic mosaics who are phenotypically more male or female, and wish to be changed. Others have no apparent genetic basis, as the individuals have normal sex chromosome complements; such persons have psychological problems in accepting their natural sex. Dr. Richards had reportedly fathered a son prior to the operation and sex change. It is very likely then that she is XY, and of course the genetic constitution of her cells is not changed.

One argument used to support the acceptance of transsexuals biologically is that hormones can change muscle tone; on the other hand, the argument may be made that transsexuals will never experience physical problems of natural females such as those associated with the menstrual cycle. At any rate, we leave it to the sociologists, psychologists, and athletes to debate the legitimacy of claims by transsexuals to compete as members of the sex they have legally, but *not* genetically, joined!

SEX-DETERMINING GENES

To return to our discussion of the determination of sex, it should be evident that much remains to be elucidated regarding the function of the sex chromosomes. To review, we do know that a balance of autosomes and X chromosomes determines sex in *Drosophila*, and that in humans the presence of the Y chromosome is the determining factor. (Some exceptional XY females and XX males have been reported, but these are very rare.) We do not know, of course, what genes are involved, or how many there might be, or whether they are on sex chromosomes or autosomes.

We do not know the answers to these questions primarily because we have found few mutations of genes determining sex. This is a very important point that must not be overlooked by students of genetics, namely, that without variations of phenotype, we cannot follow the transmission of traits. These variations depend on the presence of different alleles whose existence derives ultimately from the genetic changes we call mutation. The variations of sex development in humans have involved mainly chromosomal differences, not single gene changes.

Our lack of understanding of the Y chromosome in either *Drosophila* or humans is also due largely to our inability to find Y-linked mutations. Indeed, aside from the essential role it plays in sex determination in humans, the Y chromosome in these species has been referred to in the past as genetically inert. There appears to be little or no homology with genes on the X chromosome, and as we mentioned above there are no detectable mutations traceable to Y-linked genes. A LARGE NUMBER OF X-LINKED TRAITS HAVE BEEN FOUND, OF COURSE, IN BOTH *Drosophila* AND HUMANS, BUT NOTE AGAIN THAT NONE OF THESE TRAITS IS DIRECTLY INVOLVED WITH SEX DETERMINATION. The autosomal gene involved in testicular feminization is a rare one which is known to interfere with normal developmental processes. We shall now continue our discussion of sex determination by reviewing systems which are different from those in either *Drosophila* or humans.

OTHER SYSTEMS OF SEX DETERMINATION

In Chickens and Amphibians

Other organisms in which the female is the heterogametic sex and the male the homogametic sex include birds and amphibians. Experiments grafting testes into female amphibian embryos caused functional sex reversal in some cases. The same experiments were used to establish that females contain XY sex chromosomes, and males XX.

In such experiments, the grafted testis degenerates due to immune reactions of the sort that will be discussed in Chapter 12, and do not produce gametes. If females were homogametic, then the functional sex-reversed animal (phenotypically male, but genetically female), when mated to a normal female, should produce only female offspring. On the other hand, if females are heterogametic, such a mating should produce 3 females : 1 male since YY animals are also found to be female. Two matings of such a sex-reversed male by normal females produced 21.6% and 35.4% male offspring (see Humphrey 1945), giving strong evidence that females are heterogametic among amphibians.

Chickens have also been found to be capable of functional sex reversal. Sex chromosomes of females are referred to as ZW; of males as ZZ. Unlike the case in mammals, the development of gonads and ducts in bird embryos is not bilateral. One side remains in the undifferentiated state, while the other side continues to develop into adult structures. If the ovary and structures of the reproductive system of an adult female are removed, the rudimentary structures can develop into functional male structures and the gonad into a functional testis. Such chickens develop the appearance of roosters and have been known to father offspring. Where there is evidence that the mammalian embryo has an innate tendency toward femaleness, in birds there appears to be an innate tendency toward maleness.

In Hymenoptera

An interesting system of sex determination has been found in various species of the order Hymenoptera, which includes insects such as the honeybee and wasp. In the honeybee, sex is determined by the number of sets of chromosomes. The diploid number, which is 32, is found in female workers and in the queen bee; all males are haploid, as they arise from unfertilized eggs in a process called *parthenogenesis*. The queen bee produces haploid eggs which if fertilized develop into females, and if not, into males. The males produce sperm by mitosis rather than meiosis. In the first division, a cytoplasmic bud containing no chromosomes forms and division results in only one cell with chromosomes (see Rothenbuhler *et al.* 1968).

In wasps, it was originally thought that the system also depended on the *ploidy*, or number of sets of chromosomes. However, P. W. Whiting found that the genotype for a particular locus also was instrumental in determining sex. He found that nine alleles exist at this locus which he designated Xa, Xb, Xc, etc. Femaleness is determined by heterozygosity for any two of these alleles, whereas maleness is determined by haploidy or by homozygosity for the alleles. Males formed by homozygosity are usually sterile, and the functional males are still the haploid individuals. Since there are such a large number of alleles, insects mating at random in natural environments would be unlikely to carry the same allele, and consequently diploid males are rare.

In Plants

Cytological studies of higher plants have occasionally revealed heteromorphic pairs of chromosomes. In *Melandrium*, genetic studies have shown that these are the sex chromosomes, again designated by the letters X and Y. M. Westergaard found that no anthers develop in the absence of a Y chromosome. There is, however, an effect on sex development depending on the numbers of X and Y chromosomes. Plants with three or more X chromosomes and one or two Y chromosomes are all female.

On the other hand, a large number of plants have been shown not to have any differences in chromosomes of plants of different sex. In some plants, sex is determined by single gene pairs, as found by C. M. Rick and G. C. Hanna in the garden asparagus. They discovered that maleness is inherited as a dominant trait. Occasionally berries were produced by self pollination of rare pistils in staminate flowers. These fruits would then be the progeny of essentially a male × male cross. When germinated, the berries developed into approximately $\frac{3}{4}$ male and $\frac{1}{4}$ female plants, the ratio one would expect if the original parent plant were heterozygous for the alleles for sex determination. Further tests of the male progeny plants showed that approximately $\frac{1}{3}$ gave rise to only male plants when crossed to normal female plants, indicating that they were homozygous for the dominant allele. The 3 male : 1 female ratio, then, is simply an F_1 generation from a monohybrid cross.

SOME FURTHER THOUGHTS

The determination of sex in the various forms of living organisms is varied and fascinating. Understanding of the mechanisms involved is of interest not only from the point of view of theoretical genetics, but could be of practical value as well. For example, Rick and Hanna pointed out that staminate asparagus plants produce a higher yield of shoots, so that by carefully controlling crosses one might be able to develop all-male plantings. For purposes of seed production, homozygous males could be used to pollinate a large number of pistillate plants, whose seeds would in turn produce only staminate plants.

In domestic animals it is frequently desirable to have more animals of one sex than another, for example in dairy cattle. Perhaps someday we shall be able to separate gametes bearing the X chromosome from those carrying the Y chromosome and be able to ensure the conception of a female or male young by artificial insemination. This type of germinal choice may even be extended to humans, but only after much more investigation of poorly understood processes.

One further point can be emphasized here. It is of interest and a curious fact that in the evolution of sexually reproducing organisms, only two sexes have generally developed regardless of the mechanism underlying the determination of sex. Factors which we do not yet understand must have prevented the evolution of organisms with more than two different sexes.

PROBLEMS

1. Color-blindness is an X-linked recessive trait in humans. Is it possible for women to be color-blind? What would one have to assume about the genotype and phenotype of the father of a color-blind woman? Of the mother?

2. A color-blind man marries a normal-visioned woman. They have two normal-visioned daughters. Assuming that the girls marry normal men, what phenotypic ratios can be expected among the grandchildren? Among the grandsons?

3. Hemophilia, the "bleeder's disease," is also an X-linked trait in humans. A worker in a chemical factory sues his employer because his son was born a hemophiliac. He claims that exposure to toxic substances caused a mutation in his cells giving rise to the condition in the baby. If you were on the jury, what would your reaction be?

4. A type AB woman who is heterozygous for hemophilia marries a man who is type A and has normal blood. His father was type 0. Diagram the cross of this couple and give the genotypic and phenotypic ratios of the offspring.

5. Suppose you were maintaining a colony of wild-type *Drosophila*. A white-eyed female fly appears in a particular generation. When it is crossed with a wild-type male, the offspring are found to include only white-eyed females and wild-type males, and this is true for the F_2 generation as well. How would you explain these results?

6. Suppose the same white-eyed female, when crossed to a wild-type male, produced instead the following progeny: 32 wild-type females, 35 wild-type males, 30 white-eyed females, and 28 white-eyed males. How would you explain the results in this case?

7. In Chapter 4 we discussed the metabolic explanation for the absence of pigmentation in eyes of *Drosophila* homozygous for two recessive mutations, *scarlet* and *brown*. How might you explain the absence of pigmentation in flies due to a single X-linked mutation?

8. How can one distinguish between Y-linked transmission and sex-limited traits expressed only in males?

9. A bald man marries and fathers four sons and four daughters. Two of the sons and one daughter eventually become bald to various degrees. What is the most likely genotype of the bald man and his wife for the gene for balding (which is a sex-influenced trait in humans)?

10. Assume that we have found a male calico cat that is fertile (normally they are not). What genotypes can we expect to find among all the offspring of such a male crossed with a normal female calico cat?

11. "Superfemales" (XXX) in humans can be fertile and have children. Assume that such a female carries the recessive gene for color-blindness on two of her X chromosomes. She marries a normal male. What phenotypic ratios can we expect among their children?

12. What hypothesis can you formulate to explain the XX/X0/XY combination of cells in genetic mosaics?

13. Studies on X-inactivation have shown that in some mammals such as the kangaroo, the paternal X chromosome is preferentially inactivated in cells of females. Would this affect the expression of X-linked traits in females? Why?

14. In view of the unique system of sex determination in bees and wasps, would you expect males or females to be the more viable sex? Why?

15. If sex-reversed roosters are mated to normal hens, what ratios of males and females might we expect among their offspring?

16. Analyze the following pedigrees and give the most likely mode of inheritance (both as to dominance or recessiveness, and autosomal or sex-linkage).

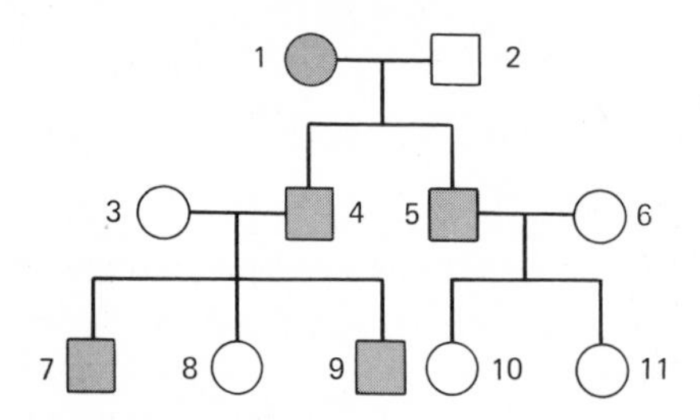

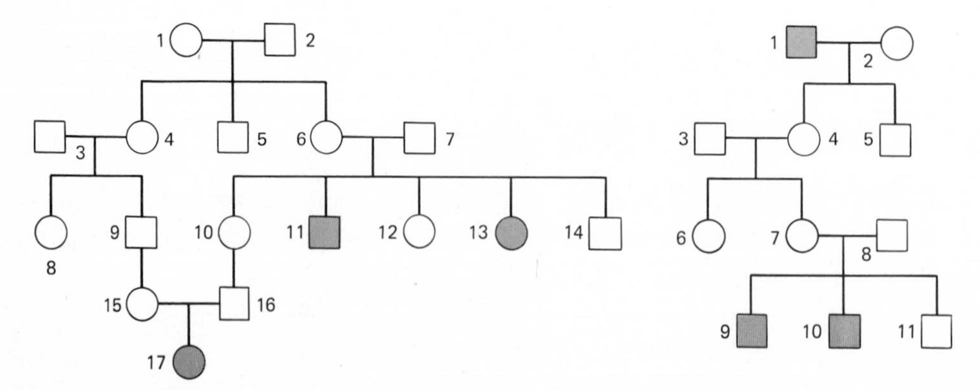

17. Using the numbers to identify them, give the most likely genotype or genotypes of the members of the three families. If the information is insufficient, designate the genotype by a ?

18. Assume that a woman is heterozygous for both albinism (an autosomal recessive trait) and hemophilia (an X-linked recessive trait). She marries a man who is heterozygous for albinism and has normal blood. Diagram the cross, giving the genotypes of the couple, and the phenotypic and genotypic ratios you would expect among their children.

19. In 1962, Gartler et al. reported that a XX/XY hermaphrodite had two populations of red blood cells for the Rh blood group: *CDe/cDE* and *CDe/cde*. Her left eye was hazel like her mother's; her right eye was brown like her father's. The mother's Rh genotype was found to be *CDe/cDE*; the father's was *cDE/cde*. What was the most likely explanation for these phenomena?

REFERENCES

BARR, M. L., AND E. G. BERTRAM, 1949. "A Morphological Distinction Between Neurones of the Male and Female, and the Behavior of the Nucleolar Satellite During Accelerated Nucleoprotein Synthesis." *Nature* 163: 676–677.

BRIDGES, C. B., 1925. "Sex in Relation to Chromosomes and Genes." Reprinted in J. A. Peters, ed., *Classic Papers in Genetics*, 1959. Prentice-Hall, Inc., Englewood Cliffs, N.J.

CHANDRA, H. S., AND S. W. BROWN, 1974. "Regulation of X-Chromosome Inactivation in Mammals." *Genetics* 78: 343–349.

DRONAMRAJU, D. R., 1965. "The Function of the Y Chromosome in Man, Animals, and Plants." *Adv. Gen.* 13: 227–310.

DAVIDSON, R. G., H. M. NITOWSKY, AND B. CHILDS, 1963. "Demonstration of Two Populations of Cells in the Human Female Heterozygous for Glucose-6-Phosphate Dehydrogenase Variants." *Proc. Nat. Acad. Sci. U.S.* 50: 481–485.

FORD, C. E., K. W. JONES, P. E. POLANI, J. C. DEALMEIDA, AND J. H. BRIGGS, 1959. "A Sex-Chromosome Anomaly in a Case of Gonadal Dysgenesis (Turner's Syndrome)." *Lancet* 1: 711–713.

GARTLER, S. M., S. H. WAXMAN, AND E. GIBLETT, 1962. "An XX/XY Human Hermaphrodite Resulting from Double Fertilization." *Genetics* 48: 332–335.

GOLDSCHMIDT, R., 1955. *Theoretical Genetics.* U. California Press, Berkeley.

HUMPHREY, R. R., 1945. "Sex Determination in Ambystimoid Salamanders." *Am. J. Anat.* 76: 33–66.

JACOBS, P. A., AND J. A. STRONG, 1959. "A Case of Human Intersexuality Having a Possible XXY Sex-Determining Mechanism." *Nature* 183: 302–303.

LARNED, D. 1976. "The Femininity Test." *Women Sports* (July, p. 9).

LYON, M. F., 1962. "Sex Chromatin and Gene Action in the Mammalian X Chromosome." *Am. J. Hum. Gen.* 14: 135–148.

———, 1971. "Possible Mechanisms of X-Chromosome Inactivation." *Nature New Biol.* 232: 229–232.

———, 1974. "Mechanisms and Evolutionary Origins of Variable X-Chromosome Activity in Mammals." *Proc. Roy. Soc. Lond.* 187: 243–268.

MIGEON, B. R., AND J. F. KENNEDY, 1975. "Evidence for the Inactivation of an X Chromosome Early in the Development of the Human Female." *Am. J. Hum. Gen.* 27: 233–239.

MONEY, J., 1973. "Intersexual Problems." *Clin. Obstet. Gynecol.* 16: 169–191.

MORGAN, L. V., 1922. "Non-Criss-Cross Inheritance in *Drosophila.*" *Biol. Bullet.* 42: 267–274.

MORGAN, T. H., 1910. "Sex Limited Inheritance in *Drosophila.*" *Science* 32: 120–122. Also reprinted in J. Peters, ed, 1959. *Classic Papers in Genetics*. Prentice-Hall, Inc., Englewood Cliffs, N.J.

OHNO, S., U. TETTENBORN, AND R. DOFUKU, 1971. "Molecular Biology of Sex Differentiation." *Hereditas* 69: 107–124.

RICK, C. M., AND G. C. HANNA, 1943. "Determination of Sex in *Asparagus Officinalis.*" *L. Am. J. Bot.* 30: 711–714. Also reprinted in L. Levine, ed., *Papers on Genetics*. C. V. Mosby, St Louis, Mo.

ROTHENBULER, W. C., J. M. KULENCEVIC, AND W. E. KERR, 1968. "Bee Genetics." *Ann. Rev. Gen.* 2: 413–438.

RUSSELL, L. B., 1961. "Genetics of Mammalian Sex Chromosomes." *Science* 133: 1795–1803.

SILVERS, W. K., AND S. S. WACHTEL, 1977. "H-Y Antigen: Behavior and Function." *Science* 195: 956–960.

STERN, C., 1973. *Principles of Human Genetics*, 3rd ed. W. H. Freeman & Company Publishers, San Francisco.

SPRATT, N. T., 1971. *Developmental Biology*. Wadsworth, Belmont, Cal.

STEWART, B. R., AND J. R. MERRIAM, 1975. "Regulation of Gene Activity of Dosage Compensation at the Chromosomal Level in *Drosophila.*" *Genetics* 79: 635–647.

WELSHONS, W. J., AND L. B. RUSSELL, 1959. "The Y Chromosome as the Bearer of Male Determining Factors in the Mouse." *Proc. Nat. Acad. Sci. U.S.* 45: 560–566.

WESTERGAARD, M., 1948. "The Relationship Between Chromosome Constitution and Sex in the Offspring of Triploid Melandrium." *Hereditas* 34: 257–279.

WHITING, P. W., 1939. "Multiple Alleles in Sex Determination of Habrobracan." *J. Morph.* 66: 323–335.

WILSON, E. B., 1925. *The Cell in Development and Heredity*, 3rd ed. Macmillan, New York.

SIX

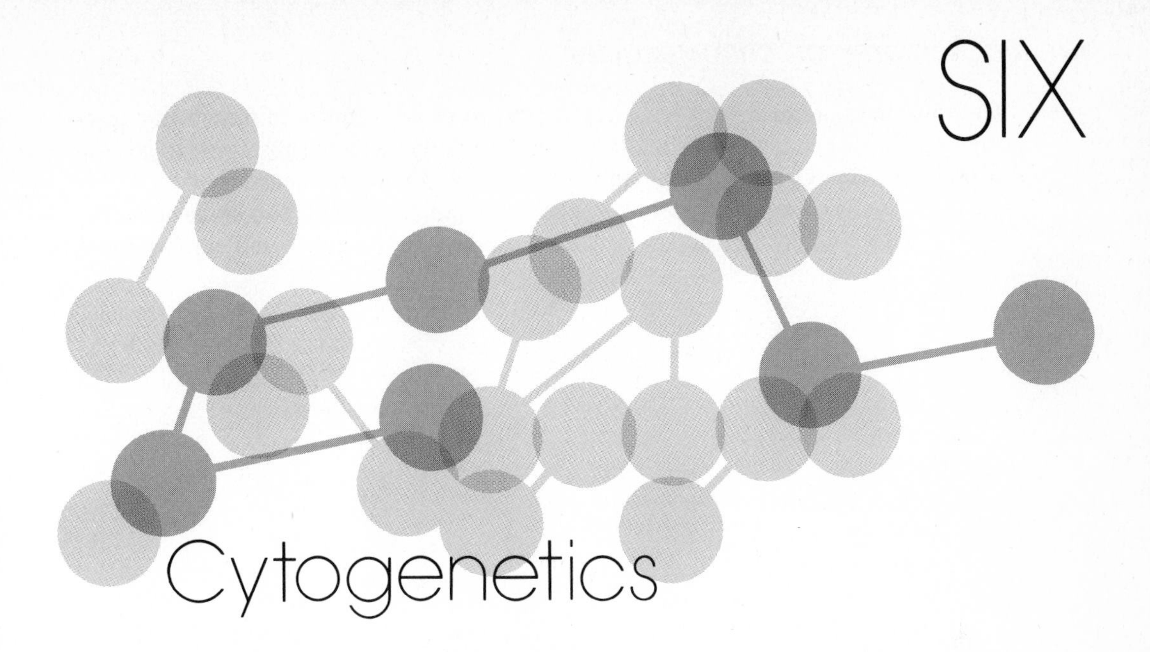

Cytogenetics

INTRODUCTION

The establishment of chromosomes as the vehicles by which genes are transmitted from parent to offspring led geneticists to focus their attention on these threadlike structures. How is the chromosome organized? Where are the genes located? What is the chemical composition of chromosomes? Scientists in many areas applied their particular expertise to the problem. Cytologists experimented with stains to bettter visualize the contents of nuclei, and biochemists opened the cells and isolated components for their analysis. Geneticists attempted to correlate these findings with their data from experimental crosses.

Accordingly, cells were stretched, squashed, and stained. As for the chromosomes themselves, Lejeune has written, "The diversity of tortures inflicted upon chromosomes in order to force them to divulge their inner information is really amazing. Boiling them, digesting them, burning them with alkali or salts, even intoxicating them—all these maltreatments have been tried in turn" (Lejeune 1974, p. 16). From all these efforts we have obtained understanding of the nature of chromosomes, and this will be our topic of discussion in the following sections. The area of genetics which deals with chromosome structure and function is known as *cytogenetics*.

THE BIOCHEMISTRY OF CHROMOSOMES

Chromosomes of eukaryotes have been found to be constructed of four major components. In this chapter, we shall limit our discussion of these substances to general aspects; later, in Chapter 9, their structure and significance for genetics on the molecular level will be explored in detail. The four components include two kinds of nucleic acids (DNA and RNA), and two classes of proteins (histones and nonhistones).

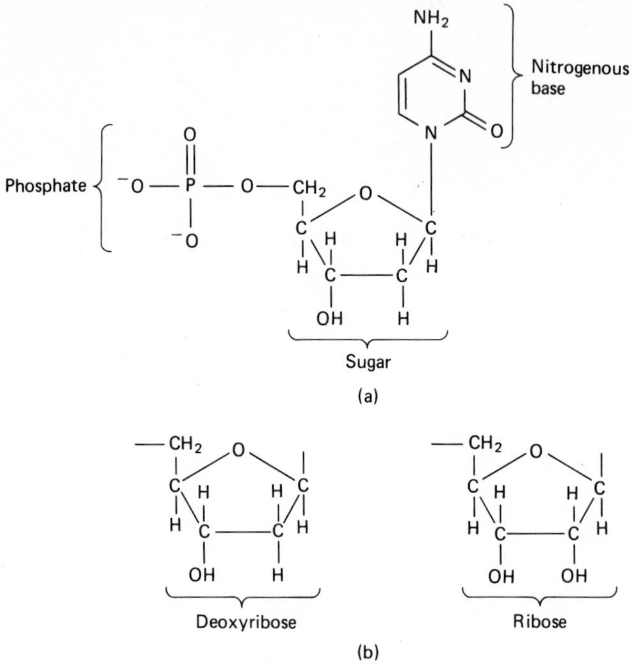

Nitrogenous base

Phosphate

Sugar

(a)

Deoxyribose

Ribose

(b)

Figure 6-1. (a) A nucleotide. (b) The sugar component of DNA (left) and RNA (right).

Nucleic Acids

Nucleic acids are organic compounds composed of carbon, nitrogen, hydrogen, oxygen, and phosphorus. They are polymers of a number of subunits bonded together. The subunits are called *nucleotides*, and in turn are composed of a molecule of a five-carbon sugar, a phosphate group, and a nitrogenous base (Fig. 6-1a). (A sugar attached to a base without the phosphate group is known as a *nucleoside*.) A number of nucleotides bonded together form a *polynucleotide*.

The two kinds of nucleic acids found in chromosomes are deoxyribonucleic acid (or, more familiarly, DNA), and ribonucleic acid (or RNA). One of the differences between these two types of nucleic acid is apparent in their names: DNA contains the five-carbon sugar deoxyribose, and RNA contains ribose, in which there is one more oxygen atom than deoxyribose (Fig. 6-1b).

Although the phosphate groups and the sugars of DNA and RNA are the same in nucleic acids of all organisms, the nitrogenous base content varies among organisms. Five different nitrogenous bases are found in nucleic acids. DNA contains the purines adenine and guanine, and the pyrimidines cytosine and thymine. A further difference between DNA and RNA is that the latter contains uracil instead of thymine (Fig. 6-2).

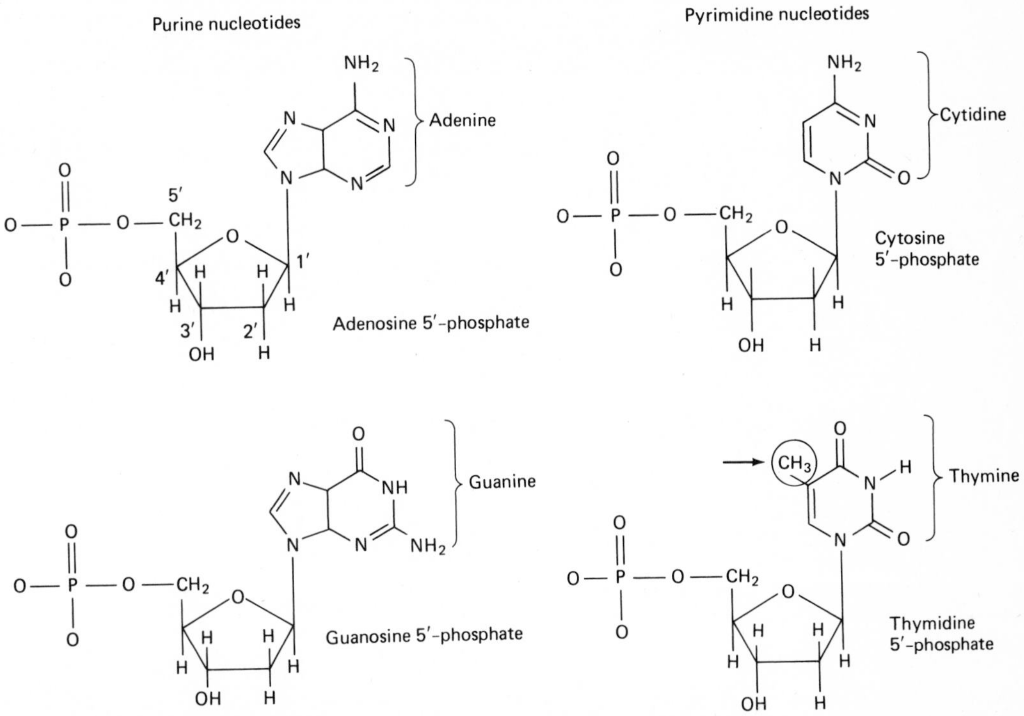

Figure 6-2. Four kinds of nucleotides found in DNA. The base uracil is found in place of thymine in RNA. Uracil has the same structure as thymine minus the methyl group which has been circled (arrow).

Proteins

Proteins are also large organic compounds composed of subunits bonded together. The subunits of protein, as you probably know, are amino acids. Basically there are 20 different amino acids (Fig. 6-3); in nature, modifications of these amino acids are found which increases the actual number. Every amino acid, however, is made of carbon, hydrogen, oxygen, and nitrogen, and each contains a COOH group and an amino group, NH_2. They are bonded together by peptide bonds formed between the COOH group of one amino acid and the NH_2 group of the adjacent amino acid (Fig. 6-4). A string of amino acids forms a *polypeptide.*

Levels of protein structure. Proteins are more than just polypeptides. Their levels of organization range from simple to complex, and to be functional, they usually attain the more complex levels. The simplest, or *primary level,* of structure refers to the sequence and number of amino acids, which distinguish one protein from

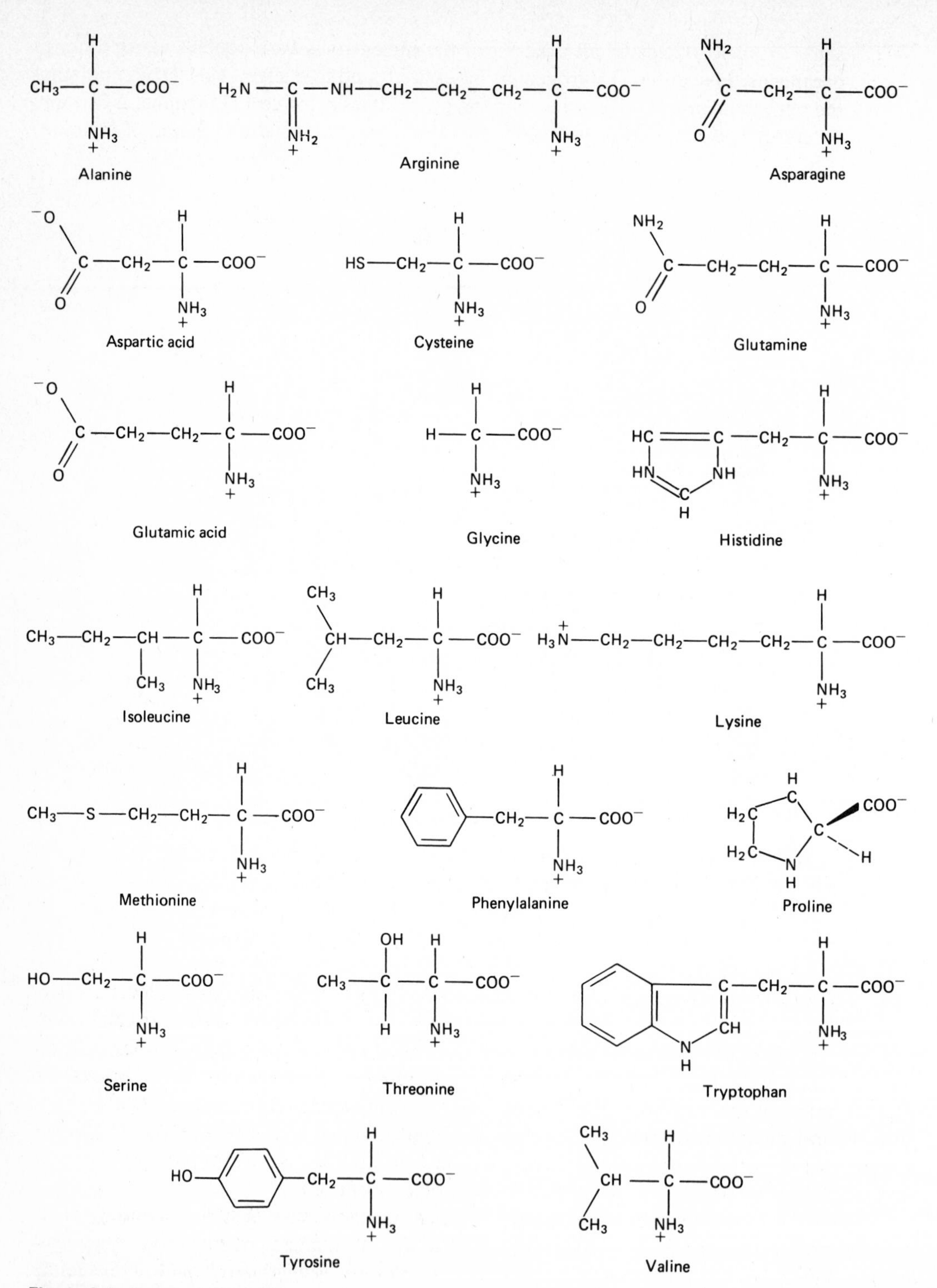

Figure 6-3. Twenty amino acids.

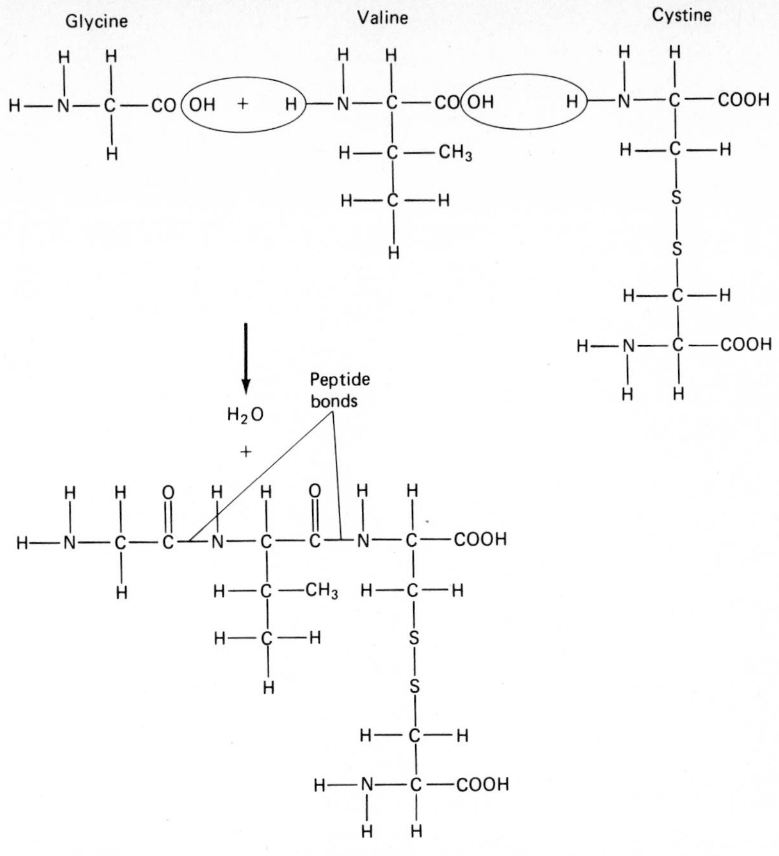

Figure 6-4. Polypeptide formation by peptide bonds joining amino acids.

another. Hydrogen bonds between different amino acids cause proteins to take on three-dimensional shapes such as the α helix discovered by Linus Pauling; this is the *secondary level* of protein structure (Fig. 6-5). The shape can be further complicated by bonds formed between different parts of the polypeptide to produce a *tertiary level.* Finally, the protein can attain the *quaternary level* of structure by different polypeptides joining together, or with other organic compounds such as lipids.

A well-known protein, which we will be discussing in detail later, that demonstrates all the levels of structure is hemoglobin, the important substance in red blood cells that allows the exchange of gases and causes the blood to be red. There are four polypeptides in the protein, or globin, part of hemoglobin. Two of these are identical and known as the *alpha polypeptides*; the two others, different from the alpha chains but identical to each other, are the *beta polypeptides* (Fig. 6-6a). These four chains are bonded together, and also to a molecule of pigment known as *heme.* Figure 6-6b shows the spatial configurations of all the components of this complex and essential protein.

Histones and nonhistones. The two classes of proteins found in chromosomes are the histone and non-histone proteins. Histones are relatively small polypeptides with molecular weights of 10,000 to 20,000 Daltons. They are basic proteins and therefore can be visualized by staining with acidic stains, or extracted from chromosome

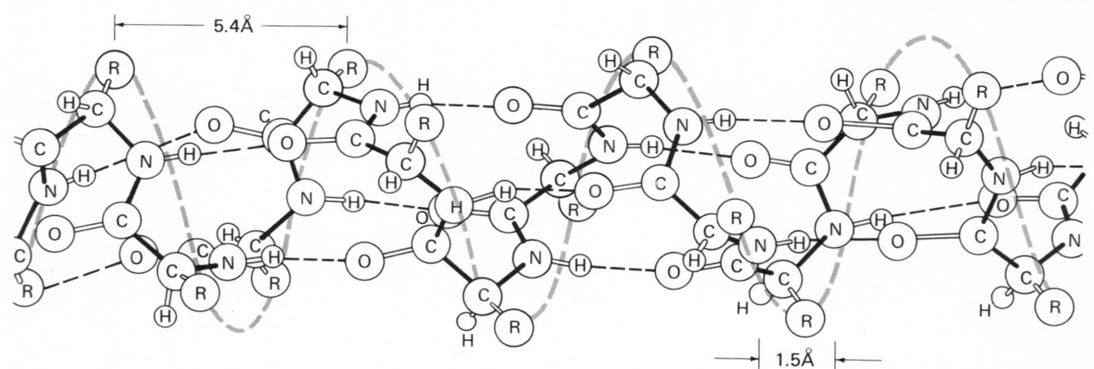

Figure 6-5. The influence of hydrogen bonding on secondary structure. In the conformation of a polypeptide chain known as α helix, there is one amino acid residue spaced every 1.5 Å along the axis of the helix. There are 3.6 amino acid residues per complete turn of the helix, a distance of 5.4 Å. The light dashed lines denote hydrogen bonds between the amide N and the carboxyl O of every third residue. The heavy dashed line traces the peptide backbone. (From D. W. Woodward and V. W. Woodward, 1977. *Concepts of Molecular Genetics*, Fig. 151, p. 239. McGraw-Hill, New York.)

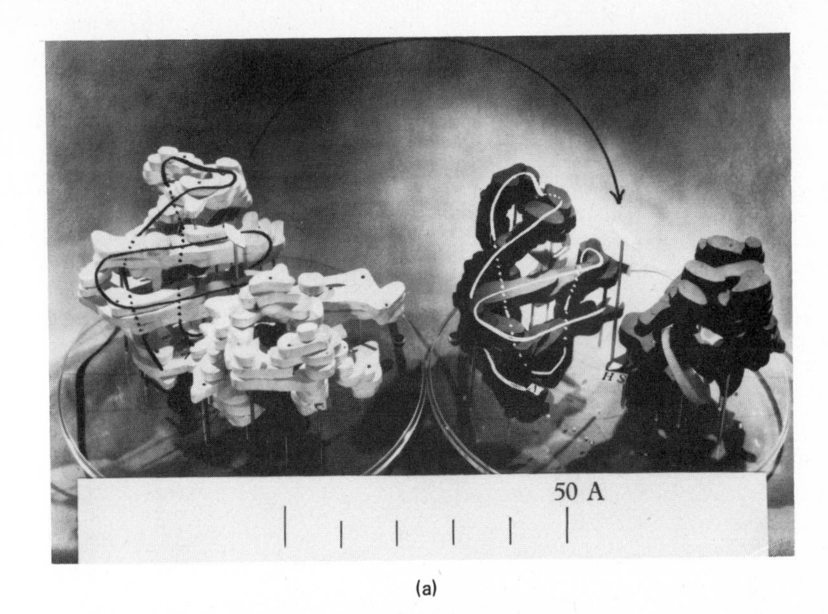

(a)

(b)

Figure 6-6. (a) The four polypeptide chains of a hemoglobin molecule. α chains are white; β chains are black. Arrow shows position of α chains. (b) Joining of α and β chains. One α chain is removed to show how well the chains fit with each other. [From H. Lehmann and R. G. Huntsman, 1974. *Man's Haemoglobins*, pp. 62 and 63. Elsevier/North-Holland Biomedical Press Amsterdam. Photographs courtesy M. F. Perutz, MRC Laboratory of Molecular Biology.]

preparations with .2N HCl and studied biochemically. There are five major groups of histones, categorized by differences such as terminal groups and the presence or absence of amino acids such as cystine. Table 6-1 shows the amino acid content for the five groups of histones found in chromosomes of cells of calf thymus.

The nonhistone proteins are the most poorly understood component of chromosomes. They are large polypeptides, with molecular weights of up to 100,000 Daltons. They are acidic in nature, with high levels of glutamic and aspartic acids. One of the reasons why they are not as well characterized as the histones is because they are much more difficult to extract from chromosomes.

Table 6-1 Amino Acid Analyses and Derived Data for the Five Main Calf Thymus Histones

	F1	*F2b*	*F2a2*	*F2a1*	*F3*
Aspartic acid A	*2.5*	5.0	**6.6**	5.2	4.2
Threonine	5.6	6.4	*3.9*	6.3	6.8
Serine	5.6	**10.4**	3.4	*2.2*	3.6
Glutamic acid A	*3.7*	8.7	9.8	6.9	**11.5**
Proline	**9.2**	4.9	4.1	*1.5*	4.6
Glycine	7.2	5.9	10.8	**14.9**	*5.4*
Alanine	**24.3**	10.8	12.9	*7.7*	13.3
Valine	5.4	7.5	6.3	**8.2**	*4.4*
Cystine/2	0.0	trace	trace	trace	**1.0**
Methionine	*0.0*	**1.5**	(trace)	1.0	1.1
Isoleucine	*1.5*	5.1	3.9	**5.7**	5.3
Leucine	*4.5*	4.9	**12.4**	8.2	9.1
Tyrosine	*0.9*	**4.0**	2.2	3.8	2.2
Phenylalanine	*0.9*	1.6	0.9	2.1	**3.1**
Lysine	**26.8**	14.1	10.2	10.2	*9.0*
ϵ-N-Methyl lysine	0.0	0.0	0.0	**1.2**	1.0
Histidine	(trace)	2.3	**3.1**	2.2	1.7
Arginine	*1.8*	6.9	9.4	12.8	**13.0**

The highest values found in all the fractions are given in boldface type and the lowest values in italics.

The amino acids were determined in the protein hydrolysates by the general method of Spackman, Stein, and Moore (1958), using Technicon Autoanalyser equipment.

The amino acids are given as moles per 100 moles of all acids measured. No corrections have been made for hydrolytic losses. Cystine is the sum of the cystine and cysteic acid found.

From J. A. V. Butler, E. W. Johns, and D. M. P. Phillips, 1968. *Prog. in Biophys. Mol. Biol.* 18: 221.

GROSS STRUCTURE OF CHROMOSOMES

All of these substances, the nucleic acids and proteins, are packaged together to form the remarkable structure known as a chromosome. Although our knowledge of the molecular structure of nucleic acids and proteins has advanced greatly, the actual manner of "packaging" of these molecules in a chromosome is still not entirely clear.

In the era of transmission genetics, knowledge of the molecular structure of these components was not available. What was available were the data from cytogeneticists who stained and studied the chromosomes from as many species as possible, and provided descriptions of the gross structure of chromosomes.

For example, studies have indicated that the usual eukaryote chromosome most likely consists of a double chain of DNA running the length of the chromosome, held together by hydrogen bonds, and that the proteins are bound in some way to the DNA. However, we have not yet actually visualized the fine structure of chromosomes. The main problem we face is that chromosomes are highly coiled at the stages of division when they become visible, and as you shall see, this renders the interpretation of their structure very difficult.

We have already mentioned some aspects of chromosome structure in Chapter One when discussing cell division. You recall that under the light microscope, chromosomes during stages of division following prophase appear to be rodlike structures with an area of constriction, the centromere. The lengths on either side are referred to as the *arms* of the chromosomes (see again Fig. 1-6).

The position of the centromere is frequently used to classify chromosomes. Those with the centromere approximately in the middle, so that the arms are equal in length, are known as *metacentric* chromosomes. Those with the centromere somewhat closer to one end are called *submetacentric*, and those chromosomes which have centromeres almost at one end are called *acrocentric*. *Telocentric* chromosomes are those whose constrictions appear to be at the very end. In addition, some chromosomes have secondary constrictions near one end which appear to result in a portion of the chromosome being separate from the rest, connected by a thin thread. These regions are referred to as *satellites* (Fig. 6-7).

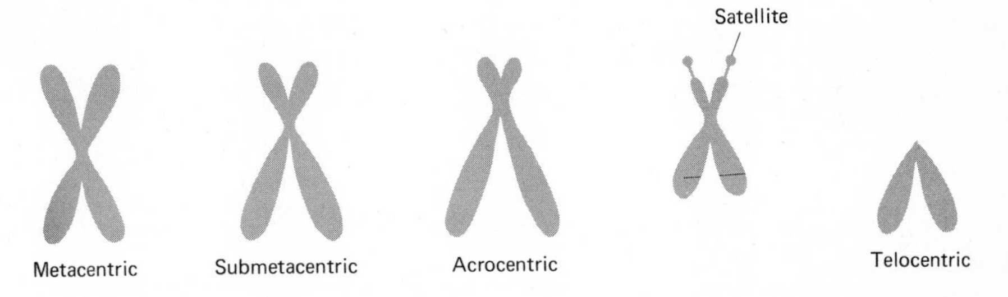

Figure 6-7. Classification of chromosomes based on gross structure.

A great deal of folding must obviously occur to condense the elongated chromatin found in interphase nuclei into the rodlike structure seen following prophase of cell division. Even in its elongated state, some folding, in the form of beadlike structures known as *nucleosomes* or *nu bodies*, is found along the length of the chromatin thread. These nucleosomes, 125 Å in diameter, represent areas of coiling of the DNA, complexed with histones (Fig. 6-8). Nucleosomes are connected by "linker" regions believed to be DNA bound by histones (Fig. 6-8). A current model is one that pictures an octomer of inner histones surrounded by DNA. It is thought that nucleo-

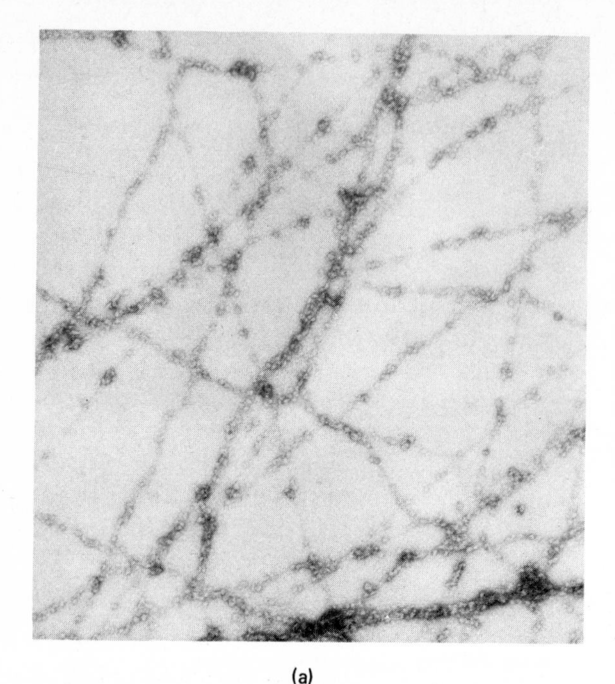

(a)

Figure 6-8. (a) Chromatin fibers showing beadlike configurations (nucleosomes) at intervals. Intermediate strands are believed to be uncoiled DNA plus H1 histone. (b) Electron micrographs of chromatin subunit particles obtained from chicken erythrocyte nuclei; (upper) monomers, (lower) dimers. (Photographs courtesy Dr. C. L. Woodcock, University of Massachusetts.)

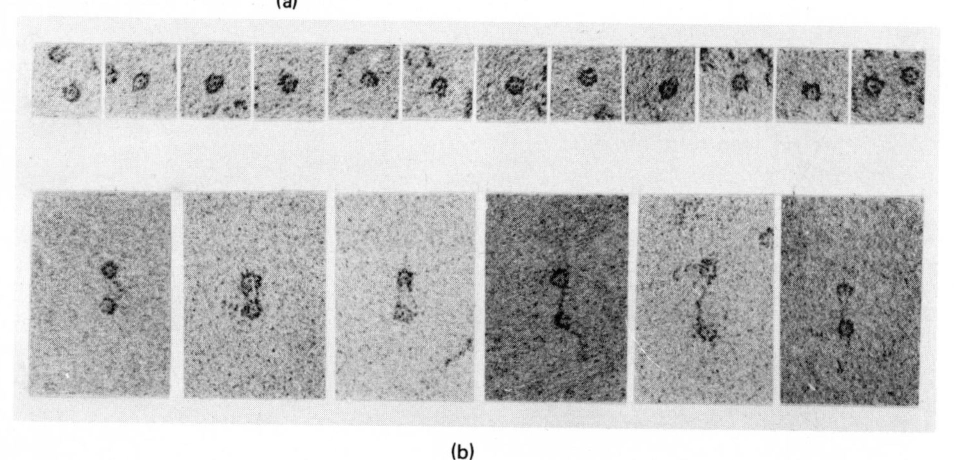

(b)

some structural alterations occur when genes become active. Details of this alteration and its control are presently being studied.

When histones are removed, the beading characteristics of chromosomes are lost. Some geneticists have proposed that there may exist a special class of nonhistone chromosome binding proteins (Mayfield and Ellison 1975), which may be responsible for the beading characteristics, but this is only speculation. As we mentioned previously, many aspects of chromosome structure are not clearly understood, and the nature of beading is one of them.

The presence of nucleosomes is largely indiscernible by metaphase. However, electron microscope studies have shown that metaphase chromosomes are not really rodlike as they appear under the light microscope; they are actually formed of folded loops of the chromosome thread. Furthermore, it is believed that within each loop there are alternating regions of heavier and lighter coiling which lead to the formation of structures known as *chromomeres*. It is the chromomeres which are believed to be related to the banding patterns of chromosomes, which we shall discuss shortly.

That there is intense folding and looping is reflected in estimates that while some of the largest human chromosomes contain DNA estimated to be 7.3 cm long, at metaphase the same chromosome measures only 2-10 $\times$ 10^{-4} cm long.

Giant Chromosomes of Drosophila

The favorite subject of study of classical geneticists, the fruitfly, also was a useful object for the analysis of chromosome structure. What made this insect so valuable was the presence of some enormous chromosomes in the salivary glands of *Drosophila* larvae that were first discovered by E. G. Balbiani in 1881 (Fig. 6-9). Later they were found in larvae of other Dipteran insects, and in tissues other than the salivary glands, such as the Malpighian tubules. The most commonly studied source of giant chromosomes, though, are those of the salivary glands.

Actually, because of their unusual size and appearance, the giant chromosomes were not recognized as chromosomes until the 1930s. At that time, T. S. Painter, among others, realized what they were and that their extraordinary size and clarity would afford a good opportunity to correlate chromosome structure with the mass of data that had been accumulated on inherited traits in Drosophila.

Polyteny. Studies have since established that the large size of the chromosomes is due to a process known as *endomitosis*. In endomitosis the genetic material duplicates as in mitosis, but the daughter strands do not separate and the cell does not divide. Consequently, as development proceeds, there is an accumulation of chromatid strands which are seen as giant chromosomes. This phenomenon is also known as *polyteny*, and the chromosomes are commonly called *polytene* chromosomes.

Polytene chromosomes can contain as many as 1000 to 4000 chromatid threads. Evidence for the number of chromatid strands has come from "squash" preparations, in which salivary gland cells are squashed between a glass slide and slip cover, and stained with acetoorcein or acetocarmine. Occasionally, the ends of polytene chromosomes so treated will be frayed, and the parallel fibers which can be seen extending from the ends can be counted. Also, biochemical studies of salivary gland cells of *Chironomus tentans*, also a Dipteran insect, have led to estimations of as much DNA (3.36 $\times$ 10^{-9} gm per haploid set) as the equivalent of 13 geometric doublings of a normal diploid chromosome set found in other somatic cells of the larvae (Daneholt and Edstrom 1967).

Because of polyteny the areas of coiling of the giant chromosomes are expanded into clearly visible banding patterns. Actually, the very existence of banding in polytene chromosomes indicates that as endomitosis proceeds, the chromatids remain together in close association, gene to gene. The more tightly coiled areas remaining side by side assume the appearance of a band across the width of the chromosome. Recall that the darker bands representing greater coiling are called *heterochromatin*, and the light bands are called *euchromatin*.

Further note that the chromosomes in Fig. 6-9 appear to be the haploid num-

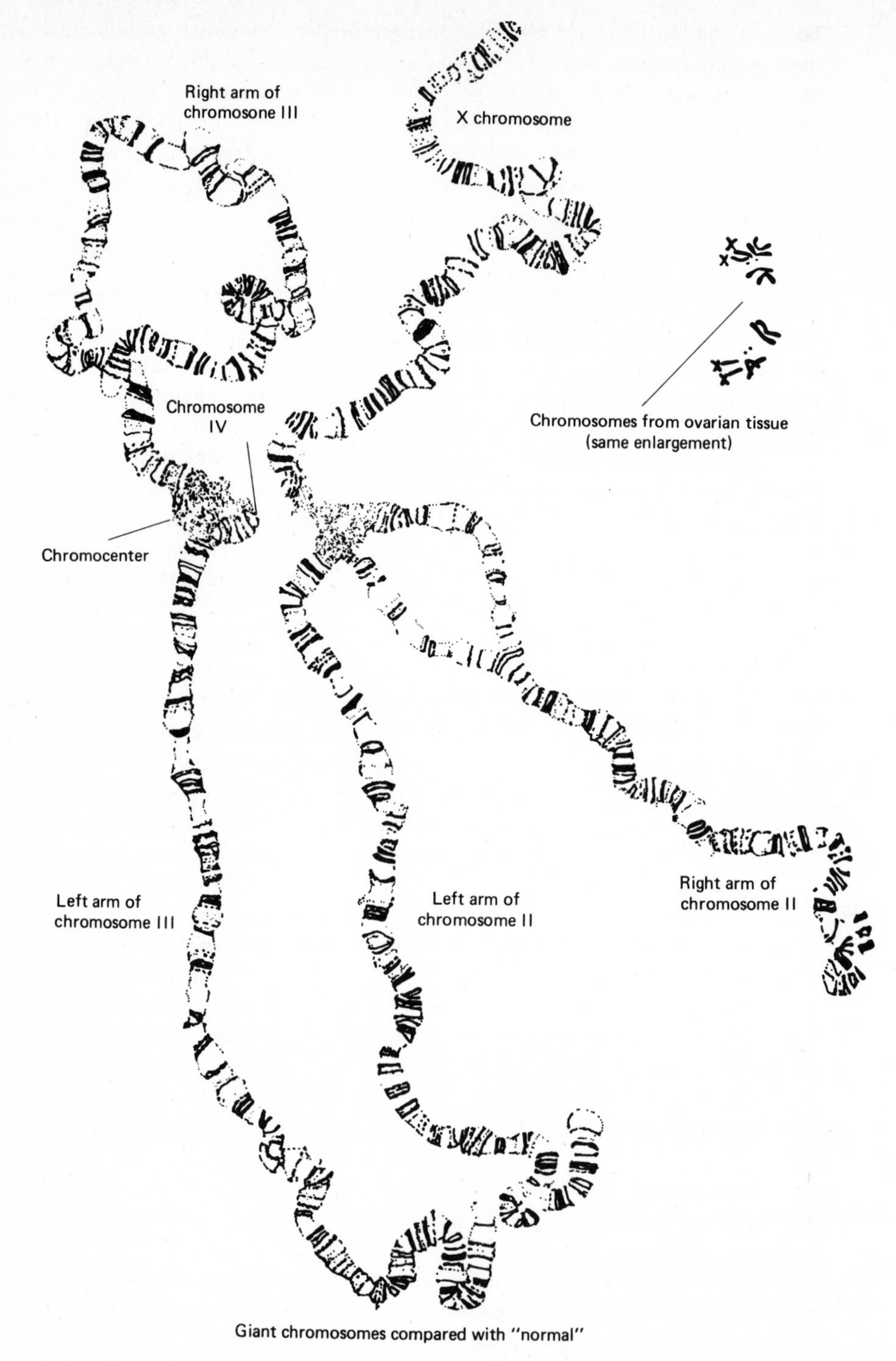

Figure 6-9. Camera lucida drawing of polytene chromosomes of a female Drosophila. (From T. S. Painter, 1934. *J. Heredity* 25: Fig. 1.)

ber, or four, for *Drosophila* , even though we are dealing with somatic cells. This is a unique characteristic of Dipteran polytene chromosomes. Not only do the chromatid strands of one homolog remain together, but the pairs of homologs remain very closely associated as in early prophase I of meiosis. We call this *somatic pairing.* In addition, all the chromosomes are in contact at one point known as the *chromocenter.* The centromere is the attachment region of each chromosome to the chromocenter.

Correlation with gene loci. While studying banding in polytene chromosomes, Painter realized that the patterns were quite consistent in salivary gland cells. He suggested that the bands might afford clues to the position of genes on the chromosomes. Although much work has been done to follow up this idea and some evidence has been obtained to relate a single gene function to the bands, the data are not always exact, and correlation with gene function has not been found for all bands (Lefevre 1974).

One of the first traits discovered that did show a relationship between a particular band and an inherited trait was the X-linked trait known as *Bar Eyes*, studied by Bridges. *Bar Eyes* is determined by a dominant gene, and affected individuals possess abnormal eyes shaped like bars (Figure 6-10a). A study of the X chromosomes of the mutants revealed that bands in a region identified as 16A are present in extra copies. In other words, the normal chromosome banding pattern has only the single area, and Bar-eyed mutants have a repeated area (Fig. 6-10d).

This information strongly indicated that the *Bar Eyes* gene and its normal allele are situated in 16A. Later, this idea received support when flies with even more seriously affected eyes were found. This condition was called *Double Bar* or *Ultra Bar*, and the X chromosome of these mutants showed the presence of a third area in 16A (Fig. 6-10d). These repeated areas are called *duplicated* regions of the chromosome, and the extra genetic information supplied by the duplicated bands is obviously detrimental to eye development.

Human Chromosomes

Unfortunately, mammals were not found to have any chromosomes like the giant Dipteran chromosomes which are so useful to cytogeneticists. Much effort has been expended to develop techniques that would allow us to uncover structural details of the smaller mammalian chromosome which might show correlation with genetic studies. The first attempts to stain for human chromosomes, for example, led to some confusion, as even the number of chromosomes in human cells could not be agreed upon. For some time in the 1940s and early 1950s, the diploid number was thought to be 48.

One of the explanations for the different chromosome counts was that the cell samples used were from patients at mental institutes as well as from other sources. A frequent consequence of an abnormal number of chromosomes in humans is mental retardation, so that quite possibly they were studying cells with aberrant chromosome numbers to begin with. Finally, in 1956, the correct diploid number was reported from two different laboratories. J. H. Tjio and A. Levan in Sweden and C. F. Ford and J. L. Hamerton in England published studies of human cells in which they found 46 chromosomes.

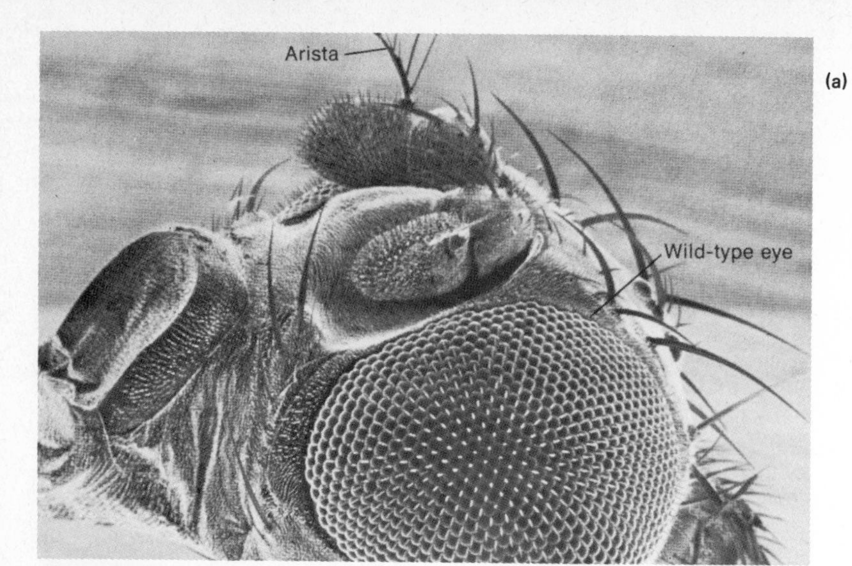

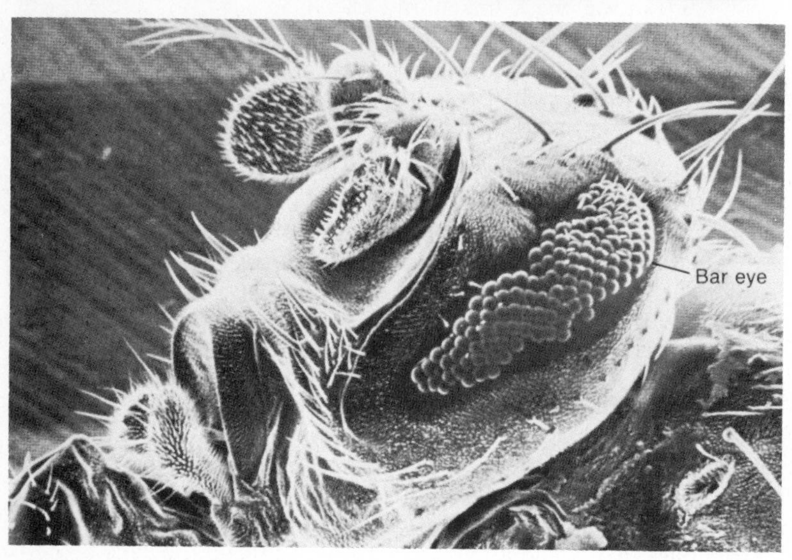

Figure 6-10. Scanning electron microscopy of (a) Drosophila wild-type eye and (b) Bar eye due to a dominant X-linked mutation. (c) Altered phenotypes of eye shape due to Bar mutations. (d) Duplications of bands in Drosophila chromosome at the Bar locus. [(a) and (b) from D. T. Suzuki and A. J. F. Griffiths, 1976. *An Introduction to Genetic Analysis,* p. 223. W. H. Freeman & Company Publishers, New York. Copyright © 1976. (c) and (d) from A. C. Pai, 1974. *Foundations of Genetics,* Fig. 11.14, p. 231. McGraw-Hill, New York. (After Morgan and Bridges)]

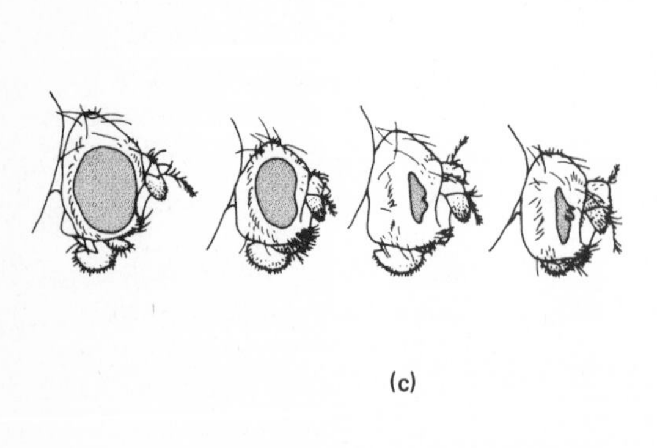

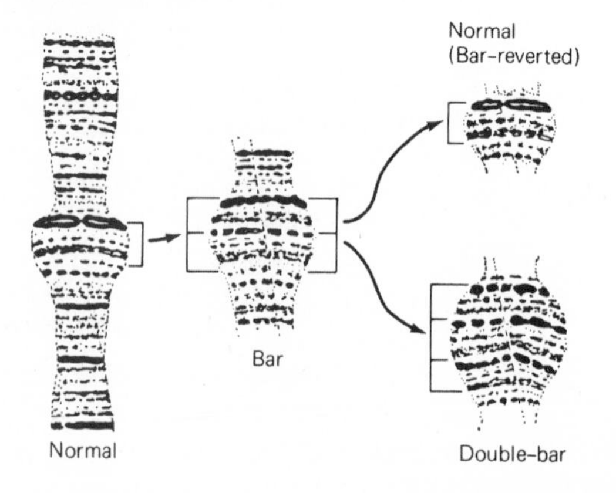

Karyotypes. The technique for isolating and staining chromosomes from a human cell is now routinely performed for diagnostic purposes by laboratory technicians, and can be easily carried out by students. Generally the procedure calls for obtaining cells from peripheral blood samples and incubating them in a nutrient medium that will encourage mitosis of the white blood cells. This is necessary, of course, to allow us to visualize the chromosomes. A drug such as colchicine is applied to arrest mitosis at the metaphase stage, and the cells are placed in a hypotonic solution which causes them to expand and the chromosomes to be spaced farther apart in the cells. The cells are fixed and stained on a microscope slide, and observed under the light microscope.

Favorable preparations are photographed, then the chromosomes are cut out of the photograph, and they are matched and grouped according to structure. The result is an orderly compilation of the entire set of chromosomes from a cell, known as a *karyotype*. Figure 6-11 illustrates karyotypes of normal human male and female cells. The chromosomes are shown both as they appear under the microscope, and after they are matched to form the karyotype.

In Fig. 6-11, we can see structures in human chromosomes that we have mentioned previously. Each has an area of primary constriction, the centromere, which is located at variable regions along the length of the chromosome. There are satellite regions, and chromosomes are grouped according to these aspects, as well as overall size. The X chromosome is considered to be in the large submetacentric C group, and the Y chromosome is similar to the acrocentric G group.

With the standardization of karyotyping techniques, diagnosis of abnormalities due to aberrant chromosome numbers was greatly facilitated. In addition, gross chromosomal defects could be detected. One of the problems encountered by geneticists, however, was that chromosomes of the different groups were so similar that the matching of chromosomes into homologous pairs was purely arbitrary. The stains used when karyotyping was first developed did not bring out any structural details in the chromosomes that would allow them to be distinguished from one another.

Banding in human chromosomes. In 1970, the first demonstration of structural details brought out by staining of chromosomes was reported by Caspersson and his coworkers. They found that fluorescent compounds, such as quinicrine mustard, stained the DNA of chromosomes in such a manner as to produce banding patterns (see again Fig. 6-11). Since then, a number of other techniques and stains, such as Giemsa staining, and denaturation-reassociation techniques, have also been found to produce banding patterns. (For papers discussing these techniques, see the reference to the Nobel Symposium on Chromosome Identification, Lejeune paper.)

Each of these methods has been found to produce a constant banding pattern on metaphase chromosomes, but the pattern differs depending on which procedure was followed in preparing the karyotype. It was apparent that a system of nomenclature needed to be developed, and a number of conferences were held to establish both an official nomenclature and a classification of human chromosomes in relation to the banding patterns.

Nomenclature and classification. The Paris Conference of 1971 was convened for the purpose of arriving at an international classification of the human chromo-

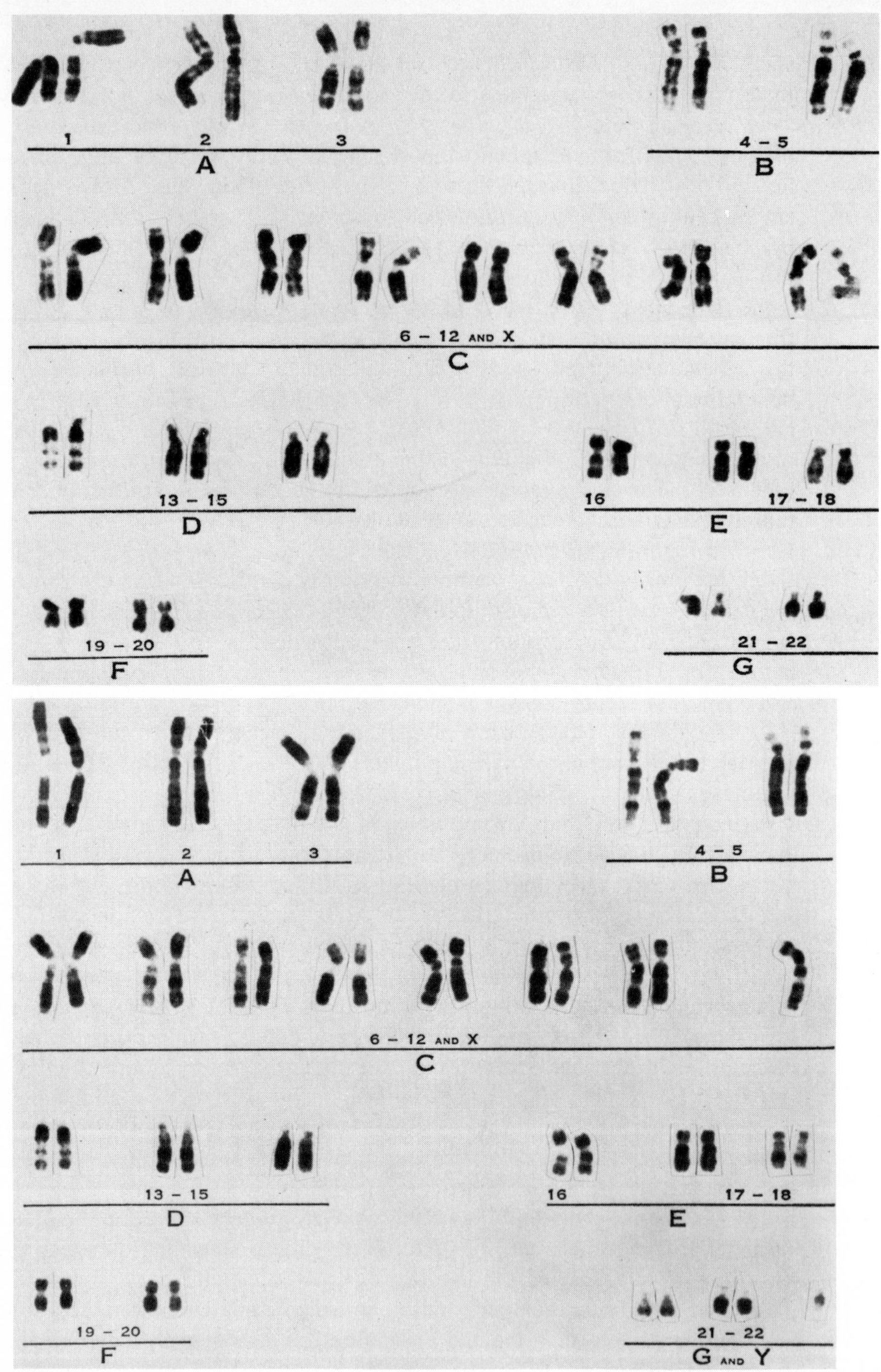

Figure 6-11. (a) Karyotype of a normal female. (b) Karyotype of a normal male. Special techniques were used to bring out banding patterns. (Photos courtesy Emilie Mules, Genetics Laboratory, John F. Kennedy Institute, Baltimore, Md.)

some complement. Figure 6-12 shows the human karyotype as it was decided upon by geneticists at the meeting.

Since the three major procedures used for staining chromosomes do result in different banding patterns (for example, the denaturation technique produces a banding pattern which is the reverse of that produced by the fluorescent stains), the different bands are represented in literature by the letters G, Q, and R. Bands brought out by Giemsa are represented by G; Q indicates the bands of the quinicrine dyes; and R the bands found with the renaturation techniques.

The nomenclature calls for assigning the letter p to the short arms and q to the long arms of the chromosomes. Bands are numbered consecutively outward from the centromere to the ends of the arms. In addition, the chromosomes are divided into "regions." A region is defined as a portion of the chromosome lying between two chromosome landmarks, which are structural features such as the centromere, the ends of the arms, and certain very consistent bands. In referring to a particular band, a geneticist must thus specify in order the number of the chromosome, which arm, the number of the region, and the band number. For example, 10q25 refers to the 10th chromosome, long arm, region 2, band 5.

This nomenclature has been extremely useful in many areas of study. It has allowed detailed studies of chromosomal abnormalities in humans. In addition, the techniques have been applied to cells of other species of vertebrates in which banding patterns have also been found. This has led to the development of an entirely new approach to the study of evolutionary relationships, as we shall discuss in Chapter 17.

There will no doubt be further modifications and improvements in techniques for the study of chromosome structure. For example, the diagram of the human karyotype in Fig. 6-12 is based on banding patterns found in metaphase chromosomes. Recent reports have shown that techniques applied to prophase chromosomes show many more bands present (Yunis 1976). It is assumed that several of these bands are lost due to the coiling of the chromosomes as they enter the metaphase stage. Figure 6-13 illustrates this point.

The more banding detail we can observe, the better we will be able to discern the relationship between chromosome structure and function. There is at present no evidence that the bands in human chromosomes represent sites of single gene function as we have discussed for the polytene chromosomes of insect larvae. In view of the variability of banding patterns, this is not surprising. These areas of ambiguity will surely be clarified as research progresses.

CHROMOSOME ABNORMALITIES

As is the case with so many genetic phenomena, our understanding of chromosome structure and its relationship to genes has been aided by analyses of abnormalities of chromosomes. We have already mentioned mutations of single genes, which are heritable changes in the genetic material that result in changes of phenotype. Scientists have also studied with interest mutations on a larger scale, namely, those affecting a considerable portion of genetic material in chromosomes, or even of numbers of chromosomes. These abnormal conditions are referred to as *gross*, or *chromosomal, mutations.*

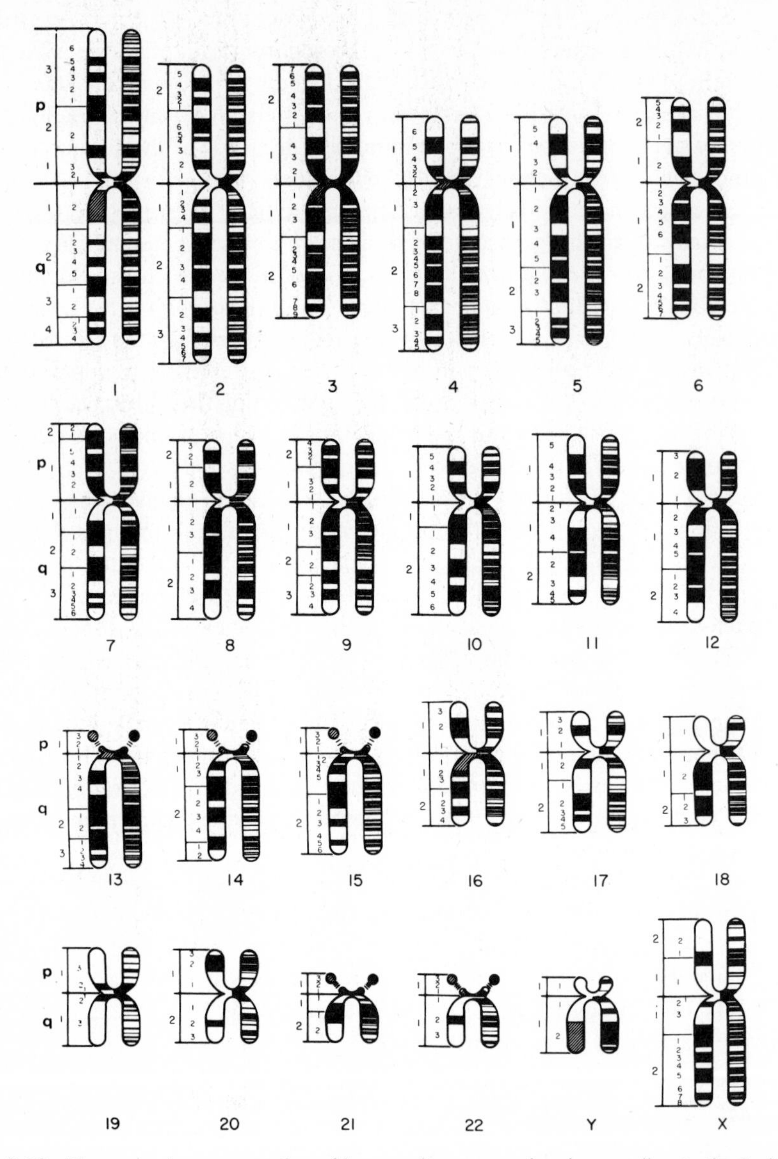

Figure 6-12. Diagrammatic representation of human chromosome bands according to the Paris Conference nomenclature. In each chromosome the left chromatid represents the banding pattern observed in mid-metaphase and the right chromatid represents the banding pattern observed in late prophase. (From J. Yunis, 1976. *Science*, p. 1269. Fig. 1; copyright 1976 by the American Association for the Advancement of Science.)

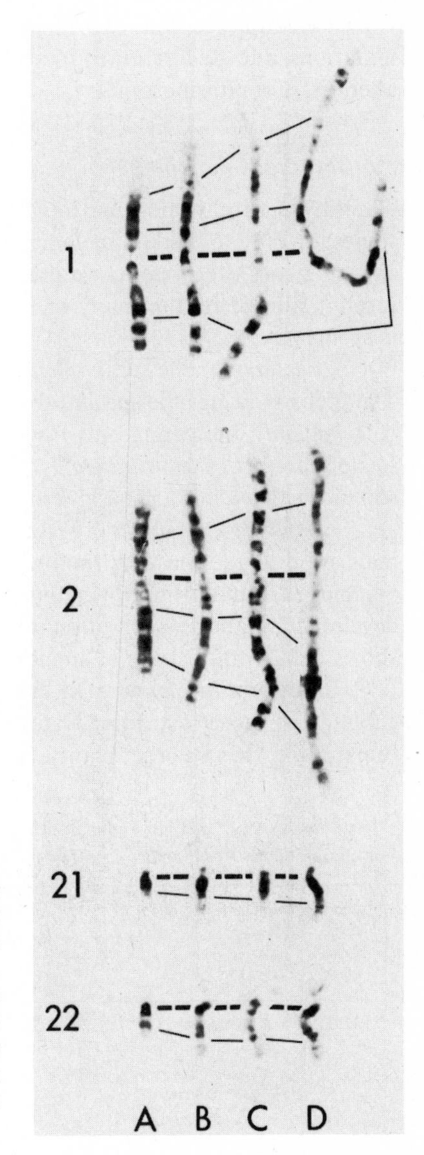

Figure 6-13. The two largest (1 and 2) and the two smallest (21 and 22) chromosomes of man at mid-metaphase (A); early metaphase (B); early prometaphase (C); and late prophase (D). Note the progressive coalescence of the multiple fine bands of late prophase into the thicker and fewer dark and light bands of metaphase. (From J. Yunis, 1976. *Science*. p. 1269, Fig. 1; copyright 1976 by the American Association for the Advancement of Science.)

Many basic concepts of genetics have been developed from the study of gross mutations. For example, Boveri studied sea urchin embryos which contained abnormal numbers of chromosomes due to dispermic fertilization (p. 2). The developmental aberrations that resulted led to our awareness that chromosomes contain different genes. Studies on abnormal chromosome numbers in *Drosophila* by Bridges led to an understanding of the role of sex chromosomes (p. 124). The *Bar* mutations indicated a relationship between banding and gene loci. These classical studies involved examples of the two major subgroups of chromosomal mutations, namely, mutations due to alteration in chromosome numbers and those due to structural alterations of chromosomes.

Changes in Chromosome Numbers

Alterations of chromosome numbers can involve either the entire set of chromosomes, or only individual members of a set. When too many or too few entire sets of chromosomes are present, we use the term *euploidy*. If the chromosome number of a cell is altered by too many or too few of individual chromosomes, the term used is *aneuploidy*.

Euploidy. The suffix *-ploidy* is used in terminology for all euploid conditions. The prefix indicates the specific abnormality. For example, *monoploidy* is synonymous with *haploidy*, indicating only n set of chromosomes. *Triploidy* would be $3n$, or three sets of chromosomes; *tetraploidy* would describe a cell with $4n$, or 4 sets of chromosomes. *Polyploidy* is a general term referring to several sets of chromosomes.

Euploidy in animals. With only a few exceptions, sexually reproducing animals are diploid, and abnormal numbers of chromosome sets are lethal. In humans, euploidy is found primarily in aborted fetuses; very rarely do euploid humans survive development, and the condition is lethal in those who do develop to term. Table 6-2 shows data from a study of spontaneous abortions.

Euploidy in plants. On the other hand, plants seem to be able to survive euploidy in more instances. Natural polyploidy has been found in mosses, grasses, tomatoes, corn, and other plants. In fact as many as 50 % of our cultivated crops may

Table 6-2 The Frequency of Different Types of Chromosome Abnormalities Found in Spontaneous Abortions

	Abortions karyotyped	*Abnormal karyotypes*	*45, X*	*Autosomal trisomics*							*Triploids*	*Tetraploids*	*Mixoploid trisomics*	*Others*
				A	*B*	*C*	*D*	*E*	*F*	*G*				
Number	1291	322	64	7	3	20	30	48	4	42	53	16	16	19
				All trisomics = 154							Polyploids = 69			
Proportion total	—	0.25	0.05	A,B,F 0.01		0.015	0.023	0.037	-	0.032	0.041	0.012	0.027	
				All trisomics 0.119							Polyploids = 0.053			
Proportion abnormal	—	1.00	0.199	A,B,F 0.043		0.062	0.093	0.149	-	0.130	0.165	0.050	0.109	
				All trisomics 0.478							Polyploids = 0.214			

From J. Hamerton, 1971. *Human Cytogenetics*, v. 2, p. 382, New York: Academic Press.

be polyploid (Briggs and Knowles, 1967). Triploidy has been found in some varieties of aspen. In fact, some of the polyploid species have been found to be more vigorous and better growers than their diploid counterparts. For example, McIntosh apple trees known to be $4n$ produce larger fruit.

Accordingly, some research efforts in agriculture and horticulture have been directed toward artificially inducing polyploidy by using drugs such as colchicine to interfere with mitosis. A triploid seedless watermelon has been developed with colchicine treatment. In this manner it is hoped that new polyploid varieties may result which will give better yield than diploid species.

In addition, polyploidization may allow the development of plants resistant to disease. For example, a cross between a virus-resistant tobacco plant and one susceptible to the virus resulted in a resistant, but sterile hybrid. Induction of polyploidy by colchicine treatment led to the production of some resistant and fertile plants. Similar experiments on more useful plants may solve nutritional and horticultural problems posed by natural pathogens. However, because of the constant genetic changes occurring in pathogens such as viruses, it is unlikely that permanent solutions to such problems will ever be found.

The difficulty in establishing polyploid species lies in problems of gametogenesis. Synapsis during meiosis may be abnormal. Triploids, for example, would pair in different combinations, as would tetraploids. The various combinations of chromosomes could produce inviable or weakened offspring. The absence of viable monoploids in most species is no doubt due to the expression of lethal genes, and the imbalance of gene products due to the absence of diploidy.

In some instances of polyploidy, the increase in chromosome sets occurs within the species. This is known as *autopolyploidy*. For example, doubling of the chromosomes without cellular division would result in tetraploidy, as in McIntosh apple trees which we mentioned earlier. Polyploidy resulting from a cross between two different species, is referred to as *allopolyploidy*.

A special form of allopolyploidy is *amphidiploidy*, in which the cells contain the complete diploid set of chromosomes from each of two different diploid organisms (Fig. 6-14). A classic experiment producing amphidiploidy was conducted by the Russian geneticist Karpechenko in 1928. He made crosses between radishes and cabbages, each of which contain a normal diploid number of 18 chromosomes. The F_1 plants were sterile for the most part, even though they possessed 18 chromosomes. The nine chromosomes from one parent were genetically different enough from the chromosomes of the other parent that normal meiosis could not take place.

Occasionally, however, some seeds were produced, and when germinated gave rise to F_2 plants. Studies of these plant cells showed the presence of 36 chromosomes, 18 from each parental species. In these amphidiploids, meiosis occurred successfully because of the pairing of homologs in each diploid set. Although it may have been the hope of Karpechenko to produce a hybrid with the leaves of a cabbage, and the roots of a radish, alas, the actual F_1—a raddage, a cabbish?—possessed leaves of a radish and roots of a cabbage!

Aneuploidy. The suffix *-somy* is used to indicate aneuploidy, and the prefix indicates whether there are too many or too few of a particular pair of chromosomes. For example, *monosomy* is the loss of one chromosome such that the individual is

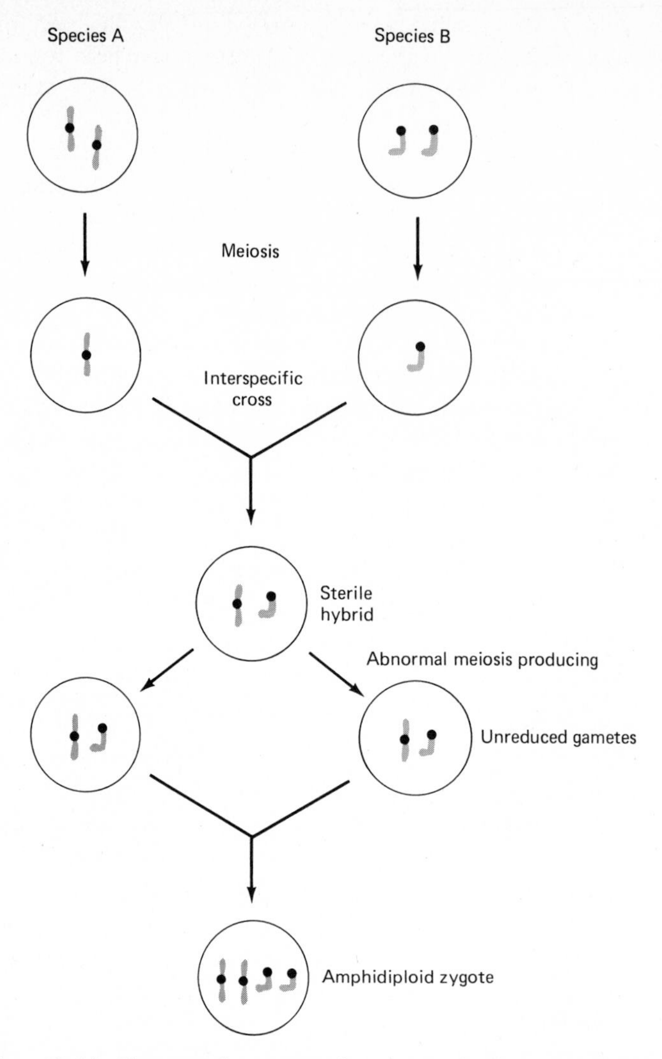

Figure 6-14. Diagrammatic representation of events leading to amphidiploidy.

$2n - 1$; *trisomy* is the presence of an extra chromosome, resulting in $2n + 1$. In each case, meiosis of germ cells in either a monosomic or trisomic individual results in gametes with different numbers of chromosomes (Fig. 6-15).

The study of aneuploid conditions began with Bridges' work on Drosophila which we discussed in Chapter 5. Generally, the addition or loss of certain chromosomes, giving rise to aneuploidy, is due to abnormalities of meiosis. This can occur naturally, for example as a result of nondisjunction, or can be induced by agents such as radiation. Radiation causes chromosomes to break and be lost, or to become slow in separating during cell division, leading to aneuploid combinations.

Aneuploidy in plants. Aneuploidy has also been found in plants. A classic study on the Jimson weed, *Datura stramonium*, revealed a number of mutations which

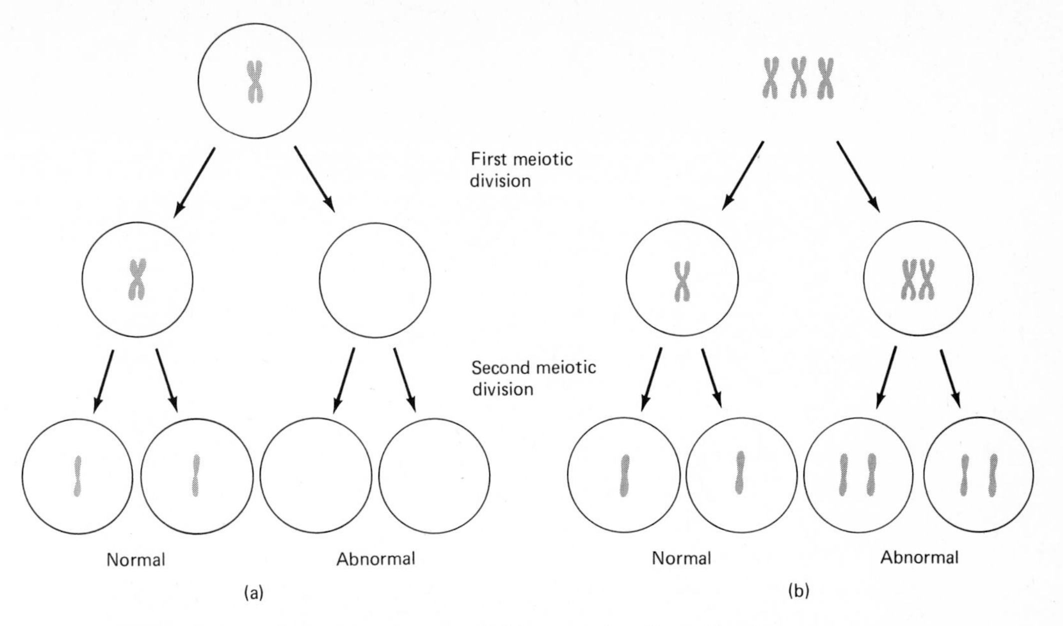

Figure 6-15. Meiosis in (a) a monosomic cell, and (b) a trisomic cell, resulting in germ cells containing too few or too many chromosomes.

were trisomic for one of each of the 12 pairs of chromosomes in *Datura* (Blakeslee *et al.* 1923).

One trisomic plant of *Datura* is Poinsettia (not the Christmas plant). Trisomy was indicated by analysis of the segregation ratios of alleles in experimental crosses. In addition there was cytological evidence for trisomy. The gene p^+ determines purple flowers in Poinsettias, while its fully recessive allele, p, determines the presence of white flowers in homozygous plants. A cross of a trisomic homozygous dominant plant to a trisomic homozygous recessive plant would yield an F_1 generation of plants that produce only purple flowers. Note, however, that the genotypes of F_1 plants are not the same. First, because of the trisomy, meiosis yields gametes with $2n$ and n numbers of chromosomes. Secondly, it has been found that male gametes with $2n$ numbers of chromosomes are nonfunctional. Figure 6-16 illustrates the expected results of this cross. Work out for yourself the expected results of an F_2 generation from a cross of F_1 plants (Problem 6-6).

Aneuploidy in humans. Humans, too, can manifest viable aneuploid conditions, some of which we mentioned in Chapter 5. The most common aneuploids derive from abnormal numbers of sex chromosomes. Figure 5-10 shows Klinefelter's syndrome patients, males whose cells contain two or more X chromosomes in addition to the Y chromosome. Such individuals are generally long-limbed, sterile due to abnormal development of the testes, and sometimes mentally retarded, with varying degrees of breast development. We should point out here that in all aneuploid human conditions there are symptoms common to each condition, but other symptoms are found to vary considerably. For example, in Klinefelter's syndrome, sterility is invari-

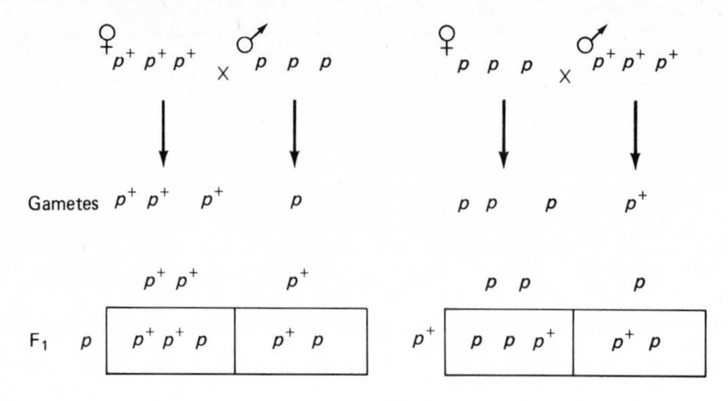

Figure 6-16. Results of crosses between plants trisomic homozygous dominant and plants trisomic homozygous recessive. Explanation in text.

able but mental development ranges from normal to distinctly mentally retarded. The greater the imbalance of sex chromosomes, the more severe the retardation.

Other aneuploidies of sex chromosomes are monosomy of the X chromosome (X0), known as Turner's syndrome (see again Fig. 5-9), XXX, and XYY. In Turner's syndrome, monosomy of the X chromosome results in the female phenotype, and the individual tends to be short. There is sterility due to abnormal ovarian development, and various other physical symptoms such as webbing of the skin in the neck region, and occasional mental impairment.

There are no physical abnormalities consistently associated with either XXX or XYY, although females trisomic, tetrasomic or pentasomic for the X chromosomes have been found to be mentally retarded, with increasingly severe effects corresponding to an increase in the number of X chromosomes.

The XYY males, although normal in phenotype, with a tendency to being taller than average, have been a subject of much controversy in recent years. Some researchers proposed that this genotype is associated with antisocial behavior. One of the main reasons for this association between XYY and the so-called "criminal syndrome" stemmed from studies of individuals in prisons and in institutions for the criminally insane. These studies seemed to indicate a higher frequency of XYY males in the institutions than the general population. Furthermore, a few men who had committed violent crimes were found to have an XYY genotype. In fact, some criminals, such as Richard Speck, murderer of eight student nurses in Chicago in the 1960's, were erroneously reported to be XYY.

A number of geneticists, however, have questioned this hypothesis. First of all, they pointed out that earlier statistical studies showing the higher frequency of XYY males in various institutions were carried out on populations presumed to have a higher frequency of this aberration. Furthermore, the study did not include adequate controls, such as XYY males in the general population, most of whom never discover their aneuploid genotype, as they are normal in phenotype.

Recently, a study was carried out on all male Danish citizens born in Copenhagen in the years 1944 through 1947 (Witkin *et al.* 1976). Only individuals, 184 centimeters or taller were studied. Sex chromosome determinations were carried out on cells from more than 4000 men, and these were correlated with the backgrounds and criminal and IQ records of each man. The data are summarized in Table 6-3.

Analysis of these data did indicate an increased incidence of crime among the XYY as well as XXY males when compared to XY males. However, lower intellectual ability also seemed to be associated with the aneuploid conditions. When compared to XY males of similar intelligence, XYYs still had a somewhat higher frequency of crime (XXYs had none), but it was much reduced. The conclusion drawn by the scientists was that there is little evidence to link XYY with abnormally aggressive behavior. They conjectured that the higher frequency of XYY males being apprehended as criminals is only a reflection of their lessened ability to escape apprehension.

Table 6-3 Crime Rates and Mean Values for Background Variable of XY's, XYY's, and XXY's. Significance Level Pertains to Comparison with the Control Group (XY) Using a Two-Sided Test. For Criminality Rate an Exact Binomial Test Was Used; for All Other Variables a *t*-Test Was Used.

	Criminality		*Army selection test (BPP)*			*Educational index*			*Parental SES*			*Height*		
Group	*Rate (%)*	*N*	*Mean*	*S.D.*	*N*	*Mean*	*S.D.*	*N*	*Mean*	*S.D.*	*N*	*Mean*	*S.D.*	*N*
XY	9.3	4096	43.7	11.4	3759	1.55	1.18	4084	3.7	1.7	4058	187.1	3.0	4096
XYY	41.7*	12	29.7†	8.2	12	0.58*	0.86	12	3.2	1.5	12	190.8†	4.6	12
XXY	18.8	16	28.4†	14.1	16	0.81‡	0.88	16	4.2	1.8	16	189.8†	3.6	16

From Witkin *et al.* 1976. *Science* 193: p. 550.

*$P < .01$. †$P < .001$. ‡$P < .05$.
SES = socieconomic status; BPP = intelligence test.

As already mentioned, all these studies suffer from inadequate controls. One must also bear in mind that if there are genes on the Y chromosome which in some way determine aggressiveness, even XY individuals could carry genes that cause them to behave criminally. In summary, there does not seem to be unequivocal scientific evidence that XYY males behave abnormally. We shall further discuss problems facing scientific analysis of behavior in Chapter 18. You will see that we do not yet have the means of testing human behavior and intelligence in a manner that would allow us to determine the genetic basis of these very interesting, but enormously complex, phenotypic traits.

One autosomal aneuploid condition in humans that has been found to be viable to varying degrees is trisomy of chromosome 21 in Down's syndrome, which we discussed in Chapter One. Figure 1-28 shows a photograph of a typical Down's syndrome patient. Facial characteristics such as epicanthal folds of the eyelids; short, blunt nose; and tendency to tongue protrusion can be seen. These patients are generally severely mentally retarded, and various other physical problems affecting the heart and vascular system shorten the life span. Other trisomies have been found such as trisomy 18 (Edwards' Syndrome Fig. 6-17), but these conditions tend to be fatal to the victims within the first year of life.

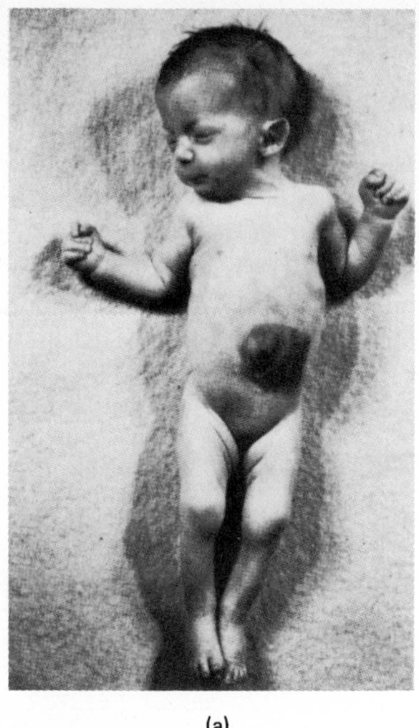

(a)

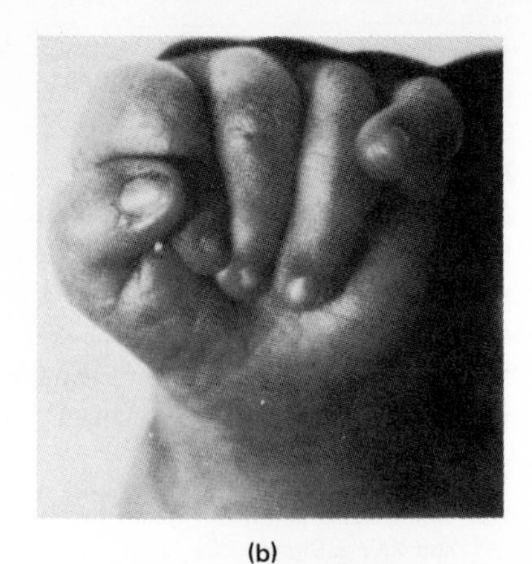

(b)

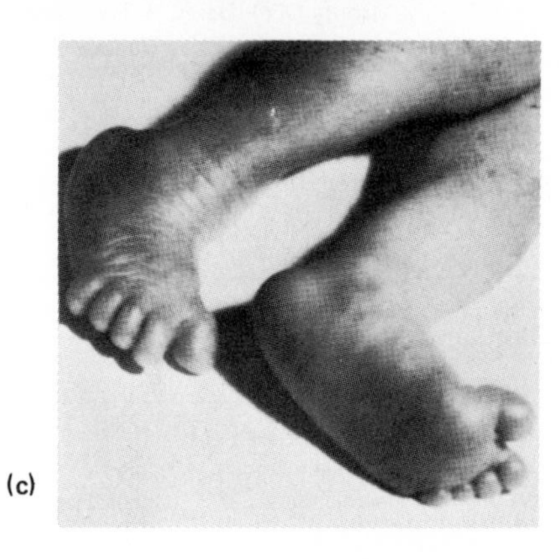

(c)

Figure 6-17. (a) Edwards' syndrome (trisomy 18). (b) Typical overlapping of fingers and (c) broad flat feet. (From J. deGrouchy and C. Turleau, 1977. *Clinical Atlas of Human Chromosomes,* p. 163. John Wiley, New York.)

Changes in Chromosome Structure

In addition to abnormal numbers of chromosomes which result in gross mutations, there are also changes of chromosomal structure which can lead to gross mutations. The four major kinds of structural alterations of chromosomes are shown in Fig. 6-18. They are deletions, duplications, inversions, and translocations.

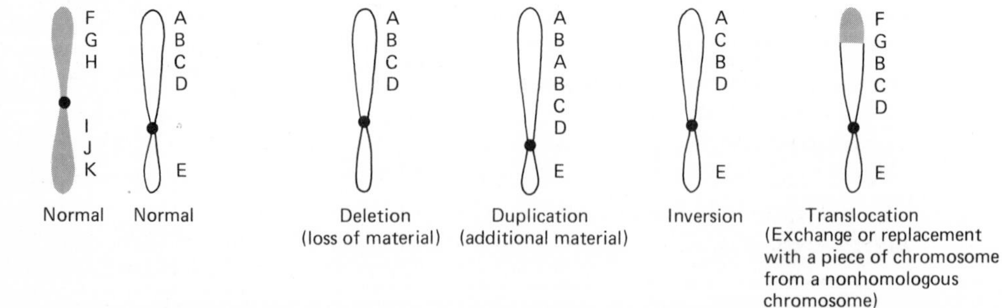

Figure 6-18. Examples of major kinds of chromosomal structural mutations.

Deletions. Deletion mutations result from the loss of a portion of the chromosome, as the term implies. Deletions at the ends of chromosomes occur as a result of a single break; deletions within a chromosome require two breaks with a reattachment of the broken ends of the chromosome. These latter deletions are referred to as *interstitial deletions*, and can be detected by using banding techniques in karyotyping. One can expect that pairing of chromosomes containing interstitial deletions during meiosis would result in the formation of loops (Fig. 6-19). In humans, the cri du chat syndrome is an example of a deletion in chromosome 5, resulting in mental retardation and early death in most cases.

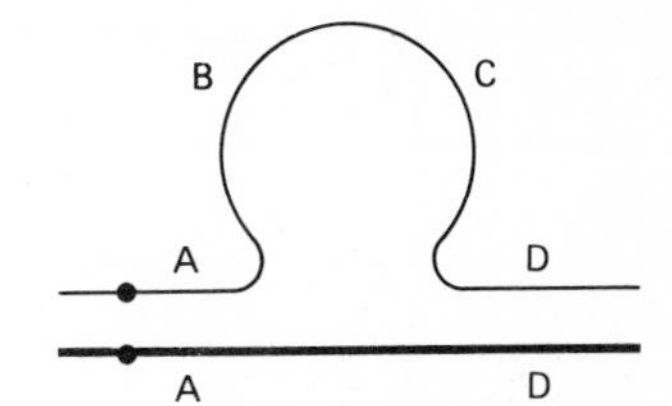

Figure 6-19. Loss of genes through deletions causes loop formation during synapsis.

A useful genetic consequence of deletions is that no further mutation at the locus or loci can be detected because of the loss of genetic material. It is known that genes which mutate from wild-type can mutate again back to wild-type, a process known as *reversion mutation.* The absence of reversion mutation at certain loci is often taken as an indication of deletion.

Duplications. Duplications are the presence of extra repeated copies of sequences within chromosomes. How an original duplication arises is not clear, but it is not an uncommon phenomenon, as genomes from most organisms show the presence of duplications. Once a duplication has taken place, unequal crossing over during meiosis can give rise to further repeats of the region. As Fig. 6-20 illustrates, this process is believed to have produced the *Ultra Bar* state in *Drosophila.*

The presence of extra portions of chromosomes can be detected by the formation of a loop during synapsis which contains the extra genetic material (Fig. 6-21). It has recently been postulated that duplication is a mechanism which protects organisms from loss of vital genetic material through mutations. The extra copies of genes ensure the presence of normal gene products.

Inversions. Chromosomal rearrangements, or inversions are regions in which the sequence of genetic material is inverted by the occurrence of two breaks, with the middle piece of chromosome swiveling 180° and reinserting into the chromosome. The inversion may involve the centromere, in which case it is referred to as a *pericentric inversion.* If the inversion does not involve the centromere, it is termed a *paracentric inversion* (Fig. 6-22).

One of the genetic effects of inversions is to interfere with crossing-over. This interference may occur because a small inversion results in abnormal pairing of genes, or in loss of pairing during meiosis. Without normal synapsis, there could be no crossing-over. Pairing can occur if the inversion is a fairly long portion of the chromosome, by the formation of an inversion loop as shown in Fig. 6-23. If crossing-

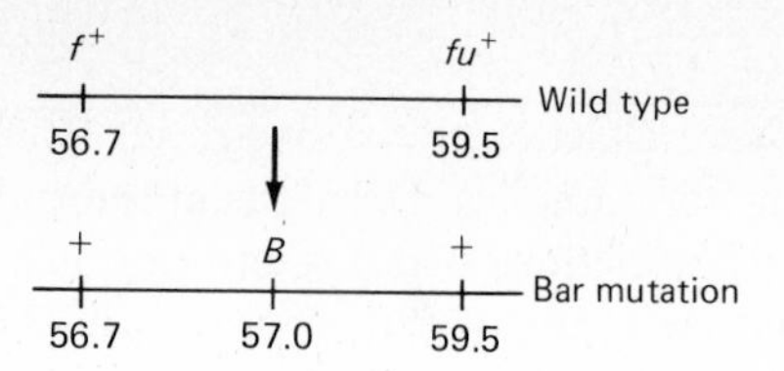

Unequal crossing over in Bar females

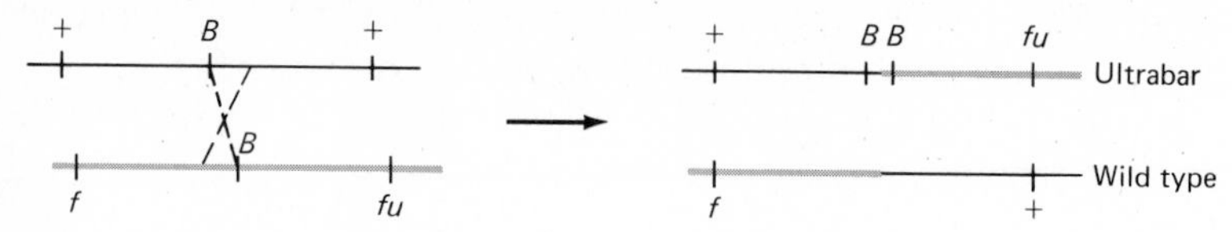

Unequal crossing over in Bar/Ultrabar females

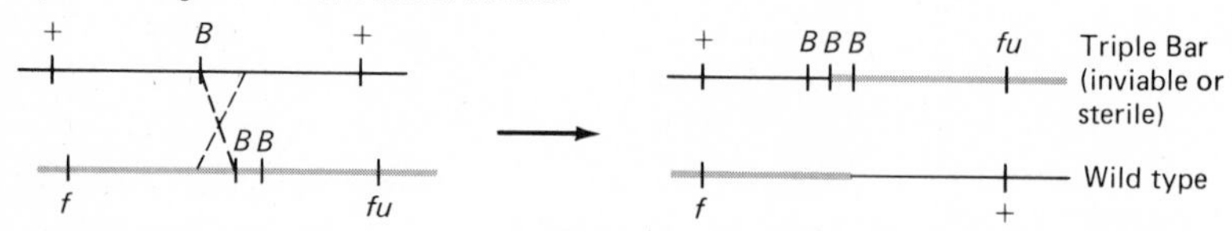

Figure 6-20. Explanation for the appearance of new Bar alleles as a result of unequal crossing over. Note that chromosomal material is added to one of the crossover products (Ultra Bar, or Triple Bar) and removed from the other (wild types). Thus both gene duplication and gene fusions may arise from the same unequal crossing over event; the former caused by repeating the sequence of an entire gene such as Bar, and the latter caused by the removal of genetic material between two formerly separate genes. (From M. Strickberger, 1976. *Genetics* 2nd ed., p. 586. Macmillan, New York. Copyright © 1976, Monroe Strickberger.)

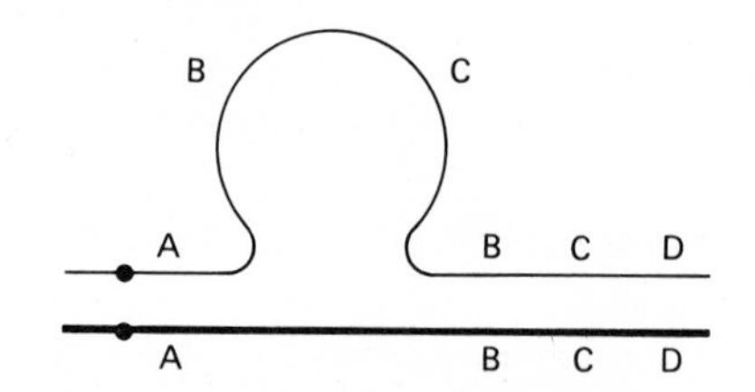

Figure 6-21. Extra genes resulting from duplication also cause loop formation during synapsis.

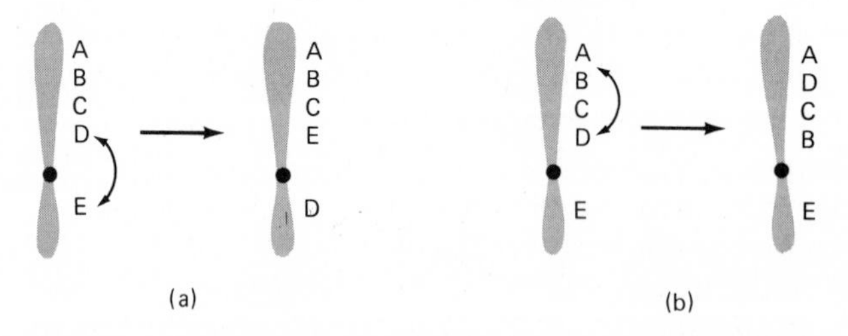

Figure 6-22. (a) Pericentric inversion; (b) paracentric inversion.

over occurs between chromatids in an inversion loop, abnormal chromosomes and acentric (without centromere) pieces of chromosomes result (Fig. 6-24), reducing the number of viable cross-over products in the gametes.

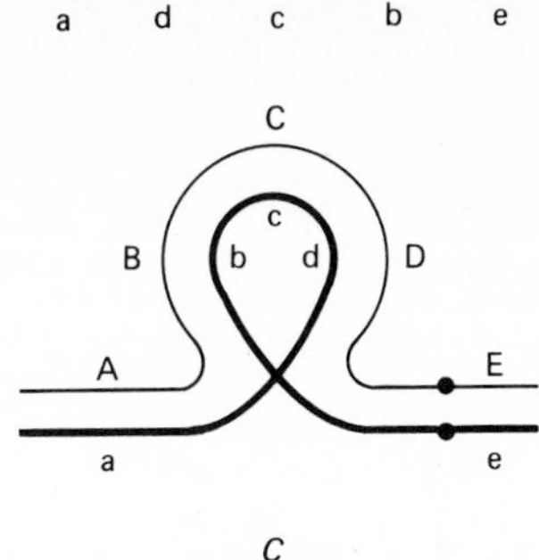

Figure 6-23. Loop formed to allow gene-to-gene pairing in synapsis of an inversion.

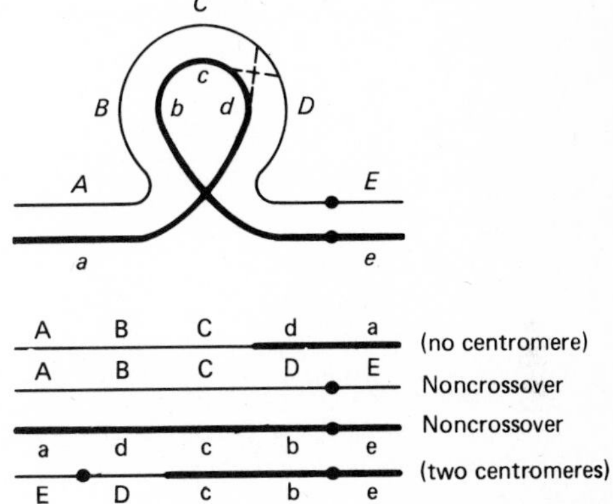

Figure 6-24. The results of crossing over within an inversion loop between two nonsister chromatids.

Translocation. Translocation occurs when a piece of one chromosome breaks away and reattaches to a nonhomologous chromosome. It can occur between two pairs of chromosomes which exchange pieces in what is known as *reciprocal translocation.* This usually does not affect the phenotype, as the full complement of genetic material remains in the cells. During synapsis, however, the two pairs of chromosomes must form a cross-shaped structure to accommodate gene-to-gene pairing (Fig. 6-25). This configuration is occasionally seen in karyotypes of various organisms.

Nonreciprocal translocations have also been found. One of the best-known examples of such a chromosomal rearrangement occurs in certain patients with Down's syndrome. These patients receive an extra copy of the 21 chromosome, but it is attached to chromosome 15. The extra-long chromosome which results is known as a 15/21 translocation (Fig. 6-26). Improved banding techniques, however, have shown that translocations of 21 to chromosomes 14 and 13 are even more frequent than to 15.

Unlike trisomy 21, which seems to be related to maternal age and spontaneous nondisjunction, the 15/21 chromosome can be transmitted to an offspring by a phenotypically normal parent of either sex. Figure 6-27a illustrates the 15 and 21 chromosomes of such a parent. Although he or she may possess only 45 chromosomes, the genetic material of all 46 chromosomes is present, and the individual is normal. Meiosis of the germ cells of such a translocation carrier results in genetically different germ cells as shown in Fig. 6-27b. Fertilization by a normal haploid germ cell theoretically results in the production of 1 normal : 1 translocation carrier : 1 Down's syndrome : 1 lethal (missing an autosome).

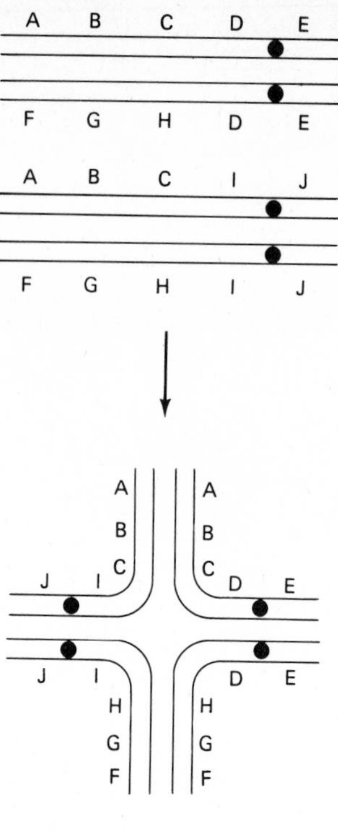

Figure 6-25. Cross-shaped configuration formed during synapsis of two pairs of chromosomes containing reciprocal translocation.

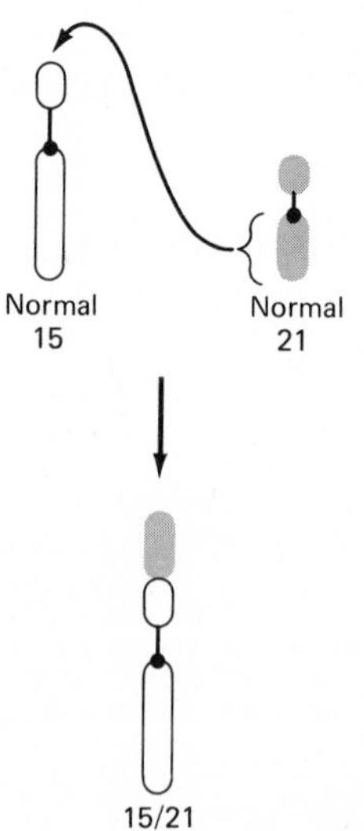

Figure 6-26. Nonreciprocal crossing over between the 15 and 21 pairs of chromosomes in humans.

Another example of translocation in humans is the Philadelphia chromosome, long associated with a particular form of cancer known as chronic myeloid leukemia. Initially thought to be a deletion of chromosome 22, the missing piece has been determined by careful banding studies to be attached to chromosome 9. What its role is in the etiology of the disease has yet to be understood.

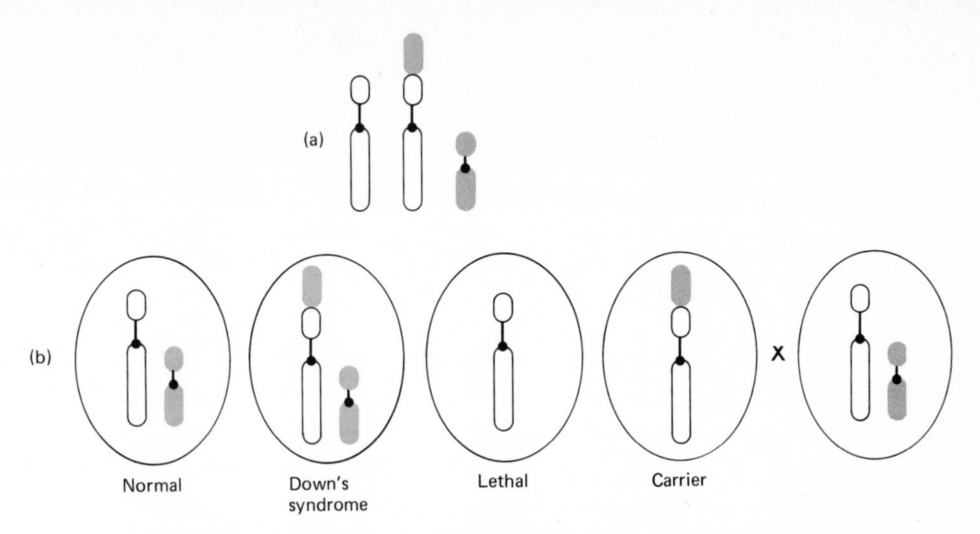

Figure 6-27. (a) The 15 and 21 chromosomes of a translocation carrier. (b) Gametes produced, which when fertilized with a normal gamete result in the four possible types of offspring.

PROBLEMS

1. Assume that a species of cultivated grass has three pairs of chromosomes as its diploid number, as does a related weedy grass. Using different colors to represent the chromosomes of each species, draw the chromosomes of an amphidiploid offspring of a cross between the plants of the two species. Then draw the various stages of meiosis to show how seed may be produced by the hybrid.

2. Assume that inversion exists in the chromosome of an individual heterozygous for five genes *A*, *B*, *C*, *D*, *E* on the chromosome:

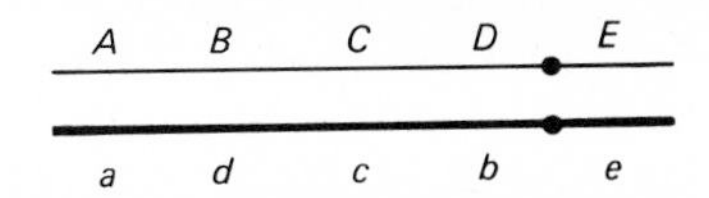

Draw the inversion loop that would allow gene-to-gene pairing during synapsis, writing in all the gene symbols on the chromatids and including the centromeres. Now, assume crossing over between the neighboring nonsister chromatids between the *B* and *C* loci. Draw the resulting combinations of genes and the positions of the centromeres on the four chromatids at the conclusion of crossing over.

3. If XXX females are fertile, diagram the sex chromosome constitutions of offspring from such a female mated to a normal male.

4. Isochromosomes are formed when a centromere splits perpendicularly to the length of a dyad so that homologous arms are now on the same isochromosome (Fig. 6-28). What would you expect to be the effect of such chromosomes when transmitted to offspring?

Figure 6-28. (Problem 4) Schematic illustration of the formation of isochromosomes. The centromere splits perpendicularly to the length of the chromosome, resulting in two chromatids whose arms are repeats of the same gene sequence.

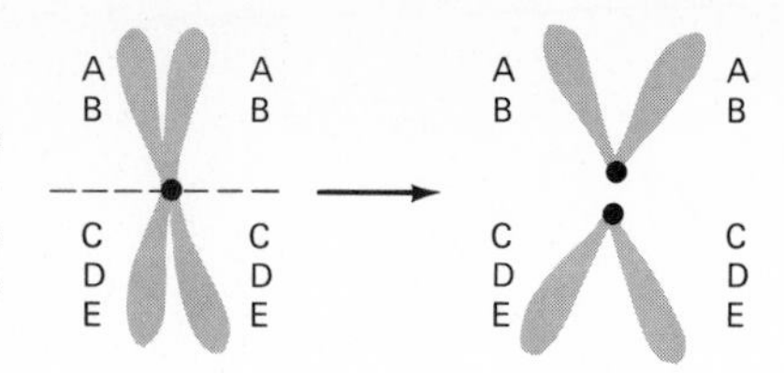

5. Suppose that you were reading an article on a human chromosome mutation, and it was reported to be 6p29. What would this mean in terms of its location?

6. Figure 6-16 illustrates the results of a cross between Poinsettia plants, which are trisomic for p^+ or p. If female F_1 plants with the genotype p^+p^+p were crossed to male plants with the genotype p^+pp, what would be the expected genotypic ratios of the F_2 offspring?

7. What would be the expected genotypic ratios of a reciprocal cross to that in problem 6?

8. Assume that resistance to a fungus in grass is due to the presence of a recessive gene, *f*; the dominant or wild-type allele *F* determines susceptibility. Assume further that a heterozygote for the alleles is induced to become autotetraploid, or *FFff*. What pairings of the homologs can you expect during synapsis of meiosis?

9. What chromosome combinations can you expect in the gametes from problem 8 (assuming all gametes to be 2n in number) and in what ratios?

10. If the autotetraploid plants are selfed, what proportion of the offspring can be expected to be resistant to fungus?

11. Assume that there is a species of New World cotten with twice the number of chromosomes as a related Old World species of cotten. Describe what you would expect to find in cells of a hybrid of these two species during Metaphase I of meiosis.

REFERENCES

The following two papers are reprinted in J. S. Peters, ed., 1959, *Classic Papers in Genetics* (Prentice-Hall, Inc. , Englewood Cliffs, N.J.):

C. B. BRIDGES, 1936. "The Bar 'Gene' a Duplication" in *Science* 83: 210–211.

T. S. PAINTER, 1933. "A New Method for the Study of Chromosome Rearrangements and Plotting of Chromosome Maps" in *Science* 78: 585–586.

BLAKESLEE, A. F., J. BELLING, AND M. E. FAIRHAM, 1923. "Inheritance in Tetraploid Daturas." *Bot. Gag.* 76: 329–373.

BRIGGS, F. N., AND P. F. KNOWLES, 1967. *Introduction to Plant Breeding*. Reinhold, N.Y.

BRIDGES, C. B., E. N. SKOOG, AND J. C. LI, 1938. "Genetical and Cytological Studies of a Deficiency (Notopleural) in the Second Chromosome of *Drosophila Melanogaster*." *Genetics* 21: 788–795.

CASPERSSON, T., L. ZECH, C. JOHANSSEN, AND E. J. MODEST, 1970. "Identification of Human Chromosomes by DNA Binding Fluorescent Agents." *Chromosoma* 30: 215–227.

DANEHOLT, B., AND J. E. EDSTROM, 1967. "The Content of DNA in Individual Polytene Chromosomes of *Chironomus Tentans*." *Cytogenetics* 6: 350–356.

DUPRAW, E. J., 1970. *DNA and Chromosomes*. Holt, Rinehart & Winston, New York.

FORD, C. F., AND J. L. HAMERTON, 1956. "The Chromosomes of Man." *Nature* 178: 1020–1023.

LEFEVRE, G., 1974. "The Relationship Between Genes and Polytene Chromosome Bands." *Ann. Rev. Gen.* 8: 51.

LEJEUNE, J., 1974. "Scientific Impact of the Study of Fine Structure of Chromatids." In T. Caspersson, and L. Zech, eds. *Chromosome Identification*, 23rd Nobel Symposium. Academic Press, New York.

MAYFIELD, J. E., AND J. R. ELLISON, 1975. "The Organization of Interphase Chromatin in Drosophilidae." *Chromosoma* 52, 37–48.

TJIO, J. H., AND A. LEVAN, 1956. "The Chromosome Number of Man." *Hereditas* 42: 106.

WITKIN, H. A., ET AL. 1976. "Criminality in XYY and XXY Men." *Science* 193: 547–555.

YUNIS, J. J., 1976. "High Resolution of Human Chromosomes." *Science* 191: 1268–1270.

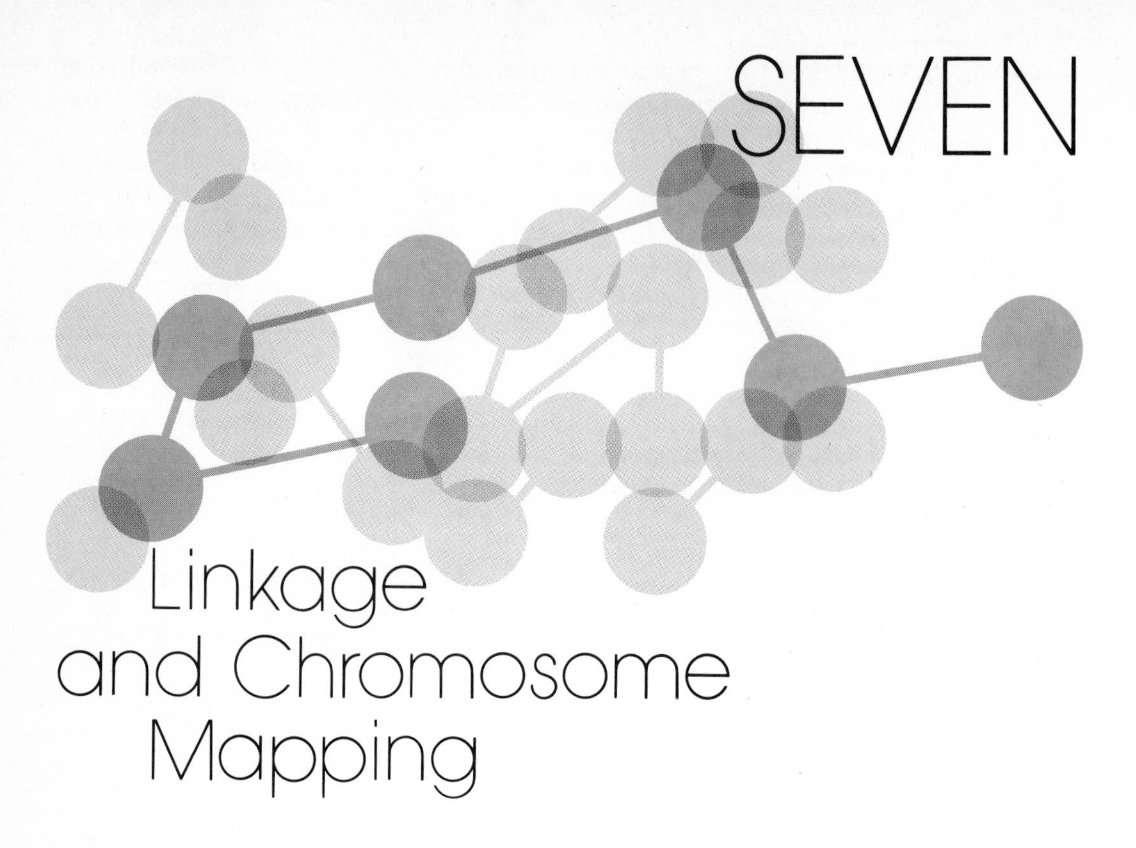

SEVEN

Linkage and Chromosome Mapping

LINKAGE

ALTHOUGH WE HAVE NOT BEEN ABLE TO RELATE THE BANDS IN HUMAN CHROMOSOMES TO GENE LOCI, THERE IS STILL NO DOUBT THAT EACH HUMAN CHROMOSOME, LIKE ALL CHROMOSOMES, MUST BE THE BEARER OF MANY GENES. FIRST PROPOSED IN THE SUTTON-BOVERI HYPOTHESIS, THE CONCEPT OF MANY GENES PER CHROMOSOME IS FIRMLY SUPPORTED BY SOME OF THE WORK WE DISCUSSED IN THE PRECEDING CHAPTER, SUCH AS STUDIES ON POLYTENE CHROMOSOMES. WE CALL THIS CONCEPT LINKAGE, AND THE GENES THAT ARE ON THE SAME CHROMOSOMES ARE REFERRED TO AS LINKED GENES. BESIDES RECOGNIZING THAT LINKAGE MUST EXIST, THE TWO PIONEERS IN THE STUDY OF THE CHROMOSOMAL BASIS OF HEREDITY ALSO REALIZED THAT IF CHROMOSOMES MAINTAIN THEIR MORPHOLOGICAL INTEGRITY DURING CELL DIVISION, THE GENES THEY CARRY MUST BE TRANSMITTED TOGETHER.

This simple conclusion holds two important implications for students of genetics. One is that if two genes are linked, there can be no independent assortment in their transmission. The progeny ratios in crosses involving traits determined by linked genes will differ greatly from the ratios we have encountered up to now, which have involved nonlinked genes. Another major implication is that studying patterns of transmission can aid geneticists in assigning genes to a particular chromosome. At a time when techniques of genetic engineering are being developed, the localization of specific genes takes on great significance.

Coupling and Repulsion

Let us address ourselves, then, to the transmission of linked genes. Figure 7-1 illustrates the difference between the transmission of a pair of linked genes and that of a pair of genes located on different pairs of chromosomes. Note that since we are using the example of homozygotes, the genetic content of gametes would be the same in either case. All gametes would receive *A* and *B* genes, although in the case of linkage, only one chromosome carries the two genes involved, and in nonlinkage, the two genes would be on different, non-homologous chromosomes.

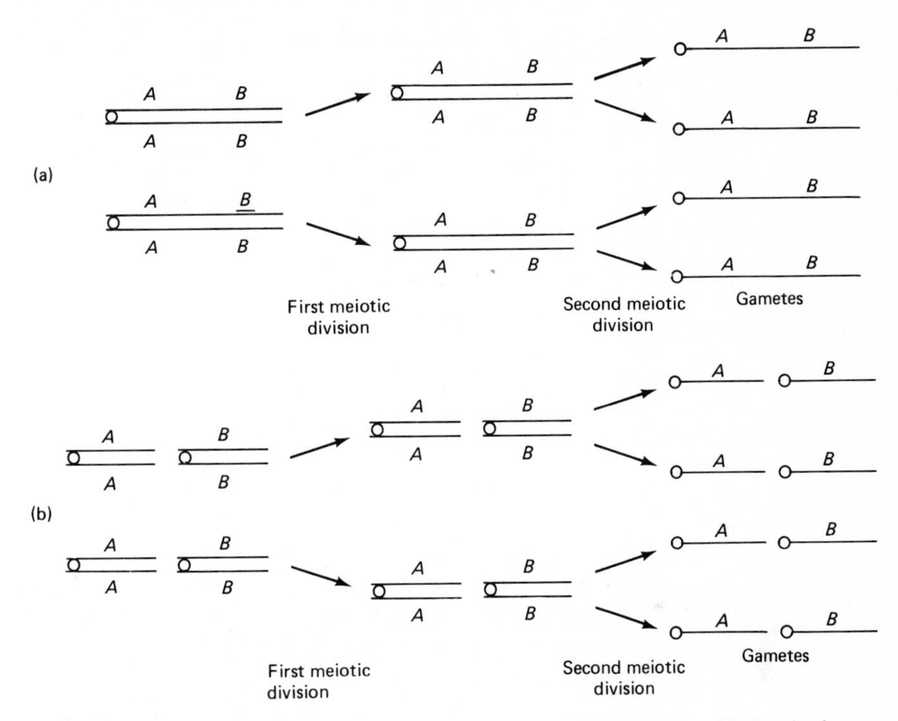

Figure 7-1. (a) Result of meiosis involving two pairs of linked genes. (b) Result of meiosis involving two pairs of unlinked genes. Each chromatid thread is represented by a line. ~~Centromeres are not shown.~~

Suppose, however, that we are dealing with heterozygotes for two pairs of genes, or dihybrids. First recognize that there are two possible ways that linked genes of a dihybrid may be situated on the pair of homologs that carry them; Fig. 7-2 illustrates this point. The two dominant genes, *A* and *B*, may be on different homologs, a situation termed *coupling phase*, or *cis position* (Fig. 7-2a). Alternatively, the dominant genes may be on the same homolog, in which case they are considered to

Figure 7-2. (a) Linked genes of a dihybrid in cis position, or coupling phase; (b) in trans position, or repulsion phase. Individuals with the different combinations are phenotypically the same.

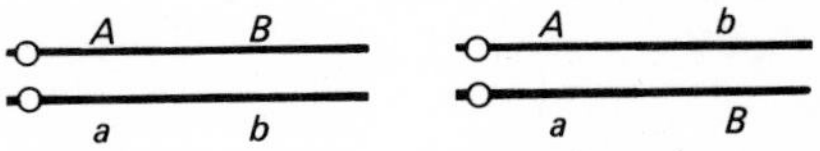

be in the *repulsion phase*, or *trans position* (Fig. 7-2b). It is important to establish whether we are dealing with repulsion or coupling when working with dihybrids of linked genes, because the expected progeny ratio varies depending on the position of the genes. This phenomenon occurs because of the combinations of genes to be found in germ cells following gametogenesis in the two different situations (Fig. 7-3).

For example, if we cross dihybrids whose linked genes are in the coupling phase, the germ cells would carry either *AB* or *ab*. The cross can be diagrammed in the following Punnett Square:

	AB	*ab*
AB	*AABB*	*AaBb*
ab	*AaBb*	*aabb*

If we assume complete dominance in the two traits, the expected phenotypic ratio of the offspring will be 3 (*A-B-*) : 1 (*aabb*).

On the other hand, a cross of dihybrids whose genes are in the repulsion phase would produce a different phenotypic ratio in the offspring:

	Ab	*aB*
Ab	*AAbb*	*AaBb*
aB	*AaBb*	*aaBB*

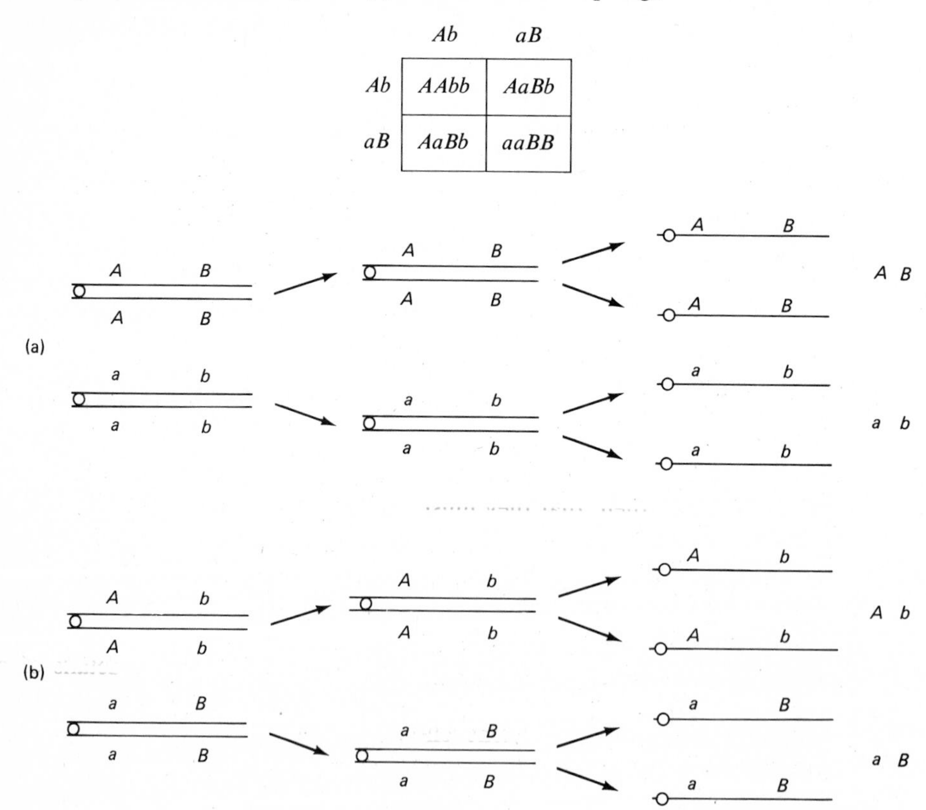

Figure 7-3. Different gene combinations to be expected following meiosis of cells from a dihybrid for linked genes whose genes are (a) in coupling and (b) in repulsion.

Such a cross would produce a 1 (*AAbb*) : 2 (*AaBb*) : 1 (*aaBB*) phenotypic ratio. Work out for yourself the expected results of a cross between a dihybrid whose genes are in the coupling phase, and a dihybrid whose genes are in repulsion. You should arrive at a 1 : 2 : 1 phenotypic ratio, but with different genotypes from the offspring of the cross between dihybrids both in repulsion.

The above results are quite obviously different from the classical 9 : 3 : 3 : 1 phenotypic ratio from dihybrid crosses with independent assortment that we have come to know. The concept of coupling and repulsion, incidentally, also applies to heterozygotes for more than two pairs of linked genes. A trihybrid in coupling, for example, would have all three dominant genes on one homolog, and all three recessive alleles on the other homolog. Note that the law of segregation still applies in that linked genes are separated from their alleles during gametogenesis by the disjoining of homologs in anaphase I.

Geneticists refer to genes transmitted together because of linkage as *linkage groups*. One would expect, then, that there would be as many linkage groups in the cells of a particular species as there are chromosome pairs, and this has been shown to be the case. In *Drosophila*, for example, there are four linkage groups. Evidence relating bands to gene loci, and other evidence from chiasma formation and genetic exchange (see Fig. 1-12) which we shall discuss in detail later in this chapter, further established that genes are situated in a linear manner along the length of the chromosomes.

The Detection of Linkage

Because of the different results to be expected when dealing with linked genes rather than independently assorting genes, it is necessary to establish whether traits under study are in fact located on the same chromosome in order to arrive at expected ratios. How do we detect linkage when dealing with inherited traits not previously studied?

Cytological evidence. There is one linkage group which can be detected by its pattern of transmission of the genes because of cytological differences between males and females. We have already spoken in detail of this group—the X-linked genes. Recall, for example, that both color-blindness and hemophilia are X-linked traits. It follows, then, that they must be linked since they are on the same chromosome, as are all X-linked genes. To date, more than 100 mutations have been assigned to the human X chromosome, some of which may be allelic, and therefore may not represent different loci. More than 40 different genes have been assigned to the *Drosophila* X chromosome.

Another line of evidence comes from purely cytological details of chromosomes, as we discussed in Chapter 6. A number of genes have been assigned to a specific chromosome in polytene chromosomes of Diptera because mutations at these loci have produced visible changes in the chromosomes, such as in size or in the banding pattern. Such assignments are referred to as *cytological maps*. The Bar mutation involving duplications of bands on the X chromosome that we spoke of earlier is one example. Bridges also found in *Drosophila* a mutation called *Noto-*

pleural which causes abnormal bristles, affects viability and fertility, and was found to be due to the loss of a portion of one arm of the second chromosome.

One of the first examples of thus assigning a gene to a particular linkage group in humans was reported by Victor McKusick. A student, Roger Donahue, was studying a karyotype of his (Donahue's) own cells and detected an unusually long chromosome of the A group which they called chromosome #1. A subsequent study of the karyotypes of the Donahue family, revealed that the transmission of this long chromosome correlated with the transmission of the Duffy blood type. Thus the gene determining this blood type is assumed to be on linkage group 1 in humans.

At the time of this study, very few other traits could be correlated with the transmission of a particular linkage group, because of the ambiguity of karyotyping. Similar family studies have identified the positions of only a few genes (McKusick and Ruddle 1977). Now that banding patterns have been established in human chromosomes, however, and as our techniques allow increasingly detailed study of the bands, we can expect to find data accumulating on the linkage groups of humans.

Genetic evidence. The first detection of linkage came from studies by transmission geneticists who obtained aberrant ratios when crossing dihybrids. For example, Bateson and Punnett were analyzing the inheritance of flower color and pollen grain shape in the sweet pea. It had already been established that the trait Purple Flowers (*P*) is completely dominant over Red Flowers (*pp*), and that Long Pollen Grains (*L*) is completely dominant over Round Grains (*ll*). They crossed homozygous dominant plants with homozygous recessive plants and obtained an F_1 generation that was uniformly dominant for the two traits. They then produced an F_2 generation, expecting to obtain a ratio of 9 purple/long : 3 purple/round : 3 red/long : 1 red/round. Their actual results were consistently similar to the following data from one of their experiments:

296 Purple/Long
19 Purple/Round
27 Red/Long
85 Red/Round

The ratios were actually closer to a 9 : 1 : 1 : 3 than the expected 9 : 3 : 3 : 1.

At the time, Bateson and Punnett did not have an explanation for this phenomenon, but it has since been established that these two genes are linked. If we ignore for the time being the two smallest classes, the 3 : 1 ratio of doubly dominant progeny to doubly recessive ones indicates that the genes of the dihybrids were in the cis position (see again p. 188).

It is also possible to arrive at this conclusion just by considering the crosses that were made. The parental cross involved homozygous dominant plants and homozygous recessive plants. The chromosomes of the parental plants can be diagrammed as follows:

A B   a b
A B   a b

Since the F_1 generation inherited a chromosome from each parent, the genes for the two traits must of necessity be in the cis position, or $\dfrac{A \quad B}{a \quad b}$. Note that when analyzing crosses involving linkage, we can no longer simply write the gene symbols, but must draw the chromosomes with the position of the alleles on them, to arrive at the expected offspring of the crosses. (The two smallest classes obtained in Bateson and Punnett's experiments were the result of crossing-over, which we shall discuss later in the chapter.)

Establishing linkage. Suppose that you wished to establish whether two autosomal genes were linked. What crosses would you make, and what results could you expect? Let us use two mutations in mice to illustrate this technique. You are already familiar with the complex *T* locus mentioned in Chapter 4. A new mutation which appeared one day in a colony of Brachyury mice was later found to be a recessive lethal mutation, *muscular dysgenesis* (*mdg*). The gene in the homozygote interferes with normal skeletal muscle development and produces other abnormalities. Among the experiments carried out with this new mutation was a series of crosses to establish whether or not *mdg* is linked to the *T* locus now established to be on linkage group 17 of mice.

We can start with the arbitrary assumption that there is linkage between the two loci. A mating of a Brachyury (short-tailed) mouse with a heterozygote for the *mdg* gene based on this assumption can be diagrammed as follows:

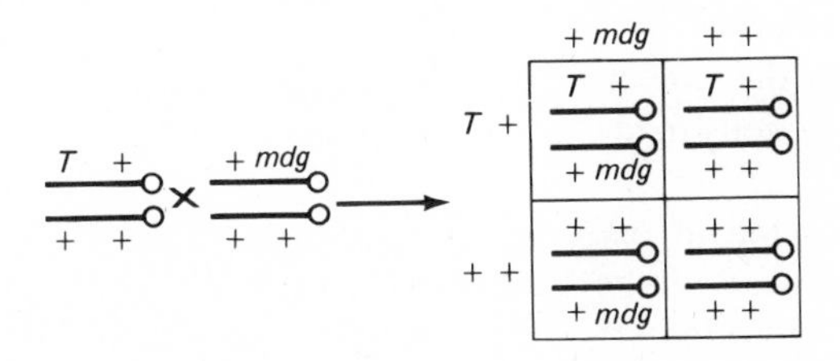

As you can see, half the progeny which are short-tailed can be expected to carry the *mdg* gene also.

We then mate the short-tailed mice to normal-tailed heterozygotes for *mdg*, to distinguish carriers of *mdg* by the presence of mutant progeny in their litters. If the Brachyury mice are also carriers of *mdg*, the cross would produce the following results:

T + / + mdg × + + / + mdg →

	T +	+ mdg
+ +	T + / + +	+ mdg / + +
+ mdg	T + / + mdg	+ mdg / + mdg

From this cross, one can expect that all *mdg* homozygotes will be normal-tailed. Conversely, all Brachyury mice will be otherwise phenotypically normal, as *mdg* is a completely recessive gene.

On the other hand, if the two genes are not linked, different results can be expected from the same crosses:

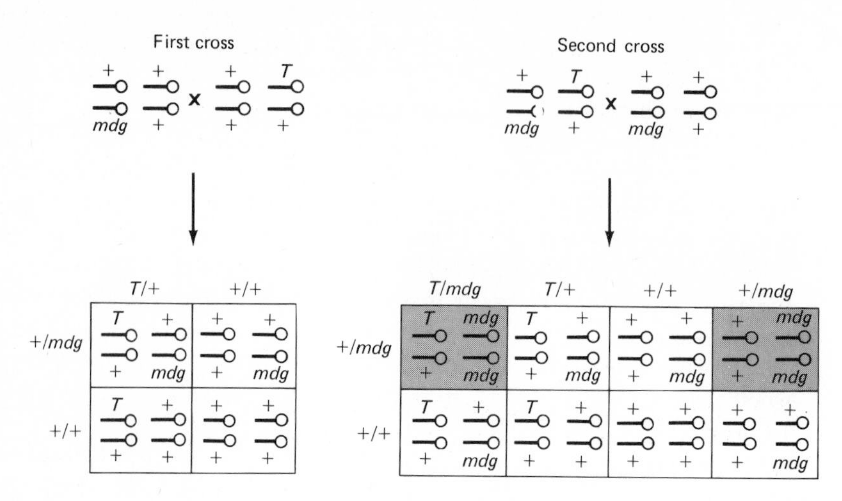

If independent assortment is occurring, we can expect both Brachyury and normal-tailed *mdg* homozygotes in the progeny of the second cross. When the actual experiments were carried out, Brachyury and normal-tailed *mdg* homozygotes were indeed observed in the expected ratios (shaded squares). If we compare our observations to the predictions made for linked and unlinked genes, we find that the loci are not linked.

This kind of reasoning is frequently used by geneticists to detect linkage or nonlinkage. One of the limitations to this technique, however, is that where nonlinkage is indicated, the experiment provides no information as to which other linkage group is involved. The preceding example was used because we know that the *T* locus is on linkage group 17. The *mdg* locus was found not to be linked; therefore, all we know is that it is not on chromosome 17. Tests with genes on other linkage groups will be necessary, but as the mouse has 20 pairs of chromosomes, you can see the amount of work needed to establish linkage of autosomal genes. There are laboratories that are equipped to do this kind of work, and they have established linkage of various genes to all the chromosomes.

Drosophila, on the other hand, are easier to work with in establishing linkage of genes because they possess only 4 pairs of chromosomes. A well-known series of crosses that can be carried out to find the specific linkage group to which a gene might belong, uses a balanced lethal system in a special strain of *Drosophila*. (The establishment of the linkage group to which a particular gene belongs is often through the discovery of a deletion or loss of a band associated with the mutant phenotype, as we mentioned previously. The gene then can serve as a marker for crosses to determine linkage with mutations for which no deletion or other cytological alterations have been found.)

This strain actually is heterozygous for four dominant mutations: *Curly* (*Cy*) and *Plum* (*Pm*) are known to be on the second chromosome, and *Dichaete* (*D*) and *Stubble* (*Sb*) are on the third pair of chromosomes. The *Curly* gene causes wing tips to curl up; *Dichaete* causes the wings to be spread wider apart than normal; *Plum* changes the color of the eyes; and *Stubble* causes scutellar bristles to be abnormally short. Because of inversions, there is also no crossing-over in chromosomes 2 and 3 in this strain of flies. The second and third chromosomes of this strain can thus be diagrammed as follows:

Cy +   *D* +
\+ *Pm*   + *Sb*

This is a balanced lethal system because a cross of flies of this strain causes death in progeny homozygous for any of the four genes. Consequently, only heterozygotes survive:

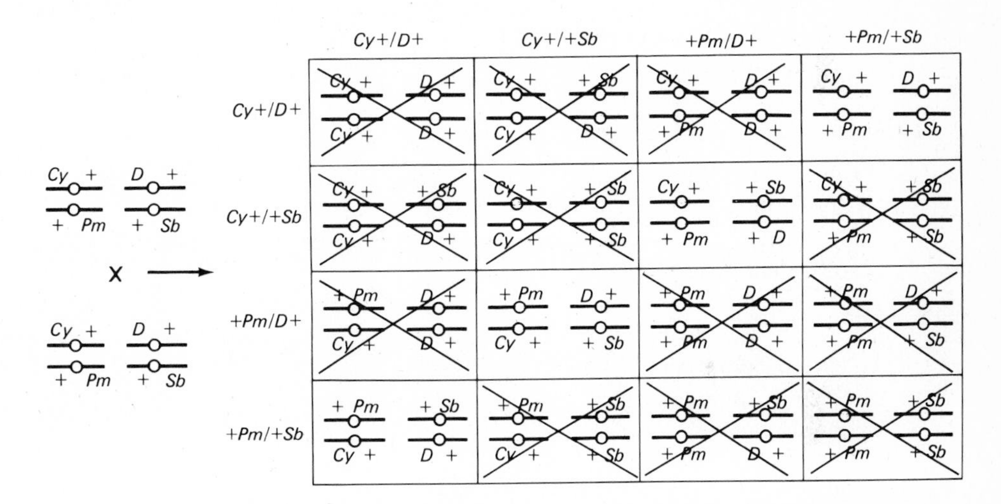

Assume that we are interested in establishing the linkage group for a recessive mutation called *ebony* (*eb*), which causes flies homozygous for the gene to have unusually dark body color. Just two experimental crosses with the *CyPm*/*DSb* strain and the mutant strain will suffice to obtain the desired information. The first cross will be *CyPm*/*DSb* males by *ebeb* females. Results of such a cross are as follows:

85 Curly/Dichaete
93 Curly/Stubble
89 Plum/Dichaete
82 Plum/Stubble

There were approximately equal numbers of males and females in the different classes. None of the progeny were ebony in phenotype.

The above result immediately eliminates one of the *Drosophila* chromosomes as the linkage group for the *ebony* gene—namely, *ebony* is not X-linked. If it were, all males of the first cross should exhibit the *ebony* trait. (It is also not Y-linked. Why?) So, *ebony* is autosomal, but the results do not tell us to which autosomal linkage group *ebony* belongs. For this, we need another cross.

The second cross is between males of any of the four progeny classes of the first cross (for example, the Curly/Stubble class) with an ebony female. You should recognize this as essentially a backcross. But to arrive at expected offspring of this cross, we have to assume linkage of the *ebony* locus with either the second or third chromosome, so let us arbitrarily assume linkage with the second chromosome.

If we assume linkage with the second chromosome, we would expect only two classes of ebony offspring: normal for the other four loci, or those with Stubble Bristles. Problem 7-6 asks you to diagram the crosses showing why you would expect these two classes of ebony offspring.

However, when the crosses were actually performed, the ebony offspring of the second cross were either normal, or had Curly Wings. If you diagram the chromosomes assuming linkage to the third chromosome (Problem 7-7), you will see that these results are compatible with those expected if ebony is located on the third chromosome.

Detection of Linkage in Humans

From the preceding discussion, it is obvious that experimental crosses with many offspring must be made in order to detect linkage of genes. This is, of course, impossible in the study of human genes, so details of linkage in humans are much less well known than in *Drosophila* or mice.

Attempts to establish linkage in humans can follow several different lines of approach. We have already described one approach, namely, the assignment of loci to chromosomes on the basis of cytological studies (p. 189). As our ability to discern the details of human chromosome structure improves, there will likely be a corresponding increase in the identification of gene function with the banding pattern.

By patterns of transmission. Human geneticists also use pedigree analysis or patterns of transmission to detect whether two loci might be linked. This approach, however, is very difficult because usually such a statistically small number of family members is involved. For example, let us assume that a study has been made of a family for the transmission of the ABO blood groups and of an abnormal dominant condition, known as Nail-Patella syndrome, which affects fingernails and kneecaps. The father has type A blood and the Nail-Patella syndrome; the mother has type B blood and is normal:

	♂		♀	
Parents	$I^A iNn$	×	$I^B inn$	
Children	$I^A iNn$	$iinn$	$I^A I^B Nn$	$I^B inn$

The pattern of transmission indicates that the *ABO* and the *N* loci are linked. Further, it appears that genes are in the cis position in the dihybrid father, because whenever the A gene is inherited, so is the N mutation. We can then draw the chromosomes of the members of this family as follows:

Parents $I^A N$ / in × $I^B n$ / in

Children $I^A N$ / in — in / in — $I^A N$ / $I^B n$ — in / $I^B n$

However, if for the same family we assume nonlinkage, the following Punnett Square shows that the four types found in the progeny can be obtained this way also:

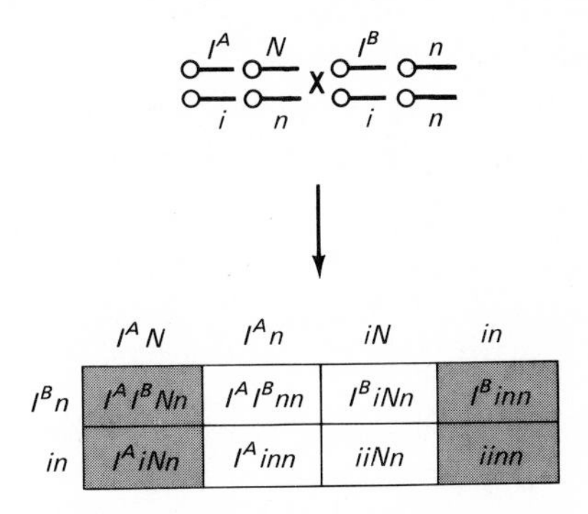

	$I^A N$	$I^A n$	iN	in
$I^B n$	$I^A I^B Nn$	$I^A I^B nn$	$I^B iNn$	$I^B inn$
in	$I^A iNn$	$I^A inn$	$iiNn$	$iinn$

The shaded boxes represent gene combinations determining the phenotypes observed in the children. Although not all the possible combinations were found in the children, remember that we are dealing with small numbers. The absence of these genotypes does not prove that they would not appear among larger numbers of offspring. In addition, one has to consider the possibility of crossing-over between linked genes. We shall discuss this phenomenon in detail shortly, but for now, recall that crossing-over results in new combinations of genes on the homologs. Therefore, what may appear as independent assortment may actually involve linked genes separated by crossing-over. Figure 7-4 illustrates the results of crossing-over in this situation.

In a small percentage of gametes of the dihybrid there would be the expected combinations of genes $I^A N$ and *in*, known as *parental* types, and new combinations iN and $I^A n$, which we refer to as *recombinant* types. The progeny would not be distinguishable from those resulting from nonlinkage. In experimental animals with large numbers of progeny the difference can be detected by analysis of data, as you will see; but in humans, with so few offspring and no way to test with experimental crosses, this is not possible.

Occasionally, however, families can be studied for the same traits and favorable pedigrees drawn up to allow fairly conclusive analysis of linkage between genes. The ABO and Nail-Patella loci, in fact, have been found to be linked. Table 7-1 lists some other linkages found in this way.

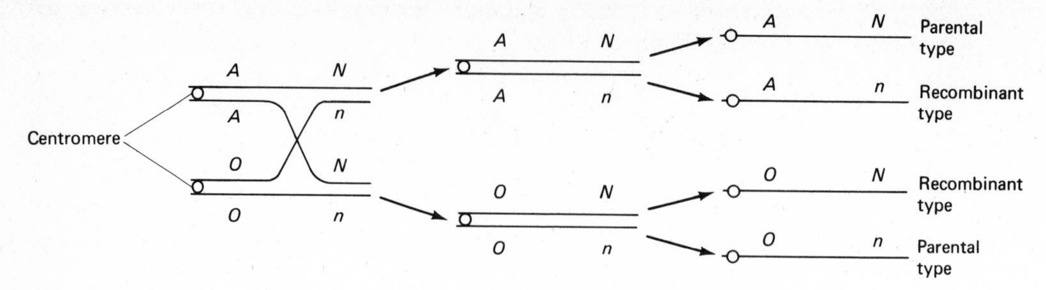

Figure 7-4. Results of crossing-over in cells of a dihybrid for linked genes. New combinations of genes, as well as the parental combinations, are found in the germ cells.

Table 7-1 Some Linked Genes in Humans

Duffy Blood Type—Cataract locus
Transfusion locus—Blood Cholinesterase locus
Albumin locus—Serum G C Protein locus
MNSs Blood Type—Sclerotyplosis (skin disease)
Beta Hemoglobin locus—Delta Hemoglobin locus
ABO locus—Nail–Patella locus—Adenylate Kinase
Rh locus—Elliptocytosis locus—6 Phosphogluconite Dehydrogenase
Lutheran Blood Group—Secretor locus—Myotonic Dystrophy

For situations in which not enough data are available to indicate linkage unambiguously, geneticists resort to mathematical tools. Described in McKusick and Ruddle (1977), these are too complex for further discussion here.

Somatic cell fusion. Obviously, linkage analysis using human pedigrees has its limitations. A technique developed in the early 1960s has, however, contributed to our studies by taking an entirely different approach. A number of laboratories (see references to Ephrussi and to Harris) found that they could cause somatic cells of mammals to fuse when grown in tissue culture. They began by using cancer cells and then found that normal somatic cells could also be fused if Sendai viruses, inactivated by exposure to ultraviolet irradiation, were introduced into the cultures. The presence of the virus apparently causes changes in the cell membrane, allowing the cells to fuse into one giant cell containing the chromosomes from both cells.

This was an important technological breakthrough, because although biologists had been studying cells cultured in vitro for many decades, the cells usually remain frozen in place, in contact but separate, a phenomenon known as *contact inhibition*. Furthermore, it was found that if hybrid cells were created by fusing cells from certain different species, such as rodent cells and human cells, for unknown reasons human chromosomes were differentially lost from the daughter cells during mitosis. Scientists seized upon this phenomenon as an opportunity to relate specific genes to specific chromosomes.

The first example of the use of this technique for genetic purposes was an experiment by Weiss and Green in 1967. They combined normal human cells with cells from a strain of mice which were known to be missing an enzyme, thymidine kinase, important in the metabolism of one of the nitrogenous bases found in DNA. They placed the cells in what has become known as HAT medium, a nutrient broth containing hypoxanthine, aminopterin, and thymidine. Ordinarily, the mutant mouse cells would not be able to survive in HAT medium because of their inability to convert thymidine into thymine; however, normal human cells could, because our cells make the same enzyme, thymidine kinase.

Although the enzymes perform the same metabolic role, mouse thymidine kinase can be distinguished from human thymidine kinase by its heat sensitivity and by a technique called *electrophoresis*. In electrophoresis, substances such as purified enzyme preparations are placed on a solid substrate such as paper or polyacrylamide gel, which has been saturated with an aqueous buffer. The substrate is then exposed to an electrical field, and the enzyme or substance under study moves toward the anode or cathode at speeds varying with the charge of molecules themselves. The speed with which a substance moves is called its *electrophoretic mobility*. The gel or paper can be stained to find the locations of the molecules and thus their comparative rate of migration toward the poles. In this manner, it was possible to determine that the enzyme produced in the hybrid cells was actually human, rather than mouse, thymidine kinase.

When the experiment was carried out, the mouse cells that did not undergo fusion with human cells were unable to survive, but the giant hybrid cells did. As division of these cells proceeded, a number of the giant cells perished, presumably because as division proceeded, they had lost the chromosome carrying the gene determining the enzyme. Some of the surviving cells were then found to have only one human chromosome left, assumed then to be the one carrying the gene determining the enzyme. Later, Migeon and Miller identified the chromosome carrying the gene for thymidine kinase to be #17. An increasing number of genes have been located on specific chromosomes as such electrophoretic studies continue.

Incidentally, there is a term frequently associated with somatic cell fusion which we can introduce here, and that is *cloning*. The cells which retained the necessary chromosome and survived would produce daughter cells which also received the same chromosome complement, resulting in a group of cells that are genetically identical, at least with respect to the chromosome in question. We refer to such a group as a *clone*. Later, we shall describe cloning experiments dealing not only with identical cells, but with genetically identical organisms.

Cytological evidence for linkage. Recently, geneticists have begun to combine somatic fusion with cytological studies to ascertain more specifically the location of a gene on the chromosome. For example, some scientists have studied abnormal human chromosomes obtained spontaneously or by irradiation of human cells. The cells were then fused with hamster cells, and the presence or absence of enzymes was correlated with the addition or loss of chromosomal material. In this way, Burgerhout and his coworkers were able to trace the approximate location of seven genes on chromosome #1.

Still other experiments utilizing translocations have been carried out to locate markers on the X chromosome of human cells (see references to Pearson *et al.*, and Risciutti and Ruddle). The specific translocation involved here was the long arm of the X chromosome translocated to an autosome. In hybrid cells formed by fusing human cells containing this abnormal chromosome to mouse cells, the presence of enzymes determined by certain X-linked genes gave good indication that these genes were located on the long arm of the X chromosome.

However, the traits determined by genes that can be studied in this manner will be limited to the various enzymes and cell products which can be detected in cell cultures. There are countless traits which cannot be ascertained in this manner, as they are found in the organism as a whole, rather than within individual cells. There would be no way to study the gene for attached earlobes, for example. Also, polygenic traits and others depending on the interaction of different genes on different chromosomes would be difficult to assign to specific chromosomes. Nonetheless, cell fusion studies have made significant contributions to the list of genes that have been identified with specific chromosomes. Figure 7-5 presents a summary of human genes that have thus far been identified with regard to linkage group and position on the chromosome.

CHROMOSOME MAPPING

Once linkage between two genes is detected, it then becomes of interest to establish the relative positions of the linked genes on the chromosomes. For example, we said that *Curly* and *Plum* were on the second pair of chromosomes in *Drosophila* but we have not indicated their position on that chromosome. It is important to know not only which chromosome the genes are on, but also how they are situated relative to each other. Incidentally, the term *synteny* is used for genes known to be linked, but whose positions have not yet been determined.

THE ASSIGNMENT OF RELATIVE POSITIONS OF GENES ON CHROMOSOMES IS KNOWN AS CHROMOSOME MAPPING. AS WE MENTIONED BEFORE, CYTOLOGICAL STUDIES HAVE LED TO THE IDENTIFICATION OF SOME GENES IN *Drosophila*. SUCH CYTOLOGICAL MAPPING HAS PROVIDED INFORMATION ON RELATIVELY FEW GENES, HOWEVER. BY FAR THE BULK OF CHROMOSOME MAPPING HAS BEEN ACCOMPLISHED BY EXPLOITING A PHENOMENON WE HAVE TOUCHED UPON SEVERAL TIMES, NAMELY, CROSSING-OVER. MAPS ESTABLISHED IN THIS MANNER ARE KNOWN AS GENETIC MAPS.

Crossing-Over

Recall that the phenomenon of crossing-over is believed to be due to an actual exchange of genetic material between chromatids of homologs in the early stages of meiosis. Because of the formation of chiasma during and after pachytene of prophase I, it was thought that the exchange occurred at that point. There is, however, no conclusive evidence for this and actual genetic exchange may occur in earlier stages.

Genetic evidence. One of the first encounters of crossing-over in genetic experiments occurred in Bateson and Punnett's studies that we discussed on p. 190. Recall that the phenotypic ratio they obtained in a dihybrid cross was closer to

9 : 1 : 1 : 3 than the expected 9 : 3 : 3 : 1. Our analysis indicated that the dihybrids were in the coupling phase, which accounted for the two largest phenotypic classes, *A-B-* and *aabb* (see again p. 181). T. H. Morgan once again supplied the theory to account for the two small classes of offspring and their unexpected phenotype combinations, namely, *A-bb* and *aaB-*. Morgan realized that the results could be explained if one assumed an exchange of genetic material during gametogenesis that resulted in new combinations of genes on the chromosomes. Figure 7-6 illustrates the results of crossing-over in the plant that Bateson and Punnett were studying.

In addition, Morgan's own studies on X-linked genes also yielded results that indicated crossing-over. In one series of experiments, for example, Morgan crossed female flies that were homozygous for the recessive X-linked trait *white eyes*, and had normal wings, to males that were red-eyed, and possessed the recessive X-linked trait *rudimentary wings*. In the F_2 generation, he found white-eyed rudimentary-winged flies. Figure 7-7 analyzes these results, which are consistent with an assumption of crossing-over between the two loci in gametogenesis of the F_1 females. (Note that the genetic exchange of which we speak is between chromatids of the two homologs, not between sister chromatids. Crossing-over between sister chromatids is difficult to detect, as they are genetically identical.)

In 1913, Sturtevant carried the analysis even further. In studies on six *X*-linked traits he found evidence that the frequency of cross-over events differs between the six loci. Some had a very low frequency of cross-over, others a far higher frequency. He postulated that if genes are situated in a linear manner, and if crossing-over is the result of an exchange of genetic material between homologs, it stands to reason that the farther apart two genes are, the more likely there is to be crossing-over between them. Conversely, the closer they are, the less likely there is to be crossing-over. One can make the analogy that there are more places to cross a long stretch of road than a short stretch of road.

Cytological evidence. In a milestone report in 1931, Harriet Creighton and Barbara McClintock gave cytological evidence correlated with genetic data to show that crossing-over is in fact due to an exchange of genetic material. They did their work on maize, and found a particular strain with a distinctly noticeable difference in the morphology of a pair of homologs. One chromosome of pair 9 possessed a knob at one end that was absent in its homolog. Furthermore, they found that an extra piece of chromosomal material (shown to be part of chromosome 8) was attached to the other end of the knobbed chromosome, so that it was also noticeably longer than the normal homolog (Fig. 7-8). They used the term *interchanged* to describe the chromosome with extra material in it. In modern terminology, however, this would be considered a translocation.

Figure 7-5. A diagrammatic synopsis of the gene map of the human chromosomes. The banding patterns and numbering are those given in the 1975 supplement to the report of the Paris Conference. An assignment is considered confirmed if found in two laboratories or several families; it is considered provisional if based on evidence from one laboratory or family. Additional assignments considered tentative (evidence not as strong as that for "provisional") or based on inconsistent results of different laboratories are not included. (From V. A. McKusick and F. H. Ruddle, 1977. *Science* 196 : 398–399; copyright 1977 by the American Association for the Advancement of Science.)

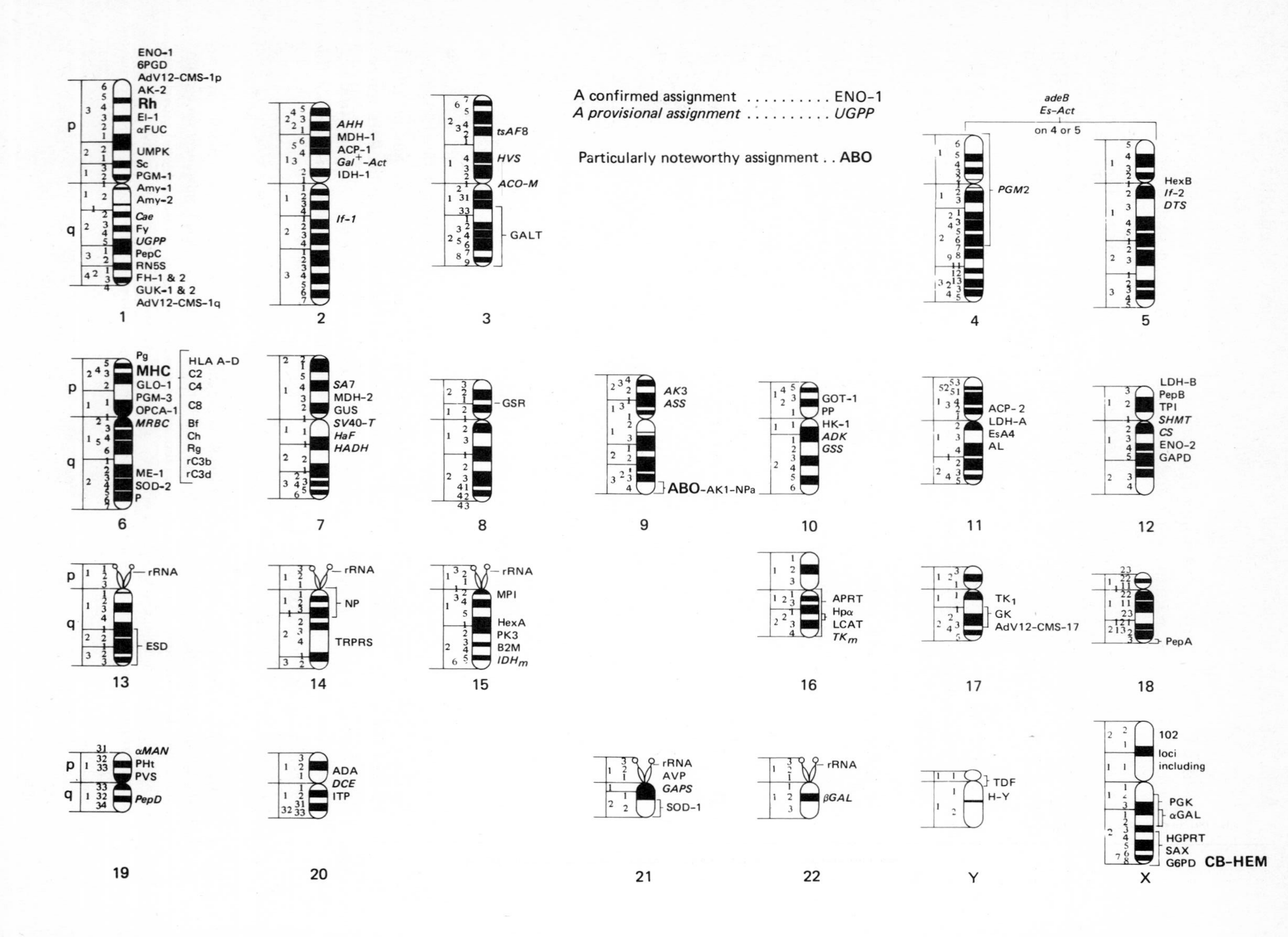

A confirmed assignment ENO-1
A provisional assignment .... UGPP
Particularly noteworthy assignment . . ABO
1
ENO-1
6PGD
AdV12-CMS-1p
AK-2
Rh
El-1
αFUC
UMPK
Sc
PGM-1
Amy-1
Amy-2
Cae
Fy
UGPP
PepC
RN5S
FH-1 & 2
GUK-1 & 2
AdV12-CMS-1q
2
AHH
MDH-1
ACP-1
Gal+-Act
IDH-1
If-1
3
tsAF8
HVS
ACO-M
GALT
4
PGM2
adeB
Es-Act
on 4 or 5
5
HexB
If-2
DTS
6
Pg
MHC
GLO-1
PGM-3
OPCA-1
MRBC
ME-1
SOD-2
P
HLA A-D
C2
C4
C8
Bf
Ch
Rg
rC3b
rC3d
7
SA7
MDH-2
GUS
SV40-T
HaF
HADH
8
GSR
9
AK3
ASS
ABO-AK1-NPa
10
GOT-1
PP
HK-1
ADK
GSS
11
ACP-2
LDH-A
EsA4
AL
12
LDH-B
PepB
TPI
SHMT
CS
ENO-2
GAPD
13
rRNA
ESD
14
rRNA
NP
TRPRS
15
rRNA
MPI
HexA
PK3
B2M
IDHm
16
APRT
Hpα
LCAT
TKm
17
TK1
GK
AdV12-CMS-17
18
PepA
19
αMAN
PHt
PVS
PepD
20
ADA
DCE
ITP
21
rRNA
AVP
GAPS
SOD-1
22
rRNA
βGAL
Y
TDF
H-Y
X
102
loci
including
PGK
αGAL
HGPRT
SAX
G6PD
CB-HEM
p
q

Human chromosome map. The symbols are explained in the following key.

ABO	ABO blood group (chr. 9)
ACO	Aconitase, mitochondrial (chr. 3)
ACO-S	Aconitase, soluble (chr.9)
ACP-1	Acid phosphatase- (chr.2)
ACP-2	Acid phosphatase-2 (chr. 11)
ADA	Adenosine deaminase (chr. 20)
adeB	FGAR amidotransferase (chr. 4 or 5)
ADK	Adenosine kinase (chr. 10)
AdV12-CMS-1p	Adenovirus-12 chromosome modification site-1p (chr. 1)
AdV12-CMS-1q	Adenovirus-12 chromosome modification site-1q (chr.1)
AdV12-CMS-17	Adenovirus-12 chromosome modification site-17 (chr. 17)
AHH	Arylhydrocarbon hydroxylase (chr. 2)
AK-1	Adenylate kinase-1 (chr. 9)
AK-2	Adenylate kinase-2 (chr. 1)
AK-3	Adenylate kinase-3 (chr. 9)
AL	Lethal antigen: 3 loci (a1, a2, a3) (chr. 11)
Amy-1	Amylase, salivary (chr. 1)
Amy-2	Amylase, pancreatic (chr. 1)
ASS	Argininosuccinate synthetase (chr. 9)
APRT	Adenine phosphoribosyltransferase (chr. 16)
AVP	Antiviral protein (chr. 21)
Bf	Properdin factor B (chr. 6)
β2M	β2-Microglobulin (chr. 15)
C2	Complement component-2 (chr. 6)
C4	Complement component-4 (chr. 6)
C8	Complement component-8 (chr. 6)
Cae	Cataract, zonular pulverulent (chr. 1)
CB	Color blindness (deutan and protan) (X chr.)
Ch	Chido blood group (chr. 6)
CS	Citrate synthase, mitochondrial (chr. 12)
DCE	Desmosterol-to-cholesterol enzyme (chr. 20)
DTS	Diphtheria toxin sensitivity (chr. 5)
El-1	Elliptocytosis-1 (chr. 1)
EMS	Echo 11 sensitivity (chr. 19)
ENO-1	Enolase-1 (chr. 1)
ENO-2	Enolase-2 (chr. 12)
Es-Act	Esterase activator (chr. 4 or 5)
EsA4	Esterase-A4 (chr. 11)
ESD	Esterase D (chr. 13)
FH-1 & 2	Fumarate hydratase-1 and 2 (S and M) (chr. 1)
αFUC	Alpha-L-fucosidase (chr. 1)
Fy	Duffy blood group (chr. 1)
Gal^+-Act	Galactose + activator (chr. 2)
αGAL	α-Galactosidase (Fabry disease) (X chr.)
βGAL	β-Galactosidase (chr. 22)
GALT	Galactose-1-phosphate uridyltransferase (chr. 3)
GAPD	Glyceraldehyde-3-phosphate deydrogenase (chr. 12)
GAPS	Phosphoribosyl glycineamide synthetase (chr. 21)
Gc	Group-specific component (chr.4)
GK	Galactokinase (chr. 17)
GLO-1	Glyoxylase I (chr. 6)
GOT-1	Glutamate oxaloacetic transaminase-1 (chr. 10)
G6PD	Glucose-6-phosphate dehydrogenase (X chr.)
GSR	Glutathione reductase (chr. 8)
GSS	Glutamate-γ-semialdehyde synthetase (chr. 10)
GUK-1 & 2	Guanylate kinase-1 & 2 (S & M) (chr. 1)
GUS	Beta-glucuronidase (chr. 7)
HADH	Hydroxyacyl-CoA dehydrogenase (chr. 7)
HaF	Hageman factor (chr.7)
HEM_A	Classic hemophilia (X chr.)
HexA	Hexosaminidase A (chr. 15)
HexB	Hexosaminidase B (chr. 5)
HGPRT	Hypoxanthine-guanine phosphoribosyltransferase (X chr.)
HK-1	Hexokinase-1 (chr. 10)
HLA	Major histocompatibility complex (chr. 6)
Hpα	Haptoglobin, alpha (chr. 16)
HVS	Herpes virus sensitivity (chr. 3)
H-Y	Y histocompatibility antigen (Y chr.)
If-1	Interferon-1 (chr. 2)
If-2	Interferon-2 (chr.5)
IDH-1	Isocitrate dehydrogenase-1 (chr. 2)
IDH_m	Isocitrate dehydrogenase, mitochondrial (chr. 15)
ITP	Inosine triphosphatase (chr. 20)
LCAT	Lecithin-cholesterol acyltransferase (chr. 16)
LDH-A	Lactate dehydrogenase A (chr. 11)
LDH-B	Lactate dehydrogenase B (chr. 12)
αMAN	Lysosomal α-D-mannosidase
MDH-1	Malate dehydrogenase-1 (chr. 2)
MDH-2	Malate dehydrogenase, mitochondrial (chr. 7)
ME-1	Malic enzyme-1 (chr. 6)
MHC	Major histocompatibility complex (chr. 6)
MPI	Mannosephosphate isomerase (chr. 15)
MRBC	B-cell receptor for monkey red cells (chr. 6)
NP	Nucleoside phosphorylase (chr. 14)
NPa	Nail-patella syndrome (chr. 9)
OPCA-I	Olivopontocerebellar atrophy I (chr. 6)
P	P blood group (chr. 6)
PepA	Peptidase A (chr. 18)
PepB	Peptidase B (chr. 12)
PepC	Peptidase C (chr. 1)
PepD	Peptidase D (chr. 19)
Pg	Pepsinogen (chr. 6)
PGK	Phosphoglycerate kinase (X chr.)
PGM-1	Phosphoglucomutase-1 (chr. 1)
PGM-2	Phosphoglucomutase-2 (chr. 4)
PGM-3	Phosphoglucomutase-3 (chr. 6)
6PGD	6-Phosphogluconate dehydrogenase (chr. 1)
PHI	Phosphohexose isomerase (chr. 19)
PK3	Pyruvate kinase-3 (chr. 15)
PP	Inorganic pyrophosphatase (chr. 10)
PVS	Polio sensitivity (chr. 19)
Rg	Rodgers blood group (chr. 6)
Rh	Rhesus blood group (chr. 1)
rRNA	Ribosomal RNA (chr. 13, 14, 15, 21, 22)
rC3b	Receptor for C3b (chr. 6)
rC3d	Receptor for C3d (chr. 6)
RN5S	5S RNA gene(s) (chr. 1)
SA7	Species antigen 7 (chr. 7)
SAX	X-linked species (or surface) antigen (X chr.)
Sc	Scianna blood group (chr. 1)
SHMT	Serine hydroxymethyltransferase (chr. 12)
SOD-1	Superoxide dismutase-1 (chr. 21)
SOD-2	Superoxide dismutase-2 (chr. 6)
SV40-T	SV40-T antigen (chr. 7)
TDF	Testis determining factor (Y chr.)
TK_m	Thymidine kinase, mitochondrial (chr. 16)
TK_s	Thymidine kinase, soluble (chr. 17)
TPI	Triosephosphate isomerase (chr. 12)
TRPRS	Tryptophanyl-tRNA synthetase (chr. 14)
tsAF8	Temperature-sensitive (AF8) complementing (chr. 3)
UGPP	Uridyl diphosphate glucose pyrophosphorylase (chr. 1)
UMPK	Uridine monophosphate kinase (chr. 1)

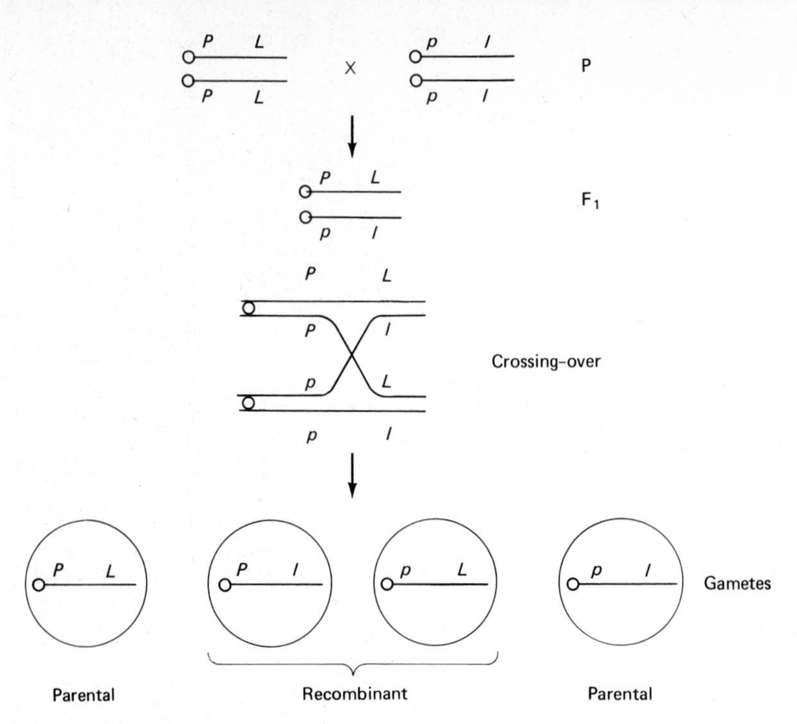

Figure 7-6. Crossing-over in Bateson and Punnett's classical experiment.

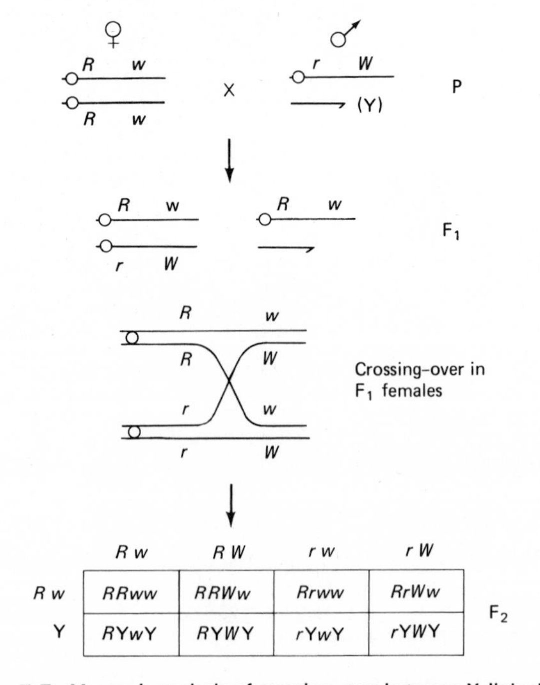

	R w	*R W*	*r w*	*r W*
R w	*RRww*	*RRWw*	*Rrww*	*RrWw*
Y	*RYwY*	*RYWY*	*rYwY*	*rYWY*

Figure 7-7. Morgan's analysis of crossing-over between X-linked genes.

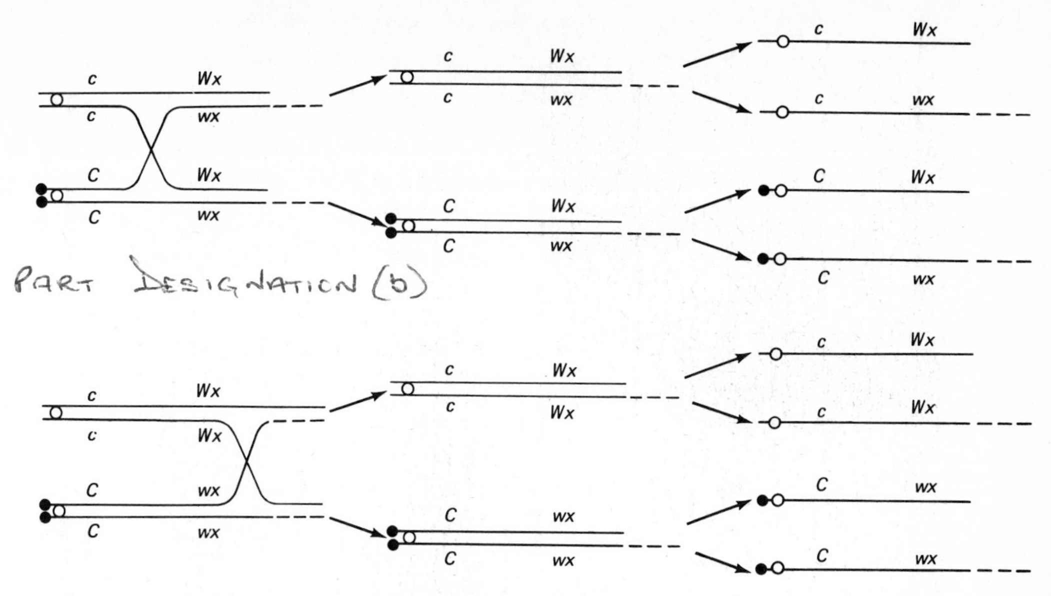

Figure 7-8. Results of crossing-over in maize. Explanation in text.

A plant was found that not only had heteromorphic ninth chromosomes, but also was heterozygous in the repulsion phase for two loci on the ninth chromosomes. These were the loci for *colored aleurone* (*C*) and *waxy endosperm* (*wx*). A cross was made between this plant (*CcWxwx*) and one possessing a normal knobless ninth chromosome whose genotype was *ccWxwx*. We can diagram the chromosomes of this cross in the following way:

c Wx / C wx × c Wx / c wx

Creighton and McClintock correctly reasoned that if crossing-over occurs between the two genes in the heteromorphic pair, recombinant gametes would be of the type illustrated in Fig. 7-8A. The recombinant offspring and the expected morphology of their chromosomes would be as listed in Table 7-2. When the chromosomes of plants with recombinant phenotypes were examined, their morphology indeed correlated with what was expected. For example, *WxWx* or *CcWxwx* progeny should have all chromosomes of normal length and should have a knobbed ninth chromosome, and they did. Note that one of the plants with *c-Wx* kernels had the interchange morphology which we can assume resulted from crossing-over between the *Wx* locus and the end of the chromosome as illustrated in Figure 7-8B.

Two-point Crosses

We will now discuss the manner in which geneticists can use the frequency of crossing-over to establish the distance between genes. Points that we have already mentioned

Table 7-2 Some Progeny of Crosses in Creighton-McClintock Experiment

$$\frac{\text{Knob-}C\text{-}wx\text{-interchanged}}{\text{Knobless-}c\text{-}Wx\text{-normal}} \times \frac{\text{Knobless-}c\text{-}Wx\text{-normal}}{\text{Knobless-}c\text{-}wx\text{-normal}}$$

Plant	*Knobbed or knobless*	*Normal or interchanged*
Class I, *C-wx* kernels		
1	Knobbed	Interchanged
2	Knobbed	Interchanged
Class II, *c-wx* kernels		
1	Knobless	Interchanged
2	Knobless	Interchanged
Class III, *C-Wx* kernels		
1	Knobbed	Normal
2	Knobbed	Normal
Class IV, *c-Wx* kernels		
1	Knobless	Normal
2	Knobless	Normal
3	Knobless	Interchanged
4	Knobless	Normal

From H. B. Creighton, and B. McClintock, 1931. "A Correlation of Cytological and Genetical Crossing-over in Zea Mays." Reprinted in Peters, J. A., ed., 1959. *Classic Papers in Genetics*, p. 159. Prentice-Hall, Inc., Englewood Cliffs. N.J.

relevant to mapping can be summarized as follows: Genes are situated on chromosomes in a linear manner; during gametogenesis there is exchange of genetic material between two homologs; the closer two genes are, the less frequently crossing-over occurs; conversely, the farther apart two genes are, the more frequently crossing-over occurs.

With these points in mind, we can now explore the technique of determining gene distances. The most straightforward way to obtain the desired information is to perform a testcross of a dihybrid with a homozygous recessive. Assume that there is a strain of *Drosophila* homozygous recessive for two mutations, *dachs legs* and *jammed wings*. Let us use the following symbols: $W =$ normal wings, $w =$ jammed wings, $L =$ normal legs, $l =$ dachs legs. If we mate flies of this strain with normal wild-type flies, all F_1 flies will be dihybrids with genes in the cis position. A back-cross of dihybrid females with homozygous males for the two mutations produces the following data:

48	wild-type
42	dachs legs, jammed wings
4	normal wings, dachs legs
6	normal legs, jammed wings
100	Total

From knowledge that the dihybrids were in the coupling phase, we know that the parental types are the wild-type and doubly recessive offspring, and that the two smaller classes are the recombinants. A percentage, called the cross-over frequency, is determined by dividing the total number of recombinant progeny by the total number of progeny. Here the total of the recombinants is 10, so that the frequency of crossing-over between the two linked genes is $\frac{10}{100}$, or 10%. This is equivalent to 10 *map units*. Map units are determined by the frequency of crossing-over, but are only relative measures of distance. It has been found that the cytological distance between bands which have been associated with specific gene functions does not correspond exactly with the calculated map distance based on crossing-over frequency between the genes involved. Also, it is known that if genes *a*, *b*, and *c* are in that order and gene *a* is 6 map units from gene *b*, and gene *b* is 10 map units from gene *c*, crossing-over between genes *a* and *c* may sometimes be more or less than the expected 16%. We shall discuss some of the factors involved later.

Distances between genes can also be determined from dihybrid crosses, though with a bit more effort. Let us return to Bateson and Punnett's classical experiments with the sweet pea plants. Recall that they obtained four classes of offspring in a 9 : 1 : 1 : 3 ratio. The actual data obtained in one cross was the following:

296 Purple/Long
19 Purple/Round
27 Red/Long
85 Red/Round
427 Total

Since both parents are dihybrids (*PpLl*), both would produce parental and recombinant gametes. As we had mentioned before, the dihybrids are again in the cis position, so that we can draw up a Punnett Square for the cross again using the following symbols: *P* = Purple, *p* = Red, *L* = Long, *l* = Round.

	PL	*pl*	*Pl*	*pL*
PL	*PPLL*	*PpLl*	*PPLl*	*PpLL*
pl	*PpLl*	*ppll*	*Ppll*	*ppLl*
Pl	*PPLl*	*Ppll*	*PPll*	*PpLl*
pL	*PpLL*	*ppLl*	*PpLl*	*ppLL*

The difficulty in using a dihybrid cross to determine cross-over frequency is that progeny of three of the phenotypic classes are derived from both parental and recombinant gametes, whereas in a backcross, the progeny are from either parental or recombinant gametes. In a dihybrid cross the only phenotype class which clearly represents progeny from only parental type gametes is the doubly recessive group, *ppll* in the above experiment.

From this one bit of information, we can now proceed to determine the cross-over frequency. The Red/Round group formed $\frac{85}{427}$ of the total offspring, or approximately 20%, .20. If *ppll* individuals are derived from *pl* × *pl* gametes, .20 = *pl* × *pl*, or pl^2. Therefore, the frequency of the *pl* gametes is equal to the square root of .20, or .45. If the frequency of one parental-type gamete is .45, that must also be the frequency of the other parental gamete because of segregation of homologs, so now we know the frequency of *PL* gametes also. Parental-type gametes total .90, or 90%. Therefore, the frequency of recombinant gametes, and the frequency of crossing-over, is 10%. We can now determine the percentage of every gamete and genotype of the cross:

		Parental		Recombinant	
		.45*PL*	.45*pl*	.05*Pl*	.05*pL*
Parental	.45*PL*	.2025*PPLL*	.2025*PpLl*	.0225*PPLl*	.0225*PpLL*
	.45*pl*	.2025*PpLl*	.2025*ppll*	.0225*Ppll*	.0225*ppLl*
Recombinant	.05*Pl*	.0225*PPLl*	.0225*Ppll*	.0025*PPll*	.0025*PpLl*
	.05*pL*	.0225*PpLL*	.0225*ppLl*	.0025*PpLl*	.0025*ppLL*

We can also determine the expected numbers of offspring of a cross, if we have prior knowledge of the distance between the two genes. Assume that two genes, *R* and *S*, are 20 map units apart. From a cross of *RRss* × *rrSS* individuals, we obtain dihybrid offspring. (Note, that the way the cross was set up, the dihybrids would be in the repulsion phase.) In a testcross of the F_1, what phenotypes, and in what numbers, would you expect from a total of 1000 offspring?

The parental types of gametes would be *Rs* and *rS*, recombinant types would be *RS* and *rs*, and in a testcross these would be mirrored in the phenotypes of the offspring. If the two loci are 20 map units apart, we would expect 20% of 1000, or 200, to be of the recombinant genotypes, *R-S-* and *rrSS*, and 800 parental types. Specifically the expected numbers would be:

400	*R-ss*
400	*rrS-*
100	*R-S-*
100	*rrss*
1000	Total

If we were to actually carry out experimental crosses such as the one above, the expected numbers could be used in a chi square test of the actual data obtained for goodness of fit.

Three-point Crosses

So now you know how to determine the distance between two genes. But a point of ambiguity still remains. Suppose you have established that gene *R* is 20 map units away from gene *S* in *Drosophila* (their actual effect on the flies is irrelevant). Suppose further that there is a third marker gene known to be linked to gene *S*, which we'll call *M*. How can we tell whether it is between *R* and *S*, or on the other side of *R* or *S* on the chromosome? (The term *marker* is sometimes used to refer to a mutant gene which serves as a guide in some way because of its distinctive features. We use it here as a marker in linkage studies.)

What we can do is to again set up a testcross of individuals trihybrid for the genes with ones homozygous recessive, and to analyze the data for the information we want. We can do this by breeding organisms to be homozygous dominant and homozygous recessive, crossing them to produce trihybrids in the coupling phase, and then making the testcross.

Figure 7-9 illustrates a very important point which you must learn in order to interpret the meaning of data when dealing with a cross of individuals trihybrid for three linked genes, or what is known as a *three-point cross*. First, there are three categories of cross-over progeny: those from a single cross-over between a gene at one end and the middle gene; those from a single crossover between the middle gene

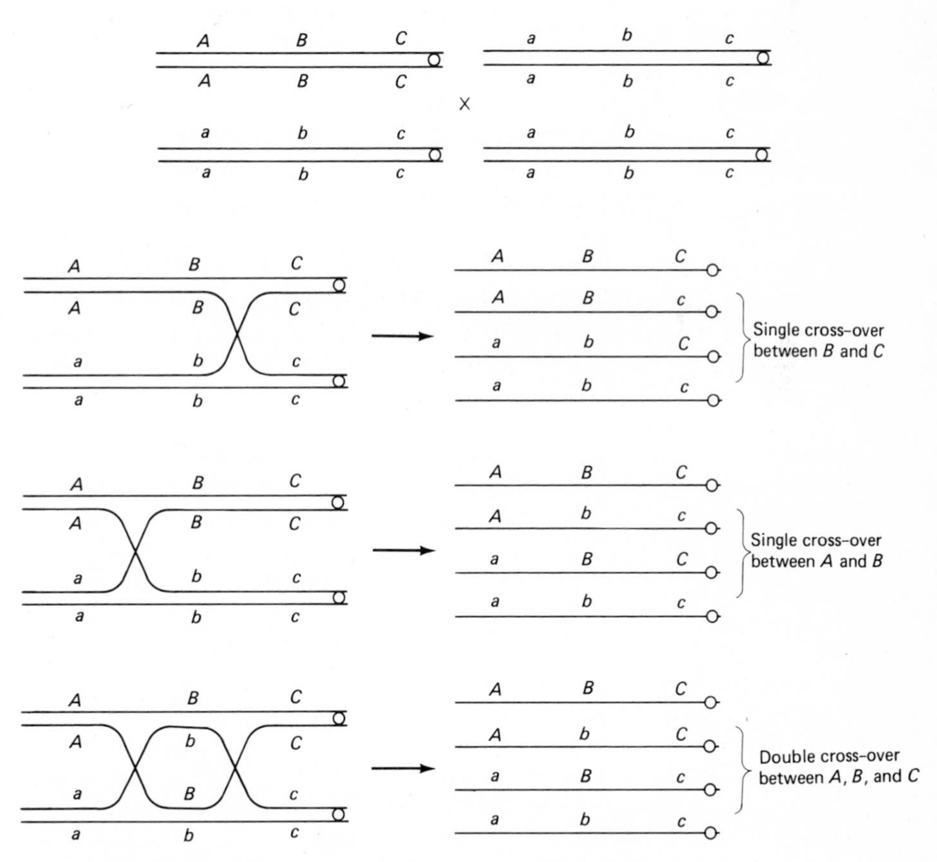

Figure 7-9. Classes of cross-over progeny to be expected when dealing with a three-point cross.

and the third gene; and finally a group of offspring that result from simultaneous crossing-over between the first and middle genes and the middle and third genes. This latter group is known as the *double cross-overs*. Let us illustrate this concept with some data. Assume the testcross yielded the following results:

Experiment 1

SrM	72
Srm	18
SRM	405
sRM	20
srm	407
srM	7
sRm	66
SRm	5
	1000 Total

In analyzing these data, the first step is to discern the different categories of offspring. The most obvious are usually the parental types and the double cross-overs, because they will be the most numerous and least numerous groups, respectively. The parental types are the most numerous, of course, because crossing-over usually occurs in less than 50% of the gametes.

The frequency of double cross-overs would be smallest, as the probability of a double cross-over is the product of the probabilities of each single cross-over. The genetic exchanges are independent events occurring at the same time, and our probability rule on p. 63 would be in effect. Thus, if crossing-over is 10% between genes *a* and *b*, and 10% between *b* and *c*, and the genes are situated *a–b–c* on the chromosome, the frequency of double cross-overs would be expected to be $.10 \times .10 = .01$. In our example above, the two parental groups are *SRM* and *srm*. (These combinations of genes should indicate coupling to you by now, even if you did not know how the matings had been set up.) The smallest groups, *srM* and *SRm*, can be assumed to be the double cross-overs.

We can now decide, on the basis of the double cross-over groups, the sequence of the genes on the chromosomes. There are three ways that genes may be situated in relation to each other. These are: *S–R–M*, or *S–M–R*, or *M–S–R*. What we now have to establish is which of these orders of the genes will give us the combinations of genes (*srM*, *SRm*) obtained in the data upon double crossing-over. Figure 7-10 illustrates the results of double crossing-over in each of the three cases, and we can see at once that the only order to yield the observed double cross-over progeny (*srM*, *SRm*) is *S–M–R*. We can assume that this is the order of the genes, and we can now tell which of the progeny must be the result of single cross-overs between genes *S* and *M*, and between *M* and *R*. This is shown in Fig. 7-11. Another method of determining gene order is to note the genes in parental types, then find the double cross-over class with two of those three genes present. The allele different from the third gene in the parental type is the gene in the middle. For example, *SRM* is one of the parental types. The double cross-over *SRm* contains two genes, *S* and *R*, in common with the parent. The different allele, *m*, is the gene in the middle.

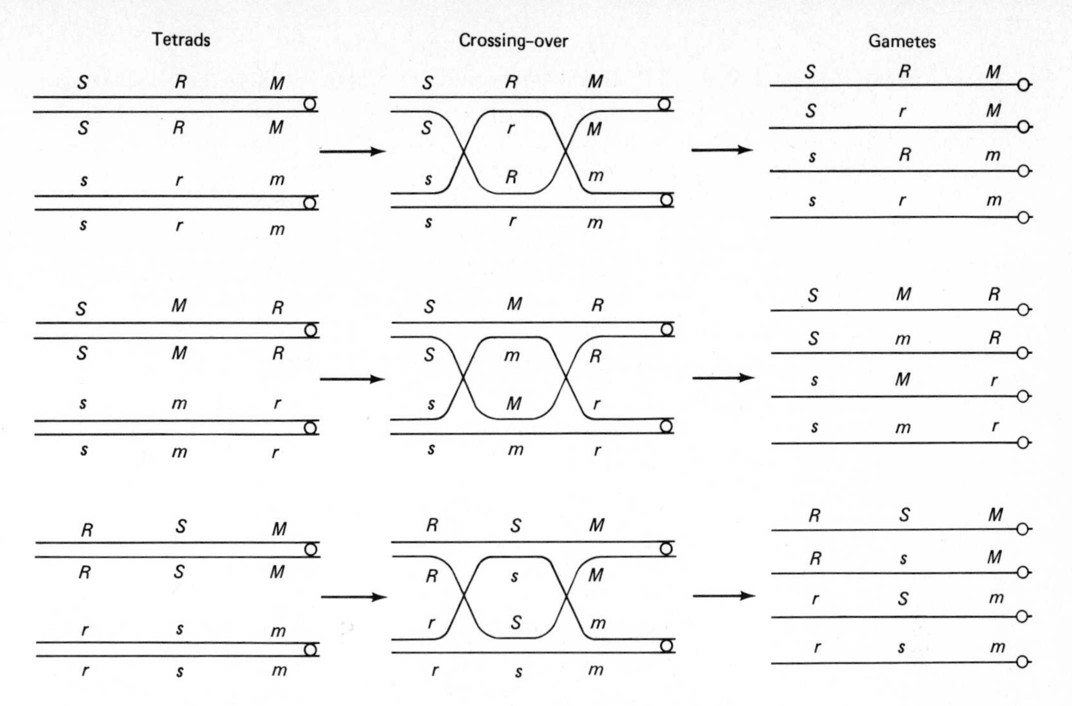

Figure 7-10. Three possible ways that three linked genes may be situated with respect to each other on the chromosome. Double cross-over products allow us to distinguish the order.

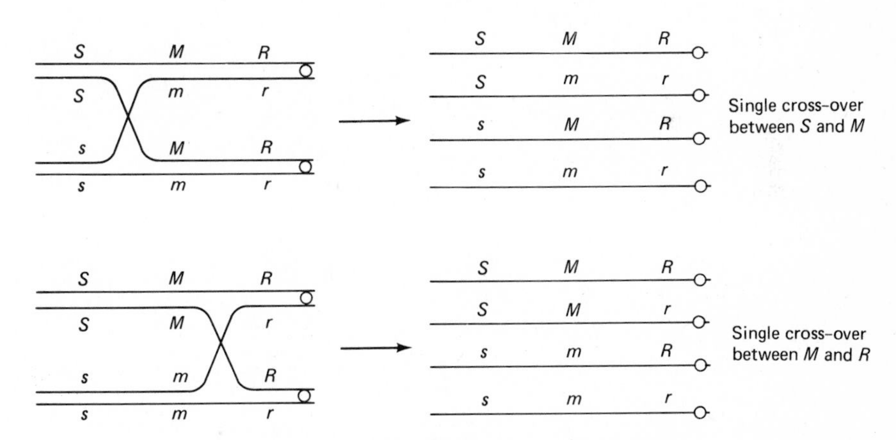

Figure 7-11. Results of single cross-over events where the order of genes is *S–M–R*.

To calculate the distance between two of the genes is now a simple matter. Totaling the number of offspring representing the single cross-overs between S and M, we find 38. To this number we must add 12, the number of double cross-overs, because each double cross-over also represents an exchange event between genes S and M. Therefore, the total number of crossing-over events between S and M is $\frac{50}{1000}$, a frequency of 5%, or 5 map units. The same must be done for crossing-over between M and R. *The total number of single cross-overs, plus the double cross-overs,* is

72 + 66 + 12 = 150, or 15% and 15 map units. Our map for these three loci is then:

$$S–5–M–15–R$$

Using the concepts of crossing-over and chromosome mapping, let us solve the following problem, which involves a slightly different approach. Since we have established the positions and distances of the three genes above, let us assume that we want to perform another series of crosses to confirm our analysis. We develop a strain of flies that is *RRmmSS* in genotype and cross them with a strain that is *rrMMss*. We obtain the F_1 generation, and testcross the F_1 with flies from a homozygous recessive strain. Out of 1000 offspring, what phenotypes would be expected and in what numbers?

Since the F_1 flies are in repulsion phase, the expected classes of progeny from the testcross would be as follows:

Experiment 2

RmS *rMs*	Parental types
RMs *rmS*	Single cross-overs between *S* and *M*
Rms *rMS*	Single cross-overs between *M* and *R*
RMS *rms*	Double cross-overs

Again, we begin our analysis by using the smallest group first, the double cross-overs. If we expect 15% crossing-over between *R* and *M*, and 5% between *M* and *S*, the expected percentage of double cross-overs is .15 × .05 = .0075. The expected numbers of double cross-overs in 1000 offspring is given by .0075 × 1000 = 7.5. Similarly, the expected numbers of single cross-overs between *M* and *S* would be 50 (because .05 × 1000 = 50) *minus* 7.5 (the expected number of double cross-overs), or 42.5. Remember that the frequency of crossing-over between two genes includes both single and double cross-overs. By the same reasoning, the single cross-overs to be expected between *R* and *M* would be .15 × 1000 − 7.5 = 150 − 7.5 = 142.5. The numbers for the parental types would of course be 1000 minus (142.5 + 42.5 + 7.5), or 807.5. We would expect 403.75 *RmS* and *rMs*. The expected offspring of the test cross can be summarized:

RmS	403.75	
rMs	403.75	
RMs	71.25	
rmS	71.25	
Rms	21.25	
rMS	21.25	
RMS	3.75	
rms	3.75	
	1000.00	Total

With the expected numbers based on the assumption that R is 15 map units from M and M is 5 map units from S, we can perform the above crosses and use the chi square test to confirm our analysis of the relationship of the three genes.

Interference and coincidence. Note that the number of double cross-over progeny in experiment 1 was different from the expected number of 7.5. OVER THE YEARS, AS GENETICISTS CONTINUED TO MAKE THREE-POINT CROSSES, IT BECAME APPARENT THAT AS OFTEN AS NOT THERE WOULD BE EITHER FEWER OR MORE DOUBLE CROSS-OVERS THAN EXPECTED, AND THE VARIATION WAS AT TIMES QUITE CONSISTENT IN STUDIES OF A PARTICULAR GROUP OF GENES. In some cases it appeared that crossing-over between the first two of the genes in some way interfered with crossing-over between the middle and the third gene, and so this was called *interference*. The occurrence of fewer double cross-overs than expected was termed *positive* interference; that of more double crossovers than expected, *negative* interference.

A measure of the degree of interference was developed, called the *coefficient of coincidence*. This is estimated quite simply by taking the ratio of the observed frequency of double cross-overs to the expected frequency. If the observed is 12 out of 1000, and we expected 7.5, the coefficient of coincidence for the loci is $\frac{12}{7.5} = 1.6$. If you think about it a bit, you will see that negative interference will always result in a coefficient of coincidence greater than 1, while positive interference will result in a coefficient less than 1. We must take the coefficient of coincidence into account when calculating expected numbers from a cross. In the preceding example, we had said that the expected number of double cross-overs should be $.0075 \times 1000 = 7.5$. However, we must now consider that previous work established negative interference, with the coefficient = 1.6. Therefore, the expected number of double cross-overs for these three loci is $.0075 \times 1000 \times 1.6 = 12.0$. We need to multiply by the coefficient of coincidence to compensate for the known interference, and to bring the numbers in line with what is expected if there is no interference. So the revised numbers for expected offspring of the trihybrid testcross would be:

RmS	406
rMs	406
RMs	69
rmS	69
Rms	19
rMS	19
RMS	6
rms	6
	1000 Total

The cause of interference cannot always be determined. In some cases, chromosomal rearrangements such as inversions have been found to suppress evidence for, or actually to suppress, crossing-over between genes. Also, there appears to be more crossing-over in females of many species compared to males, the extreme of this phenomenon being in *Drosophila*. *Drosophila* males show no crossing-over at all in sex chromosomes or autosomes. In order to carry out mapping experiments it is therefore necessary to use females as the dihybrids or trihybrids, and the males as the homozygous recessive partners. Remember that *Drosophila* males were found to have no synaptinemal complex between homologs during meiosis. It is a matter for conjecture whether this may in some way interfere with normal synapsis and crossing-over.

Our old friend, the *T* locus, also has an unusual characteristic pertaining to the lack of crossing-over. There appears to be almost complete suppression of crossing-over in heterozygotes for lethal *t* alleles over a significant region of chromosome #17, between *T* and the marker gene Tufted some 8–12 map units away. Since crossing over is normal in heterozygotes of viable *t* alleles, the reasons underlying this difference are obscure.

Other aspects of crossing-over and mapping. We can now further explore statements that we made previously. One is that map distances obtained from crossing-over frequencies are not exactly correlated to sites on the chromosomes. One of the reasons for this statement is that the farther apart two genes are, the greater the likelihood that several genetic exchanges can occur between them, some of which may negate each other so that they cannot be detected at all; Fig. 7-12 illustrates

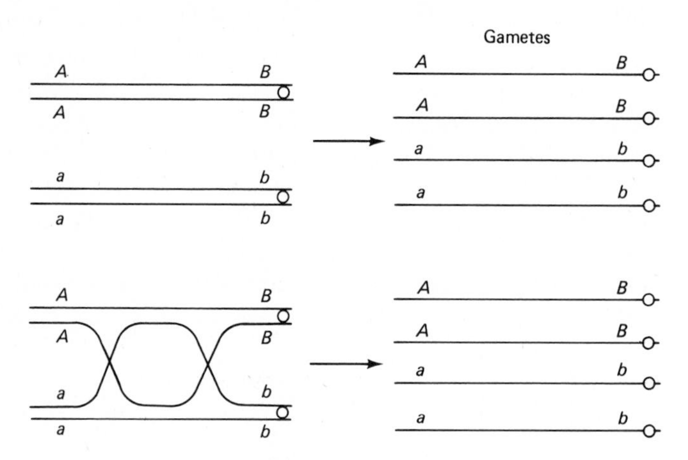

Figure 7-12. Multiple cross-overs between two loci far apart may negate each other and go undetected.

this point. Therefore, it is possible that the actual frequency of crossing-over is always higher than the observed. Furthermore, loci which are closely linked are unlikely to have multiple exchanges, and the map units obtained by cross-over frequency are more likely to be valid as a measure of the distance between the loci.

As we mentioned previously, one of the reasons that 50% crossing-over cannot be detected is that the results are indistinguishable from independent assortment. In dihybrids *AaBb*, you would expect equal numbers of the four gametes in both cases. Furthermore, mathematical analysis, which we shall not go into here, has shown that regardless of the distance between two loci, we can never observe recombination frequency greater than 50%. When cross-over frequencies approach such high levels, it becomes necessary to test the genes in question with intermediate marker genes in order to achieve any kind of accuracy.

Another aspect of crossing-over which we have not really discussed is the fact that at the stage of meiosis when genetic exchange occurs, there is tetrad formation. We have until now discussed crossing-over only between two nonsister chromatids. It is known, however, that in dealing with the tetrad as a whole, if several cross-overs occur at the same time, such as in double cross-overs, the genetic exchange does not have to involve only two of the four chromatids. In fact, all four chromatids can be involved. However, in organisms such as *Drosophila*, multiple cross-overs tend to negate each other, so that we can with some certainty proceed with the type of analysis of two-point and three-point crosses that we discussed above.

Tetrad Analysis

There are organisms in which the possibility of multiple-stranded crossing-over may play a significant role in analysis. One such organism is *Neurospora*, a fungus.

Life cycle of fungi. *Neurospora* belongs to the class of fungi known as Ascomycetes, whose life cycle is illustrated in Fig. 7-13. Fungi are mainly haploid through most of their life cycle. They are diploid only when two cells fuse, forming a *heterocaryon*, a cell with a diploid nucleus. This state of diploidy is usually only temporary, and meiosis then results in haploid spores that develop into the thread-like chains of cells called *hyphae* (sing., *hypha*), which are also haploid.

Haploidy in fungi. Since fungi are haploid, the occurrence of dominance and recessiveness is of course not possible, and the genes are all expressed; therefore, we usually do not use lower-case and capital letters as symbols to denote recessiveness and dominance. Usually + or − (with a superscript to indicate the locus) is used for wild-type, and a letter for an allele to wild-type.

Since fungi are such simple organisms, mutations generally affect some aspect of metabolism, such as the ability to produce the amino acid proline. If there has been a mutation to an inability to metabolize proline and the amino acid must be supplied in the medium in which the fungus is grown, we refer to the organism as being *auxotrophic* for proline. The wild-type, which does not need the presence of proline in the medium, is referred to as *prototrophic*.

The advantages of studying fungi are that they are haploid, which means that their genotype is reflected in their phenotype; they are small; and their life cycle is very short, allowing the study of many generations in a matter of days.

Ordered tetrads. The Ascomycetes have still another characteristic which has been extremely useful to geneticists in studying linkage of genes. When heterocaryons undergo meiosis, they do so in the confines of a saclike structure called an *ascus*, the dimensions of which cause the daughter spores to remain in place and in order. Normally, in studying diploid organisms, we have to rely on large numbers of offspring in order to discern the different combinations of genes that have resulted from linkage and crossing-over. In humans, offspring numbers are small, therefore we have no way to detect all the products of linkage and crossing-over.

In fungi such as *Neurospora*, not only do we have an organism which yields all the products of meiosis easily detectable, but the spores are held in order, so that we can analyze the kinds of cross-over events that have taken place. Because these products of meiosis and crossing-over are essentially the components of tetrads during

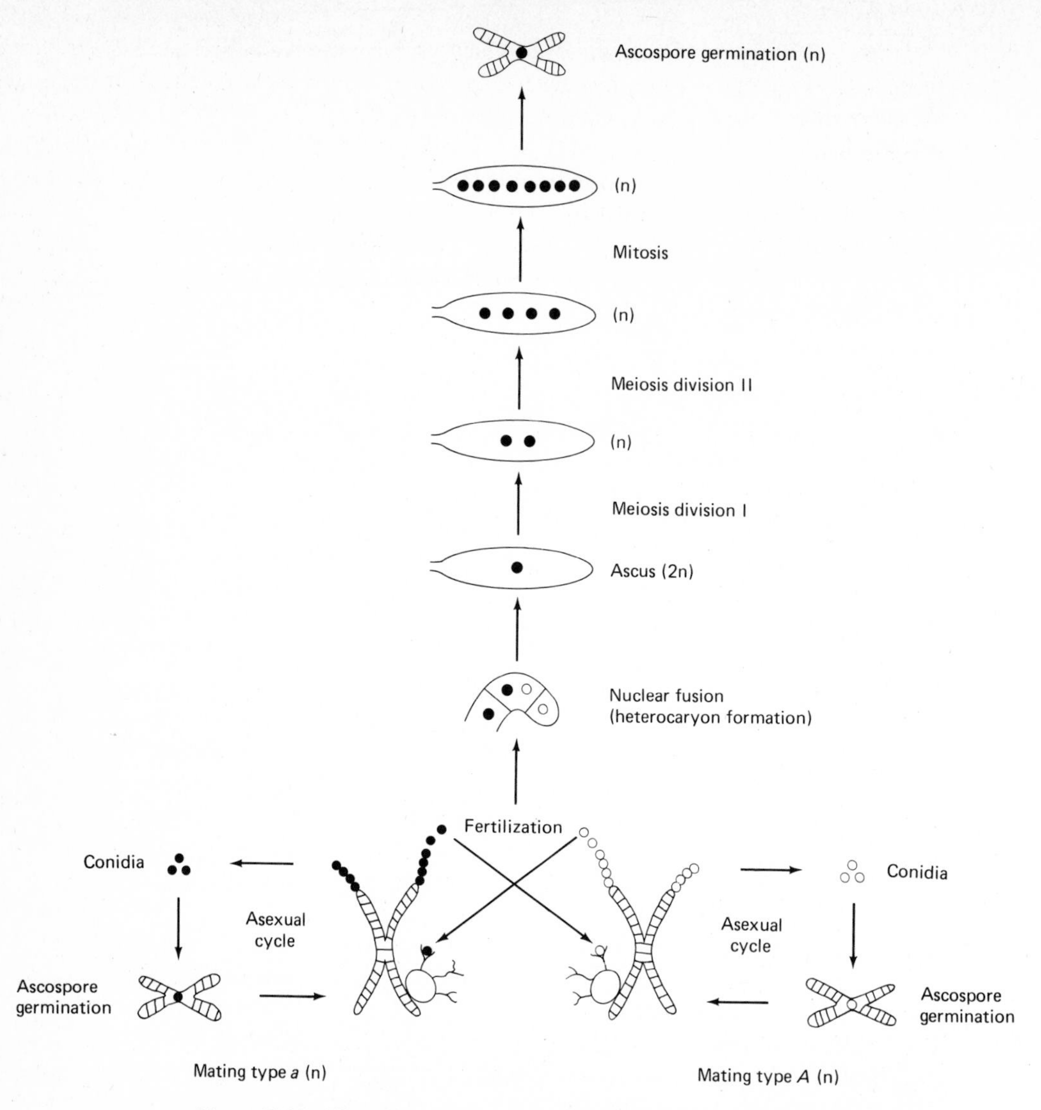

Figure 7-13. Life cycle of Neurospora. (From R. Wagner and H. Mitchell, 1955. *Genetics and Metabolism.* John Wiley, New York in U. Goodenough, 1978. *Genetics* 2nd ed., p. 83. Holt, Rinehart & Winston, New York.)

meiosis, we refer to this kind of study as *tetrad analysis*, and the products of meiosis in *Neurospora* as *ordered tetrads*. Let us use a few simple examples to illustrate the manner in which tetrad analysis can allow us to detect linkage and map the chromosomes of *Neurospora*.

The detection of linkage. Figure 7-14 shows the difference in tetrad formation, if we study the results of meiosis in a heterocaryon which is dihybrid for two linked

---

Figure 7-14. Comparison of tetrad formation in dihybrids for (a) linked genes and (b) for unlinked genes.

$+^a$ $+^b$
a b

$+^a$ $+^b$
a b

$+^a$ $+^b$
$+^a$ $+^b$
a b
a b

Tetrad formation

$+^a$ $+^b$
$+^a$ $+^b$
a b
a b

$+^a$ $+^b$
$+^a$ $+^b$

a b
a b

First meiotic division

$+^a$ $+^b$
$+^a$ $+^b$

a b
a b

or

$+^a$ b
$+^a$ b

a $+^b$
a $+^b$

$+^a$ $+^b$

$+^a$ $+^b$

a b

a b

Second meiotic division

$+^a$ $+^b$

$+^a$ $+^b$

a b

a b

or

$+^a$ b

$+^a$ b

a $+^b$

a $+^b$

$+^a$ $+^b$

$+^a$ $+^b$

$+^a$ $+^b$

$+^a$ $+^b$

a b

a b

a b

a b

Mitosis

$+^a$ $+^b$

$+^a$ $+^b$

$+^a$ $+^b$

$+^a$ $+^b$

a b

a b

a b

a b

or

$+^a$ b

$+^a$ b

$+^a$ b

$+^a$ b

a $+^b$

a $+^b$

a $+^b$

a $+^b$

(a)

(b)

genes or for two unlinked genes. The detection of linkage in fungi is essentially the same as it is in diploids, by the frequency of various classes of offspring. If genes a and b are linked, one finds only two combinations of genes in the spores assuming no crossing-over. With independent assortment, there would be four different combinations.

Note that if metabolic mutations are the subject of study, it is not possible to detect phenotype by looking at the spores. The spores must be isolated, plated on nutrient medium, and identified by their ability or inability to metabolize various substances. Occasionally a mutation causes some visible changes in the spores, such as color, but for the most part, changes from the wild-type phenotype are in the nature of metabolic variations.

Crossing-over and mapping. With these basic facts about fungal genetics in mind, let us now discuss the calculation of map distance between linked genes. Suppose that we were studying a strain of fungus auxotrophic for proline and serine (traits determined by two linked genes, p and s) and mated it to a strain that was wild-type. This would produce a heterocaryon dihybrid for the two genes in the coupling position. After spore maturation, we isolate a bunch of spores and plate them out, and record the following data:

$s\ p$	350	
$+^s\ +^p$	350	
$s\ +^p$	150	
$+^s\ p$	150	
	1000	Total

As you can see, there were 300 recombinant spores; therefore, the distance between the s and p loci is $\frac{300}{1000} = 30\%$, or 30 map units.

One further point to be made here is that if you draw the chromosomes involved in this cross as illustrated in Fig. 7-15, you can see that the order of the spores indicates that the s and $+^s$ alleles of the s locus segregated during the first meiotic division as the homologs are pulled apart into the daughter spores. This is known as *first-division segregation.* The alleles of the p locus, on the other hand, did not segregate because of crossing-over until the second meiotic division; this is called *second-division segregation.*

If mutations which cause visible differences in the appearance of the spores are being studied, we can calculate map distance between genes simply by looking at the order of phenotypes of spores in the asci. For example, let us assume that there are two linked genes which cause visible differences in spore shape and color. One locus determines the color of spores: $+^g$ = *yellow spores*; g = *gray spores*; the other locus determines smoothness of spores: $+^w$ = *smooth spores*; w = *rough spores.* Let us assume that a wild-type mating strain is crossed with a doubly mutant one. If there is no crossing-over, we can expect the appearance of spores in asci as shown in Fig. 7-16a.

If crossing-over occurs, it can involve any of the four chromatids, and Fig. 7-16b illustrates the effect on the order of spores in several different cross-over events. One can count the number of asci showing parental type spores and the number of

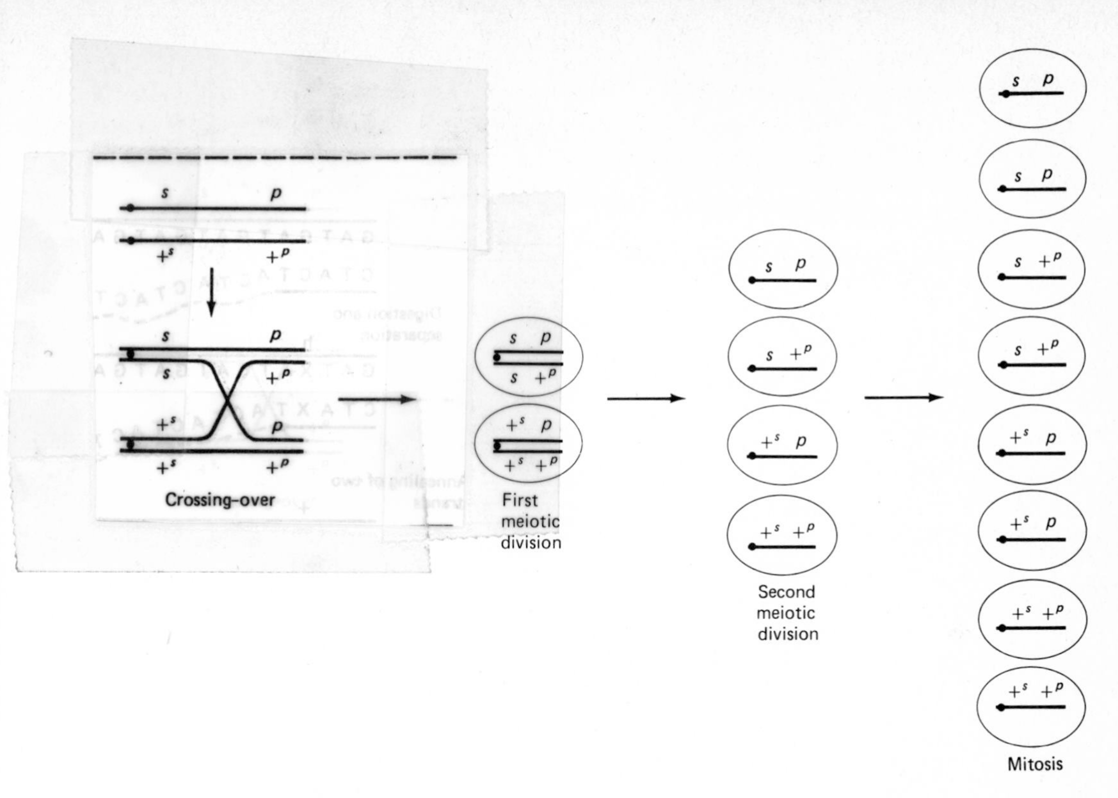

Figure 7-15. Crossing-over between linked genes in Neurospora.

asci containing recombinant type spores. If the number of asci with recombinant spores in our experiment is 30 out of a total of 100 asci studied, we might be tempted to consider the two genes 30 map units apart. However, one important factor to remember in calculating map distance in fungi is that if whole asci are used, the figure is obtained by multiplying the number of asci with recombinant spores by $\frac{1}{2}$. The reason for this is that only half the spores within each ascus are recombinant spores; the other half are parental types which have not crossed-over. If we find that 70% of all asci contain the parental combinations gw and $+^g+^w$, and 30% $g+^w$ and $+^gw$, the actual frequency of crossing-over is $.30 \times \frac{1}{2} = .15$, and the genes are 15 map units apart.

Centromere mapping. The position of the centromere can also be accounted for in tetrad analysis. If crossing-over had occurred between the centromere and the g locus, we would find second-division segregation of the g and $+^g$ alleles as well as the w and $+^w$ alleles; you should be able to work this out for yourself. Essentially, because of the effect of crossing-over between the centromere and genes on the order of spores, the centromere functions as a gene in studies of chromosome mapping in *Neurospora*.

The situation becomes rather more complicated when we take all three, the centromere and the g and w loci, into consideration. This is because we must account for all the possible types of crossing-over that can occur, single and double, and because two, three, or all four strands can be involved. Figure 7-17 illustrates the

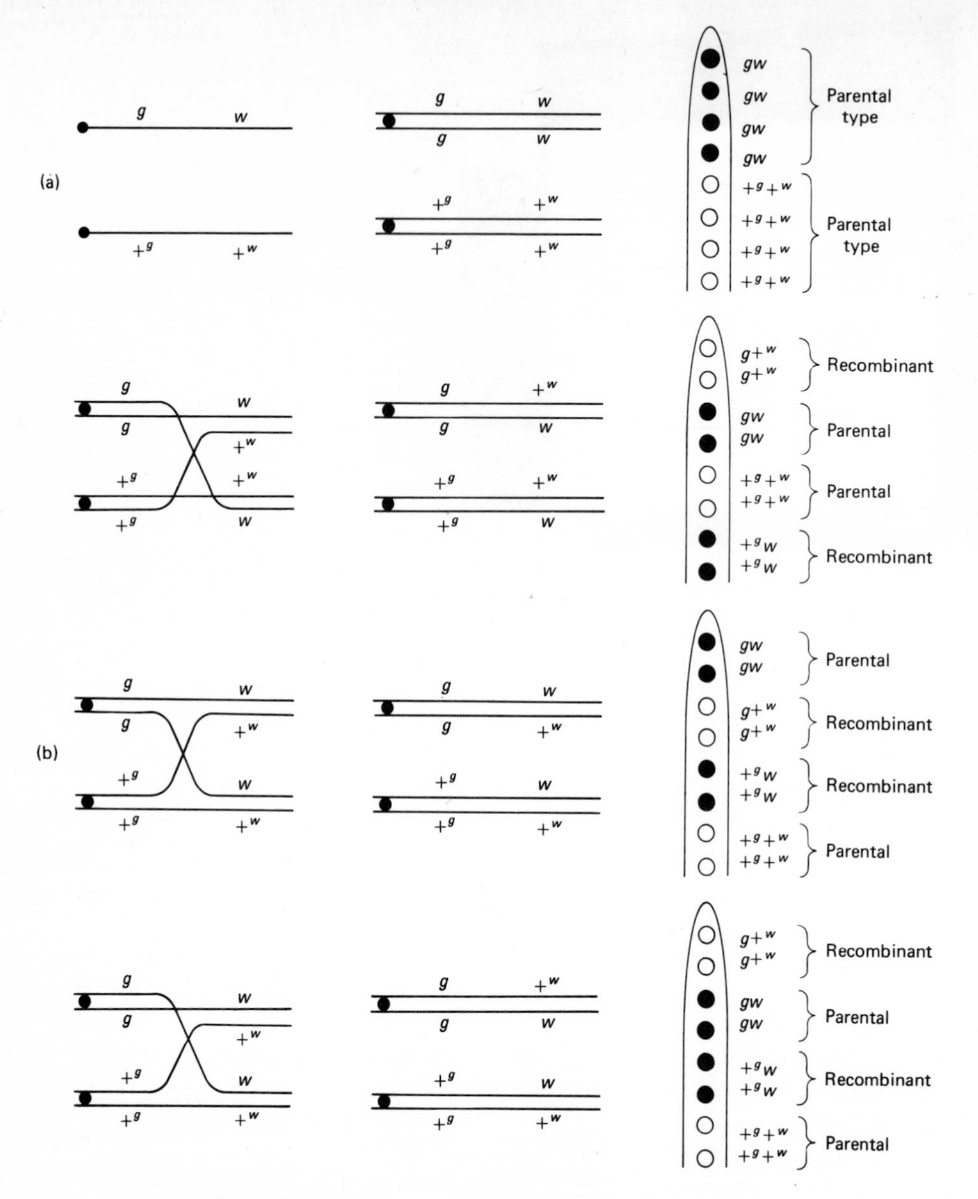

Figure 7-16. (a) No crossing-over. (b) Illustrations of different kinds of crossing-over involving all four chromatids. Explanation in text.

kinds of cross-over events, and the order of the spores in the asci following meiosis, in a cross of wild-type by *gw* fungi.

Geneticists have classified the various combinations of genes found in the spores. Asci containing only the parental-type combinations, as in cases 1 and 2 of Fig. 7-17, are called *parental ditypes* (PD). Those containing only two combinations, which, however, are not parental, are called *nonparental ditypes* (NPD), as in cases 6 and 7 of Fig. 7-17. Asci with four different combinations of genes are known as

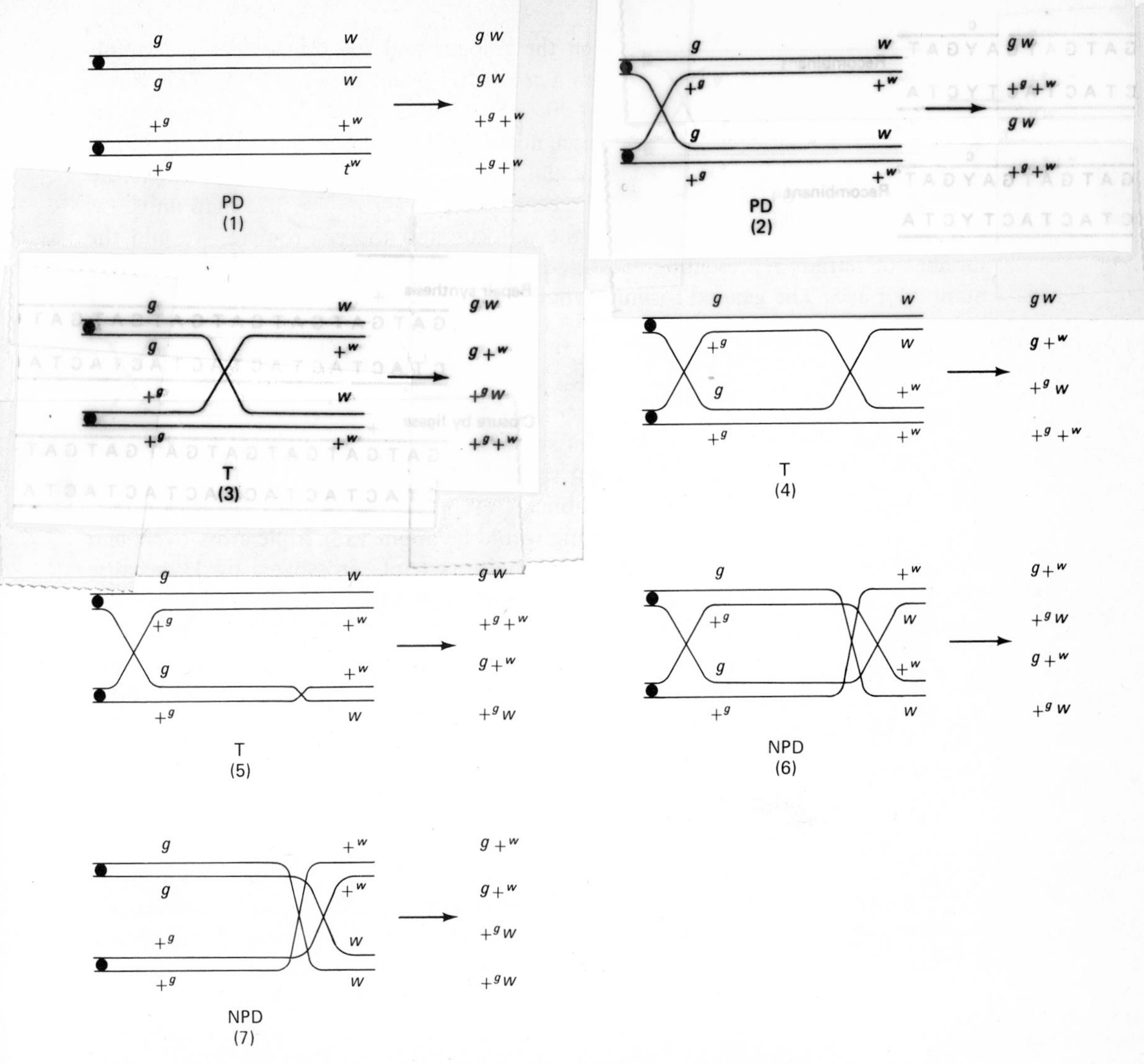

Figure 7-17. Results of different combinations of crossing-over events between the centromere and two linked genes. PD = Parental ditypes; T = Tetratypes; NPD = Nonparental ditypes. Products of mitosis and spore formation not shown for simplification.

tetratypes (T), seen in cases 3, 4, and 5 of Fig. 7-17. Let us assume that an experimental cross has been made between a *gw* strain and a $+g+w$ strain. We obtain the following data from a total of 1000 asci counted:

Classes						
1	2	3	4	5	6	7
gw	gw	gw	gw	gw	$g+^w$	$g+^w$
gw	$+^g+^w$	$g+^w$	$+^g w$	$+^g+^w$	$+^g w$	$g+^w$
$+^g+^w$	gw	$+^g w$	$g+^w$	$g+^w$	$g+^w$	$+^g w$
$+^g+^w$	$+^g+^w$	$+^g+^w$	$+^g+^w$	$+^g w$	$+^g w$	$+^g w$
705	40	180	10	20	5	40
(PD)	(PD)	(T)	(T)	(T)	(NPD)	(NPD)

To obtain the distance between the g locus and the centromere, we simply total the number of asci resulting from a cross-over event between them. This would be classes 2, 4, 5, and 6: $40 + 10 + 20 + 5 = 75$. $\frac{75}{1000} = 7.5\%$. However, remember that we are dealing here with asci, not individual spores. Since half the spores in the asci are nonrecombinant, the actual frequency of crossing-over, and therefore map units, between the g locus and the centromere is $7.5 \times \frac{1}{2} = 3.75$ map units.

To detect map units between the w locus and the centromere, we add the number of tetrads representing crossing-over in this region and divide by the total number of asci. The general formula would be:

$$\frac{\frac{1}{2}(\text{number of single cross-overs}) + 1(\text{number of double cross-overs}) + 1\frac{1}{2}(\text{number of triple cross-overs}) \text{ etc.}}{\text{total number of asci}}$$

We must multiply the number of single cross-overs by $\frac{1}{2}$ because only half the spores of such tetrads are recombinant spores. Since all the spores of double cross-overs represent the number of recombinant events in a tetrad, we take the total number of such asci. The same reasoning would be applied for triple cross-overs and higher numbers of cross-overs: We must multiply triple crossovers by $1\frac{1}{2}$ because there are again $1\frac{1}{2}$ times the number of recombination events as there are spores from a tetrad, and so forth.

For our experiment above, classes 2 and 3 represent single cross-overs; classes 4, 5, and 7 represent double cross-overs; class 6 represents a triple cross-over. Therefore, inserting the appropriate numbers into our formula:

$$\frac{(\frac{1}{2})(220) + 70 + (1\frac{1}{2})(5)}{1000} = 18.75\%$$

The distance between g and w is therefore $18.75 - 3.75 = 15$ map units.

Unordered tetrads. Some organisms that produce tetrads, such as yeast, do not have asci that allow ordered tetrad analysis. In these organisms, it is of course not possible to discern second-division segregation. However, we can still calculate the recombination value between loci by counting the numbers of parental ditypes (nonrecombinants), nonparental ditypes, and tetratypes (recombinants). Figure 7-18 illustrates these three classes.

The total number of recombinants would be the sum of all nonparental ditypes which result from recombination between all four strands, and half of the tetratypes. We halve the tetratypes because half the chromosomes are parental rather than recombinants. Then we simply divide the number of recombinants by the total number of chromosomes:

$$\frac{\text{NPD} + \frac{1}{2}\text{T}}{\text{PD} + \text{NPD} + \text{T}}$$

For example, a strain of yeast requiring substance "a" is crossed to a strain that requires substance "b." We can diagram the cross as $aB \times Ab$. Assume that the genes involved are linked. Phenotypic ratios of the tetrads produced by this cross

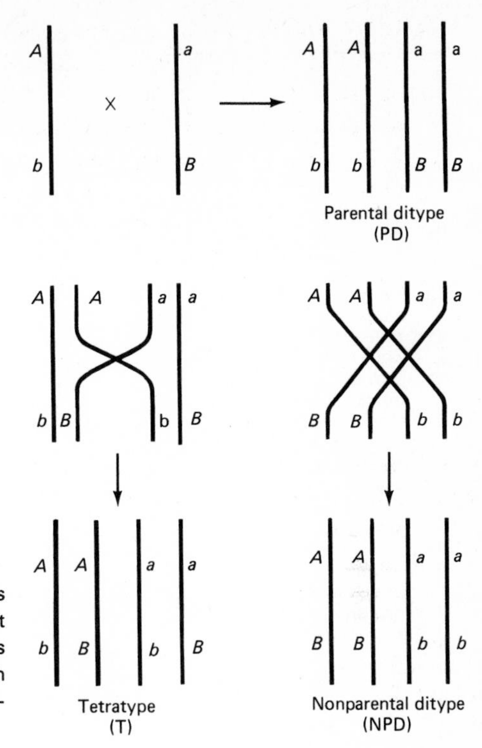

Figure 7-18. Results of a cross in organisms producing unordered tetrads. PD tetrads result from absence of recombination. Half the strands of a T tetrad represent recombination, while each strand of a NPD tetrad is the result of recombination.

were found to be:

PD (*aB* or *Ab*)	45
NPD (*ab* or *AB*)	10
T (*aB*, *Ab*, *ab*, *AB*)	15

Using our formula we obtain $\frac{10 + 7.5}{70} = .25$, or 25% recombination.

The preceding discussion covers only some of the aspects of tetrad analysis. There are many others, such as the results on tetrad order if two genes are on different sides of the centromere, or linkage between more than two gene loci and the centromere, or mitotic recombination in *Aspergillus*. We shall not go into these, however, as fungal genetics (like so many of the topics we discuss in this book) is a course in itself. What we have covered should serve as an introduction to a unique system of study in transmission genetics.

CHROMOSOME MAPS

The various techniques we have discussed in this chapter have been used to draw up chromosome maps for several species. Figure 7-19 illustrates maps for *Drosophila*, mouse, and maize.

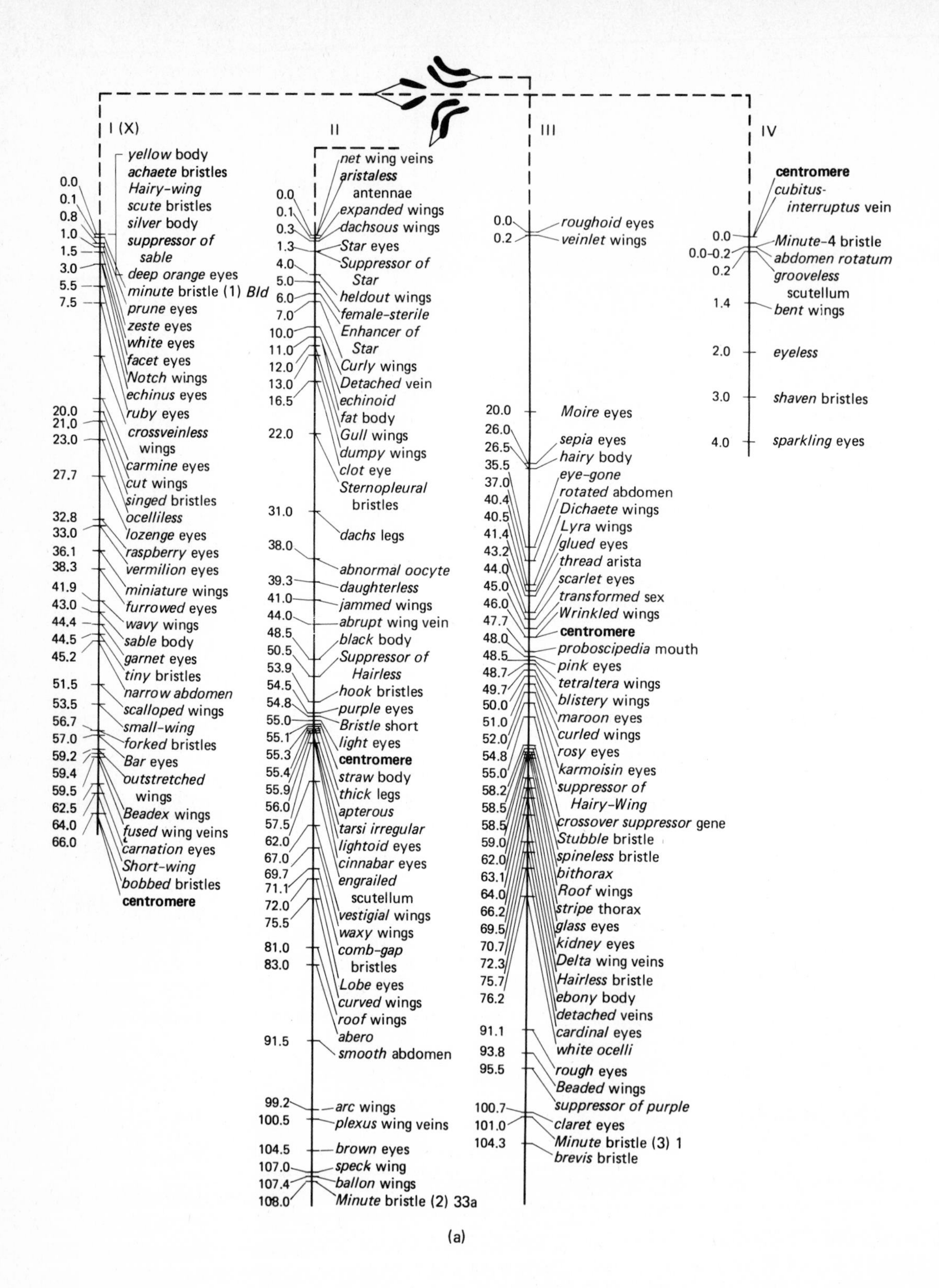

I (X)
II
III
IV
yellow body
achaete bristles
Hairy-wing
scute bristles
silver body
suppressor of sable
deep orange eyes
minute bristle (1) Bld
prune eyes
zeste eyes
white eyes
facet eyes
Notch wings
echinus eyes
ruby eyes
crossveinless wings
carmine eyes
cut wings
singed bristles
ocelliless
lozenge eyes
raspberry eyes
vermilion eyes
miniature wings
furrowed eyes
wavy wings
sable body
garnet eyes
tiny bristles
narrow abdomen
scalloped wings
small-wing
forked bristles
Bar eyes
outstretched wings
Beadex wings
fused wing veins
carnation eyes
Short-wing
bobbed bristles
centromere
0.0
0.1
0.8
1.0
1.5
3.0
5.5
7.5
20.0
21.0
23.0
27.7
32.8
33.0
36.1
38.3
41.9
43.0
44.4
44.5
45.2
51.5
53.5
56.7
57.0
59.2
59.4
59.5
62.5
64.0
66.0
net wing veins
aristaless antennae
expanded wings
dachsous wings
Star eyes
Suppressor of Star
heldout wings
female-sterile
Enhancer of Star
Curly wings
Detached vein
echinoid
fat body
Gull wings
dumpy wings
clot eye
Sternopleural bristles
dachs legs
abnormal oocyte
daughterless
jammed wings
abrupt wing vein
black body
Suppressor of Hairless
hook bristles
purple eyes
Bristle short
light eyes
centromere
straw body
thick legs
apterous
tarsi irregular
lightoid eyes
cinnabar eyes
engrailed scutellum
vestigial wings
waxy wings
comb-gap bristles
Lobe eyes
curved wings
roof wings
abero
smooth abdomen
arc wings
plexus wing veins
brown eyes
speck wing
ballon wings
Minute bristle (2) 33a
0.0
0.1
0.3
1.3
4.0
5.0
6.0
7.0
10.0
11.0
12.0
13.0
16.5
22.0
31.0
38.0
39.3
41.0
44.0
48.5
50.5
53.9
54.5
54.8
55.0
55.1
55.3
55.4
55.9
56.0
57.5
62.0
67.0
69.7
71.1
72.0
75.5
81.0
83.0
91.5
99.2
100.5
104.5
107.0
107.4
108.0
roughoid eyes
veinlet wings
Moire eyes
sepia eyes
hairy body
eye-gone
rotated abdomen
Dichaete wings
Lyra wings
glued eyes
thread arista
scarlet eyes
transformed sex
Wrinkled wings
centromere
proboscipedia mouth
pink eyes
tetraltera wings
blistery wings
maroon eyes
curled wings
rosy eyes
karmoisin eyes
suppressor of Hairy-Wing
crossover suppressor gene
Stubble bristle
spineless bristle
bithorax
Roof wings
stripe thorax
glass eyes
kidney eyes
Delta wing veins
Hairless bristle
ebony body
detached veins
cardinal eyes
white ocelli
rough eyes
Beaded wings
suppressor of purple
claret eyes
Minute bristle (3) 1
brevis bristle
0.0
0.2
20.0
26.0
26.5
35.5
37.0
40.4
40.5
41.4
43.2
44.0
45.0
46.0
47.7
48.0
48.5
48.7
49.7
50.0
51.0
52.0
54.8
55.0
58.2
58.5
58.5
59.0
62.0
63.1
64.0
66.2
69.5
70.7
72.3
75.7
76.2
91.1
93.8
95.5
100.7
101.0
104.3
centromere
cubitus-interruptus vein
Minute-4 bristle
abdomen rotatum
grooveless scutellum
bent wings
eyeless
shaven bristles
sparkling eyes
0.0
0.0-0.2
0.2
1.4
2.0
3.0
4.0

(a)

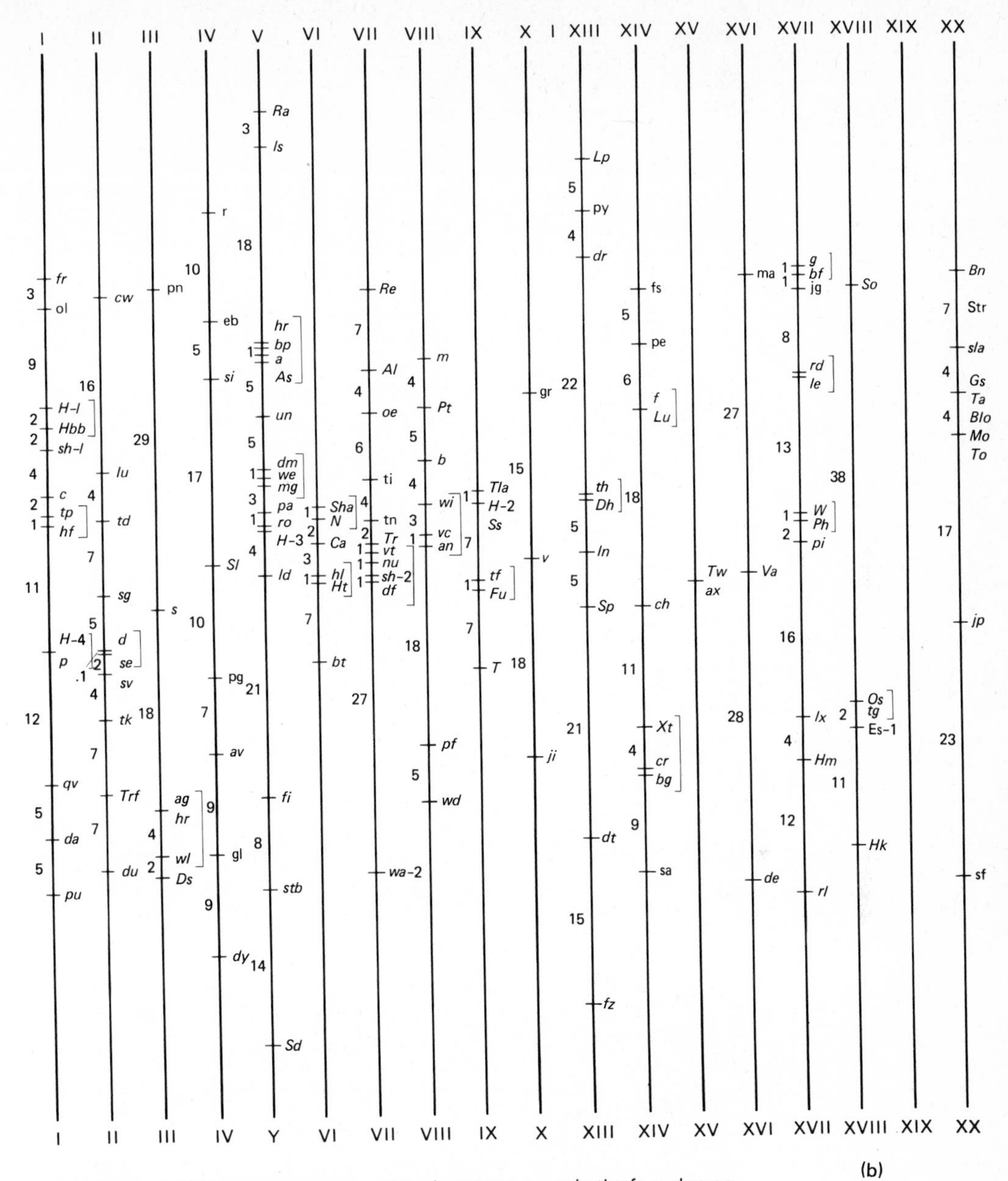

Figure 7-19. (a) Linkage map of some of the important genes in the four chromosomes of *Drosophila melanogaster*. (From M. Strickberger, 1976. *Genetics* 2nd ed., p. 341. Macmillan, New York. Copyright © 1976 by Monroe Strickberger.) (b) Linkage map of mice. Note linkage group IX is now referred to as linkage group XVII. (From E. Green, ed., 1966. *Biology of the Laboratory Mouse*, pp. 126–27. Dover Publications, Inc., New York.) (c) Linkage map of maize. Parentheses indicate probable position based on insufficient data, ○ indicates centromere position, and ● indicates organizer. (From R. C. King, ed., 1974. *Handbook of Genetics*, vol. 2, pp. 6–7. Plenum Press, New York.)

1
0 sr; 1 vp_5; (14) ag; 15 ga_6; 19 zb_4; 23 tus_{17}; 24 ts_2; 26 p; 28 zl; TB-1b; 56 as; (58) pa; Tb-1a; 64 hm; 81 br; 85 Vg; 86 f; 104 an; 106 bz_2; (108) ad; 119 Ts_3; (124) tb; 127 Kn; (128) lw; 135 gs; (154) vp_8; 158 Ts_6; 161 bm_2

2
0 ws_3; (4) al; 11 lg; rp; 30 gl_2; TB-2S; (34) d_5; 49 B, R_2; 54 gs_2; 56 sk; (60) wt; 68 fl; 74 ts; TB-2L; 83 v_4; 111 w_3; 121 Ht; 155 Ch

3
0 cr; 18 d; 26 ra_2; 31 Cg; (38) cl; 40 rt; TB-3b; (45) Rf; 46 Lg_3; 48 Rg, Rp_3; TB-3a; 49 gl_6; 50; 55 ts_4; 72 ba; (75) w_{7748}; 83 lg_2; (86) na; 111 a; 111.2 sh_2; 122 et; 128 ga_7, vp

4
0 de; 27 Rp_4; 35 Ga; (55) st; 56 Ts_5; (60) la; 63 fl_2; 66 sp; 71 su; 73 lo; TB-4a; 74 de_{16}; 84 zb_6; 86 gl_4; 107 Tu; 112 j_2; 118 gl_3; 123 c_2; 143 di

5
(0) am; (9) lu; 14 gl_{17}; 15 a_2; 18 vp_2; 19 bm; ps; 21; 22 bt; v_3; 25; TB-5a; 27 bv; (35) ga_2; (37) ac; 46 pr; lw_2; 55 ys; eg; 87 v_2

6
TB-6a; (0) rgd; 4 po; Y; 17 w^m,ms; (19) pb; ms-si; 37 pg_{11}; (43) Dt_2; 48 Pl; 49 Bh; (57) su_2; 58 sm; (59) Pt; TB-6b; 68 py

7
0 Hs; 16 o_2; 18 y_8; 20 in; 24 v_5; (25) vp_9; 32 ra; TB-7b; 36 gl; 46 Tp; 50 sl; 52 ij; 71 Bn; 109 bd; (112) Pn

8
0 v_{16}; TB-8a; 14 msg; 28 j; Clt; gl_8; sn

9
0 Dt; 7 pyd, yg_2, wd; 26 C; 29 sh; 31 bz; 40 Mr; 44 bp; 53 lo_2, ga_8; TB-9b; 59 wx; 62 d_3; (64) pg_{12}; TB-9c; (65) ar; 66 v; 67 ms_2; 69 gl_{15}; 79 bk_2; 104 Wc; 134 Bf; 138 bm_4

10
0 Rp; (12) oy; 16 Og; 24 nl; TB-10c; 28 li; TB-10b; 33 Cx; du; 35 zn; (36) sp_2; 38 l_8; TB-10a; 41; 43 T_{p2}; g; 57 E_j, R^r; 61 Lc; (63) M^{st}; 73 w_2; (92) sr_2; (99) l_2; K

(c)

SIGNIFICANCE OF LINKAGE AND CROSSING-OVER

Often in the stress of learning techniques such as analyzing data for map distance calculations, the significance of the information obtained is lost to students. It is necessary to be reminded of the potential importance of the material we have discussed in this chapter. OUR ABILITY TO MAP CHROMOSOMES IS OF MORE THAN JUST ACADEMIC INTEREST BECAUSE FROM THE PRACTICAL POINT OF VIEW WE CAN, FOR EXAMPLE, DETECT AND DIAGNOSE GENETIC CHANGES CYTOLOGICALLY. If a large area of a chromosome is affected by some mutation, we can anticipate the effect on the individual if we know the specific genes involved. Also, our ability to perform genetic engineering feats in microorganisms whose chromosomes have been thoroughly mapped (which we shall be discussing in Chapter 14) leads to speculation that similar feats may be performed on eukaryotic chromosomes in the future. We might, for example, be able to substitute a mutant gene with a normal one. But first, it is necessary to know where genes are, and that is the function of mapping.

FROM A THEORETICAL POINT OF VIEW, WE CAN REITERATE A POINT MADE IN CHAPTER ONE (P. 10), THAT CROSSING-OVER PROVIDES EUKARYOTES WITH A POWERFUL MECHANISM FOR INCREASED GENETIC VARIABILITY IN ADDITION TO INDEPENDENT ASSORTMENT. The figure that we had calculated in Chapter 2 (p. 47) for the number of genetic combinations that can be produced by any human, assuming heterozygosity at only one locus per chromosome, was already awesome. Increase that figure by crossover events that can occur between any two heterozygous loci on the chromosome (and realistically, there must be many) and the number literally as well as figuratively goes "out of sight"! The significance of variability will be more obvious to you later after we discuss mutation and evolution, but for now, we can simply state that the greater the genetic variability found in any population of living organisms, the greater the chance that population has of surviving and producing future generations.

PROBLEMS

1. Assume that genes *D* and *E* are 12 map units apart. A cross of *DDee* × *ddEE* individuals is made, and the F_1 are testcrossed to a homozygous recessive strain. Give the genotypes and numbers of offspring expected out of a total of 500 in the F_2.

2. Assume that a cross is made between two strains of rabbits homozygous for two pigment loci. One strain is homozygous dominant for the color gene (*A*), which determines that color will be expressed, and the color gene (*B*), which determines black coat color. The second strain is albino, known to be homozygous for the Brown gene, (*bb*), an allele of *B*. The F_1 progeny are all black as expected. When they were backcrossed to the recessive strain, the following data were obtained:

black	35
albino	50
brown	25
	110

Do the data indicate linkage or nonlinkage of the *A* and *B* genes? Explain.

3. Read again the classical experiment carried out by Bateson and Punnett on sweet pea flower color and pollen grain shape (p. 190). Diagram the cross if they had studied plants homozygous Red/Long and homozygous Purple/Round. What would be the appearance of the F_1 plants? Give the expected phenotypic ratios of the F_2 generation.

4. Recall that the mutation *muscular dysgenesis* (*mdg*), a recessive autosomal lethal gene, interferes with development of normal skeletal musculature. Assume that there is a dominant mutation, *shortlegs* (*Sl*), known to be on the third linkage group. Diagram the crosses you would make to determine if *mdg* is linked to *Sl*, and give the expected results if there is linkage and if there is random assortment.

5. The following is a pedigree in a family in which members have been found to be color-blind and/or have hemophilia, both recessive X-linked traits. Draw the sex chromosomes and label them with the relevant gene symbols for all numbered individuals. Is there any evidence for crossing-over? Explain.

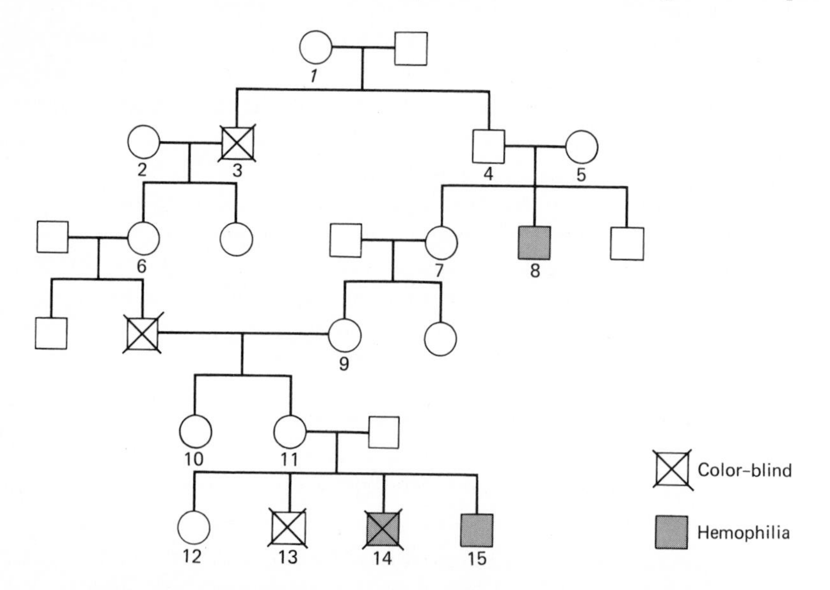

6. Diagram the chromosomes of the crosses detailed on pages 193–194. Diagram the chromosomes of the expected offspring if *ebony* is on the second chromosome of *Drosophila*.

7. Diagram the chromosomes of the crosses detailed on pages 193–194. Diagram the chromosomes of the expected offspring if *ebony* is on the third chromosome of *Drosophila*.

8. Assume that female flies trihybrid for linked genes in the repulsion phase

$$\frac{S\ m\ N}{s\ M\ n}$$

are crossed to homozygous recessive males. The *S*

locus is known to be 15 map units from *M*, and *M* is 20 map units from *N*. What genotypes and in what numbers would be expected from such a cross out of 1000 progeny if the coefficient of coincidence is known to be .6?

9. Females from a strain of flies trihybrid for the genes *A*, *B*, and *C* were mated to males homozygous recessive for all three genes. The following phenotypic data were obtained:

ABC	128
ABc	17
aBC	36
AbC	325
abc	142
abC	13
Abc	34
aBc	305
	1000

(a) Are the genes of the trihybrid in coupling or in repulsion?
(b) What is the correct sequence of genes?
(c) What is the distance between the loci?
(d) Is there interference? If so, what is the coefficient of coincidence?

10. The father of Mr. Spock, second officer of the Starship *Enterprise*, came from the planet Vulcan; his mother came from Earth. A Vulcanian has pointed ears (*P*), no adrenals (*A*), and a heart on the right side (*R*). All three genes are dominant over human alleles. The genes are autosomal and linked as follows (m. u. = map units):

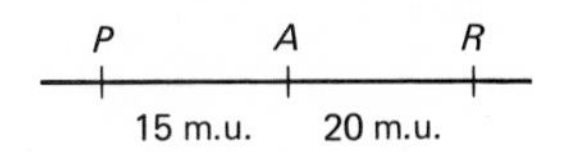

If Mr. Spock marries a human woman and there is no interference, what proportion of their children will
(a) look like Vulcans for all three characters?
(b) look like humans?
(c) have Vulcan ears and a Vulcan heart and human adrenals?
(d) have Vulcan ears and a human heart and human adrenals?
(D. Harrison, *Problems in Genetics*, © 1970, Addison-Wesley Publishing Company, Inc., page 98. Reprinted with permission.)

11. If we assume that a strain of *Neurospora* with the genotype $+^a+^b$ is mated to a strain that is *ab*, and the two genes are not linked, what would be the expected orders of genes in the tetrads (assume that the genes are very close to their respective centromeres). Now assume that the genes are linked. In the absence of crossing-over, what would be the expected orders of genes in the tetrads?

12. If a pair of genes in *Neurospora* always shows first-division segregation, what can we deduce about its distance from the centromere?

13. Assume that the two genes of problem 11 are linked and are 10 map units apart. Diagram the chromosomes and order of tetrads resulting from a single crossing-over event between a and b if the exchange occurred between

(a) the second and third strands;
(b) the first and third strands;
(c) the second and fourth strands;
(d) the first and fourth strands.

14. If two genes are linked in *Neurospora* and are 10 map units apart, and the mating is set up as in problem 11, out of 400 asci obtained, how many could be expected to show the presence of recombinant spores?

15. A cross is made of two *Neurospora* mating strains, $r+^s \times +^r s$. The r and s loci are both on one side of the centromere with r closest to the centromere. The results were as follows:

(A)	(B)	(C)	(D)	(E)	(F)	(G)
$+^r s$	$+^r s$	$+^r s$	$+^r+^s$	$+^r+^s$	$+^r s$	$+^r+^s$
$r+^s$	$+^r+^s$	rs	$+^r+^s$	rs	$+^r s$	$r+^s$
$+^r s$	rs	$+^r+^s$	rs	$+^r+^s$	$r+^s$	$+^r s$
$r+^s$	$r+^s$	$r+^s$	rs	rs	$r+^s$	rs
45	44	8	2	2	390	8

(a) Which columns represent PD (parental ditypes), NPD (nonparental ditypes), and T (tetratypes)?
(b) What is the distance between r and the centromere? What is the distance between r and s?

16. Using techniques of somatic cell hybridization, a study of four human enzymes were made in correlation with the presence of certain chromosomes in hybrid clones. The following data were obtained:

		Hybrid clones				
		A	*B*	*C*	*D*	*E*
Human enzymes	I	−	+	−	+	−
	II	+	−	−	+	−
	III	+	+	−	−	−
	IV	+	−	−	+	−
Human chromosomes	1	−	+	−	+	−
	2	+	−	−	+	−
	3	−	−	+	+	+

Are the genes determining any of these enzymes syntenic? Can any be assigned to a particular chromosome? Which do not appear to be syntenic?

17. Assume that a woman is homozygous normal for hemophilia and heterozygous for color-blindness. She is also homozygous normal for the dominant mutation Nail-Patella, and type O blood. The Nail-Patella locus and the ABO locus are linked on the same autosome. Her husband has normal blood and vision; he is heterozygous for Nail-Patella and also has type O blood. Draw their cells, showing chromosomes and genes, during:

(a) Metaphase I (b) Anaphase of Mitosis

REFERENCES

The following four papers can be found in J. S. Peters, ed., 1959, *Classic Papers in Genetics* (Prentice-Hall, Inc., Englewood Cliffs, N.J.):

BATESON, W., AND R. C. PUNNETT, 1905–8. "Experimental Studies in the Physiology of Heredity."

CREIGHTON, H., AND B. MCCLINTOCK, 1931. "A Correlation of Cytological and Genetical Crossing-Over in *Zea Mays*."

PAINTER, T. S., 1933. "A New Method for the Study of Chromosome Rearrangements and Plotting of Chromosome Maps."

STURTEVANT, A. H., 1913. "The Linear Arrangement of Six Sex-Linked Factors in *Drosophila*, as Shown by Their Mode of Association."

BURGERHOUT, W. G., A. P. M. JONGAMA, J. DEWIT, H. VAN SOMERSEN, AND P. MEERAKHAN, 1975. "Regional Mapping of Seven Enzyme Loci on Human Chromosome 1 by Use of Somatic Cell Hybridization." *Cytogen. Cell Genet.* 14 : 90–92.

EPHRUSSI, B., 1972. *Hybridization of Somatic Cells.* Princeton U. Press, Princeton, N.J.

HARRIS, H., 1965. "Behavior of Differentiated Nuclei in Heterokaryons of Animal Cells from Different Species. *Nature* 206 : 583–588.

MCKUSICK, V. A., 1971. "The Mapping of Human Chromosomes." *Sci. Am.* April : 104–113.

______ AND F. H. RUDDLE, 1977. "The Status of the Gene Map of Human Chromosomes." *Science* 196 : 390–405.

MIGEON, B. R., AND O. S. MILLER, 1968. "Human-Mouse Somatic Cell Hybrids with Single Human Chromosome (Group E): Links with Thymidine Kinase Activity." *Science* 162 : 1005–1006.

MORGAN, T. H., 1911. "Random Segregation Versus Coupling in Mendelian Inheritance." Reprinted in L. Levine, ed., 1971. *Papers on Genetics.* C. V. Mosby, St. Louis.

PEARSON, P. L., A G. J. M. VAN DER LINDEN, AND A. HAGEMEIGER, 1974. "Localization of Gene Markers to Regions of the Human X Chromosome by Segregation of X-Autosome Translocations in Somatic Cell Hybrids." *Cytogen. Cell Genet.* 13 : 136–142.

RISCIUTTI, F. C., AND F. H. RUDDLE, 1973. "Assignment of Three Gene Loci (PGK, HGPRT, G6PD) to the Long Arm of the Human X Chromosome by Somatic Cell Genetics." *Genetics* 74 : 661–678.

SHOWS, T. B., 1974. "Gene Markers for Mapping the Human Genome. The 1974 Listing." *Cytogen. Cell Genet.* p. 29.

WEISS, M. C., AND J. GREEN, 1967. "Human-Mouse Hybrid Cell Lines Containing Partial Complements of Human Chromosomes and Functioning Human Genes." *Proc. Nat. Acad. Sci. U.S.* 58 : 1104–1111.

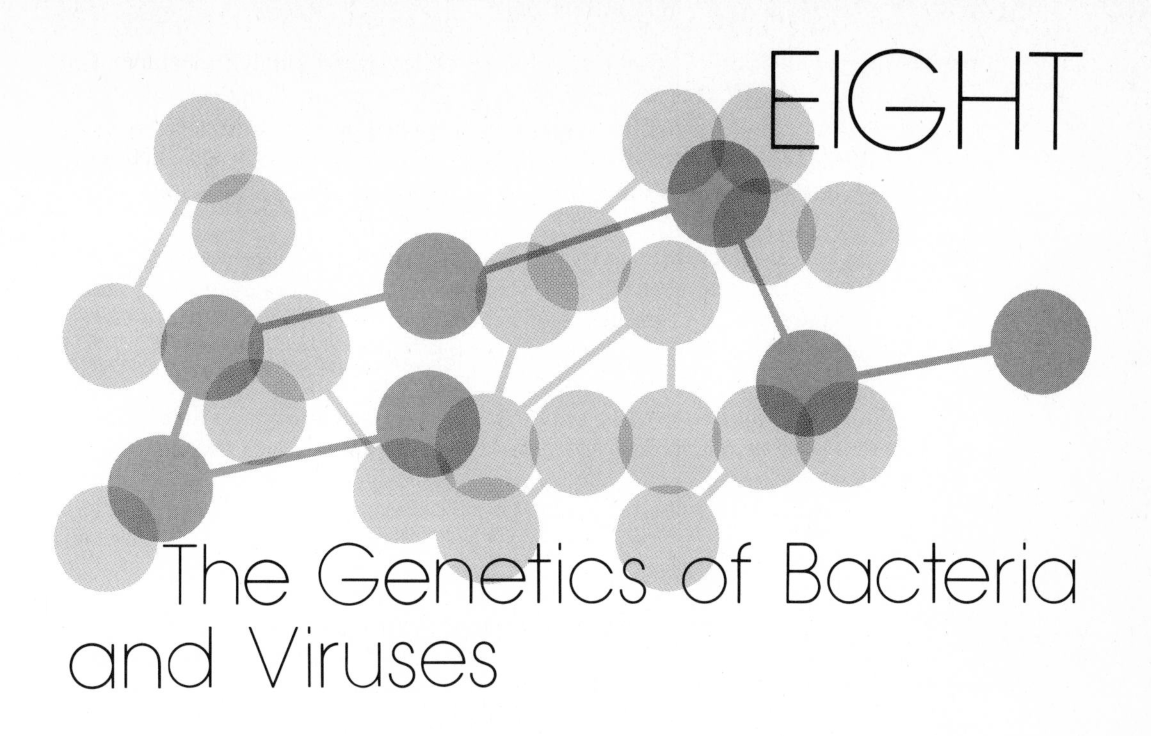

EIGHT

The Genetics of Bacteria and Viruses

INTRODUCTION

At this point in our discussion we have covered the essential aspects of classical, or transmission, genetics. We know that there are discrete units of heredity called genes which are carried on chromosomes, and which are distributed in a relatively predictable manner because of the mechanics of cell division. The genes are situated in a linear fashion along the length of the chromosome. Portions of the chromosome with their constituent genes can be exchanged in crossing-over. Since cross-overs occurred only between genes in classical studies, geneticists visualized genes on a chromosome as being similar to beads on a string, where breakage would occur only between the beads. What we have not discussed, however, is a very fundamental point: What *is* a gene?

From our discussion in Chapter 6, you know that chromosomes are composed of proteins and nucleic acids. Although the chemical nature of these organic compounds had already been elucidated by biochemists, scientists expended vast amounts of energy on genetic research in the first three decades of this century without learning whether genes were proteins or nucleic acids. Consequently, the manner in which the gene determines traits was also unknown.

You were no doubt aware, even before taking a course in genetics, that the genetic material is now known to be DNA. This fact was established by research carried out primarily in the 1940s. In the late 1930s and early 1940s, two major developments in the evolution of genetic research propelled genetics into the modern era of "molecular genetics." One was increasing interest in genes on the part of physicists and chemists, who realized that genes can be studied on the molecular as well as biological level. In the next chapter we shall be discussing some of the milestone experiments performed by researchers, who were trained as physical scientists but who contributed in an invaluable way to this most basic of all biological sciences.

The second development was in the use of new life forms in the study of heredity. As we have mentioned several times, the more frequently organisms reproduce and the larger the number of progeny per generation, the easier it is for geneticists to derive meaningful information from their data. Thus, *Drosophila* had been the organism of choice for most genetic research; however, in the early 1940s a whole new world of research opened up to geneticists when it was discovered that bacteria and viruses undergo primitive forms of sexual reproduction in which there is genetic exchange.

PROKARYOTES IN GENETICS

The existence of sexual reproduction in microorganisms is of tremendous advantage to geneticists. Generations can be studied in a matter of minutes and hours rather than the weeks required for *Drosophila* and the even longer periods for plants and mammals. In a single test tube one can store millions of bacteria and viruses, obviously reducing space and maintenance requirements.

Of equal importance is the genetic simplicity of such microorganisms. The genetic material of bacteria consists of single molecules of nucleic acid (DNA) located in the cytoplasm. Because of the absence of a nucleus, we refer to such organisms as *prokaryotes*. Their DNA appears to be generally free of histones and other proteins in contrast to chromosomes found in *eukaryotes*, organisms whose cells possess nuclei. Consequently, biologists do not consider bacteria to have true chromosomes; however, for want of a better term, chromosome is still used in publications. Of particular advantage to geneticists in the study of these microorganisms is the fact that since they are haploid, all mutations are immediately expressed.

Some Basic Facts About Bacteria

If a bacterium is so small that it can only be seen under a microscope, you may wonder how it is used in studying inherited traits. Simply described, the procedure (carried out entirely under sterile conditions) is to grow the cells in such a way that inherited morphological and physiological traits will manifest themselves. First, the cells are grown in a liquid culture medium. Drops of the medium containing cells are then pipetted onto solid agar plates (petri dishes partly filled with medium mixed with agar which is allowed to solidify). The bacteria are spread evenly over the surface of the agar, usually with a glass bar, and the cells then grow in place using the agar as a source of nutrients.

If the concentration of cells pipetted onto the agar is made low enough, individual cells can be isolated. As each bacterium undergoes binary fission (cell division), it will eventually give rise to a clump of cells that can be seen with the naked eye; these clumps are called *colonies* (Fig. 8-1). If the concentration of cells is very high, the entire surface of the agar will be covered with bacteria and present an opaque appearance. Such a layer of bacteria is sometimes referred to as a *lawn* of bacteria.

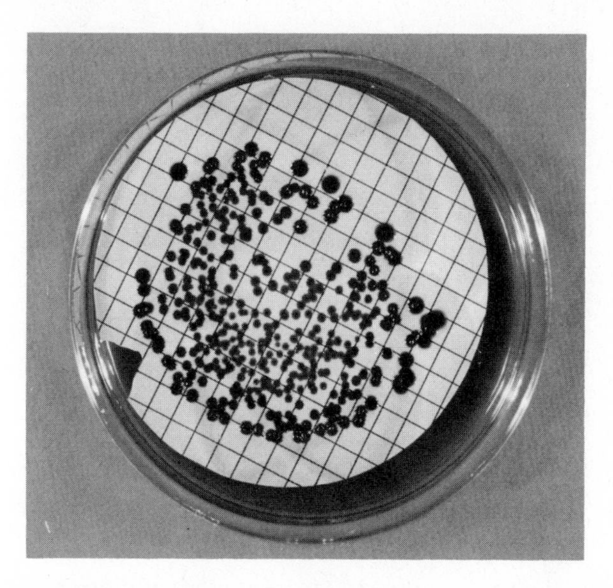

Figure 8-1. Photograph of bacterial colonies growing on agar plates. (Courtesy Dr. Leah Koditschek, Montclair State College)

Inherited traits that can be studied in bacteria include both the morphological and the physiological. Sometimes the size and shape of colonies are changed. Figure 8-2, for example, shows the difference between a wild-type strain of *Diplococcus pneumoniae* (a bacterium which causes pneumonia in mice) and a mutant strain. The wild-type strain forms large, smooth colonies, while the mutant strain forms small, rough colonies which are distinctly different from the wild-type.

Physiological traits are frequently expressed as nutritional needs, just as in the case of fungi (p. 213). The basic medium containing nutrients needed by wild-type bacteria is referred to as *minimal medium*. Often, mutations appear in bacteria which render them unable to synthesize some essential substance, such as a particular amino acid. These mutants then are auxotrophic for the substance, which must be added as a supplement to the minimal medium in order for the mutant strain to survive. One can detect such inherited traits by plating bacteria on different selective media. Remember that because bacteria are haploid, all mutations are expressed; new mutations are never masked by dominant traits.

Replica plating, a particularly ingenious technique for detecting mutations in bacteria, was devised by Joshua and Esther M. Lederberg (1952). In this method, a "master" plate of bacteria colonies is grown. A piece of wood or other stiff material (which has been cut to fit the surface of the agar plate), covered with a sterile piece of velvet cloth, is pressed to the master plate so that bacteria adhere to the

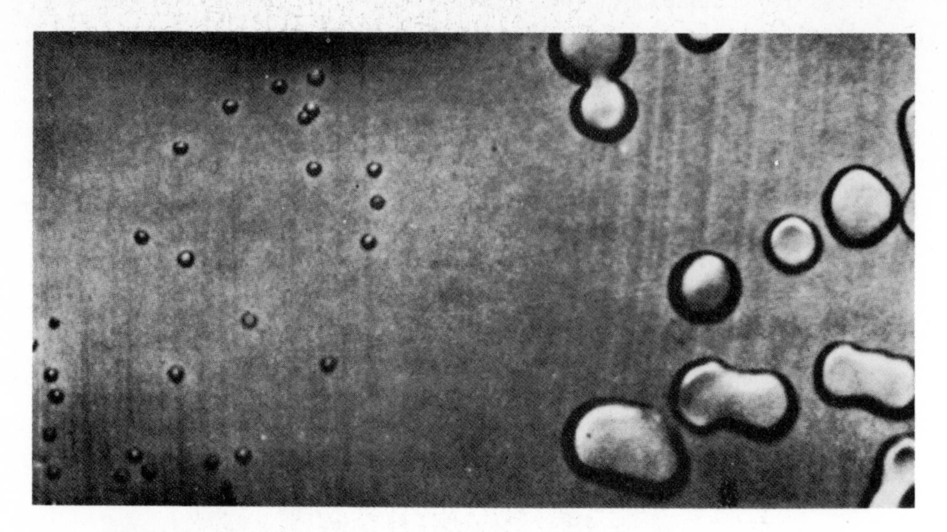

Figure 8-2. Morphological differences are easily visible in growth of small, rough mutant colonies of *Pneumococcus* (*R*, left). Wild-type bacteria form large, smooth colonies (*S*, right). (From O. T. Avery, C. M. Macleod, and M. McCarty, 1944, reprinted in J. Peters, ed., 1959. *Classic Papers in Genetics*, p. 179. Prentice-Hall, Englewood Cliffs, N.J.)

cloth. The cloth is then pressed into a number of different agar plates prepared with different media.

Colonies from cells able to survive in the different media will appear and can be transferred to a test tube and grown in liquid medium for further study. Since these plates are replicas of the master plates, the bacteria can also be identified on the master plate (Fig. 8-3). Colonies which do not survive on replica plates presumably are of cells which are mutant and auxotrophic for an essential substance.

Some Basic Facts About Viruses

Viruses are even simpler organisms than bacteria. They, too, possess only a single strand of genetic material, that is, they are haploid. Some viruses contain DNA, others RNA, but all contain very little cytoplasm. Viruses are essentially nucleic acid particles bound by a protein coat (Fig. 8-4). Because of their lack of cytoplasm, and of cytoplasmic organelles essential for metabolism and division, viruses must infect cells and depend on the host cell to provide the chemicals and machinery for reproduction. They are obligatory parasites, but unlike eukaryotic obligatory parasites, viruses cannot reproduce even if grown in a medium with every conceivable nutrient. They must be inside a cell in order to reproduce.

Viruses which parasitize bacteria are referred to as *bacteriophages*, or *phages* for short. We shall discuss viruses that parasitize animal cells in a later chapter, as they play a significant role in diseases such as polio and cancer. In the following sections we shall discuss the various developments which occur when bacteria are infected by phages.

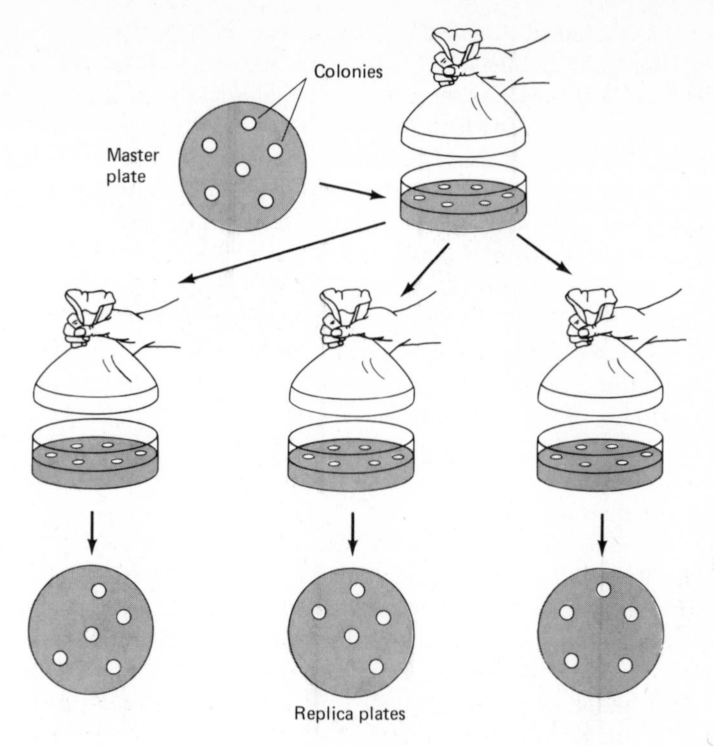

Figure 8-3. The technique of replica plating. Colonies from a master plate with complete medium are picked up by pressing a sterile piece of velvet onto the plate. The velvet is then pressed onto agar plates containing medium missing a particular nutrient. Colonies which do not appear on the replica plate indicate mutants lacking the ability to synthesize that substance. They can then be picked from the master plate for further studies.

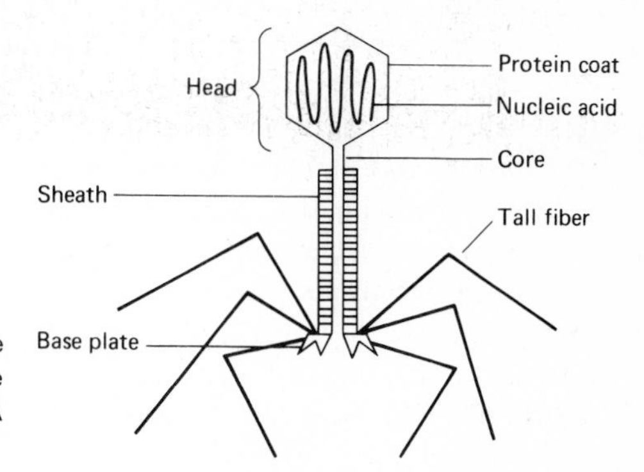

Figure 8-4. Structure of a virus, bacteriophage T2. The virus measures 2,000 Å from the tip of the head to its base plate. The head measures 650 Å in width.

The lytic phase. When a virus infects a host bacterium, one of the events that can take place is that the phage can cause the cell to produce more phage particles. Eventually the host cell breaks open and releases new viruses. This process is known as *lysis*, and results in the death of the host cell. This phase in the life cycle of viruses is known as the *lytic*, or *vegetative*, *phase*.

Lysis can be visualized as clear areas in a bacterial lawn, which result from the release of new phage particles upon lysis of a bacterium. These phages in turn infect and kill surrounding cells. The small clear area is known as a *plaque*. Some of the inherited traits which have been studied in phages involve the morphology of plaques, such as size, shape, and clearness.

It has been found that the number of plaques formed corresponds directly to the number of viruses injected into a bacterial culture. Thus it appears that each plaque originates from a single virus particle. The process of plaque formation eventually stops because of changes in bacterial metabolism on which lysis depends. Plaques are generally circular and limited in size.

The temperate phase. Another route that can be taken by some bacteriophages after infection is to remain dormant in the cell—that is, the host cell is not lysed. In the dormant state, the viruses seem not to affect the host cell. Bacteria which harbor dormant viruses are said to be *lysogenic*, and this phenomenon is referred to as *lysogeny*. Viruses in the dormant state are said to be in the *temperate phase* of their life cycle.

Studies have shown that viruses in lysogenic bacteria can exist in one of two states. They can remain autonomous in the cytoplasm of the cell. As the host cell divides, the virus can be transmitted to daughter cells. On the other hand, a daughter cell may not receive a copy of the virus; such cells are said to be *cured*.

Alternatively, the phage can become integrated into the bacterial chromosome. When a phage is thus integrated, it is known as a *prophage*. Prophages will divide along with the rest of the bacterial DNA and be transmitted to all daughter cells. Interestingly, it has been shown that the presence of a prophage confers a type of immunity to the host cell. Such a cell cannot be lysed by other viruses that might subsequently infect the cell.

Bacteriophages are also known to switch from the temperate to the vegetative state. Even when integrated as a prophage, the virus can react to various factors and break away from the bacterial chromosome in a process known as *excision*. The mechanism of excision is not fully understood. It is known, for example, that exposure of lysogenic bacteria carrying a prophage to ultraviolet light can initiate the vegetative phase. This process of causing proviruses to reproduce is known as *induction*. Figure 8-5 summarizes the life cycle of a virus.

Having discussed briefly some basic aspects of bacteria and viruses and their life cycles, let us proceed to a consideration of the forms of sexual reproduction in these organisms which result in genetic exchange.

SEXUAL REPRODUCTION IN BACTERIA

There are four major mechanisms for genetic exchange in bacteria, two of which require agents to mediate the exchange. The first to be discovered was *transformation*, the ability of certain bacteria to absorb genetic material which is in their environment, and incorporate the exogenous genes into their own chromosomes. This process can cause detectable changes in genotype of the cells.

Genetic exchange between bacteria can also occur during *conjugation*. In this

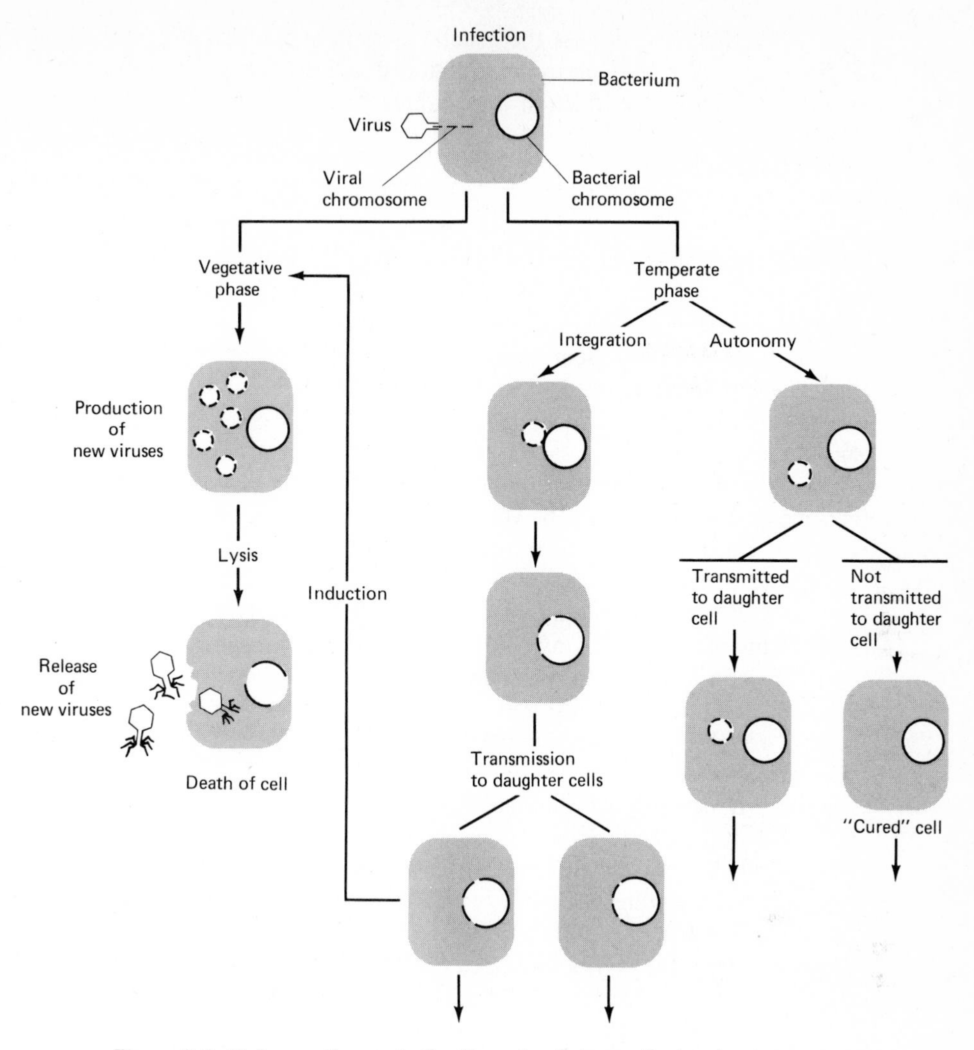

Figure 8-5. Various pathways in the life cycle of viruses. Explanation in text.

process, two bacteria actually come in contact, and genes from one cell enter the other through a "bridge" formed between the two, much as in *Paramecium* conjugation (p. 420).

Bacteriophage can effect an exchange of genes between host cells by a process known as *transduction*, in which new viral particles carry genes from one host cell into a new host cell when they infect it. There can then be an exchange of genes between the viral chromosomes and those of the second host.

The fourth process, *sexduction*, is mediated by a nonviral, extrachromosomal genetic factor in bacteria. We shall discuss such particles further in Chapter 13, but we include this process in this chapter because of the great similarity between transduction and sexduction, the results of which can be an exchange of genes between bacteria.

We shall first discuss the genetic aspects of transformation and conjugation. Following an introduction to the genetics of viruses, we will then turn our attention to transduction and sexduction. In Chapter 9 we shall explore the molecular aspects of all these mechanisms.

Transformation

As mentioned above, the first genetic exchange discovered in prokaryotes involved transformation in bacteria. Microbiologist F. Griffith (1928) reported some puzzling results from studies on *Pneumococcus*. You may recall that the wild-type strain (called *IIIS*; the *S* represents smooth) causes pneumonia in mice, and when plated on agar, produces large, smooth colonies. Griffith discovered in the course of his work a mutant strain which produces small rough colonies (called *R* strain). See again Fig. 8-2.

Experiments showed that the mutant colonies lacked the polysaccharide capsule of the wild-type cell. When injected into mice, the *R* strain *Pneumococcus* failed to cause pneumonia, and thus had lost its virulence. However, when *R* bacteria were mixed with heat-killed *IIIS* bacteria, and the mixture injected into mice, the mice died of pneumonia. Bacteria isolated from body fluids of the dead mice showed the presence of living wild-type smooth bacteria (Fig. 8-6)!

Identification of the transforming agent. Since it was unlikely that the heat-killed bacteria had been restored to life, scientists looked for some way in which the mutant rough-type bacteria could have been transformed back to wild-type. (This was why the term *transformation* was used to describe the phenomenon.) In 1944, using newly developed techniques for fractionating cells and isolating specific fractions for study, O. T. Avery, C. M. MacLeod, and M. McCarty found what was then referred to as the *transforming principle*. The only fraction of wild-type cells that could cause transformation to wild-type when mixed with cultures of *R* bacteria was that containing DNA. If the fraction was first treated with an enzyme which degrades DNA (deoxyribonuclease, or DNAase), no transformation was obtained (Fig. 8-7). On the other hand, RNAase and proteases did not affect transformation.

THIS WAS AN EXTREMELY SIGNIFICANT FINDING, FOR IT NOT ONLY CLARIFIED A METHOD IN WHICH BACTERIA CAN EXCHANGE GENES, IT STRONGLY SUGGESTED THAT THE GENETIC MATERIAL IS ACTUALLY DNA. THE PHENOMENON OF TRANSFORMATION IS A TRUE GENE EXCHANGE, FOR TRANSFORMED CELLS GIVE RISE TO PROGENY WITH THE SAME TRANSFORMED GENOTYPE.

Transformation as a genetic phenomenon. Further confirmation was needed, however, that this phenomenon was the result of genetic exchange. It was necessary to eliminate the possibility that a physiological reaction to the presence of foreign DNA might result in capsule formation. In 1949, experiments by Harriet Taylor and by R. D. Hotchkiss, among others, provided this confirmation. Taylor found a mutant strain that was extremely rough (*ER*), and found that DNA from the *R* strain could cause the transformation of *ER* to *R*. When exposed to DNA from *IIIS* strains, the transformed *R* cells were transformed a second time, to wild-type. These results are best explained by an actual transfer of genetic material.

Hotchkiss then showed that transformation of other traits is possible. He

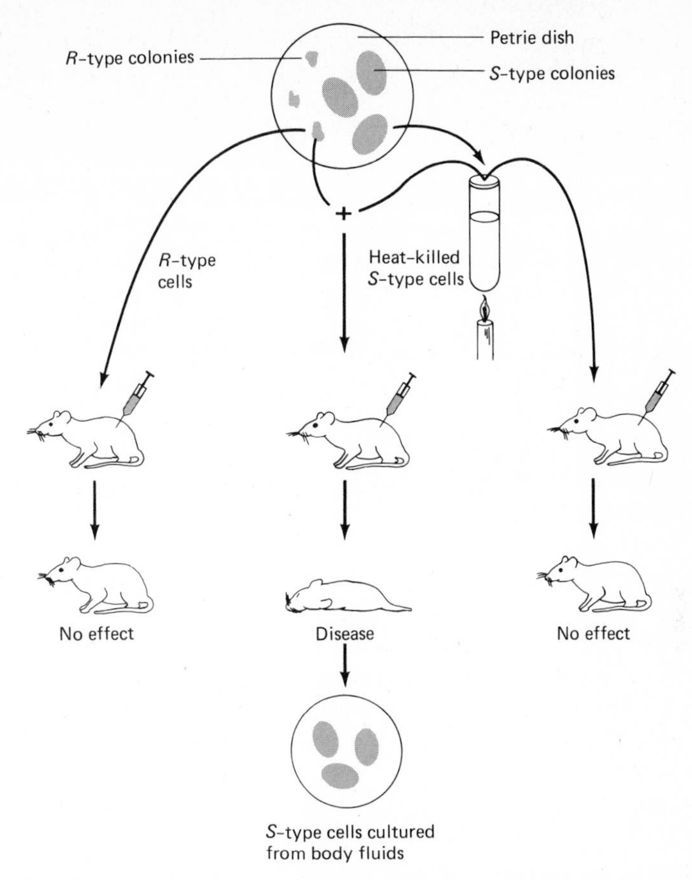

Figure 8-6. Summary of F. Griffiths classical experiment showing transformation of *Pneumococcus*. See text for explanation.

found a mutant strain of *Pneumococcus* which, though smooth-type with regard to capsule formation, is resistant to penicillin (Pen^r), unlike the wild-type which is sensitive (Pen^s). When he added DNA from the penicillin-resistant cells ($Pen^r S$) to penicillin-sensitive *R* bacteria ($Pen^s R$), he obtained transformed cells that were $Pen^s S$, $Pen^r R$, and a small number of $Pen^r S$ (Fig. 8-8). Hotchkiss concluded that transformation is the result of genetic exchange, and that there are separate genes determining the traits of penicillin resistance and capsule formation, since they can be transmitted separately.

Mapping chromosomes by transformation. Once the genetic nature of transformation was established, it was clear that transformation could provide an approach to the mapping of the bacterial chromosome. If pieces of the DNA are absorbed by bacterial cells and incorporated into the chromosome, it follows that two genes on different pieces of DNA can undergo transformation independently.

Further, the number of double transformants for the two genes should be equal to the product of single transformation events (see again our multiplication rule for independent events, p. 46). This was found to be the case. However, occa-

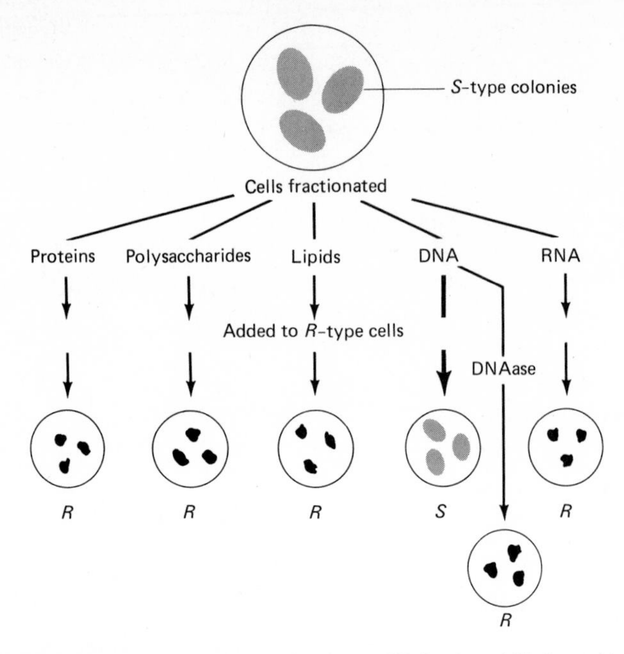

Figure 8-7. Scheme of experiments by Avery, Macleod, and McCarty identifying DNA as the transforming agent. This was the first clear indication that the genetic material of cells is DNA.

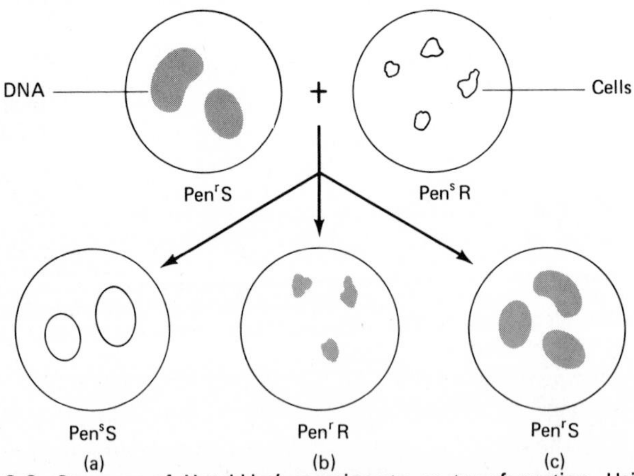

Figure 8-8. Summary of Hotchkiss' experiments on transformation. Using DNA from *Pneumococci* differing in two traits from a second strain of *Pneumococci*, the traits were found to be transmitted separately to transformed cells, as in (a) and (b), and occasionally together, as in (c). This indicated that two different genes were involved, and further confirmed DNA as the transforming agent. Cross-hatching indicates penicillin resistance.

sionally the frequency of double transformants was higher than expected, a phenomenon found to be due to very close linkage between two genes, which increased the likelihood that they would be on the same portion of the DNA that is absorbed by the transformant. The farther apart two genes are, the less likely that they will be absorbed on the same molecule of DNA.

Let us assume that we wish to determine the distance between two closely linked genes. One gene determines a nutritional requirement for leucine, and the other a nutritional requirement for threonine. We can take DNA from a strain which is wild-type for both, leu^+thr^+, and transform a strain mutant for both traits, leu^-thr^-. The same approximate number of cells are then plated on three different media: medium #1, with all nutrients except leucine; medium #2, with all nutrients except threonine; and minimal medium #3, lacking both leucine and threonine. Genotypes of cells that survive in the three media are given in Table 8-1.

Table 8-1

		Media	
	#1	*#2*	*#3*
Genotypes of surviving cells	thr^+leu^+ thr^-leu^+	thr^+leu^+ thr^+leu^-	thr^+leu^+

The percentage of single transformants would be the measure of distance between the two loci, and can be calculated by the following formula:

$$\frac{\text{number of single transformants}}{\text{number of single transformants} + \text{number of double transformants}}$$

In our example, the number of single transformants can be calculated by subtracting the number of colonies obtained in medium #3 from the number in medium #1 to obtain the number of thr^-leu^+ colonies. Similarly, by subtracting the number of medium #3 colonies from those of medium #2 we obtain the number of thr^+leu^- transformants. Applying this to the preceding formula we obtain:

$$\frac{\text{number of } thr^-leu^+ + \text{number of } thr^+leu^-}{\text{number of } thr^-leu^+ + \text{number of } thr^+leu^- + \text{number of } thr^+leu^+}$$

Although transformation did supply geneticists with information on the distance between genes, it was not the most reliable method for this kind of study. For one thing, the size of DNA fragments involved in transformation cannot be controlled, resulting in a certain amount of ambiguity. Furthermore, not all species of bacteria are transformable. In addition to *Pneumococcus*, *Bacillus subtilis* and *Hemophilus influenzae* have been found to be transformable. The bacterium which has been most commonly studied by geneticists, *Escherichia coli*, a colon bacterium found in humans, is not transformable in nature. (In Chapter 14, we will discuss a means of experimentally rendering *E. coli* cells transformable.) Conjugation, the method of genetic exchange which contributed more to genetic mapping of bacterial chromosomes than transformation, is our next topic of discussion.

Conjugation

The classical experiment which led to the discovery of conjugation in *E. coli* was carried out by Lederberg and Tatum (1946). To study genetic exchange in bacteria, they developed two strains with multiple nutritional requirements. Strain *A* required methionine (met^-) and biotin (bio^-); strain *B* required threonine (thr^-), leucine (leu^-), and thiamine (thi^-).

Cells of both strains were cultured together in complete medium and allowed to grow overnight. The cells were then washed to remove the complete medium and transferred to minimal-medium agar plates. Since the cross would be:

$$met^-bio^-thr^+leu^+thi^+ \times met^+bio^+thr^-leu^-thi^-,$$

cells of the original genotypes would not survive on minimal medium. However, a few colonies did appear on the agar plates. These had to be prototrophic for all requirements ($met^+bio^+thr^+leu^+thi^+$), and were therefore assumed to be recombinants as a result of genetic exchange.

Genetic exchange in conjugation. The difference between conjugation and transformation was revealed by experiments in which the strains of bacteria used by Lederberg and Tatum were physically separated. One such experiment (Davis 1950) utilized a U-shaped tube with a filter placed in the bend. The pores of the filter were small enough to prevent cells from passing through, but would allow medium molecules the size of DNA to pass through. Strain *A* cells were placed on one side of the filter and strain *B* cells on the other side (Fig. 8-9). No recombination occurred, thus

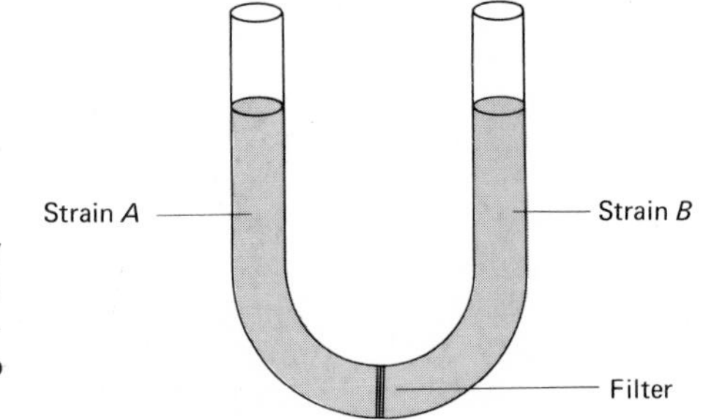

Figure 8-9. The U-shaped tube devised by Davis to separate cells of strain *A* (met−bio−) from cells of strain *B* (*thr−leu−thi−*). Prevention of physical contact between cells of the two strains resulted in absence of recombination.

physical contact was proven to be necessary for genetic recombination as observed by Lederberg and Tatum. Later, electron micrographs showed the presence of cytoplasmic bridges called *pili* (sing. *pilus*) connecting the bacteria (Fig. 8-10), through which the genetic exchange presumably occurs. The phenomenon was termed *conjugation.*

As the process of conjugation was studied further, however, a number of areas of confusion arose. The mapping of the *E. coli* chromosome by Lederberg and others using the results of conjugation experiments did not yield very clear results. Figure 8-11 shows the first genetic map of the *E. coli* chromosome. As you can see, there were problems in establishing a linear order of genes.

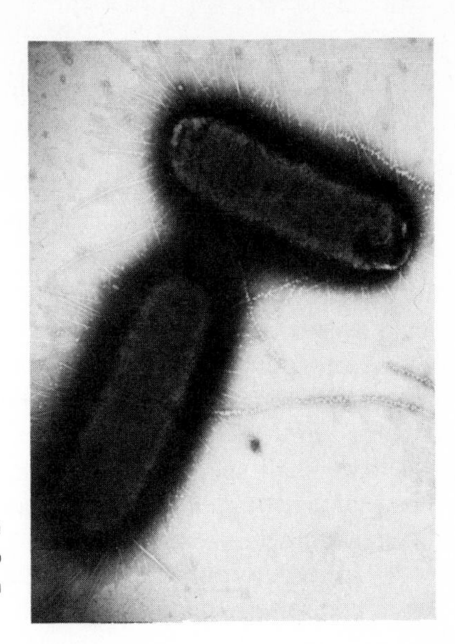

Figure 8-10. Electron micrograph of bacteria in conjugation. (Photograph courtesy Dr. Philip Silverman and Lillian Eoyang, Albert Einstein College of Medicine.)

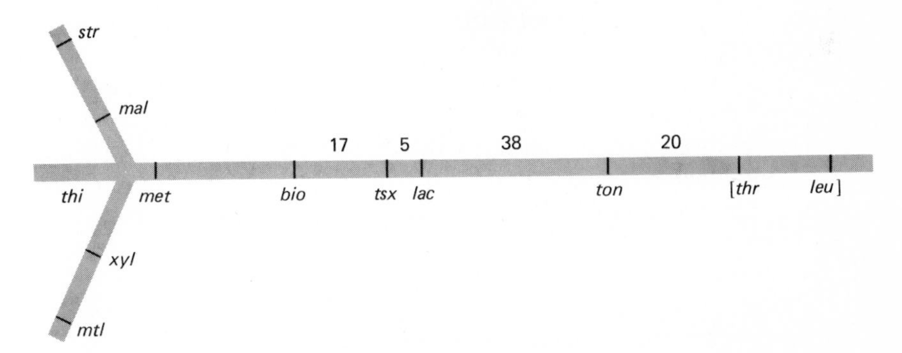

Figure 8-11. An early concept of the map of the *E. coli* chromosome proposed by Lederberg. (From G. S. Stent 1971. *Molecular Genetics*, p. 256. W. H. Freeman & Company Publishers, New York. Copyright © 1971.)

One-way transfer. One of the problems was that Lederberg and Tatum assumed that there was a *mutual* exchange of genes by the two cells undergoing conjugation. Experiments reported in 1952 by William Hayes indicated that this was not the case. Hayes found that sterilization of strain *A* cells by streptomycin did not affect the rate of recombination; however, if strain *B* cells were sterilized, no recombination occurred. Hayes concluded correctly that genetic exchange in conjugation is a one-way transfer; in this case, strain *A* is the donor strain and *B* is the recipient. Sterilization did not inhibit the ability to transfer genes, but did interfere with the actual process of recombination of genes.

The sex particle. It was also found by Hayes (1953) and others (Cavalli-Sforza *et al.* 1953) that the ability to donate genes in conjugation was itself transferable. Hayes had found a mutant strain of bacteria among the *A* cells which had lost its ability to donate genes during conjugation and had become streptomycin-resistant. Yet when these mutants were mixed with normal strain *A* cells, about one-third of the streptomycin-resistant cells became capable of being donors during conjugation. From these observations and others, it was concluded that a sex factor, called F^+ (for fertility), exists in donor bacteria and can be transferred from donor to recipient. It is absent in recipient cells which are F^-.

Subsequent studies revealed extraordinary aspects of the sex factor, and contributed much to our understanding of bacterial genetics in general. First of all, the *F* factor was shown to be constituted of DNA (Driskell and Adelberg 1961), and could thus be considered an extrachromosomal genetic element. Much like viruses, the *F* factor, or particle, was found to exist either autonomously in the cytoplasm, or integrated into the bacterial genome. However, unlike viruses, the *F* factor can only be transferred by cell contact. The *F* particle and other similar particles (of which we shall speak later in Chapter 13) were classified as *episomes*, extrachromosomal genetic elements which can exist autonomously in the cytoplasm, or in an integrated state.

In its autonomous state, the *F* particle can undergo division independently of the host chromosome. Consequently, new copies of the *F* particle can be transferred during conjugation to an F^- cell converting the F^- cell to F^+ (Fig. 8-12). The mechanism of reproduction of the *F* particle DNA is believed to be the same as that of some viruses, and we shall discuss this mechanism in detail later (p. 289, Fig. 9-24). Transfer of the episome was found to occur at a rate of 1 in 10 cells, as compared to the transfer and recombination of bacterial cells during conjugation at a rate of 1 in 10^6 cells. This was a clear indication that the *F* particle was transferred alone more frequently than not, and confirmed the autonomous state of the episome in F^+ cells.

When integrated into the host cell chromosome, the *F* particle is reproduced in synchrony with the rest of the genome. If a cell containing an integrated sex particle undergoes conjugation, the rate of recombination of bacterial genes between the cells increases a thousandfold. Such bacteria were designated *Hfr*, for high-frequency recombination (Hayes 1970).

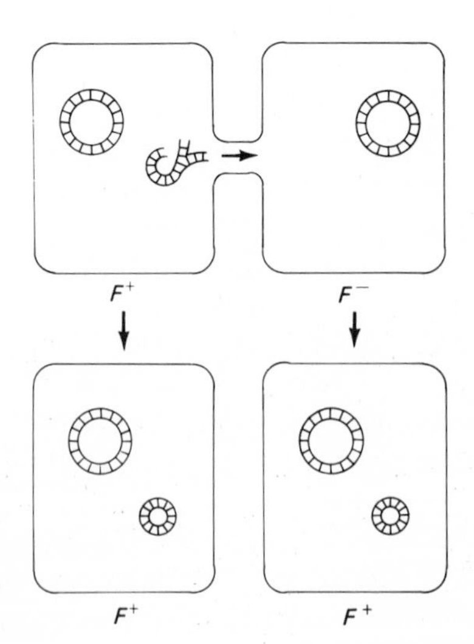

Figure 8-12. Schematic illustration of the transfer of the *F* particle from an F^+ cell to F^- recipient. At the conclusion of conjugation, the F^- cell is now F^+, and capable of being a donor cell.

The basis for this increase in recombination frequency is that when the *F* particle is integrated into the genome of the donor cell, breakage of the DNA occurs at a point within the integrated episome. The end of the DNA containing a portion of the *F* particle known as the *origin* (O) will be transferred first (Fig. 8-13). As it is transferred, the DNA is duplicated by the same mechanism as the autonomous *F* particle.

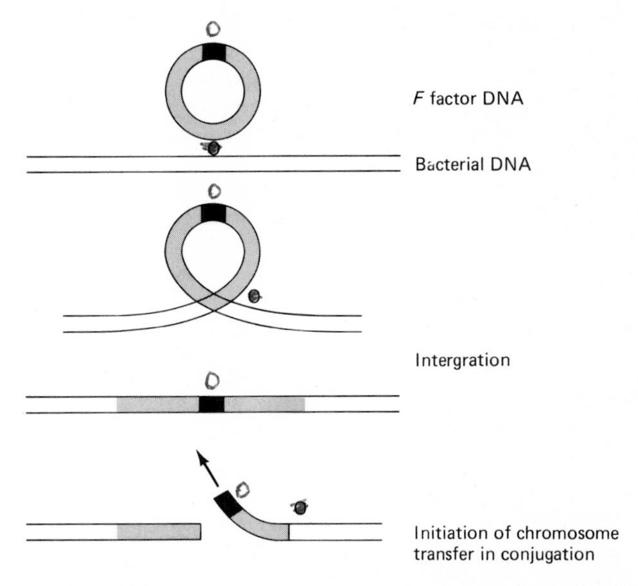

Figure 8-13. Schematic illustration of the integration of an *F* episome into the host chromosome. When conjugation occurs, the episome DNA breaks, and the first to be transferred is the origin.

If the entire new copy of DNA is transferred, the recipient cell will receive the whole integrated episome as well as the entire donor bacterial genome. The recipient F^- cell becomes *Hfr* if the *F* particle from the donor becomes integrated into its own genome by recombination. This occurs only occasionally. In the majority of cases, only a portion of the donor chromosome is transferred, and the recipient cell remains F^- since the sex particle remains in the donor cell. Nonetheless, some genetic material of the donor cell does move into the F^- bacterium and recombination between the fragment and cell genome can ensue (Fig. 8-14). There is, in other words, no way that the sex factor can be transferred when integrated unless bacterial genes are transferred in conjugation. Therein lies the explanation for the increase in recombination frequency in *Hfr* cells as compared to that in F^+ cells.

For reasons that are not understood, the sex factor in an *Hfr* cell will deintegrate. When this occurs, the cell becomes F^+. The ability of episomes to become incorporated into the host genome and to leave the genome is very similar to viruses. Episomes, however, can be transferred from one cell to another only by cell contact.

Interrupted mating experiments. The manner in which the donor chromosome fragments during conjugation had been discovered in a series of elegant experiments

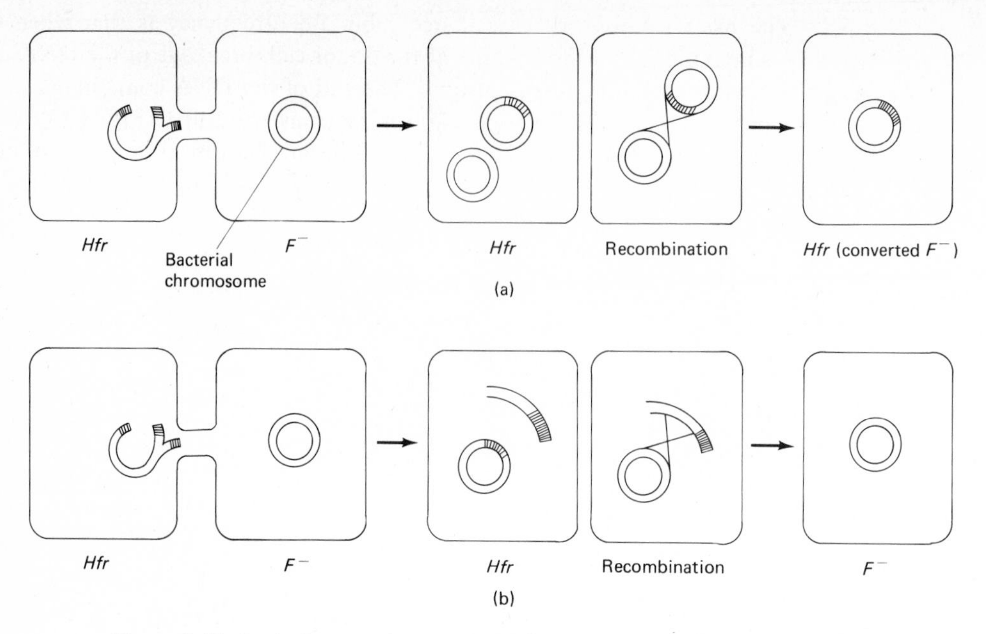

Figure 8-14. In the integrated state, the *F* episome mediates transfer of bacterial genes to recipient cells. If an entire copy of the donor genome is transferred, the recipient cell can be converted to *Hfr* (a). If only a portion of the donor genome is transferred, the recipient cell remains F^-; however, recombination of other genes can occur (b).

known as *interrupted mating experiments*, conducted by E. Wollman and F. Jacob in the 1950s. They made the following cross, $a^+b^+c^+d^+e^+f^+ \times a^-b^-c^-d^-e^-f^-$, with the wild-type cells being the donor strain. (The actual genes involved were thr^+leu^+, azi^r, Tl^r, lac^+, gal^+, but for simplicity's sake, we shall speak of *a*, *b*, *c*, *d*, *e* and *f*.) At specified times after mixing, samples were taken from the nutrient broth and processed in an electric blender, which caused conjugating cells to separate. The researchers then determined which genes had been transferred within the specified time.

They found that the a^+ gene is transferred at about 8 minutes of the experiment; b^+ and c^+ at about 9 and 10 minutes of the experiment respectively; d^+ at 17 minutes; e^+ and f^+ at 25 minutes. These results indicated that genes of the donor chromosomes are transmitted in a linear fashion. It became apparent from these results that in the majority of matings with *Hfr* strains, the chromosome is only partially transferred during conjugation. Different genes are transferred in different matings.

Other developments shed further light on the process of gene exchange during conjugation. For a sequence of genes on a chromosome, such as *abcde . . . xyz*, the sequence of transfer was found to be variable; it did not always have to begin with gene *a*. For example, it might begin with gene *e* and include the transfer of genes *xyz*. Or it might begin with *xyz* and continue with *abc*. These data led geneticists to conclude that the bacterial chromosome must be circular, as opposed to the linear chromosomes of eukaryotes. Figure 8-15 illustrates this point.

The interrupted mating experiments also established other concepts of con-

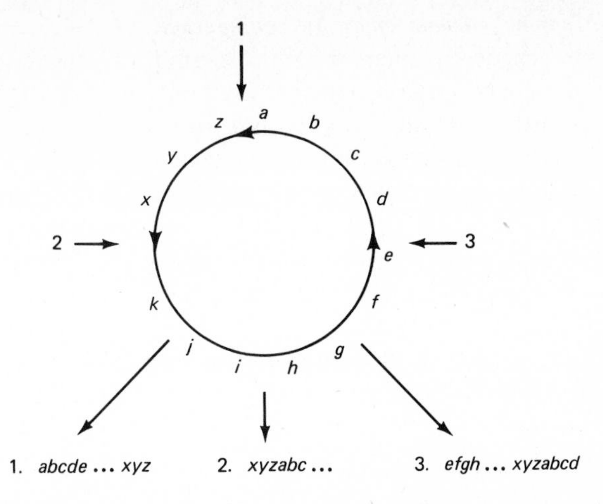

Figure 8-15. A circular chromosome, if broken in different areas (1, 2, 3), would result in the transfer of genes as written during conjugation. Arrows on the chromosome indicate direction (or polarity) of transfer.

jugation. By the randomness with which different portions of the chromosomes were transferred from donor to recipient in these experiments, it became apparent that the *F* particle can be inserted at various sites on the bacterial chromosome, and with different polarity. Thus the sequence *abcde* can be transferred either $a \longrightarrow e$ or $e \longrightarrow a$ (Fig. 8-16).

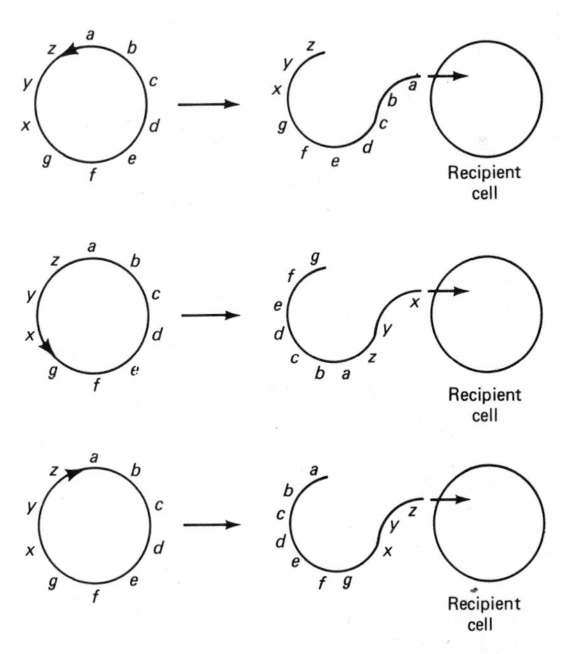

Figure 8-16. Effect of different sites of integration and orientation of the *F* particle on sequence of gene transfer during conjugation.

Mapping chromosomes by conjugation. These studies were repeated many times with genetically different strains, permitting geneticists to gain information on the linkage map of bacteria. THE SEQUENCE OF GENES COULD BE DETERMINED AS WE HAVE ALREADY DISCUSSED, AND THE TIMING OF TRANSFER OFTEN INDICATED DISTANCES BETWEEN LOCI. IN ADDITION, THE TRANSFER OF PORTIONS OF THE DONOR GENOME WHICH THEN UNDERWENT RECOMBINATION WITH HOMOLOGOUS REGIONS OF THE RECIPIENT CHROMOSOME PROVIDED MORE INFORMATION ON THE LINKAGE MAP. The piece of donor chromosome is called the *exogenote*, and the recipient chromosome is called the *endogenote*. Figure 8-17 illustrates a hypothesis for how this type of genetic exchange can occur.

There is most likely a double cross-over in order to obtain integration of the exogenote. The fragment that is the reciprocal product of this cross-over event is

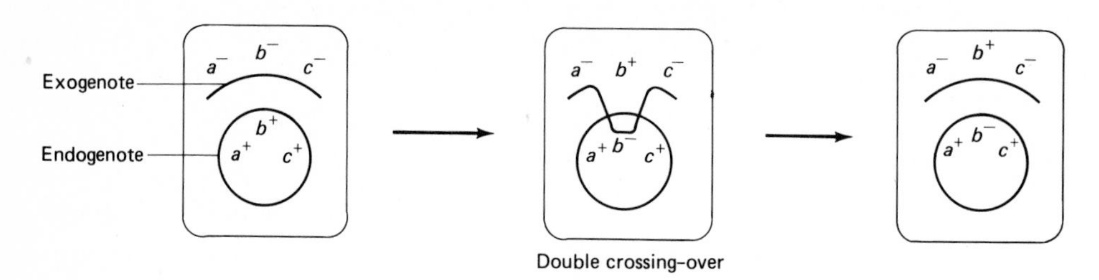

Figure 8-17. Proposed mechanism of recombination between donor DNA (exogenote) and recipient DNA (endogenote) following conjugation.

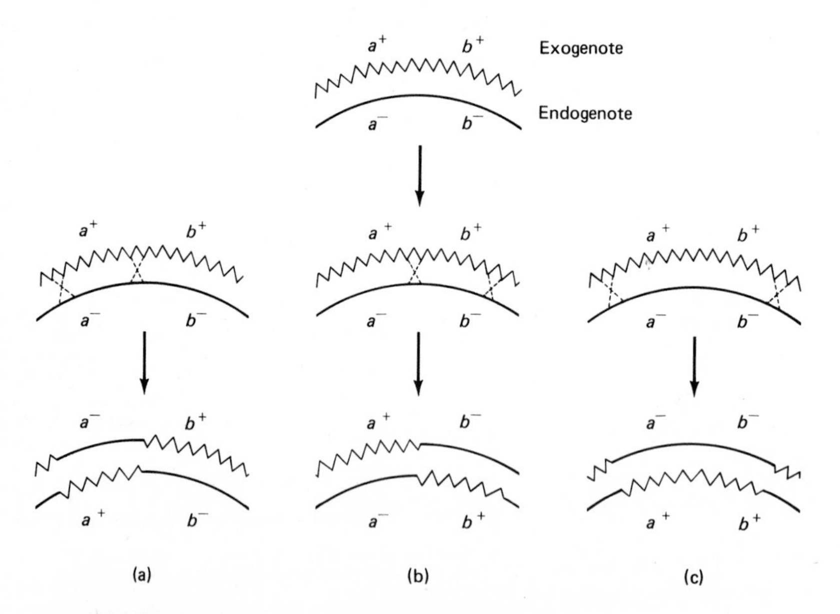

Figure 8-18. The frequency with which two loci are found transferred to the recipient chromosome is a measure of the distance between them. The closer they are, the more frequently they will be transferred together in recombination events following conjugation.

lost by enzymatic degradation. We can deduce the relative positions of genes by the frequency with which two loci are found in recombinants. For example, if we cross a donor a^+b^+ strain with a recipient a^-b^- strain, some of the a^+ recombinants will be b^+ also, and some will be b^-, as shown in Figure 8-18. Conversely, the b^+ recombinants will be either a^- or a^+. The number of a^+ and b^+ recombinants is a measure of the frequency of crossing-over and therefore of the distance between the two loci. The closer they are, the more frequently one will obtain a^+b^+ recombinants (representing no crossing-over between a and b). To determine the frequency, we use the following simple formula:

$$\frac{(\text{number of } a^+b^-) + (\text{number of } a^-b^+)}{\text{total number of colonies}}$$

Information obtained in this manner on the distance between genes has been found to correlate with the data found in interrupted mating experiments. Figure 8-19 shows the linkage map of *E. coli* as determined by studies of conjugation.

Before we can discuss transduction in bacteria, we must first familiarize ourselves with experiments on the genetics of bacterial viruses.

GENETICS OF BACTERIOPHAGE

Bacteriophages were discovered in the early years of the twentieth century, but it was not until the 1940s that they were found to be useful in genetic analysis. Their usefulness was due to the fact that study of the genetics of any organism requires inherited traits, and as we have discussed so often, inherited traits can only be detected, and their transmission studied, when variations of the trait occur. In 1946, A. D. Hershey, who was eventually to win a Nobel prize for his work in viral genetics, published reports of inherited traits in phages.

Recombination in Bacteriophages

The inherited traits which Hershey studied in phages were of two kinds: One involved the morphology of plaques formed, the other, a trait known as *host range*. The particular phage that he used was known as T2, one of a series of virulent phages that infect *E. coli*.

Wild-type T2 produces small plaques which appear to have a turbid "halo" around the perimeter. This phenomenon is apparently due to the adsorption of newly produced phage particles on bacteria that have already been infected, but have not yet lysed. The adsorption of phages onto these cells prolongs their survival for reasons not yet understood. This phenomenon is known as *lysis inhibition*, and it is a wild-type characteristic of T2, genetically determined by a gene designated r^+. Hershey had found that mutation of this gene results in large plaque formation, which he called *rapid lysis*, and the trait is designated by the letter r (Fig. 8-20).

The host range characteristic was discovered in the following way (see reference to Hayes 1970 for greater detail): if a strain of bacterium called *E. coli* B is plated with a very high concentration of T2 phage, normally lysis of all the bacteria

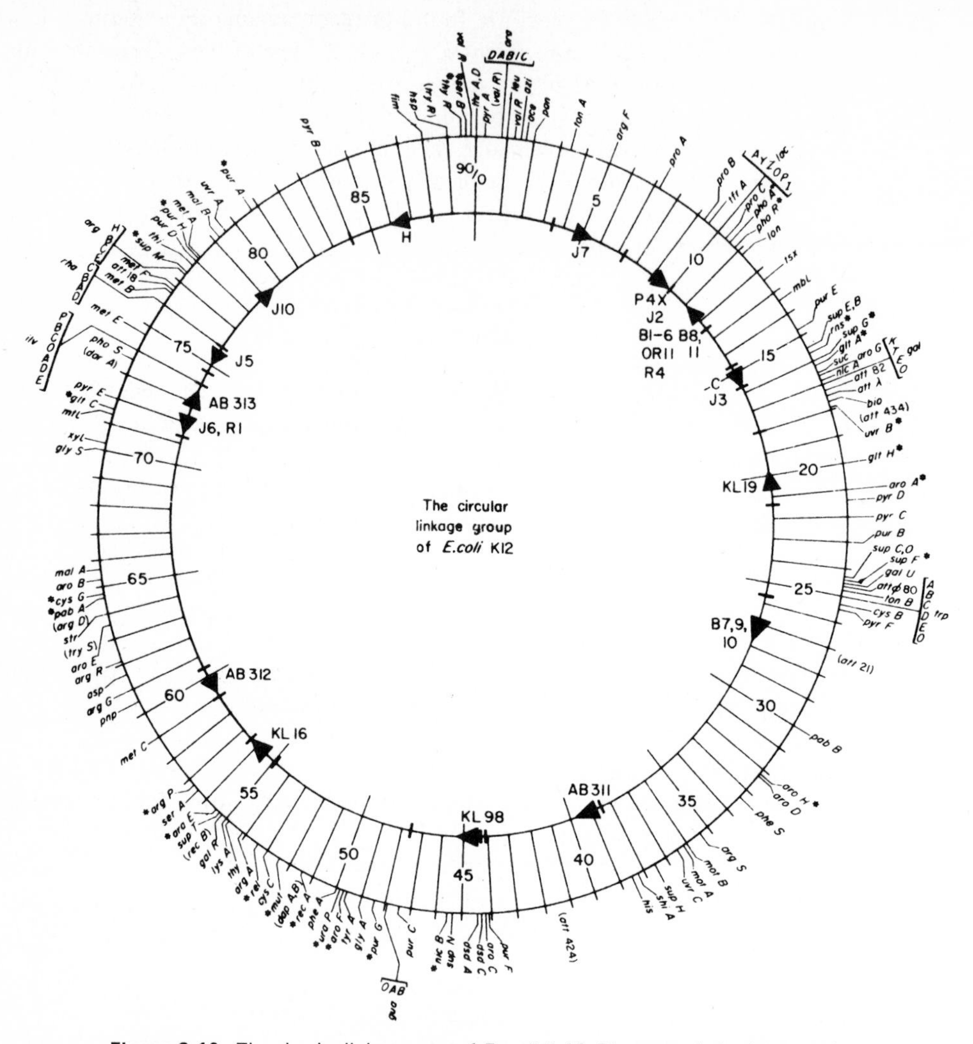

Figure 8-19. The circular linkage map of *E. coli* K-12. The outer circle shows positions of 166 loci. Inner circle indicates origins of various *Hfr* strains. Direction of arrowhead indicates polarity of transfer. The map is drawn in time units, with 90 minutes as the time required for the transfer of the entire chromosome. (From A. L. Taylor and C. D. Trotler, 1967. *Bact. Rev.* 31 : 331.)

will occur and the plaques will be confluent, resulting in a plate essentially devoid of cells. However, occasionally mutant bacteria will survive. These do not absorb T2 phage, and are called B/2 (B "bar" two). When B/2 cells are plated with T2 phages one would therefore not expect to find plaques. However, mutants of T2 can be picked up in the occasional plaque formed among B/2 cells. These mutant phages are called *host range mutants*, and the gene that determines this is designated *h*; the

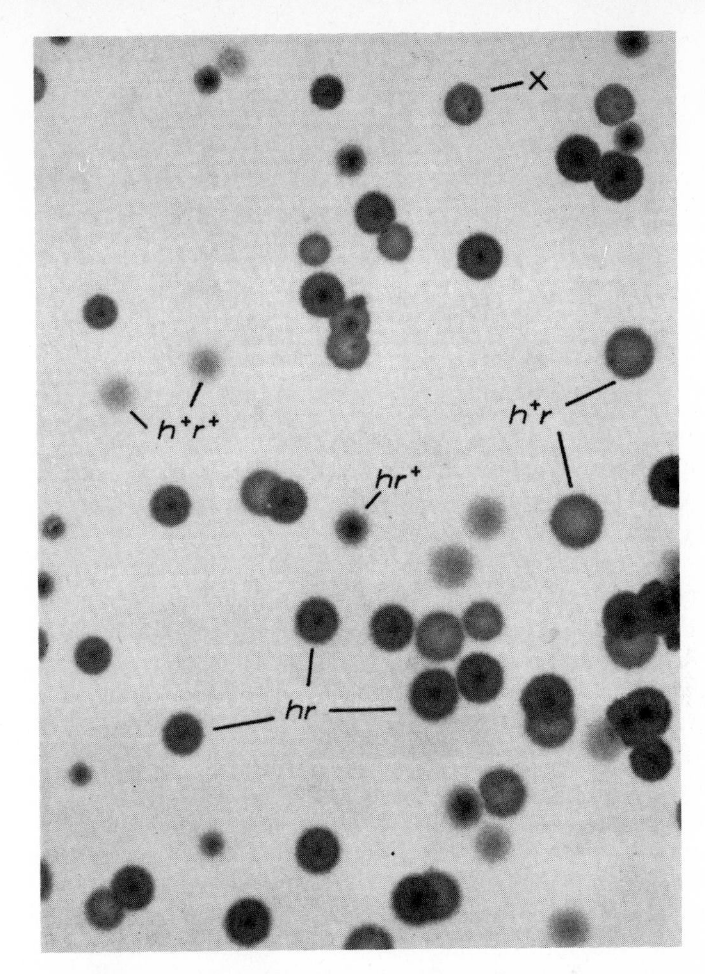

Figure 8-20. Photograph showing the morphology of plagues produced by viruses with different genotypes. h = host range; r = rapid lysis (see p. 249). (From W. Hayes, 1970. *The Genetics of Bacteria and Viruses* 2nd ed., Plate 21 facing p. 488. Blackwell Scientific Publications, Oxford and Edinburgh.)

wild-type allele is h^+ (see again Fig. 8-20). The *h* mutant phages can also infect B-strain cells.

In 1949, Hershey and Rotman infected cells with a mixture of viruses that were of different genotypes with regard to these two characteristics. In some experiments they infected cells with h^+r and hr^+ viruses; in others they infected cells with wild-type h^+r^+ and doubly mutant viruses, *hr*. They repeated the experiments using different *r* mutants which were designated by different numbers, etc. Some of their results are given in Table 8-2.

THE DATA OF MIXED-INFECTION EXPERIMENTS CLEARLY INDICATE THAT GENETIC RECOMBINATION OCCURS BETWEEN VIRUSES. In each case shown in Table 8-2, parental types and recombinant types occur among the progeny (see again Fig. 8-20, which illustrates the differences in appearance of the plaques formed by the phages of

Table 8-2 Data from Mixed Infection of Bacteria

Cross	*Percentage of progeny phenotype*			
	h^+r^+	hr	h^+r	hr^+
$h^+r7^+ \times hr7$	42	43	7	7
$hr7^+ \times h^+r7$	5.9	6.4	32	56
$h^+r1^+ \times hr1$	44	29	13	14
$hr1^+ \times h^+r1$	12	12	34	42

From A. D. Hershey and R. Rotman, 1949. Genetics 34: 44.

various genotypes). The distance between the *r* locus and the *h* locus can be calculated by dividing the number of recombinants by the total number of progeny. Using the data in Table 8-2 we find that *r7* is approximately 15-17 m.u. from *h*, and *r1* is more than 30 m.u. from *h*.

It must be pointed out here, however, that recombination in bacteriophages cannot be likened completely to that of recombination in eukaryotes. Crossing-over in eukaryotes appears to be a one-time event when it occurs, but the genetic material of viruses can apparently undergo more than one crossing-over event before it is packaged into new particles. In other words, if a recombination event has occurred between two viruses of the genotypes $a^+b^+ \times ab$ to produce a^+b and ab^+ combinations, another cross-over event may occur between the same two loci which would mask the first recombination.

With this aspect of phage behavior in mind, statistical methods were developed to calculate the frequency of recombination in phage. We shall not be using these methods here, but it is important that you be aware that recombination analysis in phage is more a statistical population analysis than the traditional methods we have used for other forms of life.

Phage geneticists soon realized that in addition to multiple exchanges between the genetic material of phages, a number of other factors can influence the frequency of recombination. For example, some phage chromosomes undergo duplication more rapidly than others and the presence of greater numbers of one type of chromosome may affect the appearance of recombinant types; therefore one may also be observing more than one generation of gene combinations.

In spite of these complexities, it was eventually possible to establish a genetic map for some of the T phages. Interestingly, the genetic material of these phages is also in the form of a circular chromosome just as in the case of bacteria. Figure 8-21 shows the linkage map of phage T4.

The Genetic Material of Viruses

In 1952, Hershey collaborated with Margaret Chase on a significant series of experiments on the mechanism of viral infection. It had been known for some time that phages adsorb onto the cell membrane of bacteria (Fig. 8-22) in order to infect the cell. What happens thereafter to cause the bacterial host to produce new viral parti-

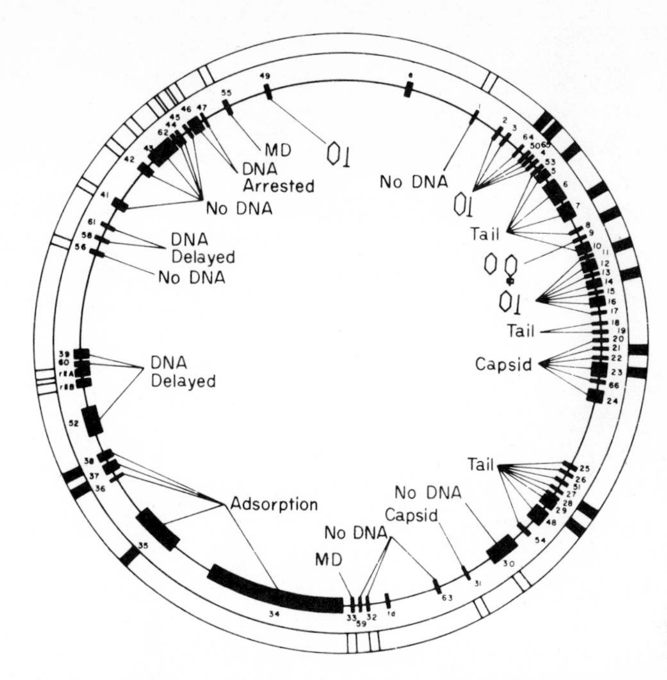

Figure 8-21. Circular linkage map of bacteriophage T4. Inner circle shows arrangement of genes and length of genes estimated by recombination frequencies. On the outer circle, white rectangles represent genes which function early after infection; black rectangles represent genes which act late after infection. (From W. Hayes, 1970. *Genetics of Bacteria and Viruses* 2nd ed., p. 491. Blackwell Scientific Publications, Oxford and Edinburgh.)

cles was not clear, however. Hershey and Chase pursued this question by growing viral particles in medium containing radioactive sulfur (^{35}S) and radioactive phosphorus (^{32}P). Recall that DNA does not contain sulfur, but proteins do. Consequently, the investigators found that ^{35}S was present only in the protein coats of viruses, and ^{32}P in the DNA of viruses.

They then infected cells with the labeled phage particles, and with the use of an electric blender, sheared the phage heads from the bacteria following adsorption to the cell wall. The phage progeny produced in the infected cell were found to contain only ^{32}P, and not ^{35}S, indicating that only the DNA of phages had entered the host cell. Proteins remained outside the cell. This was another classical experiment which clearly pointed to nucleic acids rather than proteins as the genetic material of cells.

Further confirmation that nucleic acids are the chemicals which determine the events following infection of a cell by viruses came from the studies of the tobacco mosaic virus (TMV) by Fraenkel-Conrat (1957) and others. When proteins of the TMV were separated from its nucleic acid (which, incidentally, is RNA, not DNA), the RNA was found to be the only fraction which possessed infective ability and caused the production of new viral particles. If RNA from one strain of TMV was

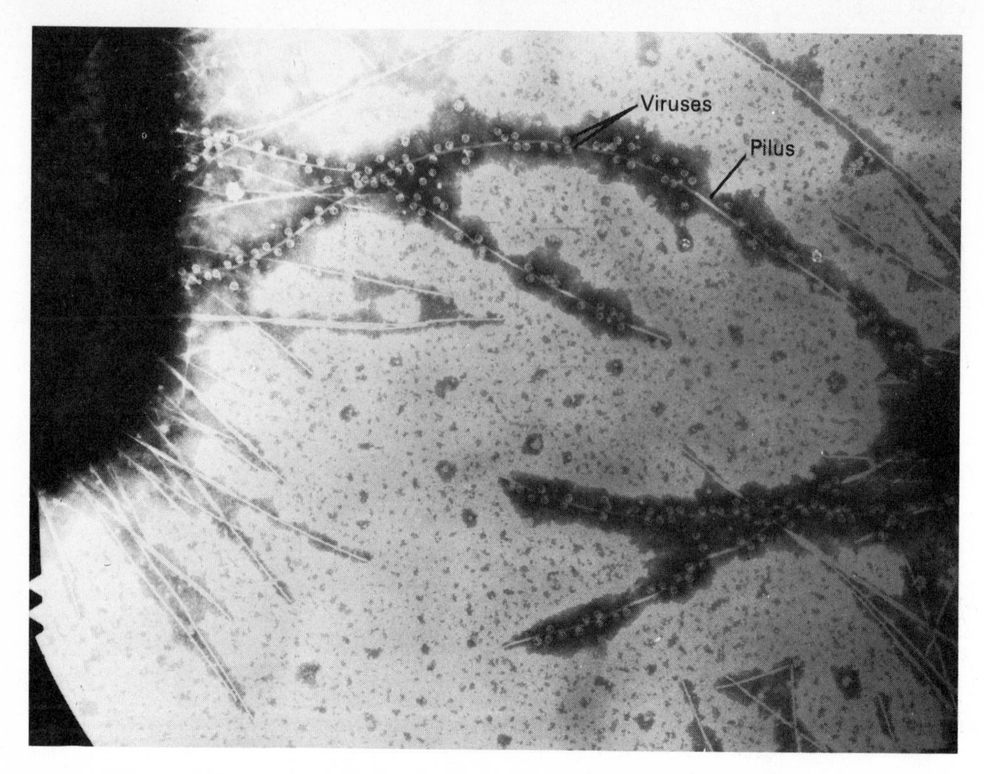

Figure 8-22. Electron micrograph showing viruses adsorbed onto pili of bacteria. (Photograph courtesy Dr. Philip Silverman and Lillian Eoyang, Albert Einstein College of Medicine.)

combined with protein from another strain, infecting cells with the "reconstituted" viruses resulted in progeny whose RNA and protein were both of the parental strain that served as the source of RNA.

Transduction

The late 1940s and early 1950s were truly the "heydays" of microbial genetics. At the same time that Hershey and Chase were reporting the results of their experiments on the role of DNA in phage, another mechanism by which viruses can cause genetic exchange was being discovered by Norton D. Zinder and Joshua Lederberg. These scientists were studying conjugation in *Salmonella typhimurium*, a bacterium which causes typhus in mice. Using strains of *Salmonella* that were auxotrophic for different nutrients, Zinder and Lederberg found that recombination of genes does occur as a result of conjugation, just as Lederberg and Tatum had established in *E. coli*.

One very important difference, however, was that when the strains were separated by a filter in a U-shaped tube, prototrophs were found among the progeny bacteria, even though the cells were too large to have passed through the pores of the filter. Furthermore, only one of the strains, let us call it strain *A*, showed the presence of prototrophs. The second strain, strain *B*, did not produce prototrophs at all. If strain *A* cells with different nutritional requirements were grown in a U tube, separated by a filter, prototrophs were obtained; on the other hand, strain *B* cells grown in a U tube still produced no prototrophic progeny.

The explanation for these observations was revealed when a filterable agent that passed from one arm of the U tube to the other was identified as a temperate phage called P22, normally associated with strain *A*. When occasionally a P22 phage became vegetative, the new phage particles would be released into the medium, and some would infect strain *B* cells. Recall that as host cells begin producing new phage particles, the bacterial chromosomes disintegrate into pieces. Apparently, some of these pieces are incorporated by accident into new phage particles. When the new phages are released upon lysis of the strain *B* host, and in turn infect *A* cells, the genes of the *B* cells are transmitted to the second host. If the portion of the strain *B* chromosome transmitted to a new strain *A* host contains wild-type genes for the nutritional requirements of strain *A*, and these genes are incorporated into the chromosome, that cell becomes prototrophic. Figure 8-23 illustrates the essential steps involved in this virus-mediated genetic exchange between bacterial cells, a process known as *transduction*.

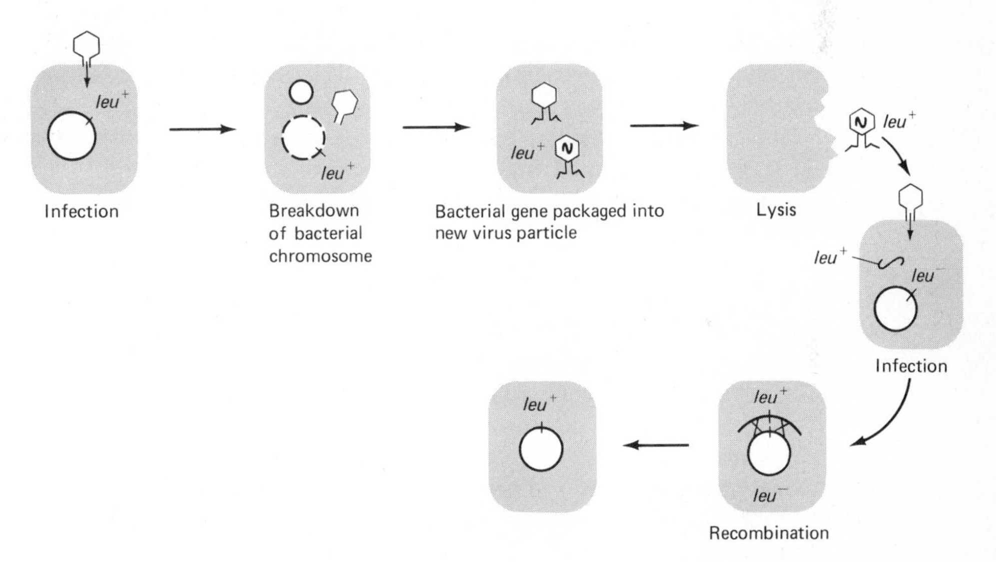

Figure 8-23. Schematic illustration of the events in generalized transduction.

Note that when a piece of bacterial chromosome is transferred from one host cell to a second host, the latter becomes diploid for those particular genes. We refer to this condition as *merodiploidy*, and the cell as a *merozygote* for those genes. The effect is only temporary, for the piece of chromosome is either exchanged for the homologous portion of the host chromosome by recombination, or is degraded by enzymes.

Mapping chromosomes by generalized transduction. GENETICISTS REALIZED THAT THE PHENOMENON OF TRANSDUCTION OFFERED YET ANOTHER MEANS OF STUDYING

LINKAGE IN BACTERIAL CHROMOSOMES. TWO GENES WHICH ARE CLOSELY LINKED WOULD TEND TO BE TRANSDUCED TOGETHER AT HIGHER FREQUENCIES THAN TWO GENES WHICH ARE FARTHER APART. IF TWO GENES ARE TRANSDUCED TOGETHER, THE RATIO OF SINGLE TRANSDUCTANTS TO THE TOTAL NUMBER OF TRANSDUCTANTS WOULD BE AN INDICATION OF THE DISTANCE BETWEEN THE TWO GENES.

For example, if genes m^+o^+ are being transduced into a cell which is m^-o^-, the distance between the two loci is determined by the formula:

$$\frac{(m^+o^-) + (m^-o^+)}{(m^+o^-) + (m^-o^+) + (m^+o^+)}$$

This is quite similar to our analysis of the distance between genes as a result of transformation. Actually, the process of genetic recombination in transformation and in transduction is probably the same, since in both cases there is a free piece of chromosome to be exchanged. The difference is of course that in transformation the piece of chromosome is absorbed from the environment, whereas in transduction, a virus injects it into the cell.

In cases such as that studied by Zinder and Lederberg, transduction is a nonspecific process. Which piece of bacterial chromosome is picked up by the phage during its packaging process is purely random—simply a matter of whichever piece happens to be close enough to be taken up. This is known as *generalized transduction*, and occurs at a frequency of about 1 in 10^5–10^7 cells.

Specialized, or restricted, transduction. A somewhat different type of transduction was discovered upon further exploration of the phenomenon. A temperate bacteriophage called *lambda*, λ, was found in *E. coli* K-12 strain by Esther Lederberg. Together with M. L. Morse, the Lederbergs carried out experiments showing that the lambda phage preferentially transduced a gene, gal^+, which determines an enzyme to break down galactose. This type of transduction, in which specific genes are transduced, and which therefore cannot be simply an accidental process, is called *specialized*, or *restricted, transduction*.

It has since been established that a special region on the *E. coli* chromosome serves as the attachment site for the lambda prophage, called *att* (see again Fig. 8-19). This site is very close to the gal^+ locus which is on one side of *att* λ. The bio^+ gene (which determines the synthesis of the vitamin biotin) is located on the other side of *att* λ. For the most part, when a virus integrates into a cell's chromosome and then is excised, the process is an exact one, that is, the particle after excision contains the same genetic material as it did when it was incorporated. However, 1 in about 100,000 prophages is excised abnormally, so that a portion of the bacterial chromosome adjacent to the phage is accidentally excised (Fig. 8-24 illustrates this process). The frequency with which the *gal* locus is transduced by λ is 1 in every 10^6 particles.

The transduced bacteria, now gal^+, were found to produce gal^- progeny fairly frequently, about 1 in 1000. This frequency implied that something other than transduction (of gal^+ to gal^-) was involved. Specifically, the investigators concluded that in specialized transduction, the piece of bacterial chromosome carrying the gal^+ gene is inserted alongside the gal^- of the host. In other words, unlike the case

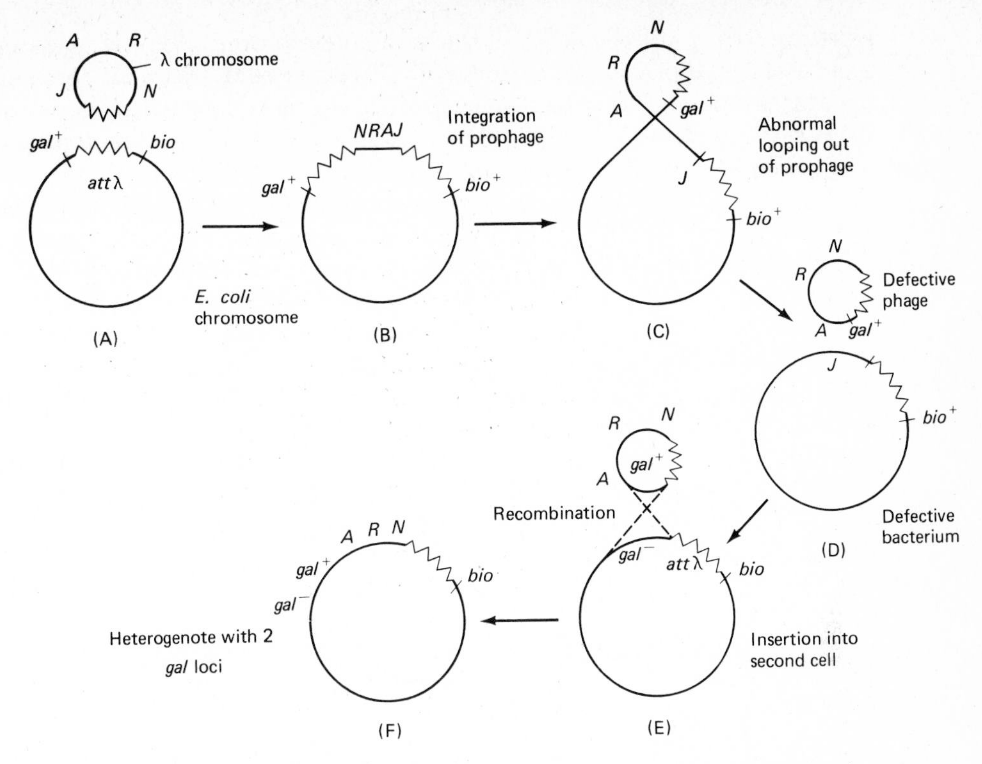

Figure 8-24. Specialized transduction *A, R, N,* and *J* are λ phage genes. Explanation in text.

with generalized transduction, this is not an exchange process; rather, the cell now has two *gal* loci (Fig. 8-24f). If the alleles are different, the cell is referred to as a *heterogenote*; if the alleles are the same (for example, if both are gal^+), the cell is a *homogenote*. The cell is also considered a *partial diploid*.

It was also discovered that the presence of the *gal* locus on the DNA of λ phages led to a loss of some of the phage genome. This resulted in some transduced cells being unable to produce new phage particles even though the λ was in its vegetative state. When examined, the λ phages were indeed found to be defective. These were then called λ*dgal*, or λ*dg*, for *lambda defective at the galactose locus*. Figure 8-24C and D illustrate how this occurs. We shall see later how the use of specialized transducing phages allowed scientists to isolate for the first time a specific gene from the *E. coli* chromosome.

SEXDUCTION

THE F ELEMENT HAS ALSO BEEN FOUND TO BE INVOLVED IN A TRANSDUCTIONLIKE PROCESS KNOWN AS SEXDUCTION. OCCASIONALLY, WHEN THE EPISOME IS EXCISED FROM THE HOST CHROMOSOME, IT PULLS AWAY WITH IT A NEIGHBORING SEGMENT OF THE BACTERIAL GENOME, IN A MANNER SIMILAR TO THE PRODUCTION OF TRANSDUCING BACTE-

RIOPHAGE (SEE AGAIN FIG. 8-24). When such an event occurs, the episome contains a piece of DNA that is homologous with a portion of the bacterial chromosome.

It stands to reason that these particles (termed F′ by Adelberg and Burns 1959) would undergo integration into a recipient bacterium's chromosome more readily and at a specific site, because of the region of homology with the bacterial chromosome, a situation in contrast to the integration at different sites of the normal *F* element. Cells that contain two copies of a particular locus are referred to as *partial diploids*, or *heterogenotes*, just as in the case of transduction. In 1959, Edward Adelberg and his coworkers found an *F′* particle which behaved in this way, and contained the *lac* locus from *E. coli.* They were able to insert the $F'lac^+$ element in F^-lac^- cells, and found the genotype of the recipients to be $F'lac^+lac^-$, with the cell being able to produce enzymes of the *lac* locus (Fig. 8-25). Isolation of *F′* particles carrying bacterial genes has provided yet another approach to geneticists studying gene recombination.

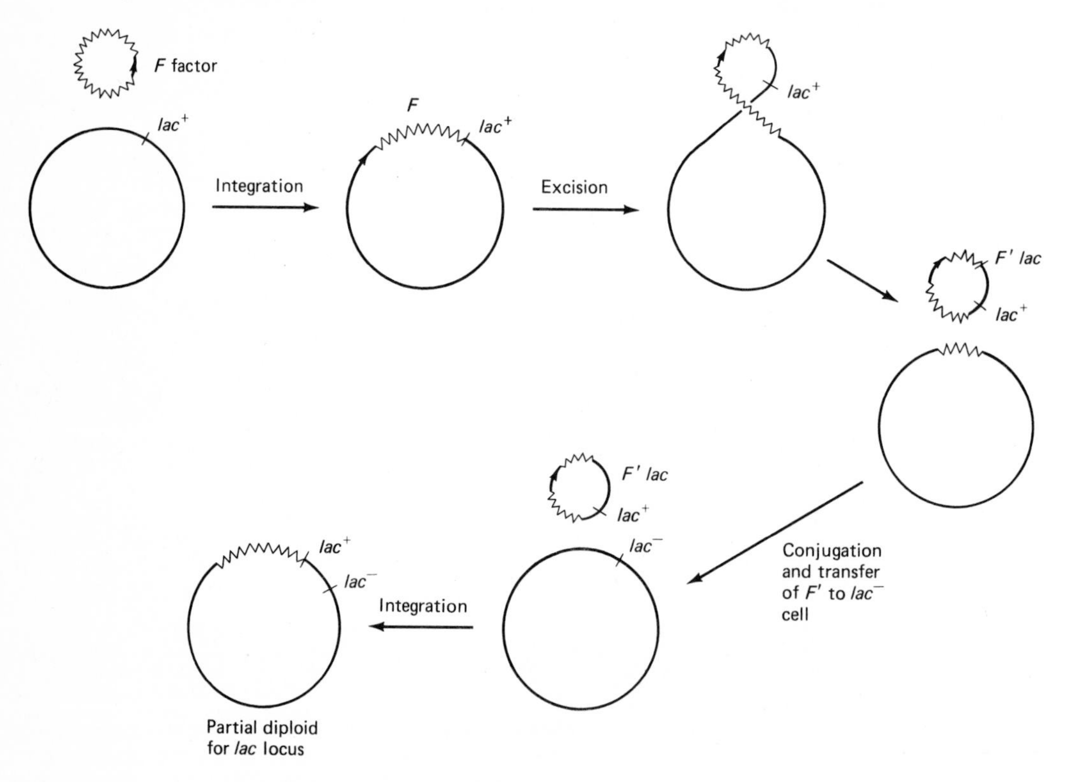

Figure 8-25. Sexduction involving the *lac* locus.

CONTRIBUTIONS OF MICROBIAL GENETICS

The material which we have covered in this chapter suggests but the bare essentials of microbial genetics. Even the essential concepts, however, should impress upon you the importance of microbial systems for geneticists. The ease of manipulation, the simplicity of their genomes, and the other favorable attributes of microbes have already been discussed. They have allowed us to examine the transmission of genetic material on a scale never before possible—the molecular level.

The following chapters, which deal with the concepts of modern genetics, essentially molecular genetics, comprise the knowledge which we have gained through the use and study of microorganisms in just a few decades. The experiments by Avery, MacLeod, and McCarty, and those by Hershey and Chase, among others, pinpointed the chemical foundation of genes to be nucleic acids. The discovery of the structure of nucleic acids, and therefore genes, set off an avalanche of research and progress. As you shall see, the precipitous nature of our gains in knowledge regarding fundamental processes of life has led to much soul-searching among nonscientists and geneticists alike. It has raised moral as well as biological questions over the manipulation of biological processes for which there are as yet no clear answers.

PROBLEMS

1. Define lysis, lysogeny, vegetative phase, temperate phase, induction, bacteriophage, virus, prophage.

2. What are the major differences between the genetic systems of prokaryotes and eukaryotes?

3. Among the first experiments on transformation were those by Hotchkiss and Marmur. They introduced DNA from cells that were resistant to streptomycin and capable of fermenting mannitol ($strp^r mtl^+$) to cells that were sensitive to streptomycin and incapable of fermenting mannitol ($strp^s mtl^-$). Describe how you should proceed to study linkage distance between the two loci.

4. If two genes being studied in transformation experiments are very far apart, how could this be discerned very quickly?

5. Let us assume that DNA from *B. subtilis* cells that are $phen^+ leu^+$ is introduced to $phen^- leu^-$ cells. The following table gives the media used and the results obtained:

Medium	*Number of colonies obtained*
#1 contains only phenylalanine	95
#2 contains only leucine	95
#3 minimal medium	90

What is the distance between the two loci?

6. What specific functions can you think of that might be determined by the genes of the *F* particle?

7. Assume that interrupted mating experiments were carried out between a donor *Hfr* strain of bacteria that were wild-type for genes, m^+, n^+, o^+, p^+, and F^- bacteria that were m^-, n^-, o^-, p^-. The following results were obtained in three different samples:

Experiment	*Genes in order of transmission*
1	n^+, p^+, m^+, o^+
2	m^+, o^+, n^+, p^+
3	p^+, m^+, o^+, n^+

(a) Give the sequence of genes on the bacterial chromosome.
(b) Draw the chromosomes, and show position and orientation of the *F* particle in each of the three experiments.

8. If an $F'lac^+$ element is inserted into a recipient F^-lac^- cell, and the latter shows lac^+ activity, how can you determine that this is due to the $F'lac^+$ gene, and not a spontaneous reversion to wild-type of the gene of the recipient cell?

9. Two different tryptophan mutations have been isolated from different *E. coli* strains. The following cross is made:

$$Hfr\ cys^+tryp_1^-str^s \times F^-cys^-tryp_2^-str^r$$

(str^s = streptomycin sensitivity; str^r = streptomycin resistance; cys = cysteine). Exconjugants plated on minimal medium containing streptomycin show the formation of colonies that are cys^+tryp^+. How do you interpret these results?

10. In most experiments such as that in Problem 9 involving conjugation between *Hfr* cells and F^- cells, the genotype of the *Hfr* bacteria includes streptomycin sensitivity, the F^- cells are streptomycin-resistant, and the exconjugants are plated out on media containing streptomycin. Why is this necessary? (Incidentally, the term used for the selection against donors is *contraselection.*)

11. In another conjugation experiment, the following cells are mated: $Hfr\ a^+b^+str^s \times F^-\ a^-b^-str^r$. The genes *a* and *b* determine nutritional requirements. Exconjugants are plated on complete medium with streptomycin. Cells from 100 colonies are tested on various media with the following results: (a) medium with substance *a* but not *b*: 40 colonies survived. (b) medium with substance *b* but not *a*: 20 colonies survived. (c) medium with neither *a* or *b*: 10 colonies survived. Give the genotypes of the surviving cells in the three media. What is the recombination frequency between genes *a* and *b*?

12. A number of mixed infections were carried out between strains of T4 phages that differed at three loci. Let us designate the genes simply as *a*, *b*, and *c*. The following data indicate the crosses made and the results obtained:

Crosses	Number of progeny			
	a^+b^+	a^-b^-	a^+b^-	a^-b^+
$a^+b^- \times a^-b^+$	18	22	160	200
	b^+c^+	b^-c^-	b^+c^-	b^-c^+
$b^+c^- \times b^-c^+$	42	48	115	95
	a^+c^+	a^-c^-	a^+c^-	a^-c^+
$a^+c^- \times a^-c^+$	11	9	35	45

Calculate the recombination frequencies and determine the order of these genes and the distance between them.

13. What reason can you think of why normally only one bacterial gene is transferred in specialized transduction?

14. In what ways might specialized transduction contribute to our understanding of gene action?

REFERENCES

ADELBERG, E. A., AND S. N. BURNS, 1959. "A Variant Sex Factor in *Escherichia coli*." *Genetics* 44: 497.

AVERY, O. T., C. M. MACLEOD, AND M. MCCARTY, 1944. "Studies on the Chemical Nature of the Substance Inducing Transformation of Pneumococcal Types. I. Induction of Transformation by a Deoxyribonucleic Acid Fraction Isolated from *Pneumococcus Type III*." *J. Exptl. Med.* 79: 137–158.

CAVALLI-SFORZA, L. L., J. LEDERBERG, AND E. M. LEDERBERG, 1953. "An Infective Factor Controlling Sex Compatibility in Bacterium coli." *J. Gen. Microbiol.* 8: 89–103.

DAVIS, B. D., 1950. "Non-filterability of the Agents of Genetic Recombination in *E. coli*." *J. Bacteriol.* 60: 507.

DELBRUCK, M., 1948. "Biochemical Mutants of Bacterial Viruses." *J. Bacteriol.* 56: 1–16.

DRISKELL, P. J., AND E. A. ADELBERG, 1961. "Inactivation of the Sex Factor of *Escherichia coli K-12* by the Decay of Incorporated Radiophosphorus." *Bacteriol. Proc.* 186.

FRAENKEL-CONRAT, H., AND B. SINGER, 1957. "Virus Reconstitution: Combination of Protein and Nucleic Acid From Different Strains." *Biochem. Biophys. Acta* 24: 541–548.

GRIFFITH, F., 1928. "The Significance of Pneumococcal Types." *J. Hygiene* 27: 113–159.

HAYES, W., 1952. "Recombination in *Bacteria coli K-12*: Undirectional Transfer of Genetic Material." *Nature* 169: 118–119.

———, 1953. "Observations on a Transmissible Agent Determining Sexual Differentiation in *Bacteria coli*." *J. Gen. Microbiol.* 8: 72–88.

———, 1970. *The Genetics of Bacteria and Viruses*, 2nd ed. Blackwell Scientific Publications, Oxford and Edinburgh.

HERSHEY, A. D., 1946. "Spontaneous Mutations in Bacterial Viruses." *Cold Spring Harbor Symp. on Quant. Biol.* 11: 67–76.

———, AND R. ROTMAN, 1949. "Genetic Recombination Between Host-Range and Plaque-Type Mutants of Bacteriophage in Single Bacterial Cells." *Genetics* 34: 44–71.

———, AND M. CHASE, 1952. "Independent Functions of Viral Protein and Nucleic Acid in Growth of Bacteriophage." *J. Gen. Physiol.* 36: 39–56.

HOTCHKISS, R. D., AND J. MARMUR, 1954. "Double Marker Transformations As Evidence of Linked Factors in Deoxyribonucleate Transforming Agents." *Proc. Nat. Acad. Sci. U. S.* 40: 55–60.

LEDERBERG, E. M., 1951. "Lysogenicity in *E. coli K-12*." *Genetics* 36: 560.

LEDERBERG, J., 1946. "Novel Genotypes in Mixed Cultures of Biochemical Mutants of Bacteria." *Cold Spring Harbor Symp. on Quant. Biol.* 11: 113–114.

———, J., AND E. M. LEDERBERG, 1952. "Replica Plating and Indirect Selection of Bacterial Mutants." *J. Bacteriol.* 63: 399–406.

———, J., AND E. L. TATUM, 1946. "Gene Recombination in *E. coli*." *Nature* 158: 558.

MORSE, M. L., E. M. LEDERBERG, AND J. LEDERBERG, 1956. "Transduction in *Escherichia coli K-12*." *Genetics* 41: 142–156.

TAYLOR, H. E., 1949. "Transformations Recipriques des formes *R* et *ER* chez le *pneumocoque*." *Comptes Rend. Acad. Sci., Paris.* 228: 1258–1259.

WOLLMAN, E. L., AND F. JACOB, 1954. "Lysogeny and Genetic Recombination in *Escherichia coli K-12*." *Comptes Rend. Acad. Sci., Paris* 239: 455–456.

ZINDER, N. D., AND J. LEDERBERG, 1952. "Genetic Exchange in *Salmonella*." *J. Bacteriol.* 64: 679–699.

NINE

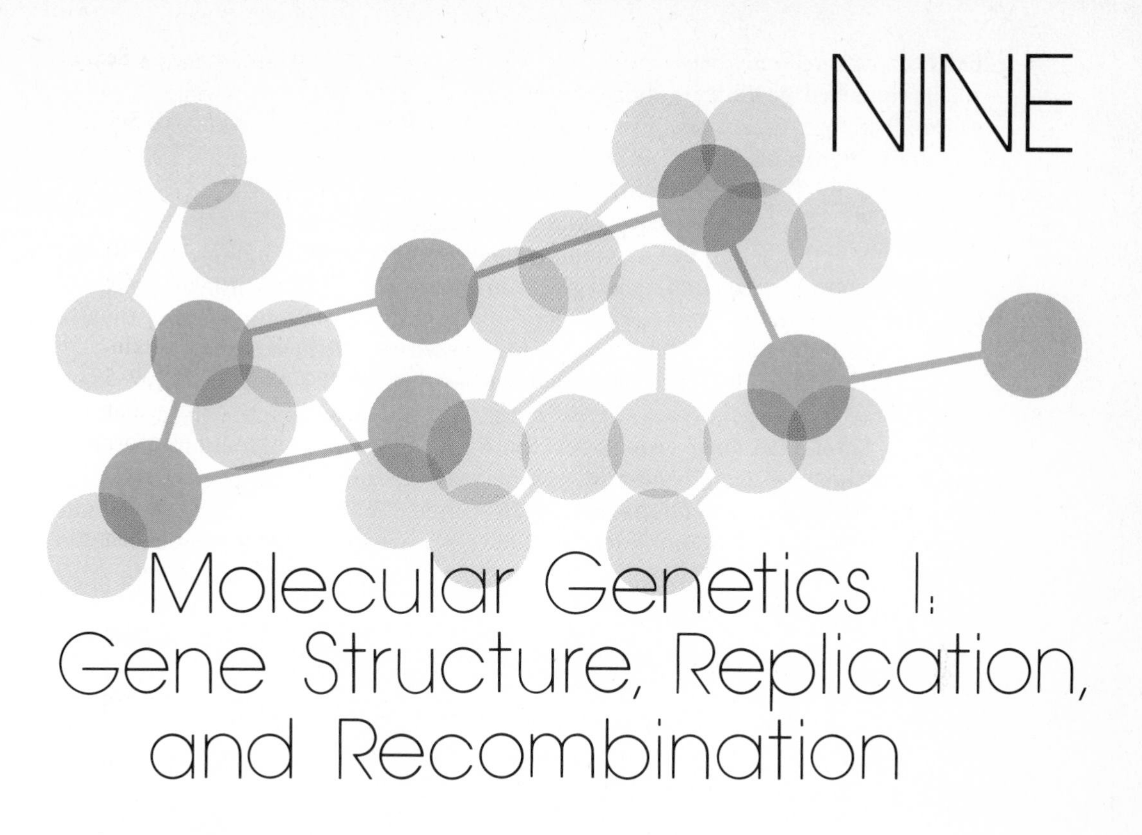

Molecular Genetics I: Gene Structure, Replication, and Recombination

INTRODUCTION

In Chapter 8 we described experiments on viruses and bacteria which indicated that the genetic material is nucleic acid. Clearly, scientists needed to direct their research toward nucleic acids in order to determine the molecular structure of genes. In this chapter we shall discuss biochemical and biophysical techniques that were adopted for the solving of this most basic biological problem.

Before we turn to the biochemical and biophysical experiments which revealed the structure of genes, however, it will be worthwhile to examine some studies of transmission genetics in *Drosophila melanogaster* which provided biologists with a clearer idea of one aspect of gene structure, namely, the continuity of genetic material in chromosomes. These studies were reported at approximately the same time that microbial genetics was making its surge to the forefront in genetic experimentation.

Pseudoalleles

In Chapter 1 we defined alleles as genes which determine the same trait or function, and are situated in the same position on homologous chromosomes. This was in keeping with the concept of "beads on a string," which visualized genes on a chromosome as being discrete, discontinuous units in a linear order. The sequence of genes is disturbed only if crossing-over occurs between genes. Since crossing-over had been detected only between genes, it was thought that the areas of chromosomes which contained genes were different from the areas of the chromosomes between genes. Therefore, no crossing-over could be expected between alleles.

The first indication that this concept was not entirely accurate came from experiments by E. B. Lewis on the *Star-Asteroid* series of "alleles" in *Drosophila* (Lewis 1950). Star (*St*) and Asteroid (*ast*) were considered allelic in the classic sense in that they both affected eye development and were located on homologs in apparently the same positions. Heterozygotes for either allele (*St*/+, *ast*/+) possessed eyes that were reduced in size, rough in outline, and with irregular facets. Heterozygotes with both alleles, *St*/*ast*, usually manifested very reduced eyes.

There was a difference between homozygotes for the two alleles, however, as *St*/*St* was a lethal genotype, whereas *ast*/*ast* was viable, although with very abnormal eyes. Lewis then made crosses between *St*/*ast* flies, and found among the progeny approximately .02% wild-type. These were interpreted to be the result of crossing-over between *St* and *ast* (Fig. 9-1a). Although both *St* and *ast* were mutants of the same gene, the mutations appear to have occurred at different locations within the gene, allowing crossing-over and recombination giving some wild-types.

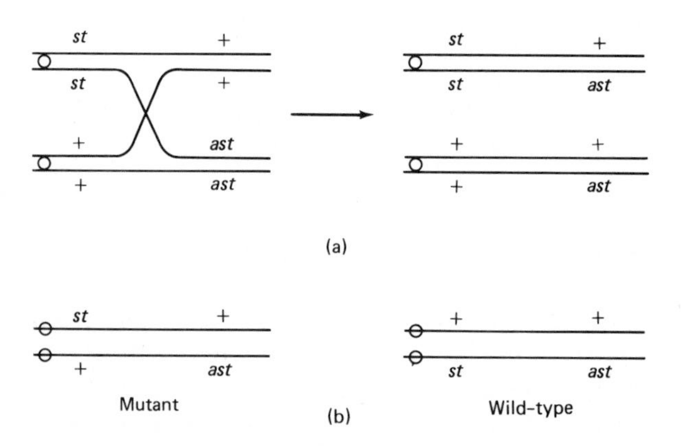

Figure 9-1. (a) Interpretation of data obtained by E. B. Lewis in studies on *St/ast* pseudoalleles. Rare wild-type progeny arose from the chromatid with wild-type genes (arrow) as a result of crossing-over. (b) The *St/ast* pseudoalleles show position effect: In the trans position the phenotype is mutant; in the cis position, the phenotype is wild-type.

Position Effect

Although crossing-over did occur between the two genes, they could not be considered two different loci in the classic sense either, because Lewis also demonstrated that a *position effect* existed, a situation in which the phenotypes of individuals differ because of different combinations of the same linked genes or because of translocation or insertion of heterochromatin between genes. In the case of *Star-Asteroid*, Lewis found that flies heterozygous for both *Star* and *Asteroid* are mutant in appearance if the genes are on different homologs, that is, in the trans position.

When crossing-over occurs so that one homolog contains both mutant genes, and the other homolog, both wild-type genes (the genes thus being in the cis position), the phenotype is normal (Fig. 9-1b).

This is of course different from the cis and trans effect of nonalleles, which we discussed in great detail in the crossing-over section in Chapter 6. Dihybrids of nonalleles are phenotypically the same regardless of cis or trans positioning (see again Fig. 7-2).

Lewis called the genes of the Star-Asteroid system *pseudoalleles* because crossing-over occurs between them. However, because of the position effect, they were also referred to as position *pseudoalleles.* Other terms to describe this effect are the *Lewis effect* and the *cis-trans effect.*

Following the publications of Lewis' research on *St/ast* pseudoalleles, similar position effects were found among other *Drosophila* genes that were thought to be allelic, such as alleles of the Vermilion locus. Two X-linked eye pigment mutations, *white* and *apricot*, had also been considered allelic until studies showed crossing-over at a frequency of 0.03% between the two genes.

BESIDES YIELDING MORE INFORMATION ON THE GENES INVOLVED, STUDIES OF PSEUDO-ALLELES INTRODUCED THE CONCEPT THAT GENES MIGHT NOT BE AS DISCONTINUOUS AS PREVIOUSLY THOUGHT, OR AS PHYSICALLY DISCRETE. SINCE CROSSING-OVER COULD OCCUR WITHIN GENES, THE AREAS OF CHROMOSOMES WHICH CONTAIN GENES MIGHT NOT BE DIFFERENT FROM AREAS OF THE CHROMOSOMES BETWEEN GENES. We use the term *intragenic crossing-over* for exchange of genetic material within genes. When the physical structure of genes was established a few years later, the accuracy of this concept was confirmed.

MOLECULAR STRUCTURE OF DNA

By 1950, the following ideas had been established concerning the actual structure of genes: Genes are situated in a linear manner on chromosomes. Some genes are separated from others in that crossing-over occurs between them; however, in some cases, genetic recombination can occur within a gene. Genes in prokaryotes were found to consist of nucleic acids, and since all chromosomes of eukaryotes contain DNA as well, it was logical to assume that DNA must also be the genetic material in eukaryotes. Obviously, the next step was to characterize the molecular structure of DNA.

As we stated before, knowledge of the structure of DNA was essential to develop a clear understanding of the manner in which genes determine the structure and function of all living cells and organisms. The expertise of men and women trained in chemical and physical sciences was applied to the problem. Never again would genetics be restricted to only the analysis of inherited traits. Although transmission genetics will always be of great importance to biology and medicine, the necessity of using techniques developed by chemists and physicists changed the emphasis in genetics research forever.

The application of these techniques to living cells and organisms gave rise to the fields of biochemistry and biophysics, and their application to genetic problems led to the development of biochemical genetics. By 1950, just a scant half-century following the rediscovery of Mendel's laws, the Age of Molecular Genetics had dawned. Let us turn now to a brief discussion of some of the most significant work which eventually led to an accurate model of the genetic material.

Biochemical Studies

In Chapter 6 we discussed the major components of DNA: sugar, phosphate groups, and nitrogenous bases. One biochemist who contributed significantly to the structural analysis of DNA was Erwin Chargaff. Chargaff used the techniques of chromatography and spectrophotometry to study the presence and distribution of nitrogenous bases in DNA from cells of different species.

The technique of chromatography allows the separation and visualization of different components of molecules. Basically it involves applying the substance being analyzed to a solid material such as absorbent paper, then dipping the material into a solvent. As the solvent migrates up the material, the substance separates into component molecules which will migrate at different rates according to their size and solubility. When separated, they can be located either by staining the preparation, or by *UV* absorption. Figure 9-2 shows a chromatograph.

Biochemists use spectrophotometry to detect the presence of various molecules by their differential absorption of light. In other words, different substances absorb light maximally at different wavelengths. Samples of the substance to be studied are placed in solution in a small container (called a cuvette), then inserted into a spectrophotometer, the components of which are illustrated in Fig. 9-3. The amount of light absorption is recorded and indicates the presence of specific components.

Chargaff established that there are four different nitrogenous bases in DNA, two purines (adenine and guanine) and two pyrimidines (thymine and cytosine). His data showed that the ratio of purines to pyrimidines always approached 1 regardless of the source of the DNA (Chargaff 1951). On the other hand, the $(A+T)/(G+C)$ ratio varied noticeably. Furthermore, the nitrogenous base composition was constant for cells from different organs of the same species and was characteristic of DNA of that species. He also noted that the amount of adenine approximately equaled the amount of thymine; similarly, the amount of guanine equaled the amount of cytosine. The question then was, how are these components put together?

Biophysical Studies

Studies utilizing techniques of biophysics contributed another approach to the determination of the molecular nature of genes. The physical structure of DNA was investigated by a number of scientists using a technique known as *X-ray crystallography*. If pure crystalline preparations of substances are irradiated by X-rays, the X-rays are diffracted by the atoms of the crystals and can be recorded as a pattern of shadows on a photographic plate. Linus Pauling, for example, had successfully studied protein structure utilizing this technique, and discovered the alpha helix of protein secondary structure (p. 157).

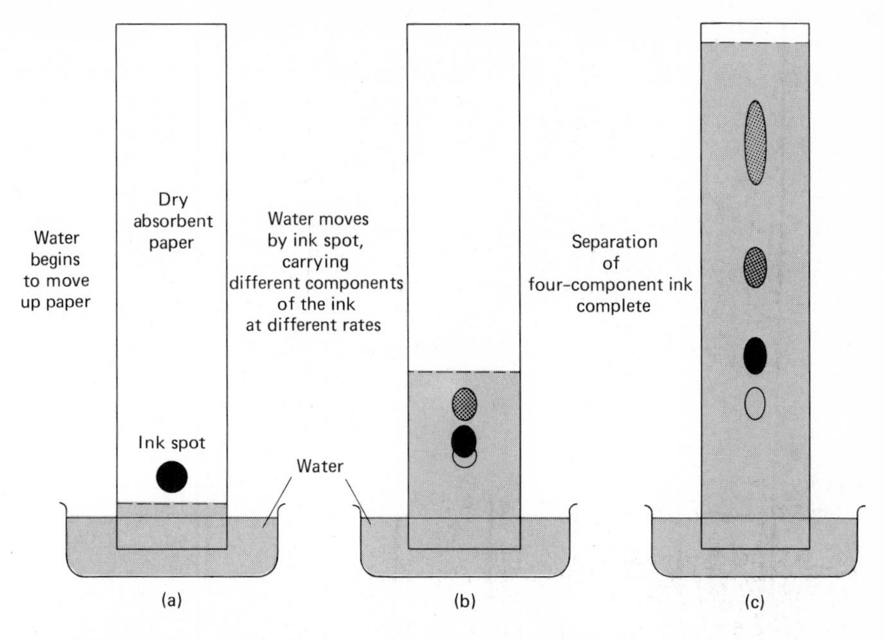

Figure 9-2. Example of a chromatograph: Paper chromatography of ink. Owing to the absorbent character of paper, water moves against gravity (a), and carries the ink dyes along its path (b). If the ink dyes move at different rates they will be separated in the developed chromatogram (c). (From M. M. Jones, J. T. Netterville, D. O. Johnston, and J. L. Wood, 1972. *Chemistry, Man and Society,* p. 9. Saunders, Philadelphia.)

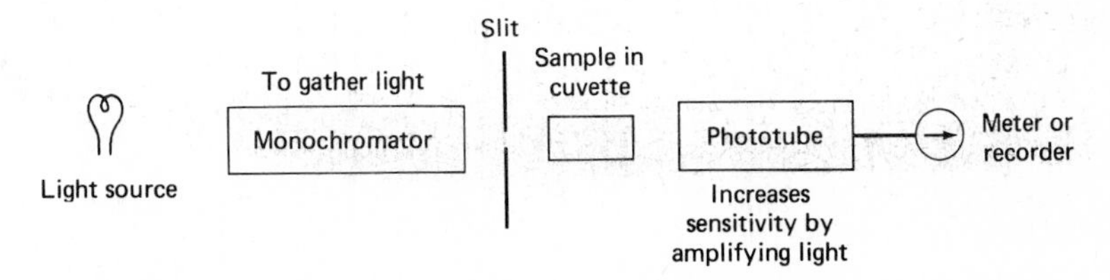

Figure 9-3. Components of a spectrophotometer. (From Brewer *et al.,* 1974. *Experimental Techniques in Biochemistry.* Prentice-Hall, Englewood Cliffs, N.J.)

Although earlier attempts had been made to determine the structure of DNA by X-ray crystallography, the most successful pictures were not obtained until the early 1950s, when especially good preparations of highly oriented DNA crystals were X-rayed and photographed by Rosalind Franklin in the laboratory of M. H. Wilkins at the University of London. Figure 9-4 shows a famous photograph published by Rosalind Franklin which gave very clear details of the molecular structure

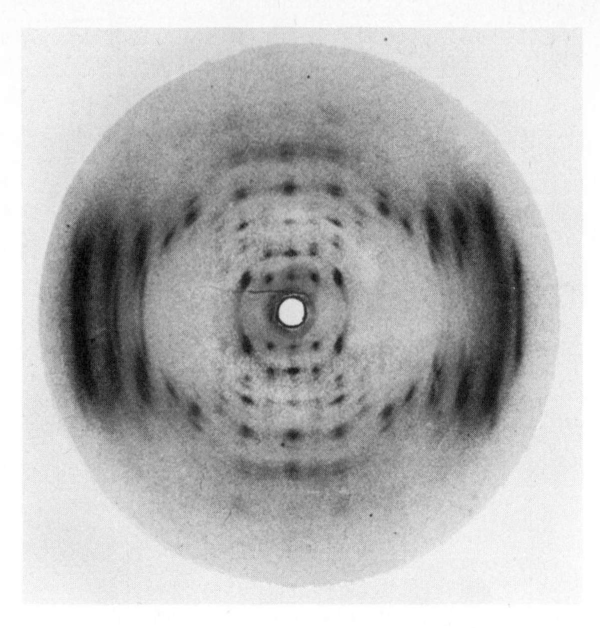

Figure 9-4. X-ray crystallograph of DNA. The darker shadows represent positions of atoms. (Used by permission of Atheneum Publishers from *The Double Helix,* by James D. Watson, p. 73. Copyright © 1968 by James D. Watson.)

of DNA that had not been previously available. To the untrained eye, the shadows present only a symmetrical picture, but to the expert, the pattern suggests a particular orientation of the molecules in relation to each other. In addition, the distance between the different components can be measured from these pictures.

The Watson-Crick Double Helix

As both scientists and the general public know, the actual structure of DNA was determined in 1953 by the brilliant Englishmen, Francis Crick, and his then 23-year-old American collaborator, James Watson. Watson had journeyed to Cambridge as a postdoctoral fellow, and there met the 35-year-old Crick. Crick had been trained as a physicist, but had, like so many physicists, become intrigued with the problems of biology (Fig. 9-5). For an entertaining (though openly one-sided) account of their search for the structure of DNA, you are referred to Watson's book.

The Double Helix

Watson and Crick combined their respective expertise in biology and physics, and their mutual interest in DNA structure, by building scale models of the molecule, based on information already provided by Chargaff, Franklin, Wilkins, and others. It was the extraordinarily clear X-ray diffraction photographs of DNA which finally provided Watson and Crick with the information they needed to arrive at the correct model.

In 1953, in a paper only two pages long, published in *Nature*, Watson and

Figure 9-5. Francis Crick (left) and James D. Watson, circa 1951. (Used by permission of Atheneum Publishers from *The Double Helix*, by James D. Watson, p. 5. Copyright © 1968 by James D. Watson.)

Crick proposed what has now been confirmed to be the structure of the genetic material, DNA. Figure 9-6 shows the basic structure without molecular details. Essentially this is the information that X-ray crystallography provided. Because of the symmetry of the patterns, and the density of the molecules, Watson and Crick concluded that the basic structure was that of a double helix. The photographs by Franklin thus were of the double helix taken from one end.

The helix has a constant diameter of 20 Å, and makes a complete turn every 34 Å. Because of the size of nucleotides, it was calculated that every turn of the helix contains 10 nucleotides. The crucial question remained: How are the nucleotides positioned in this helix?

Watson and Crick considered the fact that four nitrogenous bases exist in DNA. Since bases are the only components of DNA to vary (in amounts) from species to species, any model of DNA structure had to allow for a great variety in base composition. This was necessary to account for the enormous genetic variability found in nature.

Working with wire and cardboard facsimiles of the bases, sugar, and phosphates, WATSON AND CRICK DISCOVERED THAT ALL THE CHARACTERISTICS OF THE DNA MODEL COULD BE ADHERED TO IN A DOUBLE HELICAL CHAIN IF ONE ASSUMED WHAT IS NOW KNOWN TO BE THE CENTRAL ASPECT OF DNA STRUCTURE: COMPLEMENTARY BASE PAIRING. Watson was the first to realize that adenine and thymine held by hydrogen bonds are essentially the same size and shape as cytosine and guanine bound by hydrogen bonds (Fig. 9-7). These could then be fitted into the center of the helix, with the sugars and phosphates serving as the "backbone," as Rosalind Franklin had previously suggested. However, it was also necessary to assume an *antiparallel* orientation, in order to fit the various components within the distances dictated by the X-ray studies. Figure 9-8 shows the original model of DNA as con-

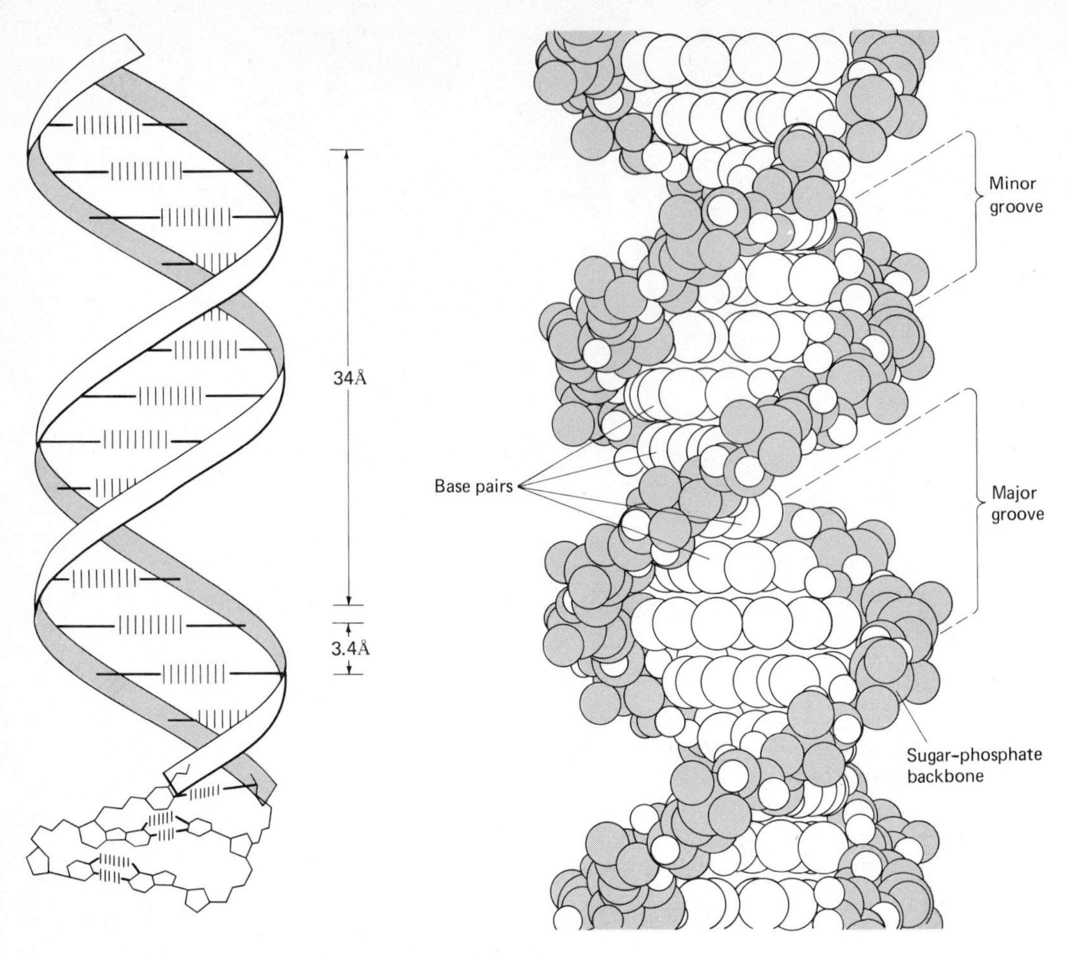

Figure 9-6. Schematic illustration of the helical nature of the DNA. (a) General dimensions of the double helix. (b) Space-filling model of the double helix. (From A. Kornberg, 1974. *DNA Synthesis*, p. 9. W. H. Freeman & Company Publishers, New York; copyright © 1974.)

structed by Watson and Crick; Figure 9-9 illustrates in detail the molecular structure of a molecule of DNA, the substance of genes.

One can visualize the untwisted DNA molecule as a ladder. The sides of the ladder are the sugar–phosphate chains; the rungs of the ladder are a pair of nitrogenous bases held together by hydrogen bonds, two between adenine and thymine, and three between cytosine and guanine. THE ESSENTIAL VARIABILITY OFFERED BY THIS MODEL LIES IN THE SEQUENCE OF BASES. NO PHYSICAL LIMITATIONS EXIST TO THE SEQUENCE IN WHICH BASES CAN BE LINKED ALONG THE LENGTH OF ONE CHAIN OF DNA. THUS THE NUMBER OF POSSIBLE DIFFERENCES IN GENETIC MATERIAL BETWEEN CELLS OF LIVING ORGANISMS IS ESSENTIALLY INFINITE.

Watson and Crick began their paper, "We wish to suggest a structure for the salt of deoxyribose nucleic acid (DNA). This structure has novel features which are

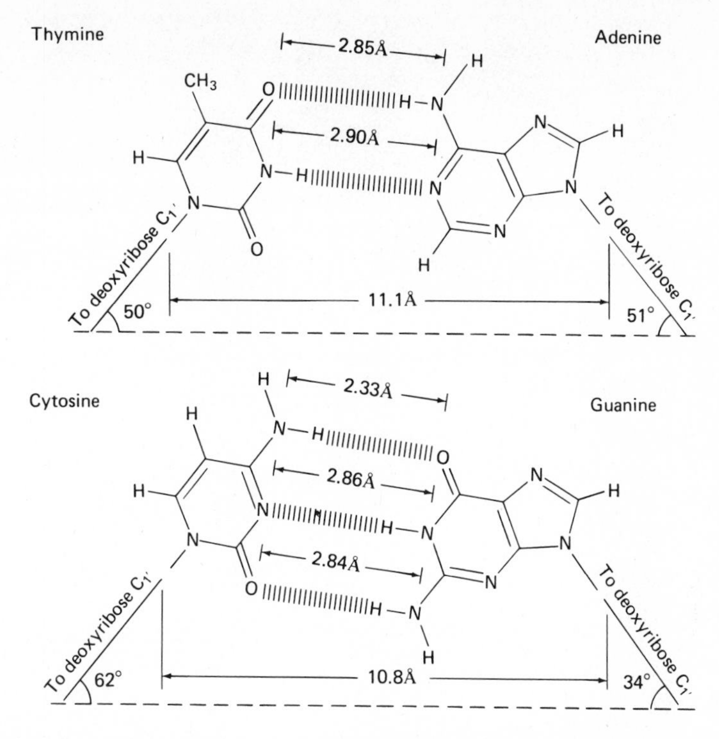

Figure 9-7. Dimensions of adenine-thymine and cytosine-guanine base pairs. Similarity allows complementary base-pairing in DNA double helix. (From A. Kornberg, 1974. *DNA Synthesis,* p. 8. W. H. Freeman & Company Publishers, New York; copyright © 1974.)

of considerable biological interest." This introduction can be best described as exquisite understatement! In 1962, together with M. H. Wilkins,[1] they were awarded the Nobel Prize for a discovery that gave biologists a very important key to understanding life.

DNA and Chromosomes

Following the publication of Watson and Crick's landmark paper on the molecular structure of DNA, the physical state of genetic material in chromosomes of cells of different species came under active scrutiny. We have already spoken of the "chromosomes" of bacteria and viruses—essentially naked strands of DNA. However, there are a great variety of forms of DNA. For the most part, the nucleic acid exists as closed circles, as in the case of *E. coli.* Often the DNAs of bacteria and viruses manifest further twisting and looping (called *supercoiling*) of the already helical molecule (Fig. 9-10).

In some viruses, such as lambda, however, the DNA is actually in the form of a rod when it is not inside its normal host, *E. coli.* Once the viral DNA has entered a

[1]Rosalind Franklin died of cancer in 1958 at the age of 37.

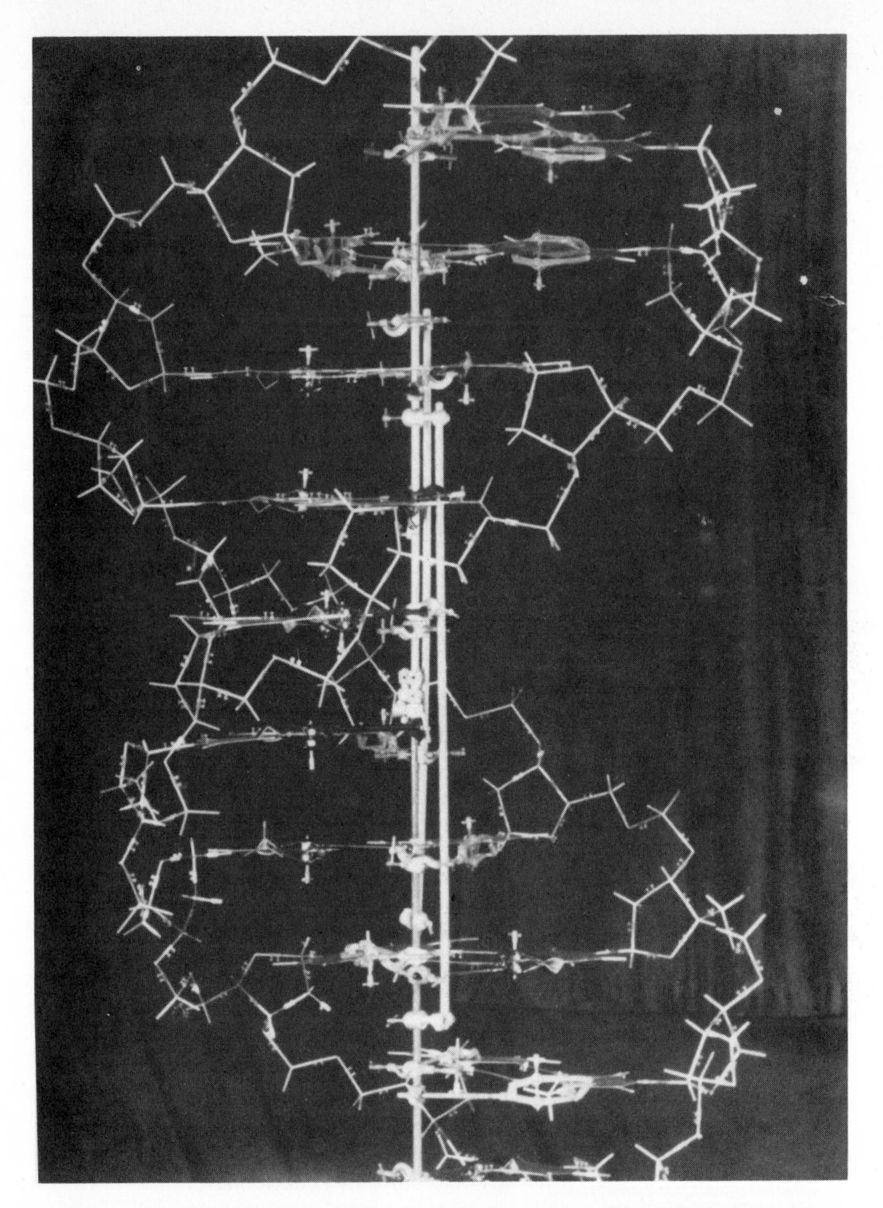

Figure 9-8. Original model of DNA double helix constructed by J. D. Watson and F. Crick at Cambridge. (Used by permission of Atheneum Publishers from *The Double Helix*, by James D. Watson, p. 208. Copyright © 1968 by James D. Watson.)

host cell, it becomes circularized and supercoiled. In yet other prokaryotes, such as T2 or T4 phages, the chromosomes are rod-shaped or linear DNA double helices even after infection of host cells.

There also exist viruses whose genetic material is not the usual double helix of DNA. Among these is the very useful bacteriophage, ϕX174, which contains only a single-stranded form of DNA. Later in this chapter we shall show how the single-

Figure 9-9. Molecular structure of DNA. (Model at the American Museum of Natural History.)

stranded nature of this virus was useful in the understanding of DNA metabolism. Some viruses, such as $Q\beta$ and tobacco mosaic viruses, contain RNA as their genetic material.

In eukaryotes, evidence indicates strongly that each chromosome normally possesses one linear double-stranded molecule of DNA. (An exception would be polytene chromosomes, which contain many strands of DNA. See again p. 164.) One difference between the eukaryotic chromosome and the genetic material found in prokaryotes is of course the histones which are bound to the DNA in eukaryotic cells. The exact molecular relationship between DNA and these proteins is not completely understood.

Another difference between eukaryotes and prokaryotes is the existence of *repetitive* sequences of bases in almost all eukaryotes (Britten and Kohne 1970;

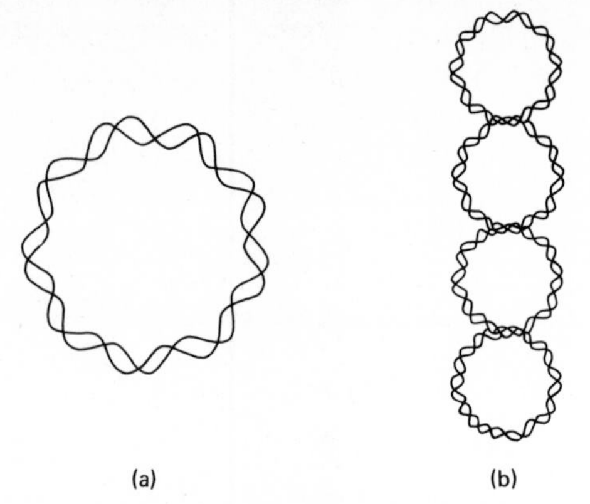

Figure 9-10. Schematic representation of (a) "relaxed" DNA and (b) supercoiled DNA.

Yunis and Yasmineh 1971). In fact, some 20 to 80 percent of the genomes of various species are comprised of these repetitive sequences. In the mouse, for example, 10 percent of the DNA consists of sequences about 300 nucleotides in length, repeated approximately 1 million times! Such highly repetitive DNA is called *satellite* DNA.

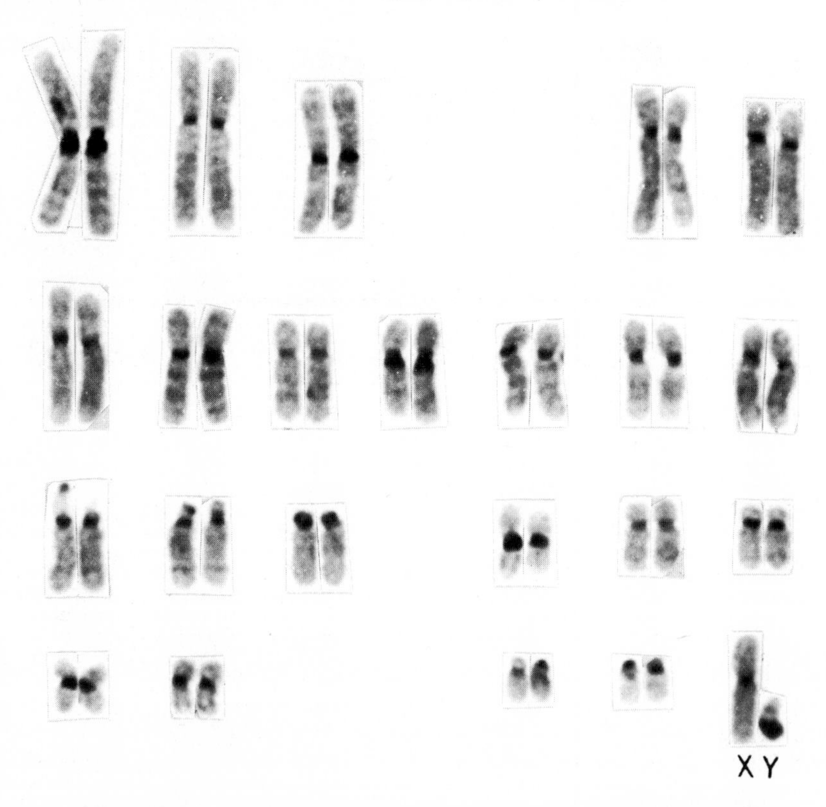

Figure 9-11. A human karyotype showing distribution of constitutive heterochromatin (C-bands). (Photograph courtesy T. C. Hsu and Sen Pathak, University of Texas M. D. Anderson Hospital and Tumor Institute, Houston, Texas.)

In addition, different sequences of intermediate repetitiveness are found 100–100,000 times in the DNA, making up about 20 percent of the genome.

In many organisms, such as amphibians and sea urchins, researchers have reported an interesting pattern of repetitive and nonrepetitive sequences of DNA. Much of the repetitive DNA is regularly interspersed among stretches of nonrepetitive, or *unique*, sequences of DNA. There are also areas of clusters of satellite DNA, and areas of uninterrupted sequences of nonrepetitive DNA. In species such as mice and humans, much of the satellite DNA is associated with the centromeric areas of the chromosomes (Fig. 9-11). We shall discuss the possible function and significance of repetitive DNA later (p. 374).

Table 9-1 gives the sizes of DNA molecules and genomes for a number of different species.

Table 9-1 Sizes of DNA molecules and genomes

	Size of Genome		
Organism	*Number of base pairs (thousands) (kb)*	*Total length (mm)*	*Shape*
	Viruses		
Polyoma, SV40	5.1	0.0017	Circular duplex
ϕX174	5.4	0.0018	Circular single strand; duplex replicative form
M13 (fd, f1)	6.0	0.0020	Circular single strand; duplex replicative form
P4	15.0	0.0051	Linear
T7	35.4	0.0120	Linear
P2, P22	40.5	0.0138	Linear
λ	46.5	0.0158	Linear
T2, T4, T6, P1	144	0.049	Linear
Vaccinia	240	0.082	Linear
	Bacteria		
Mycoplasma hominis	760	0.26	Circular
Escherichia coli	4000	1.36	Circular
	Eukaryotes		
			Number of chromosomes (haploid)
Yeast	13,000	4.6	17
Drosophila (fruit fly)	165,000	56	4
Man	2,900,000	990	23
South American lungfish	102,000,000	34,700	19

Length = (kb) (3.4×10^{-4}) mm.

From A. Kornberg, 1964. *DNA Synthesis*, p. 17, W. H. Freeman & Company Publishers, New York.

From the evolutionary viewpoint, the amount of DNA per haploid genome (referred to as the *C value*) is not directly correlated with genetic complexity. Lest we humans become too smug about our DNA constitution, note that the South American lungfish contains almost 50 times as many base pairs as humans!

BIOLOGICAL PHENOMENA REVISITED: REPLICATION

The simple yet elegant model of DNA proposed by Watson and Crick was received with great enthusiasm. Reference to past observations suggested immediately that the model was probably correct. For example, the ratios obtained by Chargaff could be explained easily on the basis of complementary base-pairing. The ratio of purines to pyrimidines of necessity equaled 1 because for every adenine there is a thymine, and for every cytosine there is a guanine. Yet, because of the variable sequence of bases, the ratio of (A+T)/(C+G) can be different, as Chargaff had found.

On the basis of the Watson-Crick model of the double helix, therefore, we can reconsider a number of biological phenomena which we have discussed previously to reach a clearer understanding of the actual mechanism involved. In the following sections we shall explore the phenomena of gene activity known as *replication* and genetic *recombination*. Replication occurs at interphase of mitosis and meiosis, and recombination is found in crossing-over in eukaryotes and in genetic exchanges in bacteria and viruses.

Semiconservative Replication

The first major application of the double helix model was in the elucidation of the manner in which DNA is exactly reproduced (*replicated*) during cell division. Watson and Crick, in fact, followed their first paper on the structure of DNA with another short paper in 1953 which proposed a likely mechanism of replication.

Watson and Crick reasoned that if the double helix were to unwind and open by the breaking of the weak hydrogen bonds which connect the bases together, the bases of the two chains would be exposed. Nucleotide monomers (single nucleotides) complementary to the exposed bases would then be attracted to and bind with the exposed bases. The process would continue until each of the original chains is completely bound to complementary bases, forming two molecularly identical chains (Fig. 9-12). THE ORIGINAL CHAINS THUS SERVE AS TEMPLATES FOR THE PRODUCTION OF NEW CHAINS. SINCE EACH HELIX THAT RESULTS FROM A REPLICATION EVENT CONTAINS ONE "OLD" CHAIN AND ONE "NEW" CHAIN, THIS TYPE OF REPLICATION IS KNOWN AS SEMICONSERVATIVE REPLICATION.

The Meselson-Stahl experiments. Evidence to confirm the hypothesis of semiconservative replication came from experiments reported by M. Meselson and F. W. Stahl in 1958. They grew a culture of *E. coli* cells in medium containing a heavy isotope of nitrogen, ^{15}N. After a few generations, all the DNA of the bacteria contained heavy nitrogen. The cells were then transferred to medium containing only the ordinary light isotope of nitrogen, ^{14}N. They were allowed to go through one replication and then DNA was extracted from the daughter cells and subjected to a technique known as *density gradient centrifugation.*

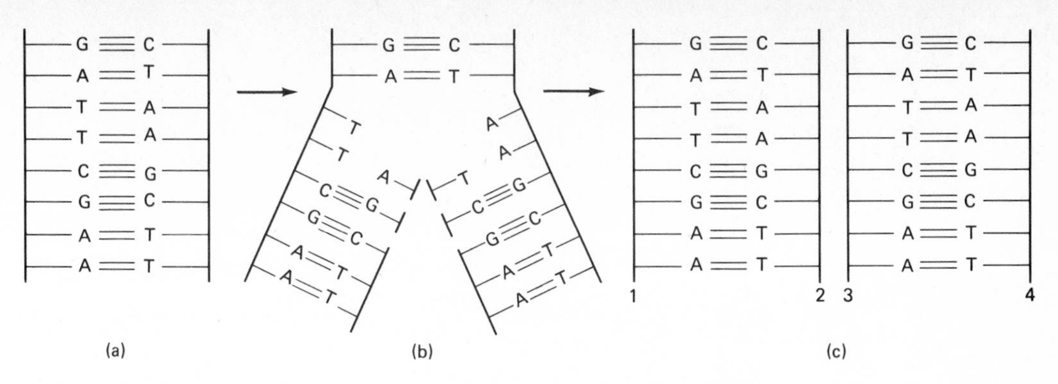

Figure 9-12. Schematic representation of semiconservative replication based on complementary base pairing. (a) Original helix. (b) Replication. (c) Daughter helices are identical to the original and to each other. Strands 1 and 4 are "old" strands; 2 and 3 are "new."

This technique involves placing the DNA in a solution of cesium chloride, CsCl. The mixture is then spun at high-speed centrifugation (around 50,000 rpm, or a force equal to 10^5 times the force of gravity). At such speeds, the Cs^+ and Cl^- ions are subjected to two different forces: the tendency to sediment toward the bottom of the tube because of the centrifugal force, and the diffusion of the ions back to the top of the tube. Eventually, an equilibrium is reached in which the movement of the ions in the two directions is balanced. At this point a gradient of CsCl exists, with the larger-sized particles and therefore the denser portions of the gradient at the bottom of the tube.

If a substance such as DNA is placed in the CsCl and subjected to centrifugation, the DNA will migrate to the region of the gradient which is of the same density as itself. The narrow band formed by the substance can be visualized by UV absorption. Meselson and Stahl found that DNA containing ^{15}N is localized in a band distinctly different in position from the band of DNA containing ^{14}N (Fig. 9-13a).

When DNA of bacteria which had been transferred from medium with heavy nitrogen to medium with light nitrogen was subjected to density gradient centrifugation, it was found that the DNA band was located precisely between the position of ^{15}N-DNA and ^{14}N-DNA. This was interpreted as indicating that after one replication the DNA molecules contained half ^{15}N-DNA ("old" strand) and half ^{14}N-DNA ("new" strand).

If bacteria so transferred were allowed to undergo two cell divisions, and their DNA again analyzed in this manner, two distinct bands appeared in the cesium chloride gradient. One corresponded to the $^{15}N/^{14}N$ DNA mentioned above, the other, to DNA with only ^{14}N. This result supported the theory of semiconservative replication, because the $^{15}N/^{14}N$ DNA would be new double helices which were formed with the ^{15}N DNA strands as template; the DNA with only light nitrogen would be new molecules formed with ^{14}N DNA molecules. As replication in ^{14}N medium continued, an increase in the number of $^{14}N/^{14}N$ molecules would be found, with no increase in the number of $^{15}N/^{14}N$ molecules. This was indeed the observation reported by Meselson and Stahl. Figure 9-13B illustrates the observations of this series of experiments.

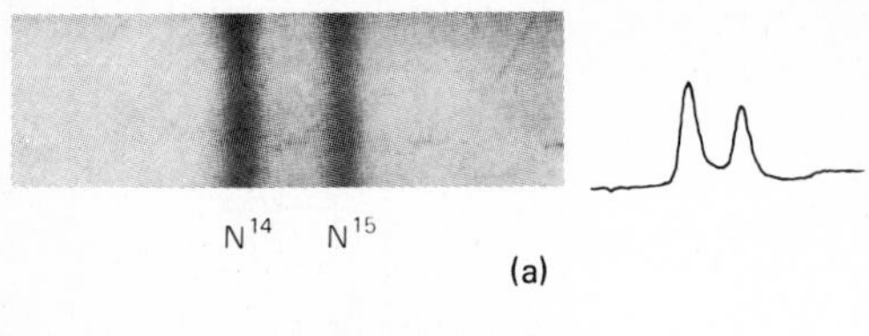

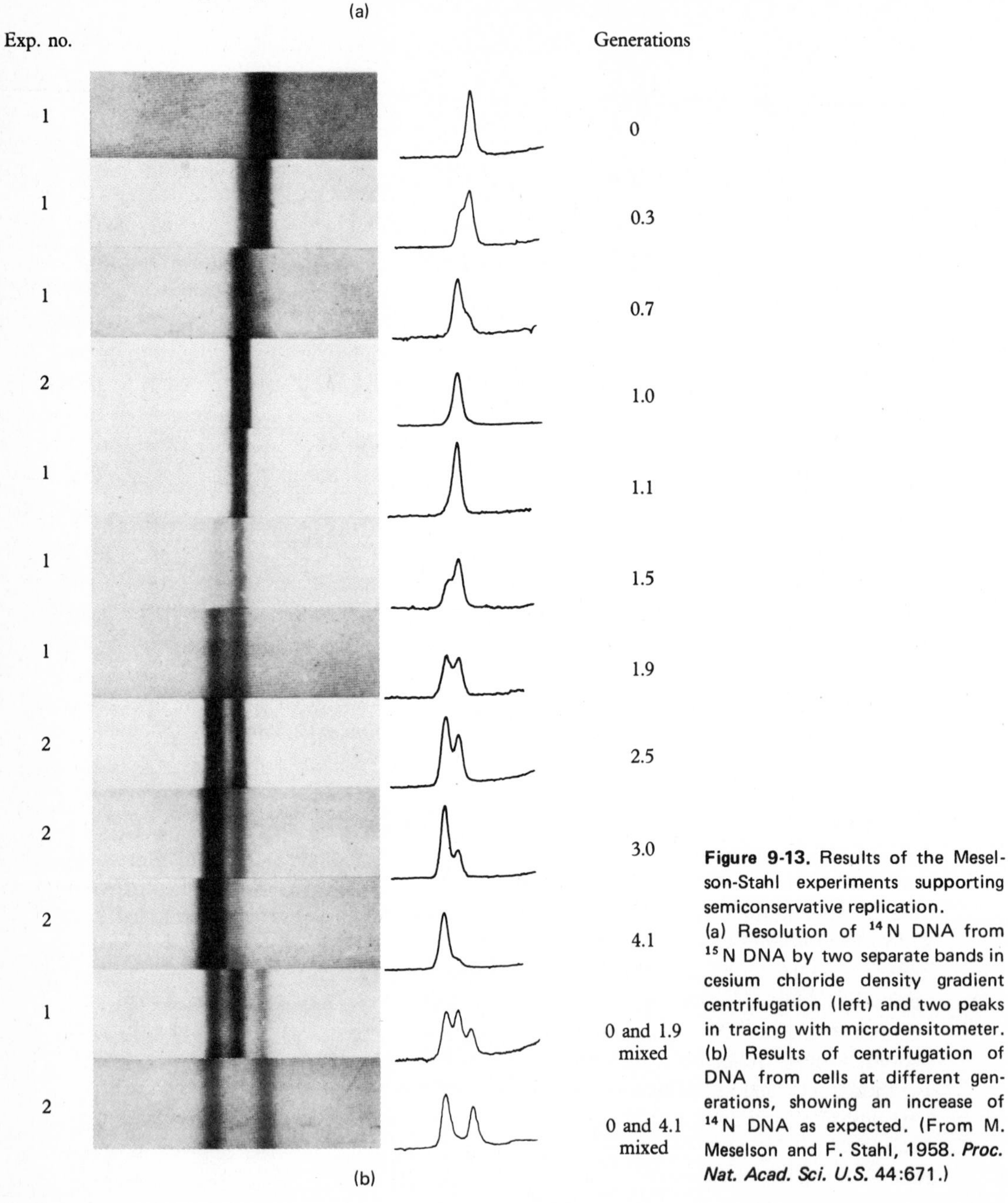

Figure 9-13. Results of the Meselson-Stahl experiments supporting semiconservative replication. (a) Resolution of ^{14}N DNA from ^{15}N DNA by two separate bands in cesium chloride density gradient centrifugation (left) and two peaks in tracing with microdensitometer. (b) Results of centrifugation of DNA from cells at different generations, showing an increase of ^{14}N DNA as expected. (From M. Meselson and F. Stahl, 1958. *Proc. Nat. Acad. Sci. U.S.* 44:671.)

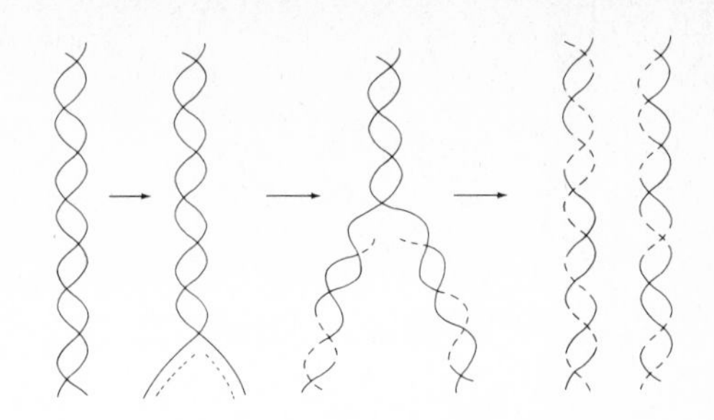

Figure 9-14. Model for dispersive replication. Breaks in the parental strands cause ends of "old" and "new" strands to join, forming the double helix whose chains contain both "old" and "new" DNA.

The results of the Meselson-Stahl experiments eliminated the possibility of *conservative* replication, in which the two template strands of the original helix remain together, giving rise to a DNA double helix with two new chains. In this mode of replication, one would expect to obtain two bands in the cesium chloride gradient after the first division. One band would represent only ^{14}N chains, and one would represent only ^{15}N chains, rather than the intermediate band ($^{14}N/^{15}N$) that was actually obtained.

Since their original report, the Meselson–Stahl technique has been applied to DNA of different species and sources, such as in eukaryotic cells, and with mitochondrial and chloroplast DNA. In each case, the results indicated semiconservative replication, which is now accepted to be the mode of duplication of genetic material.

Meselson's and Stahl's results also support the idea that the entire length of each strand of DNA serves as the template for another entire strand during replication, rather than that a double helix breaks into pieces, with each piece serving as a template. This latter concept, known as *dispersive* replication, holds that the pieces would then be joined together so that there would be "old" DNA as well as "new" DNA in a form of semiconservative replication (Fig. 9-14). If this were the case, one might expect the number of $^{14}N/^{15}N$ molecules to increase with each cell division, as well as an increase in ^{14}N DNA, and this was not observed.

Further evidence against dispersive replication was obtained when the two strands of a DNA double helix were separated by gentle heating. Under these conditions, the hydrogen bonds between the bases break, and the two chains can be studied individually. When the DNA molecules containing both ^{15}N and ^{14}N were so studied, two separate bands appeared, indicating one group of chains containing only light nitrogen, and one group with only heavy nitrogen. With dispersive replication, the chains should have contained both isotopes.

Denaturation and reannealing of DNA. DNA which has been heated so that the strands are separate is known as *denatured* DNA. If the preparation is allowed to cool slowly, single strands with complementary base sequences will reform the hydrogen bonding in a process that is known as *renaturation* or *reannealing*. If complementary base sequences do not exist between single strands, stable hydrogen bonding does not form. The greater the *homology* (a term which refers to similarity of base sequence), the more complete the renaturation. Thus the rate of reannealing in two populations of single-strand DNA is an indication of the homology of their base sequences.

The phenomenon of reannealing has been very useful in many areas of recent genetic research. It led, for example, to the discovery of *repetitive DNA* which was

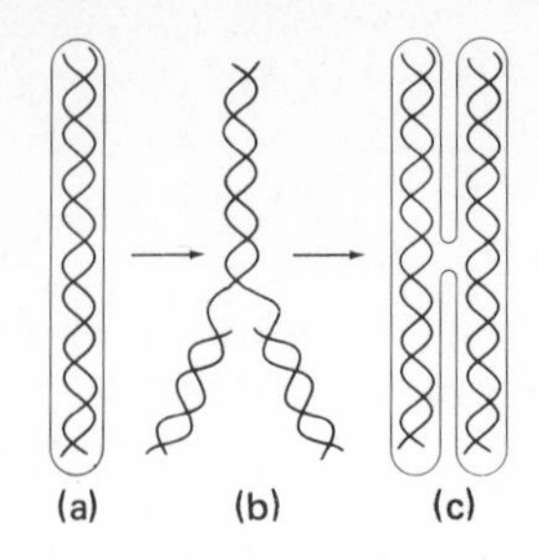

Figure 9-15. Replication in chromosomes. It is believed that each chromosome contains a single linear molecule of DNA (a). This DNA then replicates during interphase (b), resulting in sister chromatids (c) which will separate in anaphase.

revealed by the rate of reannealing of molecules with the same base sequence to many areas of different chromosomes. The greater the similarity in base sequence, the faster the reannealing.

Autoradiography experiments. A number of investigators studied the concept of semiconservative replication with autoradiography. In Chapter 1, we mentioned such studies as those by J. Taylor on cells of *Vicia faba* beans undergoing mitosis. Turn again to Fig. 1-28. The photograph is of chromosomes from cells that were grown in medium with radioactive material, transferred to nonradioactive medium, and allowed to undergo two divisions. Following two divisions, only one chromatid of each duplicated chromosome contained radioactivity in the daughter cells.

These observations indicated that semiconservative replication is the mechanism of chromosome duplication in eukaryotic cells also, and furthermore, that each chromatid most likely contains one double helix of DNA. Figure 9-15 illustrates how semiconservative replication occurs in eukaryotic cells.

Autoradiography has also been used on *E. coli* cells for studies on replication by J. Cairns. Figure 9-16 shows an autoradiograph published by Cairns which he interpreted as the circular bacterial chromosome opening up, with each strand being duplicated. It appears from the autoradiograph, and this has since been confirmed, that replication is a progressive event, rather than one in which the entire chromosome opens up at once to be duplicated.

These studies, and others too numerous to recount here, have fairly well established that genetic material is duplicated in a semiconservative type of replication that results in the two chains serving as templates for new chains down the entire length of the molecule. What of the other aspects of the double helix proposed by Watson and Crick? For example, are the two strands antiparallel in orientation; and how would this affect replication?

In the following sections we shall discuss a series of elegant experiments conducted by Arthur Kornberg and his associates which yielded information clarifying a number of points regarding not just replication, but also other aspects of DNA metabolism. In recognition of some of this work, the Nobel Prize was awarded to Dr. Kornberg in 1959.

In Vitro Synthesis of DNA

In 1957, Kornberg had isolated an enzyme from bacteria which appeared to be capable of catalyzing DNA synthesis in vitro. A mixture of the enzyme (which he termed *DNA polymerase I*), nucleotides in the form of 5′ deoxynucleoside triphosphates (in other words, dadenine triphosphate, dthymine triphosphate, etc., see Fig. 9-17), and Mg^{+} was used. In addition a piece of *primer DNA* (Fig. 9-18) was essential. A

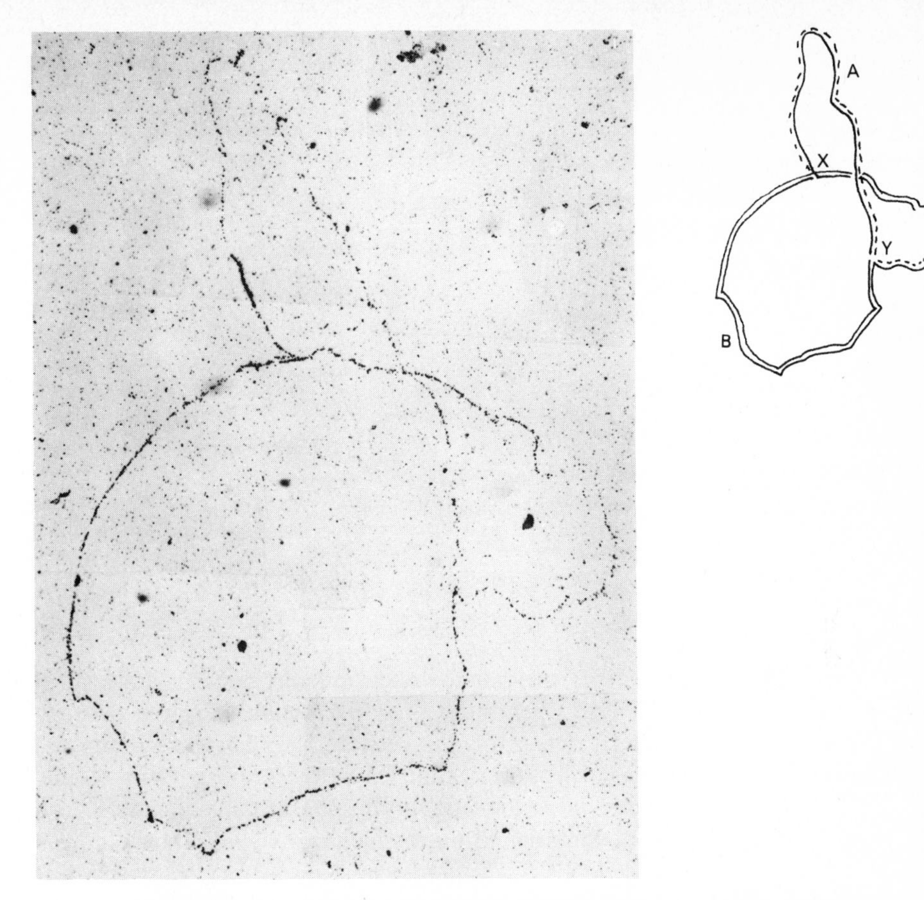

Figure 9-16. Autoradiograph of the replicating DNA of *E. coli*. Inset is a diagrammatic drawing of the autoradiograph. X and Y mark areas now called replicating forks. A and C are the replicating strands; B marks the portion of the DNA yet to be replicated. (Photo courtesy Dr. John Cairns, Mill Hill Laboratories, Imperial Cancer Research Fund, London, England.)

primer molecule, in the in vitro synthesis of new DNA molecules, is a molecule with one short strand of nucleotides onto which the deoxynucleoside triphosphates are added. The second strand of the primer DNA is long, and serves as the template for the elongating new strand.

It must be pointed out here that most in vitro DNA synthesis is not *de novo* synthesis—in other words, it is not the synthesis of DNA solely from component molecules. There usually is a piece of DNA which serves as both the primer, and the template. Such a molecule is necessary because DNA polymerase I can only add nucleotides to a preexisting chain of DNA.

Nearest neighbor analysis. Using this system for in vitro synthesis of DNA, Kornberg and his associates began to study the manner in which nucleotides are incorporated into new DNA chains. Four different preparations were made for in vitro synthesis. The only difference between the four mixtures was that in each, a different deoxynucleoside was labeled with ^{32}P attached to the 5′ carbon of the sugar.

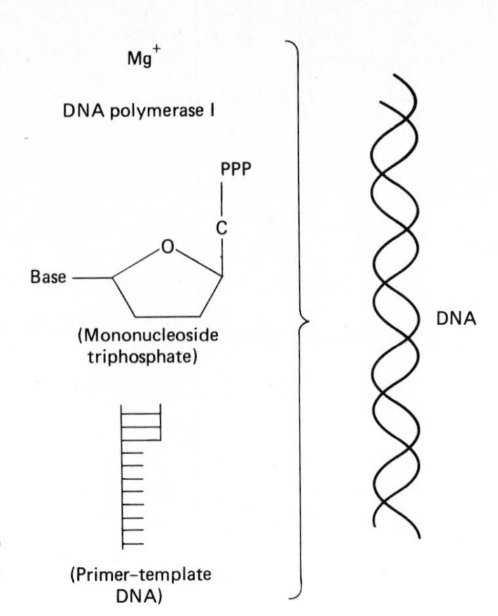

Figure 9-17. Major components for in vitro synthesis of DNA used by A. Kornberg and his associates.

The determination of whether the nucleotide monomer is added to the 3′ or 5′ end of the chain depended on the degradative action of spleen diesterase. This enzyme had been found to cause breaks between the phosphate group and the 5′ carbon in DNA. If the nucleotide is attached to the growing chain at the 3′ end, spleen diesterase would result in the labeled phosphorus remaining with the adjacent nucleotide and not with the original nucleotide (Fig. 9-19a). On the other hand, if the nucleotide is added to the 5′ end, pulse labeling (a very short exposure to the labeled nucleotides, so that only one nucleotide would be added to the primer) would show no ^{32}P in the DNA chain at all (Fig. 9-19b).

The results reported by Kornberg showed that indeed the label was attached to the nearest neighbor nucleotide, not to the original base (thus the technique came to be known as *nearest neighbor analysis*). THIS RESULT ESTABLISHED THAT SYNTHESIS OF DNA TAKES PLACE BY THE ADDITION OF NUCLEOTIDES AT THE 3′ END OF DNA CHAINS, OR IN OTHER WORDS, THAT CHAIN ELONGATION PROCEEDS FROM THE 5′ ⟶ 3′ DIRECTION. NEAREST NEIGHBOR ANALYSIS ALSO CONFIRMED TWO OTHER ASPECTS OF DNA STRUCTURE, NAMELY, COMPLEMENTARY BASE-PAIRING AND OPPOSITE POLARITY OF THE TWO NUCLEOTIDE CHAINS. Appendix A contains a detailed discussion of the nearest neighbor analysis of these two DNA characteristics.

Fidelity of base-pairing in vitro. The fidelity of complementary base-pairing during replication was also reported by Kornberg in his studies of in vitro synthesis of DNA. Base pair frequencies of the new DNA molecules were very similar to those of the primer molecules, as we mentioned previously. This was interpreted to mean that DNA polymerase I was capable of catalyzing the synthesis of complementary copies of DNA sequences, and led to the erroneous conclusion that this was the enzyme responsible for replication in vivo. (We shall speak of this later when we discuss all the enzymes and proteins involved in DNA metabolism.)

However, direct evidence for the fidelity of copying came later, when M.

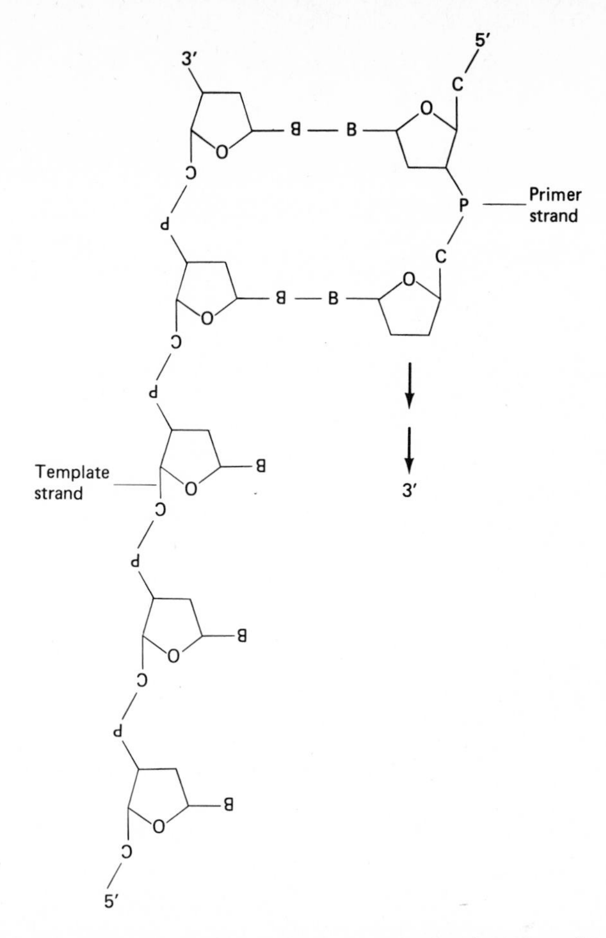

Figure 9-18. Schematic illustration of primer molecule of DNA used in in vitro synthesis. Mononucleotides will be added onto the primer strand.

Goulian, A. Kornberg, and R. L. Sinsheimer (1967) produced a biologically active molecule of DNA using viral DNA as template-primer. By *biologically active* we mean a molecule that can self-replicate. (All the earlier work resulted in the synthesis of nonactive DNA, since the template molecules did not have the same sequence as DNA of any living organism, but were made up in the laboratory.) Robert L. Sinsheimer had earlier discovered a small bacteriophage called ϕX174 (Sinsheimer 1959), whose most appealing feature to geneticists was that it contained a small *single-stranded* circular molecule of DNA, only 6000 nucleotides long.

When ϕX174 viruses infect a cell, synthesis of complementary strands of DNA ensues. The original strands are referred to as the + (plus) strands, and the complementary chains as the − (minus) strands. The new double-stranded form of the virus is known as the *replicating form*, or RF. New plus strands replicated using the minus strands as template are then packaged as new viral particles.

M. Goulian and others (1961) used the DNA from ϕX174 viruses as the primer molecule for in vitro synthesis of DNA. Minus strands formed in vitro from

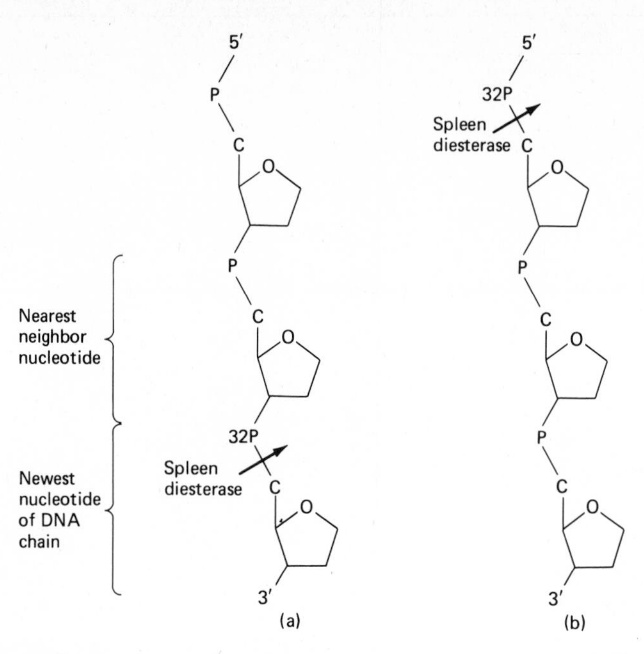

Figure 9-19. Nearest neighbor analysis developed by Arthur Kornberg. Spleen diesterase is known to break the bond between the 5′ carbon and the phosphate of a nucleotide. (a) Thus if nucleotide monomers are added to the 3′ end of a DNA chain during replication, the labeled phosphate would be found attached to the nearest neighbor nucleotide. (b) If added at the 5′ end, the label would not be found after diesterase treatment. Nucleotides were found to be added at the 3′ end.

plus strands were separated from the original DNA by gentle heating. Since these DNA chains were linear, *DNA ligase*, an enzyme capable of forming bonds between the free ends of linear DNA, was added to the preparation to obtain circular molecules. The minus strands were then used in another reaction mixture to generate plus strands. These plus strands presumably contained exactly the same sequence of bases as the original viral DNA, assuming that the in vitro synthesis of DNA is an exact process. One would therefore expect these artificially created plus-strand DNA molecules to be able to replicate in host cells. They were again separated from the minus strands by heating and circularized by ligase. When the plus strands were taken up by living bacteria, the host cells were found to produce ϕX174 virus particles (Fig. 9-20). THIS WAS CLEAR EVIDENCE THAT THE IN VITRO SYSTEM ALLOWED EXACT COPYING OF DNA SEQUENCES BY COMPLEMENTARY BASE-PAIRING.

De novo synthesis of DNA. RECENTLY, HOWEVER, BIOLOGICALLY ACTIVE MOLECULES OF DNA HAVE INDEED BEEN SYNTHESIZED DE NOVO. IN 1977, ITAKURA AND SEVERAL ASSOCIATES REPORTED THE FIRST ACHIEVEMENT OF EXPRESSION OF A GENE THAT WAS PRODUCED WITHOUT USING A NATURAL TEMPLATE IN VITRO. The gene is that which determines the hormone somatostatin. Somatostatin is found in the hypothalamus and various other tissues of vertebrates, and seems to be capable of inhibiting the secretion of insulin, glucagon, and growth hormones.

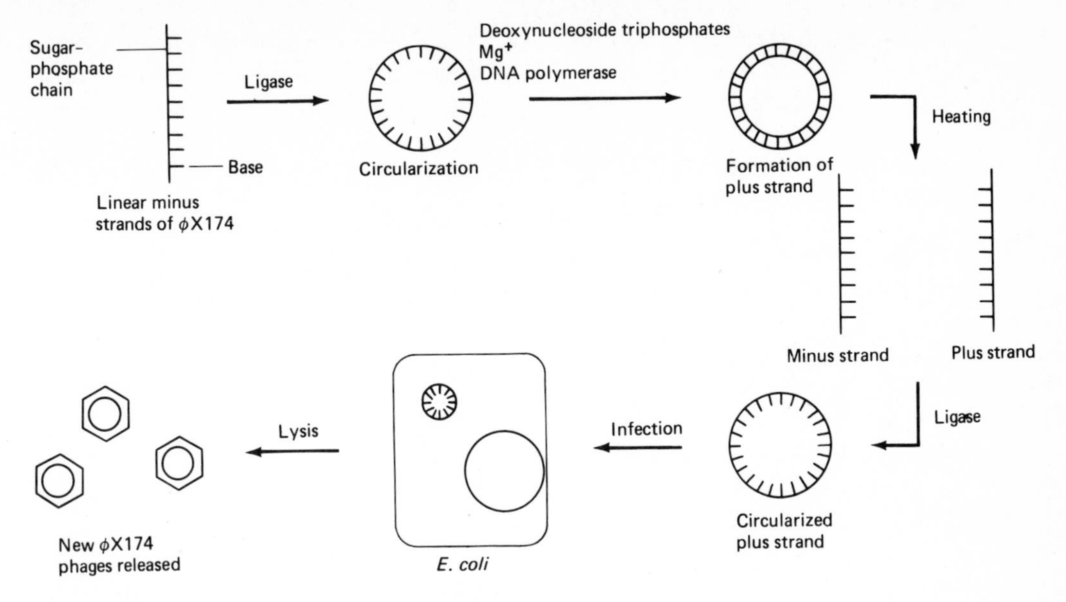

Figure 9-20. Use of ϕX174 phage DNA as primer for in vitro synthesis of DNA capable of infecting cells and causing production of new phage particles.

The characteristic of somatostatin that appealed to molecular geneticists is that it is a very small protein, composed of only 14 amino acids. After determining the amino acid sequence, the Itakura team used sophisticated biochemical techniques to produce a sequence of bases that would determine the correct sequence of amino acids. Once the 42 bases were properly linked, a complementary strand was formed, and the completed DNA molecule was inserted into an *E. coli* cell.

The insertion of the somatostatin gene into a bacterial cell was accomplished by recombinant DNA methodology. We shall discuss techniques to form recombinant DNA in detail in Chapter 14, but for our discussion here, the pertinent fact is that following the insertion of the gene, its activity was reflected in the synthesis of somatostatin by the host cell, which of course normally does not produce the hormone. Presence of somatostatin was determined by immunological tests, and the injection of extracts of bacterial cells containing somatostatin genes into experimental rats. The injections resulted in the inhibition of rat growth hormone production.

These observations indicate that the de novo synthesis of DNA can result in a faithful copy of the natural gene. The apparently normal effects derived from the bacteria-produced hormone attest to this exciting development.

Other Aspects of Replication

The experiments described above, and many others, confirmed that the basic structure of DNA was indeed as Watson and Crick had visualized, a double helix with two chains of nucleotides of opposite polarity. Complementary base-pairing is an essential feature of the double helix. Replication is semiconservative, always proceeding in the $5' \longrightarrow 3'$ direction by the addition of nucleotide monomers (single nucleotides) at the 3′ end, with the "old" chains serving as templates for new chain formation. Since these basic concepts of replication have been established, other unpredicted aspects have been brought to light, and by no means do geneticists feel that they understand the entire process.

Origin of replication. Various techniques such as autoradiography have been used to study the replicating chromosome in prokaryotes. If the molecule is circular, as it is in most bacteria and viruses, where does replication begin? IT HAS NOW BEEN ESTABLISHED THAT THERE IS INDEED A UNIQUE POINT IN THE CHROMOSOME AT WHICH REPLICATION IS INITIATED. This point is known as the *origin* of replication and has been located on the *E. coli* map, for example, at approximately 70 minutes (amount of time needed for the transfer of this locus during conjugation in a particular strain of *E. coli* p. 242).The existence of an origin of replication is apparently universal, and has been reported in eukaryotes as well (Amaldi *et al.* 1973). Additional evidence indicates that there are several origins on each eukaryotic chromosome, at which points new chains are being formed simultaneously. The various pieces are eventually linked up by ligase into a continuous polymer.

Bidirectionality of replication. From his autoradiography studies on bacterial chromosomes, Cairns originally proposed that replication was unidirectional, proceeding from the origin around the chain in one direction (Fig. 9-21a). MORE RECENT WORK, HOWEVER, HAS INDICATED THAT REPLICATION IN MANY SPECIES IS ACTUALLY BIDIRECTIONAL (FIG. 9-21b). In *E. coli*, for example, replication begins at the origin,

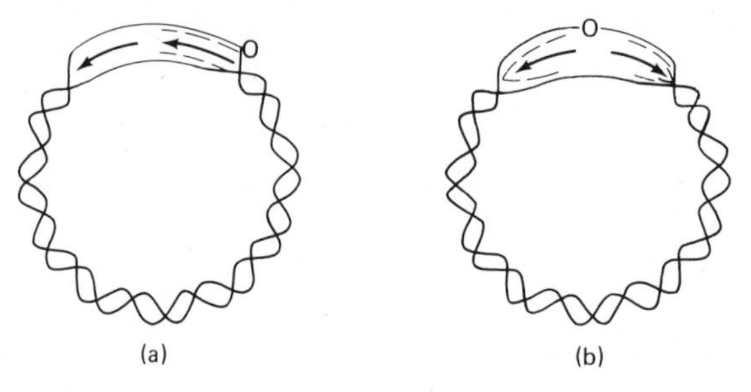

Figure 9-21. (a) Unidirectional replication as originally proposed for bacteria. (b) Evidence now indicates bidirectional replication. O = origin; dashed line = new DNA; arrows point to direction of replication. This is the so-called theta (θ) form of bacterial DNA.

proceeds at equal rates in both directions around the circular chromosome, and is completed at approximately 180° from the origin. For details of the studies which established this phenomenon, see the references to Bird *et al.* (1972) and to Prescott and Kuempel (1972).

Bidirectional replication has also been found in *Bacillus subtilis*, *Salmonella*, and lambda phage, and in several species of eukaryotes, such as *Drosophila*, yeast, and some mammals (see review by Gefter 1975). However, it is apparently not universal, as replication in some phages is unidirectional, as is replication of mitochondrial DNA in mouse cells.

Discontinuous replication. If replication is known to proceed in the 5′ to 3′ direction, and if the two chains of DNA are in opposite polarity, how does the process of new chain formation occur in opposite directions at the replication forks at the same time? This difficulty was resolved first by Okazaki and his coworkers (1968). They exposed *E. coli* to very short bursts of radioactivity (pulse-labeling), then extracted the labeled DNA, which represented new DNA replicated at the time of exposure. They discovered that the labeled DNA was in fragments about 1000–2000 base pairs long. THE INTERPRETATION OF THIS FINDING, WHICH IS NOW WIDELY ACCEPTED, IS THAT DNA REPLICATION ACTUALLY TAKES PLACE IN A DISCONTINUOUS FASHION (FIG. 9-22), AND THAT THE PIECES OF DNA CHAINS PRODUCED ARE THEN LINKED BY DNA LIGASE INTO A LONG CONTINUOUS CHAIN.

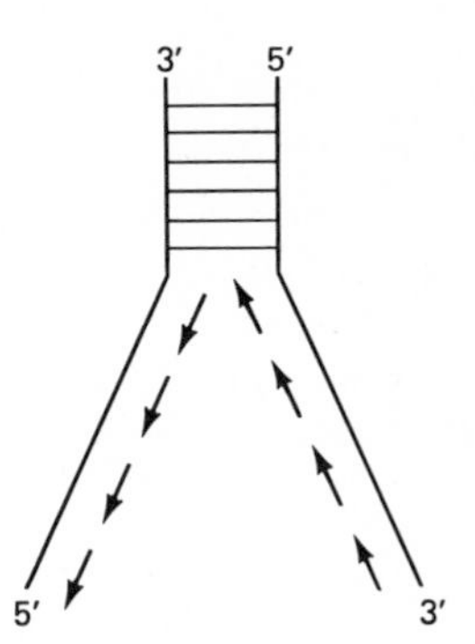

Figure 9-22. Schematic representation of "Okazaki fragments" of DNA being replicated. Arrows represent short segments of new chains that will be joined by ligase, and direction of synthesis.

Role of RNA in chain initiation. The DNA polymerase discovered by Kornberg, and other DNA polymerases (which we shall discuss shortly) that are involved in replication, have all been found to be capable of elongating only a *preexisting* chain of DNA. They have not been found to initiate new (de novo) chain formation, but need a primer molecule onto which they can add nucleotides. A search for such primer DNA molecules in vivo proved fruitless. How, then, do the DNA polymerases catalyze replication in the living cell?

IN THE EARLY 1970S, EVIDENCE WAS REPORTED BY A NUMBER OF SCIENTISTS THAT THE PRIMER FOR DNA REPLICATION IN VIVO IS ACTUALLY A SHORT SEGMENT OF RNA (LARK 1972). An enzyme, RNA polymerase, is known to be capable of causing the de novo formation of a strand of RNA using DNA as template. The RNA fragment then serves as the primer molecule in vivo, and as replication is completed, is excised from the final molecule of DNA by yet other enzymes.

Evidence to support the crucial role of RNA in DNA replication of the bacteriophage M13 was found when rifampicin was shown to prevent the formation of the replicating form of the virus. Rifampicin is an antibiotic which is known to inhibit RNA polymerase activity. The involvement of RNA in initiating DNA chain formation has also been found in a number of bacteria, phages, plasmids, and eukaryotes (Kornberg 1974).

Enzymes and Proteins in Replication

From the preceding discussion, it is quite obvious that in order to understand DNA replication we must understand the enzymes involved. Researchers have found that, in addition to enzymes, other proteins determined by genes are also necessary in order that this most basic of biological processes, self-reproduction, can be completed.

DNA polymerases. In 1957, Kornberg isolated the first DNA polymerase, appropriately called *DNA polymerase I*, which we have already mentioned. There are some 400 molecules of the enzyme in an *E. coli* cell. Although it does cause chain elongation as used in in vitro systems, DNA polymerase I is *not*, however, the primary replication enzyme in vivo. This was discovered when cells with mutations of the gene determining the enzyme were found to be viable—that is, the cells could continue to replicate in the absence of DNA polymerase I (DeLucia and Cairns 1969).

The function of DNA polymerase I appears to be as an *endonuclease* (an enzyme capable of causing a break, or *nick*, within a double-stranded DNA molecule). It also can act as an *exonuclease* (an enzyme which degrades DNA chains from an open end), capable of degrading DNA in both the $3' \longrightarrow 5'$ and $5' \longrightarrow 3'$ directions. Its primary role seems to be that of repairing "mistakes" in DNA, such as mutations, or mistakes made during replication. We shall discuss mutation and repair mechanisms further in Chapter 15.

A second polymerase, DNA polymerase II, is also believed to be involved in repairing DNA. Smaller than polymerase I, polymerase II serves as an exonuclease only in the $3' \longrightarrow 5'$ direction, and also polymerizes monomer nucleotides into chain formation.

DNA polymerase III, a product of the gene known as *dnaE*, is known to be essential for replication. The elongation of the growing chain is primarily a function of *pol III*, as DNA polymerase III is sometimes called. It is believed to complex with the primer RNA molecule. There are only 10 molecules of this enzyme in a cell of *E. coli*, and it is decidedly more unstable than either of the other polymerases. There is evidence, however, that the active form of polymerase III is a complex with at least one other cellular protein, copol III. In the literature, complexed DNA polymerase III is referred to as *DNA polymerase III**. The mechanism involved remains to be elucidated.

DNA polymerases have also been found in both the cytoplasm and the nucleus of mammalian cells, but their structure and function are not yet well understood. They do appear to differ from bacterial polymerases in that they do not have exonuclease activity.

Other proteins. We have already described the important role of DNA ligase, both in forming closed circular DNA, and in binding together Okazaki fragments. Among other proteins that have been postulated to be necessary for replication are unwinding and untwisting proteins. Recall that the DNA of the chromosome in bacteria and viruses has been found to be coiled upon itself. It is therefore necessary first to untwist the supercoils and then to unwind the double helix of the DNA preparatory to the actual duplication process. Some proteins have been found in *E. coli* which appear to affect the superstructure of DNA molecules, but details of their structure and function have not yet been characterized.

There is also evidence that replication in *E. coli* is somehow associated with the cell membrane. When the cell is fractionated, nascent (newly formed) DNA is found attached to fragments of the plasma membrane. The role of this association remains to be discovered.

Figure 9-23 shows a diagrammatic model of the events thought to occur at the replicating fork of DNA in *E. coli*. Certainly, as we learn more about the process,

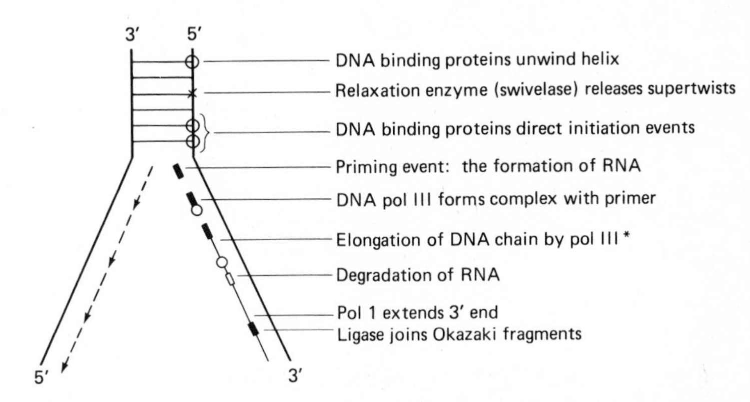

Figure 9-23. Model of events at the replicating fork of DNA in *E. coli*. Details shown of one chain are occurring on both chains. (From K. Geider, 1976. *Curr. Top. Microbiol. Immunol.* 74 : 62.)

replication is proving to be an enormously complex mechanism requiring many different proteins and enzymes. Table 9-3 lists those found in *E. coli*, some of which we have mentioned. Far from rendering basic biological phenomena, such as replication, insignificant, our increased understanding of them has, if anything, heightened the awe of scientists at the marvelously intricate machinery of a living cell.

The Rolling Circle Model of Replication

In some forms of bacteriophage such as the ϕX174, a model has been proposed for the mechanism of reproduction of the RF form of the virus (Gilbert and Dressler 1968). This is known as the *rolling circle model* (Fig. 9-24). (This is also the mechanism by which episomes such as the *F* particle are believed to reproduce and cause the transfer of DNA in conjugation discussed on p. 244.) You recall that the first step in the reproduction of this phage is the synthesis of a minus strand using the plus strand as the template. This duplex molecule is believed to be circularized and closed by ligase. An endonuclease causes a break (nick) in the plus strand, which then begins to unravel (Fig. 9-24 a). As it does so, complementary base pairing occurs on both the exposed bases of the plus and minus strands. This continues until the entire plus strand has become part of a new duplex DNA molecule which then is nicked in such a way as to cause its release from the original molecule.

The original RF or the new RF forms can produce additional RF molecules by this rolling circle mechanism. It is believed that new viral particles are formed in

Table 9-3 Genes Involved in Replication in *E. coli*

Locus	*Alternate gene symbols*	*Function in DNA synthesis*	*Location on standard chromosome map*
dnaA		Initiation at origin	73 min.
dnaB		Chain growth	81 min.
dnaC	dnaD	Initiation at origin	89 min.
dnaD	same cistron as dnaC		
dnaE	see polC		
dnaF	see nrdA		
dnaG		Chain growth	60 min.
dnaH		Initiation at origin	64 min.
dnaI		Initiation at origin	36 min.
dnaP		Membrane defect	75 min.
dnaS		(Accumulation of very short DNA)	72 min.
dnaZ	(dnaH revised)	Chain growth	approx. 11 min.
lig		DNA ligase	46 min.
nrdA	dnaF	Subunit B1 of ribonucleotide diphosphate reductase	42 min.
nrdB		Subunit B2 of ribonucleotide diphosphate reductase	42 min.
polA		DNA pol I	76 min.
polB		DNA pol II	2 min.
polC	dnaE	DNA pol III	4 min.
rif		β subunit of RNA pol, sensitive for rifampicin	77 min.

From K. Geider, 1976. *Curr. Top. Microbiol. Immunol.* 74: 55–112.

much the same way. As the plus strands unroll in some of the RF particles, instead of minus strands being polymerized on them, they are in some way prevented from serving as templates, perhaps by a binding protein. After the total plus strand has been released the molecule is packaged into new phage particles (Fig. 9-24 b).

Replication in RNA Phages

Whereas DNA constitutes the genetic material for the majority of living cells, there are a few viruses which have been found to contain not DNA but RNA, such as Qβ and R17. These viruses are very small particles, and are known to contain information for three genes, one for the coat protein, one for a maturation protein (to complete the packaging of new particles), and one for a subunit of the enzyme *RNA polymerase*, needed for the synthesis of new RNA molecules after infection of a host cell.

RNA is a single-stranded molecule containing ribose instead of deoxyribose, and uracil instead of thymine. Because it has only a single strand of nucleic acid, it is believed that the reproduction of RNA phages such as Qβ is similar to that of ϕX174, in that a plus strand is used as template for the synthesis of a minus com-

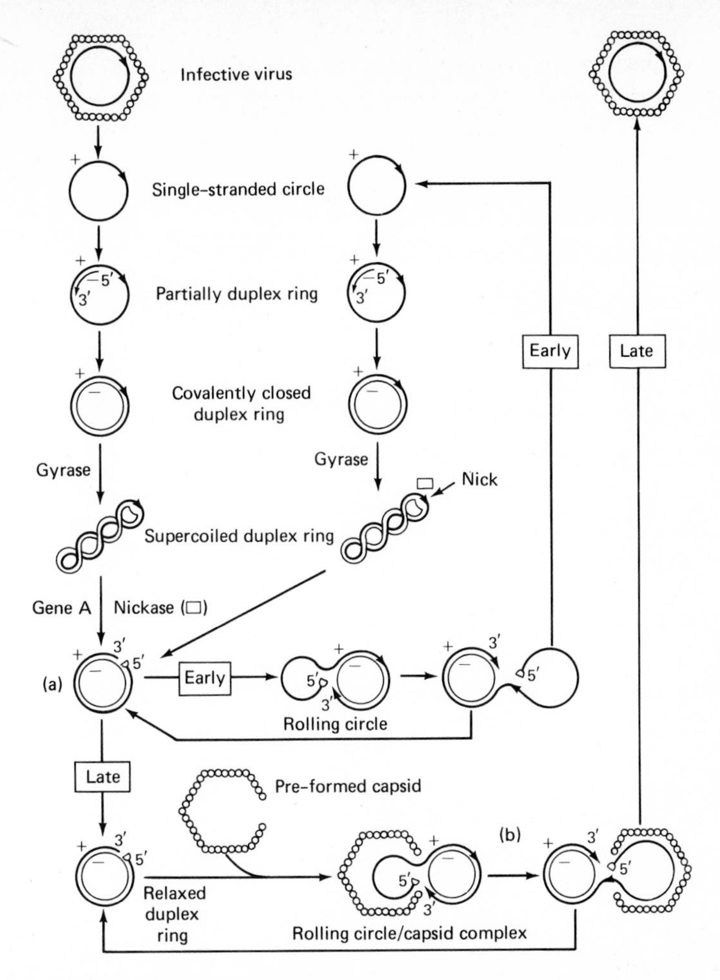

Figure 9-24. The Rolling Circle model of replication in single-stranded phages such as ϕX174. There are two synthetic events: the de novo synthesis of negative strand DNA to form duplex RF DNA, and the continuous synthesis of positive strand DNA involving a rolling circle intermediate. Further explanation in text. (From Sims, J., K. Koths and D. Dressler, 1979. CSHSQB Vol XLIII, pp. 349–365. "Single-Stranded Phage Replication: Positive- and Negative-Strand DNA Synthesis.)

plementary strand. Evidence indicates that replication is associated with the cell membrane, and that at least four cellular products are needed for successful viral reproduction. (For a review of RNA bacteriophage reproduction, see reference to Eoyang and August 1974).

MOLECULAR BASIS OF RECOMBINATION

One basic biological function of genes which has not been well elucidated by research subsequent to the Watson-Crick double helix discovery is that of recombination. The term *recombination* refers to the process by which two different strands of DNA are linked to form a new combination of genetic material. We first encountered recombination in our discussion of linkage and crossing-over. Now that we know that genes are DNA, it is apparent that new combinations of genes found in recombinant gametes must result from some process which allows the DNA of chromatids to interact during meiosis. Incidentally, since we believe DNA to be a continuous strand in eukaryotic chromosomes, one can visualize crossing-over occurring within genes as well as between genes, as we had discussed at the beginning of this chapter in the section on pseudoallelism.

In addition, recombination is involved in the exchange and retention of genes among prokaryotes. We discussed several of these phenomena in Chapter 8—for example, recombination from transformation and conjugation in bacteria, and transduction and recombination in viruses. Since these phenomena include medically significant exchanges, such as between disease viruses and animal cells, the discovery of the molecular basis of recombination holds high priority in genetic research.

Models of Recombination

Although several early models of recombination were proposed, studies of *Drosophila* and other systems yielded data that for the most part indicated a reciprocal exchange during crossing-over. Subsequent to the concept of DNA structure proposed by Watson and Crick, one "classical" model of the molecular basis for this exchange is known as *breakage-reunion*, a term actually used in the 1930s at the chromosomal level.

Breakage-reunion. Essentially the model entails the nicking of each of the DNA helices involved, the unwinding of the molecules so that bases are exposed, and then a ligation, but of strands from different DNA molecules. Any bases that may have been "lost" by this process are replaced by DNA polymerase repair, and ligase restores the molecules to continuous strands. Since recombinants in both eukaryotic and prokaryotic systems usually show little evidence of deletion or duplication of genetic material, it is believed that the nicks occur at exactly the same position in both DNA strands. Figure 9-25 illustrates how breakage-reunion may occur at the molecular level.

Copy-choice. Although some experimental evidence supports the concept of breakage-reunion, other genetic phenomena have led geneticists to postulate that other mechanisms of molecular genetic exchange may exist. Simply put, these phenomena generally involve *nonreciprocal crossing-over*. An example of nonreciprocal crossing-over would be if dihybrids of linked genes in repulsion phase are testcrossed:

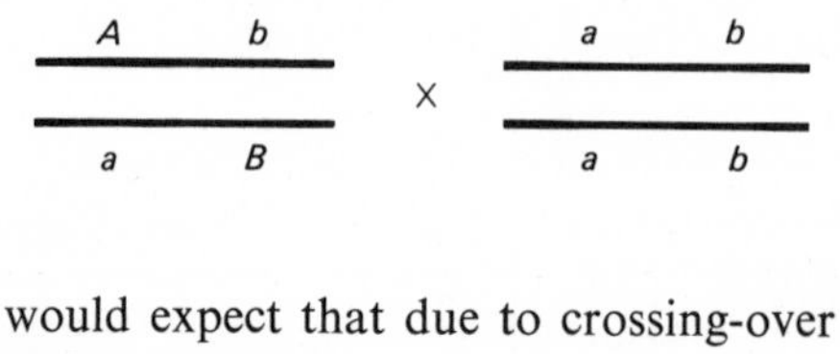

$$\frac{A \quad b}{a \quad B} \times \frac{a \quad b}{a \quad b}$$

In this situation, one would expect that due to crossing-over not only the parental types, *Ab* and *aB*, but also recombinant types should be produced. These recom-

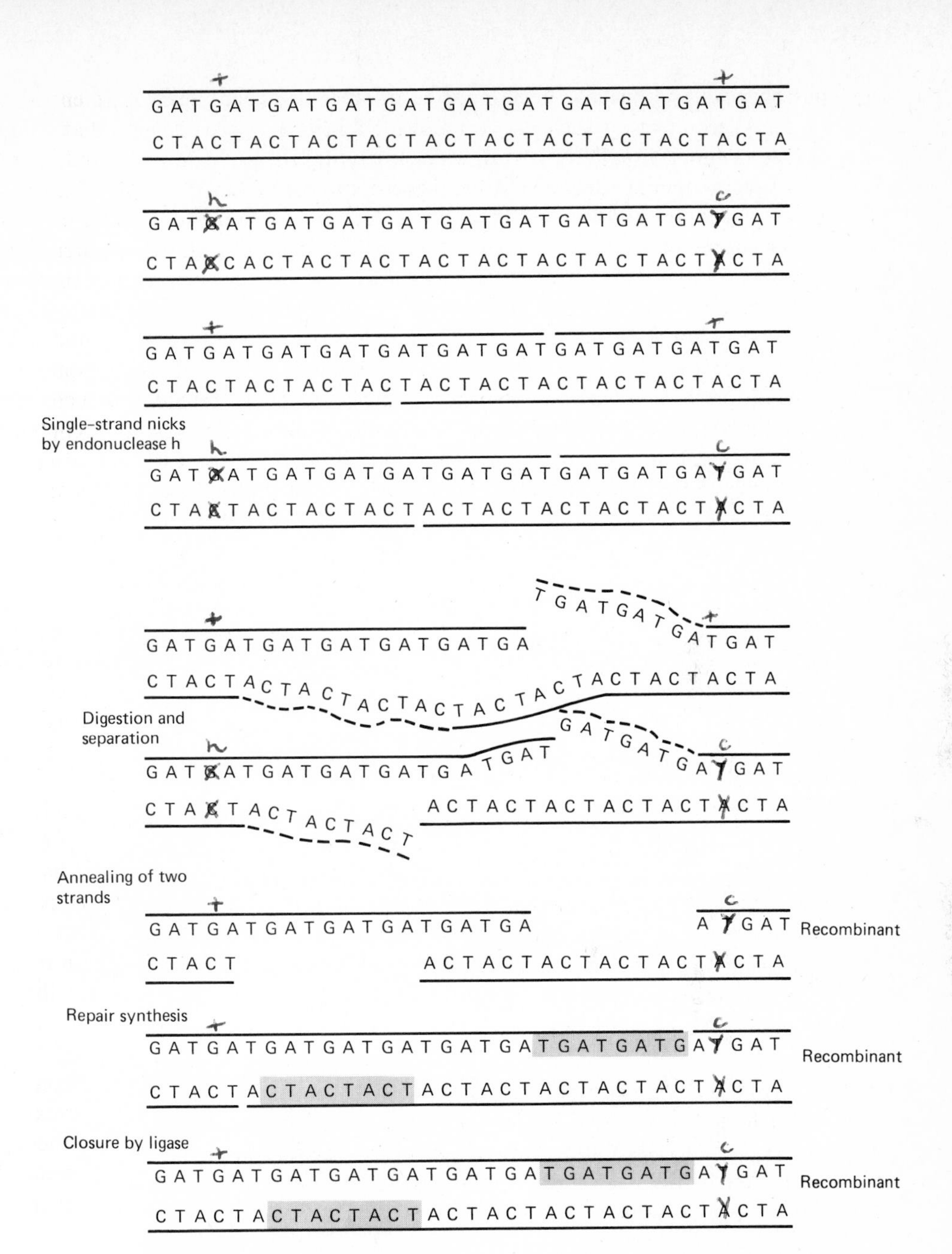

Figure 9-25. Model illustrating breakage-reunion type of recombination at the molecular level. (From D. O. Woodward and V. W. Woodward, 1977. *Concepts of Molecular Genetics,* p. 151. McGraw-Hill, New York.)

binants, *AB* and *ab*, would appear among the progeny in equal numbers. However, in some cases, only one of the recombinant types was found, an occurrence which geneticists termed *gene conversion.* Originally found by Lindegren in yeast (1953), this phenomenon has also been reported in other organisms, such as fungi (Mitchell 1955). Aberrant ratios such as 5 : 3 and 6 : 2 have also been found in recombinant tetrads in fungi, usually involving very closely linked genes.

These findings have led some geneticists to feel that mechanisms other than breakage–reunion (which should result in reciprocity) are responsible for recombination on the molecular level. The "classical" model which is most commonly proposed as an alternative is called *copy-choice*. For some reason, as a new strand of DNA is being formed, instead of proceeding along the length of one template chain, the process jumps to the other chain, and copies it.

Recent models of recombination. More recent models of recombination have generally been modifications of concepts proposed by R. Holliday in 1964. Figure 9-26 illustrates the basic scheme of the Holliday model for recombination. Breaks occur at identical sites in the DNA duplexes of homologous chromatids. Each broken chain then attaches to the homologous DNA molecule. Both DNA duplexes then form a structure similar to the shape of the Greek letter chi (χ). This structure has been called the *Holliday intermediate molecule*, or *chi form*.

There then exists an area in both duplexes which may contain mismatched base pairs. Such areas are called *heteroduplex DNA* (Fig. 9-26f). Furthermore, the heteroduplex area can change, since the region of exchange can move in either direction along the DNA molecules, a process known as *branch migration* (Fig. 9-26g). It is assumed that cellular enzymes such as DNA polymerase and ligase catalyze the steps during recombination.

Because the Holliday intermediate molecule can undergo physical rotation (Fig. 9-26h), separation of the DNA molecules at the crossing point can occur in two different planes (Fig. 9-26i). This would result in different combinations of genetic material following ligation of free ends (Fig. 9-26k,l). Molecular models have been constructed showing that this potential for rotation does indeed exist.

More recent researchers have proposed various models for recombination which are essentially modifications of the Holliday scheme. One widely accepted modification is that of M. S. Meselson and C. M. Radding (1975), who hypothesize that both asymmetric and the symmetric heteroduplex formation can be assumed in the Holliday model.

In asymmetric heteroduplex formation, a break occurs in one strand of DNA only (Fig. 9-27a). The single-strand pairs with, and induces a break in, one strand of the homologous DNA molecule, resulting in a heteroduplex region formed on only one of the DNA strands (Fig. 9-27b). If a genetic marker being studied lies in the region of the heteroduplex, dissociation of the DNA molecules will result in aberrant segregation ratios, as in gene conversion.

Isomerization can also occur. This is a change in the physical shape of the DNA so that the unbroken strands rotate around and can form a chiasma. Cleavage of the crossing strands would also result in nonreciprocal exchange (Fig. 9-27c). On the other hand, if branch migration takes place before cleavage, reciprocal or symmetric heteroduplexes can then be formed on both DNA molecules (Fig. 9-27d). Cleavage of the crossing strands would then result in "classical" crossing-over.

A number of models of the molecular basis for recombination have been proposed (see review by R. Hotchkiss 1974), many based on the Holliday scheme.

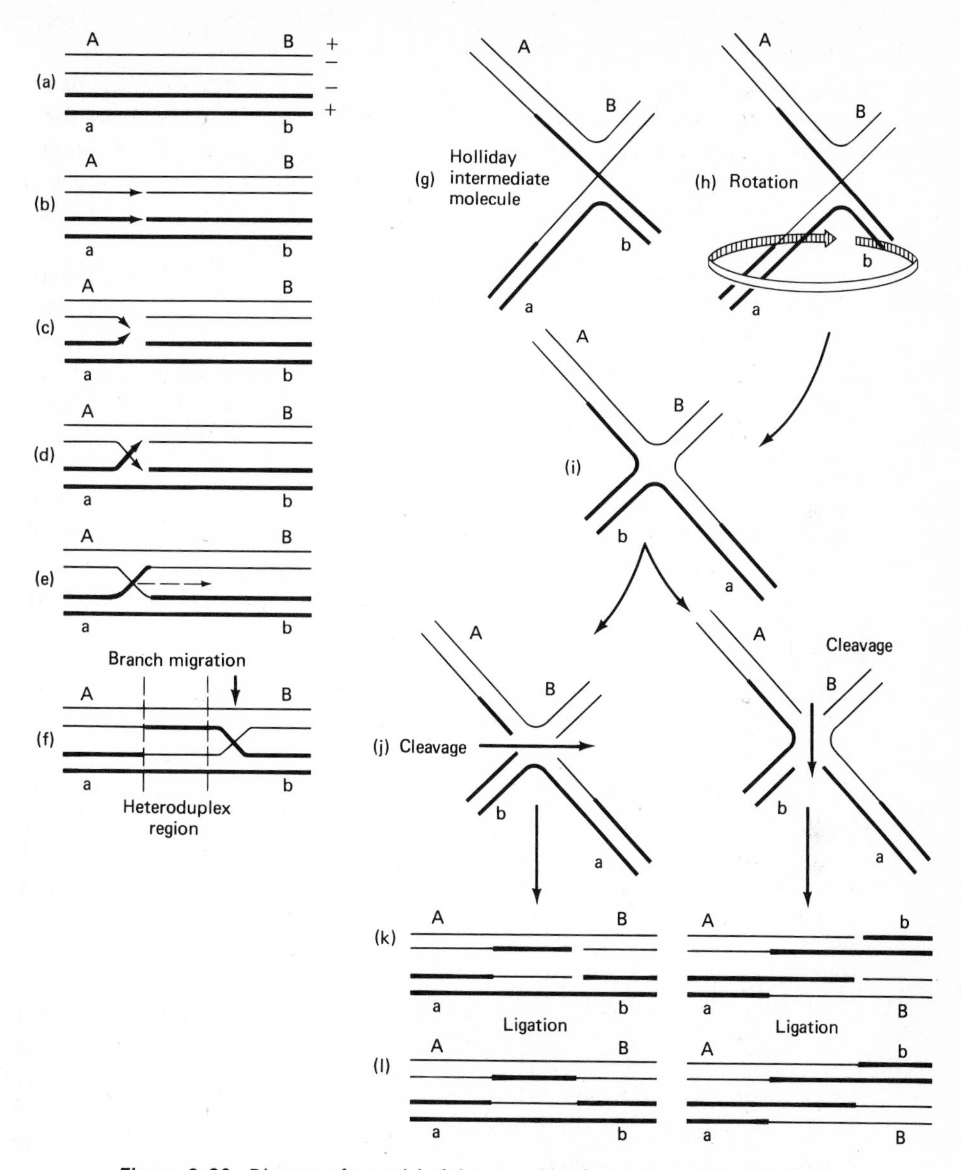

Figure 9-26. Diagram of a model of the recombination process. (a) to (f) illustrate the genetic exchange; (g) represents the Holliday intermediate molecule and region of contact; (h) to (i) postulate possible ways strands can separate, and the resulting gene combination. (From H. Potter, and D. Dressler, 1976. *Proc. Nat. Acad. Sci. U.S.* 73:3303.)

These are still hypothetical, although evidence has appeared to support certain aspects of the models. We mentioned that molecular models have been constructed which show the physical potential of DNA to undergo the scissions and rotations

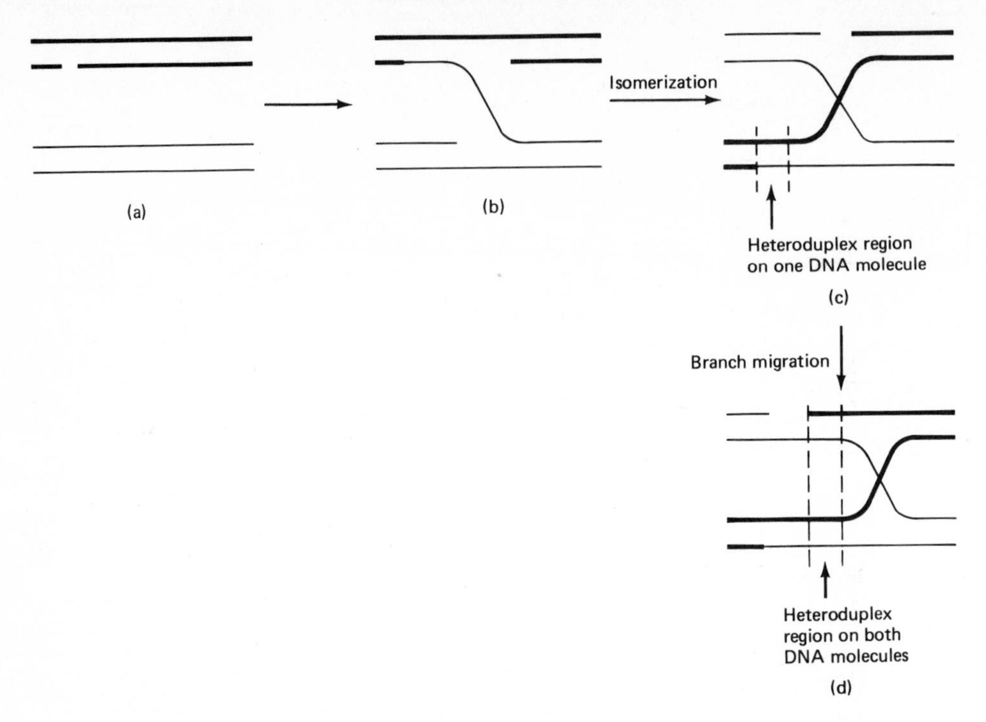

Figure 9-27. Schematic diagram of Meselson-Radding model of recombination showing asymmetric (c) and symmetric (d) formation of heteroduplex regions on DNA molecules undergoing recombination. Further explanation in text. (From M. S. Meselson, and C. M. Radding, 1975. *PNAS* 72: 359.)

that have been postulated. Recently, electron micrographs have been reported to show two DNA molecules in the Holliday intermediate form.

Other details remain to be worked out, such as the factors controlling recognition of sites on homologous DNA molecules which would allow the initiation of the events in recombination. It is proposed that the superhelical form of DNA found in prokaryotes and in nucleosomes of chromosomes in eukaryotic cells may in some way expose sequences of bases that promote recombination between chromatids.

Recombination in transformation. In the last chapter we discussed the uptake of exogenous DNA by some bacteria such as *Pneumococcus* in the process known as transformation. The incorporation of exogenous DNA into the cell's genome is essentially a recombination event. Studies have established that only double-stranded DNA is effective in being absorbed in transformation, and that cells are not always capable of absorbing DNA. We refer to cells that are capable of taking up DNA as *competent*; cells at the stage of most active division are most effectively competent.

Once the DNA is absorbed into the cells by enzyme reaction, single-stranded pieces of DNA are then integrated into the bacterial chromosome by replacing one portion of one of the strands of cellular DNA. These steps are illustrated in Fig. 9-28.

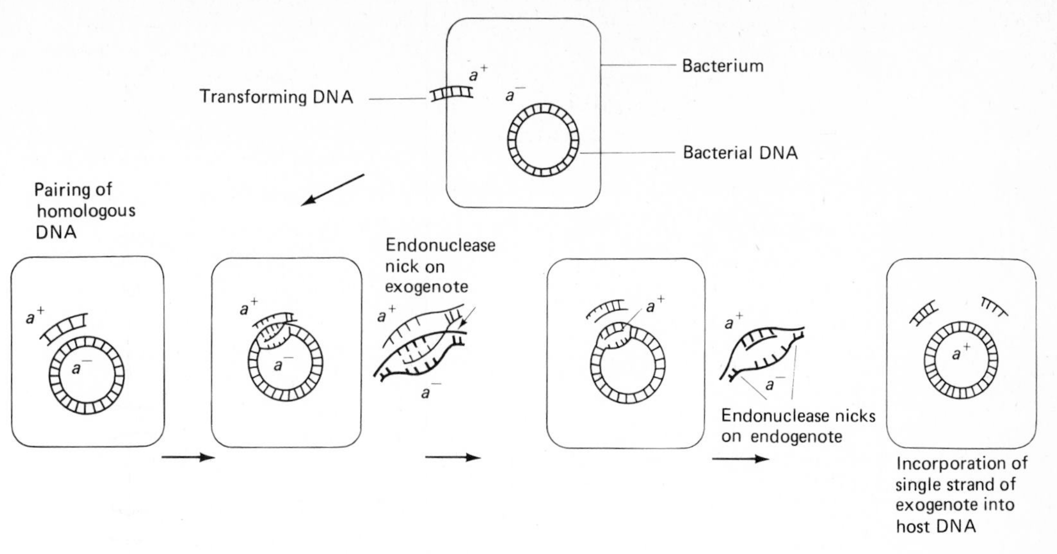

Figure 9-28. Model for genetic recombination during bacterial transformation. (After M. Fox, 1966. *J. Gen. Physiol.* 49: 193.)

Rec⁻ Mutants in Bacteria

Another area on which the study of genetic recombination has great bearing is that of the molecular basis to *repair mechanisms*. It has been found that cells possess the capacity to detect abnormal areas of DNA resulting from damage (such damage can be caused by irradiation, for example). Enzyme systems exist which excise the abnormal areas and replace them with normal sequences of bases. Repair will be discussed in more detail later (Chapter 15), but it is relevant to mention this here because several of the enzymes involved in repair are also apparently involved in recombination. This is consistent with recent models of recombination discussed in preceding sections, which assume enzyme activity in cleavage, ligation and repair of recombining molecules.

In bacteria, for example, at least three genes are known to determine various aspects of recombination. Mutations at these loci have been found, and such mutant cells are defective in both recombination and repair. Because of a lack or deficiency of recombination, these mutants cannot exchange genetic material in conjugation and transduction. They are known as $recA^-$, $recB^-$, and $recC^-$ mutations, the letters signifying the locus which has mutated.

The $recA^+$ gene determines a protein which allows reunion of DNA ends after they have been nicked in breakage–reunion (McEntee *et al.* 1976). In the absence of $recA^+$ protein, the strands undergo degradation instead of being linked to form recombinant molecules. The genes $recB^+$ and $recC^+$ determine subunits of an ATP-dependent DNAase, which seem to have both exonuclease and endonuclease activity. Its direct role in recombination has not been determined, but $recB^-$ and $recC^-$ mutants have reduced frequency of recombination.

Site-specific Recombination

In Chapter 8, we spoke of the integration of viruses into bacterial chromosomes. This process is one which involves recombination, although there is integration rather than exchange of genetic material. Recent studies carried out on the attachment sites on the DNA of both the lambda virus and its host *E. coli* have shown the reason for the specificity of attachment of the lambda prophage (Landy and Ross 1977). Within the *att* locus of the bacterium and of the virus, there is an identical

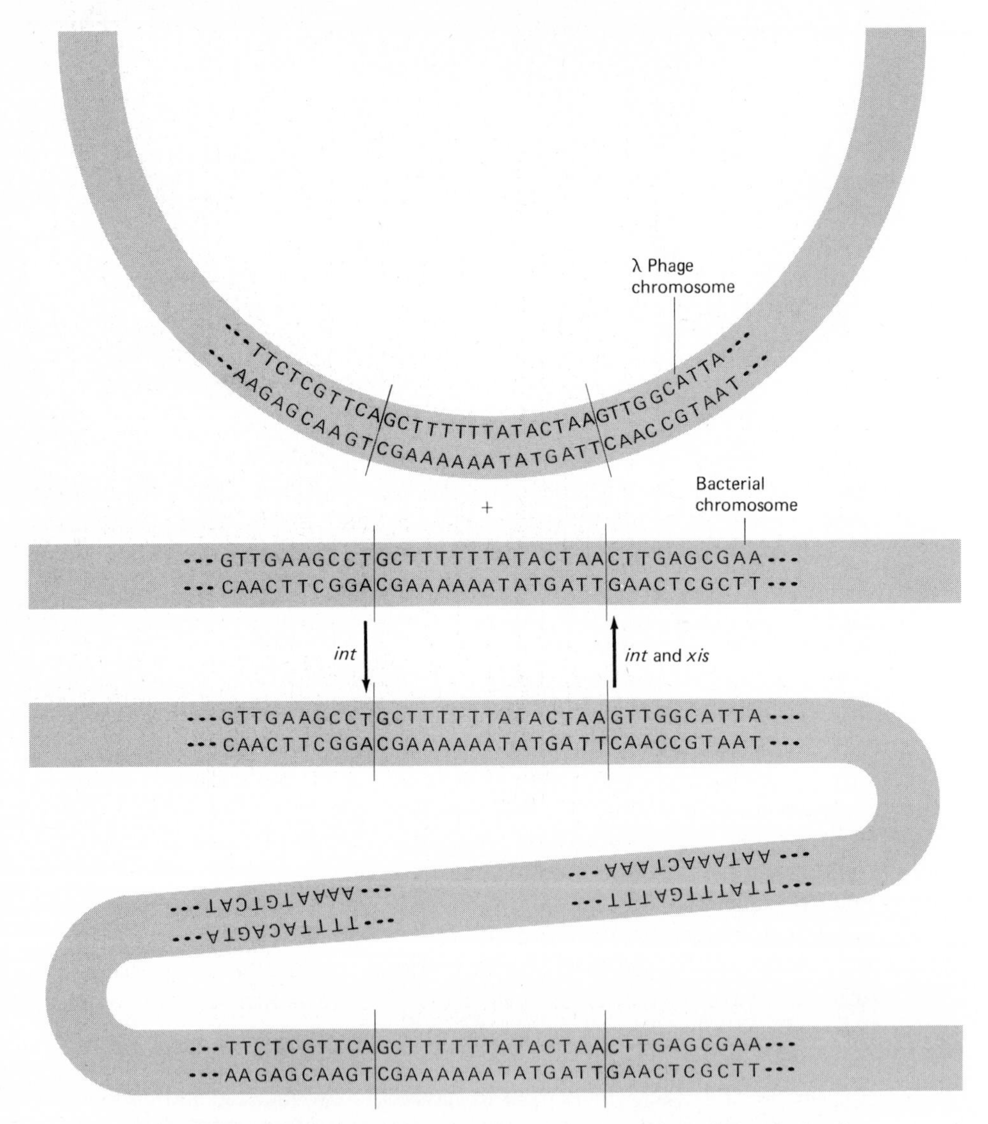

Figure 9-29. Illustration of *att* locus of λ phage and *att* locus of *E. coli*, showing identical sequence of 15 nucleotides. Recombination for integration occurs in this area. (A. Landy, and W. Ross, 1977. *Science* 197, cover illustration; copyright 1977 by the American Association for the Advancement of Science.)

sequence of 15 nucleotides called the *core region* (Fig. 9-29). Recombination of the DNA of the virus and its host DNA is believed to occur within this core region.

We now know that a number of sites on the bacterial chromosome known to be sites of attachment for "foreign" particles of DNA, such as viruses and the *F* particle in conjugation, have the same base sequence as a portion of the particle which integrates into the chromosome.

It is tempting to extrapolate from these data that perhaps all recombination events, including crossing-over in eukaryotes, involves such regions of molecular homology. Perhaps this is why recombinants show so little deletion or duplication of genetic material; if such areas exist on the same part of chromatids, exchanges would be precise.

Recombinant DNA Research

The importance of understanding the molecular basis to genetic recombination has taken on greater significance because of recently developed techniques allowing geneticists to produce specific new combinations of DNA in the laboratory, in so-called "recombinant DNA" experiments. Since the ability to do so is essentially the ability to create "new forms of life," much controversy surrounds the desirability of this type of research. We shall devote an entire chapter to the techniques, and the pros and cons, of this fascinating development in genetics in Chapter 14.

PROBLEMS

1. How many base pairs can be expected in a virus chromosome 10 microns long?

2. If the average gene is 1000 base pairs long, how many genes would the above virus contain?

3. The rate of replication of the *E. coli* chromosome is approximately 90,000 base pairs/minute. If the length of the *E. coli* chromosome is 1100 microns, approximately how many minutes are needed for one replication event in *E. coli*?

4. If the DNA sequence on one strand of a double helix is the following: 5′ ATCGGCCTT 3′, give the base sequence of an RNA strand that would be formed from it.

5. Using the techniques of the Meselson-Stahl experiments, what bands would be found in a cesium chloride gradient, if replication is conservative, after one replication following transfer of a culture of bacteria from medium with ^{14}N to medium with ^{15}N? After two replications? After five?

6. Assume that you have to analyze a primer molecule of DNA. Nearest neighbor experiments using ^{32}P thymine and ^{32}P adenine show that the label attaches to a cytosine only when ^{32}P adenine is used in a very short pulse. If ^{32}P adenine and ^{32}P thymine are used together the label is found attached to both cytosine and adenine. What is the sequence of bases of the primer DNA?

7. Nucleic acid molecules of a virus are isolated from an infected cell while they are in various stages of replication during a lytic phase. The molecules are all found to be equal to or shorter than the total length of the virus chromosome. Would this finding preclude a rolling circle form of replication? Why?

8. Devise an experiment that would allow you to distinguish between the replacement of a DNA strand and the addition of a DNA strand after genetic recombination due to transformation.

9. Show by drawing the molecules why T*–C dinucleotides are found in equal frequency to G*–A dinucleotides, but not A*–G in *Mycobacterium phlei* DNA. (This problem requires information given in Appendix A.)

REFERENCES

AMALDI, F., M. BUONGIORNO-NARDELLI, AND F. CARNEVALI, 1973. "Replicon Origins in Chinese Hamster Cell DNA, II. Reproducibility." *Exp. Cell Res.* 80: 79–87.

BIRD, R. E., J. LOUARN, J. MARTUSCELLI, AND L. CARO, 1972. "Origin and Sequence of Chromosome Replication in *Escherichia coli*." *J. Mol. Biol.* 70: 549–566.

BRITTEN, R. J., D. E. KOHNE, 1968, "Repeated Sequences in DNA." *Science* 161: 529–540.

CAIRNS, J., 1963. "The Chromosome of *Escherichia coli*." *CSHSQB* 28: 43–45.

CHARGAFF, E., 1951. "Structure and Function of Nucleic Acids and Cell Constituents." *Fed. Proc.* 10: 645–659.

DELIUS, H., AND A. WORCEL, 1973. "Electron Microscopic Studies on the Folded Chromosome of *Escherichia coli*." *CSHSQB* 38: 53–58.

DELUCIA, P., AND J. CAIRNS, 1969. "Isolation of an *E. coli* Strain with a Mutation Affecting DNA Polymerase." *Nature* 224: 1164–1166.

EOYANG, L., AND J. T. AUGUST, 1974. "Reproduction of RNA Bacteriophages." H. Fraenkel-Conrat, and R. R. Wagner, eds., in *Comprehensive Virology*, pp. 1–57. Plenum Press, New York.

GEFTER, M. L., 1975. "DNA Replication." *Ann. Rev. Biochem.* 44: 45–78.

GEIDER, K., 1976. "Molecular Aspects of DNA Replication in *Escherichia coli* Systems." *Curr. Top. Microbiol. Immunol.* 74: 55–112.

GILBERT, W., AND D. DRESSLER, 1968. "DNA Replication: The Rolling Circle Model." *CSHSQB* 33: 474–484.

GOULIAN, M., A. KORNBERG, AND R. L. SINSHEIMER, 1967. "Enzymatic Synthesis of DNA. XXIV. Synthesis of Infectious Phase ϕX174 DNA." *Proc. Nat. Acad. Sci. U.S.* 58: 2321–2328.

HOLLIDAY, R., 1964. "A Mechanism for Gene Conversion in Fungi." *Genet. Res. Camb.* 5: 282–304.

HOTCHKISS, R. D., 1974. "Models of Genetic Recombination." *Ann. Rev. Microbiol.* 28: 447–468.

ITAKURA, K., T. HIROSE, R. CREA, A. D. RIGGS, H. L. HENEKER, F. COLIVAR, AND H. W. BOYER, 1977. "Expression in *Escherichia coli* of a Chemically Synthesized Gene for the Hormone Somatostatin." *Science* 198: 1052–1063.

JOSSE, J., A. D. KAISER, AND A. KORNBERG, 1961. "Enzymatic Synthesis of Deoxyribonucleic Acid. VIII. Frequencies of Nearest Neighbor Base Sequences in Deoxyribonucleic Acid." *J. Biol. Chem.* 236: 864–875.

KORNBERG, A., 1960. "Biologic Synthesis of Deoxyribonucleic Acid." *Science* 131: 1503–1508.

______, 1974. *DNA Synthesis.* W. H. Freeman & Company Publishers, San Francisco.

LANDY, A., AND W. ROSS, 1977. "Viral Integration and Excision: Structure of the Lambda *att* Sites." *Science* 197: 1147–1160.

LARK, K. G., 1972. "Evidence for the Direct Involvement of RNA in the Initiation of DNA Replication in *Escherichia coli* 15 T^-." *J. Mol. Biol.* 64: 47–60.

LEWIS, E. B., 1950. "The Phenomenon of Position Effect." *Adv. Gen.* 3: 75–115.

LINDEGREN, C. C., 1953. "Gene Conversion in *Saccharomyces.*" *J. Genet.* 51: 625–637.

MCENTEE, K., J. E. HESSE, AND W. EPSTEIN, 1976. "Identification and Radiochemical Purification of the *recA* Protein of *Escherichia Coli K-12.*" *Proc. Nat. Acad. Sci. U.S.* 73: 3979–3983.

MESELSON, M. S., AND F. W. STAHL, 1958. "The Replication of DNA in *Escherichia coli.*" *Proc. Nat. Acad. Sci. U.S.* 44: 671–682.

MESELSON, M. S., AND J. J. WEIGLE, 1961. "Chromosome Breakage Accompanying Genetic Recombination in Bacteriophage." *Proc. Nat. Acad. Sci. U.S.* 47: 857–868.

MESELSON, M. S., AND C. M. RADDING, 1975. "A General Model for Genetic Recombination." *Proc. Nat. Acad. Sci. U.S.* 72: 358–361.

MITCHELL, M. B., 1955. "Aberrant Recombination of Pyroxidine Mutants of *Neurospora.*" *Proc. Nat. Acad. Sci. U.S.* 41: 215–220.

OKAZAKI, R. T., K. OKAZAKI, K. SUGIMOTO, K. SAKABE, AND A. SUGINO, 1968. "Mechanism of DNA Chain Growth. I. Possible Discontinuity and Unusual Secondary Structure of Newly Synthesized Chains." *Proc. Nat. Acad. Sci. U.S.* 59: 598–605.

PRESCOTT, D. M., AND P. L. KUEMPEL, 1972. "Bidirectional Replication of the Chromosome in *Escherichia Coli.*" *J. Mol. Biol.* 70: 549–566.

SINSHEIMER, R. L., 1959. "A Single-Stranded DNA from Bacteriophage ϕX174." *J. Mol. Biol.* 1: 43–53.

VISCONTI, N., AND M. DELBRUCK, 1953. "The Mechanism of Recombination in Phage." *Genetics* 38: 5–33.

WATSON, J. D., AND F. H. C. CRICK, 1953. "Molecular Structure of Nucleic Acids. A Structure for Deoxyribose Nucleic Acid." *Nature* 171: 737–738. Also reprinted in J. A. Peters, ed., 1959. *Classic Papers in Genetics.* Prentice-Hall, Inc., Englewood Cliffs, N.J.

WATSON, J. D., AND F. H. C. CRICK, 1953. "Genetical Implications of the Structure of Deoxyribonucleic Acid." *Nature* 171: 964–969. Reprinted in E. A. Adelberg, ed., *Papers on Bacterial Genetics*, 2nd ed. Little Brown, Boston.
WATSON, J. D., 1968. *The Double Helix*. Atheneum, New York.
WICKNER, W., R. SCHEKMAN, K. GEIDER, AND A. KORNBERG, 1973. "A New Form of DNA polymerase III and a Copolymerase Replicate a Long, Single-Stranded Primer-Template." *Proc. Nat. Acad. Sci. U.S.* 70: 1764–1767.
YUNIS, J. J., AND W. G. YASMINEH, 1971. "Heterochromatin, Satellite DNA and Cell Function." *Science* 174: 1200–1209.

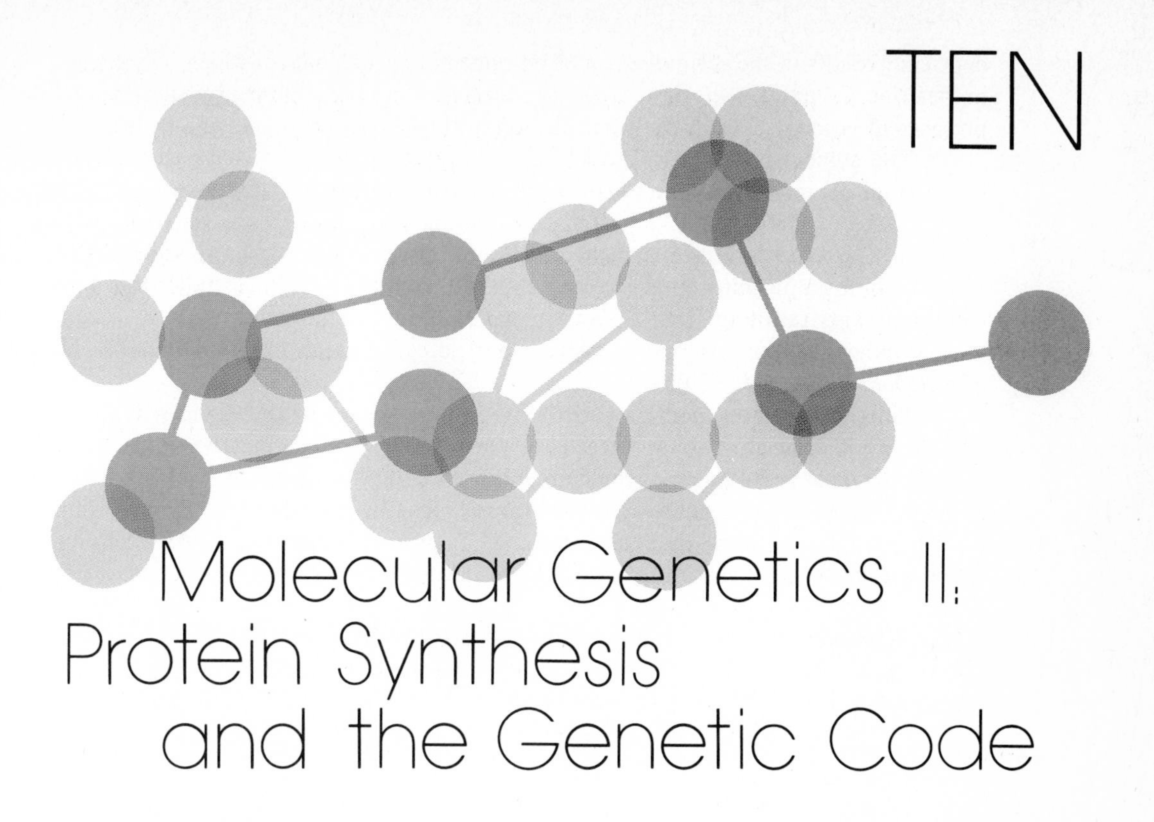

TEN

Molecular Genetics II: Protein Synthesis and the Genetic Code

INTRODUCTION

Two of the basic functions of genes are self-replication and the determination of the structure and function of cells. In Chapter 9 we discussed the mechanism of self-replication. In this chapter we will discuss how genes determine the structure and function of cells, (that is, phenotype). Once the structure of DNA was determined, the mechanism by which genes control phenotype became the subject for investigation.

Clues to the manner in which genes can determine phenotype have been compiled since the rediscovery of Mendel's laws at the turn of the twentieth century. In the following sections we will discuss some of the studies between the first decade of this century and the 1950s which have led to our present understanding of this most important aspect of gene action.

Genes and Metabolism in Humans

Shortly after the rediscovery of Mendel's laws, a British physician noticed an unusual coincidence in the background of several of his patients suffering from conditions due to a deficiency in metabolism. The physician's name was Sir Archibald Garrod. His patients were suffering from a disease known as alkaptonuria, which results from an inability of the cells to metabolize the amino acid phenylalanine. This deficiency results in the accumulation of homogentisic acid, some of which is excreted in the urine. Common symptoms of alkaptonuria are darkening of the urine (due to the presence of homogentisic acid), pigmentation in the connective tissue, and arthritis.

The coincidence which Garrod noticed was that several of his patients suffering from alkaptonuria were offspring of consanguineous marriages, suggesting a genetic basis to the disorder. Through consultation with the classical geneticist, W. Bateman, Garrod determined that the condition has a pattern of transmission characteristic of a simple autosomal recessive trait. He then expanded his studies to other conditions also resulting from defects in metabolism, and found a number which were all inherited as autosomal recessives. Garrod called these diseases *inborn errors of metabolism* (Garrod 1909).

Since then, researchers have determined that several of the diseases Garrod studied result from an inability of cells to perform one specific metabolic reaction. Among the most well-known of these are albinism and phenylketonuria (also known as PKU). These diseases are related to various blocks in phenylalanine metabolism, though at different steps than alkaptonuria. Figure 10-1 illustrates the reactions which are involved in these inborn errors. Clinically, these diseases are of concern because of their effects on homozygotes.

Albinism. Found in many species of plants and animals, albinism is well known as a condition which involves a total absence of pigmentation. The hair of albino individuals is white, and because of the lack of pigmentation in the iris and

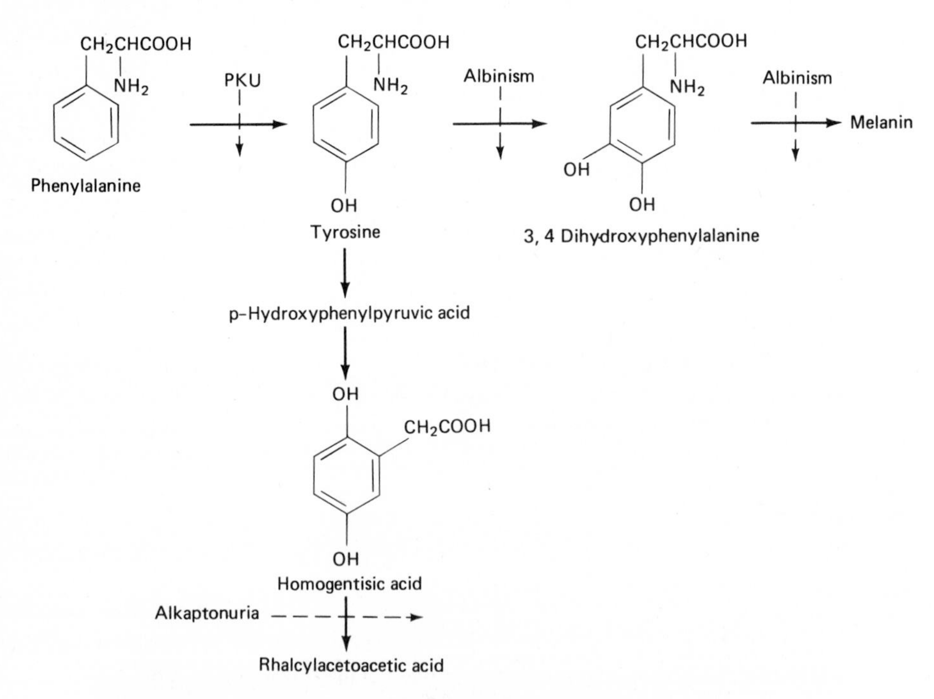

Figure 10-1. Diagram of phenylalanine metabolism, showing steps blocked in three inborn errors of metabolism: PKU (phenylketonuria), two forms of albinism, alkaptonuria.

skin, their eyes and skin are pink, reflecting the vascularization of the tissues. The unpigmented iris also causes albinos to be especially sensitive to light. Turn again to Fig. 4-9, which shows albinism in humans.

Phenylketonuria. PKU homozygotes manifest a high level of phenylalanine in the blood due to an inability to convert the amino acid into tyrosine. The excess leads to, among other defects, severe mental retardation.

Significance of inborn errors of metabolism. Although important to affected individuals and their families from the clinical standpoint, inborn errors also contributed a significant concept to geneticists, namely, that the mutation of genes can be associated with the alteration of a specific metabolic reaction. SINCE ENZYMES ARE INVOLVED IN ALL METABOLIC REACTIONS, AN INEVITABLE CONCLUSION WAS THAT GENES ARE THEREFORE SOMEHOW RELATED TO ENZYMES. Since Garrod's pioneer observations, more than 150 diseases in humans have been identified as inherited enzyme defects (see reference to Stanbury *et al.* 1972).

Genes and Metabolism in Insects

Wild-type eye pigmentation in *Drosophila* involves brown pigments called *ommochromes* (see Chapter 4). Two different eye-pigment mutations affecting ommochrome production in *Drosophila* were studied by Beadle and Ephrussi (1937). These were the autosomal recessive mutations Vermilion (*v*) and Cinnabar (*cn*). Techniques had been developed for transplanting anlage (embryonic precursors) of eye tissue from *Drosophila* larvae into the abdomen of other larvae. In this environment the cells continue to differentiate and produce pigmentation.

Beadle and Ephrussi transplanted eye primordia from *vv* larvae into Cinnabar larvae. The reciprocal experiment was also performed, in which *cncn* eye tissue was transplanted into Vermilion larvae. They observed that *vv* anlage eye cells in Cinnabar larvae were able to produce wild-type pigmentation; however, *cncn* primordia in Vermilion larvae were unable to do so. Analysis of the biochemical pathway for ommochrome synthesis indicated that these observations result from the fact that the Vermilion gene interrupts the biosynthesis of ommochromes at an earlier step than does the Cinnabar gene. Consequently, Cinnabar flies produce a substance whose synthesis is blocked in Vermilion flies; this then leads to the synthesis of ommochromes in the transplanted *v* tissue. On the other hand, Vermilion flies do not have the substance needed by cells of Cinnabar flies, and the transplanted *cn* tissue still cannot synthesize ommochromes. Figure 10-2 summarizes this important experiment.

Genes and Enzymes in Fungi

In 1941 G. Beadle and E. Tatum (Fig. 10-3) performed a series of experiments on the fungus *Neurospora crassa*. Their observations led them to formally propose, for the first time, a direct relationship between genes and enzymes. In their experiments, Beadle and Tatum exposed wild-type N*eurospora* cultures to UV radiation, which had previously been proven to cause mutation in fungus.

The irradiated cultures were then grown in complete medium, containing all the nutrients needed by cells for growth and maintenance. Samples were removed

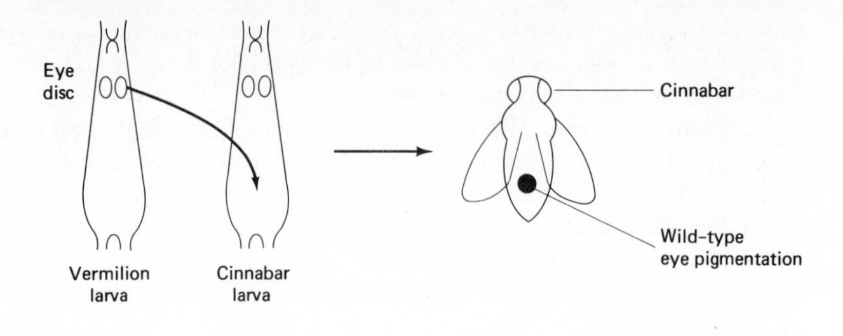

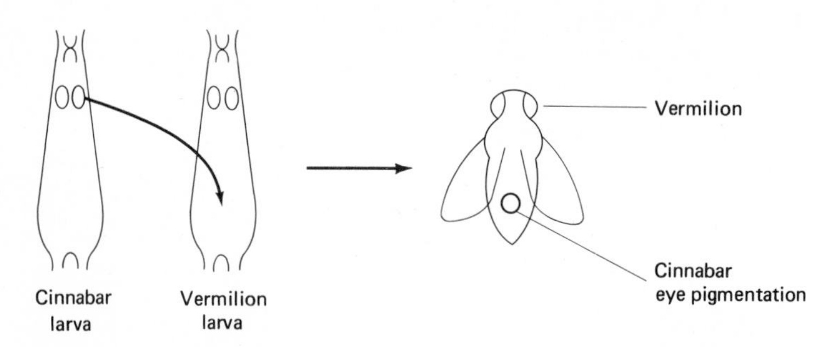

Figure 10-2. Beadle and Ephrussi's experiments on reciprocal transplants of *v* and *cn* eye primordia (eye discs) into Cinnabar and Vermilion larvae, respectively. Vermilion eyes can differentiate wild-type pigmentation in Cinnabar larvae; Cinnabar eye discs remain mutant in phenotype when transplanted into Vermilion larvae.

and transferred to minimal medium. A number of the cultures were found to be unable to survive in minimal medium, indicating mutation. Mutant cultures were then grown in selective media, which are minimal media supplemented by specific compounds such as vitamins or amino acids. It could be assumed that if the addition of a particular supplement allowed growth to occur, when other substances did not, this was the essential substance whose biosynthesis had been interfered with by the mutation (Fig. 10-4).

The association between mutation and deficient metabolism led Beadle and Tatum to reflect: "The development and functioning of an organism consist essentially of an integrated system of chemical reactions controlled in some manner by genes. It is entirely tenable to suppose that these genes which are themselves a part of the system, control or regulate specific reactions in the system either by acting directly as enzymes or by determining the specificities of enzymes."[1]

Their analysis led to the formulation of the "one gene–one enzyme" hypothesis. As you will see shortly, we now know the hypothesis to be only partially correct, but it was the first stated insight into the manner by which genes can control pheno-

[1] Beadle, G. W., and E. L. Tatum, 1941. "Genetic Control of Biochemical Reactions in *Neurospora*." *Proc. Nat. Acad. Sci. U.S.* 27: 499–506.

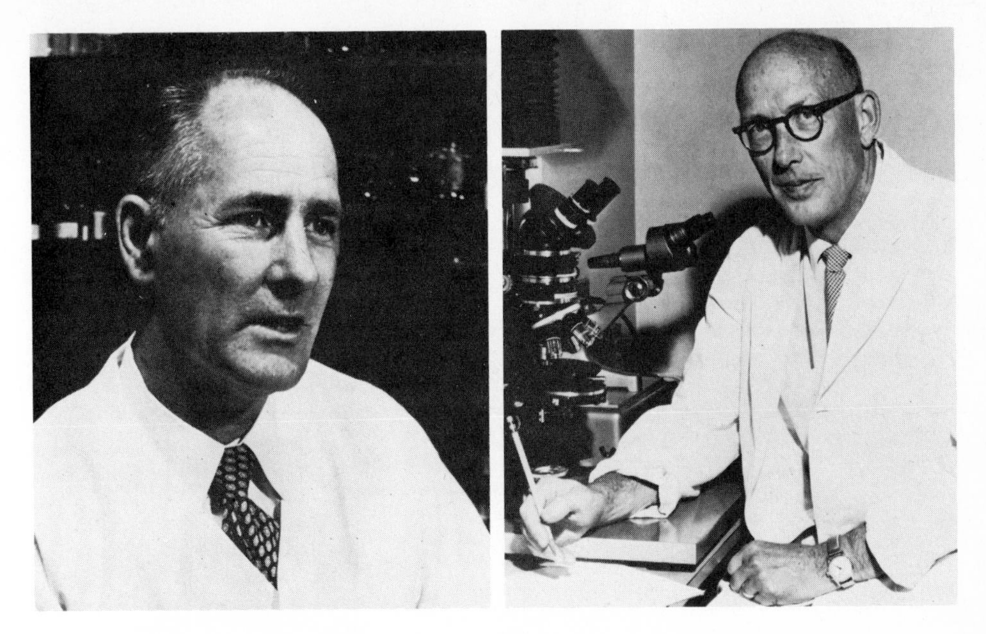

Figure 10-3. George Beadle (left) and E. Tatum. (The Bettmann Archive.)

type. For their studies in the biochemical genetics of fungi, Beadle and Tatum were awarded the Nobel Prize in 1958 along with Joshua Lederberg, whose work in microbial genetics we have already mentioned (Chapter 8).

GENES AND PROTEINS

The preceding studies indicated that genes control function by determining the ability of cells to synthesize enzymes, which are proteins. Also, many of the structures in cells, such as the membranes both surrounding the cytoplasm and within the cytoplasm, contain proteins as a major component. THESE FACTS POINTED TO AN INCONTROVERTIBLE ASSOCIATION BETWEEN GENES AND PROTEIN SYNTHESIS. The next problem to be solved was, how does the genetic material, DNA, determine protein synthesis in cells?

One of the first aspects of this problem to be resolved was that of variability of cellular proteins. Living cells are capable of synthesizing literally thousands of different proteins. What could account for this variability in the double helical structure of DNA? As we discussed in Chapter 9, the sugar-phosphate backbone of DNA in all cells is the same, and all DNA molecules are composed of the same four nitrogenous bases. However, one aspect of DNA structure did offer the prospect of unlimited variability: the *sequence* of bases.

As you will see in later sections of this chapter, many elegant experiments have confirmed that the sequence of bases in a DNA molecule contains information in the form of a code, known as the *genetic code*, which determines the synthesis of

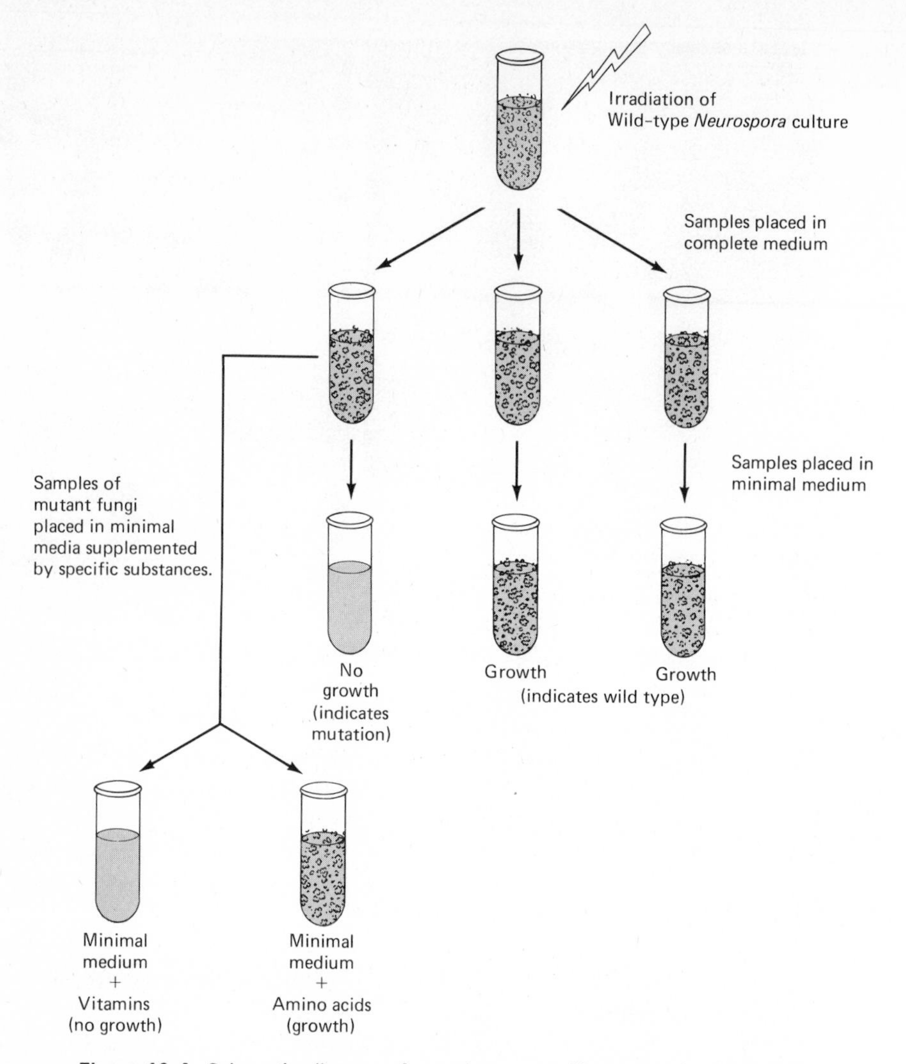

Figure 10-4. Schematic diagram of experiments on *Neurospora* by Beadle and Tatum. These results indicate the mutation under study interferes with amino acid synthesis. Additional tests with specific amino acids will indicate which one was missing or defective in mutant cells. Further explanation in text.

specific proteins by a cell. Since the 1960s, the genetic code has been completely deciphered. However, in order to understand how this was achieved, it is necessary first to explore the mechanism by which the code in the DNA is converted into protein by the cell.

Transcription

Cell biologists had earlier established that protein synthesis occurs in the cytoplasm of cells and involves cell organelles known as ribosomes. Yet the genetic determinant of protein synthesis, DNA, is located in the nucleus, which in cells of eukaryotes is separated from the ribosomes by a nuclear membrane. How, then, is the information relayed to the cytoplasm from the nucleus? EXPERIMENTS HAVE SHOWN THAT THE FIRST STEP IN PROTEIN SYNTHESIS IS THE DNA-DIRECTED SYNTHESIS OF A MESSENGER MOLECULE, SPECIFICALLY AN RNA MOLECULE, WHICH CARRIES THE INFORMATION TO RIBOSOMES. THIS MESSENGER RNA (COMMONLY CALLED mRNA) IS SYNTHESIZED IN A PROCESS KNOWN AS TRANSCRIPTION.

Studies by cytogeneticists first suggested the potential for movement of macromolecules into and out of the nucleus. Observations from vital staining experiments indicated that nucleic acid particles could move from the nucleus into the cytoplasm. The possibility of macromolecules moving through the nuclear membrane was confirmed by electron microscopy which showed that the nuclear membrane of eukaryotic cells is, in fact, full of pores.

Messenger RNA. One of the first clues leading to the identification of an actual messenger molecule came to light when a number of laboratories reported that infection of bacterial cells by phage is invariably followed by a burst of RNA production in the host cell (see references to Volkin *et al.* 1958, and Brenner *et al.* 1961). Yet the total amount of RNA in the cytoplasm of infected unlysed cells does not increase, indicating that the RNA molecules synthesized after infection are quickly degraded. Furthermore, when analyzed for nucleotide content, this class of RNA revealed base ratios very similar to those of viral DNA.

Since the infected cell was known to produce viral proteins following infection, it was thought that the short-lived RNA molecules may represent messages of the viral DNA. The very characteristic of being short-lived was consistent with what one would expect of messenger molecules, since within a cell not all proteins are produced at all times, as we mentioned at the beginning of the chapter. Thus if gene messages are degraded quickly after they have participated in protein synthesis, cells would not continuously produce these proteins. This concept has generally been proven to be true, although examples of long-lived messenger molecules have been found, which we shall discuss later.

In addition, it was not difficult to accept RNA as the relayer of information from DNA, since the sequence of bases in DNA could be reproduced in a sequence of bases of an RNA strand. In other words, DNA could serve as a template for RNA synthesis, with the exception of having uracil incorporated instead of thymine. This very neatly fit with the observation that the base ratios of RNA synthesized in newly infected cells were similar to the base ratios of viral DNA.

Since normal cells as well as infected cells synthesize protein, such a class of RNA would be expected to exist in all cells, not only in newly infected ones. Indeed, in 1961, F. Gros and his associates reported finding a class of short-lived RNA molecules with base ratios similar to those of cellular DNA in uninfected bacterial cells.

The concept of mRNA strands complementary to DNA was supported by hybridization and reannealing experiments of B. D. Hall and S. Spiegleman (1961) among others. In these experiments, RNA produced in *E. coli* cells following infection by T2 bacteriophage was isolated and mixed with dissociated viral DNA strands and with dissociated single strands of bacterial DNA. Reannealing occurred only with viral DNA. Absence of reannealing with either of the strands of bacterial DNA indicated that the sequence of bases in mRNA is complementary only to the DNA which served as a template for its production (Fig. 10-5).

DNA of viruses is mostly double-stranded. Since the nucleotide sequences of the two strands of any DNA helix are complementary, and not identical, does each strand serve as template for mRNA synthesis? Marmur and his associates (1963)

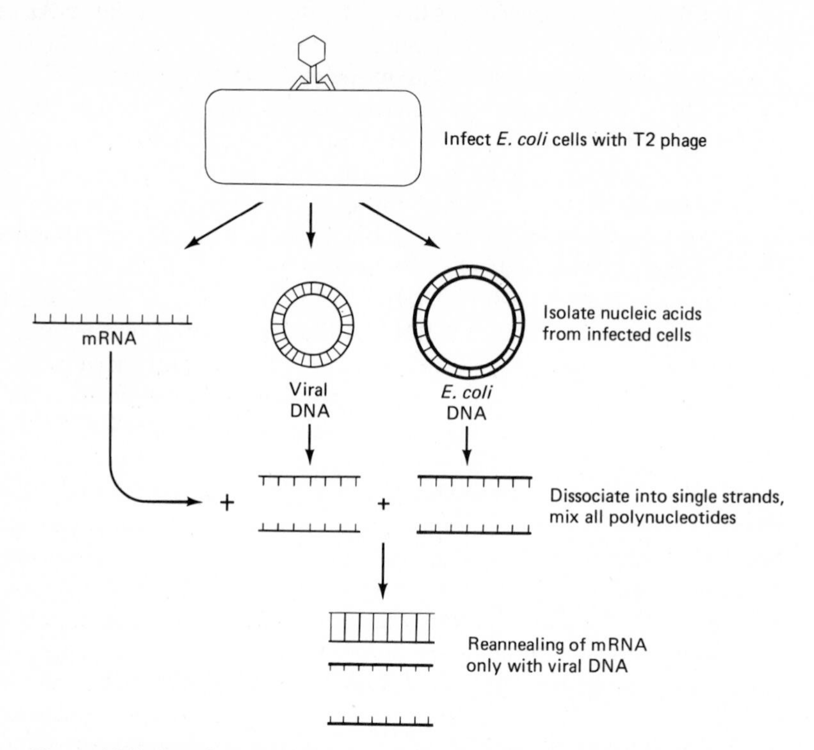

Figure 10-5. Experiment showing reannealing only between viral DNA and RNA found in bacterium following infection. Absence of reannealing between RNA and bacterial DNA indicates that viral DNA was the template for RNA synthesis, since base sequences are complementary.

reported findings that only one of the two strands of DNA serves as template for messenger RNA. They studied the bacteriophage SP8, which infects *B. subtilis* cells. Because the bouyant densities of the two strands of SP8 DNA are distinguishable, they can be separated after dissociation. When mRNA from infected cells was mixed with the heavier viral DNA strand and with the lighter strand, reannealing was achieved only between the mRNA and the heavier strand of virus DNA (Fig. 10-6).[2]

mRNA structure. As expected, mRNA nucleotide sequence is generally as varied as the proteins found in cells. However, recent analysis has revealed some

[2]Although this characteristic of transcription from only one of the two strands of a DNA double helix holds true generally, more recent findings indicate that in some cases such as viral DNA. both strands of DNA are transcribed (p. 314).

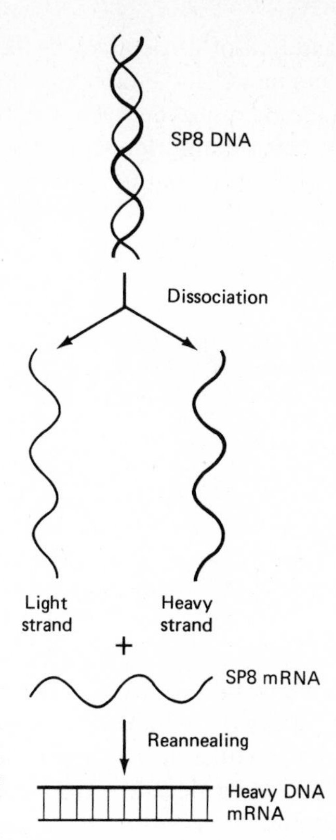

Figure 10-6. Schematic illustration of experiments showing complementarity of nucleotide sequence of viral mRNA to only one strand of DNA from which it is transcribed.

interesting characteristics common to all mRNA molecules found in eukaryotes and some viral mRNA. Methyl groups have been found at the 5′ end which are referred to as *caps*. Cap formation appears to occur shortly after initiation of transcription. In eukaryotes, this occurs before the mRNA leaves the nucleus. In vitro studies of caps in viral mRNA have indicated that they are important for binding mRNA to ribosomes, a process that we shall discuss shortly (Furuichi *et al.* 1977).

Another interesting feature found in most mRNA molecules is the presence of a polyadenylated region at the 3′ end. As many as 20–200 adenines are present in these regions. These also are added to mRNA molecules shortly after transcription from DNA. How this is done, and what its function may be, are not clear at the present time. Short repetitive sequences of bases near both the 5′ and 3′ ends of mRNA from eukaryotic cells have also been reported (Darnell 1976). These sequences do not appear to be part of the message for a protein, and their origin and role in protein synthesis is still under investigation. The general structure of mRNA molecules is illustrated in Fig. 10-7.

RNA polymerase. The synthesis of RNA molecules using DNA as template is mediated by the enzyme *RNA polymerase.* As in DNA synthesis, mRNA is formed

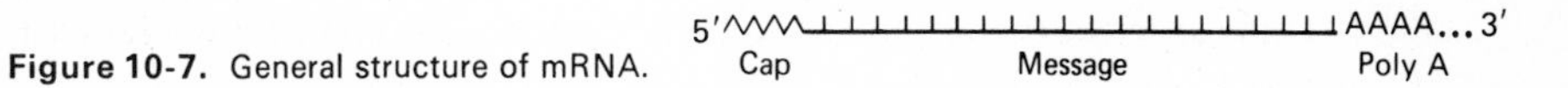

Figure 10-7. General structure of mRNA.

by the addition of nucleotides at the 3′ end of the chain. In other words, transcription occurs in a 5′ ⟶ 3′ direction (Fig. 10-8).

Like many enzymes, RNA polymerase is a complex protein composed of several subunits. One subunit, known as the *sigma factor* (or σ), appears to be involved in binding with and initiating transcription at the proper point in the DNA molecule.

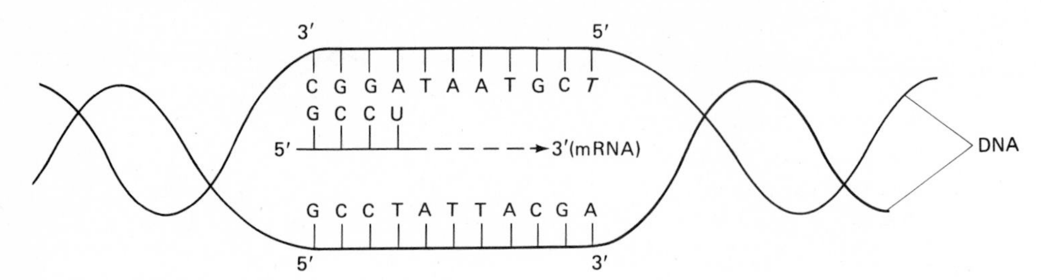

Figure 10-8. Illustration of direction of transcription. The mRNA elongates in a 5′ ⟶ 3′ direction. Note that its sequence is complementary to its template, and the same as the other strand of the DNA (with the exception of uracil).

Since we know that DNA is a continuous polymer, this is obviously an important function in assuring transcription of the complete message of a gene, rather than just a portion. In fact, it is now believed that more than the message is transcribed in eukaryotes and that the excess RNA is somehow removed before the messenger RNA is translated (Darnell 1976, Dunn and Studier 1973).

The other subunits of RNA polymerase are collectively referred to as *core proteins*; their specific function has not yet been determined. Unfortunately, RNA polymerase dissociates easily during isolation procedures, and thus it is difficult to study the enzyme in its native form. Information available so far indicates that RNA polymerase is approximately 450,000 daltons in molecular weight, and is composed of five polypeptides. In eukaryotic cells, there is evidence that at least three DNA-dependent RNA polymerases exist, which transcribe for mRNA, for rRNA, and for tRNA (to be discussed shortly), respectively.

In summary, the functions of RNA polymerase may be said to include (a) binding to the DNA template, (b) initiation of chain formation, (c) elongation of the RNA chain, and (d) termination of the RNA chain. The mechanism for termination is not yet understood, but there is some evidence that cellular proteins may be involved.

The Genetic Code

It was a logical assumption that in the sequence of bases in certain portions of the DNA must lie a kind of code which determines the sequence of specific amino acids. Much work was done in the late 1950s and 1960s which eventually led to the deciphering of the genetic code.

One aspect of the genetic code was indicated by the well-known fact that there are basically 20 amino acids which join together in different combinations to form the thousands of proteins found in living cells. If DNA and RNA contain only four bases, the code has to be longer than a doublet code, since only 16 doublets can be formed from the four bases (for example, A-A, A-T, etc.). A triplet code would provide more than enough combinations of the four bases (AAA, AAT, AAC, AAG, ATA, ATT, etc.), 64 to be exact, to code for the 20 amino acids. (This is an application of the multiplication rule discussed on page 46, that is, $4 \times 4 \times 4 = 64$.) Experiments were then designed to explore the triplet nature of the genetic code.

Frameshift mutations and a non-overlapping code. In 1961, Francis Crick and a number of associates published the results of some studies on mutations induced in T4 bacteriophage that lent support to the triplet code theory. To induce the mutations, acridine dyes were used, which cause mutations by deleting nucleotides or inserting additional nucleotides into the DNA. The group of mutations which they studied were called *rII.* These rII mutants of T4 cause rapid lysis in strain *B E. coli*, and are unable to lyse strain *K E. coli.*

When some rII mutants were exposed to acridine dyes a second time, reversions to wild-type (referred to as *revertants*) were occasionally observed, as evidenced by the ability of the revertants to once again lyse strain K bacteria. In some cases, however, the revertants were not as efficient as wild-type, although some capacity to lyse strain K was restored; these revertants were called "leaky." Analysis of the revertants indicated that the original mutation was still present, and that the change in phenotype was due to the presence of a second mutation induced by exposure to the dyes. Because of the manner in which acridine dyes can induce mutation, the conclusion was that the reversion can be accounted for if it is assumed that one of the mutations was an addition, and the second a deletion, of a base in the DNA.

If we make an analogy with a word of three letters, *eat*, and string these words together in the form *eateateat*, an insertion of an extra letter, for example a *t* at the left end, would result in *teateateat*. If we assume a triplet code, the reading of this message would become *tea tea tea t*. A second mutation in the form of a deletion of a letter, for example the second *t* from the left, would restore the sense in most of the message: *tea eat eat*. The small abnormal portion of the message which remained in the gene accounted for the leaky revertants.

To extend this line of thought, one would not expect two deletions or two additions to result in reversion to wild-type if the code is triplet. On the other hand, three additions or deletions could result in wild-type (Fig. 10-9). When Crick and his coworkers studied various combinations of addition and deletion mutations,

eateateat	Original sequence
eaeatat	Two deletions (nonsense)
eeaateateat	Two additions (nonsense)
eateat	Three deletions (sense)
eeeeateateat	Three additions (sense)

Figure 10-9. Schematic illustration of expected effects of two and three additions or deletions on the reading of a sequence if the code is triplet.

which they designated by + or − respectively, they found that the combinations behaved as expected as shown in Table 10-1. The different mutations tested are designated by FC (after a certain geneticist's initials!) and a number.

Table 10-1 Phenotypes Produced by Various Combinations of rII Mutations in T4 Phage Induced by Acridine Dyes

Addition mutations	*Deletion mutations*
FC0, FC40, FC58	FC1, FC21, FC23
Combinations of mutations	
FC0, FC1, (+, −)	FC0, FC40 (+, +)
FC0, FC21, (+, −)	FC0, FC58 (+, +)
FC40, FC1 (+, −)	FC1, FC21 (−, −)
FC58, FC1 (+, −)	FC1, FC23 (−, −)
FC0, FC40, FC58 (+, +, +)	
FC1, FC21, FC23 (−, −, −)	

From F. H. C. Crick *et al.* 1961. *Nature* 192: 1227–1232.

These mutations have been called *frameshift mutations* because each mutation shifts the reading of the message encoded in the DNA. IN ADDITION TO SUPPORTING THE CONCEPT OF A TRIPLET CODE, THESE STUDIES ALSO INDICATED THE GENETIC CODE TO BE CONTINUOUS WITHIN THE LIMITS OF A GENE. THE CODE IS THUS REFERRED TO AS COMMALESS (WITHOUT INTERRUPTIONS WITHIN THE GENE). Each sequence of three bases specifying an amino acid is called a *codon*.*

Crick's analysis of frameshift mutations also implied that a sequence *eateateat*, is read in three letter groups without any overlapping (*eat eat eat*, specifying 3 copies of the same amino acid), and not in three letter sequences with two letters previously read being read again (*eat ate tea*, which could be 3 different codes for different amino acids). THE CODE IS, IN OTHER WORDS, NONOVERLAPPING. And indeed, this nonoverlapping nature of the genetic code has been found to be the usual case. Recently, however, scientists in the laboratory of F. Sanger, who won the Nobel Prize for determining the amino acid sequence of insulin, made an astonishing discovery that in the virus ϕX174, although the sequence of bases is read in a nonoverlapping manner (Barrell *et al.* 1976), the same portion of DNA can be read in different frameshifts. For example, the sequence eateateat may be read *eat eat eat* or *e ate ate at*, with meaning to the virus.

The discovery came about when, having sequenced the nucleotides of this small virus, the scientists realized that there were not enough genes to account for the ten different proteins known to be determined by the viral DNA. When they sequenced the individual genes determining two proteins, they discovered that one gene, *E*, had the same sequence as part of another gene, *D*. Analysis showed that indeed gene *E* is located within the *D* sequence, but that it is read in a different frame-

*See page 319 for recent discoveries that modify this concept.

shift sequence. The first letter of a *D* gene codon is the last letter of an *E* gene codon (Fig. 10-10). The advantage for the virus is that optimal use is made of the small amount of DNA it contains; however, the control over which of the sequences is to be read at any one time is unclear.

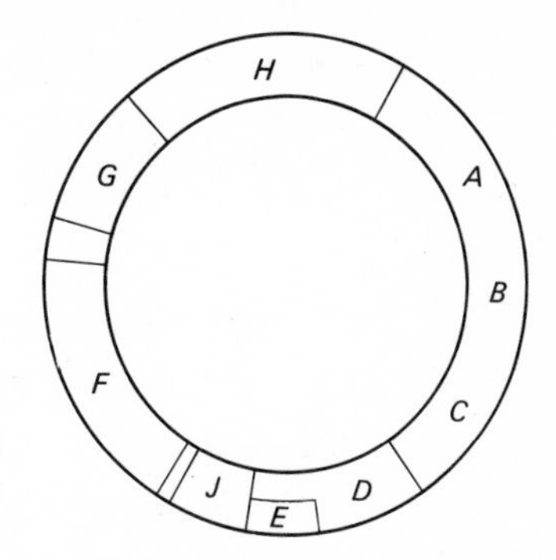

Figure 10-10. The genetic map of ϕX174 viruses. The exact boundaries of genes *A, B*, and *C* have not been determined. Gene *E* has been mapped to be within gene *D*.

Although this discovery is certainly highly significant in showing that the same sequence of bases can carry different meaning in at least one virus, it is still believed that sequences of DNA are read in a nonoverlapping manner for any one protein.

The code deciphered. As we stated above, the frameshift mutations studied by Crick and his associates established the genetic code to be triplet, commaless, and nonoverlapping. In other words, groups of nucleotides in threes specify particular amino acids to the cell. But which triplet specifies which amino acid?

In 1955, Gruneberg-Manago and S. Ochoa reported the isolation of an enzyme which catalyzes the in vitro synthesis of RNA, a key development to the deciphering of the code. The enzyme is *polynucleotide phosphorylase*. When included in a preparation containing ribonucleotide diphosphate molecules, this enzyme causes polymerization of ribonucleotides into a chain of RNA. THE SIGNIFICANCE OF THIS PROCESS IS THAT CHAIN FORMATION CAN OCCUR IN THE ABSENCE OF A TEMPLATE OR PRIMER MOLECULE. SUCH ARTIFICIAL RNA MOLECULES CAN BE USED IN COMBINATION WITH A NUMBER OF FACTORS TO DIRECT THE IN VITRO SYNTHESIS OF PROTEINS.

The other factors include ribosomes, ATP, amino acids, and two important classes of molecules for protein synthesis, tRNA and amino acyl-tRNA synthetase, which we shall discuss shortly in more detail. Because most of the experiments which led to the deciphering of the genetic code were carried out on RNA, the codons include U instead of T which is found in DNA, and are actually complimentary to the DNA sequence.

M. W. Nirenberg and J. H. Matthaei in 1961 used these systems for the first reported deciphering of a particular triplet specifying one amino acid. First they synthesized an RNA chain containing only uracil. Such a molecule, which is U–U–U–U . . . , is called *poly U*. They then used this poly U molecule for protein synthesis. The result was synthesis of a polypeptide containing only one amino acid, phenylalanine; thus, the codon for this amino acid is UUU. (The sequence of DNA determining the codon for phenylalanine in RNA would of course be AAA.) Similarly, poly C was found to code for proline, and poly A for lysine. For chemical reasons that we need not go into, use of poly G did not result in polypeptide formation.

What about the majority of codons, which are composed of combinations of different bases? Nirenberg among others (1963) designed experiments which allowed him to analyze and decode triplets containing two different bases. To synthesize mRNA molecules, two bases were added in different amounts. For example, if U and A are the two bases used, they may be added in a ratio of 5U : 1A, and so the proportions are therefore $\frac{5}{6}$ U and $\frac{1}{6}$ A. We can then calculate the expected proportions (probabilities) of the possible triplets. On pages 66–73 we presented the probabilities for a binomial experiment. Using these probabilities we get

$$P[0A, 3U] = P[UUU] = (\tfrac{5}{6})(\tfrac{5}{6})(\tfrac{5}{6}) = (\tfrac{1}{6})^0(\tfrac{5}{6})^3 = \binom{3}{0}(\tfrac{1}{6})^0(\tfrac{5}{6})^3 = 0.579$$

$$P[1A, 2U] = \left\{\begin{matrix} P[UUA] = (\tfrac{5}{6})(\tfrac{5}{6})(\tfrac{1}{6}) = (\tfrac{1}{6})^1(\tfrac{5}{6})^2 \\ P[UAU] = (\tfrac{5}{6})(\tfrac{1}{6})(\tfrac{5}{6}) = (\tfrac{1}{6})^1(\tfrac{5}{6})^2 \\ P[AUU] = (\tfrac{1}{6})(\tfrac{5}{6})(\tfrac{5}{6}) = (\tfrac{1}{6})^1(\tfrac{5}{6})^2 \end{matrix}\right\} \binom{3}{1}(\tfrac{1}{6})^1(\tfrac{5}{6})^2 = 0.347$$

$$P[2A, 1U] = \left\{\begin{matrix} P[AAU] = (\tfrac{1}{6})(\tfrac{1}{6})(\tfrac{5}{6}) = (\tfrac{1}{6})^2(\tfrac{5}{6})^1 \\ P[AUA] = (\tfrac{1}{6})(\tfrac{5}{6})(\tfrac{1}{6}) = (\tfrac{1}{6})^2(\tfrac{5}{6})^1 \\ P[UAA] = (\tfrac{5}{6})(\tfrac{1}{6})(\tfrac{1}{6}) = (\tfrac{1}{6})^2(\tfrac{5}{6})^1 \end{matrix}\right\} \binom{3}{2}(\tfrac{1}{6})^2(\tfrac{5}{6})^1 = 0.069$$

$$P[3A, 0U] = P[AAA] = (\tfrac{1}{6})(\tfrac{1}{6})(\tfrac{1}{6}) = (\tfrac{1}{6})^3(\tfrac{5}{6})^0 = \binom{3}{3}(\tfrac{1}{6})^3(\tfrac{5}{6})^3 = 0.005$$

(P = probability)

When the mRNA molecules of this experiment were used in protein synthesis, the most frequently incorporated amino acid was found to be phenylalanine as expected, and the least frequent was lysine. Other amino acids incorporated into polypeptide chains were leucine, isoleucine, asparagine, and tyrosine. It was not possible from these experiments, however, to determine exactly which codon specified each of these amino acids.

Yet another approach was developed by H. G. Khorana (1966–67), who was able to synthesize RNA molecules containing alternating bases, such as UAUAUA and ACACAC. Thus these RNA molecules can be expected to specify two amino acids, for example, UAU or AUA from the first sequence mentioned. In each case, two amino acids were found incorporated into polypeptides, but again, which codon specified which amino acid could not be determined from this experiment alone. Data from different studies, however, allowed geneticists to discern the meaning of specific codons.

A method to determine the exact meaning of codons involved the synthesis of only a single triribonucleotide—a sequence of only three bases in an RNA molecule. When such trinucleotides were used with the factors for in vitro protein synthesis, particular trinucleotides were always found to interact with tRNA molecules bound to a particular amino acid (Nirenberg and Leder 1964). For example, UUU RNA was always complexed with tRNA-phenylalanine.

All these elegant experiments led shortly to the complete deciphering of the genetic code, which is given in Fig. 10-11. For their efforts, Nirenberg, Matthaei, and Khorana joined the long list of geneticists who have been awarded the Nobel Prize.

AA- AAU, AAC: Asparagine AAA, AAG: Lysine	CA- CAU, CAC: Histidine CAA, CAG: Glutamine	GA- GAU, GAG: Aspartic acid GAG, GAG: Glutamine acid	UA- UAU, UAC: Tyrosine UAA, UAG: Nonsense
AC- ACU, ACC, ACA, ACG: Threonine	CC- CCU, CCC, CCA, CCG: Proline	GG GCU, GCC, GCA, GCG: Alanine	UC- UCU, UCC, UCA, UCG: Serine
AG- AGU, AGC: (Cysteine?) Serine AGA, AGG: Arginine	CG- CGU, CGC, CGA, CGG: Arginine	GG- GGU, GGC, GGA, GGG: Glycine	UG- UGU, UGC: Cysteine UGA: Nonsense UGG: Tryptophan
AU- AUU, AUC: Isoleucine AUA: Isoleucine? Methionine AUG: Methionine	CU- CUU, CUC, CUA, CUG: Leucine	GU- GUU, GUC, GUA, GUG: Valine	UU- UUU, UUC: Phenylalanine UUA, UUG: Leucine

Figure 10-11. The complete genetic code. Amino acids coded for are written next to their respective codons. Nonsense codons do not code for any amino acid, and terminate translation. U represents uracil, as studies leading to the deciphering of the code were carried out on RNA.

Degeneracy and nonambiguity. Note that some amino acids, such as serine, are coded for by more than one triplet, a situation referred to as degeneracy of the code. However, any triplet which codes for an amino acid codes only for that particular amino acid. For this reason, the genetic code is referred to as nonambiguous.

Nonsense codons. Note that some codons (for example, UAG and UGA) do not code for any amino acids. These are referred to as stop, or nonsense, codons because they seem to function to end the process of protein synthesis. Evidence for this was presented by Sidney Brenner and others (Brenner *et al.* 1965; Sarabhai *et al.* 1964) in a study of a number of mutants of the T4 bacteriophage which were all incapable of synthesizing the protein coat surrounding the "head" of the viruses. They were, instead, capable of producing only parts of the head protein.

Brenner and his associates analyzed the components of the mutant protein and found its amino acid sequence to be the same as that in the comparable region of the wild-type protein. The amino acids that were found in the wild-type protein immediately next to the terminal amino acid of several mutant proteins were glutamine, lysine, glutamic acid, tyrosine, tryptophan, and serine. Some of the codons coding for these amino acids are CAG, AAG, GAG, UAC, UGG, UCG. Brenner proposed that in each case, the mutation or change of only one base to a different base would result in the codon UAG. In this manner it was deduced that wherever such a codon exists in the genetic message, protein synthesis will be terminated at that point. He called this first-discovered stop codon, *amber*. Two other stop codons were subsequently identified and named: UGA (*opal*) and UAA (*ochre*). (The names have no scientific significance. For example, amber was the English translation of the German name of the mother of one of the scientists working on the mutation!)

Colinearity. Another aspect of the genetic code that was brought out by Brenner's elegant studies on the head protein mutants was that there is a direct correlation between the site of the mutation on the DNA molecule and the site of an abnormal amino acid in the protein which the DNA determines (Fig. 10-12a). THIS POINTS OUT ANOTHER CHARACTERISTIC OF THE GENETIC CODE, COLINEARITY, A DIRECT CORRELATION BETWEEN THE POSITIONS OF AMINO ACIDS IN A POLYPEPTIDE, AND THE POSITIONS OF THEIR CORRESPONDING CODON IN THE DNA.

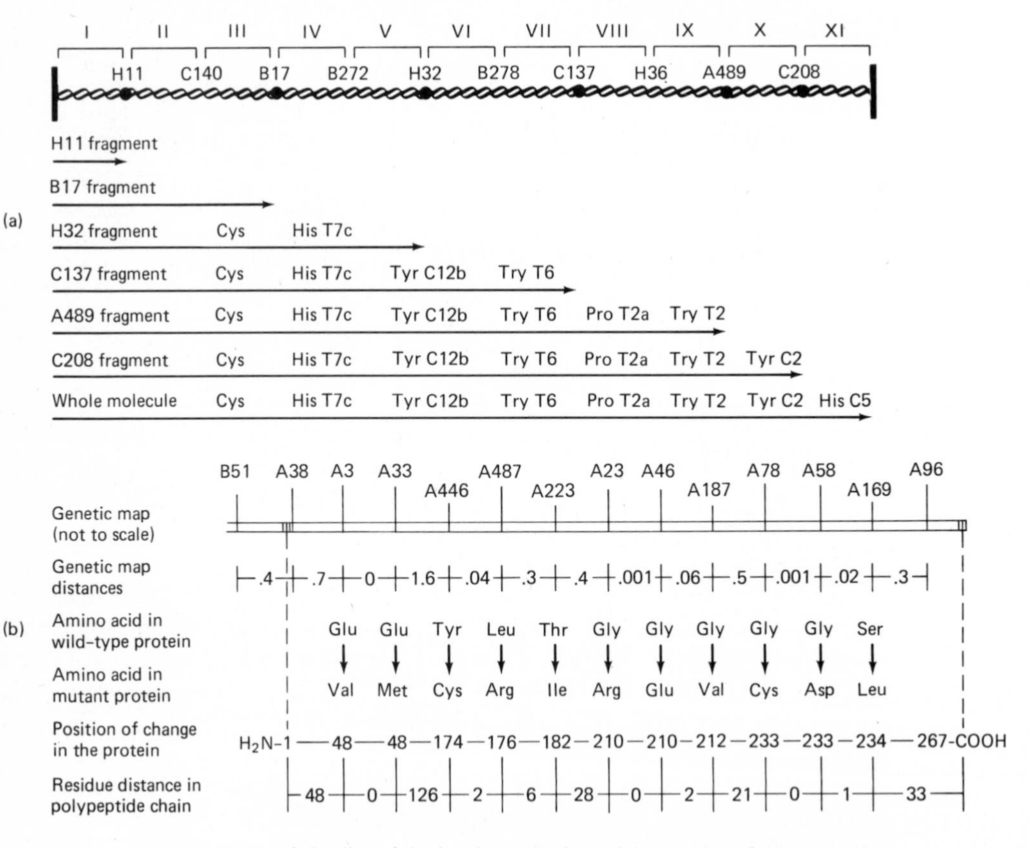

Figure 10-12. (a) Studies of the head protein showed the position of the nonzense mutations in T4 phage DNA determined a corresponding length of head protein produced. These two studies led to the concept of colinearity. (From A. S. Sarabhai *et al.*, 1964. *Nature* 201 : 13–17, Fig. 2.) (b) Studies of mutations of the tryptophan synthetase A protein. Mutation of specific codons in DNA led to mutant amino acids in corresponding part of the protein. (From C. Yanofsky *et al.*, 1967. *Proc. Nat. Acad. Sci. U.S.* 57: 296–8, Fig. 3.)

Colinearity of the genetic code with the sequence of amino acids in proteins was also reported by C. Yanofsky and his associates (1967). Yanofsky was investigating the genetic determination of the enzyme tryptophan synthetase in *E. coli.* The wild-type enzyme is composed of four polypeptides, two A chains and two B chains, determined by the *trp A* and *trp B* genes respectively. A large number of *trp A* mutations were isolated, and the position of the mutations mapped by recombinational analysis.

In each case, the A polypeptide synthesized by the mutants was studied, and again there was consistent correlation between the position of the mutation and the abnormal amino acid position on the polypeptide chain. For example, if a cell carried two mutations in the *trpA* locus, the farther apart they mapped, the more amino acids were found between the two abnormal amino acids in the protein (Fig. 10-12b).

Spacers in eukaryotic systems. In eukaryotes, the concept of colinearity of DNA structure to polypeptides has undergone some modification recently. Essentially, regions of DNA have been discovered which are not transcribed into polypeptides; these regions are referred to as *spacers.* Spacers have been found *between* genes (as in the case of ribosomal genes, which we shall discuss in more detail later in this chapter), and also *within* genes. Thus the total sequence of nucleotides within a gene is not necessarily found in the amino acid sequence of its protein. These spacers within genes are referred to as *intervening sequences* or *introns*, and apparently are removed from precursor mRNA before protein synthesis.

The gene determining β chains of hemoglobin in mice is a good illustration of such a system. A series of experiments utilizing recombinant DNA technology was reported to have resulted in the isolation of the β-globin gene (also known as $\beta G2$) from the mouse genome (Tilghman *et al.* 1977). Experiments on the isolated DNA subsequently indicated the presence of sequences not found in βG2 mRNA (Tilghman *et al.* 1978). For example, two sites of cleavage by restriction endonucleases were found to be separated by almost twice the number of bases in DNA as in mRNA molecules. (*Restriction endonucleases* are enzymes capable of recognizing and cleaving specific sequences of DNA.) The conclusion was that the extra sequences of bases in the DNA were somehow removed from the mRNA transcripts. The sequences that remain are called *exons.*

Another approach, known as *R-loop technique*, allowed scientists to visualize the excess portion of the βG2 gene. The technique calls for partially denaturing double-stranded DNA, then hybridizing a portion of the single-stranded DNA with complementary RNA. The DNA–RNA binding is more stable than DNA–DNA binding; consequently the displaced single strand opposite the DNA strand which has annealed to RNA forms an "R loop" in electron micrographs (Fig. 10-13).

When β-globin mRNA was hybridized to the denatured DNA, *two* R loops were seen instead of the expected one; furthermore, the two loops were separated by a double-stranded loop of DNA. The interpretation of this phenomenon was that βG2 mRNA hybridized to two discontinuous segments of the βG2 gene. The looped-out double-stranded region between the two segments does not complement any portion of the mRNA molecules, and thus has been excised from them. That it appears to be an excision of the excess sequence from the mRNA (called mRNA processing)

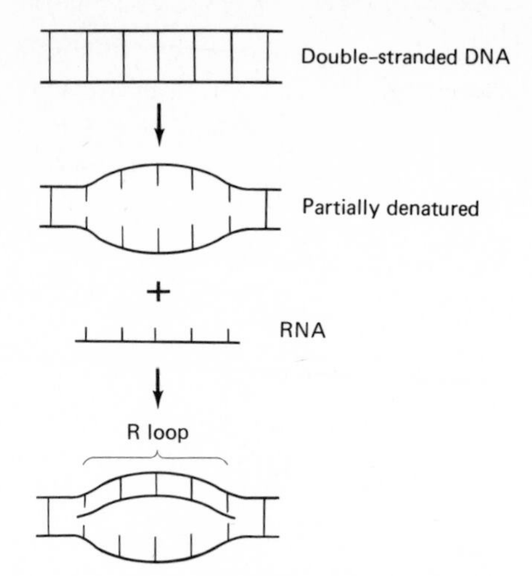

Figure 10-13. Schematic illustration of the formation of R loops.

rather than an absence of transcription of the excess sequence has been determined by the isolation of a 15S precursor mRNA for β-globin, which is then cleaved to 9S.

Nucleotide sequencing studies have also supported the idea that a sizable portion of the gene is not reflected in a comparable sequence of amino acids in the globin protein. All these studies are summarized in Fig. 10-14, which shows the map of the βG2 gene with its intervening sequences. Note that there is now believed to be a second small intervening sequence, signifying that the β-globin gene in mice is translated as three discontinuous segments.

Reports of intervening sequences in the β-globin genes in rabbits and humans have recently been published (Mears *et al.* 1978). Existence of such sequences, however, is not restricted to globin genes. There is evidence that yeast tRNA genes are longer by 14 base pairs than mature yeast tRNAs. Adenovirus and SV40 sequence analyses have also shown intervening sequences in their genomes that are not found in their viral transcripts. A large pro-insulin mRNA has been found which is known to be cleaved into smaller molecules. Chick ovalbumin genes have been found to contain at least *seven* intervening sequences which are not present in the mature mRNA (Fig. 10-15). Because of all these sequences, an astonishing 7000+ base pairs in the ovalbumin gene actually code for a mature mRNA that contains only 1859 base pairs (Dugaiczyk *et al.* 1978).

The significance of these intervening sequences is not known. Theories have been postulated that they may serve as regulatory sequences in some way. Certainly the presence of intervening sequences points to an aspect of gene function that reflects a dimension of complexity not previously anticipated. We must now assume some system of distinguishing the sites joining areas of mRNA to be translated from areas of mRNA that are not to be translated. There must then follow precise excision of the intervening sequences with equally precise ligation of the remaining pieces. In the case of the ovalbumin gene and others, there must also be ligation of a number of pieces of mRNA in the correct order. A possible role for intervening sequences in evolution will be discussed in Chapter 17.

Universality of the genetic code. All the work on the genetic code that we have cited in previous sections was accomplished using subsequent studies showed that

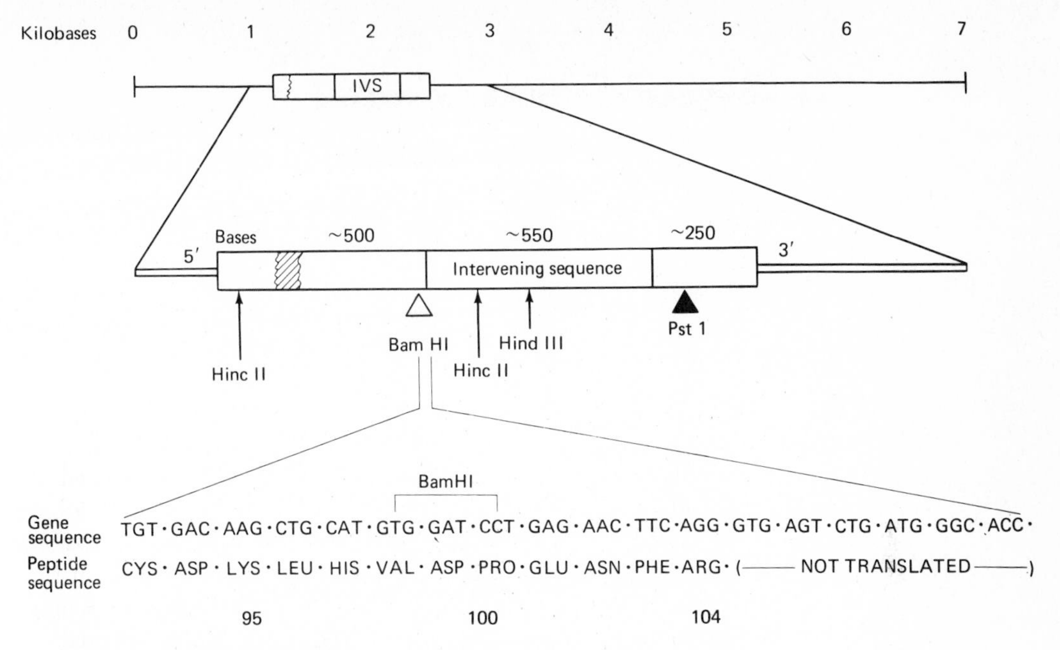

Figure 10-14. Restriction endonuclease map and nucleotide sequence of βG2 DNA. The top line represents the position of the β-globin gene within the cloned 7.0-kb mouse DNA fragment, drawn with the sequence corresponding to the 5′ end of globin mRNA at the left. The entire gene is indicated by the boxes, with the 550-base-pair intervening sequence labeled "IVS." The approximate location of the smaller 5′ interruption is drawn as a hatched box. The middle line represents an enlargement of the hybridizing sequences in βG2. The restriction sites are as indicated. The third line is the nucleotide sequence of the anti-coding strand in the region immediately adjacent to the BamHI site of βG2. The amino acid assignments for this region are drawn below. Maps were constructed from gel analyses followed by in situ hybridization to globin [^{32}P]-cDNA. (From Tilghman *et al.*, 1978. *Proc. Nat. Acad. Sci.* 75 (2): 726.)

the code is the same in eukaryotic cells. One of the experiments which gave strong evidence that the code is indeed universal was performed by H. Von Ehrenstein and G. Lipmann (1961). They isolated mRNA of rabbit red blood cells which carry the message for hemoglobin synthesis, and placed this in a preparation with tRNA and ribosomes of bacteria, and the enzymes and amino acids needed for in vitro protein synthesis. Hemoglobin protein was found following incubation of the preparation, showing that the genetic message of one species can be recognized by the protein synthesis components of an entirely different species (Fig. 10-16). NUMEROUS OTHER EXPERIMENTS HAVE ALSO CONFIRMED THAT THE GENETIC CODE IS UNIVERSAL. IN OTHER WORDS, A GIVEN CODON SPECIFIES THE SAME AMINO ACID IN CELLS OF ALL LIVING ORGANISMS. We humans are no exception, although we have had a tendency to consider our own species as special and apart from other living organisms. Molecular genetics has firmly established that the messages which direct protein synthesis in microbes are based on the same code found in our cells. We are simply more complex.

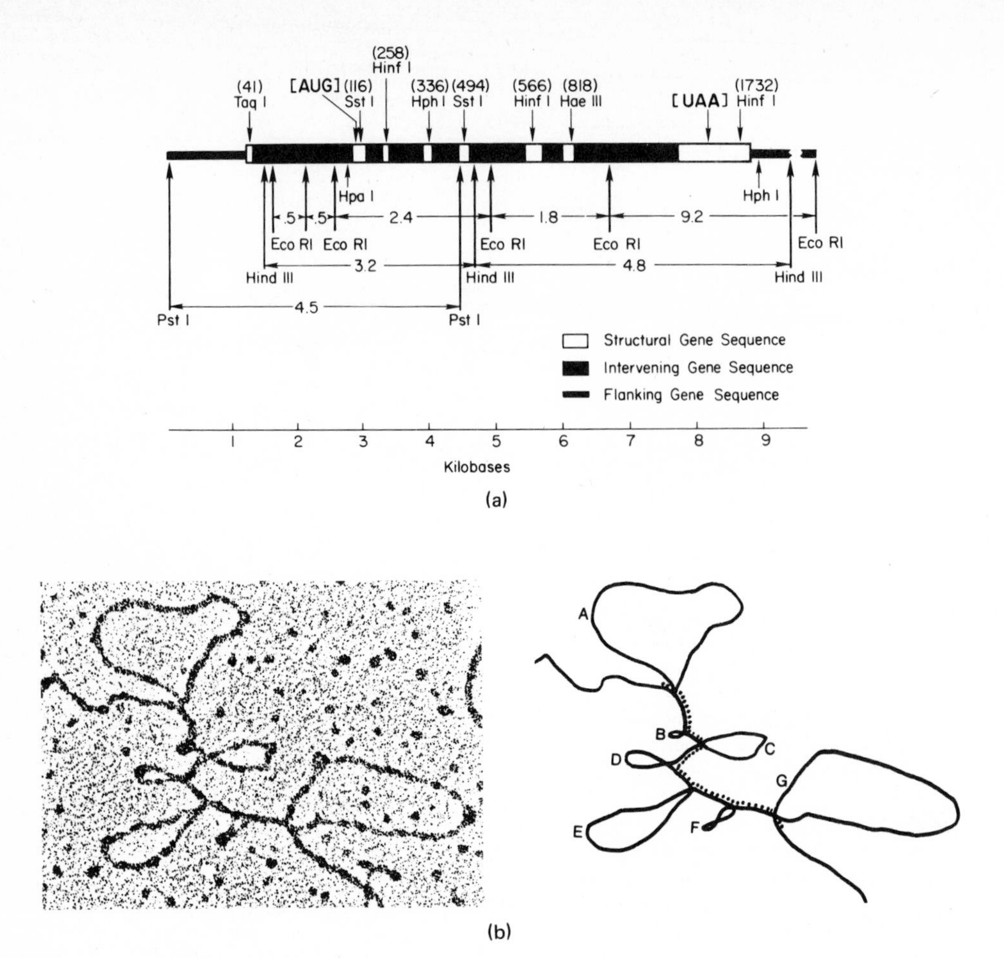

Figure 10-15. (a) Physical map of the entire natural ovalbumin gene displaying some of the key restriction cleavage sites (see Ch. 14), the locations of initiation and termination codons, and regions of interspersed structural and intervening sequences. Thin solid line, flanking sequence; thick solid line, intervening gene sequence; thick open line, structural sequence for ovalbumin. (b) Electron micrograph and drawing of the micrograph showing the loops (A to G) of DNA that have no counterpart on the ovalbumin mRNA when the two are hybridized. (Courtesy Achilles Dugaiczyk, Baylor College of Medicine.)

Summary of characteristics of the genetic code. We can now summarize the basic characteristics of the genetic code as follows:

It is triplet; each codon that specifies an amino acid is composed of a sequence of three nucleotides.

It is degenerate; more than one codon can specify the same amino acid.

It is nonambiguous; each codon specifies only one amino acid.

It is commaless; the message is a continuous one, at least in the mature mRNA.

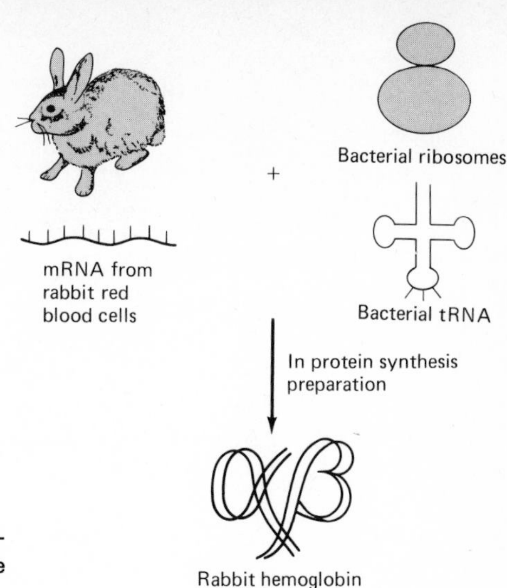

Figure 10-16. Schematic illustration of an experiment which showed the universality of the genetic code.

It is nonoverlapping for any one protein.

It is for the most part, colinear; there is a direct correlation in the position of codons on the exon DNA and the position of amino acids on the protein.

It is universal; the same codon specifies the same amino acid in cells of all living organisms.

TRANSLATION

Let us now turn our attention to the processes which convert the messages encoded in mRNA molecules into polypeptides. WE REFER TO THE COMPLEX BIOCHEMICAL EVENTS THAT LEAD TO POLYPEPTIDE FORMATION AS TRANSLATION. AS MENTIONED PREVIOUSLY, TRANSLATION OCCURS AT RIBOSOMES, WHICH ARE CELL ORGANELLES FOUND IN THE CYTOPLASM (Fig. 10-17). Much recent research has been directed toward improving our understanding of the structure of these important organelles, since such knowledge is necessary for comprehension of their function.

Ribosomes

We indicated earlier that ribosomes are quite uniform in size and structure. In all eukaryotic cells, for example, ribosomes that are not actively involved in protein synthesis dissociate into two subunits which are of different but constant size. When these particles undergo ultracentrifugation, the smaller particle sediments at 40S (S = Svedberg units, units of measurement of sedimentation rate), and the larger particle at 60S. A similar situation exists in bacteria, though the subunits are of slightly smaller sizes. *E. coli* ribosomes are composed of two particles of 50S and 30S. When engaged in protein synthesis, the subunits reassociate, and the entire ribosome in eukaryotes sediments at 80S, and in *E. coli* at 70S. (S units are not strictly additive, as these figures indicate.)

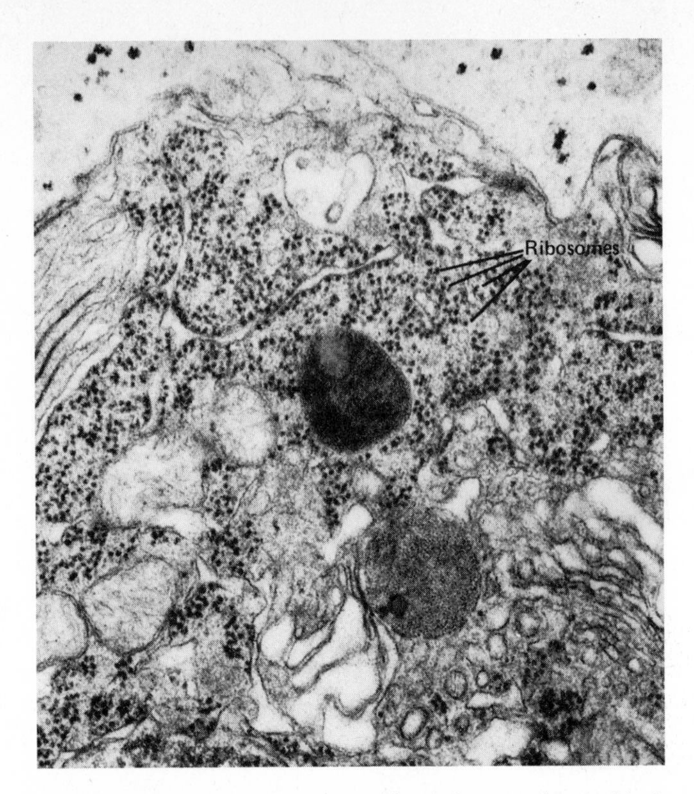

Figure 10-17. Leader points to ribosomes evident in large numbers in this electron micrograph of cytoplasm in a nerve cell. (From D. Pease, 1964. *Histological Techniques for Electron Microscopy*, p. 116. Academic Press, New York.)

Most of our present knowledge of the chemical constitution of ribosomes has come from studies on ribosomes from *E. coli*. All ribosomes are composed of approximately 40 percent RNA and 60 percent proteins. The 30S subunit of bacterial ribosomes contains one RNA molecule which has a sedimentation value of 16S and 21 different proteins. The 50S particle contains two different molecules of ribosomal RNA (which is designated *rRNA* for short), a 23S and 5S RNA, and 34 different proteins. The general structure of ribosomes is illustrated in Fig. 10-18.

M. Nomura and his associates (1973) have developed techniques to dissociate *E. coli* ribosomes into component molecules and then to reassociate them in various orders. It has been found that the various proteins and RNA molecules are incorporated into the ribosome in a special order; however, the significance of this phenomenon is not yet clear.

Generally, the organization of ribosomes in eukaryotic cells is similar to that in bacteria. RNAs in the two ribosome subunits are slightly larger than RNAs in *E. coli* ribosome subunits. The 60S subunit contains a 28S and 5S RNAs, and an unspecified number of proteins. The 40S particle contains an 18S RNA and many proteins (see again Fig. 10-18).

An interesting aspect of ribosomal protein structure has been reported follow-

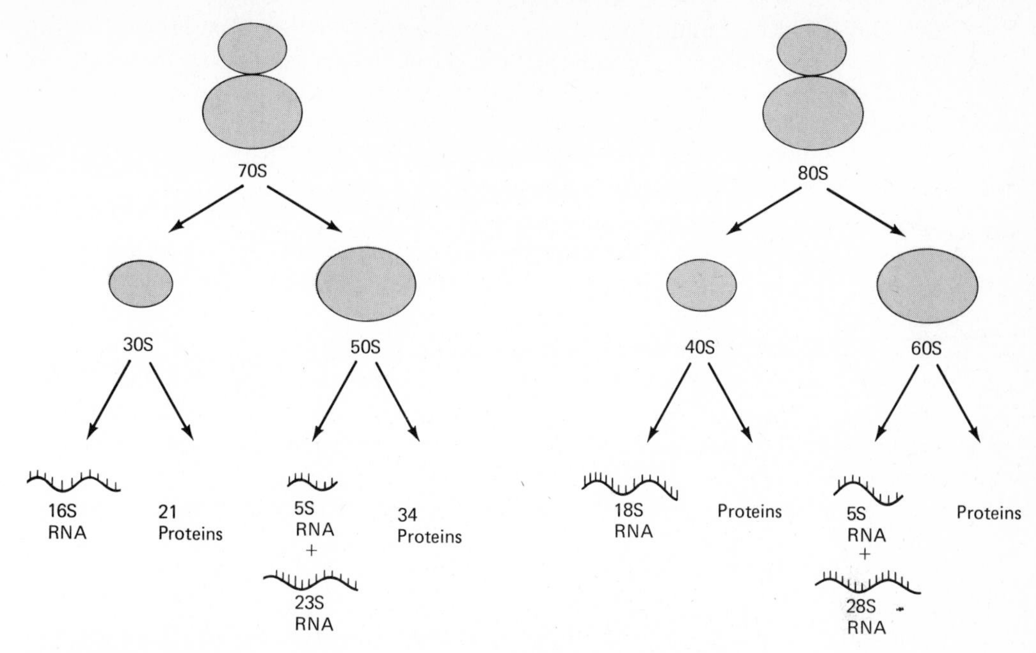

Figure 10-18. General structure and components of ribosomes from prokaryotes (a) and eukaryotes (b).

ing sequence analysis of amino acids of ribosomal proteins from different species of bacteria (Witmann 1977). It has been found that the more closely related two species are, the more homology (identical sequences) is found in their ribosomal proteins. There is no homology between amino acid sequences in proteins of bacterial and eukaryotic ribosomes. Yet ribosomes appear to have identical functions in all species.

Ribosomal RNA genes. Genetic analysis has indicated that in both prokaryotes and eukaryotes, there exist a large number of genes transcribing for rRNA. For example, the genes determining 16S and 23S RNA in bacteria are believed to be repeated as many as 40 times. Studies of cells from the amphibian *Xenopus* have led to estimates that there are about 450 genes for the 18S and 28S rRNA per set of chromosomes, and some 24,000 genes for 5S (Brown 1973). Such repeated copies of genes are believed to have arisen by a process called *duplication*, in which several transcripts of a region of DNA are incorporated into the genome; thus a sufficient supply of these essential molecules is ensured for the cell even if by chance a mutation should affect one or more of the genes.

The organization of ribosomal genes is very interesting. In bacteria, all the genes are apparently clustered in 5 to 10 regions of the chromosome. Each cluster contains genes for all three rRNAs, and is called a *transcription unit*. They are transcribed as a whole RNA molecule, and then cleaved into the appropriate sizes. As in the case of messenger RNA, many of the nucleotides of rRNA appear to have methyl groups added to them.

In *Xenopus* cells, genes determining the 28S and 18S rRNA are clustered, and genes for 5S are found on all chromosomes. Apparently, transcription of the

28S and 18S genes results in one large transcript that is then cleaved into the two separate RNA molecules. Again, methylation of the nucleotides occurs.

Furthermore, between the 28S and 18S genes, Brown and others have found regions of DNA which are not transcribed, the spacers we mentioned previously. When a comparison was made of the spacer regions of 50 different species of *Xenopus*, these spacers were found to be composed of sequences of nucleotides that are species-specific. This is in contrast to the nucleotide sequences of the rRNA genes themselves, which were similar in the different species. The function of the spacers has not yet been determined. Spacers do not appear to exist to any degree in bacterial chromosomes.

Ribosomal protein genes. Less is known about the genes determining ribosomal proteins. Studies of *E. coli* cells have indicated that many ribosomal protein genes are clustered at various regions of the bacterial chromosome. Whether these are transcribed as one large mRNA molecule or as several separate ones is not yet known.

Assembly of ribosomes in eukaryotes. Unique to eukaryotes is the manner in which the RNA and proteins of ribosomes are assembled. The actual site of assembly is in the nucleus in regions called *nucleolar organizing regions* (NOR). These actually are the sites of the 28S and 18S genes, and in some species there can be more than one such region, on different chromosomes. As transcripts for rRNA are synthesized, the protein molecules are incorporated into ribosomal particles. The cluster of ribosomal particles forms what is more familiarly known as a *nucleolus*.

Since ribosomal proteins must first be synthesized in the cytoplasm, it is believed that mRNA from genes determining ribosomal proteins is translated into polypeptides in the cytoplasm. The polypeptides migrate back into the nucleus where they become incorporated into the ribosomes in the nucleolus. The completed ribosomes are then released into the cytoplasm.

Function of ribosomes in protein synthesis. The specific function of ribosomes or their components in translation is only partially understood. We know that as the mRNA is attached to a 30S ribosome particle, a tRNA molecule carrying an amino acid is attracted to the mRNA, forming a combination known as an *initiation complex*. This complex is then joined by a 50S particle, and protein synthesis proceeds. Actually, a number of ribosomes will associate with one mRNA molecule, forming *polysomes*. Figure 10-19 shows an electron micrograph taken by O. Miller of a messenger RNA being transcribed in *E. coli* with ribosomes attached to the molecule as it is released from the DNA.

We are not yet certain what the ribosomes actually do in the process of translation. That they serve as more than just a scaffolding on which the mRNA can be supported is reflected by mutations of single proteins in ribosomes that apparently interfere with the normal reading of the message.

Nomura, for example, found a streptomycin-resistant mutation in *E. coli* which resulted in one abnormal ribosome protein, called P10. When wild-type P10 protein was incorporated into ribosomes made of all mutant RNA and other proteins, the reconstituted ribosomes caused the cells to be streptomycin sensitive, or wild-type. On the other hand, when mutant P10 protein was inserted into wild-type ribosomes from which the normal P10 protein had been removed, the cells became

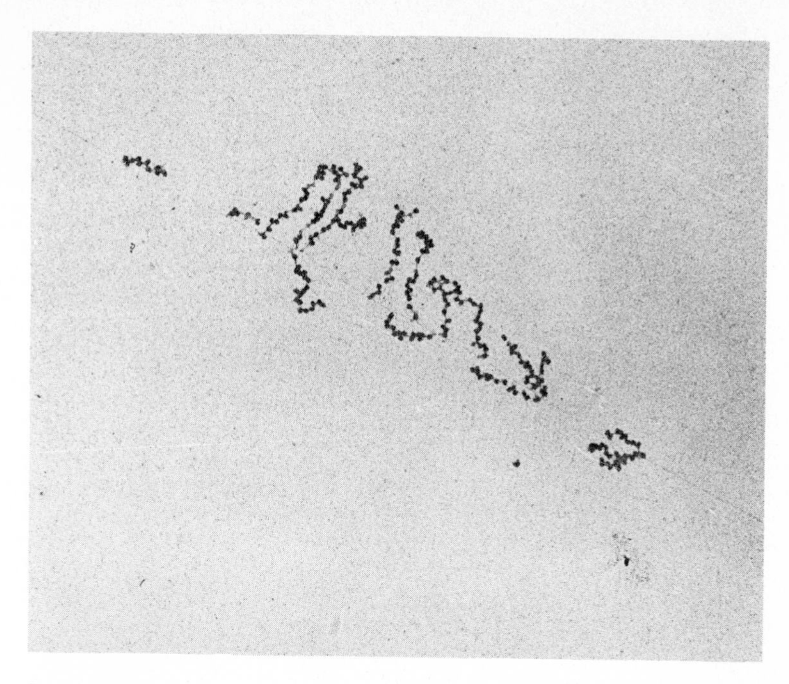

Figure 10-19. Electron micrograph of ribosomes interacting with mRNA being transcribed from one strand of DNA of a bacterium. (Photograph courtesy O. E. Miller, Jr.)

streptomycin resistant. Since the work of Davies, Gorini, and others (1964) had established that streptomycin kills bacteria by interfering with the normal reading of messenger RNA, this protein, and the ribosome itself, must be involved in the reading of messages.

Transfer RNA

There is one more class of participants in the process of translation and protein synthesis which we shall now discuss, namely, *transfer RNA*, or *tRNA*. THE MAJOR ROLE OF tRNA MOLECULES IN TRANSLATION IS TO TRANSPORT AMINO ACIDS TO MESSENGER RNA MOLECULES TO BE BONDED INTO POLYPEPTIDES. Again, to understand their function, we need to consider the structure of this class of molecules.

Structure of tRNA. Transfer RNA molecules are the smallest class of RNAs involved in the process of protein synthesis. Whereas mRNAs can contain many hundreds of nucleotides, and ribosomal RNAs consist of 120, 1500, and 3000 nucleotides for the 5S, 16S, and 23S RNAs, respectively, tRNAs are no more than 70 to 90 nucleotides long.

Multiple copies of tRNA genes have been found in both *E. coli* and eukaryotes. In *E. coli*, 60 tRNA genes have been found; in yeast, 320–400; in *Drosophila*, 750; in *Xenopus*, 8000 (Smith 1976). In *E. coli*, tRNA genes are found interspersed throughout the chromosomes, although there are clusters of 3 or 4 genes in one area. In *Xenopus*, most tRNA genes are clustered in one small part of the DNA, separated by some small sequences of DNA, while a few genes have been found scattered in other parts of the genome.

Following the transcription of tRNA, the nucleotide sequences apparently undergo considerable modification. First of all, there is evidence that tRNA transcripts are longer than the "mature" tRNA molecules, and that the transcripts are cleaved into smaller fragments, as in the case of ribosomal RNA.

Following transcription and cleavage, every tRNA molecule undergoes a number of changes. Each RNA has a sequence of three bases, 5′–CCA–3′ added at the 3′ ends of the molecules. Additional nucleotides are converted into bases not found in any other nucleic acids. For example, at specific positions some uracils are changed to dihydrouracil or pseudouracil; methyl groups are added to other bases; adenines are deaminated into inosines; the base ribothymidine is found on one of the loops. Figure 10-20 shows some of these unusual bases found in tRNA.

Dihydrouridine

Inosine

Pseudouridine

Ribothymidine

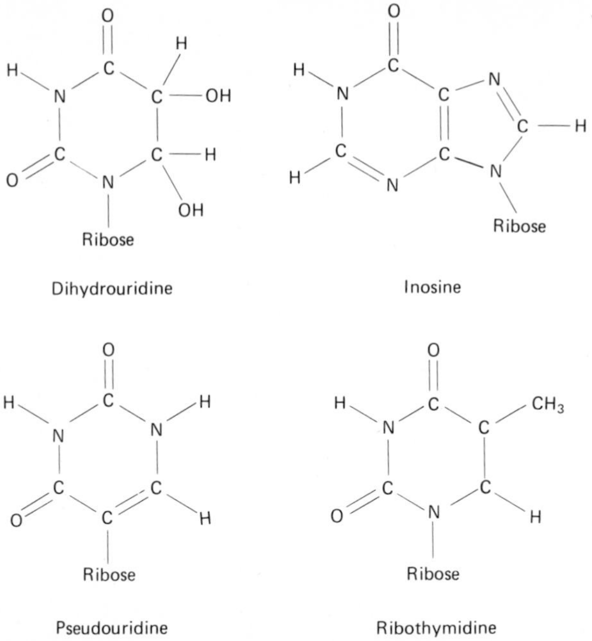

Figure 10-20. Some of the unusual bases found in tRNA.

All these changes have an effect on the final structure of tRNA. Sequences of bases in several regions of tRNA appear to be complementary to sequences in other regions, so that the single-stranded molecule will fold over and form double-stranded regions because of base-pairing. Unusual G–U pairings have also been found. However, because of the CCA and unusual bases, the 3′ end of the molecule and at least three other regions will not be able to base-pair, and therefore remain single-stranded. Thus the configuration of the tRNA seems to be one containing single-stranded loops, double-stranded areas called *stems*, and a single strand at the 3′ end. Figure 10-21 shows the "cloverleaf" two-dimensional structure of a tRNA molecule from

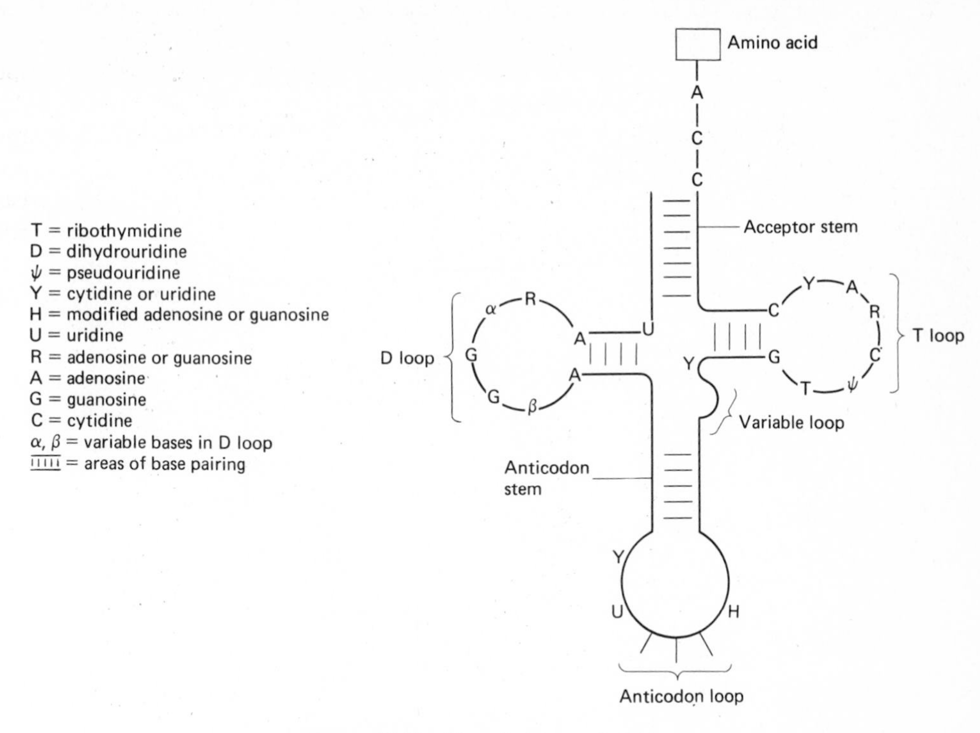

Figure 10-21. Cloverleaf form of tRNA.

yeast which was the first structure to be proposed (Holley *et al.* 1965). Figure 10-22 illustrates a different configuration for a tRNA from yeast based on X-ray crystallography analysis (Kim *et al.* 1974). The detailed skeletal model of a yeast tRNA which binds specifically to phenylalanine is shown in Figure 10-23.

Some constant features have been found in all tRNA molecules. For example, there are always five base pairs in the stem of the T loop. Furthermore, the sequences of bases in these constant regions are the same in all tRNA molecules regardless of species, a phenomenon whose significance is not known.

Attachment of tRNA to amino acid. Having discussed the structure of tRNA molecules, we can now turn our attention to their role in translation. The basic function of tRNAs is to transport amino acids to ribosomes and mRNA. In the 1950s, cell biologists discovered that amino acids must be attached to tRNA to be incorporated into protein (Hoagland *et al.* 1957). To be attached to tRNA, however, an amino acid must first interact with ATP. This interaction between amino acids and ATP is known as *amino acid activation*, and is catalyzed by enzymes generally referred to as *amino acyl-tRNA synthetase*. For each amino acid there is a specific enzyme, for example, leucyl-tRNA synthetase catalyzes the activation of leucine (Fig. 10-24 a). The same enzyme then causes the attachment of the activated amino acid to a

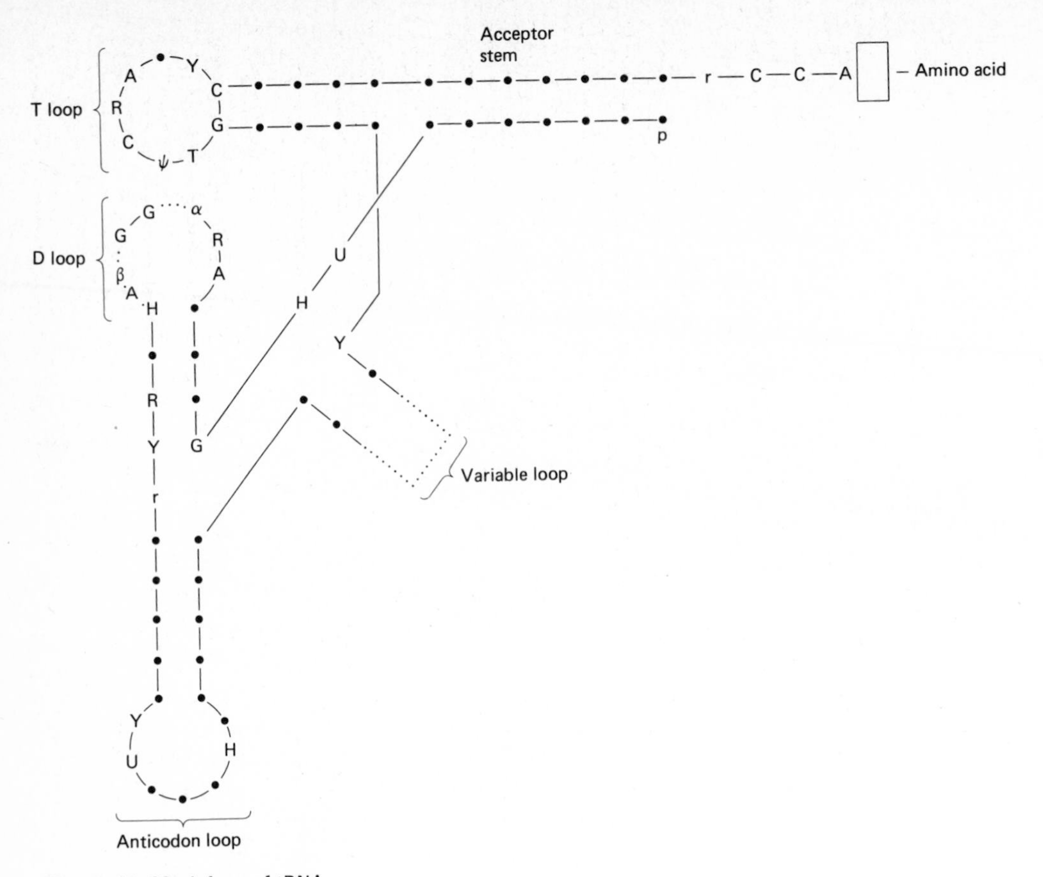

Figure 10-22. L form of tRNA.

molecule of tRNA which is specific for that amino acid (Fig. 10-24b). This reaction occurs at the 3′ end of all tRNAs.

The anticodon. There exists a triplet of bases found in one of the loops of all tRNA molecules called the *anticodon* (see again Fig. 10-22). The sequence of bases of the anticodon is generally complementary to the codon that specifies the amino acid with which the tRNA interacts. Thus when a $tRNA^{leu}$ transports its amino acid to the ribosome, the anticodon will base pair with the codon for leucine that is in the mRNA. We can designate a specific tRNA by using an abbreviation of its amino acid as a superscript.

The Wobble Hypothesis. It is necessary to state that the sequence of bases of an anticodon is only generally complementary to the bases of a codon, because in

---

Figure 10-23. Detailed skeletal model of yeast phenylalanine tRNA shows the hydrogen-bonding interactions between the nucleotide bases. It was derived in 1974 from an X-ray-crystallographic study at a resolution of three angstroms. Projection shown here was generated on a computer. Ribosephosphate backbone of the molecule is shaded a light gray; the bases are shaded a dark gray. (Courtesy Sung Hou Kim, University of California, Berkeley.)

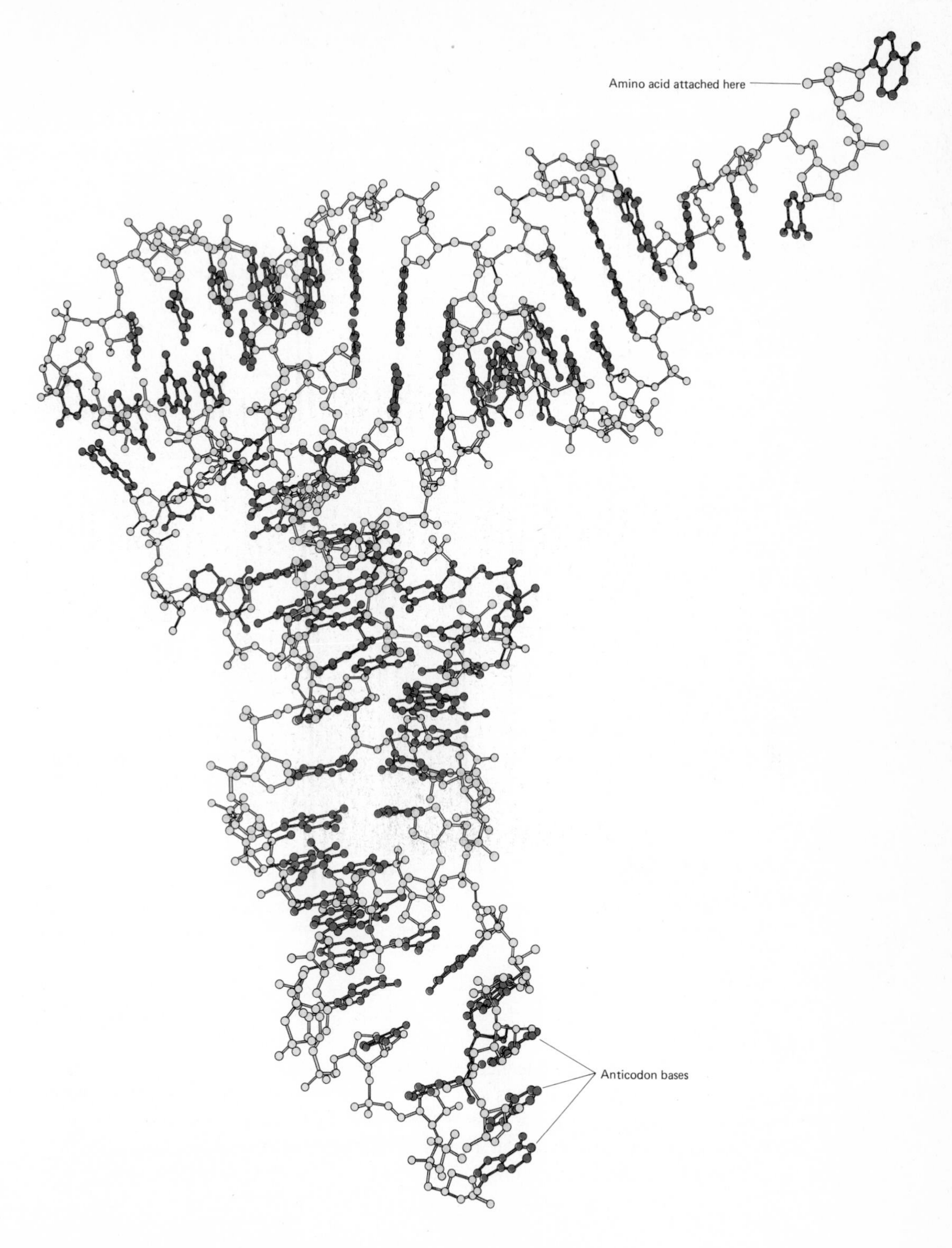
Amino acid attached here
Anticodon bases

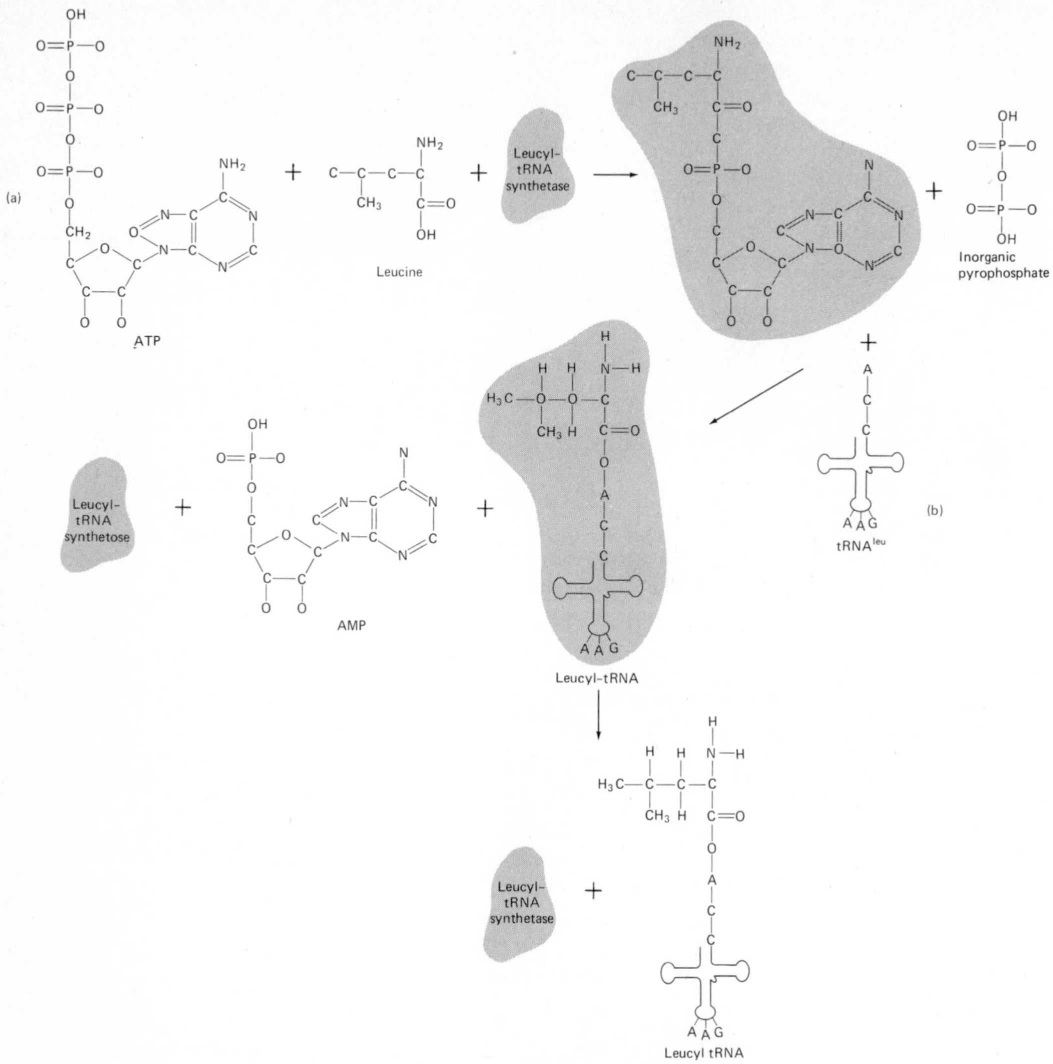

Figure 10-24. Various reactions in the activation of an amino acid, leucine (a), and its subsequent attachment to tRNA (b). Both processes are mediated by the enzyme amino acyl-tRNA synthetase.

some cases only the first two bases are actually complementary. In 1966, Francis Crick considered the base-pairing between the codons of mRNA and anticodons of tRNA, with regard to the degeneracy of the genetic code. He pointed out that in most cases of degeneracy, codons specifying the same amino acid have the same first two bases, differing only in the base at the 3′ end. For example, alanine is coded by GCU, GCC, GCA, and GCG. He hypothesized that the same tRNA molecule could bind with any of the four codons for alanine by establishing firm base-pairing with the first two bases, and having the third base form a weak, or "wobble," base pair. This has come to be known as the *Wobble Hypothesis*.

The Wobble Hypothesis also accounts for tRNA molecules that are known to have the base inosine at the 5′ end. Inosine is a derivative of adenine which has been deaminated (see again Fig. 10-20). Inosine can form base pairs with adenine, uracil, or cytosine. Uracil and guanine can base-pair with two different bases. Table 10-2 summarizes the base-pairing at the 5′ position of the anticodon (and 3′ position of the codon) as visualized by Crick.

Table 10-2 Base Pairing at the Third Position of the Codon with Wobble

Base on the anticodon	*Bases recognized on the codon*
A	U
C	G
G	U C
U	A G
I	U C A

From F. H. Crick, 1966. *J. Mol. Biol.* 19: 548–555.

This hypothesis has since been supported by experimental evidence. In some cells, though, certain codons are present more often than others which specify the same amino acid, indicating that the favored codons may be more efficient in interacting with tRNAs.

Other functions of tRNA. As important as tRNA molecules are for translation, this class of molecules is being recognized as one which appears to have a role in a large variety of cellular phenomena. For example, some tRNA molecules serve to transport amino acids to cell membranes. Some RNA tumor viruses cause the synthesis of viral DNA in infected cells by using a tRNA as a primer molecule. When we discuss mechanisms for gene regulation in Chapter 11, you will see that tRNA molecules have been implicated in the regulation of gene action as well.

Protein Synthesis

Having now been introduced to all the participants of translation, let us now run through the events of actual protein synthesis.

N-Formyl methionine. Formation of the initiation complex is the first event in protein synthesis. This involves the association of a 30S particle (in bacteria), a tRNA with its amino acid, and the mRNA. We should mention here that studies indicate that the first amino acid in polypeptides synthesized in prokaryotes is always the same one, and it is a modified form of an amino acid. It is n-formyl methionine which carries a formyl ($-\overset{H}{\overset{|}{C}}{=}O$) group on the αNH_2 end of the amino acid (Fig. 10-25).

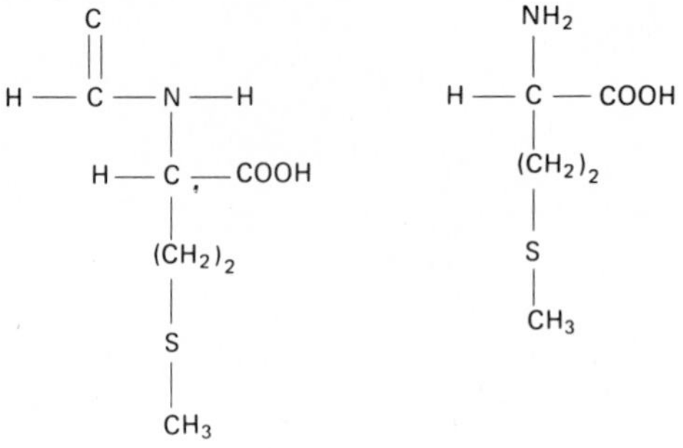

$$\text{(a)}\quad \begin{array}{c} \text{C} \\ \| \\ \text{H}-\text{C}-\text{N}-\text{H} \\ \quad\quad | \\ \quad\quad \text{H}-\text{C}-\text{COOH} \\ \quad\quad | \\ \quad\quad (\text{CH}_2)_2 \\ \quad\quad | \\ \quad\quad \text{S} \\ \quad\quad | \\ \quad\quad \text{CH}_3 \end{array} \qquad\qquad \text{(b)}\quad \begin{array}{c} \text{NH}_2 \\ | \\ \text{H}-\text{C}-\text{COOH} \\ | \\ (\text{CH}_2)_2 \\ | \\ \text{S} \\ | \\ \text{CH}_3 \end{array}$$

Figure 10-25. Comparison of the structures of (a) n-formyl methionine, the initiator amino acid, and (b) methionine, normally found within a polypeptide.

When $tRNA^{met}$, which associates with methionine, was compared to n-formyl methionine tRNA, differences in the nucleotide sequences were found. Both carry the same anticodon, however—5′ CAU 3′. Since AUG codons exist within a message, the factors which determine that n-formyl methionine tRNAs (sometimes referred to as *initiator tRNA*) will base-pair only with AUG codons at the beginning of a messenger RNA, and not with AUG within the message, have not been fully elucidated.

Polypeptide formation. With the formation of the initiation complex, the 50s portion of the ribosome joins with the 30S. Two sites on the 50S particle serve as attachment sites for tRNA molecules. One is the *aminoacyl site*, to which an incoming tRNA will attach in such a way that its anticodon will base-pair with a corresponding codon on the mRNA (Fig. 10-26a). The initiator tRNA with n-formyl methionine is then displaced to a second site on the 50S particle called the *peptidyl site* (Fig. 10-26b). A second tRNA carrying another amino acid (let us assume that it is leucine) then joins the complex at the newly vacated aminoacyl site (Fig. 10-26c). The enzyme peptidyl transferase, causes a peptide bond formation between the first and second amino acids, and the first amino acid is released from its tRNA (Fig. 10-26d). THUS PROTEIN SYNTHESIS OCCURS FROM THE AMINO END TO THE CARBOXYL END, AND TRANSLATION OF THE MESSAGE OCCURS FROM THE 5′ END TO THE 3′ END OF THE mRNA.

Figure 10-26. The sequence of events in DNA-directed polypeptide synthesis. Explanation in text.

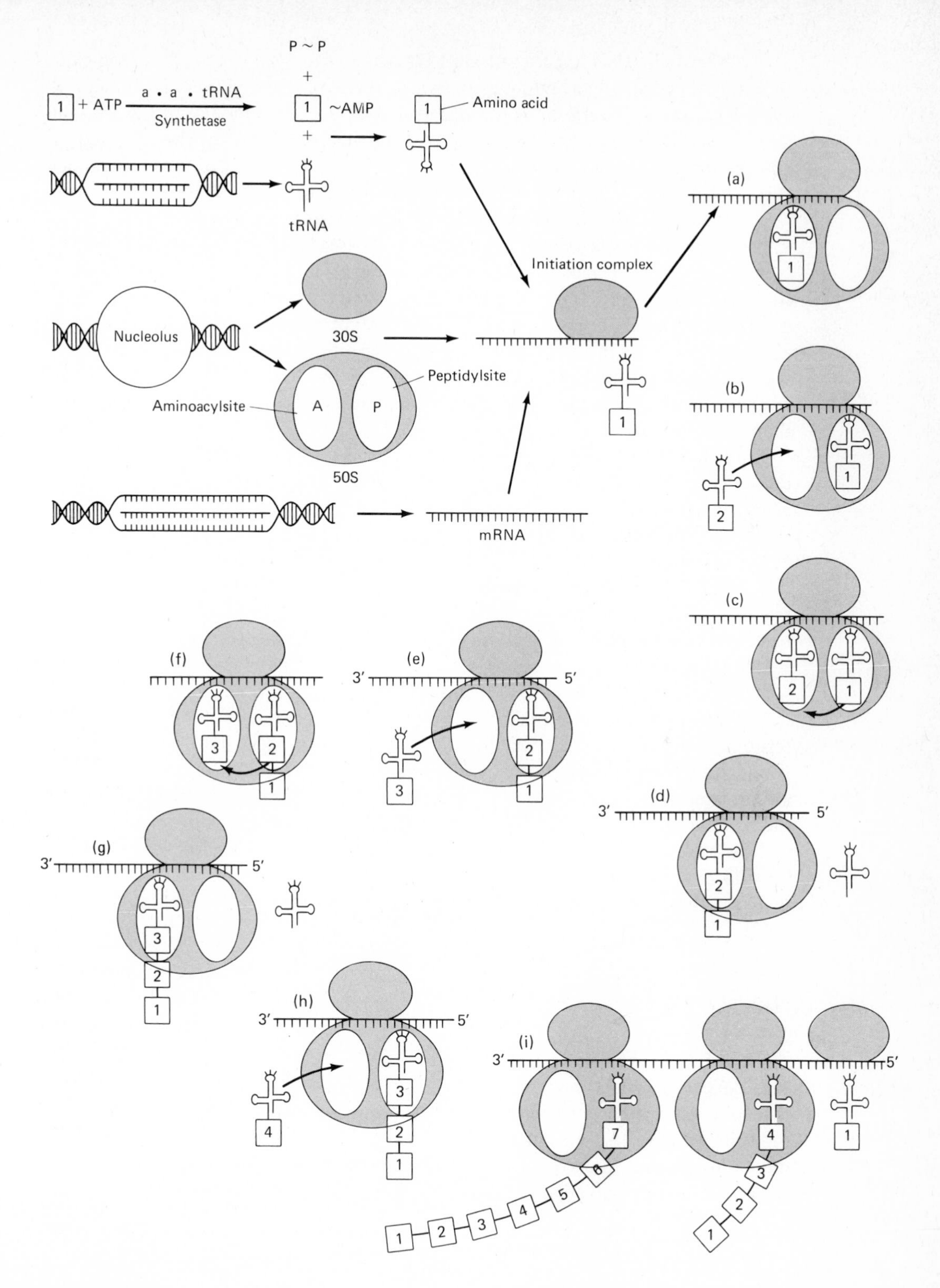
P ~ P
+
1 + ATP
a • a • tRNA
Synthetase
1 ~AMP
+
1
Amino acid
tRNA
Initiation complex
Nucleolus
30S
Peptidylsite
Aminoacylsite
A
P
50S
mRNA
(a)
(b)
(c)
(d)
(e)
(f)
(g)
(h)
(i)
3′
5′

The first tRNA then is released from the ribosome, which allows the second tRNA, now carrying two amino acids, to move into the peptidyl site (Fig. 10-26e); this event is known as *translocation*, an unfortunate example of the use of a word which already has other meanings in genetics). A third tRNA will occupy the aminoacyl site and the same series of reactions will occur, resulting in a chain of three amino acids (Fig. 10-26g). As the mRNA continued to move through or past the ribosome (Fig. 10-26f), a continuous progression of codons occurs near the aminoacyl site, thus attracting specific tRNAs with complementary bases.

Since a number of ribosomes will cluster around the same mRNA molecule, the result will be a number of polypeptides formed from the same messenger molecule (Fig. 10-26i). Termination of the process apparently occurs when a nonsense codon passes through the ribosomes. The polypeptide chains are then released from the ribosomes and undergo further modifications in form. In some cases, the initial methionine group and some subsequent amino acids are enzymatically removed from the polypeptide.

It has been estimated that, in bacteria, the rate of polypeptide synthesis is on the order of about 7 amino acids per second in the case of the tryptophan synthetase genes (Baker and Yanofsky 1972). Recent estimates of the rate of polypeptide synthesis are on the order of 20 amino acids per second in bacteria, and one amino acid per second in mammalian cells (Rich and Kim 1978).

Some of the participants in translation, such as mRNA, are degraded enzymatically. Eukaryotic mRNA has been found to be longer lasting than mRNA of bacteria. Other molecules, such as tRNA and ribosomes, will be involved in the synthesis of yet other polypeptides. As was mentioned before, the subunits of ribosomes dissociate when not engaged in the process of translation.

Other factors. As translation processes have come under ever more detailed scrutiny, additional factors which contribute to the process have been discovered. Among these are *initiation factors*, sometimes referred to as *IF* in the literature. These are proteins which contribute in some way to the dissociation of ribosomal subunits, the binding of mRNA to the 30S particle in the formation of the initiation complex, and the reassociation of ribosomal subunits. Three initiation factors have been identified in *E. coli*, and at least seven have been found in eukaryotic cells. (For a review of recent work in IF proteins, see the reference to Gruneberg-Manago and Gros 1977).

THE COMPLEXITY OF GENE-DIRECTED CELLULAR PROCESSES

It is typical of genetic research that one question leads to discovery of phenomena that raise many other questions. From our discussion in this chapter, it is quite obvious that we have learned much about gene function in a very short time. It is also quite obvious that we have much more to learn. For example, what factors regulate the maturation of polypeptide chains into functional proteins? How does a cell synthesize nonprotein substances such as steroids? In addition, since our present methods require the removal of cellular components from the cell, and therefore study of these components in an unnatural environment, scientists are aware of the need for caution in interpreting all experimental results as being completely accurate reflections of actual events in cells.

Erwin Chargaff (1976, p. 21) has recently reflected on our attempts to unravel all the complex machinations of a living cell: "Can we hope to understand the working of a cell by establishing a comprehensive catalog of its chemical abilities? I am afraid there is little chance of success, although even the complete inventory is far from being accomplished. Biochemistry, being the science of the potential, can tell us at best what may happen in the living cell, but many reactions that we know to occur seem to be mutually exclusive, and we must confess that we do not know how a cell lives."

Not all scientists are as skeptical as Chargaff about our attempts to understand gene-directed cellular functions, for, as Chargaff said, the "complete inventory" has only begun to be taken. Whether we will ever reach the point of total comprehension remains to be seen. What is certain is that each living cell is a mini-universe unto itself, with myriad molecules participating in complex interrelated reactions, each essential for the survival of the cell. What is certain is that as our knowledge expands, the living cell becomes more remarkable.

PROBLEMS

1. Assume that a mutation in strain A mice results in a condition identical to that of PKU in humans. Assume further that skin transplants can be performed successfully between this strain and an albino strain of mice. If skin from albino mice is transplanted into strain A mice, would you expect pigmentation to be produced in the transplant? Explain.

2. A culture of *Neurospora* is radiated following Beadle and Tatum's protocol (p. 306). Four nutritional mutants are isolated, which will grow if substance A is added to minimal medium. Tests on the four mutants with substances B, C, and D however give different results (+ signifies growth; – absence of growth):

Mutant strain	*A*	*B*	*C*	*D*
1	+	+	+	+
2	+	–	+	+
3	+	–	–	+
4	+	–	–	–

The results indicate that substance A is the end product of a metabolic pathway in which B, C, and D are intermediate products.

(a) Give the sequence of involvement of the substances in the pathway.
(b) Which point in the pathway is blocked by each of the four mutations?

3. A strand of DNA has the following sequence:

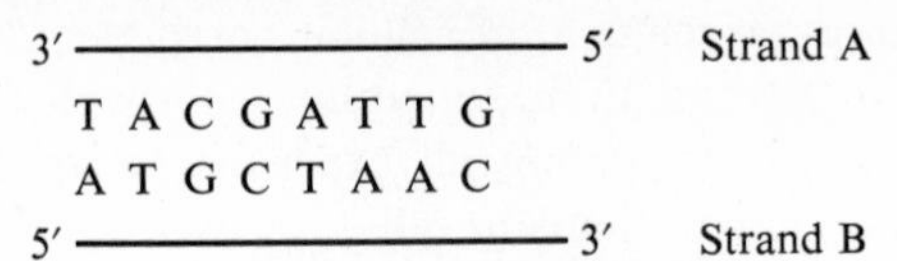

If strand A serves as the template for mRNA transcription,

(a) Give the sequence of bases of the mRNA.
(b) Which three bases of the mRNA will be transcribed first?
(c) Which three bases of the mRNA will be translated first?

4. From studies of replication in vitro as well as RNA synthesis in vitro, scientists suspected the need for a molecule of RNA to initiate replication in vivo even before the actual experimental evidence was found. Why?

5. Assume that we repeat experiments by Nirenberg on the genetic code as described on page 316, using the bases C and G in a ratio of 4C : 1G. What amino acids can be expected to be incorporated in a polypeptide using the mRNA that is synthesized? Calculate the expected proportions of each of these amino acids in the polypeptide.

6. Now assume that we prepare a mRNA that has alternating C–G bases. A polypeptide synthesized in vitro using this RNA contains alternating arginine and alanine. If the deciphered genetic code were not available, would you be able to discern the exact codons for each of these amino acids? Would you be able to discern the codons by using the data in problem 5? Explain.

7. What effect on polypeptide synthesis would you expect if mutations resulted in

(a) a stop codon at the 3′ end of the mRNA;
(b) a stop codon in the middle of the mRNA;
(c) a stop codon at the 5′ end of the mRNA?

8. What physiological advantage would there be to a cell in the degradation of mRNA molecules shortly after they have been translated into polypeptides? In humans, what cells would you expect to have mRNA that are very long-lived? Explain.

9. It is commonly known that the starvation of children results in the stunting of growth. Relate this phenomenon to gene-directed cell processes.

REFERENCES

BAKER, R., AND C. YANOFSKY, 1972. "Transcription Initiation Frequency and Translational Yield for the Tryptophan Operon of *Escherichia coli*." *J. Mol. Biol.* 69: 89–102.

BARRELL, B. G., A. M. AIR, AND C. A. HUTCHISON, 1976. "Overlapping Genes in Bacteriophage ϕX174." *Nature* 264: 34–41.

BEADLE, G. W., AND B. EPHRUSSI, 1937. "Development of Eye Colors in *Dro-*

sophila: Diffusible Substances and Their Interrelations." *Genetics* 22: 76–86.

BEADLE, G. W., AND E. L. TATUM, 1941. "Genetic Control of Biochemical Reactions in *Neurospora.*" *Proc. Nat. Acad. Sci. U.S.* 27: 499–506.

BRENNER, S., F. JACOB, AND M. MESELSON, 1961. "An Unstable Intermediate Carrying Information from Genes to Ribosomes for Protein Synthesis." *Nature* 190: 576–581.

BRENNER, S., A. O. W. STRETTON, AND S. KAPLAN, 1965. "Genetic Code: The 'Nonsense' Triplets for Chain Termination and Their Suppression." *Nature* 206: 994–998.

BROWN, D. D., 1973. "The Isolation of Genes." *Sci. Am.* 229: 20–29.

CHARGAFF, E., 1976. "Initiation of Enzymic Synthesis of Deoxyribonucleic Acid by Ribonucleic Acid Primers." *Prog. in Nuc. Ac. Res. Mol. Biol.* 16: 21.

CRICK, F. H. C., 1966. "Codon-Anticodon Pairing: The Wobble Hypothesis." *J. Mol. Biol.* 19: 548–555.

______, L. BARNETT, S. BRENNER, AND R. J. WATTS-TOBIN, 1961. "General Nature of the Genetic Code for Proteins." *Nature* 192: 227–232.

DARNELL, J. E., 1976. "mRNA Structure and Function." *Prog. Nuc. Ac. Res. and Mol. Biol.* W. E. Cohn and E. Volkin, eds. 19: 493–511.

DAVIES, J., W. GILBERT, AND L. GORINI, 1964. "Streptomycin, Suppression, and the Code." *Proc. Nat. Acad. Sci. U.S.* 51: 883–890.

DUNN, J. J., AND F. W. STUDIER, 1973. "T7 Early RNAs and *Escherichia coli* Ribosomal RNAs Are Cut from Large Precursor RNAs in Vivo by Ribonuclease." *Proc. Nat. Acad. Sci. U.S.* 70: 3296–3300.

DUGAICZYK, A., ET AL., 1978. "The Natural Ovalbumin Gene Contains Seven Intervening Sequences." *Nature* 274: 328–333.

FURUICHI, Y., A. LAFRANDRA, A. SHATKIN, 1977. "Terminal Structure and mRNA Stability." *Nature* 266: 235–9.

GARROD, A., 1909. *Inborn Errors of Metabolism.* Oxford U. Press, Oxford, England.

GROS, F., H. HIATT, W. GILBERT, C. G. KURLAND, R. W. RISEBROUGH, AND J. D. WATSON, 1961. "Unstable Ribonucleic Acid Revealed by Pulse Labeling of *Escherichia coli.*" *Nature* 190: 581–585.

GRUNEBERG-MANAGO, M., AND S. OCHOA, 1955. "Enzymatic Synthesis and Breakdown of Polynucleotides; Polynucleotide Phosphorylase." *J. Am. Chem. Soc.* 77: 3165–3166.

HALL, B. D., AND S. SPIEGLEMAN, 1961. "Sequence Complementarity of T2-DNA and T2-Specific RNA." *Proc. Nat. Acad. Sci. U.S.* 47: 137–146.

HOAGLAND, M. B., P. C. ZAMECNIK, AND M. L. STEPHENSON, 1957. "Intermediate Reactions in Protein Biosynthesis." *Biochem. Biophys. Acta* 24: 215–216.

HOLLEY, R. W., J. APGAR, G. A. EVERETT, J. T. MADISON, M. MARGINSEE, S. H. MERRILL, J. R. PENSWICK, AND A. ZAMIS, 1965. "Structure of a Ribonucleic Acid." *Science* 147: 1462–1465.

HUNT, T., T. HUNTER, AND A. MUNRO, 1969. "Control of Hemoglobin Synthesis Rate of Translation of the Messenger RNA for the Alpha and Beta Chains." *J. Mol. Biol.* 43: 123–133.

KHORANA, H. G., 1966-67. "Polynucleotide Synthesis and the Genetic Code." *Harvey Lectures* 62: 79–105.

KIM. S. H., G. L. SUDDATH, G. J. QUIGLEY, A. MCPHERSON, J. L. SUSSMAN, A. H. J. WANG, N. C. SEEMAN, AND A. RICH, 1974. "Three-Dimensional Tertiary Structure of Yeast Phenylalanine Transfer RNA." *Science* 185: 435–440.

MARMUR, J., C. M. GREENSPAN, E. PALECEK, F. M. KAHAN, J. LEVINE, AND M. MANDEL, 1963. "Specificity for the Complementary RNA Formed by *Bacilllus subtilis* Infected With Bacteriophage SP8." *CSHSQB.* 28: 191–199.

MEARS, J. G., F. RANWIERY, D. LEIBOWITZ, AND A. BANK, 1978. "Organization of Human δ- and β-globin Genes in Cellular DNA and the Presence of Intragenic Inserts." *Cell* 15: 15–23.

NIRENBERG, M., 1963. "The Genetic Code II." *Sci. Am.* 190: 80–94.

______, AND P. LEDER, 1964. "RNA Code Words and Protein Synthesis: The Effect of Trinucleotides Upon the Binding of sRNA to Ribosomes." *Science* 145: 1399–1407.

______, AND J. H. MATTHAEI, 1961. "The Dependence of Cell-Free Protein Synthesis in *E. coli* Upon Naturally Occurring or Synthetic Polyribonucleotides." *Proc. Nat. Acad. Sci. U.S.* 47: 1588–1602.

NOMURA, M., 1973. "Assembly of Bacterial Ribosomes." *Science* 179: 864–873.

______, 1976. "Organization of Bacterial Genes for Ribosomal Components: Studies Using Novel Approaches." *Cell* 9: 633–644.

RICH, A., AND S. H. KIM, 1978. "The Three-Dimensional Structure of Transfer RNA." *Sci. Am.* 78: 52–62.

SARABHAI, A. S., A. O. W. STRETTON, S. BRENNER, AND A. BOLLE, 1964. "Colinearity of the Gene With the Polypeptide Chain." *Nature* 201: 13.

SMITH, J. D., 1976. "Transcription and Processing of Transfer RNA Precursors." *Prog. Nuc. Ac. Res. Mol. Biol.* 16: 25–73.

STANBURY, J. B., J. B. WYNGAARDEN, AND D. B. FREDERICKSON, 1972. *The Metabolic Basis of Inherited Diseases.* McGraw-Hill, New York.

TILGHMAN, S. M., ET AL, 1977. "Cloning Specific Segments of the Mammalian Genome: Bacteriophage λ Containing Mouse Globin and Surrounding Gene Sequences." *Proc. Nat. Acad. Sci. U.S.* 74: 4406–4410.

TILGHMAN, S. M., ET AL, 1978. "Intervening Sequence of DNA Identified in the Structural Portion of a Mouse β-Globin Gene." *Proc. Nat. Acad. Sci. U.S.* 75: 725–729.

VOLKIN, E., L. ASTRACHAN, AND J. L. COUNTRYMAN, 1958. "Metabolism of RNA Phosphorus in *Escherichia coli* Infected with Bacteriophage T7." *Virology* 6: 545–555.

VON EHRENSTEIN, G., AND F. LIPMANN, 1961. "Experiments on Hemoglobin Synthesis." *Proc. Nat. Acad. Sci. U.S.* 47: 941–950.

WITMANN, H. G., 1977. "Structure and Function of *Escherichia coli* Ribosomes." *Fed. Proc.* 36: 2074–2080.

YANOFSKY, C., G. R. DRAPEAU, J. R. GUEST, AND B. C. CARLTON, 1967. "The Complete Amino Acid Sequence of the Tryptophan Synthetase A Protein (A Subunit) and Its Colinear Relationship With the Genetic Map of the A Gene." *Proc. Nat. Acad. Sci. U.S.* 57: 296–298.

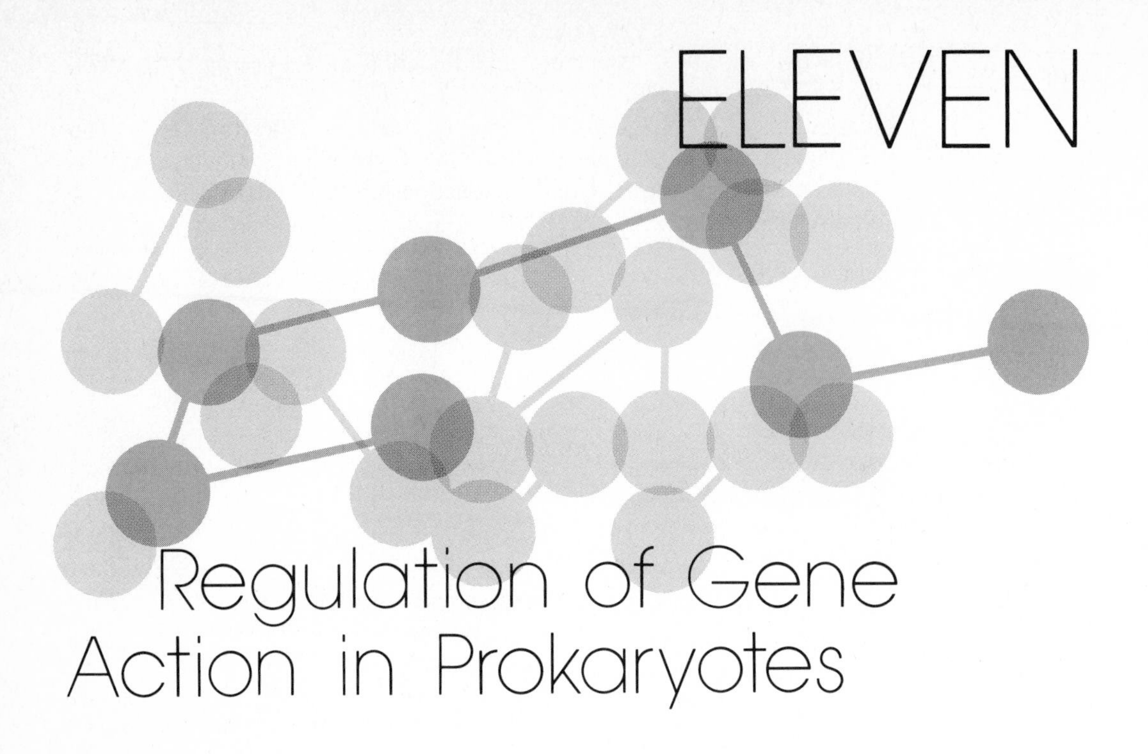

ELEVEN

Regulation of Gene Action in Prokaryotes

INTRODUCTION

In the past few chapters we have discussed the mechanisms by which genetic functions, namely, self-reproduction and the determination of phenotype, are achieved in living cells. Although many details of these mechanisms remain to be elucidated, basic concepts have been developed and accepted as universal to living organisms.

Replication, transcription, and translation have been found to occur more or less in the same manner in cells of all species. However, we have not yet discussed an extremely important aspect of gene action: how gene activity is regulated. The fact that genes are regulated becomes obvious when we consider that cells do not divide at the same rate, nor do they divide ceaselessly. For example, nerve cells in humans divide much slower than do epidermal cells. Furthermore, since there are limits to the normal size of individuals, it is also obvious that some cells do not divide continuously.

Gene regulation is also reflected in differential protein synthesis. As Stent (1971) has pointed out, an *E. coli* cell contains enough DNA to code for 3000 different polypeptides. But of the 10^7 number of protein molecules actually found in an *E. coli* cell, some polypeptides are present in very small amounts, perhaps as few as 10 molecules; others are present in large amounts, as many as 500,000 molecules. IN OTHER WORDS, SOME GENE PRODUCTS ARE CONTINUOUSLY BEING SYNTHESIZED,

REFLECTING CONTINUOUS TRANSCRIPTION OF THE RESPONSIBLE GENES, WHEREAS OTHERS ARE SYNTHESIZED SPORADICALLY, REFLECTING REGULATION OF GENE ACTIVITY.

In this chapter we shall address the concept of gene regulation as it has been studied in prokaryotes. As you will see, mechanisms of gene regulation are by no means as universal as are mechanisms of replication, transcription, and translation. In prokaryotes, a number of different regulatory systems have been found, and in fact different genes within the same cells are subject to different types of regulation. We shall discuss gene regulation in eukaryotes in the following chapter.

THE LAC OPERON

The first system of gene regulation in prokaryotes to be defined was that of the synthesis of enzymes in *E. coli* which are involved in metabolism of the sugar lactose. Normally, *E. coli* cells use glucose as their source of carbohydrates. In the presence of glucose, only very small amounts of the lactose enzymes are present. When lactose is substituted for glucose as the carbohydrate source, the synthesis of lactose enzymes is increased one thousandfold. It appears that the presence of lactose (which is converted to allolactose) induces transcription of the genes that code for enzymes for the breakdown of allolactose.

Systems such as the *lac* (short for *lactose*) system in *E. coli*, in which a substance can cause the cell to increase the activity of a particular gene or genes, resulting in the increased synthesis of certain proteins, are known as *inducible* systems. The substance that causes the increased activity of genes is known as the *effector*, or *inducer*. In the lac system, allolactose would of course be the effector.

Structural Genes

Studies showed that lactose and molecules chemically similar to lactose actually induced the concomitant increased synthesis of three different enzymes. Mapping experiments indicated that the three genes determining these enzymes are located sequentially on the chromosome. These genes were designated *z*, *y*, and *a*, and were found to determine the enzymes β-galactosidase, β-galactoside permease, and β-galactoside transacetylase, respectively (Fig. 11-1).

The enzyme β-galactosidase was found to cause the breakdown of lactose to glucose and galactose, or to allolactose (Fig. 11-2). The permease determined by gene *y* is involved in the transport of galactosides through the cell membrane. The product

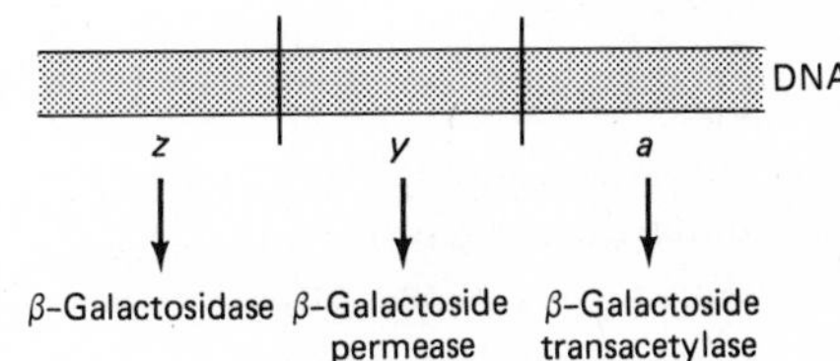

Figure 11-1. Structural genes determining enzymes for the breakdown of lactose.

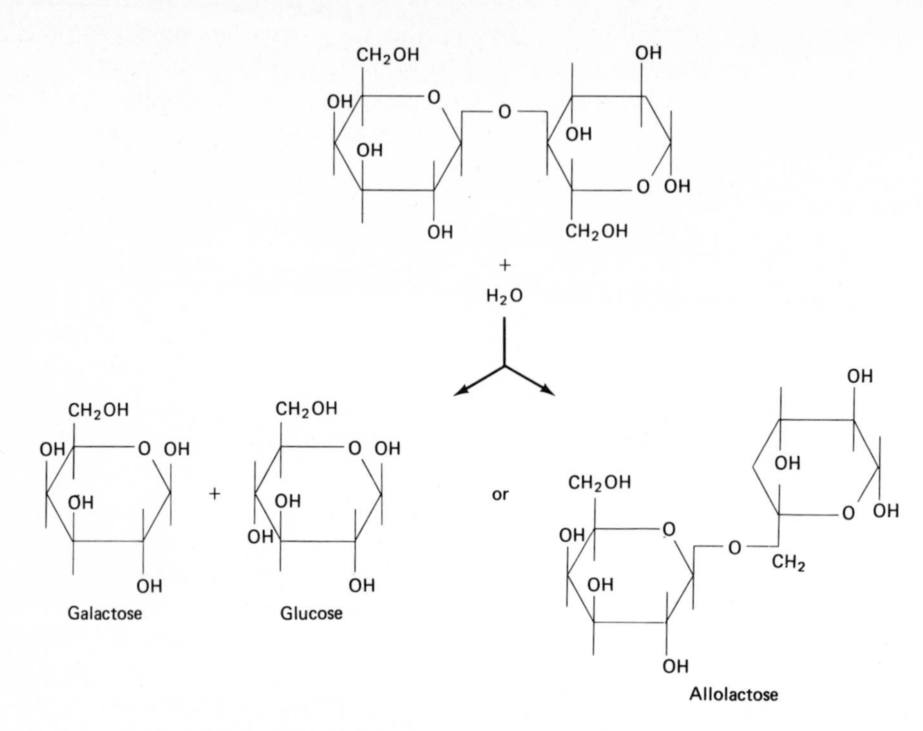

Figure 11-2. The enzyme, β-galactosidase mediates the hydrolysis of lactose into galactose and glucose, or to allolactose.

of gene *a*, β-transacetylase, is known to cause the transfer of an acetyl group from a chemical known as acetyl-CoA to galactosides, but what role this has in galactoside metabolism is not clear. Mutants lacking the transacetylase are not affected adversely by its absence.

Mutations at these loci resulted in abnormal or defective enzymes, as would be expected. However, some mutations were also found which did not affect the structure of the enzymes, but appeared to affect control over transcription of the three genes. For example, certain strains of *E. coli* cells contained a high level of each enzyme regardless of the presence or absence of lactose; these strains were termed *constitutive*, rather than inducible. Analysis of the enzymes showed that they were normal in structure and function.

Regulatory Genes and Regions

In 1961, F. Jacob and J. Monod of the Pasteur Institute in France, whose work on bacterial and viral genetics we have already discussed (p. 246), proposed that the lac system must include two different classes of genes: *structural genes*, which determine the structure of the lac enzyme polypeptides, and *regulator regions and genes*, which control the activity of the structural genes.

Operator region. Subsequent studies have confirmed that there are structural and regulatory genes (see Beckwith 1970, for review). Mapping of a number of different mutations has revealed several different regulatory sites each with its own role in the control of lactose metabolism in *E. coli.* A region situated next to the *z* locus, designated the *o*, or operator, functions to control transcription of the structural genes. In the absence of lactose, DNA in the operator region binds specifically to a protein called the *repressor*. This interaction between the repressor and the operator prevents transcription by not allowing the DNA to unwind. RNA polymerase is then blocked from moving along the structural genes for transcription. This type of control, causing the shutdown of transcription, is termed *negative* control (Fig. 11-3).

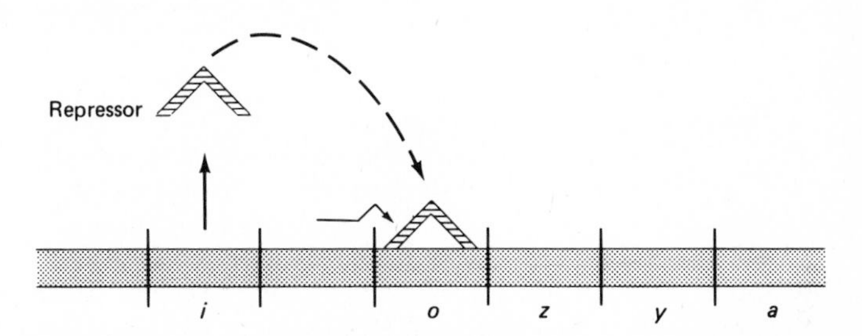

Figure 11-3. Schematic illustration depicting repressor synthesis determined by regulator gene, *i*, and repressor interaction with operator region, *o*, to shut down transcription of structural genes (arrow indicates direction of transcription).

Regulator gene. The repressor is coded for by a gene called the *i* gene, which is separated by fewer than a hundred base pairs from the *o* region. The repressor protein was identified in 1967 by W. Gilbert and B. Muller-Hill, and found to be a tetrameric protein (composed of four identical polypeptide chains) (Gilbert and Muller-Hill 1970).

If the repressor interacts with the operator to prevent transcription, Jacob and Monod reasoned that mutations at the *i* locus that affected the structure of the repressor molecules should result in i^- cells that are constitutive in their production of the lac enzymes. They would be constitutive because of the inability of abnormal repressor molecules to bind to the operator site.

A. B. Pardee, F. Jacob, and J. Monod (1959) then designed sexduction experiments which supported this hypothesis. They looked for the transfer of i^+ genes into i^- constitutive mutants. Presence of the i^+ gene restored the cell to the wild-type inducible state regardless of whether the F' episome carrying the gene was integrated into the host DNA or remained autonomous. These results confirmed that the i^+ gene determined the synthesis of a diffusible substance, the repressor, which binds to the operator of the host cell (Fig. 11-4).

Since the i^+ phenotype prevails even though the F' DNA is autonomous, geneticists refer to the i^+ gene as being *trans dominant*. In fact, since the i^+ phenotype also is expressed when the F' DNA is integrated, the i^+ gene is also considered *cis dominant* over the i^- mutation.

For the same reason, mutations at the operator locus would also be expected to result in constitutive synthesis of *lac* genes, since an abnormal operator sequence

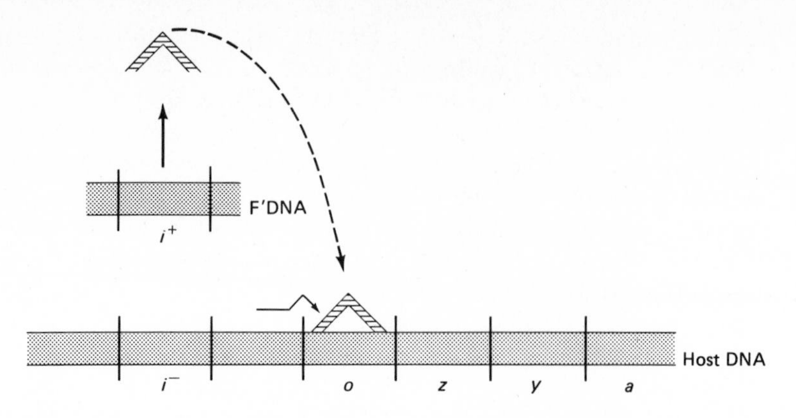

Figure 11-4. Sexduction of wild-type regulator gene into a mutant $i-$ cell (which is constitutive in its production of lac enzymes) restores the cell to wild-type (or inducible) even if the F′ DNA remains autonomous. Thus the $i+$ gene is trans dominant.

would not be able to interact with repressor molecules. Such mutations were observed and referred to as o^c mutants. However, sexduction of o^+ into o^c cells converted the host cell to the inducible state *only* if the wild-type operator gene was integrated into the host genome next to the structural genes. Cells in which the o^+ region remained in an autonomous F' episome continued to show constitutive synthesis of lac enzymes (Fig. 11-5a).

In these experiments, the o^+ region is considered trans recessive, because the mutant phenotype prevails if the wild-type operator region is on a separate piece of DNA from the *lac* structural genes. On the other hand, when the o^+ region is integrated into the bacterial chromosome by recombination, it was found to control transcription of the structural genes when bound to repressor molecules. Thus the o^+ condition is cis dominant (Fig. 11-5b).

The characteristic of o^+ being cis dominant and trans recessive was compatible with the suspected role of the operator as being regulatory in nature, rather than as determining the synthesis of a diffusible substance. This was the first observation of a DNA region which is not transcribed, and is certainly an exception to Beadle and Tatum's One Gene-One Enzyme Hypothesis.

Role of lactose. It was later determined that allolactose and other related galactosides induce the increased synthesis of lac enzymes by binding to the repressor molecules. This prevents repressors from interacting with the operator gene, and allows transcription of the structural genes to proceed (Fig. 11-6).

Accordingly, *super-repressed* (noninducible) i^s mutants were reported. In these mutant cells, presence of lactose did not increase the rate of synthesis of lac enzymes. Analysis of these data indicated that i^s mutations alter the structure of repressors so that they cannot interact with lactose, but can still bind to operator DNA. Sexduction experiments showed that i^s mutants were trans dominant over either i^+ or i^-, as would be expected (Fig. 11-7).

Promoter region. It was originally thought that the operator site functioned both to initiate transcription and to control transcription of the structural genes.

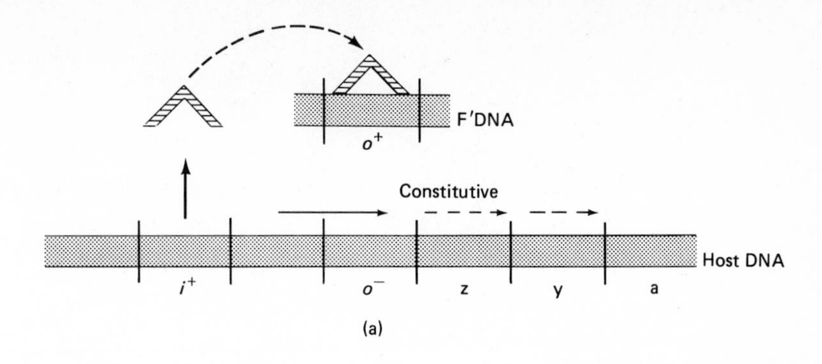

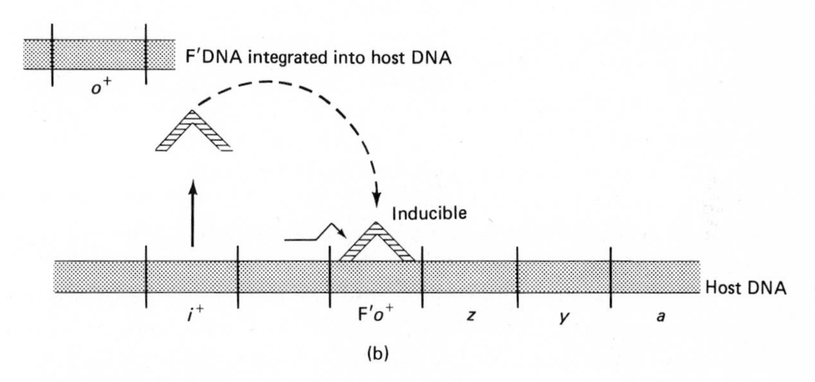

Figure 11-5. Sexduction of *O*+ into constitutive *O*− mutants does not affect constitutive production of lac enzymes in the trans position (a), but does restore wild-type inducible production when integrated into the host DNA next to structural genes (b). Thus the *O*+ gene is trans recessive and cis dominant.

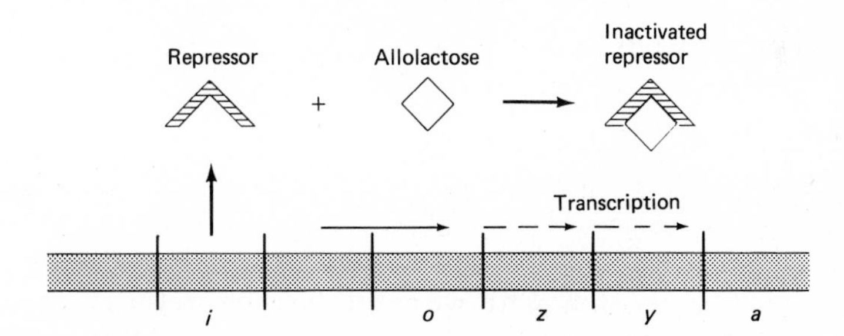

Figure 11-6. Schematic illustration of interaction between repressor and allolactose which results in the inability of repressor to bind to operator. Transcription of structural genes ensues.

However, mutations found in the region between the operator region and regulator gene also produced effects of a regulatory nature (Ippen *et al.* 1968). Mutations in this region resulted in super-repressed cells.

Further studies showed that the initiation of transcription is the function of this region, now known as the *promoter* (*p*). The portion of the promoter DNA closer

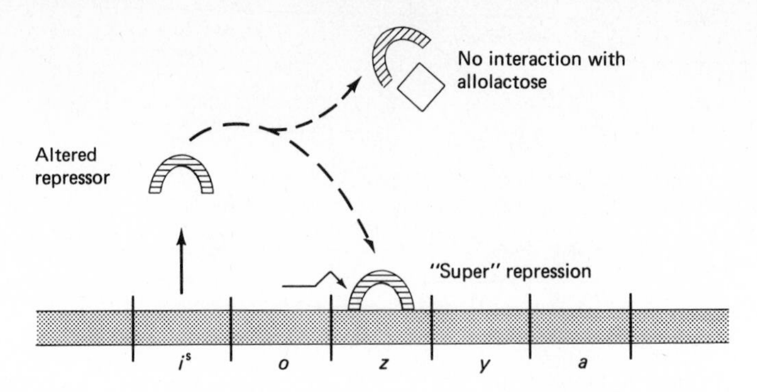

Figure 11-7. Diagrammatic representation of "super-repressed" (i^s) mutants, in which an altered repressor binds to operator, but not to effector molecules. These mutants are noninducible.

to the operator region serves as a binding site for RNA polymerase. In the absence of repressor molecules at the operator site, RNA polymerase is able to move through the promoter and operator regions to transcribe the structural genes. Mutations at this locus would of course affect binding of the RNA polymerase, and transcription would not occur at the *z*, *y*, and *a* genes.

Recently, the region of the promoter distal to the operator region has also been found to serve a regulatory function. This region binds with yet another protein, known as the *catabolite-binding protein* (or CAP). This protein must be present on the promoter to allow RNA polymerase to move through, possibly by causing a destabilization of the DNA, which then unwinds (Fig. 11-8). Studies are under way to determine the nature of the promoter, which must unwind and interact with RNA polymerase in order for the structural genes to be transcribed.

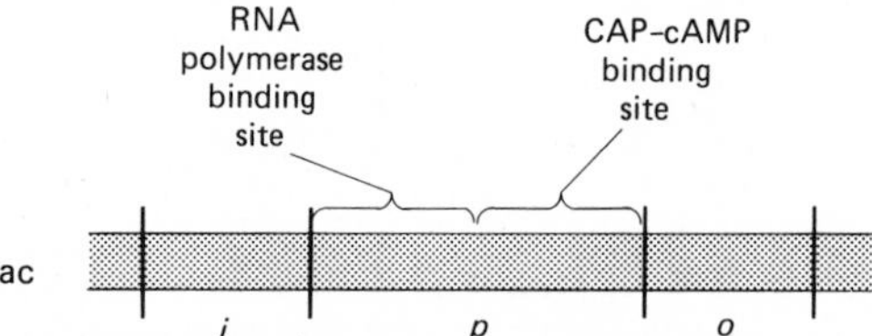

Figure 11-8. The promoter region of the lac operon.

Discovery of the CAP site and its function explained yet another phenomenon associated with the expression of *lac* genes, namely, that in the presence of both glucose and lactose, the *lac* enzymes are not synthesized. The explanation is a biochemical one, and reflects on the economy of cell metabolism. Biochemically, it is much more practical to use glucose as an energy source when it is available rather than lactose, because lactose must first be broken down to glucose, a simpler form of sugar, before it can be used by the cell.

The relationship between glucose and CAP is as follows: in order to bind to the promoter, CAP must first be bound to cyclic AMP (for a general review on the role of cyclic AMP (cAMP), see reference to Pastan and Adhya 1976). During the metabolism of glucose by the cell, the amount of intracellular AMP decreases. Without cAMP, CAP cannot attach to the promoter, and the *lac* structural genes remain repressed, in spite of the presence of lactose. It should be pointed out here that whereas the repressor effect is considered negative control, the effect of CAP is considered positive control.

Thus it appears that two conditions must be met in order for transcription of the structural genes to occur. There must be binding at the promoter locus with both RNA polymerase and cAMP–CAP, and there must be absence of the repressor–operator interaction. Induction of lac enzymes occurs only when lactose is the primary source of carbohydrate for the cell.

The Operon Concept

JACOB AND MONOD COINED THE TERM OPERON TO COVER ALL THE DIFFERENT GENES INVOLVED IN ONE FUNCTION OF THE CELL. THE FIRST OPERON TO BE DETAILED WAS OF COURSE THE LAC OPERON, IN WHICH ALL GENES WERE INVOLVED WITH THE METABOLISM OF LACTOSE. IT WAS NECESSARY TO COIN A NEW TERM BECAUSE AS WE MENTIONED BEFORE, DNA ESSENTIAL FOR THE EXPRESSION OF A FUNCTION THAT IS NOT TRANSCRIBED WAS DISCOVERED FOR THE FIRST TIME. THEREFORE THE TERM GENE, AS IT IS TRADITIONALLY USED, IS INSUFFICIENT TO DISTINGUISH REGULATORY SITES FROM STRUCTURAL GENES.

The concept of operons as first illustrated by the lac system was a very significant step toward our understanding of the regulation of gene action, but as we shall see, this concept is by no means universally applicable. In 1965, Jacob and Monod, along with A. Lwoff, who had collaborated in many landmark studies on viral and bacterial genetics, received the Nobel Prize for their outstanding contributions.

Recently, research on the lac operon has been directed toward detailing the fine structure of the operon. As was the case with DNA or any cell organelle involved in protein synthesis, we must come to understand structure before we can understand the mechanism by which various functions of the lac operon are accomplished.

Isolation of the Lac Operon DNA

In order to study the fine structure of genes, it is necessary to isolate the genes. In 1969, Shapiro and associates reported the isolation of pure lac operon DNA. By an elegant series of experiments using specific transducing phages, specific genes in pure form were obtained for the first time (Fig. 11-9).

Without going into detail, the result of many experiments with various *lac* mutations and phages was the isolation of two strains of phages carrying lac operon genes in opposing sequence. One phage, λplac5 carried the sequence *y z o p i a*; the other, ϕ80plac carried *i p o z y a* (Fig. 11-10).

The DNA of the phages was subjected to strand separation techniques. The two strands in each phage could be distinguished because in each phage one strand has more *G* and *C* residues and is therefore heavier. The heavy strands of ϕ80plac and λplac5 were then isolated and mixed for reannealment of complementary base regions. Since the only complementary regions were those containing the *lac* genes, these regions formed a duplex molecule by base-pairing (Fig. 11-11). The rest of the strands remained unpaired, since their base sequences were not complementary.

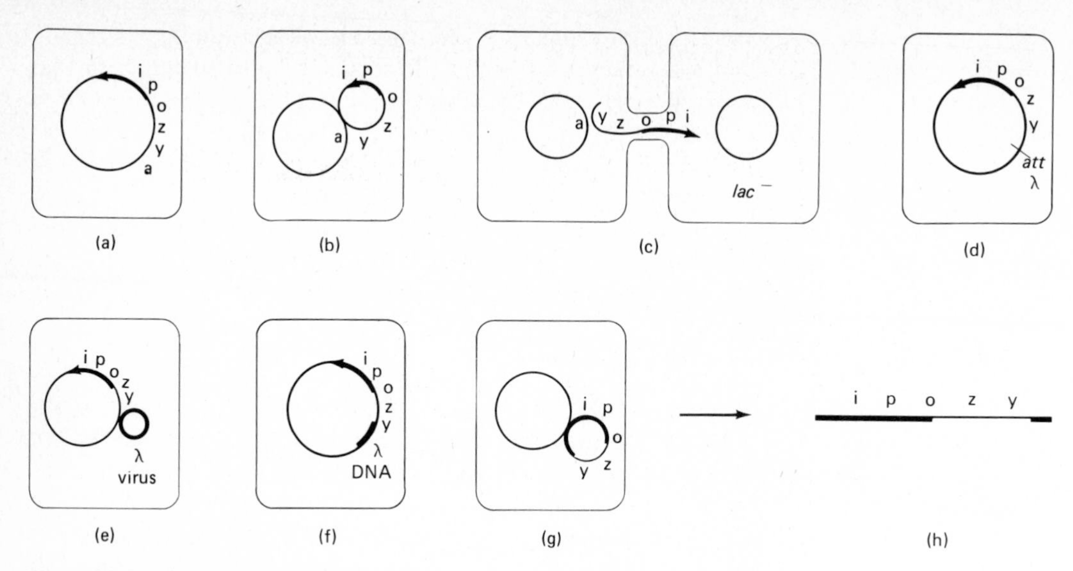

Figure 11-9. Abbreviated steps in the isolation of viruses carrying genes of the lac operon. (From Shapiro *et al.*, 1969). (a) Hfr cell containing F particle integrated into DNA adjacent to lac operon. (b) Excision of F′ containing genes *i–y*. (c) Sexduction of *lac* genes to *lac*-cell. (d) Integration of F′ and *lac* genes adjacent to attachment site of virus (in this case λ). (e) Infection of cell by virus. (f) Integration of prophage. (g) Abnormal excision of virus so that it contains *lac* genes. (h) Isolation of DNA from these viruses.

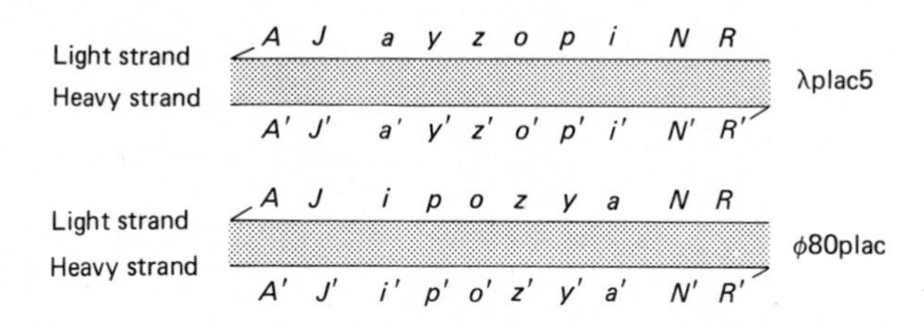

Figure 11-10. Diagram of opposite orientation of *lac* genes on light and heavy strands of λplac5 and φ80plac viruses. (From Shapiro *et al.*, 1969.)

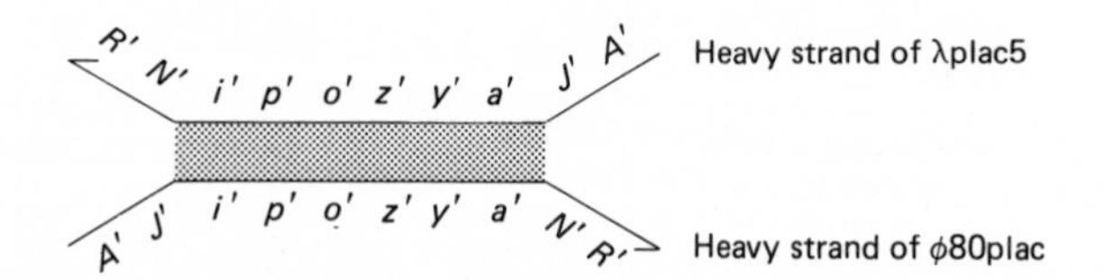

Figure 11-11. Reannealment of heavy strands from λplac5 and φ80plac resulted in base pairing of *lac* gene DNA. The single-stranded ends are digested by nuclease, leaving "pure" lac operon as the duplex molecule.

Figure 11-12 shows an electron micrograph of such a molecule. The single-stranded regions are then enzymatically digested, resulting in a molecule of "pure" *lac* operon genes.

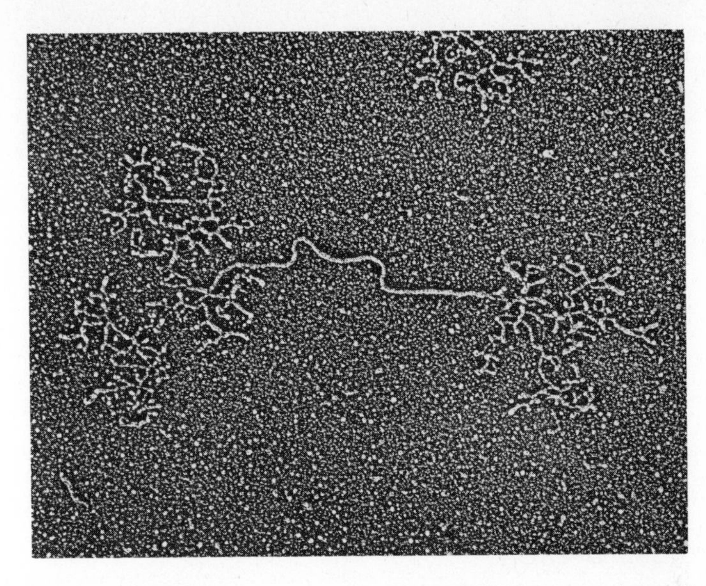

Figure 11-12. Electron micrograph of molecule resulting from reannealment of the heavy strands of λplac5 and ϕ80plac. Highly coiled regions at the ends are single DNA strands which have not annealed because of noncomplementary base pairing. (From Shapiro *et al.*, 1969. *Nature* 224: 772; photograph courtesy James Shapiro; electron micrograph by L. A. MacHattie, University of Chicago.)

This research was obviously quite complex, and required large numbers of organisms and many hours of effort to isolate the specific transducing phages with the proper sequences of lac^+ genes. Furthermore, only very small quantities of lac^+ DNA could be produced. More recently different techniques have been used to obtain lac^+ genes in larger quantities.

One of the more recent methods is to digest bacterial chromosomes with restriction endonucleases. We mentioned earlier that these endonucleases are capable of cleaving specific regions of DNA. Following fragmentation, *lac* genes can be isolated by techniques such as use of labeled repressor molecules which will bind only to those fragments containing the operator gene (Gilbert and Maxam 1973).

Yet another approach is to chemically synthesize a functional gene by using techniques developed by Khorana in 1968. (The chemistry involved is too sophisticated for our discussion.) The artificially made *lac* operator genes show binding properties to repressor molecules similar to those of natural operator sequences obtained by endonuclease digestion (Lin *et al.* 1976).

These techniques are currently being refined, and promise to afford scientists much larger quantities of specific DNA sequences for study. The sequence of bases in the promoter-operator regions of the lac operon as determined by these various approaches is shown in Figure 11-13.

Note that there are regions in both the CAP site of the promoter and in the operator gene which are areas of *rotational symmetry*. This term refers to sequences of nucleotides in DNA which are the same reading from the 3′ to the 5′ direction on both strands; another term for such regions is *palindrome*. The function of these

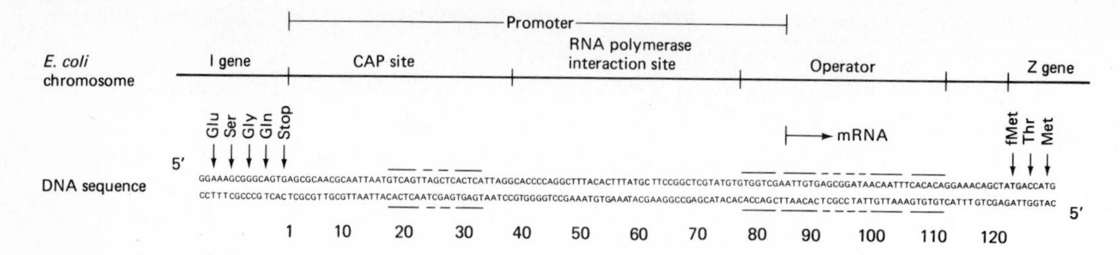

Figure 11-13. DNA sequence of the promoter-operator region of the lac operon. (From Dickson *et al.*, 1975. *Science* 187 : 32.)

regions has not yet been determined, but Dickson *et al.* (1975) conjecture that they may be recognition sites for protein-binding (such as CAP or repressor). Studies on the regulation of the *lac* operon genes continue at the present time.

THE ARABINOSE OPERON

In the late 1960s, genes in *E. coli* determining enzymes involved in the metabolism of another sugar, arabinose, were also found to be under *coordinate regulation.* This term refers to the concomitant control over transcription of several structural genes. Again, three structural genes were found to code for the enzymes that convert arabinose to D-xylulase 5-phosphate, a simpler pentose sugar. The genes are known as *araD*, *araA*, and *araB*, and determine the synthesis of arabinose epimerase, arabinose isomerase, and arabinose kinase, respectively (Englesberg *et al.* 1965).

As with the lac operon, it was found that *ara* genes are not active in the absence of arabinose. Addition of arabinose to the medium, in the absence of glucose, results in the induction of ara enzyme formation. As you might now expect, regulatory genes as well as structural genes were discovered (Schleif 1971).

Genes of the Ara Operon

Figure 11-14 illustrates the arabinose operon. One of the differences between this system and the lac system is the existence of two structural genes which are not in

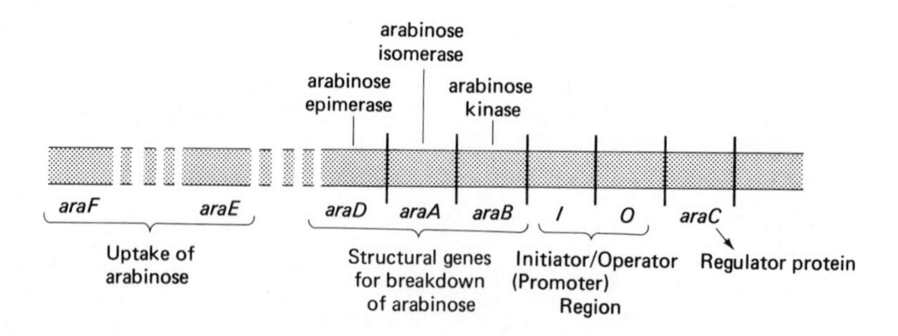

Figure 11-14. Genes of the arabinose operon.

the same cluster as *araD*, *araA*, and *araB*. These genes, *araF* and *araE*, which determine enzymes involved in the uptake of arabinose by the cell, are nonetheless under the same control as the other structural genes. How this is accomplished is not yet understood.

Another difference is that the promoter, or initiator, region is situated next to the structural genes, *araD*, *araA*, and *araB*, and the operator region is next to the promoter region. On the other side of the operator from the promoter is *araC*, the locus of the regulator gene. A promoter site for *araC* is located between *araC* and the operator for *araBAD*.

Regulation of Ara Genes

Regulation of transcription of *ara* structural genes has been found to be under dual control, very similar to the case with the lac operon. A repressor protein known as P1 is the product of *araC*. In the absence of arabinose, P1 binds to the operator site and transcription is repressed—again, negative control over transcription. In the presence of arabinose, P1 is converted to P2 by interaction with arabinose. P2 then acts as an inducer by interacting with the initiator region to initiate transcription. Thus, arabinose serves as an effector for this inducible operon (Fig. 11-15).

A further similarity to the lac operon was discovered when mutants for arabinose enzyme synthesis that are noninducible were found to respond to the presence

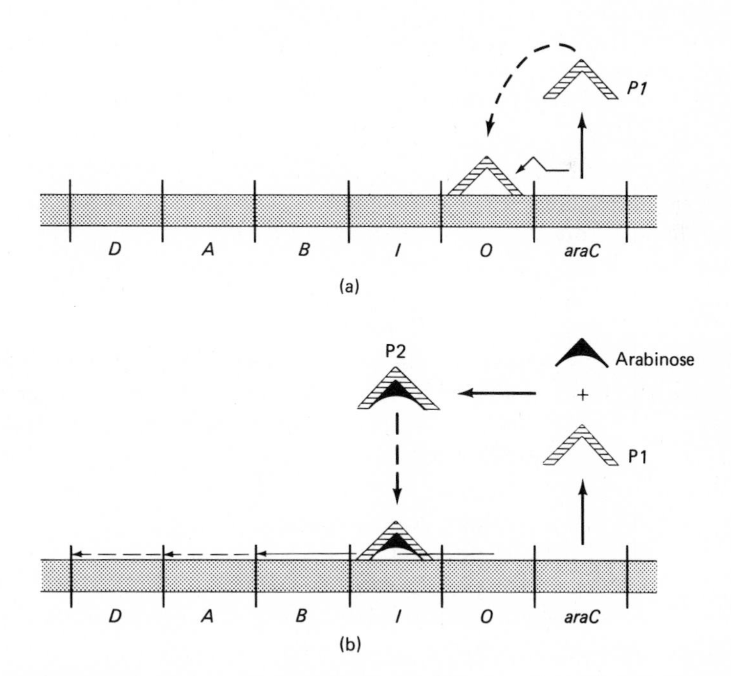

Figure 11-15. (a) The protein, P1, determined by *araC* acts as a repressor by binding to the operator and preventing transcription of *ara* structural genes. (b) In the presence of arabinose, P1 is converted to P2 which interacts with the initiator (promoter) site to allow transcription of the structural genes.

of cyclic AMP by synthesizing ara enzymes (Perlman and Pastan 1969). This led to the identification of a cAMP–CAP binding region in the arabinose promoter region. As in the lac operon, CAP must be bound to the promoter in order for RNA polymerase to transcribe the structural genes. These *ara*⁻ mutants were actually mutant for a gene, *cya*, which codes for the enzyme adenylate cyclase, that converts ATP to cAMP. In the absence of cAMP, the CAP is unable to interact with the promoter, and hence the mutants were super-repressed.

Interaction of Ara Promoters

A unique aspect of the ara operon has recently been reported. Hirsh and Schleif (1977) found that separate promoter sites exist for *araC* and *araBAD*, each located in the region contiguous with the respective genes. Figure 11-16 illustrates the positions of

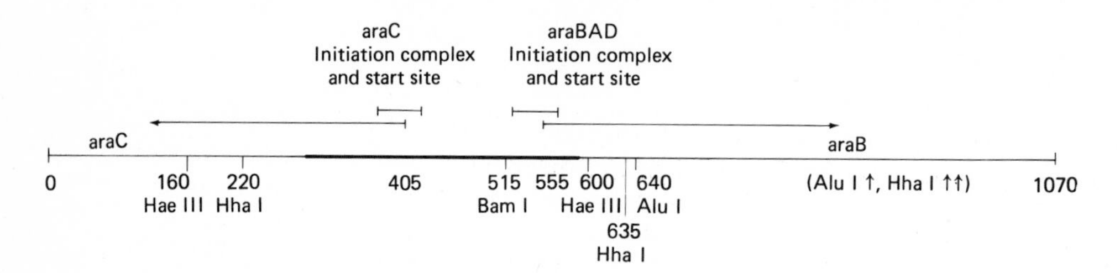

Figure 11-16. Regulatory region of the ara operon. Hae, Bam, Hha, Alu are abbreviations of endonucleases used to fragment the DNA. (From Hirsh and Schlief, 1977. *Cell* 11 : 545–550, Fig. 2; copyright held by M.I.T. Press.)

the two promoter regions. As you can see, they are separated by approximately 100 base pairs, including the operator region. Furthermore, the polarity of transcription is in opposite directions in *araC* from *araBAD* genes. Since RNA polymerase functions to produce RNA molecules in the 5′ to 3′ direction, transcription of *araC* and *araBAD* must be off opposite strands of the DNA.

Although the two promoters are some distance apart, Hirsh and Schleif made the interesting observation that the two regions appear to "communicate." When RNA polymerase is bound to one promoter, the other promoter is inhibited from binding with RNA polymerase at the same time. Evidence to show that this is controlled at the level of DNA resulted from experiments that cleaved the DNA between the two promoter sites, thus separating them. Once separated, both promoters can bind with RNA polymerase at the same time. The *araBAD* promoter requires CAP–cAMP; *araC* promoter does not.

Analysis of the phenomena associated with both the lac and ara operons continues today. Both are good illustrations of how the presence of metabolites in the cytoplasm of cells activates genes, leading to synthesis of gene products. However, as we mentioned before, systems of gene regulation are not universal in their mechanisms. In the next few sections, we shall explore systems of gene regulation in prokaryotes which are somewhat different from either the ara or the lac operons.

OPERATOR-ATTENUATOR SYSTEMS OF REGULATION

As studies on the lac operon contributed to greater understanding of the system of gene regulation, geneticists were optimistic that the system would apply to all genes. After all, the genetic code, replication, transcription, translation, and a number of other genetic phenomena had been found to be universal in their essential details. However, further studies of gene regulation have indicated that no such universality exists. In the following sections we shall discuss operons in which dual control systems regulate transcription of structural genes.

The Tryptophan Operon

In the 1960s, studies of genetic systems determining enzymes for the biosythesis of some amino acids in *E. coli* indicated that regulation of gene action involved a form of *feedback inhibition*. This term refers to the biochemical phenomenon in which production of a substance eventually results in the inhibition of the biochemical reactions leading to the synthesis of that substance. This inhibition generally is caused by an interaction between the end product and an enzyme determining one of the reactions. Presence of tryptophan in *E. coli* cells resulted in repression of its synthesis, a phenomenon initially interpreted as feedback inhibition.

Genes of the trp operon. F. Imamoto and his associates (1965) showed that enzymes involved in tryptophan synthesis are produced by the cell from *polycistronic messenger RNA* molecules. Such mRNA molecules are transcribed as one continuous molecule from a number of contiguous structural genes. There are known to be five tryptophan structural genes, designated *trpE*, *trpD*, *trpC*, *trpB*, and *trpA*, linked in that order. The *trp* operon genes are shown in Figure 11-17.

Early studies showed that the *trp* structural genes are controlled by a promoter and operator region linked closely to them (Morse and Yanofsky 1969). Studies of

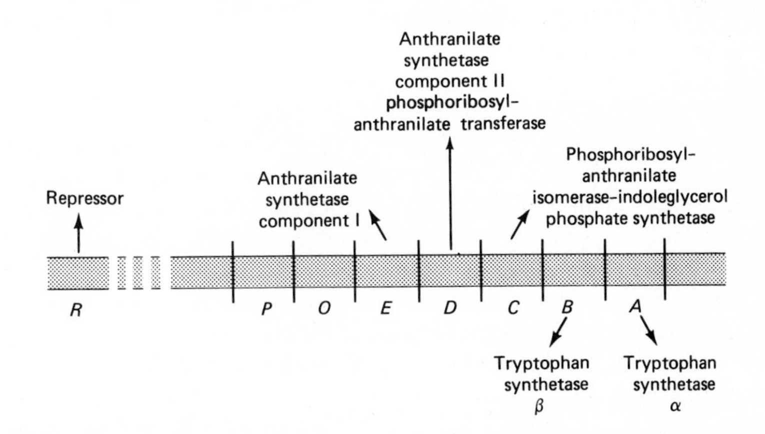

Figure 11-17. Genes of the tryptophan operon as determined by Imamoto *et al.*, 1965, and Morse and Yanofsky, 1969.

mutants which produced the enzymes constitutively even in the presence of tryptophan showed that a regulatory gene, *trpR*, determines a repressor molecule. Normally the repressor interacts with tryptophan, and the complex then binds to the operator site, shutting down transcription of the structural genes. (This, then, is not feedback inhibition as we defined it above, but transcription repression.) The *trpR* gene is not, however, closely linked to the cluster of structural genes.

It is important to digress briefly here to point out a possibly misleading use of terms. Microbial geneticists frequently refer to two genes as being *linked* or *unlinked*. Remember, though, that since bacteria and viruses have but one "chromosome," all their genes must be linked. The term *unlinked*, in microbial genetics, simply means that the two genes are not situated next to one another in a contiguous manner.

To return to the trp operon, the above-mentioned studies all indicated a negative form of control. In other words, the operon is shut off by a repressor, and derepression results from the absence of the amino acid. There is a promoter region next to the operator, as in the lac operon, which serves as a binding site for RNA polymerase. More recent studies have revealed a number of novel features about the regulatory mechanisms of the tryptophan operon.

Novel Features of the Trp Operon

Figure 11-18 shows the various regulatory and structural genes of the trp operon as it has been recently detailed. The gene *trpR* is not included in the diagram. First note that there are actually two promoter sites, the principal one in what is now called the

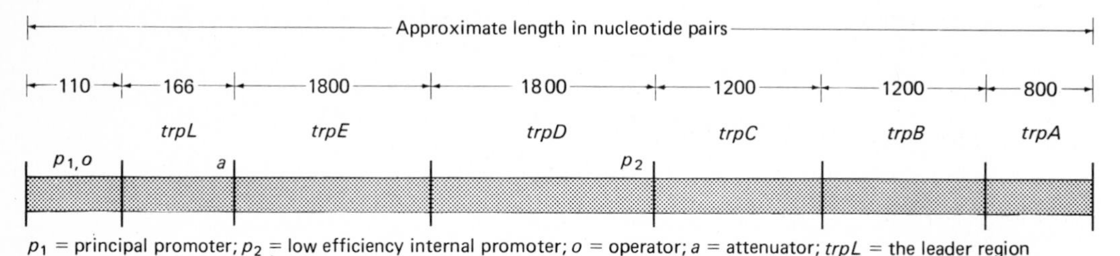

Figure 11-18. Structural features and regulatory sites of the tryptophan operon of *E. coli.* (From C. Yanofsky, 1976.)

promoter-operator region, and a minor promoter in the distal third of the *trpD* gene. We shall not dwell on the minor promoter, as its biological function is not known other than that it has a low level of interaction with RNA polymerase. It has been found that the repressor-tryptophan molecule represses transcription by actually competing with RNA polymerase for binding at the *p-o* region.

The original concept of the promoter-operator region of the trp operon being similar to that in the lac operon became subject to revision when Jackson and Yanofsky reported in 1973 on the effects of a number of deletion mutations that occurred in the DNA just before *trpE*, the first structural gene. These mutations resulted in an increased expression of the structural genes, even though it could be shown that repressor binding to the operator site was unaffected.

Leader region. It was then suggested by Yanofsky and his coworkers, and since supported with experimental evidence (Bertrand *et al.* 1975), that the region preceding *trpE* has a regulatory function that is separate from that of the promoter–operator region. This portion of the *p–o* DNA is now designated the *leader region,* and given the symbol *trpL.* Studies have shown that even when repression of *trp* structural genes exists, RNA transcripts which correspond to the first 130–140 nucleotides of *trpL* can be found. This indicates that transcription of *trpL* does escape the effects of the repressor, but that a second mechanism must exist to terminate transcription just before it reaches the structural genes.

Attenuator region. The mechanism that terminates transcription has been assigned to the 30 nucleotides of *trpL* just preceding *trpE.* This sequence of bases has been designated the *attenuator region,* or *a.* Recent studies (Zurawski *et al.* 1978) have shown that the leader region contains tandem codons for tryptophan. When tryptophan is available to the cell, the leader transcript is translated past the codons for tryptophan, at which point the ribosome already translating the mRNA encounters a stop codon in the leader sequence (Fig. 11-19a). Translation then terminates and the RNA polymerase which precedes the ribosome ceases transcription.

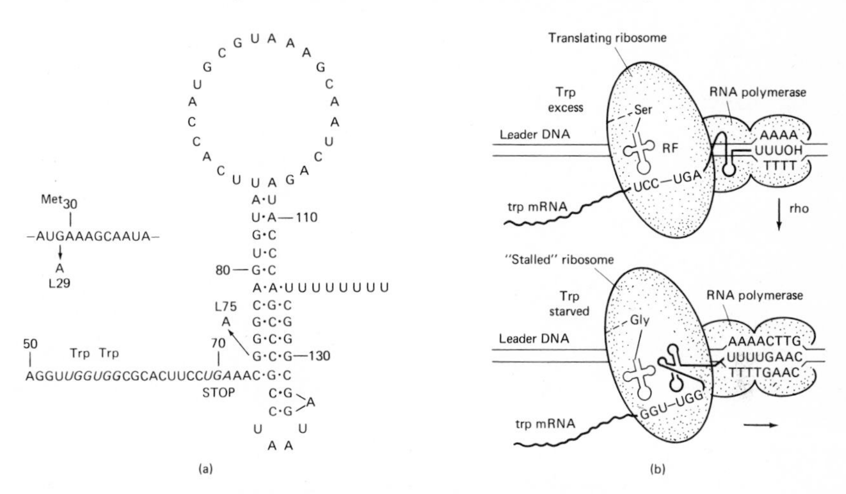

Figure 11-19. (a) The nucleotide sequence of two portions of the leader transcript from the *E. coli trp* operon. Note the tandem codons for tryptophan and the stop codon. (b) A proposed model for the regulation of transcription termination at the attenuator of the *E. coli trp* operon. Under conditions of tryptophan excess the first ribosome translating *trp* leader RNA follows closely behind the transcribing polymerase and synthesizes the entire *trp* leader peptide. At the stage shown the ribosome is reading the translation stop codon, UGA, and the polypeptide release factor (RF) is about to release the polypeptide. The second RNA stem and loop has formed, signaling the polymerase to cease synthesis. Rho factor (30) then will promote dissociation of the termination complex. Under conditions of tryptophan starvation the first ribosome stalls at the trp codons. In this configuration, interactions between the trp codons or their immediately distal region and a more distal segment of the transcript result in the formation of an alternative RNA secondary structure. The existence of this secondary structure prevents transcription termination, allowing the polymerase to continue transcription into the first structural gene, *trpE,* and beyond. (From: G. Zurawski, *et al.,* 1978. *Proc. Nat. Acad. Sci. U.S.* 75 (12) 5989, Fig. 1.)

On the other hand, absence of tryptophan in the cell is believed to cause the ribosome to "stall" in its translation of the leader region of the attenuator at the tryptophan codon. A second structure of the mRNA is formed which allows the RNA polymerase to proceed in its transcription of the structural gene. Figure 11-19b illustrates the proposed model of regulation of transcription of the tryp operon.

Two regions in the leader nucleotide sequence have been found to have special affinity for ribosome binding in vitro, possibly indicating that the in vivo function of RNA transcripts of the leader region may be to bind with ribosomes. It is thought that such binding with ribosomes could serve to protect the mRNA from degradation by cellular nucleases. In addition, the attenuator can be seen to possess a GC-rich sequence followed by an AT-rich sequence, each of which shows some symmetry of base sequence. How these areas function in gene regulation remains to be determined.

Let us turn our attention now to an operon which has been found in the genome of the bacterium *Salmonella typhimurium*. The control of this operon has some similarities to the operator–attenuator system of gene control of the trp operon. We shall then discuss briefly a proposed relationship between attenuators and cellular activity under deprivation or starvation conditions.

THE HISTIDINE OPERON IN SALMONELLA

In 1963, Ames and Hartman reported that genes determining synthesis of the amino acid histidine in *Salmonella typhimurium* are clustered in one region of the chromosome. The transcription of these genes is controlled by an operator region which interacts with a repressor substance (Fig. 11-20).

Early studies seemed to point to either histidine or an enzyme, phosphoribosyl transferase (determined by the *hisG* gene), as the repressor. However, regulatory mutants were found which mapped outside the his operon. Furthermore, some *hisG* mutations did not affect regulation. These data indicated that other factors controlled genes of the his operon.

Multiple regulatory genes. No fewer than five different regulatory genes determining the synthesis of repressor substances have now been found, all located outside the his operon, scattered through the DNA of *Salmonella*. Each of these genes was found to determine either histidyl-tRNA or histidyl-tRNA synthetase (Brenner and Ames 1972). Furthermore, it has been reported (Singer and Smith 1972) that $tRNA^{his}$ is effective as a repressor only if the modified base, pseudouridine, is present in the anticodon loop of the tRNA molecules.

Attenuator in the his operon. In 1975, the effects of a deletion mutation in the *hisO* gene were reported by Artz and Broach to indicate a possible operator-attenator system of control. The mutation increases transcription of *his* genes, even though there is repressor interaction with part of the operator region. Thus, a dual control over transcription, operator-repressor and attenuator, has been postulated for the his operon. The attenuator region has been proposed to be the region of the *hisO* gene adjacent to *hisG*.

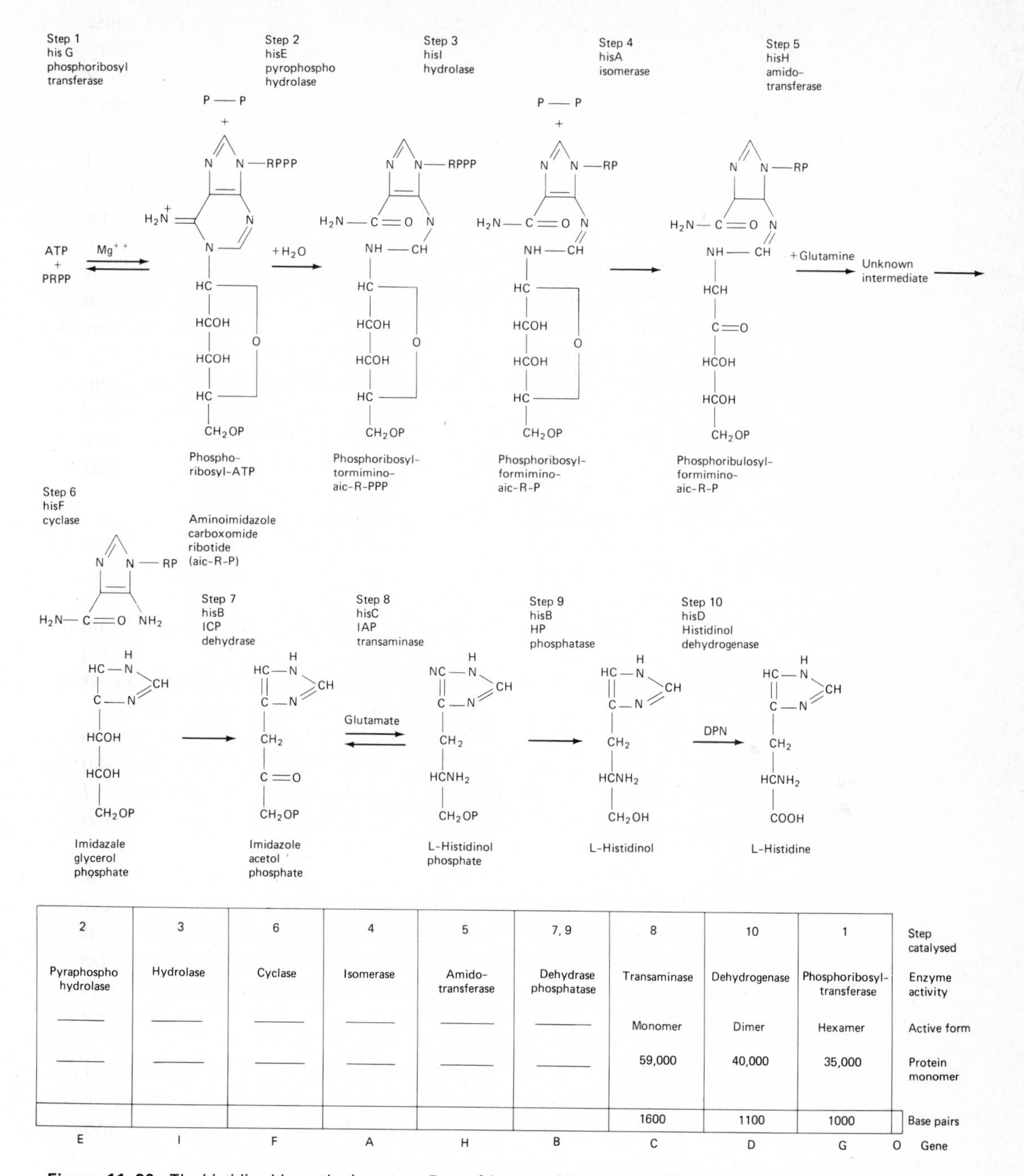

Step catalysed	2	3	6	4	5	7, 9	8	10	1	
Enzyme activity	Pyraphospho hydrolase	Hydrolase	Cyclase	Isomerase	Amido-transferase	Dehydrase phosphatase	Transaminase	Dehydrogenase	Phosphoribosyl-transferase	
Active form	———	———	———	———	———	———	Monomer	Dimer	Hexamer	
Protein monomer	———	———	———	———	———	———	59,000	40,000	35,000	
Base pairs							1600	1100	1000	
Gene	E	I	F	A	H	B	C	D	G	O

Figure 11-20. The histidine biosynthetic system. Data of Ames and Hartman (1963), Roth *et al.* (1966). (From B. Lewin, 1977. *Gene Expression* 1 : 318–319. John Wiley and Sons Ltd., London; copyright 1974, B. Lewin. Reprinted by permission of John Wiley & Sons Ltd.)

Possible Role of the Attenuator

A clue to the possible function of the attenuator regions of the trp and his operons was found in studies of cells grown in media without amino acids. In such amino-acid-poor media, tRNA synthesis ceases. The control over this is unclear, but of interest here is the fact that such deprived cells show roughly a fivefold increase in the amount of tryptophan mRNA over that found in cells in rich medium.

Concomitant with the decrease in tRNA molecules in a tryptophan-deprived or histidine-deprived cell is the appearance of two unusual nucleotides, guanosine tetraphosphate (ppGpp, with one diphosphate at the 5′ end, and one at the 3′ position) and guanosine pentophosphate (ppGppp, with the triphosphate group at the 5′ end of the molecule) (see reference to Cashel and Gallant 1969). It is thought that ppGpp may in some way relieve transcription termination by the attenuator in the trp and his operons. This would lead to increased synthesis of amino acids, which could compensate for the deficient medium. Such a substance, if it indeed acted to increase transcription, would be considered positive control over these operons.

Furthermore, ppGpp has been found to accumulate only in the presence of normally functioning ribosomes. Addition of chloremphenicol, which inhibits ribosome function, leads to rapid decay of ppGpp. Coincidentally, F. Imamoto reported that antibiotics which affect polysome formation decrease trp mRNA transcription (Imamoto 1973), and suggested that the presence of normal translational machinery may be a regulatory factor in the transcription of *trp* genes.

At this point in time, studies of systems involving the biosynthesis of amino acids in several bacterial species all indicate an attenuator region which plays a significant role in transcription termination. It may be proposed, then, that the presence of an attenuator could be a general characteristic of the regulation of biosynthetic systems.

REGULATION IN LAMBDA BACTERIOPHAGE

All the systems of gene regulation which we have thus far explored deal with individual cellular functions, namely, the breakdown or synthesis of specific substances. The complexity inherent in these processes must be magnified many degrees if one contemplates regulation and coordination of *all* genetically controlled cellular processes. Because of this complexity, we do not yet have models for regulation at the cellular level. Therefore, we turn to simpler forms of life, such as viruses. By attempting to relate the various life cycles of the bacteriophage lambda to specific gene activity and regulation, we hope to gain some insight into the coordination of cellular activity.

To review briefly, bacteriophages were found to have two major routes of activity following infection of a host cell (see again p. 235). The virus can enter the lytic, or vegetative, state during which its nucleic acids are replicated, viral proteins are synthesized, and new viruses released with the death of the cell. Alternatively, the virus can enter the dormant, or temperate, stage, during which it remains inactive in the cytoplasm, or becomes integrated into the host genome as a prophage. Viral particles can be induced to excise from the host genome and enter the lytic cycle under the influence of agents such as UV irradation (see again Fig. 8-5).

Bacteriophage Lambda

Lambda was discovered by Esther Lederberg in 1951, as a prophage in *E. coli* K12 strain. Later studies proved it to be a specialized transducing phage, indicating that it attaches during lysogenization to a specific site (now designated *att*) in the *E. coli* DNA. Recent reports show the fine structure of the attachment sites on both the host and the phage to have an identical sequence of 15 base pairs (Landy and Ross 1977). It has been hypothesized that these are the areas in which recombination between the phage and bacterial DNA must occur, leading to integration of the λDNA (see again Fig. 8-24).

What is the genetic control over the life cycle a virus will enter when it infects a cell? Studies of the bacteriophage lambda have revealed a number of different regulatory pathways which lead to the vegetative or the temperate phase. As we shall point out, these studies have also revealed unique aspects of transcription in lambda viruses, such as transcription in opposite directions. This feature is similar to that of the ara operon we discussed above; however, researchers have also found that overlapping transcription occurs in opposite directions of the *same* portion of the DNA. The situation is complex, but remember that we are dealing here with the entire scope of life functions of the virus, not just one process.

Genes Involved in Lysis

The genome of λ phages is a double-stranded linear molecule of DNA of 32×10^6 daltons in molecular weight (for a review, see Herskowitz 1973). Estimates propose enough DNA for the synthesis of about 50 polypeptides. Following the infection of a cell, the linear DNA of lambda viruses circularizes because it contains single-stranded cohesive ends which come together because of complementary base-pairing. The first genes to be transcribed (by host RNA polymerase) are known as gene *N* and gene *cro* (sometimes also referred to as *tof*). The promoter-operator region which controls the transcription of *N* lies to its right on the linear map of the λDNA. Since transcription of the *N* locus proceeds in a leftward direction, the promoter for *N* is designated *pL*, and the operator, *oL*. The promoter-operator region controlling *cro* is to its left, and transcription occurs in a rightward direction; thus, these genes are known as *pR* and *oR*. Accordingly, the DNA strand which serves as the "sense" strand for *N* gene transcription is referred to as the *l* strand, and the "sense" strand for *cro* is the *r* strand (Fig. 11-21).

The product of *N* is apparently needed for the expression of two sets of genes (operons). How the *N* protein functions is not known, as it is very unstable and has not been isolated. One set is the group including *cIII* through *int*, which are transcribed in the same direction as gene *N* (Fig. 11-22). These genes are involved in the process of lysogeny, and their activity is repressed by the product of *cro* during the lytic life cycle.

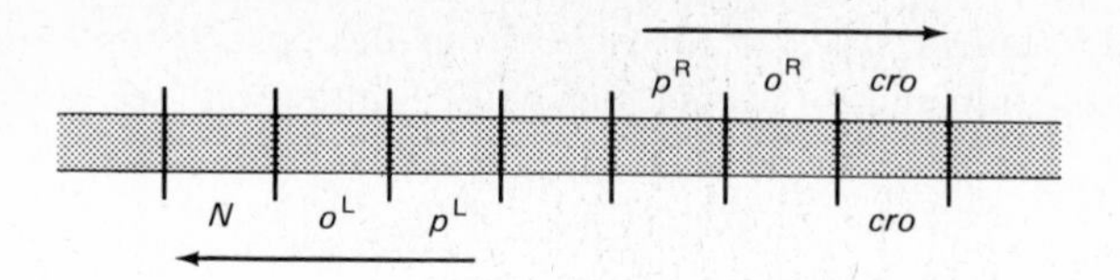

Figure 11-21. Genes expressed immediately following infection of a cell by λ viruses are *N* and *cro,* each controlled by separate operator/promoter regions. Transcription is in different directions on opposite DNA strands.

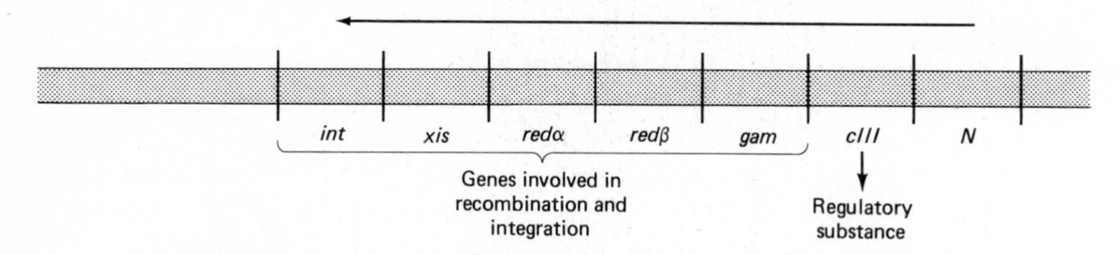

Figure 11-22. Diagram of genes adjacent to *N* gene which are expressed in the presence of *N* gene product.

The second set of genes dependent on *N* gene product for their expression is the group containing *O* and *P*. These genes have been found necessary for phage DNA replication, which is of course an inherent part of lysis. Recent reports (Moore *et al.* 1977) indicate that genes *O* and *P* determine the synthesis of an *O* protein, which functions as a specific initiator of replication by interacting with the replicon, *ori*, of lambda. Mapping studies have shown that the λ replicon lies in an area between *pR* and the middle of gene *O* (Furth *et al.* 1977).

The onset of replication then enhances transcription of gene *Q*. The product of *Q* has been found to be essential to promote transcription of genes determining lysis (*S* and *R*), as well as head and tail protein structural genes and packaging genes (*A–J*). Transcription of this set of genes occurs in a direction opposite to that of *N–int*. The complete genetic map of lambda is given in Fig. 11-23.

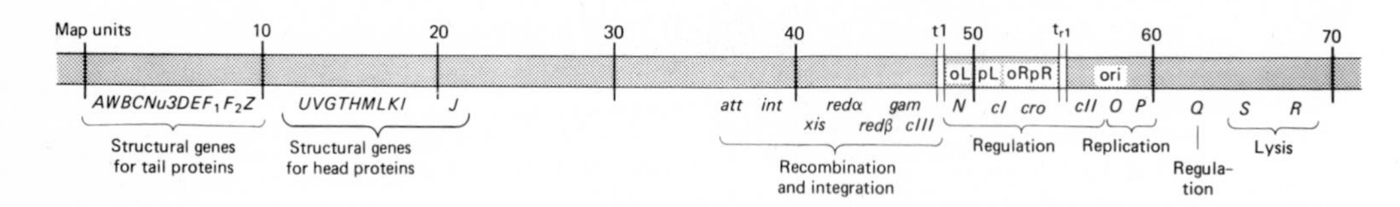

Figure 11-23. The complete genome of bacteriophage λ.

Since genes *N* and *cro* are transcribed immediately after infection of the host cell, they are among the genes classified as *early* genes. The two sets of genes stimulated by *N* gene product are considered *delayed early* genes, and the genes for lysis, head and tail proteins, and packaging are called *late* genes. The development of new phage particles during lysis thus depends on an orderly sequential expression of λ genes.

Genes Involved in Lysogeny

In the temperate phase of the lambda life cycle, transcription beyond gene *N* leads to expression of genes such as *int*, which determines the ability of the viral DNA to integrate into the host genome at the attachment site. At the same time, genes *cII* and *cIII* are also transcribed (see again Fig. 11-23). These two genes promote the efficiency of the expression of a third gene, *cI*. The product of gene *cI* acts as a general repressor of all lambda genes (Ptashne 1967) by interacting with both operators, *oR* and *oL*, to shut off transcription of all lambda genes, including those that would lead to lysis.

Studies on the fine structure of the operators of lambda, made possible by fragmentation with restriction endonucleases and isolation methods to be discussed in Chapter 14, have shown that there exist similar base sequences in both *oR* and *oL*. There are three of these sequences in *oR*, called oR^1, oR^2, oR^3, and three in *oL*, oL^1, oL^2, oL^3 (Fig. 11-24). Each of these sites is a repressor binding site. It has been speculated that three binding sites constitute stricter control over transcription, and decrease the possibility of inadvertent λ gene expression, which would lead to lysis and cell death (Maniatis and Ptashne 1976).

Control of the *cI* gene is unique, in that there appear to be two different promoter sites for the *cI* locus. One promoter initiates transcription of *cI* and is known as *pre* (*p*romoter for *r*epressor *e*stablishment), and is located to the right of the *cro* locus. A second promoter which maps between the *cro* and *cI* loci (possibly overlapping with *oR*) is responsible for the continuance of repressor production. This promoter, known as *prm* (for *p*romoter of *r*epressor *m*aintenance) appears to be

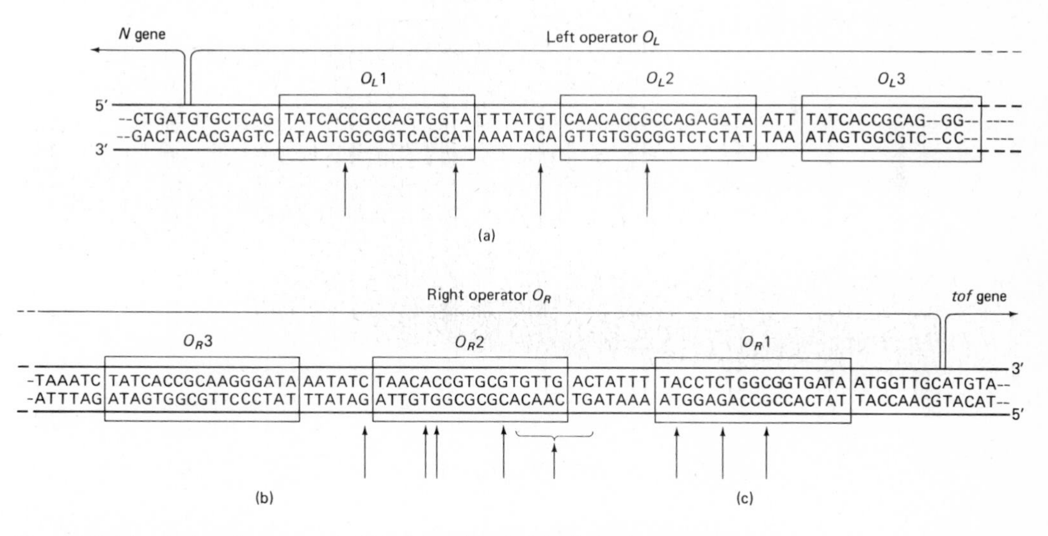

Figure 11-24. Nucleotide sequence of left (above) and right (below) operators of lambda phage. Rectangles mark areas of presumed binding to repressor. Within each site there exists considerable symmetry on either side of the central base pair. Arrows point to regions of mutation in different lambda strains. Each mutation caused a loss of affinity of repressor to that site. (From T. Maniatis, and M. Ptashne, 1976. *Sci. Am.* 234 : 64–76 ; copyright 1976 by Scientific American. All rights reserved.)

activated by the repressor protein itself. Thus *cI* is responsible for the establishment *and* maintenance of lysogeny.

Note that the area of DNA containing *pre* and *cro* is one in which the direction of movement of RNA polymerase on the two strands overlaps in opposite directions (Fig. 11-25). There is some evidence that transcription in different directions may interfere with the efficiency of either transcription or translation (perhaps by the formation of mRNA molecules which, having complementary sequences, may form duplex molecules) (Herskowitz 1973).

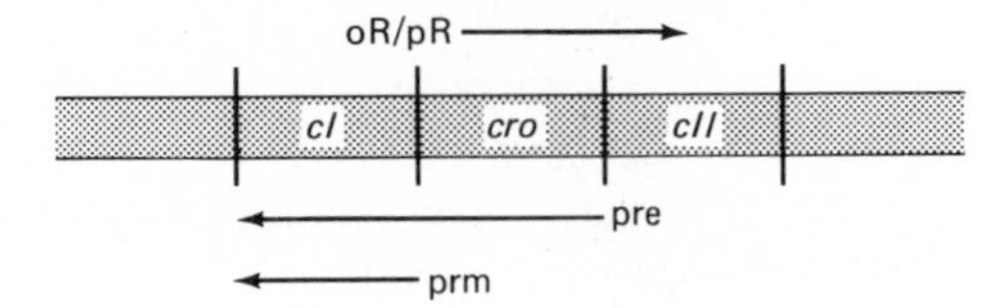

Figure 11-25. Three different promoter regions of λ map in overlapping regions near the *cro* gene. Since the direction of transcription of pR is opposite to pre and prm, both strands of the DNA in this region are transcribed. (From L. Reichardt and A. D. Kaiser, 1971. *DNAS* 68: 2185–2189.)

Lysis vs. Lysogeny

In summary, when a lambda virus enters a host cell, there is immediate transcription of the *N* and *cro* genes. Thus, in the beginning, genes are expressed which are involved in both lysis (*cro, O, P, Q*) and lysogeny (*int, cII, cIII*). Yet these are, by definition, mutually exclusive processes: The cell dies in lysis, and it lives in lysogeny. The factors which determine how far the cell proceeds in one or the other direction are not yet understood. Nor is induction fully understood, although clearly it must involve the activation of λ genes, and therefore the inactivation of repressor molecules.

For the normal functioning of the virus, then, there exist a number of regulators. Besides the operators and promoters, there is negative regulation (*cI* repressor), positive regulation (*N, cII, cIII, Q* gene products), and *autogenous* regulation. The latter can be defined as the ability of a substance to affect the control of its own production, and again, the *cI* repressor would illustrate this phenomenon.

Transcription of both strands of DNA, in some areas in an overlapping manner, is a feature of λ gene activity rarely seen in other genetic systems. To transcribe effectively in this way would be a very conservative use of DNA. Perhaps this is again compensation for the relatively minute amounts of genetic material found in viruses.

SIMILARITIES AND DIFFERENCES IN PROKARYOTIC SYSTEMS

The systems of gene regulation that we have discussed here are representative of control over gene activity in prokaryotes. As we said at the outset of this chapter, the systems share some similarities and also exhibit differences. There are promoter-initiator regions in all which interact with RNA polymerase to begin transcription. There are operator regions which repress transcription of structural genes by interaction with various repressors. Differences include separate control systems, such as the attenuator found in some systems but not in others. Many laboratories continue to explore gene regulation in systems too numerous even to mention here. Gene regulation is, in short, one of the most perplexing and challenging aspects of current genetic research.

What similarities exist between gene regulation in viruses and cellular forms of life, especially in eukaryotes, remains for the future to determine. It is logical to assume, however, that our understanding of regulation in genetically simpler systems is a necessary first step toward eventual understanding of the more complex genetic systems. In the next chapter we shall explore gene regulation in eukaryotic cells, including cancer, which is essentially a disease of regulation.

PROBLEMS

1. Define the following terms: positive regulation, negative regulation, autogenous regulation, promoter, operator, attenuator, repressor.

2. What would you expect to happen if a mutation of the promoter gene occurred in the lac operon? Explain.

3. Sexduction of an i^s mutant gene of the lac operon into a wild-type cell resulted in super-repression. Would you expect the i^s gene to be trans dominant or trans recessive? Cis dominant or cis recessive? Explain.

4. A super-repressed strain of *E. coli* was found to contain all wild-type genes for the lac operon. Give a possible explanation for this observation.

5. Summarize the effects on the lac and ara operons in *E. coli* if glucose is added to the culture medium.

6. Distinguish the differences between autogenous regulation and feedback inhibition.

7. What advantage would you expect the presence of transcription terminators to confer on a cell or virus?

8. Summarize the sequence of lambda gene expression in the process of lysis; of lysogeny.

9. What would be the effects of mutation in the *N* gene of phage lambda? of mutation in the *Q* gene?

10. Mutation in which gene would account for an increase in the expression of λ genes *cIII–int*? Explain.

11. Studies of λ mutations of *cI*, *cII*, and *cIII* genes yielded the following observations: cI^- mutants are essentially unable to lysogenize (less than 10^{-6} infected cells show lysogeny); cII^- mutants lysogenize at a frequency of 10^{-4}, and $cIII^-$ mutants lysogenize at a frequency of 10^{-2}. Lysogens of cII^- and $cIII^-$ viruses are stable. From these observations, what can you conclude about the roles of these three genes in lysogeny?

12. If *E. coli* cells are infected simultaneously with cII^- and $cIII^-$ viruses which are wild-type in all other genes, would you expect any lysogeny to occur? Explain.

REFERENCES

AMES, B. N., AND P. E. HARTMAN, 1963. "The Histidine Operon." *CSHSQB* 28: 349–356.

ARTZ, S. W., AND J. R. BROACH, 1975. "Histidine Regulation in *Salmonella Typhimurium:* An Activator–Attenuator Model of Gene Regulation." *Proc. Nat. Acad. Sci. U.S.* 72: 3453–7.

BECKWITH, J. R., 1970. "*Lac:* The Genetic System." In J. R. Beckwith, and D. Upser, eds. *The Lactose Operon*, pp. 5–26. Cold Spring Harbor Laboratory, New York.

BERTRAND, K., ET AL., 1975. "New Features of the Regulation of the Tryptophan Operon." *Science* 189: 22–26.

BRENNER, M., AND B. N. AMES, 1972. "Histidine Regulation in *S. Typhimurium.* IX. Histidine tRNA of the Regulatory Mutants." *J. Biol. Chem.* 247: 1080–1088.

CASHEL, M., AND J. GALLANT, 1969. "Two Compounds Implicated in the Function of the RC Gene of *E. coli.*" *Nature* 221: 838–841.

DICKSON, R. C., J. ABELSON, W. M. BARNES, and W. S. REGNIKOFF, 1975. "Genetic Regulation: The *Lac* Control Region." *Science* 187: 27–85.

ENGLESBERG, E., J. IRR, J. POWER, AND N. LEE, 1965. "Positive Control of Enzyme Synthesis by Gene *C* in the L-Arabinose System." *J. Bact.* 90: 946–957.

FURTH, M. E., F. R. BLATTNER, C. MCLEESTER, AND W. F. DOVE, 1977. "Genetic Structure of the Replication Origin of Bacteriophage Lambda." *Science* 198: 1046–1051.

GILBERT, W., AND B. MULLER-HILL, 1970. "The Lactose Repressor." In J. R. Beckwith, and D. Zypser, eds., *The Lactose Operon*, pp. 93–110. Cold Spring Harbor Laboratory, New York.

GILBERT, W., AND A. MAXAM, 1973. "The Nucleotide Sequence of the Lac Operator." *Proc. Nat. Acad. Sci. U. S.* 70: 3581.

HERSKOWITZ, I., 1973. "Control of Gene Expression in Bacteriophage Lambda." *Ann. Rev. Gen.* 7: 289–324.

HIRSH, J., AND R. SCHLEIF, 1977. "The *araC* Promoter: Transcription, Mapping and Interaction with the *araBAD* Promoter." *Cell* 11: 545–550.

HOZUMI, N., S. TONEGAWA, 1976. "Evidence for Somatic Rearrangement of Immunoglobulin Genes Coding for Variable and Constant Regions." *Proc. Nat. Acad. Sci. U.S.* 73 (10) 3628–32.

IMAMOTO, F., N. MORIKAWA, AND K. SATO, 1965. "On the Transcription of the Tryptophan Operon: III. Multicistronic Messenger RNA and Polarity for Transcription." *J. Mol. Biol.* 13: 169–182.

IMAMOTO, F., 1973. "Diversity of Regulation of Genetic Transcription." *J. Mol. Biol.* 74: 113–136.

IPPEN, K., J. H. MILLER, J. SCAIFE, AND J. BECKWITH, 1968. "New Controlling Element in the Lac Operon of *E. coli.*" *Nature* 217: 825–826.

JACKSON, E. N., AND C. YANOFSKY, 1973. "The Region Between the Operator and First Structural Gene of the Tryptophan Operon of *Escherichia Coli* May Have a Regulatory Function." *J. Mol. Biol.* 76: 89–101.

JACOB, F., AND J. MONOD, 1961. "Genetic Regulatory Mechanisms in the Synthesis of Proteins." *J. Mol. Biol.* 3: 318–356.

KHORANA, H. G., 1968. "Nucleic Acid Synthesis." *Pure Appl. Chem.* 17: 349–380.

LANDY, A., AND W. ROSS, 1977. "Viral Integration and Excision: Structure of the Lambda Att. Sites." *Science* 197: 1147–1160.

LEDERBERG, E. M., 1951. "Lysogenicity in *E. Coli* K-12." *Genetics* 36: 560.

LIN, S., ET AL., 1976. "The Interaction of Chemically Synthesized 21 Base Pair Lac Operator with the Lac Repressor." In D. P. Nierlich, W. J. Rutter, and C. Fred Fox, eds., *Molecular Mechanisms in the Control of Gene Expression*, pp. 143–158. Academic Press, New York.

MANIATIS, T., AND M. PTASHNE, 1976. "A DNA Operator–Repressor System." *Sci. Am.* 234: 64–76.

MOORE, D. D., K. DENNISTON-THOMPSON, M. E. FURTH, B. G. WILLIAMS, AND F. R. BLATTNER, 1977. "Construction of Chimeric Phages and Plasmids Containing the Origin of Replication of Bacteriophage Lambda." *Science* 198: 1041–1046.

MORSE, D. E., AND C. YANOFSKY, 1969. "Amber Mutants of the *trpR* Regulatory Gene." *J. Mol. Biol.* 44: 185–193.

PARDEE, A. B., F. JACOB, AND J. MONOD, 1959. "The Genetic Control and Cytoplasmic Expression of 'Inducibility' in the Synthesis of β-Galactosidase by *E. Coli.*" *J. Mol. Biol.* 1: 165–180.

PASTAN, I., AND S. ADHYA, 1976. "Cyclic Adenosine 5′ Monophosphate in *Escherichia coli.*" *Bact. Rev.* 40: 527–551.

PERLMAN, R. L., AND I. PASTAN, 1969. "Pleiotropic Deficiency of Carbohydrate Utilization in an Adenyl Cyclase Deficient Mutant of *Escherichia coli.*" *Biochem. Biophys. Res. Commun.* 37: 151–157.

PTASHNE, M., 1967. "Isolation of the λ Phage Repressor." *Proc. Nat. Acad. Sci. U.S.* 57: 306–312.

SCHLEIF, R., 1971. "L-arabinose operon messenger of *E. coli.* Its Inducibility and Translation Efficiency Relative to Lactose Operon Messenger." *J. Mol. Biol.* 61: 275–280.

SHAPIRO, J., L. MACHATTIE, L. ERON, G. SHLER, K. IPPEA, AND J. BECKWITH, 1969. "Isolation of Pure *lac* Operon DNA." *Nature* 224: 768–774.

SINGER, C. E., AND G. R. SMITH, 1972. "Nucleotide Sequence of Histidine tRNA." *J. Biol. Chem.* 247: 2989–3000.

STENT, G. S., 1971. *Molecular Genetics.* W. H. Freeman & Company Publishers, San Francisco.

YANOFSKY, C., 1976. "Regulation of Transcription Initiation and Termination in the Control of Expression of the Tryptophan Operon of *E. coli.*" In

D. P. Nierlich, W. J. Rutter, and C. Fred Fox, eds., *Molecular Mechanisms in the Control of Gene Expression.* Academic Press, New York.

———, AND L. SOLL, 1977. "Mutations Affecting $tRNA^{trp}$ and Its Charging and Their Effect on Regulation of Transcription Termination at the Attenuator of the Tryptophan Operon." *J. Mol. Biol.* 113: 663–677.

ZURAWSKI, G., D. ELSEVIERS, G. V. STAUFFER, C. YANOFSKY, 1978. "Translational Control of Transcriptional Termination at the Attenuator of the *Escherichia coli* Tryptophan Operon." *Proc. Nat. Acad. Sci. U.S.* 75(12): 5988–92.

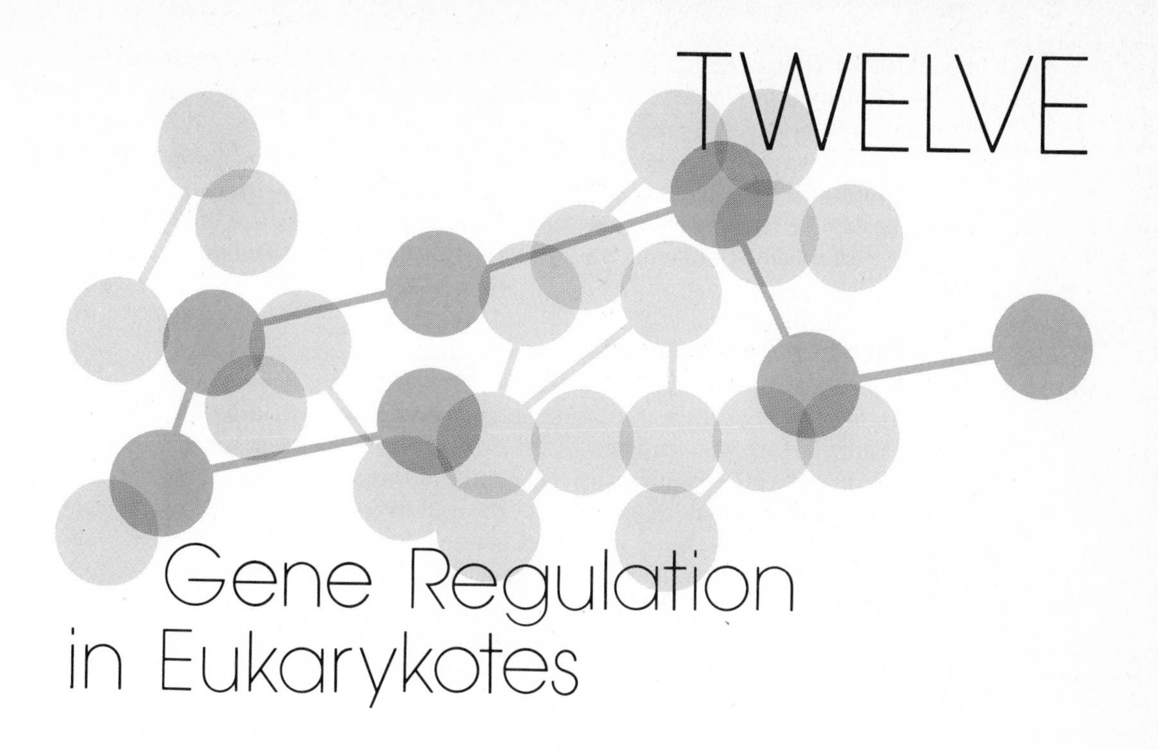

TWELVE

Gene Regulation in Eukarykotes

INTRODUCTION

Having looked into gene regulation in prokaryotes in some detail, we shall now turn to the topic of gene regulation in eukaryotes. This is a fascinating area of genetics which commands much attention in research laboratories. It is also an area which has yielded less information than almost any other which we have discussed, because the complexity inherent in genomes of eukaryotic systems prevents easy access to their molecular details.

To understand gene regulation in eukaryotes is ultimately to understand how all living organisms come to be as they are. As we have mentioned before, most living cells in an organism, whether that organism be prokaryote or complex, are known to be similar in the great majority of structural genes. This has been determined by the presence of the same cellular proteins controlling essentially similar metabolism in all living cells. Therefore, the multitude of differences between species most probably result from differences in regulatory, rather than structural, genes.

Interest in gene regulation in eukaryotes is also heightened by increasing recognition that many disease conditions in humans probably stem from alteration of regulatory mechanisms. Examples of such diseases are immune diseases, cancer, and some forms of circulatory diseases. Achieving our goal of understanding these diseases will not only enable us to prevent or cure such conditions, but also will undoubtedly shed light on normal control processes. On the other hand, basic research on gene regulation in normal cells may also help us to understand and treat aberrant cell behavior in diseased organisms.

In this chapter we discuss examples of gene regulation in eukaryotes. Because of the complexities inherent in diploidy and intercellular interactions in multicellular organisms, progress in understanding regulation in eukaryotes has been slow—we have yet to formulate actual models of regulatory mechanisms as we have in bacteria.

GENE REGULATION IN DEVELOPMENT

The study of gene regulation in complex organisms can be traced back many centuries to a time before we even knew of the existence of cells, much less of genes. As far back as the time of Aristotle, scientists pondered the mysteries of embryogenesis, the process by which a multicellular organism with diverse histotypes develops from a single fertilized egg cell. Since histotypes are cellular phenotypes, and phenotypes are the result of gene action, it follows that the events in differentiation and development are the end result of gene activity. Embryology and genetics have thus been intrinsically interrelated fields of biology.

Along this line, recall that it was an embryologist, Theodor Boveri, who in 1903 formulated the chromosomal theory of heredity, simultaneously with the pioneer geneticist, Sutton. Boveri and other embryologists in the late nineteenth and early twentieth centuries established that the zygote first undergoes a stage of development known as *cleavage*, during which time the egg cell divides into two identical *blastomeres* (daughter cells which arise by mitosis of a fertilized egg cell). These two blastomeres undergo mitosis, as do their daughter cells, until the embryo is a multicellular cluster of genetically identical cells (Fig. 12-1).

During the stages of development following cleavage, there is cell movement leading to the formation of basic embryonic structures. Cells begin to look noticeably different from each other. For example, in the embryo, some become rounded and lose their nuclei—these are the red blood cells. Other cells elongate and fuse, eventually forming skeletal muscle cells. Still others produce, and become imbedded in a substance that hardens and gives rise to bone. Yet because all cells arose by mitosis from one cell, these very different-looking cells are genetically identical.

THUS THIS GRADUAL, PROGRESSIVE PROCESS THROUGH WHICH CELLS WHICH ARE GENETICALLY IDENTICAL ATTAIN DIFFERENT PHENOTYPES (DIFFERENTIATION), MUST BE THE RESULT OF DIFFERENTIAL GENE REGULATION. For example, it is assumed that genes determining the production of hemoglobin in red blood cells are not active in skeletal muscle cells; similarly, genes determining the production of actin and myosin in skeletal muscle cells are active only in these cells and not in red blood cells.

Questions with which we are still confronted address many aspects of differentiation. For example, when does the process of differentiation begin? Is the process the same for all cells? Once differentiation has begun, is it irreversible? And of course, what are the mechanisms of gene regulation leading to differentiation? In the following sections we will discuss some experiments which bear on these very important questions.

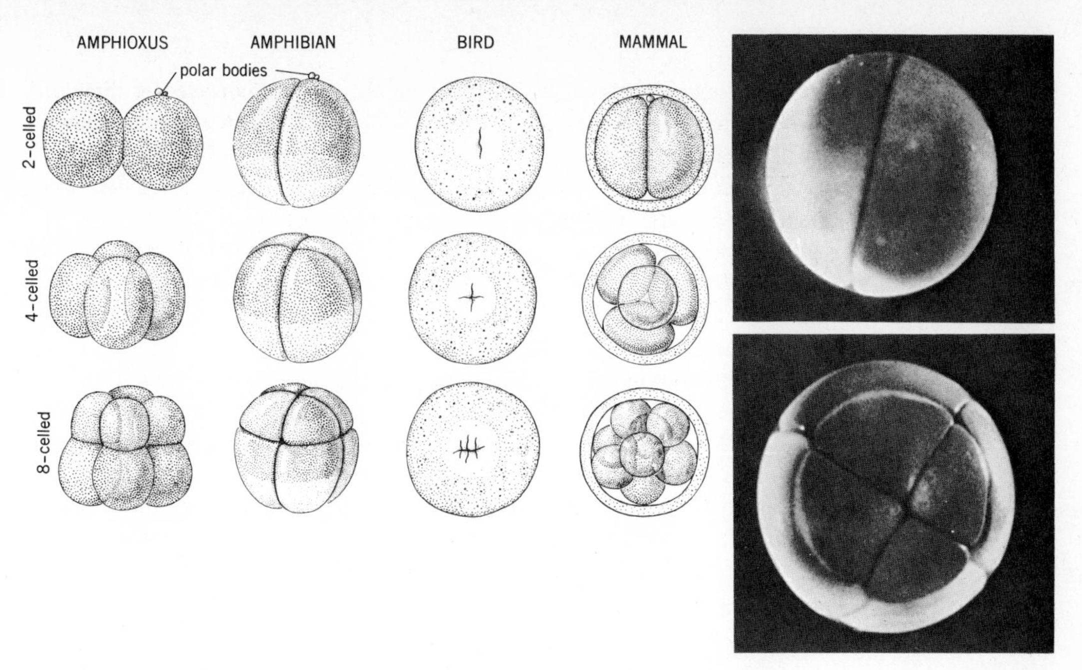

Figure 12-1. Cleavage in chordate embryos. (From N. J. Berrill, and G. Karp, 1976. *Development*, p. 133. McGraw-Hill, New York.)

Determination in Development

Experiments by classical embryologists have shown that onset of a direction of differentiation for an embryonic cell differs among cells and among species. For example, the developmental fate of cells in different areas of the amphibian embryo was followed using vital stains. The consistency with which cells of certain areas of the early amphibian embryo developed into specific structures allowed scientists to draw *fate maps* of the embryos, as shown in Fig. 12-2.

The great German embryologists H. Spemann and O. Mangold developed techniques for the transplantation of cells from one area of the blastula into another

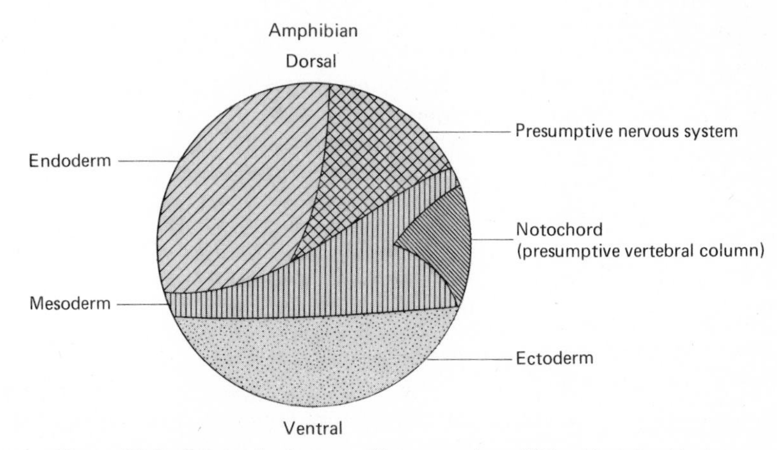

Figure 12-2. Schematic diagram of fate map of amphibian blastula, sideview.

area. They found that even in the early gastrula stages, cells which are destined to become neural tissue, for example, still possess the potential to form other ectoderm or mesoderm structures if implanted in those regions of the embryo. We refer to this as *totipotency;* in genetic terms, it means that gene regulation has not yet become established, and that the cells are capable of responding to different environments and attaining phenotypes other than those which would be their normal fate.

However, at about the time that differentiation normally becomes visible through changes in the appearance and behavior of cells, namely, during late gastrulation, the developmental potential of embryonic cells becomes restricted. This process of restriction of developmental potential is known as *determination.* In the amphibian embryo, if presumptive neural cells of late gastrula stages are transplanted to other regions, they will no longer adapt to their new surroundings, and will still form neural structures. Figure 12-3 shows an embryo with two neural tubes (forerunners of the spinal cord) resulting from the transplantation after gastrulation of additional presumptive neural cells into the side of the embryo.

Regulative vs. Mosaic Development

Embryonic systems, such as the amphibian, which show developmental flexibility until the late gastrula stages, have been traditionally referred to as having *regulative* development. Echinoderm development is also highly regulative. On the other hand, E. B. Wilson, in his classical studies on molluscs, showed that in this class determination occurs very early, even by the first cleavage of the fertilized egg. Embryonic systems with very early determination are referred to as having *mosaic* development.

Do these studies indicate that there are actually different systems of regulation governing mosaic and regulative development? The answer is, probably not. Even though the amphibian embryo has been categorized as regulative, experiments have shown that its development has mosaic aspects as well. For one thing, the amphibian egg contains different cytoplasms, one light and one dark. In frogs, the first two blastomeres are separated after the first division, each forms a normal individual. If after the second division the blastomeres are separated "vertically," regulative development again results in two normal frogs. However, if the blastomeres are separated "horizontally," the half containing the portion of the egg with lighter-color cytoplasm (known as the *grey crescent*) develops normally, whereas the half which contains only darker cytoplasm fails to gastrulate (Fig. 12-4).

It thus appears that some of the differences observed between mosaic and regulative systems may be due to the experimental approaches used. Since even regulative systems show determination eventually, one might also conjecture that the differences may only be ones of time. At any rate, such studies indicate that the cytoplasm of the egg cell plays a very significant role. In the next section, we shall explore some of the factors, discovered in recent studies of gene activity in oogenesis, which are believed to play a role in the events of early development.

Gene Activity in Oogenesis

In Chapter 1 in our discussion of oogenesis, we pointed out that one of the differences between maturation of eggs and sperms is the retention of most of the cytoplasm

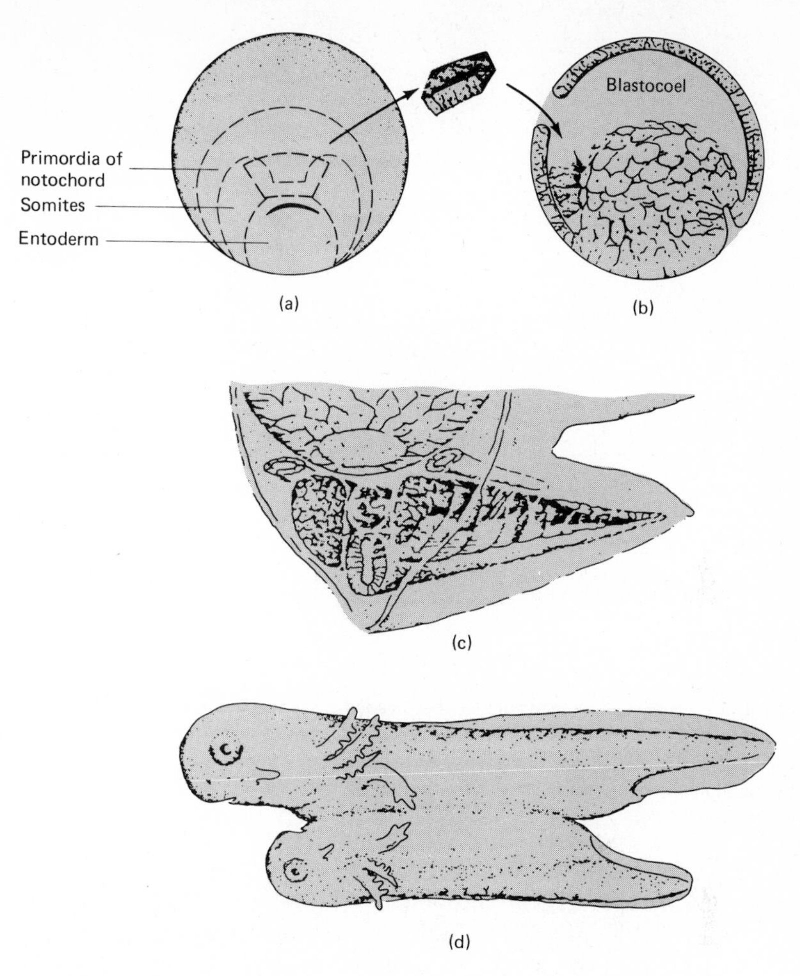

Figure 12-3. Diagram showing transplantation of presumptive notochord (a) into a second embryo (b). In (c) the tissues of the graft are white; contribution to secondary embryo by host tissue is in black. (d) Gross appearance of secondary embryo. (From N. T. Spratt, 1971. *Developmental Biology*, p. 447. Wadsworth, New York.)

by the ovum and the loss of most cytoplasm by the sperm. The events of cleavage are apparently determined by genetic messages stored in the egg cytoplasm, and transcription of most nuclear genes is not detectable until later stages. Evidence that gene activity occurs in oogenesis directed toward the production of genetic messages to be stored in the cytoplasm has been gathered by both cytologists and biochemists. Cytologically, chromosomal structural changes in maturing oocytes have been detected. Biochemically, evidence for transcription during oogenesis, and the presence of long-lived mRNA in the egg cytoplasm, have been reported. Following are a few examples.

Lampbrush chromosomes. During diplotene of prophase I of meiosis, regions of bivalents assume a configuration of loops which look like brushes, and accordingly

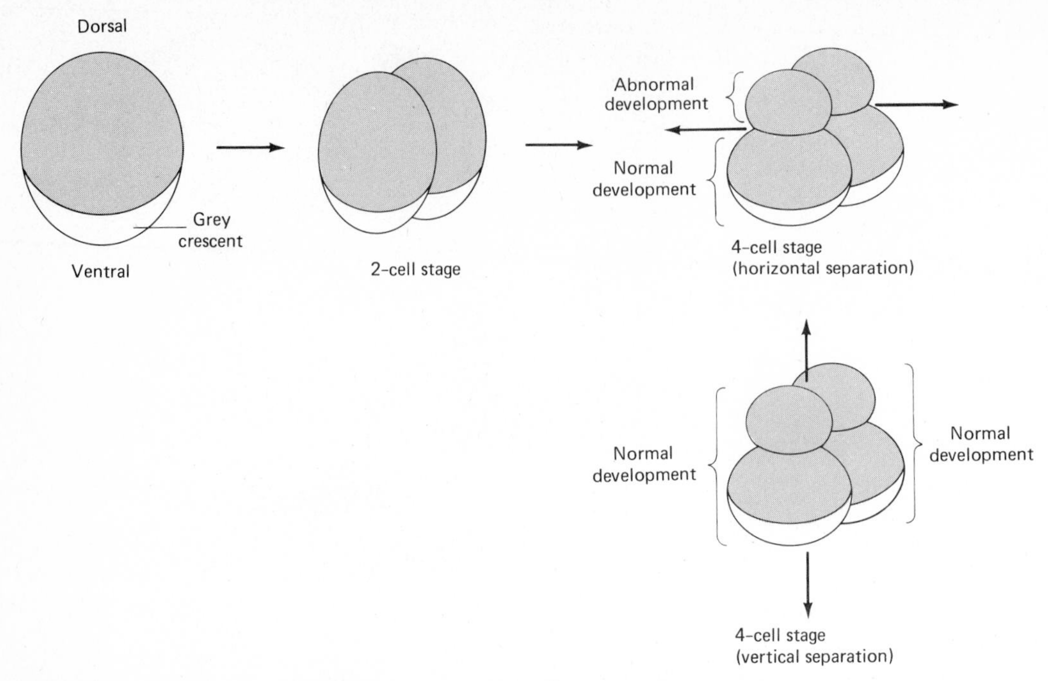

Figure 12-4. Cleavage in amphibian egg. Double-headed arrows represent plane of separation of blastomeres. Horizontal separation at the 4-cell stage results in abnormal development of dorsal blastomeres and normal development of ventral blastomeres. Vertical separation leads to development of two normal larvae.

have been designated *lampbrush chromosomes* (Fig. 12-5). First studied in amphibian oocytes, lampbrush chromosomes have now been found in mammals, insects, and worms. The loops are constituted of a strand of DNA. Approximately 5% to 10% of the DNA of the oocyte is involved in loop formation.

Edstrom and Gall reported in 1963 that the base composition of the RNA particles found on the loops is very similar to the base composition of the DNA. To further support the concept that the loops are regions of transcription, the RNA has been found to contain a polyadenylic region (recall that mRNA molecules contain poly *A* regions,). Davidson (1976) has reported that these RNA molcules are very large, more than 10^5 nucleotides long. He has also shown the presence of repeated sequences of bases. Davidson feels that all these characteristics classify the RNA associated with lampbrush loops as *heterogeneous nuclear RNA* (*hnRNA*) rather than mature mRNA. The function of hnRNA is not clear, although some molecules are believed to be mRNA precursors.

The function of RNA molecules synthesized by the lampbrush loops is still open to speculation. There is certainly the possibility that the RNA is packaged with nucleoproteins, and stored in the cytoplasm to serve the blastomeres during early stages of development. Furthermore, since both repetitive and nonrepetitive, or unique sequences are found in these molecules (see again p. 275), they may serve either structural or regulatory functions—or both.

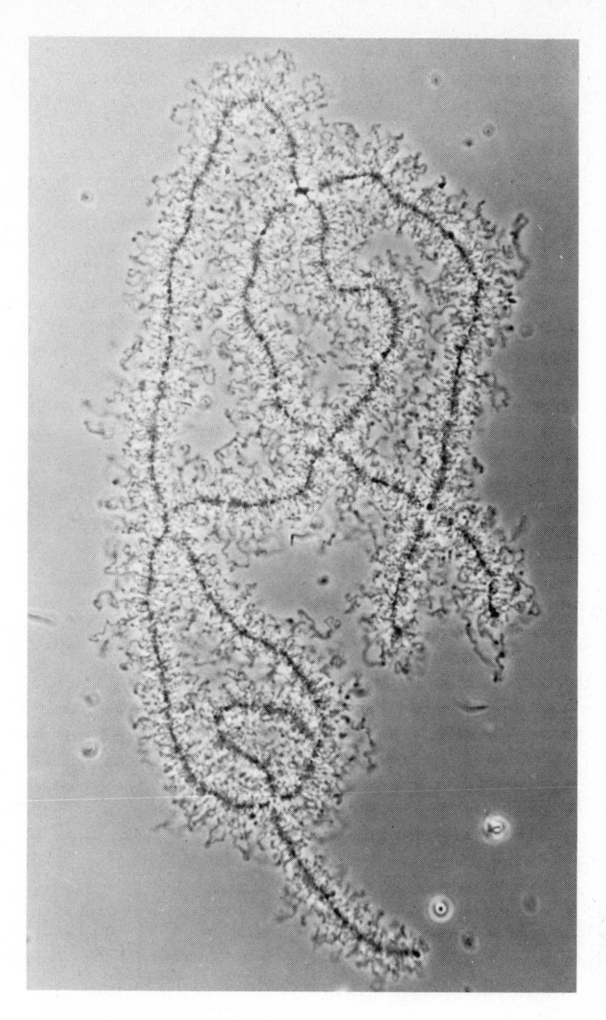

Figure 12-5. Photomicrograph of lampbrush chromosomes. (Photograph courtesy J. G. Gall, Yale University.)

Gene amplification in amphibian oocytes. Classical embryologists had noted that nuclei of amphibian eggs during early oogenesis were extremely large; hence they were called *germinal vesicles*. Embryologists found one explanation for the enlargement when they discovered that the number of nucleoli in these egg cells was much greater than expected. Somatic cells of amphibians normally contain one nucleolus per haploid set of chromosomes, therefore one would expect to find four nucleoli during prophase I when the cells are temporarily tetraploid. Instead, in oocytes of different species of amphibians, 600 to more than 1000 nucleoli have been reported (Brown and Dawid 1968)!

This phenomenon has been termed *gene amplification* (Fig. 12-6), and results from the increased replication of the genes for 18S and 28S ribosomal RNA particles (see again p. 326), primarily during pachytene stage. Perkowska *et al.* (1968) estimated

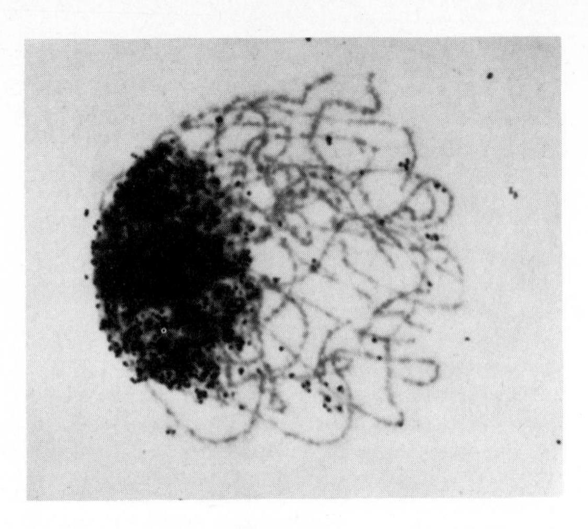

Figure 12-6. Autoradiograph of Xenopus oocyte nucleus. "Cap" of stained material represents high concentration of RNA in nucleoli. (Photograph courtesy Dr. Joseph G. Gall, Yale University.)

that based on 1500 nucleoli in *Xenopus* oocytes, the total number of ribosomal gene sets would be 3–5 $\times$ 10^3, and the total number of these genes would be 1.5–2.5 $\times$ 10^6! At the end of oogenesis, an amphibian egg may have as many as 1.1 $\times$ 10^{12} ribosome particles in its cytoplasm. It has been estimated that without amplification of the genes, it would take an amphibian cell with just one ribosome gene about 500 years to synthesize this number of ribosomes. By the end of oogenesis, the extra nucleoli no longer exist. The ribosome particles are distributed in blastomeres as cleavage proceeds.

Amplification is achieved by the synthesis of rDNA from the chromosomal genes. This DNA remains extrachromosomal, and Hourcade and his associates (1974) have proposed that the bulk of gene amplification takes place by way of the rolling circle model of gene replication (p. 289). Ribosomal RNA molecules are transcribed from the extrachromosomal DNA and pass through nuclear pores into the cytoplasm. The addition of nucleic acid molecules, as well as components of yolk material (which originates elsewhere in the body of the female and is then taken into the egg cell), accounts for the enormous growth of the egg cell during maturation. Oocyte volume increases by as much as 100,000 times in *Rana pipiens*.

Both formation of lampbrush chromosomes and gene amplification, and their eventual cessation, must be under some form of control, but the nature of this control has not yet been determined. As is the case for many aspects of development, our present knowledge of these two phenomena is primarily descriptive.

Evidence of Selective Gene Activity During Development

If the progressive changes that occur during embryogenesis reflect differential gene activity, we should be able to detect specific examples of genes becoming active, or being inactivated, during development. In this section we shall discuss briefly a number of such examples at the chromosomal and biochemical levels.

Before we do so, however, a very important point must first be established. Thus far in our discussion we have been assuming that blastomeres are genetically identical because they arise from mitosis. The expression of certain genes in some cell types and not in others has been assumed to be due to gene regulation. Scientists realized that it was necessary, however, to establish that the basis for this phenomenon is indeed regulation, and not some other mechanism such as loss of genes during development.

Nuclear transplantation and cloning. In the early 1950s, R. Briggs and T. J. King, and later J. B. Gurdon, developed the technique of nuclear transplantation to address the question of whether cells retain the entire genome as differentiation proceeds. Working with the relatively large and accessible frog egg, researchers used micropipettes to remove the nucleus from a number of mature, unfertilized egg cells. Somatic endoderm cells from *Rana* gastrulas and blastulas were similarly enucleated, and the diploid nuclei transplanted into enucleated egg cells. Briggs and King reported that a number of the egg cells developed into normal frogs (Fig. 12-7). THESE RESULTS INDICATE THAT CELLS WHICH HAVE ALREADY UNDERGONE DIFFERENTIATION DO INDEED RETAIN ALL THEIR GENES, OR ELSE NORMAL DEVELOPMENT COULD NOT HAVE TAKEN PLACE.

Gurdon later extended these nuclear transplantation studies and also found that, generally, the more specialized the cell type used as the source of nuclei, and the older the donor, the less able are the nuclei to support normal development. For example, nuclei from endoderm cells of early embryos gave more successful results than nuclei from endoderm and gut cells of hatched and swimming tadpoles (Gurdon 1962). Gurdon and Laskey also reported (1970) that nuclei from adults resulted in little normal development; however, if the cells were first grown in vitro, and the nuclei then removed, normal frogs were obtained. Cells grown in tissue culture are known to undergo phenotypic changes resembling dedifferentiation. Although it might be argued that these observations do not prove that differentiated cells retain their original genes, since nuclei from adult cells do not support development, it is generally accepted that the data do support this conclusion. Research with plant cells shows that "adult cells" do contain all the genes for an organism (see p. 385).

The individuals which develop from nuclei taken from cells of the same frog are genetically identical, and are referred to as *clones*. In the years since this research first was reported, many words have been expended on extrapolating these data to the possibility of someday cloning humans. The vision of being able to duplicate, as it were, some famous (or infamous) people has captured the imagination of the public in novels and movies. Therefore, it might be worthwhile to comment briefly here on the present state of the art of cloning humans.

It is unlikely that the cloning of humans, using nuclei from adult cells, can be achieved at the present time. First, as we mentioned above, Gurdon's work on frogs has shown irreversible gene regulation which is unaffected by transfer of the nucleus from *adult* cells into the cytoplasm of the egg cell. Second, it is one thing to work with frogs' eggs, which are approximately $1\frac{1}{2}$ millimeters in diameter, and therefore easily visible to the naked eye; it is quite another to work with a human egg cell which is on the order of 100 μ, visible only with a microscope. Third, it must be emphasized that humans, being mammals, need the uterine environment for develop-

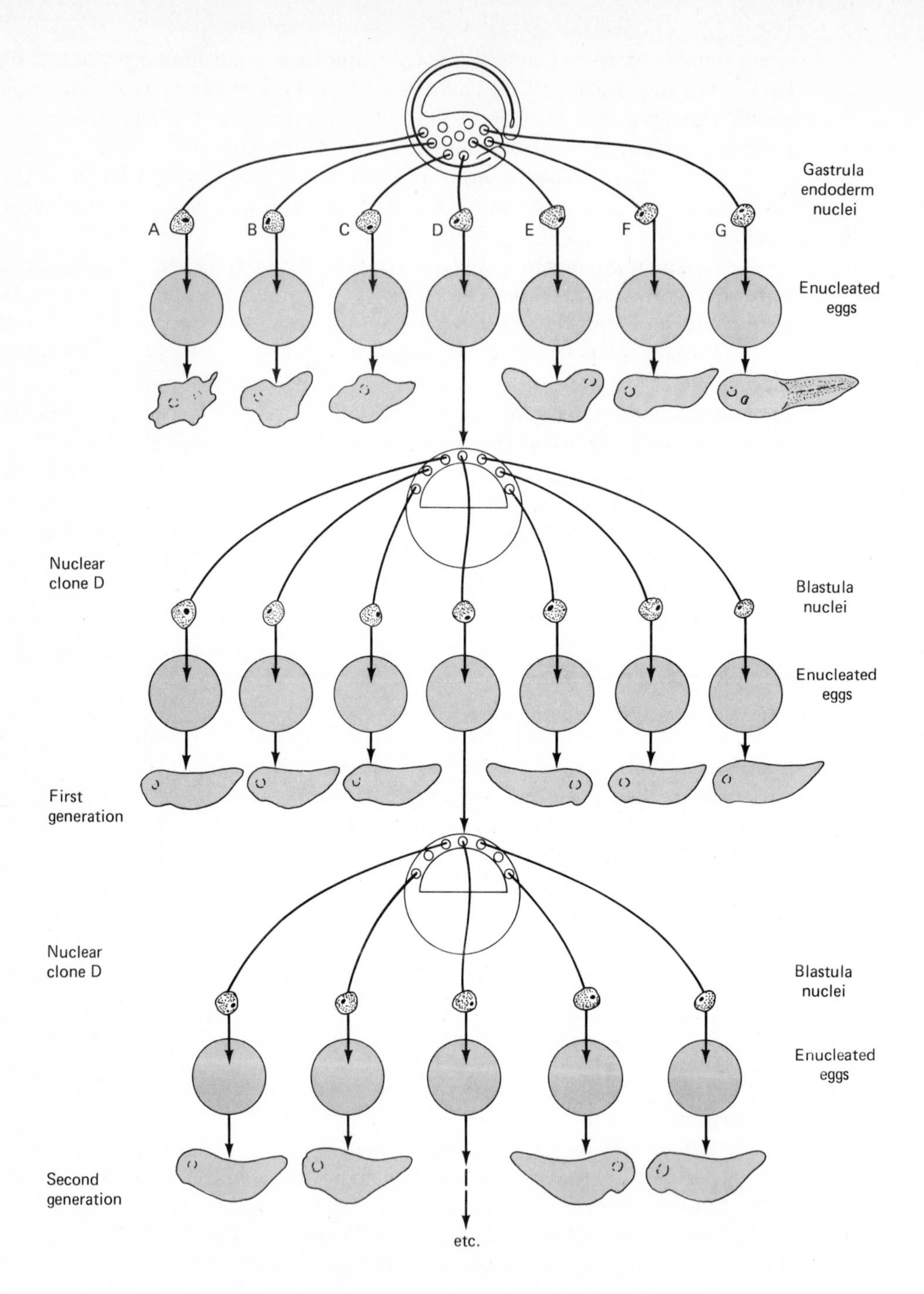

Figure 12-7. Diagrammatic depiction of nuclear transplant experiments. Explanation in text. (From N. T. Spratt, 1971. *Developmental Biology*, p. 435. Wadsworth, New York.)

ment, unlike oviparous organisms such as the amphibians. A human egg cell would have to be implanted into the uterus of a "foster" mother. Although technically possible (as shown by the "testube baby" born in England in July 1978), this requirement would certainly restrict any large-scale cloning.

Given that a way may be found around all the technical obstacles in the future, one might in addition question the desirability of cloning humans, which would only add to the problems of overpopulation. Remember, too, that the environment plays a large role in determining whether any individual reaches his or her genetic potential. To wish to have oneself cloned would seem to require motivation from an enormous ego, accompanied by an equally profound biological ignorance.

Balbiani rings. At about the same time that nuclear transplantation experiments were being carried out, other experiments were reported which showed structural changes in polytene chromosomes of dipteran larvae to correlate with developmental events. The changes involved areas of the chromosomes becoming diffuse and enlarged (Fig. 12-8). Named for a cytologist who had observed them earlier, these

Figure 12-8. Photograph of chromosome puff at one band of *Drosophila* salivary gland chromosome (arrow). (Photo from Poulson and Metz, 1938. *J. Morph.* 63 : 389 courtesy D. F. Poulson, Yale University.)

regions are known as *Balbiani rings*, or *chromosome puffs*. W. Beermann (1964) showed that the appearance of puffs was consistent at specific areas of chromosomes during specific periods of development. Furthermore, the pattern of puffing was tissue specific.

Two experiments clearly supported the relationship between puffs and development. First, when salivary glands from larvae of one stage were transplanted to larvae of a different stage, the chromosomes of the transplanted glands assumed the pattern of puffing of the host larva (Becker 1962). Another experiment involved a hormone, ecdysone, which is produced by larvae prior to the third molting stage. Normally, two puffs appear in conjunction with this stage. Injection of the hormone into younger larvae caused formation of the same two puffs, even though the larvae were not at the proper stage in development (Clever and Karlson, 1960). Whether the hormone directly causes the puffing, or interacts with substances in the cell in a way which then results in puffing, has not been determined.

Evidence that puffing is related to gene activity accumulated rapidly. Besides their correlation with developmental stages, the puffs appeared to be situated at single bands on the polytene chromosomes. Recall that the bands are believed to represent single gene loci (p. 165). Furthermore, when radioactive precursors of RNA were introduced, labeling occurred only in the region of puff formation. A number of questions remain to be answered, such as whether the RNA transcribed functions as mRNA or hnRNA. We also do not know if the RNA molecules represent transcription of the entire DNA of the bands involved, or of only partial areas. Nonetheless it is widely accepted that the puffs represent regions of DNA which uncoil in order to allow transcription to occur.

Inactivation of the X chromosome. In Chapter 5 we discussed the phenomenon of X chromosome inactivation first proposed by Mary Lyon (p. 144). As we pointed out then, this regulation of gene activity occurs in early development, by the late blastocyst stage. Its mechanism has not been determined.

Interestingly, the property of being inactivated is one which can spread from an inactivated X to autosomal genes translocated onto parts of the inactivated X chromosome, a phenomenon reported by Russell *et al.* (1963). The fact that inactivation of the translocated autosome varied in degree depending on its position in the X chromosome led Russell to suggest that the X chromosome may contain centers of inactivation. The closer an autosome is translocated to these centers, the more effective the inactivation.

It is doubtful, however, that the spreading effect of the inactivated X is unique to the X chromosome; more likely it is a property of heterochromatin in general. As far back as 1950, E. B. Lewis had reported that a mutation for white eyes seemed to be inactivated when translocated near heterochromatin. Flies heterozygous for the gene showed variegation in the pigmentation of the eye. Lewis termed this phenomenon *position effect.*

Regulation of Protein Synthesis in Development

The preceding discussion demonstrated that development results from selective activation or inactivation of genes during embryogenesis. One would therefore expect to encounter evidence for this phenomenon in proteins synthesized during differentiation of cells, as well as in visible changes in the activity of portions of chromosomes. In fact, differentiation may be defined as the ability of cells to synthesize special proteins. Although examples of differential protein synthesis abound in embryological studies, we shall discuss only two of the most widely known examples, namely, changes in the synthesis of hemoglobin and of the enzyme lactate dehydrogenase (or LDH).

Hemoglobin synthesis. Perhaps the best-characterized of all proteins synthesized by eukaryotic cells is the globin portion of the hemoglobin found in red blood cells. The reasons for this are, of course, accessibility and constant supply. Hemoglobin is found in a readily replaceable tissue, and one which can be removed without permanent injury to the donor.

Hemoglobin is a large molecule containing one molecule of the pigment heme, and four polypetide chains in the globin component. There are two chains of each of two different polypeptides in the tetrameric protein. The structure of hemoglobin is illustrated in Fig. 6-6.

For students of gene regulation in eukaryotes, hemoglobin is a desirable object of research. Its polypeptides have been studied in great detail, and a number of ontogenic (developmental) changes in the polypeptides have been established. Half of the polypeptide chains are different in embryonic, fetal, and adult hemoglobins. Since we assume that different polypeptides are determined by different genes, the pattern of changes in hemoglobin reflects patterns of gene regulation.

Table 12-2 compares the embryonic and adult hemoglobin polypeptides in a number of vertebrate species. Capital letters represent the class of hemoglobin, and Greek letters in parentheses represent the particular polypeptides.

Table 12-2 Patterns of Hemoglobin Synthesis Compared with Development

	Cell Lines			
	Embryonic		*Definitive (adult)*	
Species	*Site*	*Hemoglobin*	*Site*	*Hemoglobin*
Mouse	Yolk sac	E_I (xy) E_{II} (αy) E_{III} (αz)	Liver and bone marrow	(Fast A?) and A ($\alpha\beta$)
Man	Yolk sac (?)	Gower ($\alpha\epsilon$)	Liver and bone marrow	F ($\alpha\gamma$), A ($\alpha\beta$) and A_2 ($\alpha\delta$)
Chicken	Yolk sac	E P	Yolk sac and bone marrow	A, D
Frog	Liver	Type I Type II t	Liver and bone marrow	Frog I and Frog II

From P. A. Marks, and R. A. Rifkind, 1972. *Science* 175: 956.

In man, for example, there is one form of embryonic hemoglobin which contains alpha (α) and sigma (ϵ) polypeptides. During later development, sigma chains are no longer produced but are replaced by gamma (γ) chains, forming fetal hemoglobin (type F). Finally, in the adult, gamma chains are replaced by beta (β) chains in the vast majority (97%) of red blood cells. This form in the adult is known as hemoglobin A. The remainder of red blood cells are type A_2, with both alpha and delta (δ) chains.

Furthermore, note that cells which manufacture hemoglobin are derived from different sources at different stages in development. In chickens, mice, and probably humans, the yolk sac is the source of red blood cells in embryos; in adults, the liver and bone marrow assume this function. Whether the cells of the yolk sac migrate to the liver and bone marrow to serve as stem cells for erithropoiesis (red-blood-cell formation) is not clear. What is clear is that the gene determining alpha chains is continuously active at all stages of development, and that there is regulation of the nonalpha genes.

Evidence has now been accumulated that nonalpha genes (γ, δ, β) are linked in the order in which they are expressed in development, and on a different linkage group from the *alpha* locus. An interesting abnormality of the beta chain in some humans, which leads to a type of anemia known as *Lepore hemoglobin*, is cited as evidence of linkage of the *delta* and *beta* genes. Lepore hemoglobin has been observed to be deltalike in amino acid sequence at the amino terminal end, and betalike at the carboxyl terminal end (Kabat 1972), a situation believed to have resulted from unequal crossing-over or translocation between the β and δ loci. A mild anemia results because the delta gene mRNA is thought to be transcribed at a slower rate than the beta chain, and there is a deficiency in the rate of hemoglobin production.

The fact that the rate of hemoglobin synthesis is under some form of control is also reflected in another inherited anemia, known as *thalassemia*, or *Cooley's anemia*. This is one of the so-called *ethnic diseases*, which are found significantly more frequently in certain populations of a common origin. Thalassemia is associated with people whose ancestry traces from Mediterranean countries such as Italy, Greece, and Spain. The disease is transmitted as an autosomal recessive, and homozygotes suffer from thalassemia major, a lethal anemia; heterozygotes are anemic, but theirs is a nonfatal condition known as thalassemia minor. The defect is in an abnormally low rate of beta chain production. The hemoglobin which is produced is structurally normal, but there simply is not enough to sustain normal body functions.

Some insight into the mechanism of regulation of nonalpha genes has been derived from studies of humans manifesting a hereditary condition known as *hereditary persistence of fetal hemoglobin*, or *HPFH*. In these individuals, only normal fetal hemoglobin of the composition $2\alpha 2\gamma$ is produced; the *beta* gene is completely suppressed. However, studies of individuals heterozygous for HPFH and sickle-cell anemia in both the cis and trans positions were found to express the *beta* gene only if it is in the trans position from the HPFH mutation. The fact that the HPFH gene appears to suppress beta gene activity only in the cis position indicates a regulatory function similar to a promoter or operator gene rather than one which determines a regulatory substance.

Since individuals with fetal hemoglobin appear to be normal, it would be of great interest to be able to understand and perhaps someday control the production of the various hemoglobin chains. In sickle-cell anemia patients, one would encourage the production of gamma chains (which are not affected by the sickle mutation) in adults. A natural form of this type of compensation is known to exist in sheep rendered anemic experimentally. In such sheep, a different *beta* gene is found to be switched on by the production of the β^c polypeptide normally found only for a few months in newborn sheep (Kabat 1972).

Lactate dehydrogenase. Another tetrameric protein which has been studied in great detail and has shown ontogenic changes is the enzyme *lactate dehydrogenase*, or *LDH*. LDH catalyzes the interconversion of pyruvic acid and lactic acid in glycolysis, the biochemical pathway involving the breakdown of glucose. Actually, several different forms of LDH have been found in every vertebrate studied. The different forms all catalyze the same reaction, but show differences in molecular structure. Enzymes with these characteristics are referred to as *isozymes*, or *isoenzymes*.

Most of the LDH isozymes found in vertebrates have combinations of two different polypeptide chains, A and B. Dissociation and electrophoresis of the isozymes (Markert 1963) showed that there are five forms, called LDH1, LDH2, LDH3, LDH4, and LDH5. They contain, respectively, BBBB, BBBA, BBAA, BAAA, and AAAA. Although all five forms can be found in the cells of many tissues, one of the forms frequently is present in greater quantity than the others, as shown in Table 12-3.

As you can see, LDH1 predominates in heart cells, and LDH5 in skeletal muscle cells, possibly because slightly different biochemical properties of the different isozymes cause one to be more effective in one cell type than the other.

More pertinent to our discussion here are the changes in LDH forms that C. L. Markert and others have found in tissues of mouse embryos at different stages of development. For example, adult mouse heart tissue contains mostly LDH1 and LDH2, as shown in Table 12-3. However, studies of mouse embryos at about the

Table 12-3 A Comparison of LDH in Different Tissues of Vertebrates

		Percent Distribution of LDH Isozymes				
Species	*Tissue*	LDH1	LDH2	LDH3	LDH4	LDH5
Human	Heart	67	29	4	1	1
	Brain	25	25	34	15	1
	Kidney	42	48	9	1	
	Skeletal muscle	1	4	8	9	78
	Liver	2	2	3	12	80
Rat	Heart	50	30	12	6	2
	Skeletal muscle	1	1	3	5	90
Rabbit	Heart	94	2	1	3	1
	Skeletal muscle	1	1	1	3	95

From Maclean, N., 1976. *Control of Gene Expression*. Academic Press, New York. Portions of Table I.

tenth day of gestation show mostly LDH4 and LDH5. As development continues, there is a noticeable shift to LDH1 and LDH2, as illustrated in Figure 12-9. Here again is a clear example of one gene, the A gene, being active in early development then being suppressed in later development, while the B gene becomes more active.

Two other subunits of LDH have also been found, but they are produced only briefly and/or in special tissues of some vertebrates. The LDH subunit known as *C* has been found to be expressed only in primary spermatocytes of pigeons, and its synthesis is suppressed as soon as the cells have divided to form secondary sper-

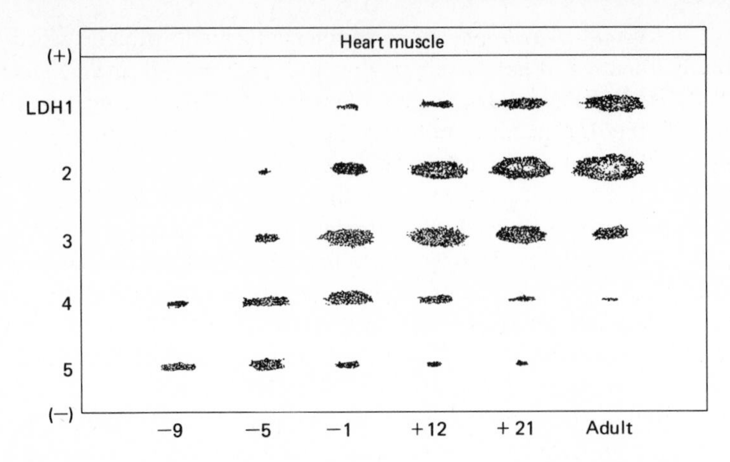

Figure 12-9. Changing LDH isozyme pattern in mice, showing a shift from LDH5 to LDH1 and 2. (From C. L. Markert and H. Ursprung, 1971. *Developmental Genetics*, p. 44. Prentice-Hall, Inc., Englewood Cliffs, N.J.)

matocytes. In nerve cells of adult fish retinas, another subunit, E, has been reported. Both C and E subunits can form tetramers with A or B (Markert and Ursprung 1971). We shall discuss the evolutionary significance of this phenomenon in Chapter 17.

As we mentioned earlier, the genetic determination of the protein components of hemoglobin and LDH are but two examples of regulation of protein synthesis during development. Countless other systems have been studied; however, no specific information yet exists that would allow us to formulate models for mechanisms of control which can be tested as in prokaryotes.

Reversibility of Regulation in Development

Since this book cannot discuss more than a few examples of the mass of work being carried out on developmental systems, the preceding discussion must suffice. Before we turn our attention to two other areas involving gene regulation of cells in adult eukaryotes, however, it is necessary to make a few general comments on concepts of regulation during development.

Again, our knowledge of development is based primarily on descriptive studies. We can see that in all eukaryotes, the zygote is a totipotent cell, capable of giving rise to daughter cells of many different phenotypes. The loss of this potency in daughter cells is progressive during development, often likened to the available pathways which a ball in a pinball machine may travel. When it begins its journey, many channels are open, but as the ball travels down the machine, fewer and fewer options are available to it, until finally there is but one way to move toward the final pocket.

What does this progressive determination of phenotype mean genetically? There is at present a controversy over whether it represents a series of activations and repressions of different loci, or whether all loci are at first transcribable, with irreversible repression then resulting in differentiation. The first theory implies that repression of genes would be reversible; the latter, that repression is irreversible.

The latter concept is supported by scientists who point to the presence of a highly diverse population of mRNA in early developmental cells (Caplan and Ordahl 1978). They propose that this diversity indicates that all genes are being transcribed, to different degrees. Some will be repressed in early development, others in later stages. Still others, called *housekeeping* genes, which determine basic metabolic reactions such as glycolysis, are never repressed.

Our main problem is to find models for testing either theory regarding the possible reversibility of gene regulation. For example, one might point to the regenerative ability of certain animal species as worthy of study. An elementary and fascinating lab exercise in basic biology courses is to bisect organisms such as the flatworm *Planaria*, and observe its regeneration of the missing halves to form two whole individuals.

Even some vertebrates show regeneration; the newt, for example, can regenerate an amputated limb. However, our analysis is clouded by the fact that we cannot distinguish the source of the cells that take part in the regeneration process. Do the already differentiated cartilage and muscle cells of a newt dedifferentiate, and then redifferentiate in a second developmental process, or are there mesenchymal cells which have never differentiated, and remain in an embryonic state? These questions remain to be answered.

An example of great genetic flexibility of cells is well known to botanists and anyone who has ever propagated plants. For example, it is common knowledge that a stem cut from a flowering plant, if given the right conditions, will develop roots. This phenomenon certainly reflects the existence of cells retaining totipotency even in an adult organism. In fact, in different species, cuttings of roots and even leaves can develop roots and eventually the entire structure of the mature plant. It would not be an exaggeration to represent certain tissues of the plant as being in a permanent state of embryogenesis (For a review, see Steeves and Sussex 1972).

An extreme example of the ability of adult plant cells to differentiate into different tissues is the work by Steward and his coworkers (1958), who were able to dissociate cells from phloem tissue of carrot roots, and then obtain the growth of individual cells into nodules. They reported the fascinating discovery that these nodules produced roots, and when transferred to a growth medium solidified with agar, developed into an entire plant. Since the dissociated cells were from presumably differentiated cells, one would certainly have to interpret these results as indicating the ability of repressed genes to become active again.

It is quite clear, however, that regulation in plants and animals is by no means the same. Not all plants can produce the same results as the carrot root cells; certainly, no single somatic cell from an adult animal has ever been able to regenerate an entire organ, much less a whole individual. IT IS FAR MORE LIKELY THAT EACH SYSTEM OF GENE REGULATION IN DEVELOPMENT WILL HAVE TO BE RECOGNIZED AS UNIQUE.

Some general characteristic may prove to be ascribable to all systems, for we are, after all, dealing with DNA in all cases. However, an understanding of gene regulation has so far eluded us because of the staggering complexities and differences in eukaryotic cells and organisms. For example, different species show distinct differences in the time of onset of embryonic gene activation. Amphibians and echinoderms show expression of paternal genes at the gastrulation stage. On the other hand, recent evidence has shown that in the mouse embryo, and most probably other mammalian embryos as well, there is expression of gene activity during the early cleavage stages, possibly even in the two-cell stage. It is to be hoped that new techniques which will enable us to dissect out portions of the genome for study, simplifying the systems, will come to our aid in this very important area of genetic research.

Also, the use of new systems of development for study, such as imaginal discs in *Drosophila* larvae, may yield more information than classical systems have. *Imaginal discs* are single-layered clusters of cells which are determined in early development to give rise to specific structures of the integument. They can easily be removed at stages prior to visible differentiation and subjected to experimentation. For example, they can be removed and transplanted to flies that are genetically different, or grown in vitro individually or in mosaic combinations. (For a compilation of the kinds of studies being carried out on imaginal discs we refer you to vol. 17, no. 3, of the journal *American Zoologist*.)

Let us turn our attention now to two topics which are both interesting in terms of gene regulation, and of great importance to medical research: immune reactions, and the group of plant and animal diseases known collectively as cancer.

GENE REGULATION IN IMMUNE REACTIONS

We have stated above that events in development result from regulation of gene activity. REGULATION OF COURSE IS NOT CONFINED SOLELY TO DEVELOPMENTAL STAGES, BUT CONTINUES DURING THE ENTIRE LIFE SPAN OF THE INDIVIDUAL. AS WE DISCUSSED, ALTHOUGH THERE ARE "HOUSEKEEPING" GENES WHICH ARE NEVER REGULATED, MANY GENES ARE QUITE OBVIOUSLY TURNED ON AND OFF.

Genes involved in immune reactions are among those which have received much attention because of the diversity of cell products and functions which they determine in cells of the immune system. Such diversity among genetically identical cells could result only from gene regulation. Before discussing the gene activity involved, we will first briefly describe basic characteristics of the immune system found in higher animals.

Basic Characteristics of Immune Reactions

Research in the field of immunology has expanded with ever-accelerating speed in the past two decades, thanks to technical breakthroughs in molecular biology. The results of this research have depicted an enormously complex picture of cellular differentiation, cellular interaction, and regulation of gene-determined cellular products. We shall attempt to present here only major concepts of immunology related to the genetic control of immune phenomena.

The immune system has been recognized as a bodily system of defense against foreign agents called *antigens*, which lead to disease. There are two principal categories of immune reactions. One is the *humoral*, or *circulating*, type of reaction, dependent on the production of protein molecules by a special class of lymphocytes known as *plasma cells*. These protein molecules, known as *antibodies*, circulate in the serum of blood, and can react to antigens to destroy them in a number of ways. Antibody reactions with antigens on cell surfaces, for example, can result in agglutination or lysis of the cells, as in incompatible blood transfusions, or bacterial infections.

The second category of immune reaction is the *cell-mediated* form of reaction, which is dependent on the recognition of antigens by other lymphocytes which do not produce antibodies. Instead these lymphocytes together with macrophages destroy the foreign agent or cells by phagocytosis. Rejection of incompatible tissue or organ transplants is a prime example of a cell-mediated immune reaction.

B Cells in humoral immune reactions. One of the best examples of enormous genetic flexibility, and therefore gene regulation, is the ability of some lymphocytes to produce antibodies (human cells are known to produce millions of different antibodies). It has been established that each cell that produces antibodies produces only one type of antibody. Since antibodies are proteins, within each human individual there must be millions of cells transcribing different genes determining antibody structure.

Antibody-secreting cells are known collectively as *B cells.* They were first discovered to originate in a structure in the lower intestine of chickens called the *Bursa of Fabricius* (hence, the B). Humans and other mammals, however, have no Bursa, instead producing B cells in bone marrow. Whether we possess a structure equivalent to the Bursa is a matter of conjecture which has not been experimentally decided one way or the other.

In humans and other mammals, development of immunity to most antigens occurs postnatally. Newborn babies are therefore relatively susceptible to infections, but can derive some immunity from antibodies transmitted by the milk of nursing mothers. However, an exception to this general lack of antibodies in the newborn is the presence of antibodies to antigens of the ABO blood groups: At birth, type B individuals have anti-A antibodies in their blood serum, and type A people have anti-B antibodies. Type O people have both anti-A and anti-B antibodies. Why cells are stimulated to produce these antibodies in the apparent absence of antigens is not understood. Anti-Rh antibodies, on the other hand, are not produced by Rh^- people unless they are exposed to the Rh antigen as in transfusion or pregnancy (p. 92).

Plasma cells, or B cells, "recognize" a specific antigen because they have receptor antibody molecules on their cell surface which can react with the specific configuration of a particular antigen. When such a cell encounters its antigen, it can be stimulated to divide actively, and each daughter cell will respond to the same antigen by producing the same antibody (Fig. 12-10). For reasons that are not yet clear, plasma cells may be paralyzed rather than stimulated in the presence of large amounts of the antigen, or in the continuous presence of small amounts of antigen below the level necessary to stimulate antibody production.

B-cell and T-cell interactions. An interesting phenomenon which has only recently been recognized is that plasma cells cannot respond directly to all antigens against which they eventually form antibodies. On this basis, antigens have been separated into two different classes. Some antigens stimulate B cells directly. In others, the stimulatory effects are mediated by another set of immune cells, the *T cells,* which previously had been thought to be involved solely with cell-mediated immune

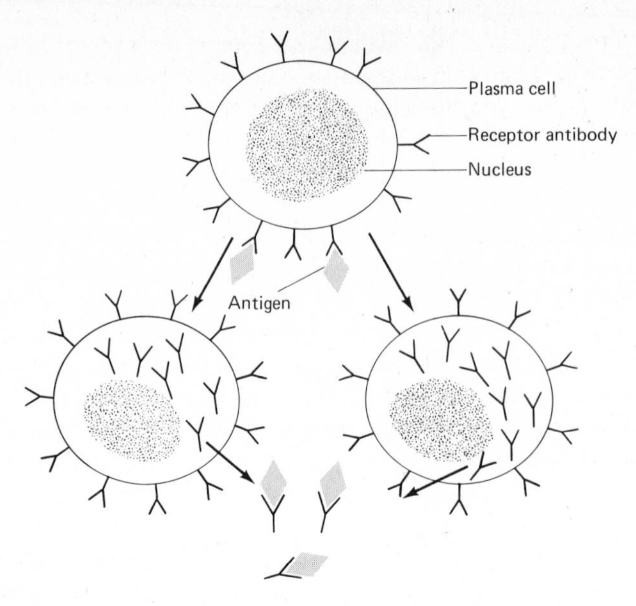

Figure 12-10. Schematic representation of B-cell reaction to antigen. Presence of an antigen recognized by receptor antibody molecules on the surface of a plasma cell leads to division and production of large amounts of antibody to attack the antigen.

reactions. Because T cells are lymphocytes which mature in the thymus gland, the antigens they recognize are known as *thymus-dependent* antigens. (Accordingly, antigens which are recognized directly by B cells are called *thymus-independent*.) How the T cells transmit information to B cells to stimulate antibody production, certainly a form of gene regulation, remains obscure.

As we indicated, T cells migrate during development of the organism from the bone marrow to the thymus, where they mature into *immunocompetent* cells, that is, cells capable of taking part in immune reactions. They then migrate from the thymus to sites such as the lymph nodes.

Recent studies show this class of lymphocytes to have diverse capabilities (Paul and Benacerraf 1977). The T cells which interact with B cells to cause antibody production against certain antigens are called *helper* T cells. There are also *suppressor* T cells, which apparently interact with B cells to cause suppression of antibody production. Some controversy exists as to whether these are in fact two different populations of T cells, or whether the same cell can act as helper on one occasion, and suppressor on another. There is evidence now that T cells may secrete substances (know simply as *T-cell factors*) which can stimulate or suppress B-cell activity (Taussig and Monroe 1975).

THUS, IT IS GENERALLY RECOGNIZED THAT T CELLS ACT TO REGULATE B CELLS IN IMMUNE REACTIONS. IN ADDITION, T CELLS ACT DIRECTLY IN CELL-MEDIATED REACTIONS. We shall discuss cell-mediated immunity in the following sections.

T Cells in cell-mediated immune reactions. Simply stated, cell-mediated immune reactions are those in which no antibody production is elicited. T cells recognize antigens on the surface of cells known as *histocompatibility* antigens. If the antigens are "foreign," that is, not the same as those on the cells of the individual, then certain T cells known as *killer* T cells will be stimulated to divide, and to migrate to the source of the foreign antigens and destroy them by phagocytosis. This reaction also appears to involve T-cell interaction with other cells, namely, macrophages (Shevach 1976).

This form of immune reaction occurs in transplantation of tissue and organs from one individual to another. It has long been established that a transplant of skin between mice of an inbred strain will be accepted, whereas a transplant between mice of different strains will be rejected (Billingham and Silvers 1971). More recently, much publicity has surrounded attempts of doctors to transplant essential organs from deceased donors to recipients with chronic ailments such as heart defects.

On occasion, transplants are accepted and the recipient enjoys the benefit of prolonged life span. More frequently, however, the transplants are rejected, or even cause a graft vs. host reaction because of incompatibility of the cellular antigens. The *graft vs. host reaction* is one in which lymphocytes of the graft attack cells of the host, causing wasting and disease. This type of reaction occurs when bone marrow is grafted into individuals suffering from an inherited deficiency of immunocompetent cells known as *agammaglobulinemia*. In such cases, even grafts between siblings usually fail, as lymphocytes of the graft attack the recipients own cells.

The reason for transplant rejections and graft vs. host reactions lies in genetic differences between donor and host. Just as inbred mice of the same strain can accept grafts, so can identical human siblings accept grafts from one another, such as in kidney transplants. The more alike two people are in genes determining immune reactions, the more likely it is that transplants between then will be accepted. In the following discussion we shall focus on genes involved in immune reactions. These genes determine antibody production, T-cell reactions, and interactions of different immunocompetent cells. Although much remains to be learned about the genetic basis of immune reactions, significant advances have been achieved in recent years toward the understanding of this extremely complex system.

Genetic Basis of Immune Reactions

From the preceding brief introduction to some immune phenomena it should be quite obvious that genes play a large role in the normal functioning of this critical body-defense system. Gene regulation governs the differentiation of lymphocytes into B cells, helper T cells, and so forth. We have no information, however, about which genes are involved, nor about what the factors may be that determine which developmental pathway a particular lymphocyte will take.

Our current understanding of genes involved in immune reactions centers around those determining antibody structure, and around genes of the major histocompatibility complex which determine histocompatibility antigens and other immune phenomena. These phenomena generally involve gene products, which we can isolate and analyze for molecular structure, as well as explore through linkage studies in experimental animals. The study of gene regulation is especially difficult because to date researchers have found no gene products related to regulatory genes that can be similarly studied.

Genetics of antibody production. All antibody molecules are tetrameric proteins. In 1973, Gerald Edelman and Rodney R. Porter (Edelman 1973; Porter 1973) jointly received the Nobel Prize for their work in establishing the molecular structure of antibody molecules (Fig. 12-11). These molecules are constructed of two identical polypeptides of about 50,000 molecular weight called the *heavy chains*, and two identical smaller polypeptides of about 23,000 molecular weight called the *light chains;* disulphide bonds join the four chains (Fig. 12-11a).

Comparison of different antibody molecules has revealed that certain portions of the heavy and light chains remain relatively constant, whereas other portions differ considerably. These are termed, respectively, the *constant* and the *variable* regions of the heavy and light chains. When the variable regions are cleaved from the constant regions by enzymes, only the variable portions of the polypeptides bind to antibodies. The fragment containing the light variable and heavy variable region is known as the *Fab* fragment (for *f*ragment, *a*ntigen *b*inding).

Further studies showed that the constant heavy chain contains three segments similar to the constant light chain in amino acid sequence and structure. Edelman suggested that these segments, designated *domains* (Fig. 12-11b), are folded in a specific manner to give the antibody molecule its final three-dimensional configuration; this interpretation has recently been confirmed by X-ray crystallography. The significance of the similarity in amino acid sequence of the domains, and thus of the DNA determining them, with regard to the evolution of immune genes will be discussed later (p. 394).

In humans, antibody molecules are classified into five major groups depending on the structure of the heavy constant chains. These are IgG (immunoglobulin gamma), IgA (alpha), IgM (mu), IgD (delta), and IgE (epsilon). Each class includes

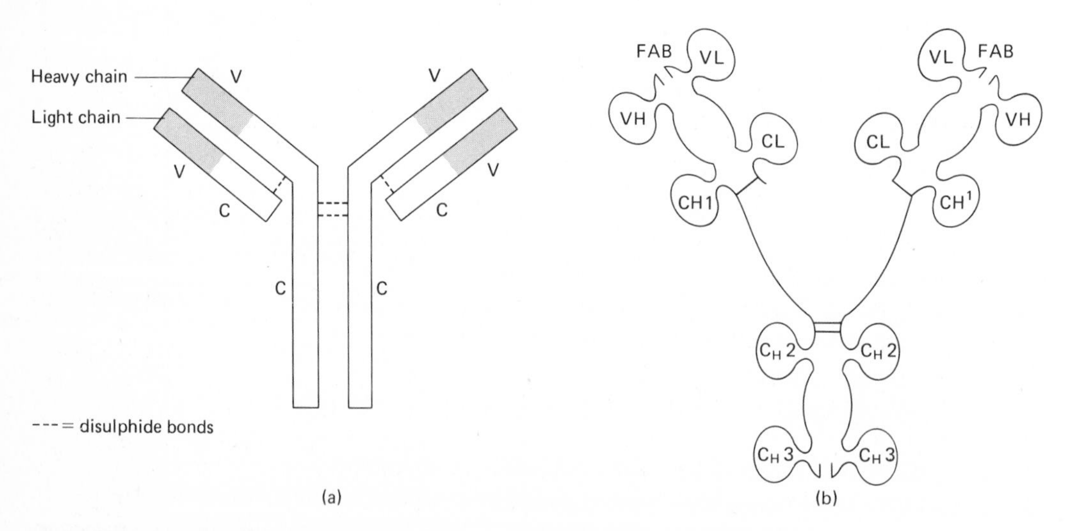

Figure 12-11. (a) Simplified diagram of antibody structure. (b) Schematic diagram showing folding of polypeptides. FAB = antigen binding fragment; VH = variable heavy; VL = variable light; CH = constant heavy, CL = constant light; C_H1, C_H2, C_H3 = constant domains. (From N. K. Kabat, 1972.)

subgroups based on certain amino acid sequences shared by areas of the constant heavy chains. The term *allotype* is used to refer to different forms of an immunoglobulin.

The different classes of antibodies are thought to play somewhat different roles in immune reactions. For example, IgE antibodies are associated with allergic reactions. During disease infection reactions, IgM antibodies are first produced by B cells, and then IgG antibodies are produced which are the major class of antibodies in defense against infections.

The production of IgM and then IgG antibodies again reflects a pattern of gene regulation. The IgM molecules are mostly incorporated into the cell membrane of B cells, where they serve as receptor molecules for a particular antigen. There is evidence that this occurs in chickens while B cells are differentiating in the Bursa, and that as the cells migrate from the Bursa to the bone marrow, the IgM gene is turned off and the IgG gene is turned on.

Light chains have been found to belong to one of two groups, again based on similarities of amino acid sequences. There are *kappa* variable and *kappa* constant chains, and *lambda* variable and *lambda* constant chains. Unlike heavy chains, in which the different kinds of constant chains can be associated with various variable chains, kappa constant can only join with kappa variable, and the same is true of lambda.

It may have struck you by now that we are treating the constant and variable regions of antibody polypeptides as separate entities, even though they are part of one polypeptide molecule. We have, until now, assumed that each polypeptide is determined by one gene, and all polypeptides determined by a particular gene would be invariable in amino acid sequence because of colinearity with the sequence of nucleotides of the DNA. Yet in antibody polypeptides there is variability *within* regions of the molecules.

This phenomenon initially puzzled biologists, but the explanation is now well accepted. Simply put, each of the polypeptides that constitute antibody molecules is determined by *two* different genes. IN OTHER WORDS, AT LEAST TWO GENES DETERMINE ONE POLYPEPTIDE IN THE CASE OF IMMUNOGLOBULINS. We shall not dwell on the evidence supporting this extraordinary situation, but certain references will provide further details (Hood 1972). It is mind-boggling to consider how these genes are regulated, since a stimulated plasma cell can produce up to 2000 antibody molecules per second! Table 12-4 summarizes present concepts of genes involved in the determination of immunoglobulins in humans and mice.

Genetic studies indicate that in humans, the kappa genes, lambda genes, and heavy-chain genes are all on different linkage groups. The variable genes for each type of chain (kappa, lambda, and heavy) are closely linked to their respective constant genes, but are not in a continuous cluster. A distinct region separates the cluster of variable genes from the cluster of constant genes.

Recent studies of mouse embryonic cells have shown that the V and C regions are widely separate, but that during differentiation of lymphocytes into B cells, the regions are brought close together to form a single transcriptional unit (Hozumi and Tonegawa 1976). It has been proposed that this occurs by a somatic recombination event, coupled with excision of the intermediate regions.

Table 12-4 Genes Involved in the Determination of Antibody Chains

Man		V-genes	C-genes
Light chains	κ	Ia Ib II III	—
	λ	I II III IV	Arg Lys Gly
Heavy chains		I II III	y4 y2 y3 y1 α1 α2 μ2 μ1 σ ϵ
Mouse		**V-genes**	**C-genes**
Light chains	κ	I II III IV V VI VII VIII etc.	—
	λ	I [II]	I II

V = variable chain
C = constant chain

From Milstein *et al.*, 1976. In *The Immune System*, p. 75. F. Melchers, and K. Rajewsky, eds. Springer-Verlag, New York.

Furthermore, sequence studies have shown that there are 2 segments in the V DNA. One codes for the variable chain, and one codes for intermediate polypeptides in the B cells. Because the intermediate polypeptides serve to join the variable and constant portions of the antibody chain, they are now known as joiner DNA or J DNA.

The J DNA is still separated from the C DNA by approximately 1.2 kilobases, an intervening sequence which is believed to be spliced out of the mRNA. Thus in the processing of genes and mRNA which leads to antibody production, two complex events are believed to occur: somatic recombination with excision, and mRNA splicing (Sakano *et al.* 1979). Figure 12-12 illustrates the various steps proposed in the maturation of B cells and the production of functional antibody molecules.

An area of controversy concerning regulatory mechanisms in cells of the immune system is the tremendous diversity that exists in the immunoglobulins produced by the cells, a phenomenon alluded to earlier. There must be millions of different molecules to respond to antigenic substances. Since any one lymphocyte produces only one antibody molecule, there must then be millions of lymphocytes manifesting functional genetic diversity. How does this diversity arise? There are two major lines of thought concerning this question: the *germ-line theory* and the *somatic cell theory* (Jerne 1973).

The germ-line theory proposes that all lymphocytes are genetically identical, and that reaction with the antigens or other immune cells in some way establishes certain combinations of the genes present to result in a unique protein produced by a particular cell. This cell then gives rise to daughter cells similarly constituted to produce the same antibody molecule.

The somatic cell theory proposes that lymphocytes are genetically different, so that cells can produce only a particular antibody. How the genetic diversity arises is another matter of controversy. Some feel that the genes determining antibody polypeptides may be especially mutable, and that they constantly undergo somatic

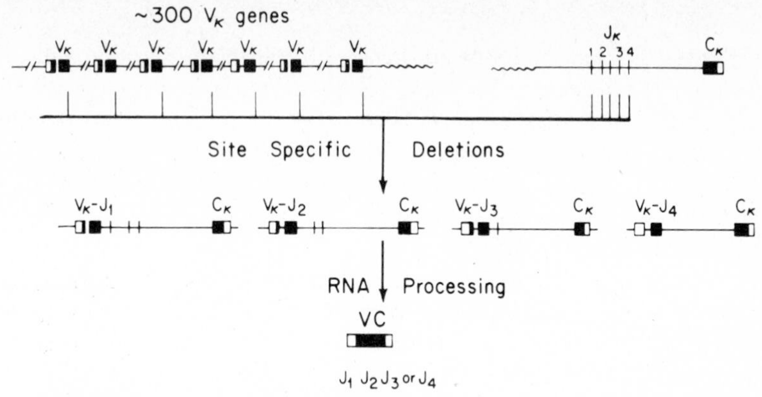

Figure 12-12. Schematic illustration of present theories regarding processing of antibody genes during lymphocyte maturation which results in site specific deletions to bring variable, joiner and constant region DNA together. This is then followed by mRNA splicing resulting in the final antibody structure. (Diagram courtesy Robert P. Perry, The Institute for Cancer Research, Fox Chase, Philadelphia.)

mutations. The question is still under study, although there seems to be more support for the germ-line theory than for the somatic theory.

Histocompatibility loci. Recognition of a genetic basis to transplantation reactions stems from work begun in 1916 by Clarence C. Little and his coworkers at the Jackson Laboratory in Bar Harbor, Maine.[1] The key to their work was to develop inbred strains of mice which they then used in controlled experiments for skin transplantation and the grafting of tumors between and within strains. They reported the now well-known phenomenon that grafts between genetically different organisms were rejected, while grafts within strains were not. In addition to conducting research activities and offering research training programs, the Jackson Laboratory remains today a center for the development of inbred strains of mice.

In subsequent years, scientists in many laboratories have determined that rejection is a form of immune reaction, and that numerous histocompatibility genes are involved in transplant rejection. Some genes have been found to determine relatively minor reactions to incompatible transplants. In both humans and mice there exists a cluster of genes called the major histocompatibility complex (or MHC) which determines the strongest immune reactions.

In humans the MHC is called *HLA* and is found on the 6th chromosome; in mice it is known as *H*-2 and is found on the 17th chromosome, not far from the *T* locus. More details are known about the H-2 locus because of experiments on inbred mouse strains differing with regard to the MHC genes. In the following discussion we shall be speaking primarily of the mouse MHC; however, studies on human genes of the MHC have revealed a large degree of similarity with H-2.

Tissue antigens originally studied in transplantation experiments are determined by genes *H*-2*D* and *H*-2*K* in mice and *HLA*-*A*, *B*, and *C* in humans. These antigens are glycoproteins and are found on the surface of all nucleated cells and platelets. They are also referred to as *serologically defined antigens*, or *SD*. If these

[1] Dr. George D. Snell of the Jackson Laboratory shared the Nobel Prize in Medicine in 1980 with Drs. Baruj Benacerraf and Jean Dausset for their work in immunogenetics, transplantation, and diseases.

genes are different in donor and recipient, transplant rejection will occur by stimulation of killer T cells.

Recent studies have revealed some interesting features of the SD antigens (Henning *et al.* 1976). Depending on the method used in extracting them from the cell membrane, the antigens have been found to be either a tetrameric protein composed of two heavy chains and two light chains, or a single heavy chain with a single light chain. Possibly both forms exist in the natural state.

The light chain of H-2 and HLA antigens has been identified as beta-2-microglobulin, a polypeptide found to be produced by all cells, not just cells of the immune system. What is fascinating is that β-2-microglobulin contains sequences of amino acids very similar to those of the constant domains of antibody molecules (Smithies and Poulik 1972).

Furthermore, amino acid sequences of both *H-2K* and *H-2D* are quite similar. Such discoveries have led to the proposal that, the gene for β-2-microglobulin (and possibly other genes of the MHC) arose during evolution by duplication of existing genes. Their relationship with constant domains determined by antibody genes brings to mind that antibody genes also exist in clusters in the genome, and duplication may also be the process by which antibody genes arose. In fact, it is frequently conjectured that MHC and immunoglobulin genes may have derived from the same ancestral gene(s).

Initially thought to determine histocompatibility antigens primarily, the MHCs of both mice and men are now known also to contain genes that determine a wide variety of immune reactions, including humoral immune reactions (see Appendix B). As Fig. 12-13 illustrates, the MHC regions contain genes determining different functions mapping in an overlapping manner (Paul and Benacerraf 1977; Albert 1976).

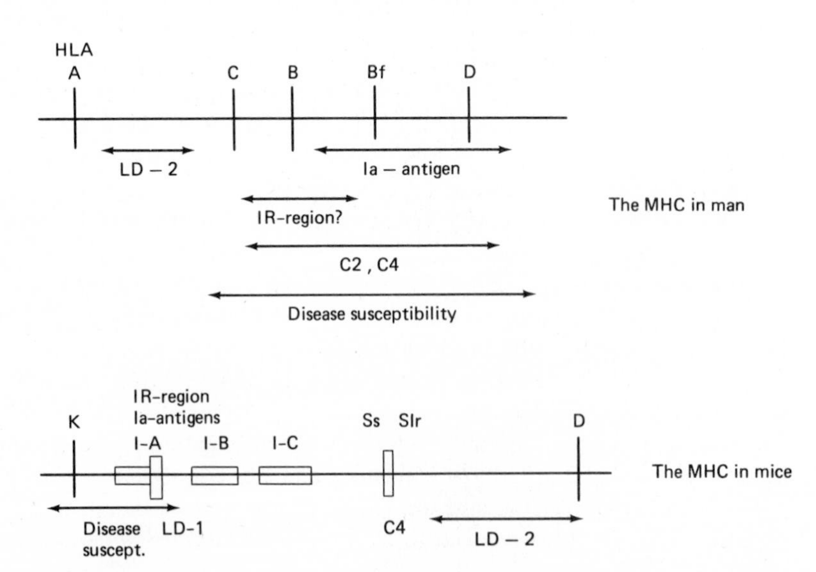

Figure 12-13. Diagrammatic representation of the MHC in man and mouse. (From E. D. Albert, 1976. *The Immune System*, F. Melchers and K. Rajewsky, eds., p. 136. Springer-Verlag, New York.)

GENE REGULATION IN CANCER

Another area of intensive study in both genetics and medicine has involved the disease known as cancer. Actually, cancer is not one disease, but a group of diseases in which cells which become abnormal share certain characteristic changes in structure and function. Since the structure and the function of cells are determined by genes, it is increasingly obvious that our eventual understanding of cancer will depend upon our uncovering genetic changes which occur in the cells that become cancerous. A number of these genetic changes are of a regulatory nature.

In our discussion we will again stress those studies most relevant to genetic phenomena, and restrict our exploration of clinical aspects to only the general characteristics of *neoplasia*, another term for cancer (*malignancy* is also synonymous with cancer).

General Characteristics of Cancer Cells

When cells become cancerous, some characteristic changes occur in basic cell behavior. The process by which a cell becomes cancerous is termed *transformation*. Recall that we have also encountered this term in an entirely different context, that of genetic transformation in bacteria (p. 238).

In transformation to malignancy the cell and its daughter cells become *hyperplastic*, or in other words, they divide ceaselessly. Transformed cells in vitro will also divide infinitely, as opposed to normal cells that have a finite time span of division. There also occur a number of cellular changes which are believed to be associated with the cell membrane in some ways; for example, transformed cells acquire the ability to migrate from one site to another in vivo, a phenomenon known as *metastasis*.

Cell division. Our finite size and shape reveal the obvious fact that normally there is a limit to the growth of our tissues by division. Whatever the limiting mechanisms are, they are either altered or bypassed in cancer. The existence of tumors and the predominance of certain white blood cells in leukemia, for example, all indicate that unrestrained division is a factor in cancer. This characteristic of infinite growth also has been affirmed in vitro: normal cells will survive and divide for a few generations in culture, but then no matter how they are cared for, will die. On the other hand, transformed cells will continue to divide and survive in vitro far past the life span of their normal counterparts.

Some cancer cell lines, such as the *Hela cell*, have been cultured for decades. Hela cells are so named after the woman (*He*nrietta *La*cks) from whom a cervical tumor was removed which served as the original source cell line. Hela cells are generally quite stable, with characteristic patterns of enzyme production by which they can be identified. They are such hardy cells that laboratories all over the world have recently encountered problems of contamination of other cultures by Hela cells.

Metastasis. The ability of malignant cells to migrate is the characteristic that distinguishes malignant growths from benign growths in vivo. Cells of a benign tumor remain in one site, therefore when the tumor is removed, the patient is considered cured. Malignant tumors, however, can "seed" tumors in other parts of the body as cells leave the original tumor and travel to begin growth elsewhere.

In vitro, this behavior may be mirrored by the loss of a normal characteristic known as *contact inhibition*. When a normal cell migrates over the substratum of a culture dish and encounters another cell, the two cells will cease to move upon contact. This results in a monolayer of cells which frequently shows a fairly regular pattern of cell position. Transformed cells show no contact inhibition, and continue to move over one another, resulting in a multilayered mass of cells (Fig. 12-14).

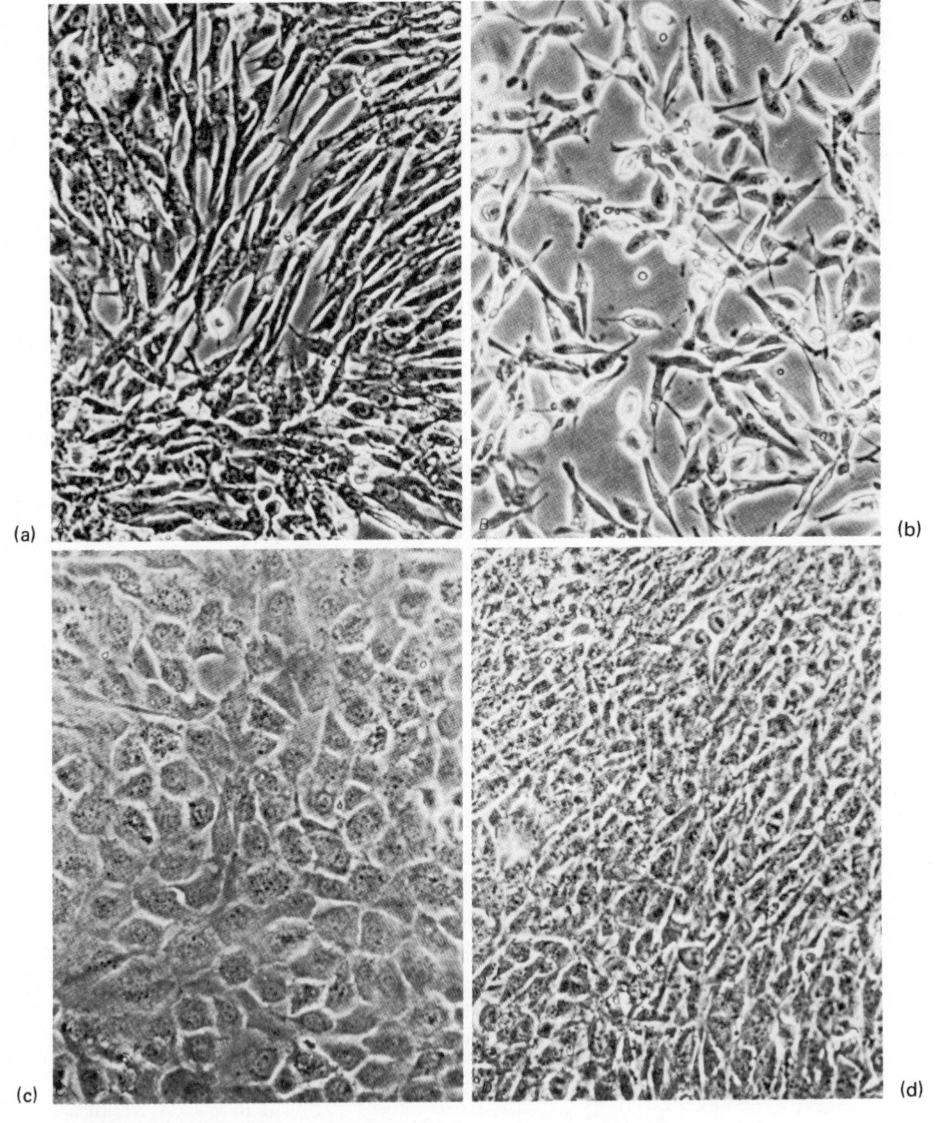

Figure 12-14. (a) and (c) Photographs of normal cells in culture; (b) and (d), of cancer cells. Note the better organization of the normal cell cultures. (Photos courtesy Dr. Renata Dulbecco, Imperial Cancer Research Fund Laboratories, London.)

Biochemical changes. When cells become malignant, biochemical changes accompany the other changes in cell division and movement described above. For example, some specialized proteins whose synthesis represents the differentiated state of the cell are sometimes no longer produced. At the same time, proteins not normally produced by cells are found when they become cancerous.

The presence of unusual substances produced by cancer cells is, however, difficult to interpret from the point of view of causation. For example, G. Abelev of the Soviet Union and others (1974) have reported finding an antigen called alpha-fetoprotein in transformed cells of both mouse and human. The interesting point here is that this is an antigen which is normally produced only during fetal stages of development, and ceases to be produced after birth. Thus a gene active during embryogenesis and then repressed is once again activated in transformed cells. It is not clear whether the expression of such genes or the repression of others is a cause or effect of transformation.

The production of alpha-fetoprotein is further evidence for a striking number of similarities noted between transformed cells and embryonic cells. Both show unregulated cell division; both are capable of migration; transformed cells by their loss of specialized gene activity resemble undifferentiated cells. It is not surprising, then, that several cancer specialists have proposed that among the genetic changes in transformation is the altered regulation of gene activity, so that normally active genes are repressed or altered; or normally repressed genes are reactivated. This would account for biochemical changes associated with transformation, some of which we have mentioned above.

What are the factors which result in this abnormal regulation of cells? We shall explore this question in the following sections.

Causal Factors in Transformation to Malignancy

The high cancer death rate in human populations (almost 20 percent in the United States) has prompted increased funding for research on the cause of the diseases. Optimism in the 1960s that science was on the brink of a breakthrough in understanding cancer has proved to be premature. We are now aware that this goal is not yet in sight. For one thing, gene regulation is involved in transformation, and you know by now that our understanding of regulation in eukaryotic cells is incomplete to say the least.

Furthermore, it is apparent that genetic regulation of the malignant cell is only a part of the story. Viruses, environmental factors, and genes determining susceptibility to cancer all play a role in the onset of this group of diseases. All these factors add up to an extremely complex and difficult situation. Indeed, where we hope that basic research into gene activity may enable us to understand what has gone wrong in a cancer cell, many researchers feel that such knowledge about the cancer cell will also give us the needed clues to understand normal cellular functions.

The solution is so elusive that one scientist likened cancer research to staring at a small light in the dark. The more you look directly at the light, the dimmer it becomes. We have to look to the periphery of the light to see it more clearly. In a manner of speaking many scientists are now looking at peripheral aspects of cancer in their attempts to understand its mechanisms, or to find some means of controlling the disease.

Viruses in cancer. That viruses are a factor in the development of cancer has been suspected since the early years of this century. In 1910, Peyton Rous isolated a cancer-causing virus in chickens known now as the *Rous sarcoma virus* (*RSV*). However, because the RSV could not induce cancer in mammals, the significance of viruses in general in the etiology of human cancers was generally disregarded. Support for a viral role in neoplasia increased in later decades, and was established by research on animal cancers in the 1960s. However, no virus has yet been isolated from human cancer which has been proven capable of inducing cancer when injected into normal human cells.

Let us first briefly discuss some of the similarities and differences in the behavior of bacteriophage and animal-cell viruses upon infection of host cells. Bacteriophage, you recall, enters either a vegetative life cycle resulting in lysis of the host cell, or a temperate phase during which the virus remains dormant in the host cell without having any adverse effects on it.

Viruses which infect animal cells can also follow these two lines of behavior. As with bacteria, animal cells can also lose viruses and be "cured" of them, a process known as an *abortive infection*.

In addition, new viral particles can be produced by the cell without lysis. Animal cell viruses can replicate their nucleic acids and incorporate a small portion of the host cell membrane as their protein coat. They are then released from the cell by a form of budding (Fig. 12-15). This process, which is known as *productive infection*, results in new virus particles without causing death of the host cell.

Yet another route which an animal virus can take following infection (which is not seen in bacteriophage) is that resulting in transformation of the host cell to cancer. Although both RNA and DNA viruses have been linked to cancer in a wide variety of animal species, most naturally recurring virus-induced cancers are the result of RNA virus infection. These viruses are also known as *retroviruses*, or *oncornaviruses*. Examples of oncornaviruses are the Rous sarcoma virus, the Gross leukemia virus in mice, and the Bittner mammary tumor virus, also in mice.

RNA tumor viruses. Retroviruses are also classified according to gross structure. Some are called type C particles because of their centrally positioned nucleoid (genetic material). Type B particles have an acentrically located nucleiod. Type A particles are smaller than either type B or C, and have a double-layered shell surrounding the nucleoid (see again Fig. 12-15). Generally, retroviruses contain two to four identical nucleotide chains of about 3×10^6 m. w. The term *virogene* is sometimes used to refer to the complete virus genome.

One of the major achievements derived from studies on RNA tumor viruses stemmed from studies on the mechanism by which they are reproduced, and on the observation that DNA sequences complementary to viral RNA were found incorporated in whole or in part into the genome of the host cell. Furthermore, efforts had failed to find RNA-dependent RNA polymerase in uninfected cells. In 1964, Howard Temin of Wisconsin proposed that a DNA intermediate is involved in RSV infection

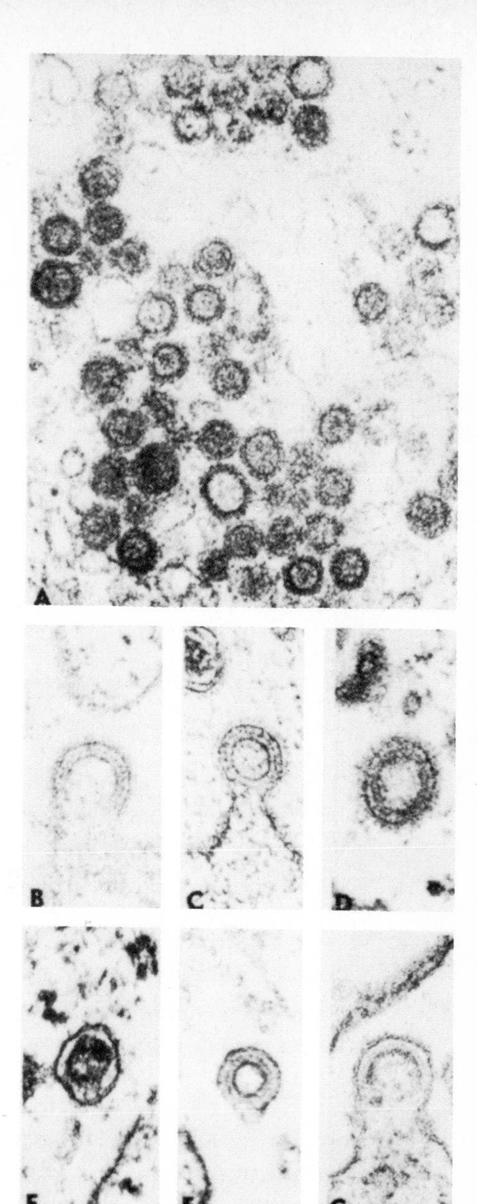

Figure 12-15. Electron micrographs of oncornaviruses [magnifications: (A) ~ ×100,000; (B to G) ~ ×140,000]. (A) Group of intracytoplasmic type A particles in a mouse mammary tumor. (B) Type B particle budding from a mouse mammary tumor cell. (C) Late bud of type B particle from mouse mammary tumor. (D) Free, immature, type B particle with spikes on surface, electronlucent center in the nucleoid, and spokes radiating from outer surface of the nucleoid to inner surface of the envelope. (E) Type C particle in extracellular space. (F) Late bud of type C particle in tissue of human embryo kidney cells infected with a strain of Rauscher leukemia virus. (G) Type C particle budding from human embryo lung cell in tissue culture infected with feline leukemia virus. [Source: Albert J. Dalton. National Cancer Institute. (From T. Maugh, 1974. *Science* 183: 1183.)]

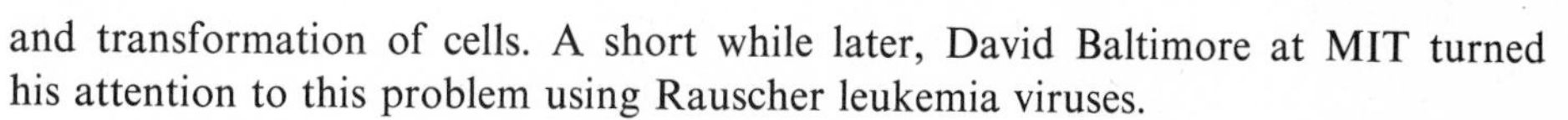

and transformation of cells. A short while later, David Baltimore at MIT turned his attention to this problem using Rauscher leukemia viruses.

In 1970, both men reported the isolation of an RNA-dependent DNA polymerase in cells infected with RSV and the Rauscher leukemia virus. The enzyme, commonly known as reverse transcriptase, is determined by one of the genes of the viruses, now called gene pol. This was the first discovery of an RNA ⟶ DNA pathway.

It is now accepted that the following sequence of events occurs when an oncornavirus infects a cell: During the first few hours, reverse transcriptase is produced. These enzyme molecules then catalyze the synthesis of DNA complementary

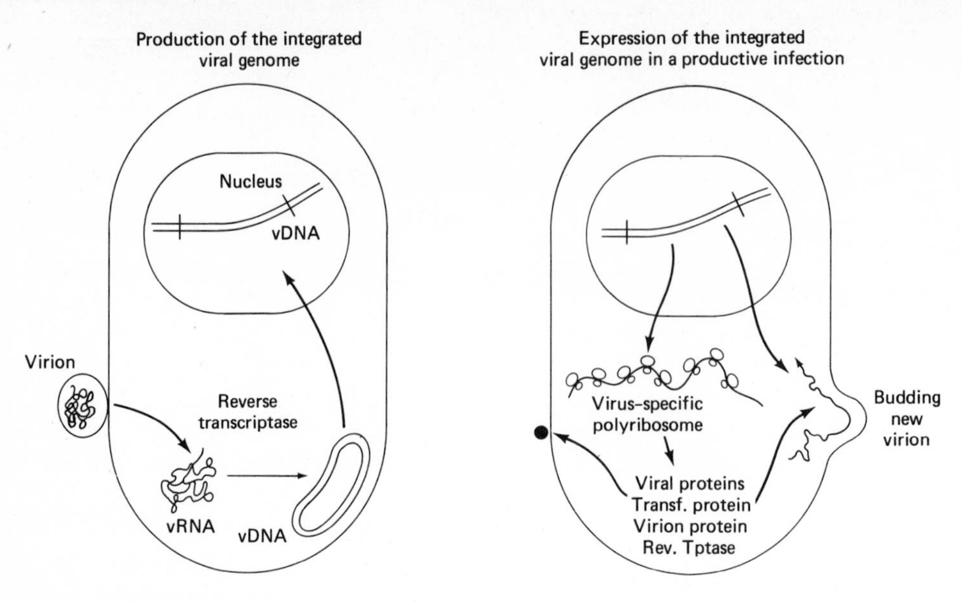

Figure 12-16. Life cycles of oncornaviruses. Explanation in text. (From D. Baltimore, 1976. *Science* 192: 634; copyright 1976 by the American Association for the Advancement of Science.)

to viral RNA. Tryptophan tRNA is believed to serve as the primer molecule for RSV; proline tRNA serves as the primer in murine viruses. The DNA then integrates into the cell genome. The second half of the life cycle then involves the production of viral proteins from mRNA transcribed from the provirus DNA. These proteins direct virus reproduction and/or transformation (Fig. 12-16).

Isolation of reverse transcriptase provided scientists with a powerful tool to study gene structure and function. In the years subsequent to its first isolation, experiments have used messenger RNA molecules isolated from cells (such as mRNA for hemoglobin and mRNA for insulin) to serve as templates for the synthesis of DNA using reverse transcriptase. In this way researchers have been able to produce the nucleotide sequence of individual genes. Once such individual genes are available, experiments can be designed to study their molecular structure, and to alter them enzymatically to study the effects on gene function. We shall discuss this work further in Chapter 14. For their discovery, Temin and Baltimore received the Nobel Prize in medicine together with Renata Dulbecco, a pioneer in the study of cancer viruses, in 1976.

Studies of RNA tumor viruses have led to a generalized concept of their genome. Besides the gene *pol* for reverse transcriptase, they are thought to possess at least three other loci. One is designated *gag*, and determines structural proteins for viral reproduction. The gene *env* determines the formation of the membrane envelope surrounding new viral particles. Finally, the gene *onc* is believed to determine a transforming protein that in some way initiates the cellular changes leading to malignancy of the host cell.

One of the disturbing revelations to come from research on retroviruses is that cells of most, if not all, animal species contain DNA related to retroviruses in their genome (Todaro 1974). Since these nucleotide sequences have been found in germ cells as well as somatic cells, they are transmitted vertically (from one generation to another) as well as horizontally (from one individual to another within the same generation). Whether the genes are triggered to initiate transformation depends on the genome of the individual and the presence of environmental factors which affect the activity of cancer genes.

What might be the origin of these virus-related gene sequences? We can only speculate on the answer to this question. Temin has proposed that there exist gene sequences which under certain circumstances can become the precursors of retroviruses. In other words, cellular DNA "gone wrong" is the source of new viruses. This has been called the *protovirus theory*. Todaro and his associates have postulated that at some point in our evolutionary past, viral genes were integrated into the DNA of cells of our ancestors, and have been transmitted over the centuries as part of our genome. This is the *oncogene theory*. Still others, such as Baltimore, support the idea that retroviruses inserted into cellular DNA can at any time deintegrate and give rise to new viral particles which can then reattach. Over the years, this may result in genetically different particles through the accumulation of mutations.

What evidence exists that the viral genes causing transformation to cancer can be passed from one individual to another like other infectious agents? One of the earliest studies to indicate this possibility was done on the transmission of the Bittner mammary tumor virus in mice. Inbred strains of mice were developed which differed greatly in incidence of mammary tumor development. When high-incidence females served as foster mothers to low-incidence newborn mice, the foster offspring developed a much higher incidence of mammary tumors. The reason was discovered to be transmission of the mammary tumor virus through the nursing females' milk to the young, which then were infected. More recently, researchers have reported evidence that feline leukemia virus is also infectious.

DNA tumor viruses. Among the DNA viruses that have been found to cause transformation in various species of animals are *papova* viruses (so named because they cause a variety of malignancies—*pa*pilloma, *po*lyoma, *va*ccuolating tumor viruses, and herpes viruses). Since their genetic material is DNA, these viruses can integrate directly into the genome of host cells.

The DNA viruses that are best characterized genetically are two very small ones, polyoma (a naturally occurring virus in mice) and simian virus 40 (SV40). Both induce tumors when injected into newborn hamsters and mice, but neither has been shown to affect human cells. The entire nucleotide sequence of the SV40 virus has recently been determined (Reddy *et al.* 1978). Figure 12-17 shows a schematic representation of some of the principal features of the SV40 genome. There are sequences for the determination of at least four proteins—VP1, VP2, VP3, and a locus determining the T antigen (also known as the A protein) produced early in infection in all cells transformed by SV40. How it functions is unknown, but the presence of T antigen is necessary for transformation.

Among some of the interesting features of the SV40 genome is the fact that some 23 percent of the DNA is not involved in synthesis of the known proteins. This

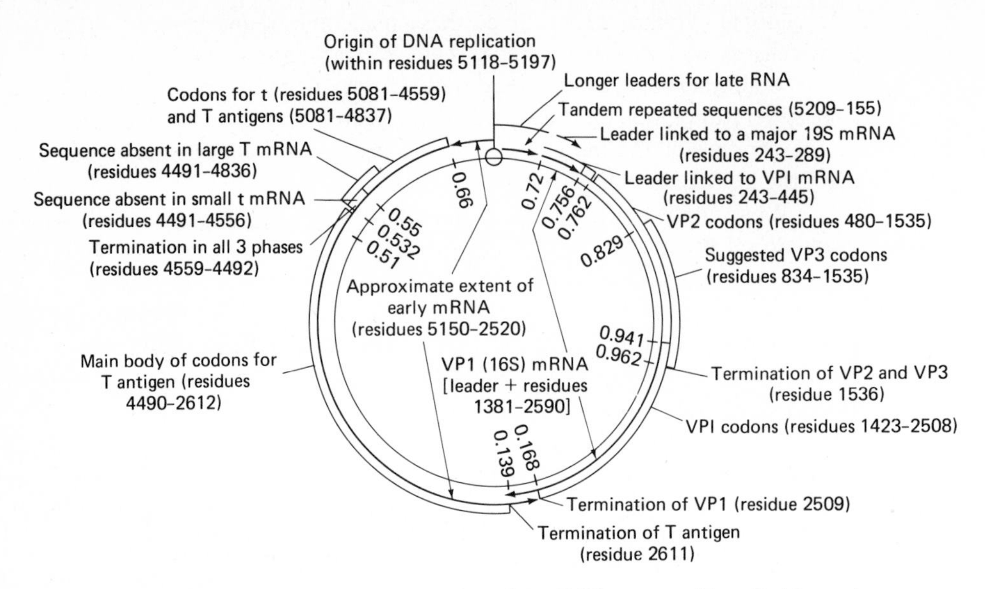

Figure 12-17. Schematic representation of the SV40 genome. (From Reddy *et al.*, 1978. *Science* 200: 500; copyright 1978 by the American Association for the Advancement of Science.)

DNA contains areas of rotational symmetry, continuous A–T sequences, and repetitive sequences. There are also regions of the DNA which contain termination codons, and so could conceivably direct the synthesis of small polypeptides.

Furthermore, analysis of the region believed to code for the T antigen reveals a number of termination codons, such that the largest sequence of sense codons is much smaller than would be expected for a protein the size of the T antigen. This situation may be similar to that for the genes determining antibody structure: a cluster of separated genes which give rise to one continuous polypeptide.

Tumor viruses in humans. Although the role of viruses in animal cell tumors has unquestionably been established, their role in human tumors is still uncertain. There have been sporadic reports of the isolation of viral particles from human cells. In 1975, for example, Gallagher and Gallo reported finding viruses in human acute myelogenous leukemia cells. Later, however, these viruses were found to be indistinguishable from woolly monkey virus and a baboon virus (Klein and Smith 1977). The question that cannot be answered is how these viruses came to reside in human cells (possibly by horizontal transmission?).

Still, there are reports which link human cancers to viruslike nucleic acids, even though no complete virus particle has yet been isolated from tumor cells which, when injected into normal cells, causes them to be transformed. One study (Spiegleman *et al.* 1972) reported finding nucleic acid sequences related to mouse mammary tumor virus in cells from human breast carcinoma, using reannealing techniques. These sequences were absent in normal mammary cells.

Recent studies have revealed that at least three sites in different human chromosomes serve as integration sites for the SV40 virus. This phenomenon may be interpreted as the presence of sequences in our genomes related in some way to the SV40 virus.

The two viruses which are most frequently spoken of in conjunction with human cancers are the herpes virus 2, associated with cervical carcinomas, and the Epstein-Barr virus, found in Burkitt's lymphoma. Both are DNA viruses, but unlike the proven oncogenic viruses, when transmitted horizontally are found to be benign. The question of virus involvement in human cancers thus remains unresolved.

Immunological aspects of cancer. Awareness that we are dealing with viruses and various cell-surface changes in neoplasia, including the production of foreign cell-surface antigens in cancer cells, has led to application of immunological principles and techniques in an attempt to find a means to control malignancies. SEVERAL LINES OF EVIDENCE INDICATE THAT THE IMMUNE SYSTEM IS IN FACT ACTIVATED AND RESPONDS TO THE PRESENCE OF BOTH CANCER VIRUSES AND CANCER CELLS.

Response to the polyoma virus in mice is seen in the inability of such viruses to induce tumor formation in adult mice. However, tumors can be induced if the polyoma virus is injected into newborn mice (whose immunological reactions are not yet developed), or into adult mice whose immune system has been artificially weakened.

In both humans and experimental animals infection with bacterial vaccines seems to increase resistance to tumors. Apparently the bacteria stimulate cells of the immune system to great activity, and accordingly allow recognition and rejection of the tumor cells. Factors which can stimulate the immune system to react more effectively against other antigens are known as *immunopotentiators*.

Given this very clear evidence that the immune system can react to cancer viruses and cells, one would have expected success in developing vaccines or immunological therapy that would cure patients of the disease. In the 1960s there was great optimism that this goal could be accomplished within a very short time, and the government announced support for a crusade against cancer that would lead to a "cure" within years.

Quite obviously, the increasing numbers of cancer deaths reveal that we have not reached this goal. For one thing, as far as developing a *single* vaccine for cancer is concerned, there are too many different viruses involved for this to be possible. More than 100 oncornaviruses have been isolated from animal cells. Furthermore, no human cancer virus has been found. One obviously cannot develop a vaccine unless the virus can be isolated. Even chemically induced tumors show great variability. The experiments on methylcholanthrene-induced tumors that elicit an immune reaction showed each tumor to be immunologically different from every other tumor even though they were induced by the same chemical in the same inbred strain.

But what of the tumor cells known to have certain specific antigens, such as the T antigen of SV40, and others associated with polyoma and leukemia viruses? Although one would expect the immune system to be able to detect and defend against specific antigens, the development of cancer from these viruses indicates otherwise. It appears that cancer cells have different "tricks" to escape immunological surveillance. For example, TL, a specific cell surface antigen in leukemia cells, disappears from cancer cells in animals immunized to the antigen (Old 1977). Yet the same cells produce the TL antigen again when they are transplanted into nonimmunized animals. This behavior reflects a form of gene regulation which enables the cancer cells to survive immune attack. Remember, too, that cancer cells still have many self antigens which may confuse recognition by lymphocytes.

In addition to immune reactions which are of a specific nature, that is, in which antibodies may be produced against a specific antigen, animal cells also have a nonspecific response to viral infection. This nonspecific response involves the production of a glycoprotein known as *interferon* (for a review, see Burke 1977). The infection of a cell by virus elicits interferon production. In humans there are two genes known to determine interferon production, which are found on chromosomes #2 and #5. Interferon produced by one cell can spread to surrounding cells by binding to a receptor on the cell determined by a gene on chromosome #21. Protection from further viral infection may then be provided by antiviral substances produced by the cells, possibly in response to the presence of the interferon.

Interferon appears to act only against viruses, as cellular processes are not affected at all. It seems to have no specificity of effect, as interferon protects against all viruses. Initial studies had indicated that specificity did exist, however, in that interferon produced by cells of a particular species could protect cells of only that species and no other. Recently findings have indicated that interferon from one species may in fact protect cells from another species; for example, human interferon appears to protect bovine cells from viral infection.

If so, this may be a powerful tool in preventing viral infections, as interferon is so powerful that a few molecules are sufficient to protect cells. However, cells produce interferon in such extremely small amounts that it has been very difficult to collect enough material to purify and analyze, much less enough to perform a significant test on cancer patients. The small amounts that have been administered to terminal patients have produced negative and ambiguous results. The first steps to solve this problem of scarcity of material were taken in 1980, when several laboratories succeeded in isolating the structural genes for human interferon using techniques of genetic engineering. These genes are now being produced in great quantities, and it is hoped that the availability of interferon for testing will be increased substantially. More details of this work will be given in Chapter 14.

Other factors in cancer etiology. To complicate the situation even more, still other factors are implicated in the onset of cancer. From the genetic point of view, there is the association of cancer with aging cells, familial incidence of cancer in humans, genetic susceptibility and resistance in experimental animals, and the presence of cytogenetic abnormalities in cancer cells.

The relationship between incidence of cancer and aging is very clear in humans (Cairns 1975). Most cancers develop in older people. Cancer of the large intestine, for example, increases about a thousandfold between the ages of 20 and 80. The increase in somatic mutations as well as the decreasing ability of aging cells to repair mutation (which we shall discuss further in Chapter 15) are believed to be related to this increase in frequency.

In humans, the genetic basis of susceptibility to cancer is not at all clear. Where there are reports of familial clustering of cancer, there is not evidence in a Mendelian sense that onset is genetically determined. Certainly, there must be a genetic factor. For one thing, our immune reaction is genetically determined, and we have already seen that the immune response can be a factor in development of malignancies. Unfortunately, at the present time, there is no way to predict in a Mendelian way the likelihood of any individual developing cancer.

Research on inbred strains of mice has given clear evidence that susceptibility or resistance to some cancers is genetically determined: Some strains have a high incidence, some have a low incidence. Crosses have indicated this trait is Mendelian. The AKR strain, for example, is known to have a high incidence of leukemia. Two loci have been found to determine inducibility of murine leukemia virus by chemicals in cells in tissue culture. One locus is linked to the first linkage group. How these loci determine inducibility is not known.

That cytogenetic abnormalities are associated with both aging and neoplasia has been known for many years. For example, the well-known Philadelphia chromosome is found in some patients with chronic myeloid leukemia (p. 182). Evidence that this is a somatic mutation is the fact that only affected white blood cells show this aberration; all other cells of the patients have normal #22 chromosomes. With the development of banding techniques, the presence of both numerical and structural chromosome abnormalities in neoplastic cells has become well established (Mark 1977). It is again a problem, however, to interpret the significance of these abnormalities. They may very well be the result of cellular changes due to transformation, rather than the cause of transformation.

Finally, of critical importance to epidemiologists is the effect of environmental factors in the etiology of cancer. Studies on substances such as cigarette tars and asbestos (which are specifically linked with lung cancers and abdominal cancers, respectively) are well known. Such cancer-causing factors are known as *carcinogens.* How carcinogens are able to activate genes involved in transformation is the subject of much interest, but again of little understanding. Even more perplexing is the fact that years may ensue following an encounter with a carcinogen before the onset of transformation. Asbestos inhaled by workers in shipyards in the 1940s was linked to an increasing number of cases—appearing in the 1970s—of mesothelioma, a formerly rare cancer of the lining of the abdomen.

From our brief look at various aspects of cancer, you can see that the situation still poses many problems. Although no "cure" is in sight, we can expect continued progress in the understanding of some specific forms of cancer, as there has been in recent years. Most scientists have always been aware that there is no single explanation for transformation, and no single cure for cancer.

FACTORS IN GENE REGULATION

Our discussion of gene regulation in eukaryotes has essentially only catalogued known systems in which gene regulation plays a part. The actual mechanisms of regulation were not described, since they are as yet unknown. There are, however, some factors known to be involved that may some day be incorporated into an overall scheme, and we shall discuss a few of these now.

If we define promoter, or initiator, genes as being areas of DNA to which

RNA polymerase binds for the initiation of transcription, no doubt there are promoter genes in eukaryotic cells. There is transcription and RNA polymerase in our cells, and it would appear that there must be areas of the DNA for which the enzyme has affinity. We just have not been able to identify a specific promoter gene.

We also know that there are many areas of the DNA in eukaryotic cells which apparently are not transcribed at all, therefore resembling operator and promoter genes in prokaryotes, which also are not transcribed. In Chapter 7 we mentioned that these areas, known as *satellite* or *repetitive* DNA, are present in all eukaryotic cells. It has been conjectured by many geneticists that they may serve some regulatory function, since far more DNA is available than is needed to account for the structure of cellular proteins. It is believed that this excess, much of it repetitive, may be regulatory in nature.

DNA and Chromosome Structure in Regulation

Related to the possible regulatory function of repetitive DNA is the proposal by some scientists (Wells *et al.* 1977) that inherent in the structure of DNA itself may be areas that serve as regulatory sites. They point out that though the overall base composition of two regions of DNA may be the same, differences in the sequence of bases can result in distinct physical differences, such as bouyant density in CsCl gradients. Differences in sensitivity to various nucleases, presence of long blocks of AT pairs or GC pairs, and the existence of repetitive sequences, or palindromes, are all clues to possible different functions of regions of the genome.

The ability of heterochromatin to suppress gene expression (p. 380) has been discussed previously. Although it is not difficult to visualize the lack of expression of genes within a tightly coiled portion of the chromosome, it is difficult to explain the ability of heterochromatin to suppress gene expression in neighboring sites (such as revealed in the experiments translocating autosomal genes to the inactivated X chromosome). Perhaps the physical nature of heterochromatin is such that DNA located next to it becomes distorted enough to prevent interaction with RNA polymerase molecules.

Chromosomal proteins in regulation. One component of chromosomes that has been associated with both structure and gene regulation are the proteins, both histones and nonhistones. In 1962, Huang and Bonner published the results of studies on cell-free preparations for protein synthesis, showing that addition of histones sharply reduced transcription of genes. Histones, as you may recall, are known to be complexed with DNA to form nucleosomes (p. 161). However, the effect is apparently nonspecific, in that regardless of gene activity, the histones themselves are unchanged from cell to cell, and from species to species.

Recently, evidence has been reported that nonhistone proteins play an important role in gene regulation in eukaryotic cells (Stein *et al.* 1974). Nonhistones, as we mentioned earlier (p. 157) are extremely diverse, believed to contain both structural and functional proteins, such as enzymes. Furthermore, different nonhistone proteins are found in different cells, in different species, and even in cells at different stages of differentiation, leading to speculation that the nonhistones function in gene regulation.

If nonhistones are added to cell-free systems in which histones have been added to repress transcription, protein synthesis increases. This phenomenon has been found to be associated with phosphorylation of the nonhistone proteins. Stein and his coworkers believe that when nonhistone proteins bind to histones, and become phosphorylated, the result is a negative charge of the complex. Since DNA also is negatively charged, there may ensue a repelling of the proteins and freeing of the DNA for transcription.

If this concept of gene regulation is correct, we must assume that genes determining the synthesis of nonhistones are activated at certain times in order for the proteins to regulate genes as the cell differentiates and metabolizes. This assumption still leaves unanswered a very fundamental question, namely, what turns on the nonhistone protein genes?

Chemical regulators in eukaryotic systems. The answer may someday be related to the many chemical changes which we know can stimulate genes to activity. In our discussion on Balbiani rings for example, we mentioned that the hormone ecdysone can stimulate specific puffs on polytene chromosomes of *Drosophila* larvae. Hormone stimulation of various cells in the human body is certainly well established.

Different Levels of Regulation in Eukaryotes

The difficulties we have encountered in our search for mechanisms of gene regulation in eukaryotes stem from both genetic complexity and from different levels of organization which are not found in prokaryotes. Aspects of the genomes of higher plants and animals which cause us problems include diploidy, polygeny, nonlinked genes determining one trait, the amount of genetic material, and its interaction with proteins.

In addition, we must contend with the communication between different cells. In fact there is communication at many levels: Cells communicate with each other within an organ; cells of different organs communicate with each other for proper functioning of the organism as a whole. Sometimes, as in the case of hormones, this communication occurs over quite a distance.

A number of models have been proposed to account for many of the systems of gene regulation we have discussed in this chapter. For details of two of these, we refer you to papers by Britten and Davidson (1969) and Tomkins *et al.* (1969). No experimental tests exist, however, to substantiate these models. Perhaps the development of new approaches to gene regulation in eukaryotes will lead to verification, or to new models which can be tested experimentally.

New Approaches to the Study of Gene Regulation

Somatic cell fusion. We discussed earlier the technique of somatic cell fusion with regard to the identification of linkage groups of specific human genes (p. 196). The same technique has been used to study regulation in different systems. By fusing cells of different phenotypes, scientists can detect the effects of fusion in the hybrid cell.

Studies have been reported on the effect of fusion between malignant and nonmalignant cells. Although results have been somewhat variable, the general observation has been that fusion of cancerous and noncancerous cells results in a suppression of malignancy. For example, fusions of normal mouse x tumor mouse cells resulted in tumorous growth in vitro in only 21 percent of the hybrid cells, if such hybrid cells retain their full complement of chromosomes. Hybrid cells which showed full expression of tumorigenicity were generally found to have extensive chromosomal alterations.

Suppression of malignancy has also been reported in human cell hybrids. Hela cells were fused with nonmalignant fibroblasts and then injected into either immunosuppressed mice, or a special strain of genetically athymic mice. The athymic mice of course are unable to carry out cell-mediated immune reactions, and incidentally, are also hairless and often referred to as *nude mice*. In these hosts, hybrid cells are nonmalignant, whereas the Hela cells alone would give rise to tumor growth. For a review on such studies, see reference to Ozer and Jha (1977).

Another approach with cell fusion has been to form hybrids between normal and malignant cells of different species. Recall that if human and mouse cells are fused, there is preferential loss of human chromosomes, thereby enabling us to correlate certain human gene activity with a specific chromosome that remains in the hybrid cells (p. 197). Similar studies have been made to attempt to correlate expression of malignancy with a particular chromosome or combination of chromosomes. For example, Croce and his coworkers (1975) fused SV40-transformed human cells and a variety of mouse cell types, and found that expression of the SV40 T antigen depends on the presence of human chromosome #7. We can thus assume that there is a site on chromosome #7 which is the site of integration of SV40 DNA. More recently, evidence has been provided that there is also a site for SV40 integration on human chromosome #17 (Croce 1977), and perhaps a third site.

This very interesting application of somatic cell fusion also extends to the study of other systems involving gene regulation. For example, Gretchen Darlington of Cornell Medical School has fused mouse hepatoma (liver tumor) cells with human fetal liver cells and found that human liver functions are expressed in hybrid cells. If mouse hepatoma cells are fused with human leucocytes (white blood cells), human albumin (normally produced in the liver) is produced by the hybrid cells. This fascinating observation must reflect the ability of mouse hepatoma genes or gene products to derepress albumin genes of the human cells. Similar work continues today studying the effects of fusing aged cells with young cells and many other different systems which will no doubt yield important clues in our search for understanding gene regulation.

Recombinant DNA. Finally, we can mention again here the technique that we shall discuss in detail in Chapter 14, which is being used in many laboratories in studies of gene regulation and in other areas of genetic research. It is a technique which uses restriction endonucleases to break DNA into specific fragments to be isolated and studied alone, or in combination with other fragments. This general approach is called *recombinant DNA* or *gene splicing* technology.

This technique allows us to break up the genome of the SV40 virus, for example, and to make different combinations to detect what controls the expression of the gene for the T antigen. Furthermore, we can recombine DNA fragments of cells from different species to explore whether regulatory genes from prokaryotes have any effect on the expression of structural genes from eukaryotes. In Chapter 14 we shall describe the kinds of experiments that have been done and the information they have provided. This approach is not without controversy, however, and we shall also discuss arguments both criticizing and supporting the continuation of such research.

PROBLEMS

1. Define the terms ontogenic, oncogenic, induction in development, induction in prokaryotic regulation, transformation in bacteria, transformation in cancer.

2. What are the differences between gene amplification and gene duplication?

3. Design an experiment which would show the transmission of chemical factors (messengers?) between different components of an embryonic induction system (such as the lens and the epidermis).

4. The structure of antibody molecules was determined with the aid of studies on immunoglobulins obtained from patients with myeloma, a cancer of antibody-producing cells. In such patients, the antibodies produced were predominantly of one molecular type, though not identical in different patients. These uniform proteins are called the *Bence Jones proteins*. What would be the advantages of studying antibody structure from Bence Jones proteins from such an individual rather than antibodies from a normal individual?

5. Summarize the reasons why it is so difficult to find immunologically compatible tissues for transplantation in humans.

6. Many organisms do not possess immune systems as we know them. Speculate on possible mechanisms by which such organisms can resist infection by pathogens such as viruses and bacteria.

7. Explain in genetic terms why scientists are puzzled by and actively pursuing the basis for the ability of female mammals to retain embryos implanted in the uterus during normal pregnancies.

8. What is the evidence which has caused scientists to propose a definite relationship between cancer virus nucleic acids and cellular nucleic acids?

9. Studies by Temin and his coworkers showed that if DNA made from the genome of RNA tumor viruses is injected into nontransformed cells, the RNA viruses can eventually be produced in these cells. This process is known as *transfection*. Give an explanation for these observations.

10. There is a rare autosomal recessive disease known as *progeria* (page 496). Its effect is to cause premature aging in patients, who show various symptoms of old age such as wrinkling, and death from atherosclerosis (in their teens). Like naturally aging cells, progeria cells recover poorly from exposure to X-rays (we shall explain why in the next chapter). The following cell fusion experiments were performed, and results reported:

(a) young normal cells $\times$ progeria cells $=$ improved recovery from X-rays
(b) old normal cells $\times$ progeria cells $=$ no improvement in recovery

(c) progeria[1] cells × progeria[2] cells = improved recovery.
(1 and 2 refer to different cases.)

How would you explain these results?

REFERENCES

ABELEV, G. I., 1974. "Alpha-fetoprotein as a Marker of Embryo-Specific Differentiations in Normal and Tumor Tissues." *Transplant. Rev.* 20: 3–37.

ALBERT, E. D., 1976. "The Major Histocompatibility Complex and Its Biological Function." In F. Mechers, and K. Rajewsky, eds., *The Immune System.* Springer-Verlag, New York.

BALTIMORE, D., 1976. "Viruses, Polymerases, and Cancer." *Science* 192: 632–636.

BECKER, H. J., 1962. "Stadienspezifische Genativierung in Speicheldrusen nach Transplantation bei *D. Melanogaster.*" *Zoo. Anz. Suppl.* 25: 92–101.

BEERMANN, W., 1964. "Control of Differentiation at the Chromosomal Level." *J. Exp. Zool.* 157: 49–62.

BENACERRAF, B., AND M. O. MCDEVITT, 1972. "Histocompatibility-linked Immune Response Genes." *Science* 175: 273–279.

BILLINGHAM, R., AND W. SILVERS, 1971. *The Immunobiology of Transplantation.* Prentice-Hall, Inc., Englewood Cliffs, N. J.

BRIGGS, R., AND T. KING, 1952. "Transplantation of Living Nuclei From Blastula Cells Into Enucleated Frogs Eggs." *Proc. Nat. Acad. Sci. U. S.* 38: 455–463.

BRITTEN, R. J., AND E. H. DAVIDSON, 1969. "Gene Regulation for Higher Cells: A Theory." *Science* 165: 349–359.

BROWN, D. D., AND I. B. DAWID, 1968. "Specific Gene Amplification in Oocytes. Oocyte Nuclei Contain Extrachromosomal Replicas of the Genes for Ribosomal RNA." *Science* 160: 272–50.

BURKE, D. C., 1977. "The Status of Interferon." *Sci. Am.* 236(4): 42–50.

CAIRNS, J., 1975. "The Cancer Problem." *Sci. Am.* 233: 64–78.

CAPLAN, A. I., AND C. P. ORDAHL, 1978. "Irreversible Gene Repression Model for Control of Development." *Science* 210: 120–130.

CLEVER, U., AND P. KARLSON, 1960. "Induction of Puff Changes in the Salivary Gland Chromosomes of *Chironomus Tentans* by Ecdysone." *Exp. Cell Res.* 20: 123–6.

CROCE, C. M., E. ADEN, AND H. KOPROWSKI, 1975. "Somatic Cell Hybrids Between Mouse Peritoneal Macrophages and Simian Virus-40 Transformed Human Cells: II Presence of Human Chromosome 7 Carrying Simian Virus 40 Genome in Cells of Tumors Induced by Hybrid Cells." *Proc. Nat. Acad. Sci. U. S.* 72: 1397.

CROCE, C. M., 1977. "Assignment of Integration Site for Simian Virus 40 to Chromosome 17 in GN54VA, A Human Cell Line Transformed by Simian Virus 40." *Proc. Nat. Acad. Sci. U. S.* 574: 315–318.

DARLINGTON, G. J., H. P. BERNARD, AND F. H. RUDDLE, 1974. "Human Serum Albumin Phenotype Activation in Mouse Hepatoma—Human Leucocyte Cell Hybrids." *Science* 185: 859–62.

DAVIDSON, E. H., 1976. *Gene Activity in Early Development*, 2nd ed. Academic Press, New York.

EDELMAN, G. H., 1973. "Antibody Structure and Molecular Immunology." *Science* 180: 830–839.

EDSTROM, J. E., AND J. G. GALL, 1963. "The Base Composition of Ribonucleic Acid in Lampbrush Chromosomes, Nucleoli, Nuclear Sap, and Cytoplasm of *Triturus* Oocytes." *J. Cell Bio.* 19: 279–289.

GALLAGHER, R. E., AND R. C. GALLO, 1975. "Type C RNA Tumor Viruses Isolated From Cultured Human Acute Myelogenous Leukemia Cells." *Science* 187: 350–353.

GURDON, J. B., 1962. "The Developmental Capacity of Nuclei Taken from Intestinal Epithelian Cells of Feeding Tadpoles." *J. Emb. Exp. Morph.* 10: 622–640.

GURDON, J. B., AND R. A. LASKEY, 1970. "The Transplantation of Nuclei from Single Cultured Cells into Enucleated Frogs' Eggs." *J. Emb. Exp. Morph.* 24: 227–248.

HENNING, R., J. W. SCHRADER, R. J. MILNER, K. RESKE, J. A. ZIFFER, B. A. CUNNINGHAM, AND G. M. EDELMAN, 1976. "Chemical Structure and Biological Activities of Murine Histocompatibility Antigens," pp. 235–248 in *The Immune System*, eds., F. Melchers and K. Rajewsky. Springer-Verlag, N.Y.

HOOD, L. E., 1972. "Two Genes, One Polypeptide Chain—Fact or Fiction?" *Fed. Proc.* 31: 177–188.

HOOD, L. E., AND P. TALMADGE, 1970. "Mechanism of Antibody Diversity. Germ Line Basis for Variability." *Science* 168: 325–34.

HOURCADE, D., D. DRESSLER, AND J. WOLFSON, 1974. "The Nucleolus and the Rolling Circle." *CSHSQB* 38: 537.

HOZUMI, N., S. TONEGAWA, 1976. "Evidence for Somatic Rearrangement of Immunoglobulin Genes Coding for Variable and Constant Regions," *Proc. Nat. Acad. Sci.* U.S. 73: 3628–3632.

HUANG, R. C., AND J. BONNER, 1962. "Histone, A Suppressor of Chromosomal DNA Synthesis." *Proc. Nat. Acad. Sci. U.S.* 48: 1216–22.

JERNE, N. K., 1973. "The Immune System." *Sci. Am.*: 52–60.

KABAT, D., 1972. "Gene Selection in Hemoglobin and in Antibody Synthesizing Cells." *Science* 175: 134–140.

KLEIN, P. A., AND R. T. SMITH, 1977. "The Role of Oncongenic Viruses in Neoplasia." *Ann. Rev. Med.* 28: 311–27.

LEWIS, E. B., 1950. "The Phenomenon of Position Effect." *Adv. Genetic* 3: 75–115.

MARK, J., 1977. "Chromosomal Abnormalities and Their Specificity in Human Neoplasms. An Assessment of Recent Observations by Banding Techniques." *Adv. Canc. Res.* 24: 165–222.

MARKERT, C. L., 1963. "Lactate Dehydrogenase Isoenzymes, Dissociation and Recombination of Subunits." *Science* 140: 1329–30.

MARKERT, C. L., AND H. URSPRUNG, 1971. *Developmental Genetics*. Prentice-Hall, Inc., Englewood Cliffs, N.J.

OLD, L. J., 1977. "Cancer Immunology." *Sci. Am.* 236(5): 62–70.

OZER, H. L., AND K. K. JHA, 1977. "Malignancy and Transformation: Expression in Somatic Cell Hybrids and Variants." *Adv. Cancer Res.* 25: 53–93.

PAUL, W. E., AND B. BENACERRAF, 1977. "Functional Specificity of Thymus Dependent Lymphocytes." *Science* 195: 1293–1300.

PERKOWSKA, E., H. C. MACGREGOR, AND M. L. BERNSTIEL, 1968. "Gene Amplification in the Oocyte Nucleus of Mutant and Wild-Type Xenopus Laevis." *Nature* 217: 649.

PORTER, R. R., 1973. "Structural Studies of Immunoglobulins." *Science* 150: 713–716.

REDDY, V. B., ET AL, 1978. "The Genome of Simian Virus 40." *Science* 200: 494–502.

RUSSELL, L. B., 1963. "Mammalian X Chromosome Action: Inactivation Limited in Spread and Region of Origin." *Sci.* 140: 976–978.

SAKANO, H., K. HUPPI, G. HEINRICH, AND S. TONEGAWA, 1979. "Sequences at the Somatic Recombination Sites of Immunoglobulin Light-Chain Genes." *Nature* 280: 288–294.

SHEVACH, E. M., 1976. "The Role of the Macrophage in Genetic Control of the Immune Response." *Fed. Proc.* 35: 2048–52.

SMITHIES, O., M. D. POULIK, 1972. "Dog Homologue of Human 2-Microglobulin." *Proc. Nat. Acad. Sci. U.S.* 69: 2914–2917.

SPIEGELMAN, S., R. AXEL, AND J. SCHLOM, 1972. "Virus Related RNA in Human and Mouse Mammary Tumors." *J. Natl. Cancer Inst.* 48: 1205–22.

STEEVES, T. A., I. M. SUSSEX, 1972. *Patterns in Plant Development.* Prentice-Hall, Inc., Englewood Cliffs, N.J.

STEIN, G. S., T. C. SPELSBERG, AND L. J. KLEINSMITH, 1974. "Nonhistone Chromosomal Proteins and Gene Regulation." *Science* 183: 817–824.

STEWARD, F. C., M. O. MAPES, AND K. MEARS, 1958. "Growth and Organized Development of Cultured Cells. II. Organization in Cultures Grown from Freely Suspended Cells." *Am. J. Botany* 45: 705–708.

TAUSSIG, M., AND A. MONROE, 1975. "Antigen-Specific T-Cell Factor in Cell Cooperation and Genetic Control of the Immune Response." *Fed. Proc.* 35: 2061–2066.

TEMIN, H. M., 1977. "RNA Viruses and Cancer." *Cancer* 39: 422–428.

TODARO, G. L., 1974. "Endogenous Primate and Feline Type C Viruses." *CSHSQB* 39(2): 1159–68.

TOMKINS, G. M., ET AL. 1969. "Control of Specific Gene Expression in Higher Organisms." *Science* 166: 1474–1480.

WELLS, R. D., *et al.* 1977. "The Role of DNA Structure in Genetic Regulation." *CRC Critical Rev. in Biochem.* 4: 305–341.

WILSON, E. B., 1926. *The Cell in Development and Heredity*, 3rd ed. Macmillan Co., New York.

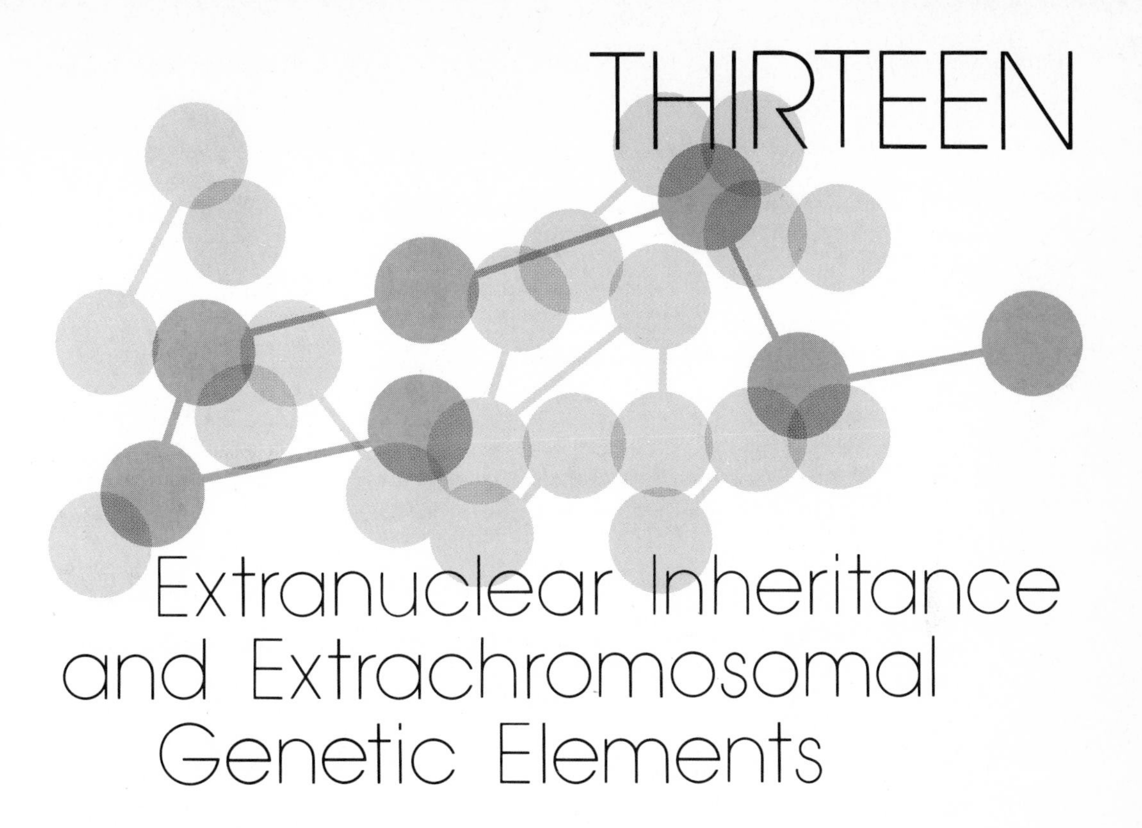

THIRTEEN

Extranuclear Inheritance and Extrachromosomal Genetic Elements

INTRODUCTION

Most of our discussion thus far has centered on the structure, function, and transmission of genetic material found in the nuclei of eukaryotes and the "chromosomes" of prokaryotes. Much interest has focused lately on genetic material which is descriptively referred to as *extranuclear*, or *extrachromosomal, genetic elements* (ECEs). As the term indicates, these are genetic determinants which are found in the cytoplasm of cells, rather than in the nuclei or chromosomes.

Included in this class of ECEs are viruses, which we have already discussed, and a number of other factors which we will touch upon in this chapter. As you will see, they have been found to play a significant role in determining the structure and function of cells in a variety of different organisms.

MATERNAL EFFECTS AND CYTOPLASMIC INHERITANCE IN EUKARYOTES

For many decades, geneticists have known of the existence of cytoplasmic genetic factors which influence phenotype in various eukaryotic organisms. Although not completely understood, several of these systems are well documented and should be familiar to students of genetics. Geneticists refer to these systems as *maternal effect* and *cytoplasmic inheritance*.

Maternal Effect in Snails

One of the best-known examples of a trait that appears to be due to maternal influence was that of the direction of coiling of the shell in the water snail *Limnaea.* In these organisms, the shell can coil to the right (dextral coiling) or to the left (sinistral coiling). Although the dextral trait appeared to be dominant to sinistral, it was found that the phenotype of the offspring was dependent on the *mother's* genotype, not on its own. Figure 13-1 illustrates this interesting system.

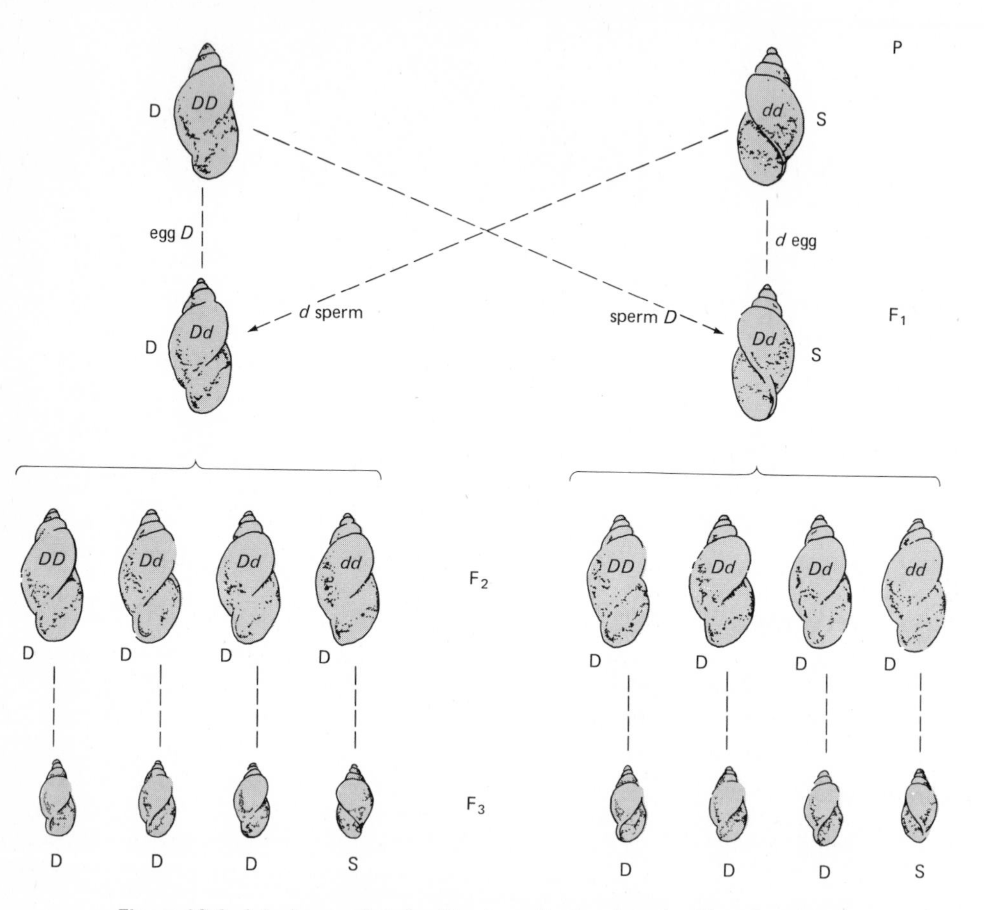

Figure 13-1. Inheritance of shell coiling in snails. D = dextral coiling; S = sinistral coiling. Genes are designated *D, d.* (From E. W. Sinnott, L. C. Dunn, and T. Dobzhansky, 1958. *Principles of Genetics,* p. 361. McGraw-Hill, New York.)

Note that the original cross is $DD \times dd$, with the mother being the dextral parent. All offspring are *Dd*, and therefore dextral as well. When $Dd \times Dd$ crosses are made, the F_2 generation is entirely dextral in phenotype because of the mother's genotype even though $\frac{1}{4}$ of the offspring are *dd.* If the F_2 are allowed to self-fertilize to produce an F_3 generation, $\frac{1}{4}$ of the F_3 progeny are sinistral in shell coiling, because of the *dd* genotype of the F_2 parent from which they derive.

This system, however, is not actually cytoplasmic inheritance because the effect is apparently due to nuclear genes which affect the early cleavage patterns of the fertilized egg (Conklin 1922). Information derived from maternal genes is present in oocyte cytoplasm and causes mitotic spindles of the zygote and daughter cells to form in a certain pattern (Fig. 13-2). For this reason, the phenomenon determining shell coiling in snails is referred to as *maternal effect* rather than cytoplasmic inheritance.

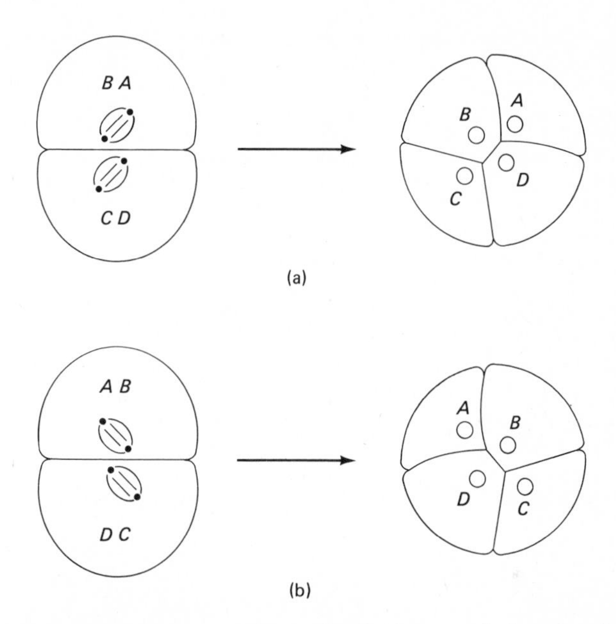

Figure 13-2. Formation of mitotic spindles leading to (a) dextral coiling, and (b) sinistral coiling. (From E. W. Sinnott, L. C. Dunn, and T. Dobzhansky, 1958. *Principles of Genetics*, p. 362. McGraw-Hill, New York.)

Streptomycin Resistance in Chlamydomonas

More recently, studies by Ruth Sager and her coworkers on the green alga *Chlamydomonas* revealed a system that appears to be an example of true *cytoplasmic inheritance*. As Sager has written: "Classically, the principal criteria for identification of cytoplasmic genes have been (a) differences in the results of reciprocal crosses; (b) non-Mendelian ratios; (c) extensive somatic segregation during vegetative or clonal growth; (d) infectivity in mycelial grafts in fungi; and (e) independence of nuclear and cytoplasmic gene assortment in suitable systems, e.g., heterokaryons." (Sager 1972 pp. 47–48). The life cycle of *Chlamydomonas* is shown in Fig. 13-3. There are two mating types, designated mt^+ and mt^-, determined by nuclear genes.

A mutant strain of *Chlamydomonas* was isolated which was highly resistant to streptomycin (sm^r). When crossed to a wild-type strain, the progeny were all resistant; none were streptomycin sensitive (sm^s). When backcrossed to wild-type, mt^+ parents transmitted resistance, but mt^- did not. Further tests showed that which-

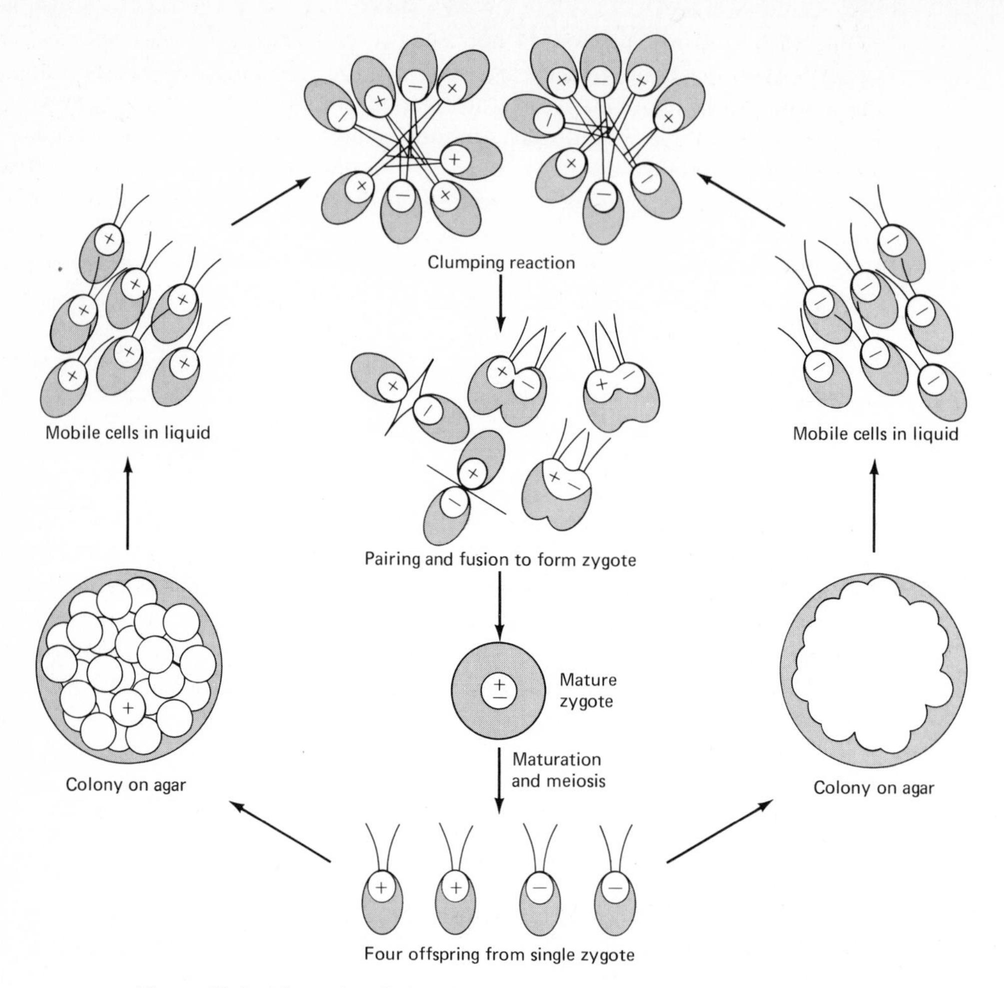

Figure 13-3. Life cycle of the alga *Chlamydomonas.* (From R. Sager, 1972. *Cytoplasmic Genes and Organelles*, p. 52. Academic Press, New York.)

ever allele was carried by the mt^+ parent was transmitted. Later, other mutations were found that exhibited the same patterns of transmission. One big difference between this situation and that of shell coiling in snails is that in the alga, the trait persisted over several generations, whereas in snails, the effect is on only one generation. Sager found that even four generations of backcrosses of sm^r cells to sm^s cells produced no individuals of the sm^-s phenotype.

The *sm* gene is believed to be present in the chloroplast, a cell organelle that is known to possess DNA. Chloroplast DNA has since been found in the form of a closed circle (Manning *et al.* 1971). Thus when we speak of cytoplasmic inheritance, we are referring to the function of genes in the cytoplasm of a cell, sometimes in association with cell organelles.

Since the genes found in cell organelles are not bound to nuclear chromosomes, their replication is of course not governed by the cell cycle, and segregation of alleles is not dependent on meiosis. However, it has been found that if mating occurs so as to produce heterozygosity for a pair of alleles, such colonies grown from single zoospores show a 1: 1 segregation of the cytoplasmic alleles. These studies are greatly complicated by recombination events and mutations of alleles of chloroplast genes, which cannot be detected by conventional transmission genetics methods.

A number of mutations have been found in *Chlamydomonas* which are also transmitted in the same manner as the *sm* mutation, namely, by one mating type, mt^+. Incidentally, probably by tradition, this mating type is also considered to be the maternal parent. It was found that the maternal effect is significantly affected by UV irradiation. Although spontaneous mutations at about 1 percent were found in which the transmission of streptomycin resistance becomes either paternal or biparental, UV treatment greatly increased this incidence.

Since there is fusion of cells during mating events in *Chlamydomonas*, it was unusual that there should be any maternal effects at all. Both parent cells must contribute an equal number of cell organelles. In isotope labeling experiments, however, differences have been observed in the fate of maternal compared to paternal chloroplasts. There is a shift of maternal chloroplasts to a lighter density following fusion. What causes this shift and what relationship it has to the lack of expression of paternal chloroplast DNA is not yet known. There may be a modification of maternal DNA in such a way as to prevent its degradation by enzymes of the zygote, which then proceed to degrade paternal chloroplast DNA which has not undergone modification (Sager and Romanis 1974).

Mitochondrial Mutations in Yeast

Saccaromyces cerevisiae, better known as baker's yeast, was found in the 1930s to be capable of sexual reproduction (the life cycle of yeast is shown in Fig. 13-4). As in *Chlamydomonas*, this species has two mating types determined by a pair of chromosomal alleles, *a* and *α*. Haploid cells of different mating types can fuse and form a diploid zygote. These diploid cells either divide mitotically to form a clone of diploid cells, or they can undergo meiosis and sporulation to form haploid ascospores in an unordered tetrad. The spores can reproduce and give rise to a haploid clone.

In the late 1940s, Ephrussi and his coworkers reported that they had observed a mutation in yeast which resulted in the absence of certain respiratory enzymes. This in turn led to the formation of unusually small colonies, therefore the mutation was termed *Petite*. The mutant yeast cells were also incapable of forming spores. When wild-type cells of either mating type were crossed with Petite mutants, the progeny were all wild-type. Backcrosses of the F_1 cells to Petite cells still produced virtually no Petite offspring. Since respiratory enzymes are associated with the mitochondria, the discovery of DNA and replicating properties of mitochondria eventually led to studies that indicated the Petite genes to be mitochondrial, that is, cytoplasmic.

In contrast to chloroplast DNA, which has been found to be relatively uni-

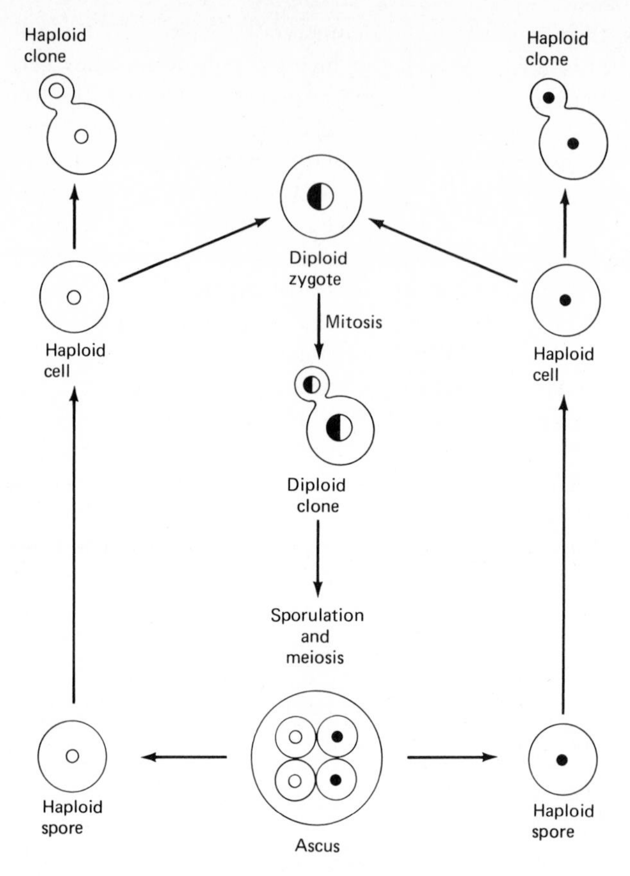

Figure 13-4. Life cycle of yeast. Different shades of nuclei represent different mating types.

form in size from species to species, mitochondrial DNA varies, being larger in plants and prokaryotic microorganisms than in animal cells. It is circular. Table 13-1 includes data on the size of mitochondrial DNA in several species.

A few years later, Ephrussi (1953) reported the discovery of a Petite mutation which again demonstrated cytoplasmic inheritance, but in this case, the cross of wild-type by Petite yielded only Petite offspring. To distinguish between the original Petite mutation and the later one, Ephrussi termed the earlier one Neutral and the later one Suppressive Petite. That these mutations were of mitochondrial genes was supported by evidence of physical alteration of mitochondrial DNA in mutant cells, such as changes in density gradients and nucleotide sequences. Furthermore, mitochondrial genes determining resistance to antibiotics have been found, and these are occasionally affected by the presence of a Petite mutation (Saunders *et al.* 1971).

There also appears to be a nuclear gene determining respiratory functions. In 1950, Chen and his coworkers reported studies showing Mendelian segregation

Table 13-1 Size and Form of Mitochondrial DNAs

Species	*Form*	*Contour length* (μ)
Man	Circular	4.8–5.3
Monkey	Circular	5.5
Mouse Liver	Circular	5.0–5.1
Chick	Circular	5.1–5.4
Frog	Circular	5.9
Fly	Circular	5.2
Sea urchin	Circular	4.6–4.9
Yeast	Circular	25.0

From R. Sager, 1972. Cytoplasmic Genes and Organelles, p. 26. Academic Press, New York.

ratios in progeny of crosses between wild-type and a Petite mutation discovered in their laboratory. The mutation is apparently recessive to wild-type. In crosses of the nuclear Petite to cytoplasmic Petites, complementation is observed, as would be expected of two different mutations, and the diploid zygotes manifest wild-type phenotype.

Male Sterility in Plants

Cytoplasmic inheritance has also been reported in a variety of plants. One of the best-known examples is male sterility in corn. M. M. Rhoades (1933) discovered a strain of corn in which pollen formation was defective. Normally, corn produces both pollen (male gametes) and ovules. The defective plants thus were male sterile, although female gametes were normal. When these were crossed to wild-type, or male-fertile, plants, the offspring were found to be male sterile. A series of back-crosses resulting in offspring possessing all the chromosomes of the male-fertile plants nonetheless yielded only male-sterile plants. Rhoades interpreted this phenomenon as being cytoplasmic inheritance, since in each case the ovule was from a male-sterile plant.

Cytoplasmic male sterility has been reported in many other plant species, such as rice, broad bean, sorghum, and petunia (see Grun 1976). In some cases, including corn, nuclear genes can influence the expression of sterility. In corn, for example, a dominant nuclear gene is able to reverse male sterility in strains that have the cytoplasmic elements for sterility.

Male sterility in plants can be of commercial importance, for it is a natural mechanism by which plants are prevented from self-pollinating. Through selected crosses of particular strains, the genetic makeup of plants can easily be manipulated. However, there is also a risk involved. In the United States, male-sterile lines of corn containing a specific type of cytoplasm, called Texas cytoplasm, were used to a large extent in producing hybrid corn seed in 1970. It was discovered too late that the Texas line was also very sensitive to the leaf blight *Helminthosporium maydis*. Some 15 percent of the corn crop was lost, and breeders have reverted to male-fertile lines.

A microorganism known to have a sexual cycle of reproduction is the protozoan *Paramecium.* This unicellular eukaryote (which possesses two micronuclei and a macronucleus) can normally reproduce in several ways. One is by binary fission, in which its two micronuclei divide mitotically, and the macronucleus apparently splits into two halves. If cells of two different mating types are brought together, a second type of reproduction occurs, in which they form a cytoplasmic bridge and undergo conjugation. Figure 13-5 illustrates this interesting phenomenon.

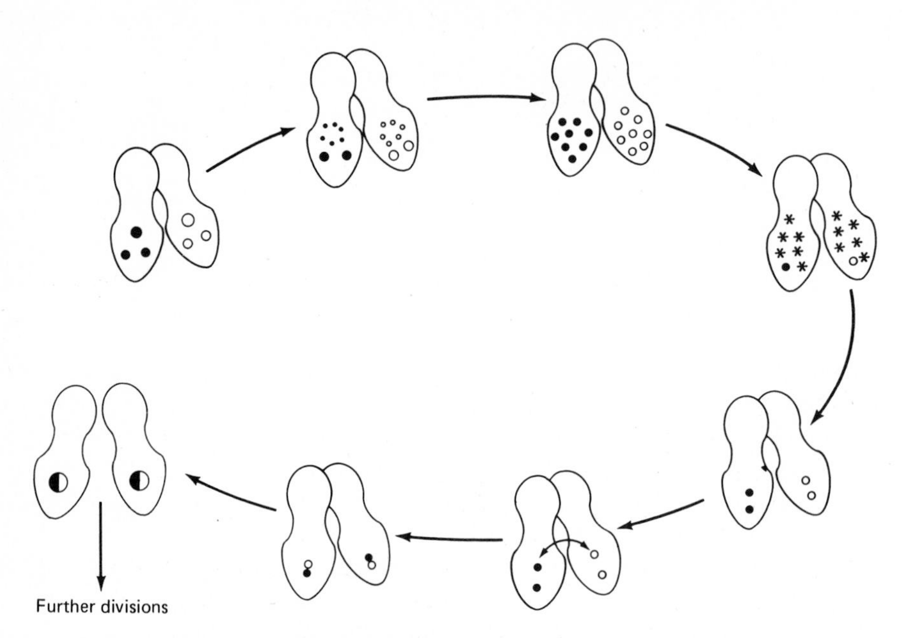

Figure 13-5. Conjugation in *Paramecium.* Explanation in text.

In conjugation the macronucleus fragments and the micronuclei undergo meiosis. Of the eight haploid nuclei that result, seven degenerate, and the remaining nucleus divides mitotically. One copy remains in the original cell, while the other crosses the cytoplasmic bridge to fuse with the recipient cell's nucleus, which now is diploid. The two cells separate and are then called *exconjugants.* Two mitotic divisions in each exconjugant result in four nuclei, of which two expand to become macronuclei. When binary fission occurs, the other two nuclei divide, and one macronucleus is transferred to each daughter cell, which gives it the normal complement of two micronuclei and one macronucleus.

Finally, the process of *autogamy* can occur, which is very much like conjugation in the division and disintegration of nuclei. The one big difference is that there is no partner cell. The two micronuclei within a cell fuse to form a diploid zygote in what is essentially self-fertilization. What mechanism exists to trigger each of these specific reproductive methods is yet to be discovered.

In 1943, T. M. Sonneborn found that a strain of *Paramecium* was able to secrete a toxic substance named *paramecin* which killed cells that were then termed *sensitive* (see Sonneborn, 1964). He was able to induce matings between "killer" cells and sensitive cells, however, and found that the exconjugants gave rise to equal clones of killer and sensitive cells. In normal conjugation very little cytoplasm is exchanged. If prolonged conjugation occurs in which cytoplasm as well as nuclei are transferred, the killer characteristic is transferred and the sensitive cells in turn become killer cells.

The correlation between transfer of the killer trait and transfer of cytoplasm in conjugation led scientists to the discovery that the ability to produce paramecin was determined by a cytoplasmic particle known as *Kappa* (Preer *et al.* 1974). As you probably have guessed, the particle was found to contain DNA. That this DNA is different from *Paramecium* nuclear genes was reflected in the fact that respiratory enzymes of killer strains are different from those normally found in *Paramecium*, but similar to those in some bacteria. Furthermore, the ribosomal RNA of Kappa particles is typical of bacteria and will hybridize with DNA from *E. coli*, but not with DNA from *Paramecium*. (However, it was found that a dominant nuclear gene *k* was necessary for the maintenance of the Kappa particles and therefore the killer characteristics.)

Actually Kappa is one of many endosymbionts found in *Paramecium*. Figure 13-6 shows a cell with hundreds of endosymbionts. It is believed that independently existing bacteria were the progenitors of the Kappa particles in *Paramecium*, much as it has been theorized that mitochondria and chloroplasts also developed at one time from bacteria. We shall discuss the evolutionary significance of this possibility in Chapter 17. Further evidence for the similarity between Kappa and bacteria has been found by Preer (Preer *et al.* 1974), who reported the discovery of viruses from Kappa particles which may actually be the determinants of paramecin secretion (Fig. 13-7). He likens these viruses to bacteriophage; in fact, the presence of these refractile particles in Kappa (which he refers to as R bodies) is a criterion for distinguishing Kappa from other endosymbionts. Preer has indicated that these R bodies also bear similarities to colicin particles of bacteria that produce toxins against bacteria (p. 430).

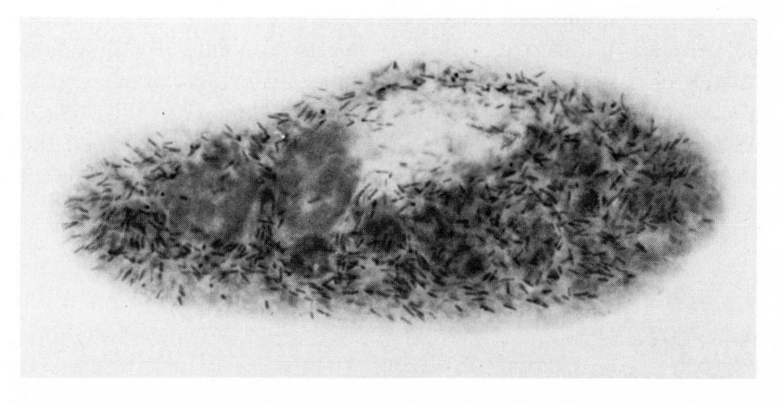

Figure 13-6. Osmium-lacto-orcein preparation of a *Paramecium*. Dark rods are endosymbionts (×500). (From J. R. Preer *et al.*, 1974. *Bacteriol. Rev.*, 38:116; photograph courtesy John R. Preer, University of Indiana.)

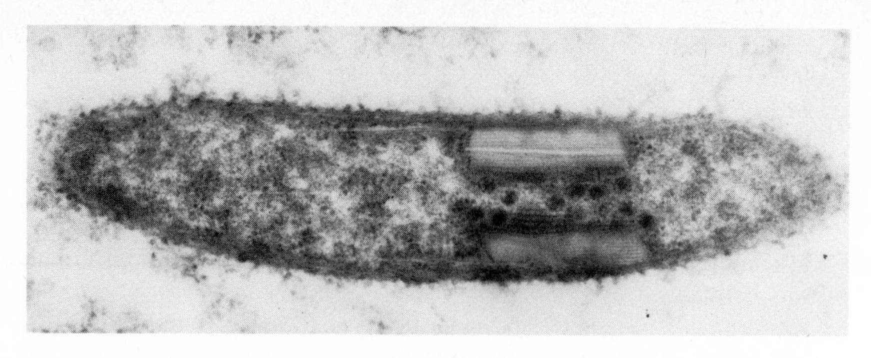

Figure 13-7. Electron micrograph of *Paramecium* with a bright Kappa particle containing spherical R bodies (×60,000). (From J. R. Preer, *et al.*, 1974. *Bacteriol. Rev.*, 38:116; photograph courtesy John R. Preer, University of Indiana.)

Other Systems of Infective Inheritance

Two other examples of the transmission of traits through cytoplasmic or maternal inheritance deserve a brief mention here. We spoke earlier of studies of the transmission of murine mammary cancer virus through the milk of nursing mice (p. 401). When babies of noncancerous mice were foster-nursed by females of the cancerous strain, the incidence of mammary tumors among fostered female young equaled the incidence in the cancerous strain.

In *Drosophila*, a strain of flies studied by L'Heritier (1958) appeared to be very sensitive to exposure to CO_2. These flies did not recover from exposure at levels which are nonlethal to normally resistant flies. Sensitive mothers always produced sensitive young no matter what the genotype of the father was; resistant mothers produced resistant young. That an infective particle was involved was indicated by the fact that extracts of sensitive flies, when injected into resistant flies, converted them to the sensitive state. Although the particle has not yet been characterized, it has been called Sigma and is believed to be a small viruslike structure.

There are a number of other examples of maternal and cytoplasmic inheritance which we shall not discuss here, but the systems described clearly demonstrate the existence of particles in the cytoplasm of germinal and somatic cells which are capable of conferring traits to the organism. The presence of nonchromosomal genetic elements has received even greater attention in recent years, as they have been discovered ubiquitously in prokaryotes. In fact, their effect on the genetics of prokaryotes is in many ways more significant than the examples we have given above of extrachromosomal inheritance in eukaryotes.

EXTRACHROMOSOMAL GENETIC ELEMENTS IN PROKARYOTES

In their studies on the sexual reproduction of bacteria, geneticists discovered the existence of very small pieces of DNA in the cytoplasm of bacterial cells. We have already discussed one class of these, namely, the viruses. But more recently, much excitement has been generated by the discovery of another class of nonnuclear genetic particles which are collectively referred to as *extrachromosomal genetic elements*, or ECEs.

These, like viruses, are genetic particles, which differ significantly from viruses in both structure and function. As we shall see, their presence has been established in virtually every species of pro- and eukaryote. Genetics today can be likened to the days when prokaryotes came into their own as tools of genetic experimentation. With the awareness of the ECEs, and the development of techniques to manipulate their genes, scientists are both electrified and deeply concerned by the potential for great progress in understanding gene action and for the manipulation of genes. In the following sections we shall explore some of the basic characteristics of these extrachromosomal genetic elements, and in the subsequent chapter, dwell on the use of these systems in genetic research and engineering.

The F Particle and Conjugation in Bacteria

The first discovery of these nonviral ECEs came from studies on conjugation in bacteria, a subject we covered in Chapter 8. Recall that genes which cause the formation of pili between an F^+ and an F^- bacteria are actually carried by an episome which was called the *F* particle. Besides functioning to cause the transfer of chromosomes from one cell to another when integrated into the bacterial chromosome, the *F* particle is also capable of carrying individual bacterial genes as an autonomous molecule during sexduction (p. 244).

F particles carrying bacterial genes are referred to as F', and recent techniques have allowed us to visualize these molecules under the electron microscope. If F' strands of DNA are allowed to reanneal with F^+ DNA strands, base-pairing will occur over most of the length of the DNA. The exception will be the region of the bacterial gene in the F', which will form a single-stranded loop. This structure is illustrated in Fig. 13-8a, and actual electron micrographs are shown in Figure 13-8b. We term these composite molecules *heteroduplex* DNA. The study of heteroduplex molecules has been very useful in recent genetic engineering research, as we shall see in the next chapter.

Not long after the initial discovery that the *F* particle exists and can mediate genetic exchange between bacteria, evidence for other episomes in bacteria was reported. One class of these is of great concern not only to geneticists, but to health authorities as well.

Antibiotic Resistance Factors in Bacteria

In 1956, an unusual dysentery epidemic was reported in Japan (Watanabe 1963). Patients treated with several antibiotics did not respond in the usual manner, indicating that the bacteria causing the disease, *Shigella flexneri*, had become resistant to all the antibiotics used.

In itself, antibiotic resistance is not uncommon in bacteria. Physicians are well aware of the need to increase dosage or to switch from one treatment to another as a patient responds less and less to treatment when receiving a particular drug

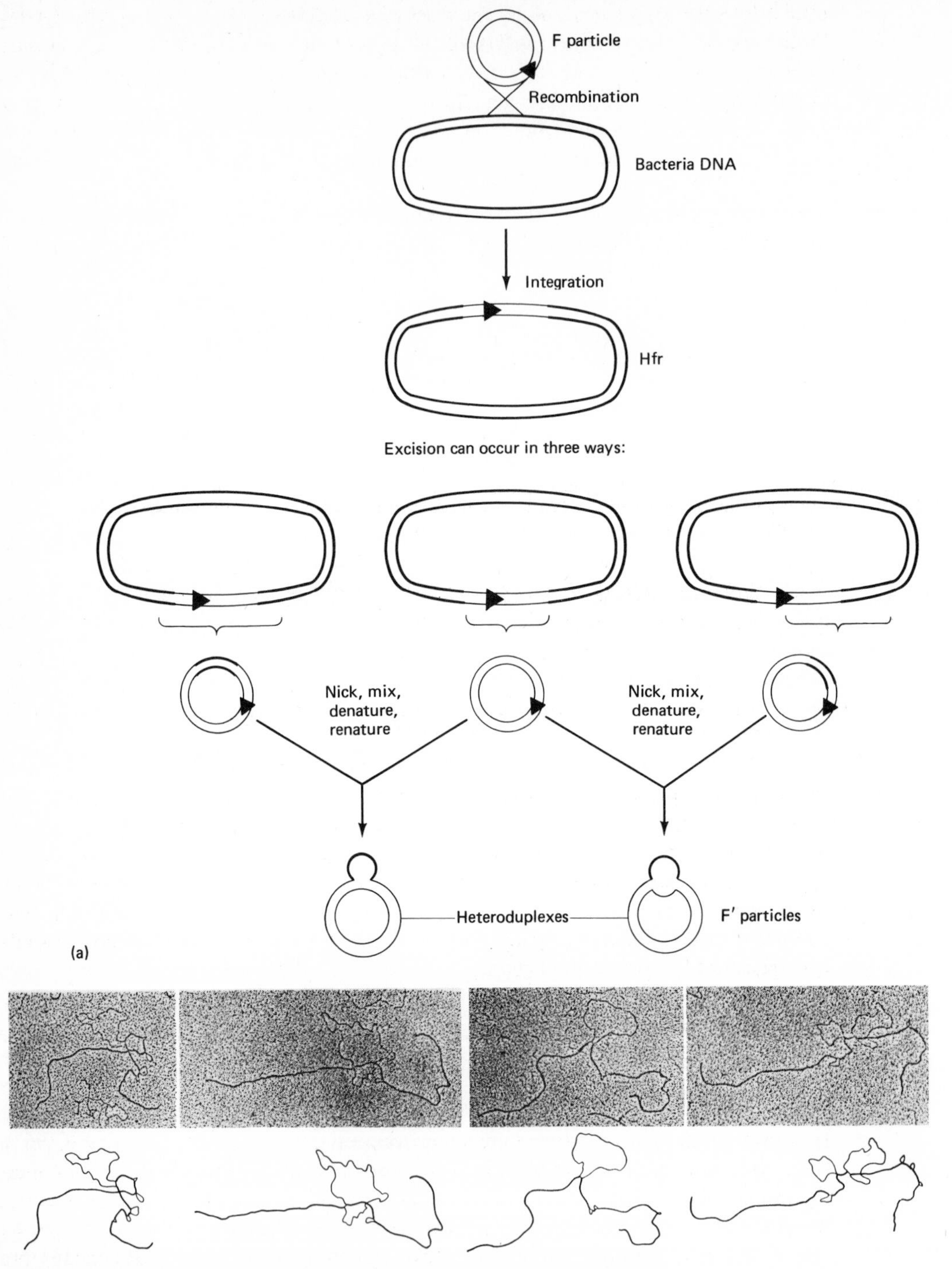

Figure 13-8. (a) Processes leading to formation of F′ particles. Presence of bacterial genes results in heteroduplex molecules seen as loops. (b) Electron micrographs and tracings of heteroduplex molecules. These were obtained by reannealing rDNA which were of different lengths from *Xenopus laevis*. [With permission from P. K. Wellauer *et al.*, 1976, *J. Mol. Biol.*, 105 : 473 ; copyright by Academic Press, Inc. (London) Ltd.]

over a long period of time. The reason for this of course is that the presence of antibiotics would select against a sensitive strain and in favor of bacteria which were resistant. The resistant strains then reproduce until they dominate the bacterial population (Fig. 13-9).

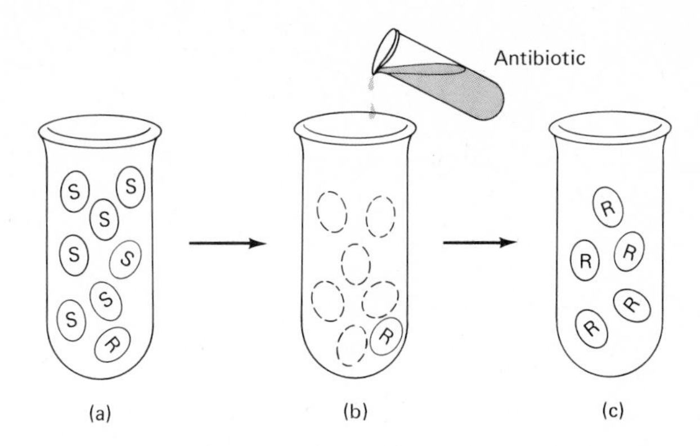

Figure 13-9. (a) A population of predominantly sensitive cells, (b) when exposed to an antibiotic, (c) becomes predominantly resistant as sensitive cells are killed.

Resistance to antibiotics had been assumed to be due to mutation. For example, resistance to streptomycin was found by Gorini and others to be due to a mutation which caused a change in the ribosome structure. Normally, streptomycin kills bacteria by interacting with ribosomes, thus interfering with protein synthesis. With the alteration of the ribosome by the streptomycin-resistant mutation, the antibiotic can no longer interfere with protein synthesis.

What was unusual about the *Shigella flexneri* cells isolated from patients of the dysentery epidemic in Japan was that the bacteria were at once resistant to four different antibiotics: chloramphenical, streptomycin, sulphonamide, and tetracycline. That this was due to mutation was highly unlikely since such mutations occur at a very low frequency, about one in several hundred million. The probability of the occurrence of four mutations in one cell is so low that it can essentially be discounted. Shortly afterward, in the United Kingdom, a similar epidemic due to multiply-resistant *Salmonella* was reported.

The unusual bacteria were isolated and carefully studied. It was discovered that the resistance to many antibiotics was a transferable characteristic: If the multiply-resistant cells were cultured with sensitive cells, within hours the sensitive cells were all multiply resistant. It was then found that the transfer occurs during conjugation. Further research indicated that sometimes the transfer occurred as a result of transduction, *independently of chromosomal genes.*

The similarity between these phenomena and the transfer of the F^+ episome in conjugation was inescapable, and a search for an extrachromosomal genetic element ensued. DNA extracts of resistant cells were analyzed by the technique of cesium chloride density-gradient centrifugation (see p. 277). As you may recall, this technique causes DNA fractions of different densities to migrate to positions in the tube where the density of the cesium chloride is the same as their own. Addition of the dye ethidium bromide aids in distinguishing different fractions of DNA. Identification of the fractions can be achieved by measuring either the amount of ultraviolet light absorption, or (if the bacteria had been grown in culture with radioactive thymidine), the level of radioactivity. As can be seen in Fig. 13-10, two distinct classes of DNA were found: cellular and noncellular.

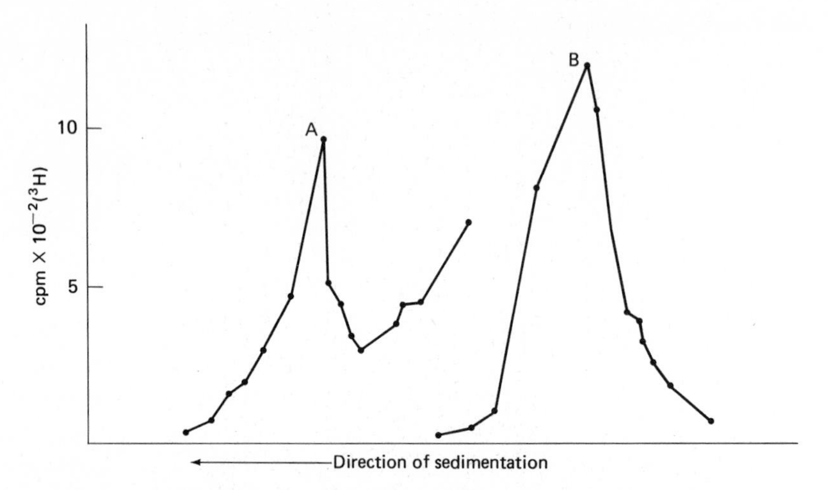

Figure 13-10. Centrifugal analysis of ^{3}H-labeled DNA extracted from R^+ bacteria. Equilibrium conditions after centrifugation in a CsCl + ethdium bromide solution for 36 hours at 22,000 g. Peak designated A represents extrachromosomal DNA. Peak B represents DNA of bacteria chromosome. (From A. B. Stone, 1975. *Sci. Prog. Oxf.* 62: 93.)

Structure of resistance factors. The noncellular fraction of DNA (the denser band) was then isolated and prepared for electron microscopy. Figure 13-11 shows what the geneticists found: an ECE in the form of closed circles. As the micrograph shows, the ECE assumed two forms: a loop or cycle, and a supercoiled form in which the circle is twisted on itself. It was determined later that the supercoiled condition is converted into the loop form by a nick in one of the strands of DNA (Fig. 13-12).

It is in fact the coiling of the ECE which causes it to be denser than cellular DNA in centrifugation experiments. The much larger bacterial chromosome tends to fragment and thus migrate to less dense areas of the cesium chloride. The dye ethidium bromide also attaches to the linear fragments, which decreases their buoyant density due to the fact that intercalation of the dye into DNA causes a further unwinding of the helix. As the dye has less affinity at high concentrations for the coiled DNA of the ECE, separation of the two classes of DNA can be achieved fairly easily.

Interspecies transfer of resistance factors. The potential danger to public health in the spread of multiple resistance led to intensive study of bacterial ECEs, which were then referred to as *RTFs*, or *resistance transfer factors*. The urgency of this task increased when it was discovered that these factors could be transferred

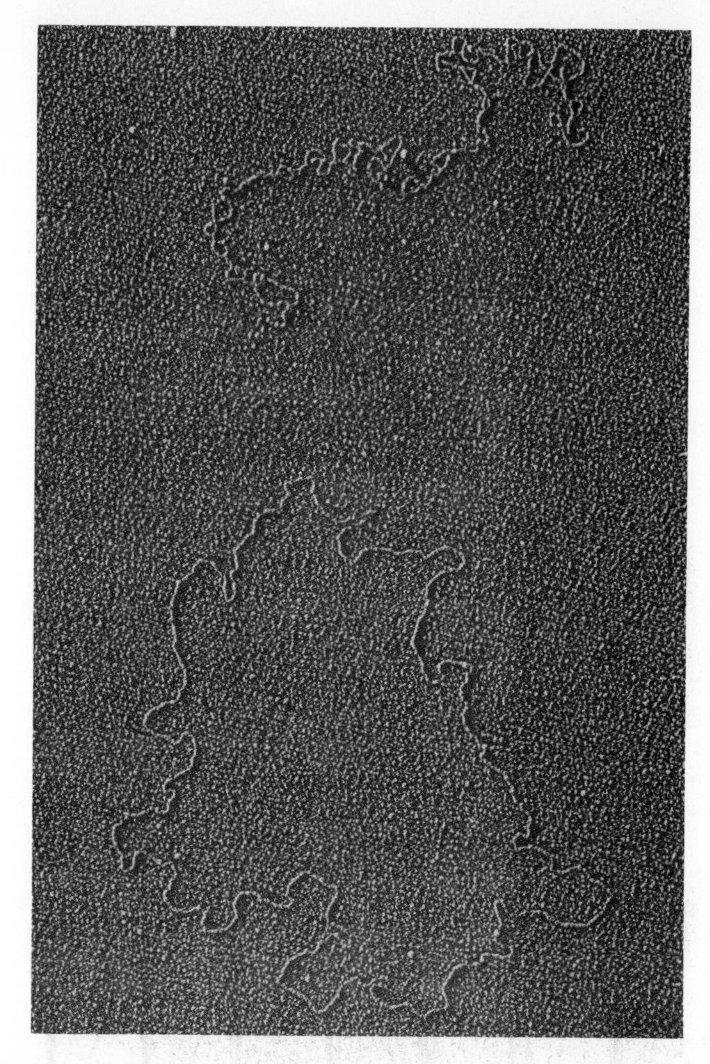

Figure 13-11. R-factor plasmids in an electron micrograph, enlarged 62,000×. The top molecule is in the supercoiled form; the bottom molecule is a "nicked duplex loop." (From R. C. Clowes, 1973. "The Molecule of Infectious Drug Resistance." *Sci. Am.* 228: 18; copyright 1973 by Scientific American. All rights reserved.)

not only within a species of bacterium, but also between species; for example, RTF from *E. coli* could be transmitted to *Salmonella*, *Shigella*, and others. Generally, the gram-negative bacteria are the ones involved in multiple resistance; however, multiple resistance has recently been found in *Streptococcus* and *Streptomyces*, both of which are gram positive.

Antibiotic-resistant ECEs have also been discovered in *Staphylococcus aureus* (Lacey 1973). The genetic analysis of plasmids in these cells is complicated by the fact that unlike *E. coli*, staphylococci are difficult to grow on simple media, and lysing cells causes denaturation of the DNA, so that normal extraction procedures

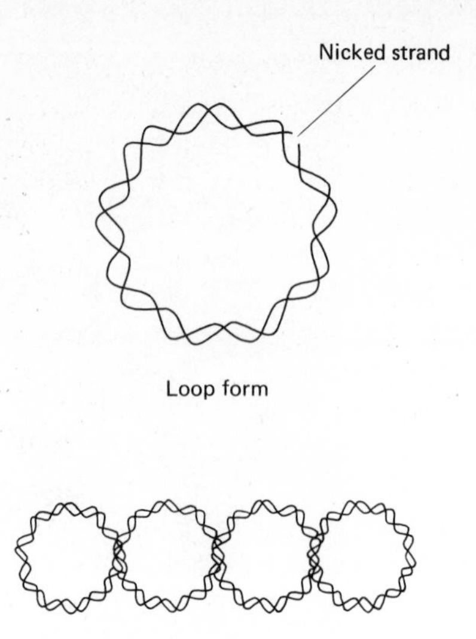

Figure 13-12. Two forms of plasmid DNA.

are not effective. Also, there is no conjugation in these bacteria, and gene transfer must be by transduction.

Distribution of resistance factors. ALTHOUGH RTFS WERE RARELY OBSERVED BEFORE THE 1950S, STUDIES HAVE SINCE ESTABLISHED THEIR PRESENCE IN BACTERIA POPULATIONS ALL OVER THE WORLD—IN HOSPITALS, IN DOMESTIC ANIMALS, IN THE SEWAGE OF URBAN RIVERS. THEY, NOT MUTATION AS PREVIOUSLY THOUGHT, APPEAR NOW TO BE THE PRIMARY SOURCE OF ANTIBIOTIC RESISTANCE IN BACTERIA.

In 1974, for example, Tanaka and associates surveyed 338 strains of *E. coli* from clinical sources for resistance to four drugs: tetracycline, chloramphenicol, streptomycin, and sulfanilimide. They found that 20.3 percent were singly resistant, 19 percent doubly resistant, 7.7 percent triply resistant, and 18.3 percent were resistant to all four drugs. Resistance factors were found in 29.2 percent of the resistant strains most commonly in the quadrupally resistant cells. They also found a small number of cells resistant to kanamycin (1.5 percent) and ampicillin (4.7 percent). These showed 100 percent and 76.9 percent presence of RTFs, respectively, indicating that RTFs are the primary cause of resistance to these drugs. The implications of these ECEs for clinical bacteriology are obvious and of great concern to health authorities.

Since the resistance transfer factors were originally thought to be cytoplasmic genetic particles capable of autonomous replication, but *not* capable of integration into the host chromosome, they were not considered episomes. A new term was used: *plasmid.* However, later studies have shown that occasionally plasmids can undergo an exceptional integration into cellular DNA (Stone 1975). Terminology referring to different ECEs is thus still somewhat ambiguous. As one geneticist stated, "What's one person's episome is another person's plasmid!" The tendency now is to use the term plasmid rather than episome.

Composite nature of resistance factors. A number of different laboratories found that in several instances the resistance transfer factor was actually a composite of two or more different elements. One, now called the *r*, or *R*, *determinant factor*, is the source of antibiotic resistance. Some RTF particles contain several R factors, each of which determines resistance to a particular drug. The other component is the RTF particle, as it contains the genes necessary for the formation of a pilus to allow conjugation, similar in this function to the *F* factor.

Clear evidence for this was obtained by Cohen (Cohen and Kopecko 1976) and others who analyzed the DNA of plasmids from *Proteus mirabilis*, known as 222/R plasmids. The DNA was originally found to be around 70 megadaltons in size, but extracts also contained DNA of 12 and 58 megadaltons. This suggested the ability of the larger particle to dissociate into two smaller particles, which were later found to be capable of autonomous replication.

The idea of composite plasmids was further supported by observations that following conjugation, segregation of the genes determining resistance to various drugs could be found in the daughter cells. Also found was segregation of the ability to transfer resistance with the actual resistance to drugs. In other words, when multiply resistant cells underwent conjugation with sensitive cells, the results of gene transfer included cells that were resistant to some of the drugs but not others; cells that were resistant, but could not transfer the resistance (containing R, but not RTF); cells that could determine conjugation, but were no longer resistant to antibiotics (containing RTF, but not R).

If resistant cells incapable of conjugation were mixed with sensitive cells that could conjugate, daughter cells were found which once again exhibited infectious drug resistance. In these cells, the R and RTF particles reassociate. When genes for resistance and for transfer are found in the same particle, the plasmids are called *cointegrates*. Figure 13-13 illustrates some of these interactions.

THE DISSOCIATION AND RECOMBINATION OF THE VARIOUS COMPONENT PLASMIDS OF A COINTEGRATE PLASMID HAS BEEN FOUND TO INVOLVE GENETICALLY SPECIFIC SITES ON THE DNA. This was determined when independently isolated RTF plasmids were found to be identical in molecular structure.

The term *nonconjugative* plasmid has been coined for ECEs which cannot be transferred by conjugational events (such as the R plasmids). Plasmids which can cause conjugation and transfer of DNA, such as RTF particles, and cointegrate plasmids, are known as *conjugative* plasmids. Generally speaking, nonconjugative plasmids have been found to be of relatively low molecular weight and are present in multiple copies in a host cell. Conjugative plasmids are of higher molecular weight and are present in only one or two copies, indicating that some form of regulation exists to coordinate the replication of the plasmid with the replication of the bacterial chromosome.

In Cohen's studies on the 222/R plasmids, isolation of DNA fractions by centrifugation showed that the transfer plasmid, called *delta*, was a single loop of DNA 60 megadaltons in size. This was found in the cells that were nonresistant, but capable of causing transfer. In cells that were noninfectious, but resistant to either streptomycin or ampicillin, two components were found, one 60 megadaltons and one 6 megadaltons.

THIS ABILITY OF PLASMIDS TO DISSOCIATE AND REASSOCIATE CONFERS SELECTIVE ADVANTAGE ON THE PLASMID DNA BY INCREASING THE GENETIC VARIABILITY OF THE

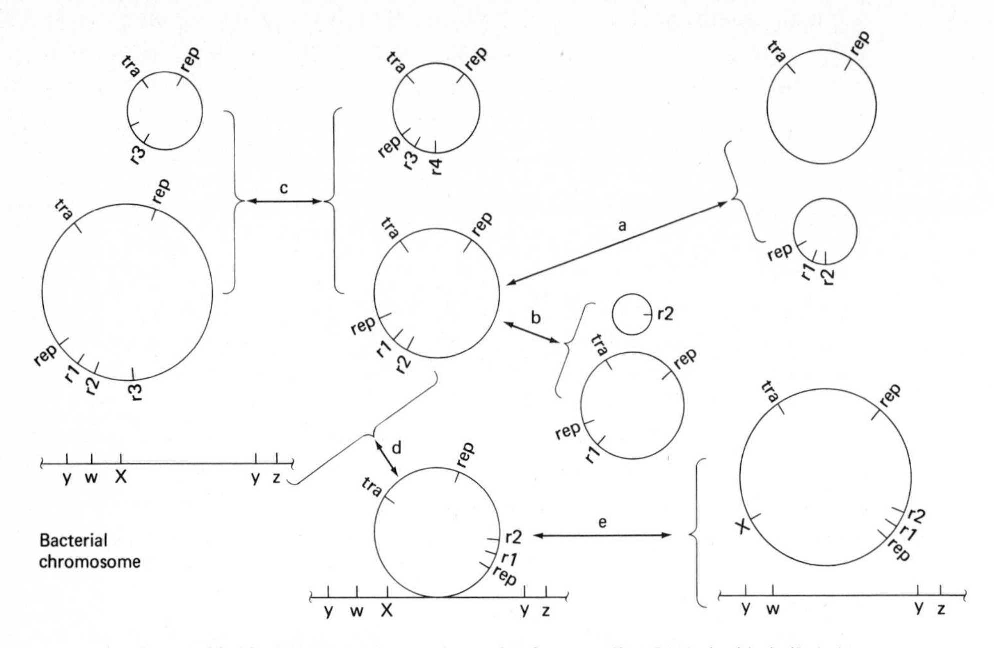

Figure 13-13. DNA-DNA interactions of R factors. (The DNA double helix is here represented by a single line for clarity.) (a) Dissociation, giving a transfer factor and a non-transmissible resistance factor; (b) loss of one resistance determinant; (c) recombination between two R factors, resulting, in this particular case, in the acquisition by one factor of a resistance determinant lost by the other; (d) insertion into the bacterial chromosome; (e) pick-up of a chromosomal region during excission from the bacterial DNA. *rep*, DNA replication genes; *tra*, plasmid transfer genes; r1, r2, r3, r4, plasmid genes determining resistance to four antibiotics; v, w, x, y and z, genes on the bacterial chromosome. (From A. B. Stone, 1975. *Sci. Prog. Oxf.*, 62:96.)

PARTICLES. THE RESULTING GENETIC VARIABILITY OF THE HOST CELLS POSES GREATER PROBLEMS FOR THE CLINICIAN.

It has been found, however, that not all plasmids can coexist in a particular host cell. Some plasmids are incompatible so that after a few cell generations, only one of the two remains. The mechanism underlying this phenomenon, known as *superinfection immunity*, has not yet been determined. Furthermore, it has also been found that a resident plasmid can prevent another plasmid from entering its host cell, or from replication if it enters. Known as *exclusion*, this is another phenomenon whose mechanism is not understood.

Colicin Factors in Bacteria

The ability of bacteria to produce toxins that kill other bacteria cells has been known for many decades. In the 1950s, one class of bacteriotoxins was found to be determined by plasmids harbored by *E. coli* cells. These were called *colicins*, and were found to be macromolecules sensitive to protein-degrading enzymes such as trypsin.

There are many different colicins, and they kill strains of bacteria related to the host cell, but not unrelated strains. One is reminded here of the Kappa particles and the R bodies associated with the killer effect of *Parameciun.*

In the 1950s, Frederic discovered that when *Col*$^+$ bacteria were mixed with *Col*$^-$ cells, the ability to produce the toxin spread to the *Col*$^-$ cells upon conjugation. This of course indicated the presence of an ECE, which was later found to be present in the form of circular DNA molecules. The plasmids which contain genes determining colicins are known as *Col* plasmids. It has been estimated that 25–40 percent of all *E. coli* strains carry Col plasmids. (Helinski 1976).

Col factors are referred to by capital letters, and their mode of action has been found to fall into three groups. Colicins A, B, E1, K, Ia, and Ib belong to one group which kills bacteria by affecting energy metabolism of the cell. Colicin E2 inhibits DNA metabolism, and colicins D and E3 damage the ribosomes and inhibit protein synthesis (Davies and Reeves 1975).

Each colicin is determined by a single plasmid gene, and only one molecule of colicin is required to produce the killing effect. Apparently the colicin molecules attach to receptor sites on the outer layer of the bacterial cell wall. In the laboratory of Salvador Luria, (1975), recent studies on ColE1 and ColK showed that the initial attachment of a colicin molecule is not lethal, and that in fact the cells can be "rescued" by treating the preparation with trypsin to destroy the colicin. The lethal stage finds the synthesis of all macromolecules in the victim cell at a standstill: no synthesis of proteins, nucleic acids, or glycogen occurs.

Using normal cells as well as a strain of mutant cells which are unable to produce ATPase, Luria and others now believe that the effect of the colicin is to interfere with normal utilization of membrane energy. This affects the active transport of essential ions such as potassium and magnesium, and the generation of ATP for essential metabolic reactions. Further studies should not only elucidate the mechanism of action of colicins but also yield information on the function of cell membranes.

Insertion Elements

In the late 1960s, several researchers found mutations in bacteria strains which possessed characteristics different from the usual mutations. For one thing, they occurred at much higher frequencies than normal. Whereas normally mutations in bacteria are detected at about 1 in 10^6, these mutations were found as frequently as 1 in 10^2, and they were strongly polar in nature. For example, the first such mutation was found by Starlinger and Saedler (1972) in the gal operon of *E. coli.* Although the mutation mapped outside of the structural gene for galactose kinase, the synthesis of the enzyme was reduced to less than 1 percent. In other words, the mutation was found to have a repressive effect on neighboring genes.

Furthermore, the mutations reverted to wild-type spontaneously, but this could not be induced with chemicals which were known to cause DNA alterations, and which were known to be capable of inducing reverse mutation back to wild-type. When the mutation was transferred to a lambda dg transducing phage, density gradient studies showed that the phage was carrying an extra piece of DNA. The reversion to wild-type was discovered to coincide with the loss of the extra piece of DNA.

These studies implied an important aspect of the "mutant" DNA, namely, that loss of the DNA is not random, for if it were, one would expect that in some cases only part of the mutation would be removed, or that in other cases, some of the neighboring genes may be lost also. Yet the reversion mutants were always wild-type. There was also site-specific insertion; in other words, insertion into the bacterial chromosome did not occur randomly. For this reason, these mutations were referred to as *insertion mutations*. Further, the extra DNA was not present because of recombination events, as they were detected in *rec*$^-$ mutant bacteria.

STUDIES ON A NUMBER OF SUCH INSERTION MUTATIONS INDICATED THAT THEY WERE CAUSED BY A PREVIOUSLY UNKNOWN CLASS OF DNA MOLECULES WITH THE ABILITY TO BE TRANSLOCATED READILY FROM ONE POSITION TO ANOTHER. THEY HAVE SINCE BEEN REFERRED TO AS INSERTION ELEMENTS (IS), OR TRANSPOSONS (TN). Four classes of IS have been detected, called IS1–IS4. They are all small molecules of DNA of 700–1400 base pairs in size. Hybridization experiments have shown there to be complete homology of nucleotide sequence between molecules of a particular class.

Role of IS elements in microbial genetics. Of great interest to geneticists was the discovery that IS DNA exists not only in chromosomes of bacteria, but also in plasmids. In fact, as studies on these interesting elements proceeded, it became apparent that they serve essential roles in the function of plasmids. For example, IS2 elements were found in the F^+ episome. Researchers also discovered several sites on the *E. coli* chromosome which are homologous to the IS2 sequence, and are probably IS2 elements that have been inserted. It then stands to reason that when the F^+ particle is transferred during conjugation, the areas of homology on the bacterial chromosome serve for recombination with the episome, leading to integration of the F^+, or donor, DNA (in the case of *Hfr*) into the host chromosome.

In a study of ampicillin-resistant plasmids, Heffron, Falkow, and their coworkers (1975) found that in 35 plasmids studied, which were of different sizes, all had in common a segment of DNA about 3×10^6 daltons in size. This piece of DNA, called *Transposon A*, or TnA, was found to produce *polar mutations* (mutations affecting genes to one side of the mutation) when inserted into bacterial DNA. It was found to be larger than insertion elements, because it contained not only sequences similar to IS, but also the gene for ampicillin resistance. They found that the latter was actually bounded by palindromes, or inverted complementary sequences.

The presence of inverted complementary sequences on either side of a gene, such as the one for antibiotic resistance, can be detected when the strands are separated. Intrastrand reannealing will occur, leading to the formation of "hairpin" loops which can be detected in the electron microscope. Figure 13-14a diagrams how the bases of the sequences pair to form the stalk or stem of the hairpin, and how the loop contains the R gene which will remain single-stranded for lack of complementarity of the bases. Figure 13-14b shows an actual micrograph of such a loop.

Kanamycin and tetracycline resistance are also bounded by inverted repeat sequences, and in tetracycline resistance, the sequences have been identified as IS3.

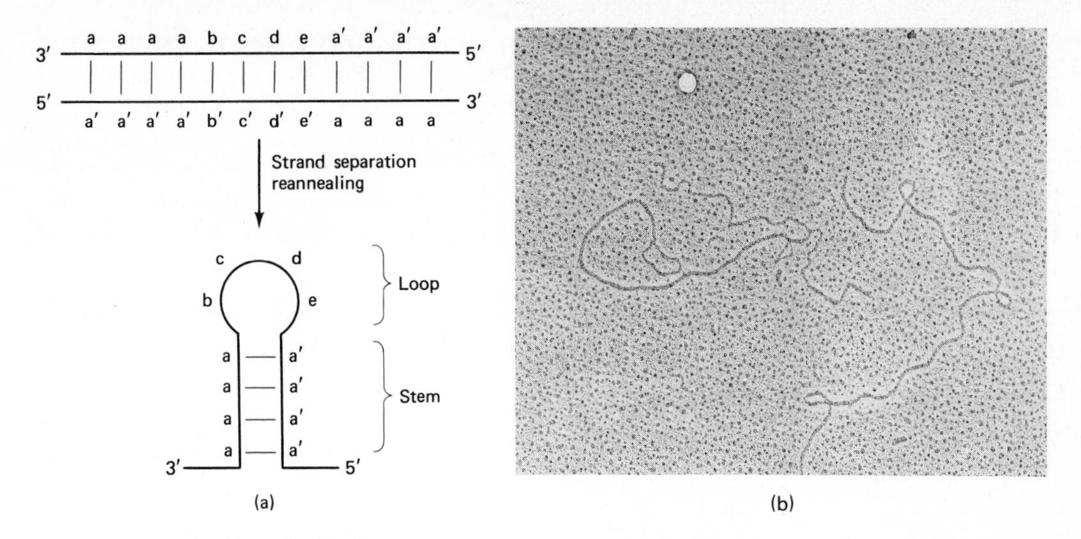

Figure 13-14. (a) Diagram of "hairpin" loop formation; *aaaa* and *a′a′a′a′* represent bases of a palindrome; *bcde* represent bases of a gene for antibiotic resistance. (b) Actual photomicrograph of hairpin loops. (Photo courtesy J. Manning, University of California, Irvine, Calif.)

Insertion elements have also been found at the boundary of R and RTF factors in cointegrate antibiotic-resistant plasmids.

The ability of these molecules to regulate gene function was demonstrated in plasmids also. Sharp and his associates (Sharp *et al.* 1973) found a spontaneous variant of a tetracycline plasmid R6 which no longer conferred resistance to tetracycline. This variant, R6-5, was found to have an extra insertion element which repressed the gene for tetracycline resistance (Fig. 13-15).

Significance of insertion elements. Although the mechanism of translocation remains unclear, the ubiquity of IS sequences has led to an awareness of their poten-

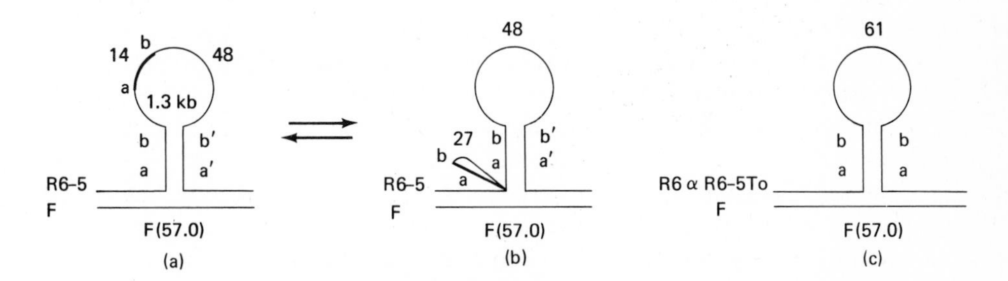

Figure 13-15. Schematic illustration of the effect of inverted repeat sequences in the R6-5/F heteroduplex. The molecule can assume two configurations if there is duplication of the sequence a–b either of which can pair with the complementary sequence a′–b′ as in (a) and (b). Only one configuration results if the sequence is not duplicated (c). [With permission from P. A. Sharp *et al.*, 1973. *J. Mol. Bio.* 75: 246, Fig. 6; copyright by Academic Press, Inc. (London) Ltd.]

tial significance for microbial genetics. It has been postulated, as we mentioned above, that ISs play important roles in conjugation, in gene regulation (by their repressive effect on neighboring genes), and in the transmission of plasmid DNA from one particle to another or between plasmid and chromosomal DNA. Their existence also helps to explain the transferability of plasmids between species of bacteria. Their presence at the boundary between the R and RTF elements of a cointegrate R plasmid explains the readiness of association and dissociation of these elements and the precision with which this occurs.

POSSIBLE RELATIONSHIPS BETWEEN ECEs

The possible role of ISs in the evolution of different ECEs has not been lost on geneticists. Heffron and associates have written: "The genes specifying many phenotypic characteristics of plasmids seem to have arisen independently from the transfer and replication region, and it is tempting to speculate that a large proportion of the 'non-essential' plasmid-mediated determinant such as antibiotic resistant determinant of R factors, which confer transient selective advantages, may be attributed to a limited pool of translocatable DNA segments" (Heffron *et al.* 1975, p. 255). Figure 13-16 shows a model proposed by Cohen and Kopecko for the role of ISs in association-dissociation events in R plasmids, and the formation of polygenic plasmids.

INSERTION ELEMENTS IN EUKARYOTES

In studies that were decades ahead of their time, Barbara McClintock reported in 1950 on a gene, which she called a *controlling element*, that was able to move from one chromosome site to another. The element, *Ds*, can be detected by its ability to induce chromatid breaks or to modify the action of neighboring genes. These effects are dependent on the presence of another gene, *Ac*, which was also found capable of translocation and of causing mutations. The exact nature of these genes is not known, but McClintock has postulated that they may be similar to control systems in bacteria (McClintock 1961).

More recently, repeated sequences in eukaryotic chromosomes have been postulated to play a regulatory role, as they bear some resemblance to the IS regions discovered in microbial DNA. Indeed, palindrome sequences have been reported by D. A. Wilson and C. A. Thomas (1974), in DNA from Hela cells, *Drosophila* cells, and mouse cells on the basis of heteroduplex studies. Most of the regions appeared to be 300–1200 nucleotides in length, and are in clusters of 2–4, separated by a region of approximately 2000 base pairs. The authors estimate there to be "thousands to hundreds of thousands of these clusters in the eukaryotic DNA studied." A genetically unstable mutation in the X chromosome of *Drosophila* has been found to be associated with a tandem duplication of the region containing the White locus. This was found to be highly mutable, showing spontaneous reversion to wild-type upon loss of the extra DNA. It has been postulated that this may represent an insertion mutation, although the nature of the extra genetic material has not been determined (Green 1975).

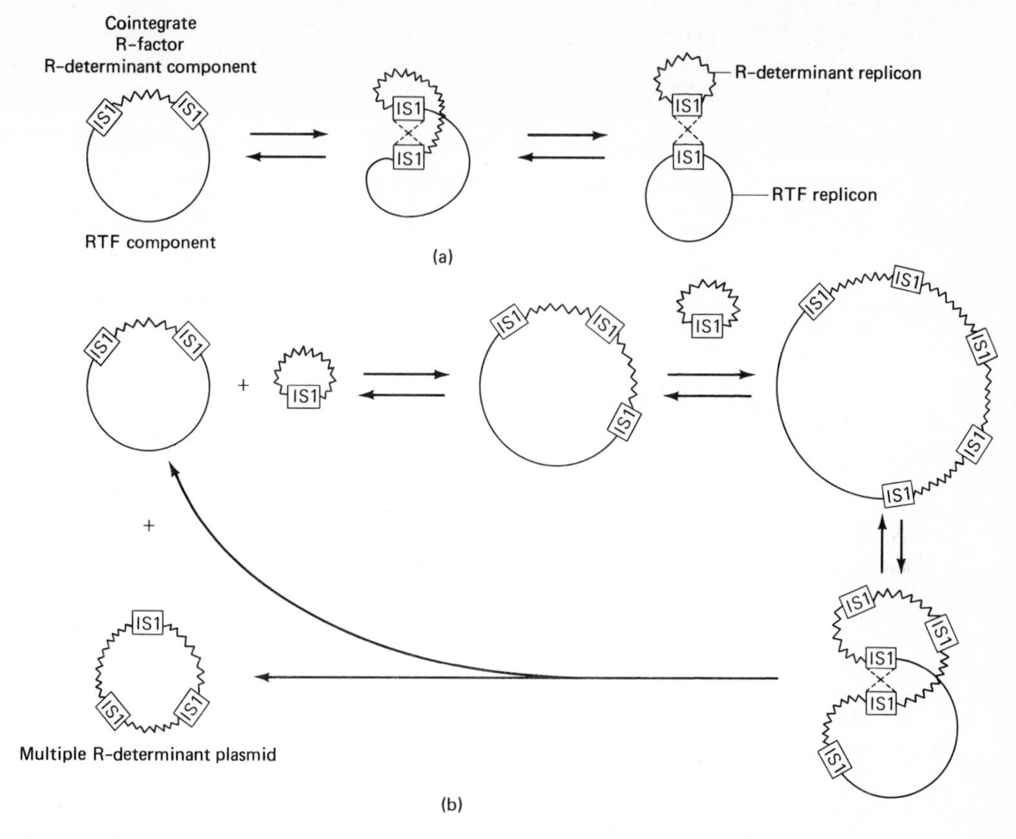

Figure 13-16. Proposed mechanism for reversible dissociation of cointegrate R-plasmids at sites of IS1 insertions. The model provides for the formation of independent RTF and R-determinant replicons (a) and for the production of polygenic R-determinant replicons (b). (From S. N. Cohen and K. J. Kopecko, 1976. *Fed. Proc.* 35: 2033.)

Although it is of unquestionable importance to the genetics of microorganisms, the presence in eukaryotes of a class of DNA molecules with the same function and importance as the insertion elements must not be assumed. Indeed, the very selective advantage IS molecules confer upon their host cells by allowing greater genetic flexibility through association-dissociation of DNA would seem to be a decided selective *disadvantage* in eukaryotes.

The primary reasoning is that the microorganism is, for the most part, unicellular and independent of other cells for survival. Its own genotype determines whether it will live or die; therefore, genetic flexibility is highly desirable. In multicellular eukaryotic organisms, on the other hand, the interaction of different cells and tissues is precisely defined, with the organism able to function as a whole only if every one of its parts fulfills its particular role. It would be difficult to envision the advantage to eukaryotes, therefore, to have genetic elements which can cause the rapid change of genotype associated with transposons. On the other hand, one might venture the speculation that as multicellularity evolved, ISs, or DNA similar to them, were modified in some ways, perhaps retaining their regulatory activity to some extent, but losing their capability of movement in the eukaryotic genome. Having considered the importance of ECEs in nature, we shall in the next chapter discuss their significance in the laboratory.

PROBLEMS

1. Assume that you are making the following cross with *Limnaea:*

$$\male DD \times \female dd$$

(a) Give the expected genotypes and phenotypes of the F_1.
(b) If the F_1 are selfed, what would you expect in the F_2 generation?

2. What results would you expect from matings of *Chlamydomonas* cells as diagramed below if there is streptomycin in the culture medium?
(a) $mt^+sm^r \times mt^-sm^s$
(b) $mt^+sm^s \times mt^+sm^r$

3. Why is observation of non-Mendelian ratios one of the criteria for cytoplasmic inheritance?

4. Another criterion for cytoplasmic inheritance is differences in results from reciprocal crosses. Is this ever found in Mendelian nuclear inheritance? Explain.

5. The nuclear gene, *K*, is dominant and necessary in order for *Paramecium* containing Kappa particles to maintain the particles and the killer characteristic. If prolonged conjugation occurs between two killer *Paramecium* of the genotype *Kk*, what offspring would you expect with regard to the killer trait?

6. Some killer cells can be caused to undergo many divisions by binary fission in rapid succession. Eventually, they are found to be sensitive and free of Kappa particles. What explanation can you give for this observation?

7. Assume that you have discovered a strain of bacteria resistant to mercury and penicillin. Give an account of the experiments you would carry out to determine if a plasmid is involved.

REFERENCES

ADELBERG, E. K., AND S. N. BURNS, 1959. "A Variant Sex Factor in *Escherichia coli.*" *Genetics* 44: 497.

CHEN, S. Y., B. EPHRUSSI, AND H. HOTTINGUER, 1950. "Nature genetiques des mutants a deficience respiratorie de la souche B-II de la levure de boulangerie." *Heredity* 4: 337–351.

COHEN, S. N., AND K. J. KOPECKO, 1976. "Structural Evolution of Bacterial Plasmids: Role of Translocating Genetic Elements and DNA Sequence Insertions." *Fed. Proc.* 35: 2031–2036.

CONKLIN, E. G., 1922. *Heredity and Environment.* Princeton U. Press, Princeton, N. J.

DAVIES, J. K., AND P. REEVES, 1975. "Genetics of Resistance to Colicins in *Escherichia coli* K-12: Cross-Resistance Among Colicins of Group A." *J. Bact.* 123: 102–117.

EPHRUSSI, B., 1953. *Nucleocytoplasmic Relations in Micro-Organisms.* Clarendon Press, Oxford.

FREDERICQ, P., 1958. "Colicins and Colicinogenic Factors." *Symp. Soc. Exp. Biol.* 12: 104–122.

GORINI, L., 1966. "Antibiotics and the Genetic Code." *Sci. Am.* 24: 102–109.

GREEN, E. L., ed., 1966. *Biology of the Laboratory Mouse.* McGraw-Hill, New York.

GREEN, M. M., 1975. "Genetic Instability in *Drosophila Melanogaster*: Mutable Miniature (*mu*). *Mut. Res.* 29: 77–84.

GRUN, P., 1976. *Cytoplasmic Genetics and Evolution.* Columbia Univ. Press, New York.

HAYES, W., 1970. *The Genetics of Bacteria and Their Viruses.* Blackwell Scientific Publications, Oxford.

HEFFRON, F., C. RUBENS, AND S. FALKOW, 1975. "Translocation of a Plasmid DNA Sequence Which Mediates Ampicillin Resistance: Molecular Nature and Specificity of Insertion." *Proc. Nat. Acad. Sci. U.S.* 72: 3623–3627.

HEFFRON, F., R. SUBLETT, R. W. HEDGES, A. JACOB, AND S. FALKOW, 1975. "Origin of the TEM-Beta-Lactamase Gene Found on Plasmids." *J. Bact.* 122: 250–256.

HELINSKI, D. R., 1976. "Plasmids." *Fed. Proc.* 35:2024.

KOLATA, C. B., 1976. "Jumping Genes: A Common Occurrence in Cells." *Science* 197: 392–394.

LACEY, R. W., 1973. "Antibiotic Resistance Plasmids of *Staphylococcus Aureus* and Their Clinical Importance." *Bact. Rev.* 39: 1–32.

LEDERBERG, J., 1952. "Cell Genetics and Hereditary Symbiosis." *Physiol. Rev.* 32: 403–430.

L'HERITIER, P. H., 1958. "The Hereditary Virus of *Drosophila*." *Adv. Vir. Res.* 5:195–245.

LURIA, S. E., 1975. "Colicins and the Energetics of Cell Membranes." *Sci. Am.* 233: 30–37.

MANNING, J. E., D. R. WOLSTENHOLME, R. S. RYAN, J. A. HUNTER, AND O. C. RICHARDS, 1971. "Circular Chloroplast DNA from *Euglena Gracilis*." *Proc. Nat. Acad. Sci. U.S.* 68: 1169–1173.

MCCLINTOCK, B., 1961. Some Parallels Between Gene Control Systems in Maize and Bacteria. *Am. Nat.* 95: 265–277.

PREER, J. B., L. B. PREER, AND A. JURAND, 1974. Kappa and Other Endosymbionts in *Paramecia Aurelia*." *Bact. Rev.* 38: 113–163.

RHOADES, M. M., 1933. "The Cytoplasmic Inheritance of Male Sterility in *Zea Mays*." *J. Genetics.* 27: 71–93.

SAGER, R., 1972. *Cytoplasmic Genes and Organelles.* Academic Press, New York.

______, AND X. ROMANIS, 1974. "Mutations That Alter the Transmission of Chloroplast Genes in Chlamydomonas." *Proc. Nat. Acad. Sci. U.S.* 71: 4698–4702.

SAUNDERS, G. W., B. ELLIOT, M. K. TREMBATH, H. B. LUKUS, AND A. W. LUNNANE, 1971. "Mitochondrial Genetics in Yeast." In *Autonomy and Biogenesis of Mitochondria and Chloroplasts*. Aust. Acad. Sci. Symp. North-Holland Publ., Amsterdam.

SAEDLER, H., AND B. HEISS, 1973. "Multiple Copies of the Insertion DNA Sequences IS1 and IS2 in the Chromosome of *E. Coli* K-12." *Mol. Gen. Gen.* 122: 267–277.

SHARP, P. A., S. N. COHEN, AND N. DAVIDSON, 1973. "Electron Microscope Heteroduplex Studies of Sequence Relations Among Plasmids of *Escherichia coli.* II, Structure of Drug Resistance (R) Factors and F Factors." *J. Mol. Biol.* 75: 235–255.

SONNEBORN, T. M., 1969. "The Gene and Cell Differentiation." *Proc. Nat. Acad. Sci. U.S.* 46: 149–165.

STARLINGER, P., AND H. SAEDLER, 1972. "Insertion Mutations in Microorganisms." *Biochimie* 54: 177–185.

STONE, A. B., 1975. "R Factors: Plasmids Conferring Resistance to Antibacterial Agents." *Sci. Prog. Oxf.* 62: 89–101.

TANAKA, T., A. KOBAYASHI, K. IKEMURA, H. HASHIMOTO, AND S. MITSUHASHI, 1974. "Drug Resistance and Distribution of R Factors Among *Escherichia coli* Strains." *Jap. J. Microbiol.* 18: 343–347.

WATANABE, T., 1963. "Infectious Heredity of Multiple Drug Resistance in Bacteria." *Bact. Rev.* 27: 87–115.

WILSON, D. A., AND C. A. THOMAS, 1974. "Palindromes in Chromosomes." *J. Mol. Biol.* 84: 115–144.

WOLLMAN, E. L., AND F. JACOB, 1958. "Sur les Processus de conjugaison et de recombinaison chez *E. coli.* v. Le Mechanisme du transfer de material genetique." *Ann. Inst. Pasteur* 95: 641.

FOURTEEN

Recombinant DNA and Gene Manipulation

INTRODUCTION

By the late 1960s, geneticists had gained an enormous amount of insight into the genetics and physiology of microorganisms. Viruses and bacteria had contributed instrumentally to our understanding of the gene and its ability to direct the structure and function of cells. The genetically simple chromosomes of bacteria had been mapped; various means of gene exchange between bacteria, between viruses, and between bacteria and viruses had been discovered. Techniques had been developed to allow scientists to transfer specific genes from one cell to another, or to isolate specific genes, by exploiting the phenomenon of transduction.

ALONG WITH GREATER DEPTH OF UNDERSTANDING OF NATURAL PROCESSES COMES THE ABILITY TO CONTROL THEM AND EVEN TO MANIPULATE THEM. GENETICS HAD ARRIVED AT THE POINT WHERE SCIENTISTS COULD LOOK BEYOND JUST STUDYING WHAT EXISTS IN NATURE, AND COULD THINK OF ACTUALLY ALTERING GENOMES TO SUIT EXPERIMENTAL SPECIFICATIONS. The fact that this would be tantamount to creating new forms of life was not lost on the scientific community, but to refrain from doing so would be counter to human intellectual instincts. From the enormous progress which has been made in the 1970s (to be discussed in this chapter), it is obvious that scientists have found it difficult to resist following their instincts.

RECOMBINANT DNA

The manipulation of the genomes of bacteria and viruses was made possible by the isolation and study of enzymes involved in the metabolism of DNA. Endonucleases and exonucleases which break DNA molecules in various regions have already been discussed in preceding chapters. Ligases were found that bond different segments of DNA together. With the understanding of the role of these enzymes, the possibility of being able to break DNA into fragments and then join different fragments together presented itself to scientists.

Along this line, another enzyme, terminal transferase, was found to be capable of catalyzing the addition of nucleotides to 3′ ends of DNA molecules which had been exposed by treatment with exonuclease. Through the addition of adenines to the ends of some molecules and thymines to the ends of others, base-pairing would occur in the regions of poly A and poly T if the two kinds of DNA were mixed together. DNA polymerase and ligase could then be used to complete the bonding of the two molecules, and the result would be a duplex DNA, actually a combination of what were disparate DNA fragments. Such new combinations are referred to as *recombinant DNA* (Figure 14-1).

To many scientists, including some actively working in this area of molecular genetics, the development of methodology to produce new gene combinations (that is new genomes) is both enormously exciting and startling. The recombinant molecules which are now being produced are fully capable of self-reproduction, mutation, and directing metabolism of cells. We shall discuss the implications of this research in later sections of this chapter.

One of the first experiments using terminal transferase to form a recombinant DNA molecule involved the linking of genes from the tumor virus SV40 to genes of λ phage and *E. coli* (Jackson *et al.* 1972). Since all the DNA used in this experiment exists as closed circles, an endonuclease called *Eco RI* was used to open the circles. In order to fully understand the steps of this and similar experiments, we must digress a moment to discuss the nature of enzymes such as Eco RI. These are enzymes which have played an extremely important role in recombinant technology, although the scientists who first performed the experiment were not aware of their potential at the time.

Restriction Endonucleases

Enzymes such as Eco RI were actually discovered in the 1950s, when geneticists studied the phenomenon of a prophage protecting its host cell upon subsequent infection of the cell by certain viruses. You recall that *r* mutants of T2 phages are unable to infect and lyse *E. coli* strain B. It was found that this inability to affect strain B was due to the presence of a λ prophage in the strain B cells which directed the synthesis of an enzyme that specifically cleaves the double-stranded DNA of the *r* mutant viruses. This enzyme was termed a *restriction endonuclease.*

Specificity of cleavage. The first restriction endonuclease isolated was nonspecific in its site of cleavage (Linn and Arber, 1968; Meselson and Yuan 1968). Only two years later (Smith and Wilcox 1970), there came a report of a RE which cleaved the DNA at specific sites. This was an enzyme determined by a gene of an influenza

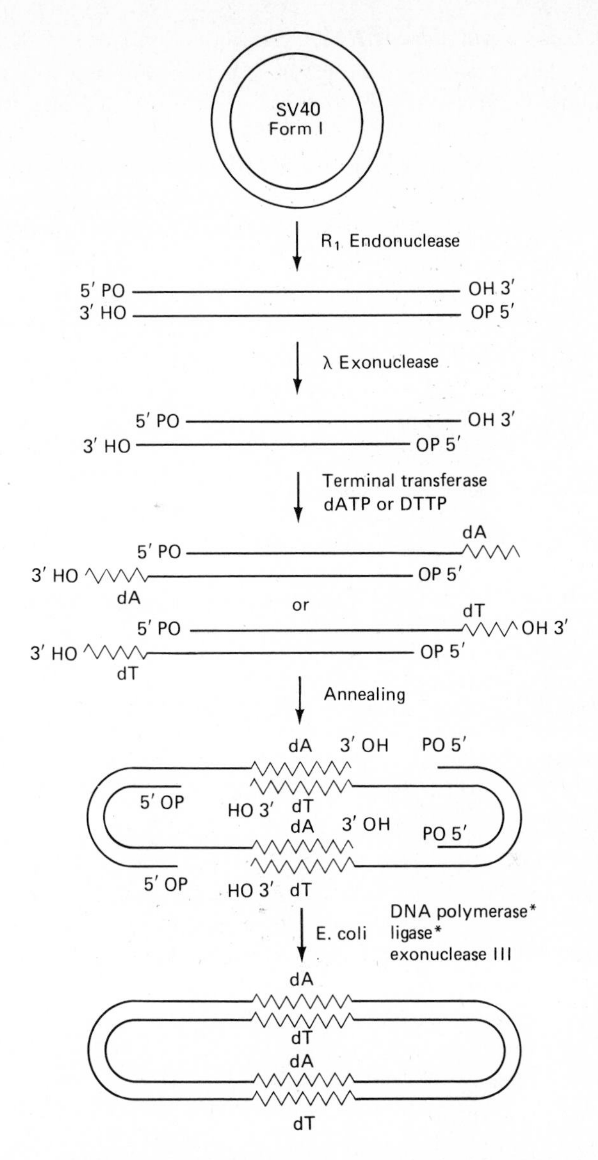

Figure 14-1. General protocol for producing covalently closed SV40 dimer circles from SV40(I) DNA, using terminal transferase. *The four deoxynucleoside triphosphates and DPN are also present for the DNA polymerase and ligase reactions, respectively. (From D. A. Jackson *et al.*, 1972. *Proc. Nat. Acad. Sci. U.S.* 69: 2905.)

virus, *Hemophilus influenza* strain Rd. The enzyme was designated Hind II, and it was found to cleave DNA at sites containing the following sequence of bases: G–T–Pyrimidine–Purine–A–C. The point of cleavage is between the pyrimidine and the purine. (In 1978, Drs. Smith, Arber, and D. Nathans, whose use of REs in the analysis of the SV40 virus DNA will be discussed later, were awarded the Nobel Prize in Physiology or Medicine for their discoveries.)

Class I and class II REs. Since the discovery of the first restriction endonucleases, literally dozens of others have been found in a variety of organisms (Nathans and Smith 1975). Generally they have been classified into two groups. The REs that cleave DNA at nonspecific sites are class I enzymes. They are relatively large molecules of about 300,000 molecular weight, and are composed of nonidentical subunits. Site-specific enzymes, which are more useful in recombinant research, are class II enzymes. These are smaller, ranging from 20,000 to 100,000 molecular daltons. The best-studied of the class II enzymes is Eco RI, the enzyme used by Paul Berg and his colleagues at Stanford (see again Jackson *et al.* 1972). Eco RI is known to have two identical subunits. Whether this is true of all class II enzymes remains to be established. Table 14-1 lists some of the known class II REs, their site specificity (arrow), number of sites in viruses (Ad 2 = adenovirus 2), and source (species of bacterium known to produce the enzyme). More than 200 REs have been found to date.

Table 14-1 Restriction Endonucleases

Enzyme	*Sequence*	*Number of cleavage sites*			*Microorganism*
		λ	Ad2	SV40	
Alu I	AG↓CT	50	50	32	*Arthrobacter luteus*
Bam HI	GGA↓TCC	5	3	1	*Brevibacterium albidum*
Eco RI	G↓AATTC	5	5	1	*Escherichia coli PI*
Hae II	PuGC↓GCPy	30	30	1	*Haemophilus aegyptius*
Hha I	GC↓GC	50	50	2	*Haemophilus hemolyticus*
Hap II	CC↓GG	50	50	1	*Haemophilus aphrophilus*
Hind II	GTPy↓PuAC	34	20	7	*Haemophilus influenza Rd*
Hpa II	CC↓GG	50	50	1	*Haemophilus parainfluenzae*
Sma I	CCC↓GGG	3	12	0	*Serratia maraescens Sb*
Xma I	CCC↓GGG	3	12	0	*Xanthomonas malvacearum*

Note that different REs can attack the same sequence, and in the case of Hap II and Hpa II, the cleavage occurs in the same place. Different REs which act on the same sequence of bases are called *isoschizomers.*

As you can see, the cleavage sites of class II enzymes are generally four to six base pairs in length. All involve areas of rotational symmetry (palindromes). In some cases, such as Eco RI, the cleavage point is more toward one end of the sequence than the other, resulting in staggered cuts which leave complementary bases at the termini, or "tails" (Fig. 14-2a). In other cases, such as Hind II and Sma I, the cleavage leaves even ends (Fig. 14-2b). DNA molecules cleaved by these enzymes can then be bonded at the cut ends to other DNA fragments to form recombinant molecules.

Use of RE-induced Fragments for the Study of Genes

IDENTIFICATION, ISOLATION, AND DISCOVERY OF THE SPECIFICITY OF ACTION OF REs WAS OF GREAT INTEREST FOR A NUMBER OF REASONS. FOR ONE THING, THEY SERVED AS OBVIOUS TOOLS FOR THE CHOPPING UP OF DNA MOLECULES INTO SPECIFIC SEGMENTS, WHICH WOULD OF COURSE ALLOW GENETICISTS TO STUDY THE STRUCTURE OF SPECIFIC AREAS OF THE GENOME OF A CELL. Let us now continue our discussion of recombinant technology.

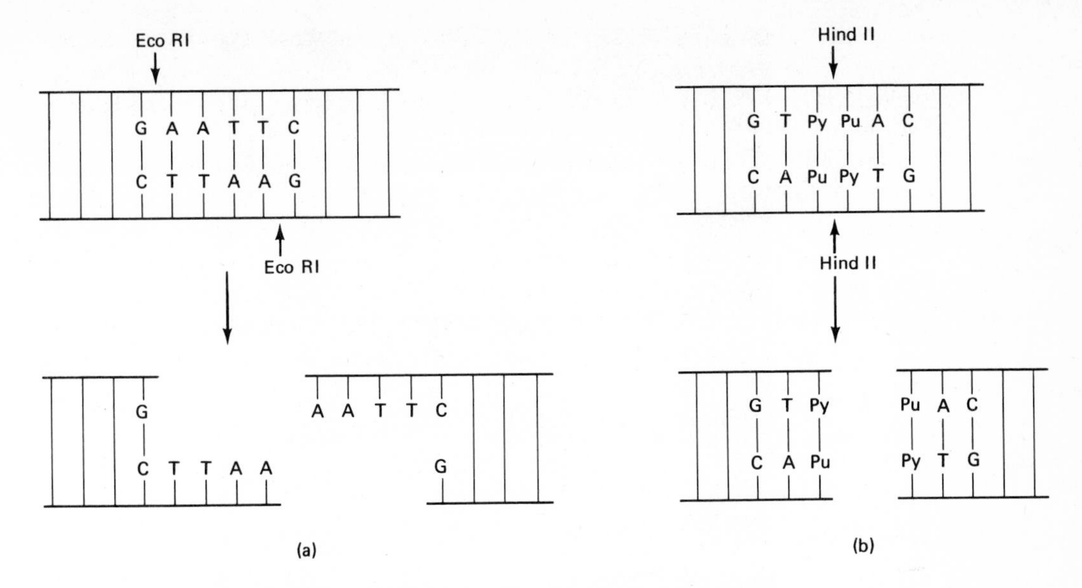

Figure 14-2. (a) Some restriction endonucleases produce uneven termini. (b) Other enzymes produce even-ended molecules.

Isolation of specific DNA fragments. The isolation of specific fragments begins with cleavage of DNA by selected restriction endonucleases. When subjected to gel electrophoresis, the fragments which result migrate at different speeds depending on their size and composition, and can be visualized as bands on the gels using dyes such as ethidium bromide (Fig. 14-3). DNA of the different bands has been found to be homogeneous by methods such as heteroduplex analysis. Fragments can be eluted from various bands and used in further experiments.

Identification of DNA fragments. One of the new procedures designed to identify the portion of the genome represented in fragments of a particular band is known as the *marker rescue* experiment. Wild-type virus DNA is fragmented with RE, and the fragments separated by electrophoresis. Fragments from a particular band are recovered and reannealed with DNA from viruses mutant for a particular gene. Heteroduplexes are formed from the wild-type single-stranded fragments and mutant viral DNA. A host is infected with the *chimera* DNA (another term for recombinant DNA) and mutant viruses, and if the virus wild-type activity is restored, it is assumed that the fragment from wild-type DNA carries the gene that is defective in the mutant virus (Fig. 14-4). Should the researcher wish to study the gene further, he can repeat the fragmentation and recover DNA from the same band following electrophoresis.

Another approach used to identify genes present in DNA fragments is to expose the DNA to various regulatory proteins and other enzymes such as RNA polymerases. For example, we know that the promoter region of the lac operon interacts with the protein CAP as well as with RNA polymerase. If DNA fragments from bacteria are mixed with the proteins on a membrane filter, only those DNA molecules containing the promoter region of the lac operon will be bound to CAP and RNA polymerase.

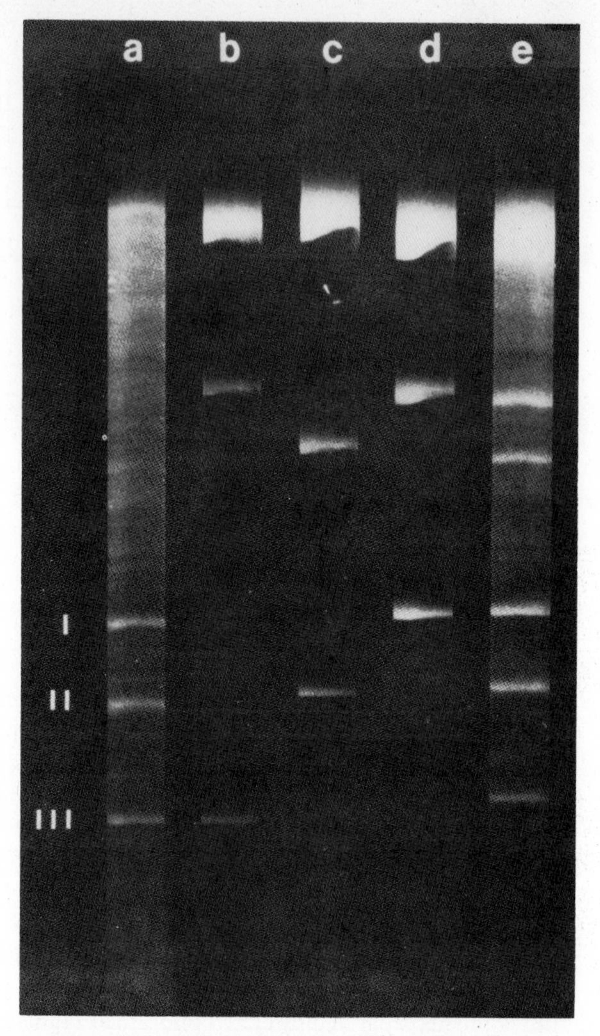

Figure 14-3. An example of banding on an electrophoresis gel. Each band represents DNA fragments of a particular size. Column (a) contains DNA from yeast; (b) (c) and (d) contain DNA from different plasmids; (e) represents a mixture of DNA from all four sources. The DNAs were cleaved with Eco RI. (From R. Kramer *et al.*, 1976. *Cell* 8: 230, Fig. 3; copyright held by M.I.T. Press.)

Use of certain stable RNA transcripts has also been reported for the identification of genes on restriction fragments. The RNA can be labeled with radioactive isotopes and when applied to DNA fragments, hybridization will reveal the presence of the genes involved.

Formation of specific recombinant DNA. Having established the specificity of cleavage sites of class II REs, scientists immediately exploited the advantages of being able to produce single-stranded DNA molecules with complementary bases at the termini from different organisms. Techniques for the identification of genes in

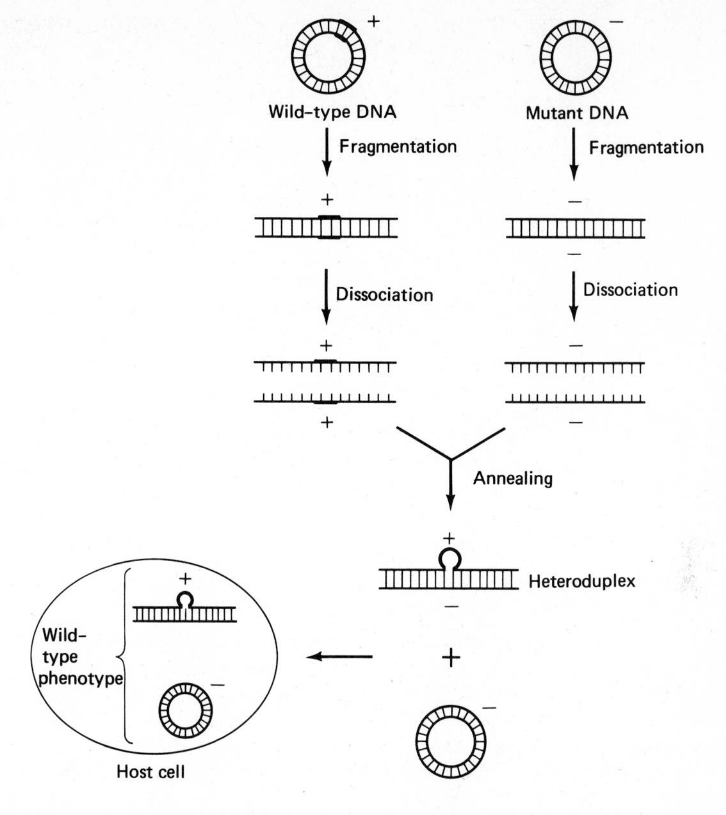

Figure 14-4. General scheme of marker rescue experiments. Presence of wild-type gene in chimera molecule complements DNA mutant for that gene, resulting in wild-type DNA function.

fragments made it possible to produce at will specific combinations of genes from different organisms. Essentially, four factors are involved: a restriction enzyme to break the DNA chain, exonucleases to remove some bases, DNA polymerase to fill in gaps when pairing between two molecules occurs, and ligase to complete the covalent bonding (Fig. 14-5).

Limitations of studies on eukaryotic DNA. Application of the technology described above to eukaryotic genomes has not yielded the same degree of success. Repetitive DNA in eukaryotes is known to have spaced cleavage sites. For example, Hind II fragments calf thymus DNA into 12 bands; mouse satellite DNA gives approximately 20+ bands with integral multiples of 240 base pairs in the repeats (Nathans and Smith 1975). However, unique DNA cannot be studied in this manner at present, since with gel electrophoresis, a million or more bands appear following exposure of the genome to restriction enzymes. This of course renders it impractical to attempt to isolate fragments for research. Nonetheless, the opportunity to isolate specific areas of DNA using recombinant methodology has been seized upon by many laboratories, and we shall describe some of the progress being made later in the chapter.

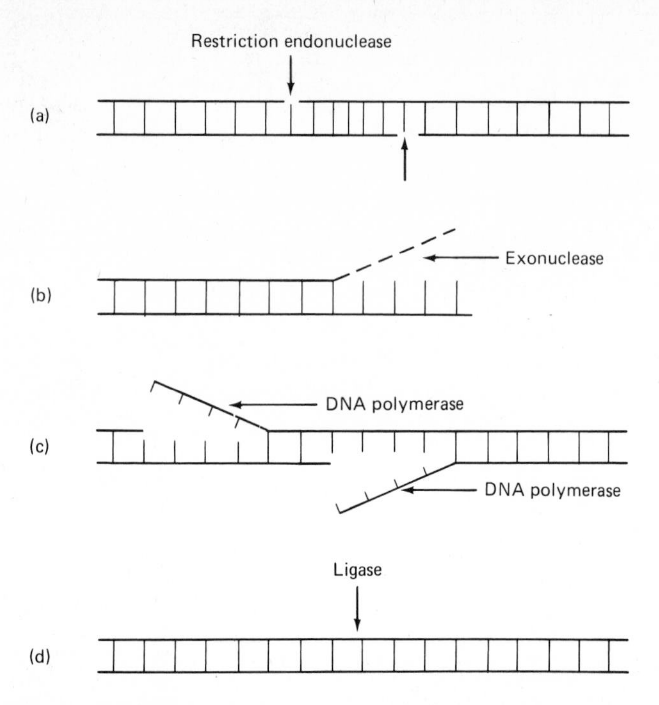

Figure 14-5. Schematic summary of major steps and enzymes involved in recombining DNA fragments: (a) Fragmentation. (b) Removal of some bases. (c) Reannealing of ends of two fragments and gap filling. (d) Ligation.

TRANSMISSION OF RECOMBINANT DNA

Once scientists constructed recombinant DNA molecules, it was natural to envision the next step, namely, insertion of the molecules into host cells in order to study their activity. It was natural for geneticists to think of using *E. coli* as the host organism because they had studied it so thoroughly; however, recall that *E. coli* was one bacterium found to be nontransformable. In other words, *E. coli* cells do not absorb exogenous DNA molecules. Therefore, a method was needed to render *E. coli* transformable, and indeed, such a technique was developed in a short time.

Transformation in E. coli

In the early 1970s, researchers found that treating *E. coli* cells with calcium salts changed the permeability of the cell membrane in such a way as to allow the uptake of noninfectious DNA. In 1972, Stanley Cohen, Annie Chang, and Leslie Hsu at Stanford reported that when *E. coli* cells were treated with calcium chloride, the cells were able to absorb purified R factor DNA. Recall that R factor DNA is normally noninfective without the RTF genes. Thus, by chemical manipulation, it became possible to transform *E. coli.* As we shall see, this was an important technological breakthrough in technology.

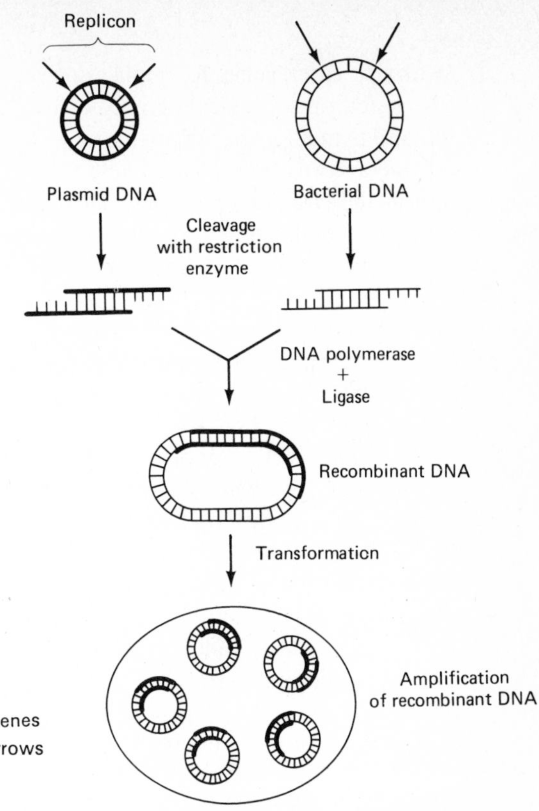

Figure 14-6. Steps in cloning cellular genes using plasmids as sources of replicons. Arrows indicate sites of cleavage.

Use of RTF Plasmids as Vectors

It was inevitable that as our knowledge of the biochemistry of DNA developed, geneticists would realize the exciting possibilities of combining what had been learned about plasmids with what could now be accomplished with restriction enzymes and fragments. THE ADDED ADVANTAGE OF JOINING DNA FROM A CELL TO DNA OF A PLASMID RESIDES IN THE ABILITY OF PLASMIDS TO REPLICATE AUTONOMOUSLY IN HOST CELLS. ONE COULD THEREFORE INTRODUCE SPECIFIC GENES INTO HOST CELLS, AND BECAUSE THEY WERE LINKED TO REPLICONS OF PLASMIDS, THERE WOULD BE REPLICATION AND THE PRODUCTION OF LARGE QUANTITIES OF THE GENES UNDER STUDY (FIG. 14-6).

The terms *cloning* and *gene amplification* are widely used to describe experiments in which genes are inserted into host bacteria, to be replicated along with plasmid genes to which they covalently bond. Through this process many copies of the genes would be produced. One obvious benefit of this technology would be to supply scientists with a source of specific DNA for their research. For example, in order to fully understand the reaction between the repressor and the operator of the lac operon, a large amount of pure lac operator gene must be available. A number of laboratories have accordingly set about to essentially manufacture the gene with recombinant DNA technology.

Biological activity of recombinant DNA. The Stanford group had in their laboratory a plasmid designated pSC101 which had been derived from a larger plasmid that conferred multiple antibiotic resistance. The smaller plasmid, pSC101, was known to carry genes for replication and a gene for resistance to tetracycline. They first tested for biological activity of recombinant DNA by exposing plasmids to Eco RI, and then allowing fragments to join at the regions of complementary bases. DNA polymerase was used to complete the double-stranded state, then ligase was used to reform the circular molecules. These reconstituted plasmids were found to be fully capable of replicating and conferring resistance to tetracycline to host cells, thereby proving that genes combined in this manner retained full biological activity (Cohen *et al.* 1973).

Following this experiment they proceeded to splice together genes from different plasmids of *E. coli*, and then genes from plasmids of different species. A plasmid from *Staphylococcus aureus*, pI258, containing a gene for penicillin resistance, was cleaved with Eco RI. Fragments were reannealed to DNA from pSC101 also treated with Eco RI. Duplex DNA was formed with appropriate enzymes, and when the chimera DNA was inserted into *E. coli* cells, the transformed hosts were found to be resistant to both tetracycline and penicillin (Cohen 1975).

The splicing of DNA from different species in vitro had actually been reported earlier by Paul Berg and his associates. Recall that they had constructed a recombinant molecule from DNA of the tumor virus SV40 with genes from λ virus and the galactose operon of *E. coli* (p. 440). However, at the time, the recombinant molecule was not used to infect host cells. The success reported by Cohen's laboratory and others in obtaining expression of gene activity from recombinant DNA molecules generated an avalanche of research which continues to build momentum today.

In 1974, Morrow and associates published the first report of the application of these techniques to the study of eukaryotic genes. They isolated amplified ribosomal genes from *Xenopus laevis* and cleaved the DNA with Eco RI. The fragments were then separated through gel electrophoresis and mixed with Eco RI-generated fragments of pSC101 plasmids. A number of different studies were carried out, such as heteroduplex analysis, density gradient centrifugation, and electrophoresis, to determine if recombinant molecules had formed. All results were positive, indicating that indeed they had produced chimera *Xenopus*-plasmid molecules.

The experiments described earlier involving drug resistance genes allowed gene activity of recombinant molecules to be detected easily. Only the presence of the drug in the culture media and survival of bacterial colonies was necessary to indicate gene activity. Activity of *Xenopus* ribosomal DNA, on the other hand, required somewhat different methods of analysis. For evidence of gene activity it was necessary to be able to detect the presence of ribosomal RNA transcripts.

One of the techniques used to identify *Xenopus* RNA transcripts was to incubate transformed cells in medium containing ^3H. Labeled RNA from these cells was applied to purified *Xenopus* rDNA which had been immobilized on nitrocellulose membranes. Any hybridization between labeled RNA and the DNA could be detected by autoradiography (Fig. 14-7), and in fact was observed.

Another technique used an unexplained phenomenon in some strains of *E. coli*. In these strains, a budding occurs, in which spheres of membrane-contained cytoplasm (*minicells*) are pinched off. Minicells containing recombinant DNA were isolated. In the absence of cellular DNA, all RNA transcripts can be assumed to be from the chimera molecules. Successful hybridization between RNA from minicells

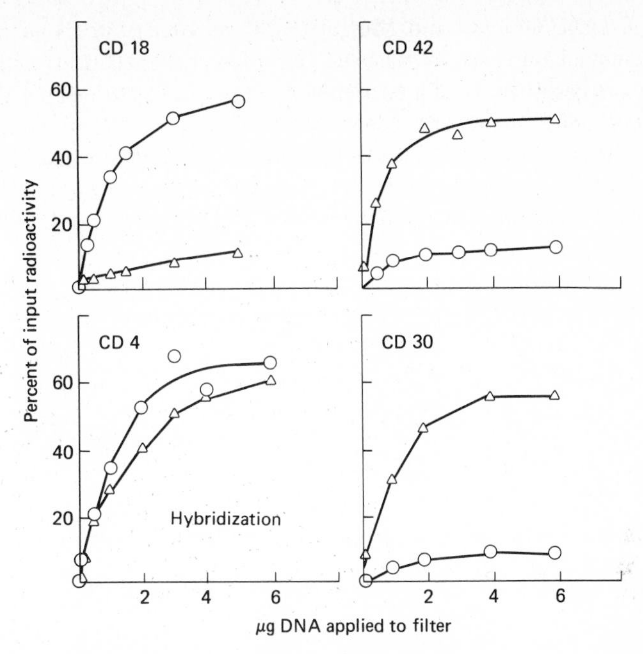

Figure 14-7. Hybridization of ³²P-labeled rRNA of *Xenopus laevis* with DNA of recombinant *X. laevis–E. coli* plasmids. CD 18, 42, 4, and 30 are different plasmids used. Δ = 18S rRNA O = 28S rRNA. (From J. F. Morrow *et al.*, 1974. *Proc. Nat. Acad. Sci. U.S.* 71 : 1746.)

with rDNA of *Xenopus* again indicated that eukaryotic genes can retain biological activity when covalently bonded to prokaryotic genetic material.

Other Vectors for Recombinant DNA

Shortly after it was shown that recombinant molecules containing eukaryotic genes retain biological activity, a number of laboratories reported success in inserting such hybrid DNA molecules into host cells using other forms of extrachromosomal genetic elements as the *vector*. The term *vector* is applied to extrachromosomal genetic particles which contain genes for replication in host cells that are spliced to foreign DNA. Without vectors, cloning would not be possible.

Colicin plasmid. One of the new vectors was a colicin plasmid, ColEI. ColEI was found capable of recombining with DNA from ϕ80 pt phage carrying typtophan genes, and kanamycin-resistance gene from pSC101 (Hershfield *et al.* 1974). Cells transformed by such molecules were found to be kanamycin resistant and had increased levels of synthesis of tryptophan enzymes. The ColEI plasmid does not contain the gene for colicin production, making it useful for experiments with bacteria sensitive to colicin as well. Also, in the presence of chloramphenicol, bacterial DNA replication is inhibited, whereas ColEI DNA can duplicate. Within 12 to 16 hours, 1000 to 3000 copies of the plasmid accumulate in each cell.

λ Phage. Viruses are obvious potential vectors since they are known to be capable of infection and autonomous replication. Several laboratories (Thomas *et al.* 1974, Murray and Murray 1974) have exploited one of the characteristics of phage packaging which we have described earlier (p. 256)—the fact that the DNA of new phages must be of a particular length in order to be encapsidated. Viable deletion mutants of λ phage were used which still contained the essential genes for replication and other functions. These mutants were obtained when phage DNA was cleaved into three fragments by Eco RI. The left and right fragments were joined to form a molecule missing a considerable portion of the genome. Foreign DNA which had also been cleaved by Eco RI was then inserted into the deletion areas, producing DNA of the proper size to be encapsidated and thus is formed an infectious particle (Fig. 14-8).

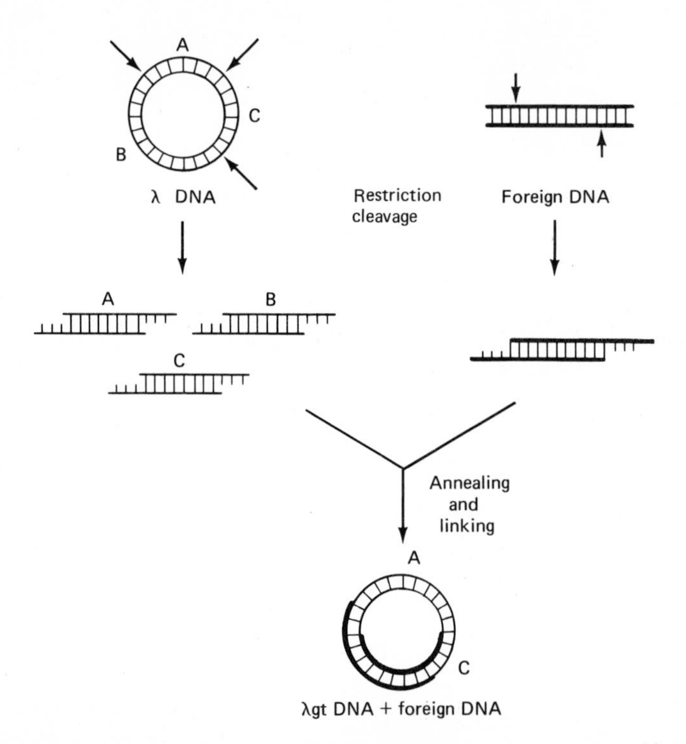

Figure 14-8. General scheme for use of *λgt* phages as vectors of cellular DNA. One portion of the phage genome (B) is removed and replaced by a piece of foreign DNA.

Following infection of host cells by the recombinant viruses, the presence of foreign DNA can be determined by various tests such as hybridization, electrophoresis, and complementation tests. Proponents of using λ viruses as vectors point out that one advantage is that if the gene *N* and the promoter of λ are included in the vector, they would allow RNA polymerase to read right through transcript stop signals. Thus foreign DNA spliced into a λ virus genome can be expressed and identified even though a stop signal may have been incorporated which would have rendered the gene silent with another vector.

Animal viruses. Recent studies have shown that DNA from SV40 viruses can be used as a vector when joined to DNA fragments from *E. coli* or bacteriophage. Of interest to scientists here would be the opportunity to study prokaryotic gene expression in a eukaryotic environment (Hamer 1977).

A somewhat different approach has also been developed in the study of animal viruses as potential vectors for the transmission of genes. Portions of SV40 genomes, for example, have been joined to deletion mutants of adenovirus 2 (Sambrook 1977). The adenovirus serves the same function in the transformation of monkey cells as the λ deletions mentioned before.

This work is in its preliminary stages, and one of the differences between it and other experiments dealing with new vectors is that the reaction is somehow mediated by coinfecting monkey cells with both viruses; it is not due to exposure of the viral DNAs to specific restriction enzymes. The viral DNA molecules are broken up in monkey cells (no doubt through enzymatic reaction), and fragments are joined together into covalently bonded circular DNA molecules.

NEW DIRECTIONS FOR MOLECULAR GENETICS

The potential usefulness of DNA fragments obtained by cleavage with REs, and recombinant DNA molecules that can be formed from them, has led to their study in a rapidly increasing number of laboratories all over the world. Most such work is being done in the United States. THE POSSIBLE APPLICATIONS OF THIS NEW METHODOLOGY TOWARD FURTHERING OUR UNDERSTANDING OF THE GENE AND GENE FUNCTIONS INCLUDE RESTRICTION MAPPING; NUCLEOTIDE SEQUENCING OF GENES; ELUCIDATION OF BASIC GENETIC PHENOMENA SUCH AS RECOMBINATION, MUTAGENESIS, AND TRANSFORMATION; CLONING AND AMPLIFICATION OF SPECIFIC GENES; AND DE NOVO SYNTHESIS OF DNA. We shall briefly discuss examples of each of these in the following sections.

Restriction Mapping

Because REs recognize specific sequences at specific sites on the chromosomes, scientists can produce specific fragments of DNA. Furthermore, sequential use of two or more restriction enzymes can result in overlapping fragments. Isolation of the fragments by electrophoresis and analysis of their genetic function can yield information on the organization of the genome (Fig. 14-9).

In the case of several viruses, this approach to mapping the position of fragments has been most fruitful. For example, Eco RI, Hpa I, and Hind III have been used sequentially to study and map the chromosome of the animal virus SV40 (Danna *et al.* 1973, Lebowitz *et al.* 1974) Figure 14-10 shows the map of the SV40 virus as determined by treatment with restriction enzymes. Such maps are logically referred to as *cleavage*, or *restriction*, *maps*. Similar work has been done on the polyoma virus (Fig. 14-11). Once cleavage maps have been established, specific segments of the DNA can be isolated, spliced to vectors, and inserted into host cells, where their function can be analyzed.

This method is not yet applicable to eukaryotic chromosomes, which are far

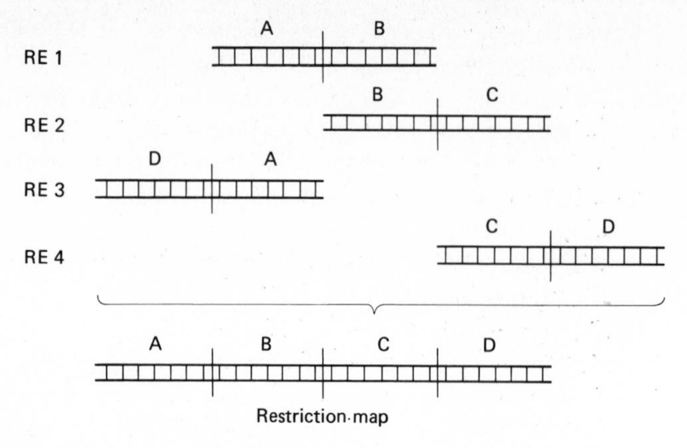

Figure 14-9. Schematic diagram of restriction mapping with overlapping fragments produced by use of different restriction enzymes (RE1 to RE4). Letters A to D represent gene activities associated with those portions of the genome.

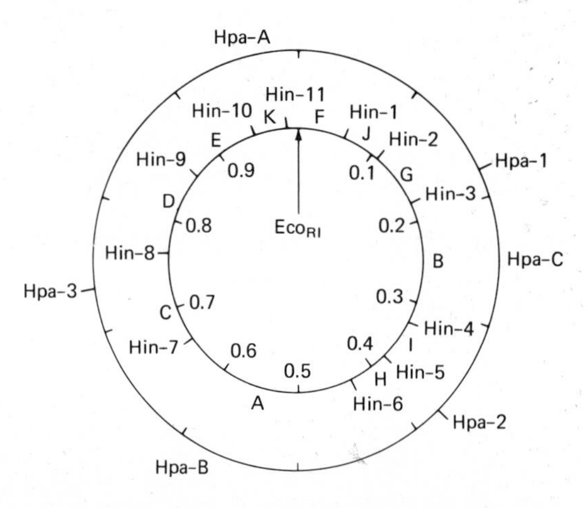

Figure 14-10. A cleavage map of the SV40 genome, as determined by restriction enzymes, Eco RI, Hpa, and Hin. Map units are given as:

$$\frac{\text{Distance from Eco RI site}}{\text{Length of SV40 genome}}$$

[With permission from K. J. Danna, *et al.*, 1973. *J. Mol. Biol.* 78: 374, Fig. 3; copyright by Academic Press, Inc. (London) Ltd.]

more complex, and would yield too many fragments to allow study of overlapping sequences. Nonetheless, a beginning has been made in the studies of repetitive DNA of calf thymus mentioned earlier, and in work such as that on bovine satellite DNA.

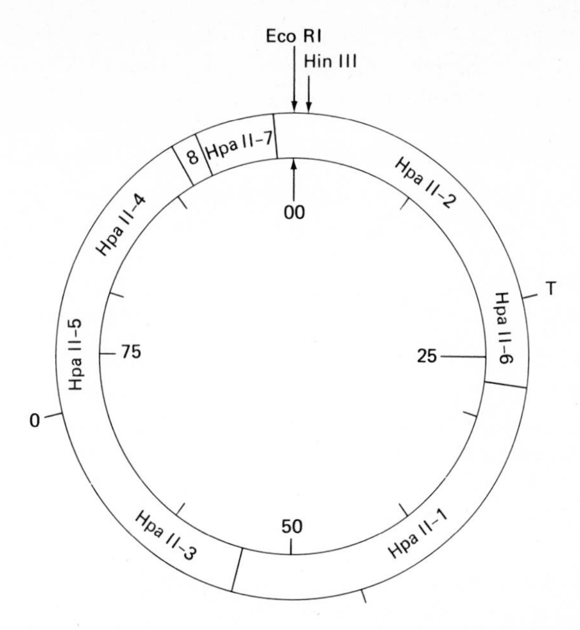

Figure 14-11. A physical map of polyoma DNA. The map, divided into 100 units, is based primarily on the eight fragments obtained when polyoma DNA was digested with Hpa II. The cleavage sites of Eco RI and Hind III are shown. Assignments of the origin (*O*) and termination (*T*) of DNA replication are indicated. (From B. E. Griffin, *et al.,* 1974. *Proc. Nat. Acad. Sci. U.S.* 71 : 2080, Fig. 5.)

Cleavage with Hin enzyme (a combination of Hind II and III) has yielded 12 fragments. Following this, exposure to Hae has resulted in 30 fragments. Mapping of these fragments is now being undertaken (Mowbray *et al.* 1975).

Nucleotide Sequencing

Recent experiments by Maniatis and his coworkers (1975) can serve as an illustration of how DNA fragments can be analyzed for their molecular structure. In studying the action of Hph on λ DNA, Maniatis found that a fragment containing most of the right-side operator gene was excised. This piece of DNA was 75 bp (base pairs) in length, and could be cleaved further into pieces 40 and 35 bp long by the enzyme Hin.

The DNA fragments were labeled with ^{32}P at the 5′ends, and then digested at the 3′ end with exonuclease. This process produced fragments of varying lengths which were then partially separated by gel electrophoresis. The duplex molecules were dissociated into single strands, and the pieces then subjected to a relatively new technique known as *homochromatography*. In this process, DNA molecules are transferred to the edge of a glass plate coated with a thin layer of ionic resin. This plate is then dipped into a solution containing 7 percent RNA fragments of various sizes. The RNA molecules will compete with the DNA for binding sites in the resin as the molecules migrate up the plate. The DNA will be pushed up the plate by the RNA, and the shorter the DNA the farther it will move; the longer the fragment, the less it will move. The result is the separation of fragments which differ by as little as one base. Finally, the preparation is exposed to film, on which the labeled DNA fragments will be visualized in their final positions.

The angle of displacement of the molecules on the chromatograph further allowed the scientists to surmise which base had been removed from each of the DNA fragments (Fig. 14-12). This in turn led to the entire sequencing of bases for the *OR* gene of λ phage, as shown in Fig. 14-13.

Two recent commonly used methods of nucleotide sequencing devised by Walter Gilbert and his coworkers, and by Frederick Sanger and his associates, are similar in approach. Restriction endonuclease is used to fragment the DNA into specific pieces which are dissociated into single strands. One end of each strand is radioactively labeled. Gilbert's technique then calls for the exposure of the DNA fragments to four different enzymes, each of which destroys the DNA fragments at the site of one of the four bases. Sanger's method uses enzymes to copy the single strand up to one of each of the four bases. In each case, pieces of DNA of different sizes ending at a specific base are created. These are then separated by electrophoresis. The shorter the molecule, the farther the piece migrates down the gel, resulting in a ladder-like gel that allows scientists to read the sequence of bases directly from the gel (see reference to Gilbert and Villa-Komaroff 1980 for a review of both techniques). Drs. Gilbert and Sanger shared the Nobel Prize for chemistry in 1980 for developing these techniques, along with Dr. Paul Berg, whose work we mentioned earlier in this chapter.

Other approaches to the sequencing of DNA fragments have been reported (see again the review by Nathans and Smith 1975), and no doubt more will be developed.

The Study of Gene Function with Recombinant DNA

Because of the prominent role of viruses in recombinant DNA experiments, it stands to reason that the study of physiological genetics of viruses would benefit most immediately from the new technology.

Cancer viruses and transformation. One of the most active areas of research involves the study of oncogenic viruses. Most of the work that has been done involves cleaving viral DNA into fragments and attempting to discern the role of genes in various fragments in the process of transformation of cells to cancer. A number of laboratories have concentrated on the SV40 virus. We have already discussed the cleavage map of SV40; now attempts are being made to assign functions to various fragments.

For example, it is known that a viral gene *A* causes the formation of an antigen very soon after cells are infected with SV40 viruses. This antigen, called *T antigen*, is necessary not only for the initiation of viral DNA replication, but also for the maintenance of transformation to cancer of the host cell. Jessel and his associates (1976) have cleaved SV40 DNA with Eco RI and Hpa II, and found that the T antigen binds to specific regions of the fragments, including the fragment containing the origin of DNA replication.

The significance of the binding is not yet understood, but scientists are in the process of cleaving the SV40 DNA into smaller fragments to pinpoint the sites of binding. Then, using techniques we have described for the sequencing of bases, they may obtain from the fine structure of the regions of affinity for the T antigen some clues as to the genetic control of T antigen production. This in turn may elucidate the physiological function of the binding. There is some speculation that the presence in infected cells of the T antigen, which binds with portions of the SV40 genome, may serve some regulatory purpose in transformation. Others are presently seeking to establish the position of genes responsible for the production of the T antigen (Berg 1977).

The pattern of transcription of the SV40 genome has also been analysed using fragments obtained with cleavage by Eco RI and Hpa I (Sambrook *et al.* 1973). When SV40 viruses infect permissive cells, two classes of stable RNA have been found in the cytoplasm of infected cells. By labeling the 3′ ends of DNA and hybridizing the fragments with RNA found in infected cells, it has been determined that one class of RNA is complementary to 30–35 percent of one strand of the viral genome and is present both early and late in infection, whereas the other class is complementary to 65–70 percent of the other strand and is present only early in infection. The pattern of hybridization showed transcription to be clockwise on the one strand and counterclockwise on the other. Further, transformation to cancer can occur in the absence of late genes (Botcham *et al.* 1976).

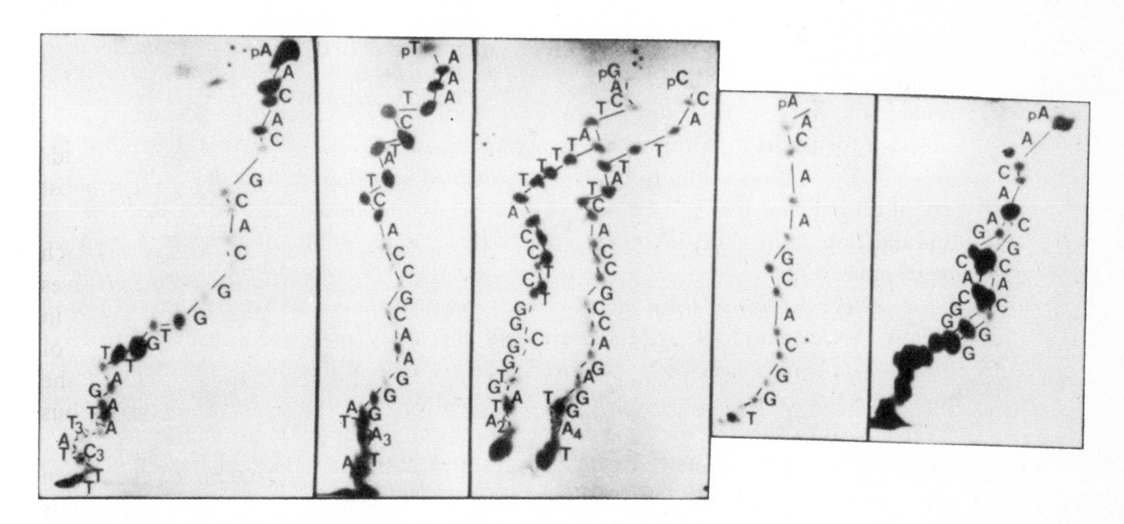

Figure 14-12. Examples of nucleotide sequences obtained by homochromatography following cleavage and gel electrophoresis of DNA. (From T. Maniatis *et al.*, 1975. *Proc. Nat. Acad. Sci. U.S.* 72: 1185.)

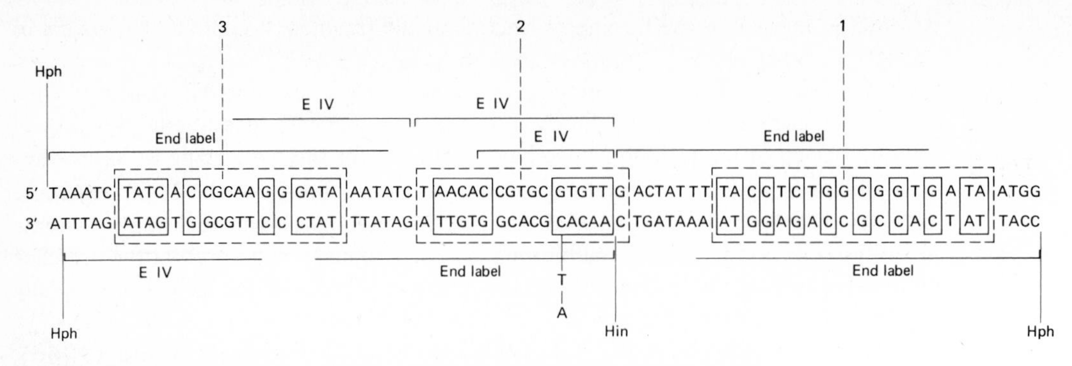

Figure 14-13. Sequence of 74 base pairs of OR. The position of the Hind II and the two Hph cutting sites are indicated. The lines labeled End Label cover those sequences determined by end labeling and partial nuclease digestion, and the solid lines labeled E IV cover bases sequenced by analysis of endonuclease IV digestion products. The vertical broken lines indicate the centers of the three 17 base-pair sequences believed to be the repressor binding sites. The bases in boxes are arranged symmetrically about these axes. (From T. Maniatis *et al.*, 1975. *Proc. Nat. Acad. Sci. U.S.* 72: 1186, Fig. 7.)

The polyoma virus is an oncogenic virus known to cause lysis of mouse cells (though occasionally transformation occurs) and transformation of hamster and rat cells; its cleavage map is shown in Fig. 14-11. A class of mutants called hr-t polyoma mutants which have lost the capacity to transform were studied by infecting cells with the mutant viruses as well as with fragments of wild-type DNA, which served as rescue markers (Benjamin 1977). A particular segment called *fragment 5* was found to be the DNA which most efficiently complemented the hr-t mutants. This fragment included the origin of replication. Thus a gene or genes proximal to the origin of replication, in the early transcription portion of the DNA, determines some essential function for transformation by the polyoma virus.

Our eventual ability to control the diseases known as cancer depends to a large extent on whether we shall be able to uncover the molecular nature of the structure and function of the genes involved. Toward this end, recombinant DNA studies should be of great value.

Other aspects of DNA metabolism. Because of the relative newness of recombinant DNA methodology, its usefulness for study of various aspects of DNA metabolism is only now being explored. For example, it has been suggested that our understanding of mutagenesis may be advanced by inserting base analogs into DNA fragments. The effect of the base analog on the activity of genes in the fragments may give us clues to the function of the genes and basic processes such as replication (Weissmann *et al.* 1977).

Another application of recombinant DNA methodology may be in resolving the molecular aspects of recombination. The structure of DNA molecules of colicin plasmids apparently undergoing recombination has been observed in electron micrographs (Potter and Dressler 1976). Analysis of the "figure 8" structures has been aided by the cleaving action of Eco RI, showing that there actually are two plasmid circular molecules attached at a presumed site of homology. The configuration is thought to be evidence supportive of the Holliday model of recombination (see again p. 294).

The Cloning of Eukaryotic Genes

We have previously described techniques for the cloning and amplification of genes. In this section we shall discuss a few examples of this type of work, especially experiments dealing with eukaryotic genes.

As we have stated, it is not feasible to isolate specific unique DNA fragments from eukaryotic chromosomes for analysis, not only because of the genetic complexity of eukaryotic organisms, but also because of the polygenic nature of many traits. If the interaction of ten genes located on different chromosomes is necessary for the expression of a genetic characteristic, isolating one or two fragments containing some of the genes will not result in expression of the trait, and detection of the genes will therefore be impossible.

There are, however, some genes which are known to determine a particular cell function without interaction, such as histone genes, or genes determining ribosomal RNA. Since their expression is detectable at the cellular level, such genes would of course be likely candidates for study with recombinant technology. But how can one find the fragment or fragments that contain the desired gene? Only in amphibians does amplification of ribosomal genes occur during oogenesis. Suppose we wish to study ribosomal genes of *Drosophila*? The procedure used in such cases—the *shotgun experiment*—has generated much controversy. We shall deal with the controversy later, but for now, let us discuss only the procedure.

Shotgun experiments. Simply described, a shotgun experiment is one in which the total genome of an organism such as a sea urchin or *Drosophila* is removed from cells and exposed to restriction endonucleases. As we mentioned before, this procedure results in a large number of DNA fragments. The fragments are mixed with DNA from a vector molecule, and random ligation is permitted to take place. Recombinant molecules are then inserted into host cells, and those cells with chimeras containing the desired genes are isolated by various methods. These recombinant molecules can then be recovered and cloned in other host cells (Fig. 14-14).

Histone genes of sea urchin cells have been isolated in this manner (Kedes *et al.* 1975). DNA from sea urchin cells was cleaved with Eco RI, as were DNA from pSC101 plasmids. The fragments were mixed together and treated with ligase. Three classes of recombinant molecules were found: recombinant plasmid DNA, recombinant molecules containing nonhistone sea urchin genes, and recombinant molecules containing histone genes. The last class, the ones to be studied, were detected by hybridization to labeled histone mRNA. The fragments could now be subjected to further analysis for the fine structure and function of the histone genes.

Polytene chromosomes of *Drosophila* have also been cleaved with Eco RI, and the fragments mixed with fragments of pSC101 plasmids and ligase. The largest hybrid molecules were eliminated by a special sucrose gradient technique, for they were believed to contain more than one *Drosophila* fragment. Using only those recombinant molecules that were tested to have only one *Drosophila* fragment, in situ hybridization

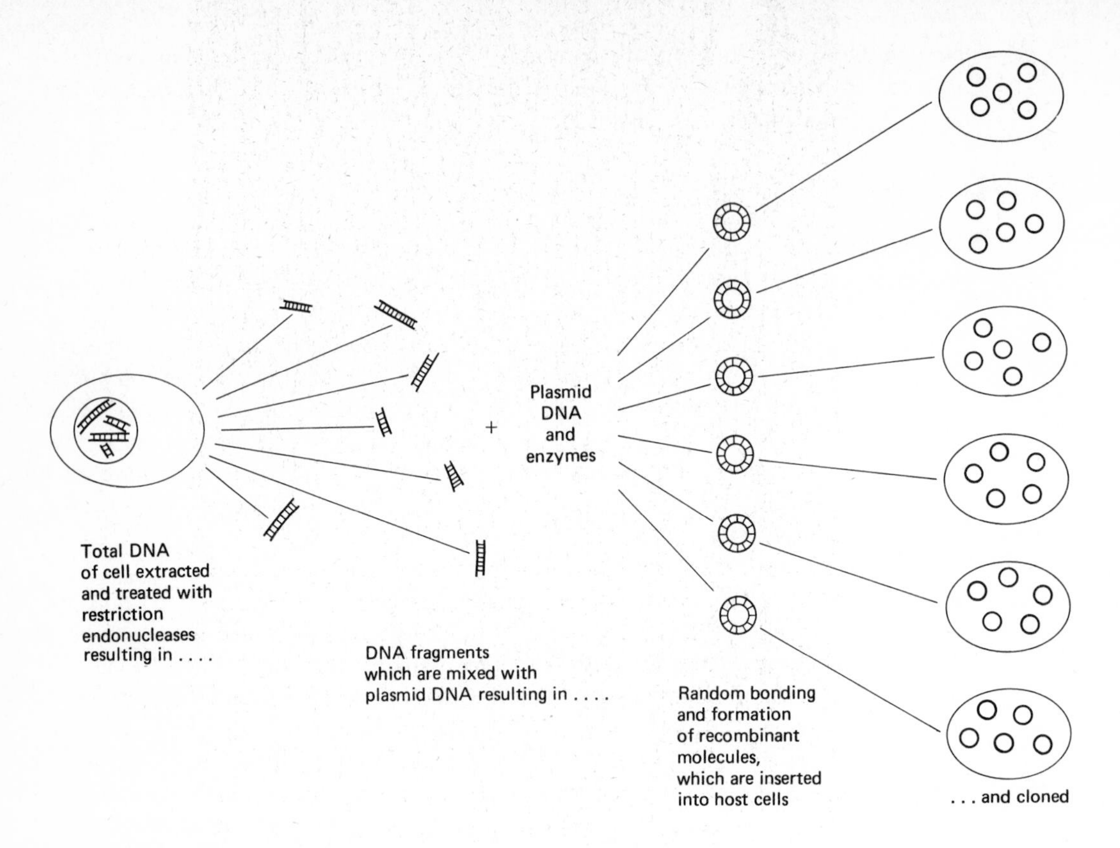

Figure 14-14. The "shotgun" experiment.

onto polytene chromosomes was carried out with labeled fragments. Six different fragments were identified; four of them were associated with single chromomeric regions, one with the Y chromosome, and one with the nucleolar organizer.

The same procedure of cleaving DNA with Eco RI, and then bonding of fragments to pSC101 fragments with ligase, was used to study the spacer regions between the genes for ribosomal RNA of *Xenopus laevis*. It has recently been reported that two regions in the spacer DNA are variable in length because of the number of internally repetitious sequences (called *subrepeats*) of less than 50 bp. The two variable regions are separated by another sequence which is constant in length. The constant region has no subrepeats (Wellauer *et al.* 1976). It remains for future studies to show the function of the subrepeats, and indeed, of the spacer regions themselves.

Application of the shotgun technique led to the isolation of interferon genes in 1980. This development evoked much interest for two reasons. One was of course the potential to clone the genes in great numbers, and to have bacteria producing interferon in quantities as large as hundreds of liters. Research heretofore restricted by the lack of material will of course be able to proceed unhindered in the future. The second point of interest is that the cloning of interferon genes now promises to be of great commercial value, as does the cloning of any substances of medical value. We shall speak more on this aspect of modern biology later.

Experiments such as these promise to help us understand more fully the organization of eukaryotic genomes. As techniques are refined in the future, no doubt we shall be capable of more precise study of the DNA of complex organisms. We still need to exercise caution in the interpretation of results obtained with recombinant technology as being an exact reflection of the genetic material in vivo. For example, it is possible that the treatment of DNA with various enzymes may eliminate portions of the genome in some undetected way. Also, we cannot be sure that hybridization experiments actually represent areas of total homology; small heterologous areas may not always be detectable with present techniques.

Nitrogen fixation genes. One area in which recombinant DNA research may prove to be of great practical value is in the study of nitrogen fixation in plants. Let us digress a moment to recall some basic aspects of nitrogen fixation. Large quantities of nitrogen exist in our atmosphere; however, it cannot be used by living organisms for biosynthesis until it has been reacted with hydrogen or oxygen, a process known as *nitrogen fixation*. We need say no more about its importance to living systems other than to point out that there is nitrogen in every amino acid.

There are two main sources of nitrogen fixation, one natural and the other industrial, both requiring large amounts of energy to convert N_2 to forms usable by cells. The natural process is dependent on the presence of nitrogenase, an enzyme that catalyzes fixation in various prokaryotes. Nitrogenase consists of two proteins determined by genes known as *nif* genes. The result of their activity is the production of ammonia, NH_3. Some of the prokaryotes capable of nitrogen fixation (there are no such eukaryotes) are free-living forms such as the bacterium *Klebsiella pneumoniae*. Others have developed symbiotic relationships with higher plants, such as *Rhizobium*, which is found in root nodules of legumes.

The industrial form of nitrogen fixation occurs in chemical reactions during the production of fertilizers. Here nitrogen is reacted with hydrogen under high pressure and temperature to produce ammonia. Anyone who has had to buy fertilizer for farming, gardening, or a lawn has to be aware of sharply rising prices for fertilizer. The reason is that the source of hydrogen in the industrial reaction is fossil fuels such as petroleum. In 1975, the total amount of energy required for the worldwide production of ammonium fertilizers was reported to equal 2 million barrels of oil per day (Shanmugam and Valentine 1975). The current shortage of fossil fuels plus the increasing demand for agricultural fertilizers suggest that there will be continuing food shortages. We have indeed already witnessed serious famine in many parts of the world.

In 1971, scientists at the University of California, La Jolla, and in England discovered and mapped the *nif* genes, which are important in nitrogen fixation. In *Klebsiella* they were found to be clustered near the histidine operon (Shanmugam and Valentine 1975). Because *Klebsiella pneumoniae* is genetically similar to *E. coli*, it was possible to transduce the *nif* genes to *E. coli*, which normally do not possess nitrogen-fixing capabilities.

Of perhaps more interest to our discussion here is the discovery that conjugation can occur between *K. pneumoniae* cells and *E. coli*, with the transfer of *nif* genes to nif^- *E. coli* recipients. Subsequently, a plasmid-carrying *nif* genes has been found, which is noninfectious, but can be transferred if attached to an RTF or an F' plasmid.

The plasmid was reported by Cannon and his coworkers in 1974. It is a covalently closed circle with a molecular weight of 9.5×10^6. Whether the *nif* genes found in the chromosome of bacteria are plasmid genes which have been incorporated remains to be established.

With these facts in mind, it is not difficult to understand present attempts by scientists to apply recombinant technology to the cloning of *nif* genes for transfer to plant cells which normally do not fix nitrogen. A number of problems must be resolved, however, before we can reach this desirable goal. One is that the actual number of genes involved is not known. Elucidation of the fine structure of the *nif* genes is the most likely immediate contribution of recombinant technology.

There is also evidence that nitrogen fixation in bacteria is under some form of regulation. The presence of fixed nitrogen slows the process of nitrogen fixation in bacteria such as *Rhizobium*. The enzyme glutamine synthetase is also known to be necessary. Whether *nif* genes can function in foreign cells without genes involved in their regulation remains to be seen; however, once the regulatory mechanisms are known, one approach to increasing the supply of fixed nitrogen may be to develop strains of bacteria that constitutively fix nitrogen.

It must also be borne in mind that simple transfer of all the *nif* genes to other plant cells is not in itself sufficient to cause nitrogen fixation to take place. One of the major reasons is that nitrogenase is irreversibly denatured by contact with oxygen, which is why nitrogen-fixing bacteria are normally found deep in the soil under almost anaerobic conditions. Thus if *nif* genes were introduced into corn, for example, the structure and metabolism of the corn plant would have to be extensively modified (Brill 1977). Still, with our modern techniques, vectors and plants that would serve as suitable hosts may be discovered in the near future—to the benefit of starving people and our dwindling supply of fossil fuels.

De novo synthesis of a functional gene. On page 284, we mentioned that K. Itakura and coworkers reported the synthesis of a biologically active gene de novo, that is, without a template. The technique that they used was a modification of one developed by Khorana in 1968.

Although we shall not dwell on the sophisticated biochemistry involved, essentially the procedure called for the synthesis of overlapping fragments of the gene for the hormone somatostatin. T4 DNA ligase was then used to join the fragments. Once the entire sequence of 42 nucleotides was constructed, the gene was inserted into a plasmid carrying various parts of the lac operon. These included the lac promoter, CAP binding region, operator, and the first seven amino codons of the *z* structural gene. In addition, the plasmid pBR322 also carried genes for resistance to tetracycline and ampicillin.

Because the somatostatin gene sequence was inserted next to the *z* gene, transcription was initiated by the *lac* promoter gene and carried through the *z* gene, and through the gene for somatostatin. Cells transformed with this chimera DNA could be detected simply by growing them in selective media containing tetracycline and ampicillin. As we pointed out in Chapter 9, the somatostatin isolated from transformed cells was tested to be the same as the hormone produced in vivo.

Other approaches. There are countless other interesting projects involving cloning of genes which will eventually lead to our understanding of many biological processes which are now obscure. One technique for making available specific eukaryotic genes for study uses mRNA for a specific polypeptide to generate cDNA with reverse transcriptase. Single-stranded cDNA is then used as a template-primer for the synthesis of a complementary strand (designated sDNA) with *E. coli* DNA polymerase I. The duplex gene can then be used in recombinant molecules and cloned for further studies. An example of this is the work being done by Efstradiadis *et al.* (1976) on the gene determining the β-globin chain of rabbit red-blood cells (Fig. 14-15).

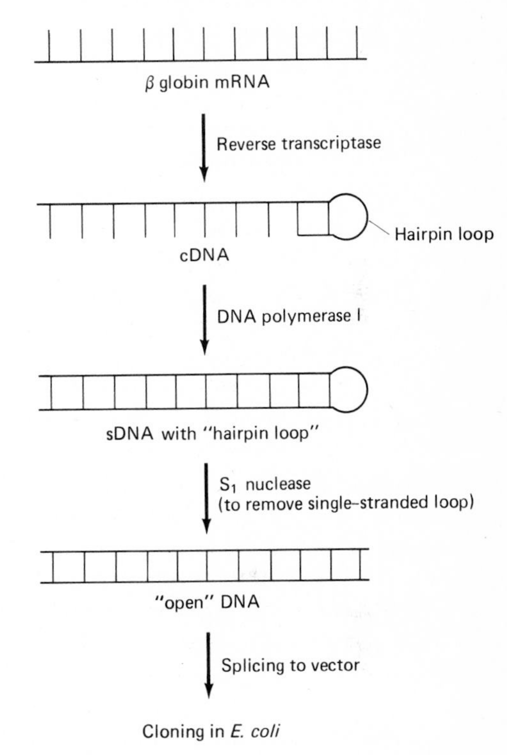

Figure 14-15. Schematic illustration of cloning specific genes using mRNA to generate the DNA, which eventually is joined to a vector and cloned.

Immunological light chain genes from mice have also recently been formed in this manner. The difficulty in this study has been in obtaining sufficient quantities of a pure form of mRNA. Nonetheless such studies may eventually provide information on the number of *Ig* genes. Sequencing studies on the *Ig* genes may then lead to our understanding of antibody variability (Mach *et al.* 1977).

The cloning of yeast genes using different vectors such as Col EI or λgt phage has been used to study whether eukaryotic genes can be expressed with fidelity in a prokaryotic environment. Preliminary results have shown that yeast genes are expressed, as complementation of deletion mutations of *E. coli* host cells has been achieved (Carbon and Ratzkin 1977). This type of research will contribute toward the development of techniques for the cloning of eukaryotic genes determining a desired cellular product, such as insulin in bacteria. The bacterial host can then serve as a true factory for the production of the substance in large quantities.

Much of this work is in its preliminary stages, but it should be quite clear to you that recombinant technology offers geneticists myriad opportunities to explore gene structure and function on a scale never before attainable. It is not feasible in an introductory course to discuss all the fascinating experiments being reported in journals, but the preceding discussion should give you an idea of some of the work currently underway (the September 19, 1980 issue of *Science* is devoted to reports of research in this field).

THE GREAT DEBATE

In July 1974, an unprecedented letter was published in *Science* magazine, signed by some of the foremost practitioners of the new techniques of recombinant DNA. They included Paul Berg, Herbert W. Boyer, Stanley N. Cohen, David S. Hogness, and Daniel Nathans—all of whose work we have already discussed—and other prominent scientists such as Nobel Laureates David Baltimore and James Watson. THIS LETTER WAS UNPRECEDENTED BECAUSE IT CALLED FOR A VOLUNTARY MORATORIUM ON CERTAIN KINDS OF GENE-SPLICING EXPERIMENTS. FOR THE FIRST TIME, SCIENTISTS WERE THEMSELVES SERIOUSLY QUESTIONING THEIR OWN WORK.

The idea of a moratorium was instituted by Paul Berg, who had earlier been asked about the possible consequences of this kind of genetic manipulation. Concern was also voiced by others at the 1973 Gordon Research Conference. In view of the potential good to be derived from the work we have discussed in this chapter, such concern may surprise you. It is a feeling probably not unlike that of Berg and others, whose enthusiasm and imagination had been kindled by the promise of their research, when first confronted with the problem.

Basically the question underlying the call for a moratorium was one of safety: What would be the consequences if new combinations of genes, which are introduced into infectious organisms, escaped from the confines of the laboratory? It is possible that such infected organisms might pose a threat to the health and existence of living organisms in nature, including, of course, humans. Such molecules could accidentally escape via an infected or contaminated technician, or insects with access to the laboratories, or in contaminated waste which is improperly treated and disposed of.

The letter to *Science* contained four proposals. One was that scientists voluntarily defer work on two types of experiments. Type 1 was the construction of recombinant molecules carrying antibiotic resistance genes or toxin genes to be transferred to cells of species which have not been found to contain these capabilities in nature. Type 2 was the linking of oncogenic viral genes or genes from other animal viruses which if released from the laboratory may cause an increase in the incidence of human diseases such as cancer.

Although no moratorium was asked for a third type of research, the Berg letter called for caution in the use of animal DNA in recombinant research, since sequences of bases in some animal cell DNA have been found in RNA tumor viruses. The third proposal called for the National Institutes of Health to establish a committee to oversee research, develop procedures to minimize the spread of experimental DNA molecules to the environment, and set up guidelines for research. Finally, they asked for an international meeting to discuss the problems.

Pros and Cons of Recombinant DNA Research

A number of meetings were held, including the first international meeting at Asilomar, California, in 1975 (Berg *et al.* 1975). During all the negotiations, salient points supporting the research and moving arguments against it were raised. The debates have raged and are raging in many facets of society, from articles in respected scientific journals to meetings of town councils, by scientists and nonscientists alike. It is important to consider major points of both sides of the debate, because as students of genetics you should be aware of them to formulate your own opinions, and because such debate carries with it even greater implications of intellectual freedom, and the role of science in our modern society.

Pros. All the potential gains that we have already pointed out in the earlier sections of this chapter constitute the best arguments in favor of continuing recombinant DNA research. Its promise of elucidating basic gene structure and function, such as that of eukaryotic genes and chromosomes, would allow inroads into areas in which progress has been blocked for lack of a feasible technical approach. Applications of gene-splicing techniques may also lead to the understanding and therefore control of a number of diseases such as cancer. Societal problems such as food shortages may be alleviated with the construction of new plants capable of nitrogen fixation, and so on.

Cons. The Berg letter already pointed out some of the misgivings on the part of even some researchers working on recombinant DNA, namely, that new combinations of detrimental genes may be placed in cells that could escape from the laboratory and cause widespread disease. For example, the presence in nature of antibiotic-resistant bacteria is enough of a problem without artificially adding to their numbers.

Furthermore, critics of recombinant research, who include some prominent scientists such as Robert Sinsheimer of Cal Tech, Erwin Chargaff of Columbia, and George Wald of Harvard, also point out that combining of prokaryotic and eukaryotic genes does not occur in nature, and therefore may result in undesirable consequences that we cannot foresee. For example, how would eukaryotic genes respond to regulatory systems of prokaryotes? As we mentioned earlier, the presence of the *N* protein from λ phage could cause transcription to occur right through normal stop signals. The absence of stop signals may be desirable in some experiments, such as the continuous production of a desirable cell product, but suppose undesirable genes also continue to be transcribed?

Shotgun experiments are of particular concern from this point of view, because it is known that some detrimental recessive genes carried in the genomes of complex organisms are not expressed because of diploidy. By random insertion of DNA fragments into infectious particles, it is feared that such detrimental genes may be cloned.

There are doubts over the use of *E. coli* as the host organism of choice by geneticists in recombinant research. It is the choice of course because of the detailed knowledge of its genetic makeup gained in past decades. However, critics point out that the use of *E. coli* may be undesirable because it is in fact a natural inhabitant of the human intestine. If cells escape from the laboratory carrying detrimental genes, humans would be among those most likely to be infected.

Sinsheimer and Wald are also worried that we are essentially tampering with evolution by crossing species lines in the recombination of genes from different organisms. As Wald has written: "Up to now living organisms have evolved very slowly, and new forms have had plenty of time to settle in. It has taken from 4 to 20 million years for a single mutation . . . to establish itself as the species norm. Now whole proteins will be transposed overnight into wholly new associations, with consequences no one can foretell, either for the host organism or their neighbors. The results will be essentially new organisms, self-perpetuating and hence permanent. Once created, they cannot be recalled." (Wald 1976, p. 7).

Sinsheimer, because of his prominence as a molecular geneticist, has perhaps been the critic of recombinant DNA research receiving most attention. He does not feel that the guidelines are stringent enough, nor could they be since they were essentially drawn up by those active in the field to solve their own experimental problems. He further questions the right to total freedom of inquiry that has been so zealously guarded by scientists since humans evolved the intelligence to question.

To quote from one of his recent talks: "Truth is necessary but not sufficient. . . . Scientific inquiry, the revealer of truth, needs be coupled with wisdom if our object is to advance the human condition. . . . Our thrusts of inquiry should not too far exceed our perception of their consequence. There are time constants and moments in human affairs. We need to recognize that the great forces we now wield might—just might—drive us too swiftly toward some unseen chasm." (Reported in Wade 1976, p. 304)

Counterarguments. For every point that critics have brought up, a counterargument has been produced by those in favor of the research. With regard to tampering with evolution, it has been pointed out that transformable bacteria probably absorb DNA of other species quite commonly, as in the decay of dead eukaryotes.

As for countering the "evolutionary wisdom of the ages," Cohen has written: "It is this so-called evolutionary wisdom that gave us the gene combinations for bubonic plague, yellow fever, typhoid, polio, diabetes, and cancer. . . . The acquisition and use of all biological and medical knowledge constitutes an intentional and continuing assault on evolutionary wisdom. Is this the 'warfare against nature' that some critics fear from recombinant DNA?" (Cohen 1977, p. 655).

Proponents of the new technology are confident of their knowledge of *E. coli*, and feel control can be maintained to minimize if not eliminate any dangers to public health. Along this line, they note that known pathogens are being investigated in hospitals and other laboratories, and have been for decades, without significant hazard to communities in which they are located.

They agree that scientists do not have the right to carry out any experimentation they wish regardless of consequences, and point to the fact that the initial call for cessation of certain experiments came from within the science. As we mentioned before, this kind of societal consciousness is really unique in the history of science.

Nonetheless, nagging doubts remain which cannot be refuted. We do *not* know what the effect of all combinations of DNA will be. There will always be the possibility of human error, regardless of containment facilities; humans will be conducting the experiments. Sinsheimer and others are worried that the proliferation of institutes active in genetic engineering may include those whose personnel have not had appropriate training and background to safeguard against mistakes.

In addition, once a laboratory has been approved for this work, there is no practical mechanism for policing the work done. In other words, there will be no way to safeguard against intentional abuse of the techniques, either by an individual or a government. And if we worry about sabotage of nuclear plants, what of sabotage of genetics laboratories?

NIH Guidelines for Recombinant Research

The guidelines which were finally issued in 1976 were the culmination of many meetings and revisions. Many of the points were based on suggestions put forth at the Asilomar meetings. As the guidelines state, the regulations are not final. As new techniques are developed, and more information obtained on the potential hazard or safety of certain types of research, the guidelines will have to be revised accordingly. The guidelines can be obtained upon request from the Office of Recombinant DNA Activities, National Institutes of Health, Bethesda, Md., 20014.

ESSENTIALLY THE GUIDELINES DOCUMENT TWO AREAS OF CONTAINMENT: PHYSICAL CONTAINMENT AND BIOLOGICAL. THE MORE HAZARDOUS THE EXPERIMENT, THE HIGHER THE LEVEL OF BOTH TYPES OF CONTAINMENT IS REQUIRED. We shall describe very generally the various levels, and give examples of combinations of physical and biological containment for various types of experiments required at the present time.

Levels of physical containment. There are four levels of physical containment, P1–P4. The P1 level requires only the usual measures taken in a clinical microbiological laboratory. These include no smoking, eating, or drinking in the lab; wearing lab coats; using cotton-plugged pipettes; and disinfecting contaminated materials. The next level, P2, adds features such as limiting access to the labs, with warning signs when experiments are in progress; also, no mouthpipetting, and use of biological safety cabinets if techniques may create aerosols which would disseminate recombinant molecules. P3 adds more protective clothing for personnel, and more stringent facilities such as special air filters and negative air pressure. The P4 level requires in addition to all the above features, isolation of the laboratory by airlocks, showering upon leaving the lab, and treatment of air and wastes to inactivate or destroy any DNA molecules or carriers of biologically active molecules.

The most stringent levels, P3 and P4, which are to be used in the most hazardous types of experiments, require hundreds of thousands of dollars in construction and equipment. P3 laboratories already exist, and several others are being built. Sinsheimer feels that the type of work necessitating P4 labs should be done in one laboratory (such as Fort Detrick in Maryland, where biological warfare experiments have been conducted), since the restrictions of this level would be difficult to maintain in a university environment.

Levels of biological containment. The three levels of biological containment have been designated EK because it was assumed at the start that the host to be used in most recombinant experiments would be strain K12 of *E. coli*, which we have already described. Since it is believed to be unable to colonize in the human intestine, K12 is the accepted host for EK1-level experiments. EK2 requires the use of K12 strains which include mutations that would render the cells so inviable in the natural environment, that only 1 of 10^8 cells would survive. EK3 would require the use of K12 strains with the mutations of EK2, but which had actually been tested and proven to be ineffective when fed to experimental animals.

Vectors to be used also fall under the three levels of biological containment. EK1 vectors, including nonconjugative plasmids such as pSC101, Col EI, and variants of phage EK2 vectors, again must contain mutations that would render their survival unlikely if they should escape from the laboratory; for example, if phages are used, they must contain mutations rendering no more than 1 in 10^8 particles able to survive. EK3 vectors will be those EK2 vectors actually tested in experimental animals and shown to have the 1 in 10^8 survival rate.

Following are some examples of experiments to be carried out under various levels of containment. Experiments using primate DNA recombinants from shotgun experiments with *E. coli* should be carried out with either P3 + EK3 or P4 + EK2 containment; use of invertebrates and lower plants in shotgun experiments is deemed less hazardous, and can be done with P2 + EK1; prokaryotes such as the enterobacteriae which normally exchange genes with *E. coli* can be used in recombinant research at P1 + EK1; however, pathogenic prokaryotes can be used only under at least P3 + EK2 levels of containment.

The guidelines also prohibit certain types of experiments altogether, such as those dealing with genes for the biosynthesis of extremely potent toxins such as botulinum, use of DNA from the most hazardous pathogenic organisms, transfer of drug resistance to organisms not known to acquire it naturally, and genes for plant pathogens in recombinant molecules that would increase their incidence in nature.

Although the requirements for different combinations of physical and biological containment are certainly debatable—for who can be sure if a particular combination is actually sufficient, or unnecessarily costly—they were developed to the best ability of those most knowledgeable. Most scientists, even those in disagreement who feel they are too stringent, appear to be willing to comply.

Development of Safe Hosts and Vectors

Because of the deep concern over the safety of some proposed experiments, some laboratories have turned their attention and energies to the development of vectors and hosts that would be efficient tools in gene manipulation, but would be considered safe under the limits of survival set by the NIH guidelines. One of the first breakthroughs along this line was achieved in the laboratories of Roy Curtiss at the University of Alabama. He and his associates have developed a mutant strain of *E. coli* to serve as host for EK2 and EK3 level experiments.

Like most of the work we have discussed in this chapter, Curtiss' research is still in progress, but a brief recounting of what has been accomplished to date will give you an idea of the approach. After two years of work, a strain of *E. coli* K12 was developed and accepted by NIH. Patriotically named $_{\chi}1776$, some of its useful characteristics include:

It is restrictionless; that is, there are no sites in the DNA which can be acted on by RE.

It is easy to transform with plasmid or recombinant DNA.

It is easy to lyse, and therefore easy to recover recombinant molecules from it.

It produces minicells to allow study of gene action.

It requires thymine or thymidine and undergoes degradation of DNA in the absence of these chemicals.

It is sensitive to bile, and therefore cannot survive passage through the human intestines.

It is sensitive to detergents, antibiotics, drugs, and a number of chemicals, so that it is less likely to survive if it escapes into sewers.

It is incapable of light or dark repair, and so would be sensitive to sunlight.

These and other characteristics bred into the $_{\chi}1776$ strain have endowed it with a formidable genotype: *F^- tonA53 dapD8 minA1 supE42* $\Delta 40$ [*gal^-uvrB*] λ^- *minB2 rfb^-2 nalA25 oms^-2 thyA57 metC65 oms^-1* $\Delta 29$ [*$bioH^-asd$*] *cycB2 cycA1 dsdR2.* And the work is not done. For example, since this strain has a very low rate of reversion mutation to *thy^+* (10^{-9}), researchers are now attempting to introduce nonrevertant mutations into the strain.

At the same time, work is progressing in the construction of safe vectors for recombinant research. Table 4-2 lists the safety features of λ gt variants developed by Leder and his associates at NIH. Frederick Blattner of the University of Wisconsin has constructed a series of λ phages, some of which have been certified as EK2 vectors by NIH (Blattner *et al.* 1977). He has designated the strains *Charon* phages, in an allegorical reference to the boatsman of the river Styx in Greek mythology. As Blattner and others have pointed out, there are natural aspects of containment of λ phages. For one thing, permissive strains of bacteria that serve as hosts are very rare. Of more than 2000 strains of *E. coli* isolated from nature or from hospital patients, none was found to be sensitive. Also, λ DNA degrades at levels of acidity as low as the pH normally found in human stomachs. In addition, Amber mutations have been introduced into Charon phages which reduce the likelihood of their survival. For example, there are Amber mutations in capsid genes, some of which can be suppressed only by a transfer RNA not usually found in wild-type *E. coli.*

Mutations have also been introduced which enhance the effectiveness of Charon phages as vectors. Different-sized DNA can be incorporated by different strains of Charon phages. Some genes have been introduced into the phages by recombination which produce easily discernible plaque characteristics, such as colorless plaques, that indicate cloning of certain genes.

Other laboratories have also been engaged in the search for safe vectors. Table

14-2 lists the safety features of λ gt phage variants developed by Leder and his associates at NIH (Leder *et al.* 1977).

Table 14-2 Safety Features of $\lambda gtWES - \lambda C$ or $\lambda gtWES - \lambda B$

1. Three conditional lethal mutations (amber)
   *E*am 1100; no phage heads and no cleavage of DNA to form cohesive ends
   *W*am 403; no joining of phage heads and tails
   *S*am 100; no lysis or phage release
   *sup*F suppresses all three mutations
   *sup*E does not suppress Sam 100 efficiently
2. *c*Its857
   Temperature-sensitive repressor; at 37°C, λ cIts857 will not form a stable lysogen.
3. *nin*5 deletion
   Removes a transcription stop signal; greatly reduces probability of plasmid formation.
4. Recombinants are recombination defective (*red*$^-$)
   λ *red*$^-$ phage grow less well than *red*$^+$ phage, conferring a selective disadvantage.
5. Under permissive conditions, recombinants are capable of high-titer growth in small, easily manipulated volumes.
6. Restriction barrier
   Phage can be propagated in nonmodifying hosts; unmodified DNA is destroyed on infection of a restricting host.

From P. Leder, *et al.*, 1977. *Science* 196: 176.

RECOMBINANT DNA IN PROSPECTIVE

And so the work continues, and the argument rages and no doubt will for some time. Arguments without clear answers have a tendency to be drawn out. But in view of all the research which has taken place, and that which continues today, the prospect for recombinant DNA experiments seems clear: *The work will go on.* Short of legislation to ban research in recombinant DNA, there is little that will be able to slow the momentum of this area of genetics which generates such excitement (and so many grants for research). Even industry is involved, as pharmaceutical companies are constructing P3 facilities. They and even some academic institutions have begun applying for patents on techniques for the production of various substances in bacteria through cloning.

On June 16, 1980, a unique page in the history of biological research was written when the Supreme Court of the United States voted 5 to 4 that new forms of bacteria produced in the laboratory are eligible for patents. The case involved a bacterium capable of "digesting" oil, but the decision has much farther-reaching implications: All bacteria containing chimera DNA would fall within the category of new forms of bacteria. To what extent patent applications would restrict use of the basic techniques of gene-splicing remains for future decisions on patents pending to determine. An interesting legal development, to say the least!

There is no question, however, that scientists will be influenced in the progress and even direction of their work by more than their own intellectual interests. Already we have seen governing bodies of towns and cities all over the United States meeting in serious debate over proposed research in laboratories within their borders (Wade 1977b).

Cambridge, Massachusetts, is a prime example (Wade 1977a). The town council met in July 1976 and voted for a three-month moratorium on certain recombinant research activities at Harvard and MIT. A committee of both scientists and nonscientists was established to look into the matter. The moratorium was extended another three months to the exasperation of the geneticists involved. In January 1977, the committee recommended that all research be allowed to continue, except for the most hazardous type. On these experiments they imposed even more stringent requirements for containment than the NIH guidelines.

What effects if any the recommendations by the committee will have on research at Harvard and MIT remains for the future to determine. But this episode, which has probably not seen its conclusion, certainly emphasizes that genetics will be subject to public scrutiny unique in the history of scientific endeavor. We have, indeed, entered a new era in the role of science in society.

REFERENCES

ARBER, W., D. DUSSOIX, 1962. "Host Specificity of DNA Produced by *E. coli* I. Host-controlled Modification of Bacteriophage Lambda." *J. Mol. Biol.* 5: 18–36.

BENJAMIN, T. L., 1977. "HR-T Mutants of Polyoma Virus." In J. Schultz and Z. Brada, eds., *Genetic Manipulation as It Affects the Cancer Problem*, pp. 73–85. Ninth Miami Winter Symposium. Academic Press, New York.

BERG, P., D. BALTIMORE, H. W. BOYER, S. N. COHEN, R. W. DAVIS, D. S. HOGNESS, D. NATHANS, R. ROBLIN, J. D. WATSON, S. WEISSMANN, AND N. D. ZINDER, 1975. "Potential Biohazards of Recombinant DNA Molecules." *Science* 188: 991–994. (Asilomar Conference)

BERG, P., 1977. "Biochemical Pastimes . . . and Future Times." In W. A. Scott, and R. Werner, eds., *Molecular Cloning of Recombinant DNA*, pp. 1–34. Ninth Miami Winter Symposium. Academic Press, New York.

BLATTNER, F. R., B. G. WILLIAMS, A. E. BLECHL, K. DENNISTON-THOMPSON, H. E. FABER, L. A. FURLONG, D. J. GRUNWALD, D. O. KIEFER, D. D. MOORE, J. W. SCHUMM, E. L. SHELDON, AND O. SMITHIES, 1977. "Charon Phages: Safer Derivatives of Bacteriophage Lambda for DNA Cloning." *Science* 196: 161–169.

BOTCHAM, M., W. TOPP, AND J. SAMBROOK, 1976. "The Arrangement of Simian Virus 40 Sequences in the DNA of Transformed Cells." *Cell* 9: 269–287.

BRILL, W. J., 1977. "Biological Nitrogen Fixation." *Sci. Am.* 263: 68–81.

CANNON, F. C., R. A. DIXON, J. R. POSTGATE, AND S. B. PRIMROSE, 1974. "Chromosomal Integration of *Klebsiella* Nitrogen Fixation Genes in *Escherichia Coli*." *J. Gen. Microbiol.* 80: 227–239.

CARBON, J. C., B. RATZKIN, L. CLARKE, AND D. RICHARDSON, 1977. "The Expression of Eukaryotic DNA Segments in *Escherichia coli*." In W. A. Scott, and R. Werner, eds., *Molecular Cloning of Recombinant DNA*, pp. 59–72. Ninth Miami Winter Symposium. Academic Press, New York.

COHEN, S. N., A. C. Y. CHANG, AND L. HSU, 1972. "Nonchromosomal Antibiotic

Resistance in Bacteria: Genetic Transformation of *E. Coli* by R Factor DNA." *Proc. Nat. Acad. Sci. U.S.* 69: 2110–2114.

COHEN, S. N., A. C. Y. CHANG, H. W. BOYER, AND R. B. HELLING, 1973. "Construction of Biologically Functional Bacterial Plasmids in Vitro." *Proc. Nat. Acad. Sci. U.S.* 70: 3240–3244.

COHEN, S. N., 1975. "The Manipulation of Genes." *Sci. Am.* 233: 25–37.

______, 1977. "Recombinant DNA: Fact and Fiction." *Science* 195: 654–657.

CURTISS, R., III, M. INOUE, D. PEREIRA, J. C. HSU, AND L. ALEXANDER, 1977. "Construction and Testing of Safer Host Strains to Facilitate Recombinant DNA Molecule Research." In W. A. Scott and R. Werner, eds., *Molecular Cloning of Recombinant DNA*, pp. 99–114. Ninth Miami Winter Symposium. Academic Press, New York.

DANNA, K. J., G. H. SACK, JR., AND D. NATHANS, 1973. "Studies of Simian Virus 40 DNA. VII. A Cleavage Map of the SV40 Genome." *J. Mol. Biol.* 78: 363–376.

EFSTRATIADIS, A., F. C. KAFATOS, A. M. MAXAM, AND T. MANIATIS, 1976. "Enzymatic In Vitro Synthesis of Globin Genes." *Cell* 7: 279–288.

GILBERT, W., AND L. VILLA-KOMAROFF, 1980. "Useful Proteins from Recombinant Bacteria." *Sci. Am.*, 242 (April), 74–94.

GLOVER, D. M., R. L. WHITE, D. J. FUNEGAN, AND D. S. HOGNESS, 1975. "Characterization of Six Cloned DNAs from *Drosophila Melanogaster*. Including One That Contains the Genes for rRNA." *Cell* 5: 149–157.

HAMER, D. H., 1977. "The Construction and Characterization of a Eukaryotic Transducing Virus." In J. Schultz and Z. Brada, eds., *Genetic Manipulation as It Affects the Cancer Problem*, pp. 37–47. Ninth Miami Winter Symposium. Academic Press, N.Y.

HERSHFIELD, V., H. W. BOYER, C. YANOFSKY, M. A. LOVETT, AND D. R. HELINSKI, 1974. "Plasmid ColEI as a Molecular Vehicle for Cloning and Amplification of DNA." *Proc. Nat. Acad. Sci. U.S.* 71: 3455–3459.

JACKSON, D. A., R. H. SYMONS, AND P. BERG, 1972. "Biochemical Method for Inserting New Genetic Information Into DNA of Simian Virus 40: Circular SV40 DNA Molecules Containing Lambda Phage Genes and the Galactose Operon of *Escherichia coli*." *Proc. Nat. Acad. Sci. U.S.* 69: 2904–2909.

JESSEL, D., T. LANDAU, J. HUDSON, T. LALOR, D. TEMIN, AND D. M. LIVINGSTON, 1976. "Identification of Regions of the SV40 Genome Which Contained Preferred SV40 T-Antigen-Binding Sites." *Cell* 8: 535–545.

KEDES, L. J., A. C. Y. CHANG, D. HOUSEMAN, AND S. N. COHEN, 1975. "Isolation of Histone Genes from Unfractionated Sea Urchin DNA by Subculturing Cloning in *E. Coli*." *Nature* 255: 533–537.

LEBOWITZ, P. T., J. KELLY, JR., D. NATHANS, T. N. H. LEE, AND A. M. LEWIS, JR., 1974. "A Colinear Map Relating the Simian Virus 40 (SV40) DNA Segments of Six Adenovirus-SV40 Hybrids to the DNA Fragments Produced by Restriction Endonuclease Cleavage of SV40 DNA." *Proc. Nat. Acad. Sci. U.S.* 71: 441–445.

LEDER, P., D. TIEMEIER, AND L. ENQUIST, 1977. "EK2 Derivatives of Bacteriophage Lambda Useful in Cloning of DNA from Higher Organisms: The *λ gtWES* System." *Science* 196: 175–177.

LINN, S., AND W. ARBER, 1968. "Host Specificity of DNA Produced by *E. coli*, X. In Vitro Restriction of Phage fd Replicative Form." *Proc. Nat. Acad. Sci. U.S.* 59: 1300–1306.

MACH, B., F. ROUGEON, S. LONGAERE, AND M. F. AELLEN, 1977. "DNA Cloning in Bacteria as a Tool for the Study of Immunoglobulin Genes." In W. A. Scott, and R. Werner, eds., *Molecular Cloning of Recombinant DNA*, pp. 219–236. Ninth Miami Winter Symposium. Academic Press, New York.

MANIATIS, T., A. JEFFREY, AND D. G. KLEID, 1975. "Nucleotide Sequence of the Rightward Operator of Phage λ." *Proc. Nat. Acad. Sci. U.S.* 72: 1184–1188.

MESELSON, M., AND R. YUAN, 1968. "DNA Restriction Enzyme from *E. coli*." *Nature* 217: 1110–1114.

MORROW, J. F., S. N. COHEN, A. C. Y. CHANG, H. W. BOYER, H. M. GOODMAN, AND R. B. BILLINGS, 1974. "Replication and Transcription of Eukaryotic DNA in *Escherichia coli*." *Proc. Nat. Acad. Sci. U.S.* 71: 1743–1747.

MOWBRAY, S. L., S. A. GERBI, AND A. LANDY, 1975. "Interdigitated Repeated Sequences in Bovine Satellite DNA." *Nature* 253: 367–370.

MURRAY, N. E., AND K. MURRAY, 1974. "Manipulation of Restriction Targets in Phage λ to Form Receptor Chromosomes for DNA Fragments." *Nature* 251: 476–481.

NATHANS, D., AND H. O. SMITH, 1975. "Restriction Endonucleases in the Analysis and Restructuring of DNA Molecules." *Ann. Rev. Biochem.* 44: 273–293.

POTTER, H., AND D. DRESSLER, 1976. "On the Mechanism of Genetic Recombination: Electron Microscope Observation of Recombination Intermediates." *Proc. Nat. Acad. Sci. U.S.* 73: 3000–3004.

PTASHNE, M., 1976. "The Defense Doesn't Rest." *The Sciences* 16: 11–12.

SAMBROOK, J., B. SUGDEN, W. KELLER, AND P. A. SHARP, 1973. "Transcription of Simian Virus 40 III Mapping of 'Early' and 'Late' Species of RNA." *Proc. Nat. Acad. Sci. U.S.* 70: 3711–3715.

SAMBROOK, J., 1977. "SV40-Adenovirus Hybrids." In J. Schultz and Z. Brada, eds., *Genetic Manipulation as It Affects the Cancer Problem*, pp. 139–160. Ninth Miami Winter Symposium. Academic Press, New York.

SHANMUGAM, K. T., AND R. C. VALENTINE, 1975. "Molecular Biology of Nitrogen Fixation." *Science* 187: 919–924.

SMITH, H. O., AND K. W. WILCOX, 1970. "A Restriction Enzyme From *Hemophilus influenzae*." *J. Mol. Biol.* 51: 379–91.

THOMAS, M., J. R. CAMERON, AND R. W. DAVIS, 1974. "Viable Molecular Hybrids of Bacteriophage Lambda and Eukaryotic DNA." *Proc. Nat. Acad. Sci. U.S.* 71: 4579–4583.

WADE, N., 1976. "Recombinant DNA: A Critic Questions the Right to Free Inquiry." *Science* 194: 303–306.

______, 1977a. "Gene Splicing: Cambridge Citizens OK Research But Want More Safety." *Science* 195: 268–269.

______, 1977b. "Gene-Splicing: At Grass Roots Level a Hundred Flowers Bloom." *Science* 195: 558–559.

WALD, G., 1976. "The Case Against Genetic Engineering." *The Sciences* 16: 6–11.

WEISSMANN, C., E. DOMINGO, AND T. TAYAGUCHI, 1977. "Site Directed Mutagenesis." In J. Schultz and Z. Brada, eds., *Genetic Manipulation as It Affects the Cancer Problem*, Ninth Miami Winter Symposium. Academic Press, New York.

WELLAUER, P. K., I. B. DAWID, D. D. BROWN, AND R. H. REEDER, 1976. "The Molecular Basis for Length Heterogeneity in Ribosomal DNA from *Xenopus Laevis*." *J. Mol. Biol.* 105: 461–486.

WELLAUER, P. K., R. H. REEDER, I. B. DAWID, AND D. D. BROWN, 1976. "The Arrangement of Length Heterogeneity in Repeating Units of Amplified and Chromosomal Ribosomal DNA from *Xenopus Laevis*." *J. Mol. Biol.* 105: 487–505.

FIFTEEN

Mutation, Mutagenesis, and Repair

INTRODUCTION

In this chapter we will deal more directly with aspects of mutation. We will classify mutations, point out their significance, and show that cells possess an innate mechanism that repairs damage which would otherwise lead to mutation.

The term *mutation* can be simply defined as a change or alteration of the genetic material of a cell or organism. For the layman, this may evoke images of monsters or aberrations, but without the genetic variability which is the consequence of mutation, there would be little life as we know it. (Without mutation there would in fact be no science of genetics, which depends on variability as its basis for analysis.)

Although living organisms depend on mutation for genetic variability, it is an unfortunate fact that most mutant traits are detrimental. We need, therefore, to understand the sources of *mutagens* (factors which cause mutation) and how they may be controlled. In this chapter we shall also discuss environmental factors which cause mutation. This is another area of genetics which extends beyond the laboratory, because cells of all living organisms, including our own, are exposed to environmental agents that cause mutational changes.

CLASSIFICATION OF MUTATIONS

There are a number of ways to classify mutations. One classification is based on the type of cell which undergoes mutation. For example, *somatic mutations* are mutations of body cells, and would not be transmissible to the next generation. *Germinal mutations*, on the other hand, occur in germ cells, and are transmitted from one generation to the next, or "vertically." Somatic mutations may affect the individual with the mutation; germinal mutations will not be expressed in the individual, but in the individual's descendants.

One can also classify mutations on the basis of their effect on the expression of other genes. Some do not determine phenotype directly, but act to suppress the expression of genes that do; these are known as *suppressor* mutations. Certain other genes increase the mutability of neighboring genes; these are called *mutator* genes.

Some mutations are distinguished by their effect on the gene product. *Missense* mutations result in an aberrant gene product, whereas *nonsense* mutations result in the absence of a gene product owing to termination of transcription.

Most of these terms refer to changes of single genes, and are known as *point*, or *intragenic*, mutations, in contrast to *gross* mutations, which we discussed in Chapter 6. It is the various aspects of point mutations which we shall consider in this chapter.

POINT, OR INTRAGENIC, MUTATIONS

Point mutations are far more frequently observed than are gross mutations, possibly because of increased viability of conditions effected by single gene changes. We have already pointed out many studies of point mutations which have contributed to our knowledge of genes and gene action. In the following section we shall briefly discuss some classic experiments which used point mutations to elucidate gene structure.

Cistrons, Recons, and Mutons

One of the landmark studies which led to clarification of a number of aspects of molecular genetics was carried out on T4 bacteriophage by Seymour Benzer in 1955. Recall that a series of mutations of T4, known as *rII* (p. 313), produce large plaques through rapid lysis of host bacteria. Certain *rII* variants were found which were incapable of lysing *E. coli* strain K; however, infecting strain K bacteria with viruses containing different *rII* mutations did occasionally produce lysis indicating that the mutations could complement each other. This in turn indicated that the two mutations occurred in DNA controlling different functions. If two *rII* mutants did not produce lysis, it was assumed that both mutations occurred in the portion of DNA controlling the same function.

Occasionally, recombination between the DNA of two noncomplementing mutants led to lysis. This phenomenon was interpreted as indicating that recombination restored one of the DNA molecules to the *rII* state, while leaving the reciprocal molecule doubly mutant. Since the mutations produce the mutant phenotype only in the trans position and rapid lysis in the cis position, this kind of complementation test (known as the *cis-trans test*) indicates whether two mutations determine the same function. Figure 15-1 summarizes the important aspects of this study.

All the *rII* mutants could be grouped into one of two complementation groups. Benzer therefore concluded that there are two regions of DNA each determining a polypeptide that plays a role in the lysis of *E. coli* strain K. He termed these two regions A and B. Since these two regions of DNA were involved in determining different polypeptides, Benzer coined the term *cistron* to designate a region of DNA determining a particular function. All mutations within a single cistron could not complement each other. *Cistron* is now considered synonymous with *structural gene.*

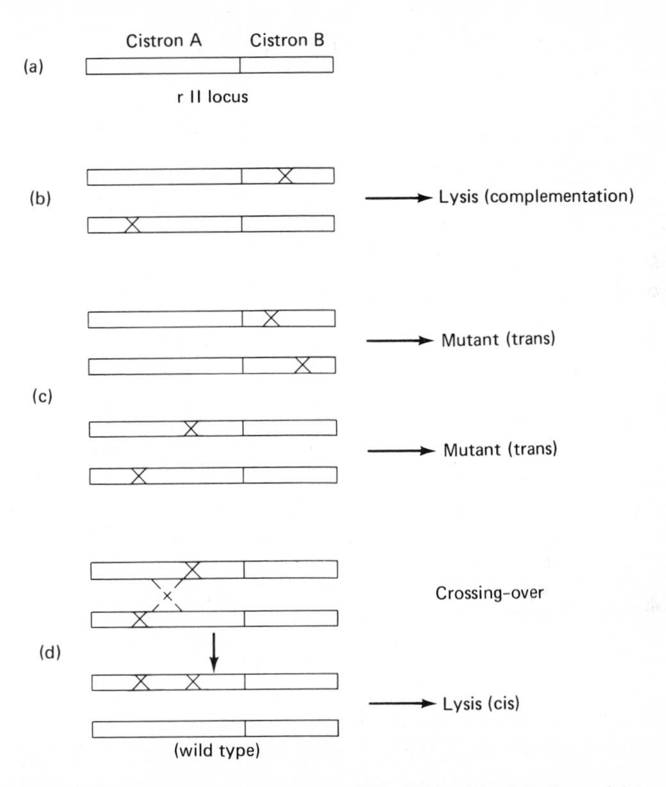

Figure 15-1. (a) Represents the *rII* locus of T4 phage. A and B cistrons determine different proteins necessary for lysis. (b) Infection of a cell with mutants defective in each cistron results in production of both proteins by normal cistrons and lysis. (c) Infection of a cell with mutants defective in the same cistron results in no lysis, unless (d) crossing-over occurs to produce a DNA with normal genes in both cistrons.

In all, Benzer found some 2000 mutations to map at about 308 sites in the *rII* region. He found some sites to be highly mutable in comparison to other sites. Those having a high frequency of mutations were called *hotspots* (Fig. 15-2). Benzer also estimated the smallest units at which mutation or recombination can occur within a gene, calling these units *muton* and *recon*, respectively. In each case, it was estimated that the recon and muton could be as small as, but no smaller than, a single base pair. Through these studies, Benzer revealed the fine structure of a particular

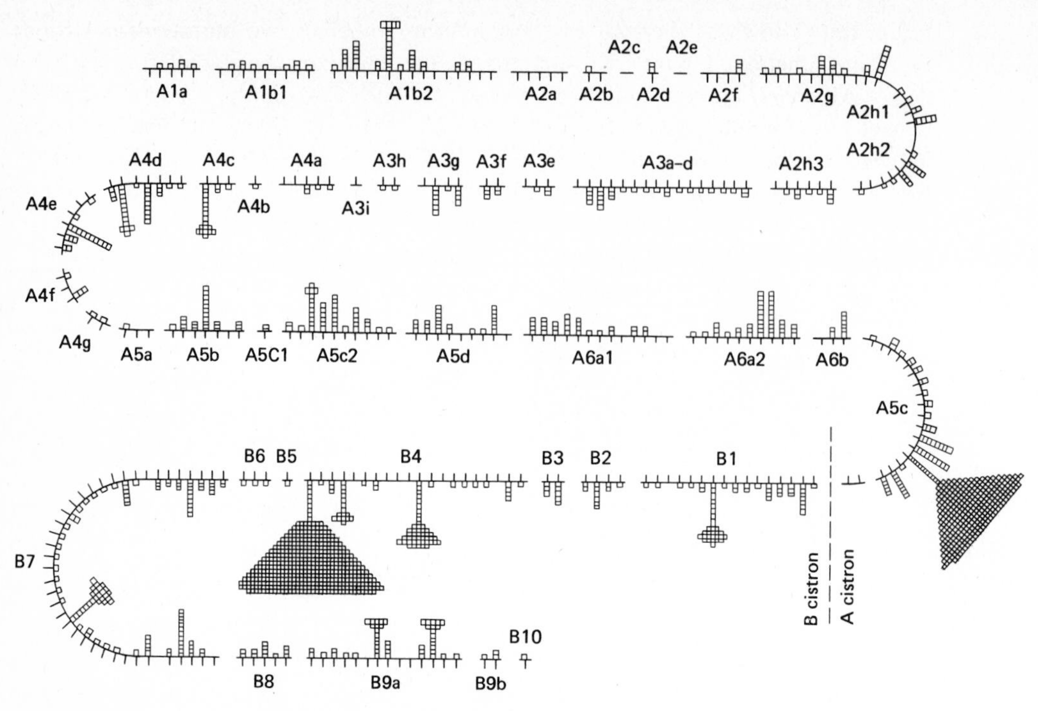

Figure 15-2. Map of the *rII* region. Each square represents a mutation event. Some areas, called *hotspots*, show high mutability. (From S. Benzer, 1961. *PNAS* 47 : 410.)

region of the T4 chromosome, and showed that changes of even a single base pair can result in mutation.

Rearrangements of Base Pairs

Since the 1950s, thanks partly to the data from Benzer's studies, geneticists have been aware that changes of base pairs within genes are the molecular basis for point mutations. What kinds of changes can occur to cause mutation? Since a gene "makes sense" to a cell through the sequence of its bases, it stands to reason that rearrangements of bases would cause either a change in meaning (a missense mutation), or a loss of meaning (nonsense mutation). Several terms referring to these rearrangements of bases are similar to those applied to rearrangements of a chromosome in gross mutations which we have already mentioned: *deletion*, *addition* (either of existing base sequences or new ones), and *inversion*. In addition, there can occur *substitutions* of base pairs which also lead to a mutant phenotype.

We have already discussed deletions and additions in our treatment of frameshift mutations (see again p. 313). These change the number of bases in a sequence and cause a shift in the reading of the codons. Inversion of a sequence of bases would cause the insertion of the wrong sequence of amino acids in an abnormal protein, assuming that the inverted sequence has meaning for the cell. For example, 5′-GGA-CAU-3′ would be translated to glycine-histidine; however, 5′-UACAGG-3′ would be translated to tyrosine-arginine.

Substitutions of bases are further classified as transitions or transversions. *Transitions* are substitutions of one pyrimidine for another, or of one purine for another. *Transversions* are substitutions of a purine for a pyrimidine, or vice-versa. One chemical change which can lead to substitutions is a shift in the position of hydrogen in a base. This process is known as *tautomerization*, and can result in abnormal base pairing as shown in Fig. 15-3.

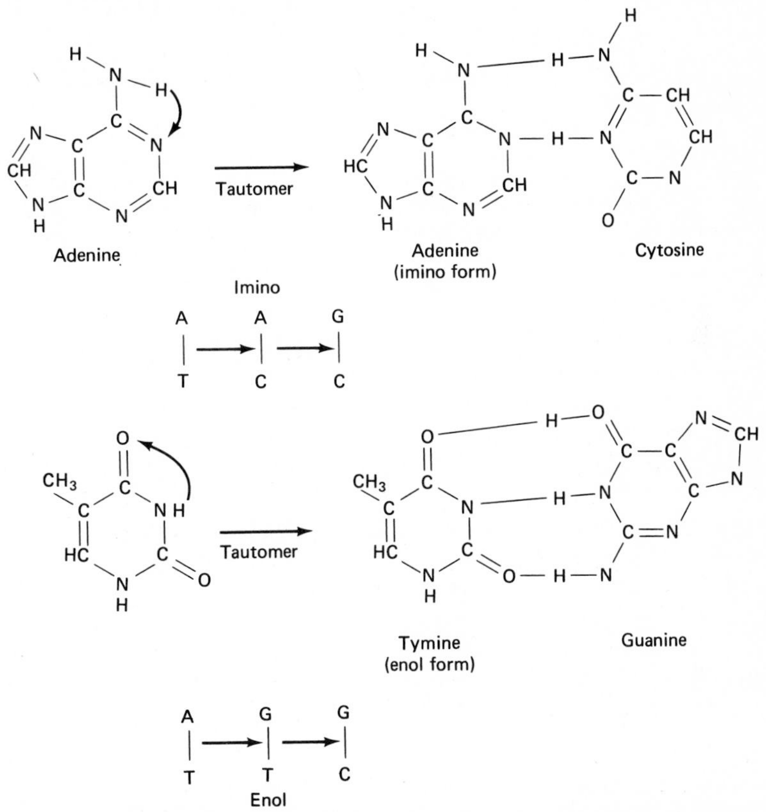

Adenine

Tautomer

Adenine (imino form)

Cytosine

Imino

A–T → A–C → G–C

Tautomer

Tymine (enol form)

Guanine

A–T → G–T → G–C

Enol

Figure 15-3. Tautomers of bases are formed by a shift in the position of hydrogen, leading to mispairing of bases in replication. This eventually leads to substitution of an adenine-thymine pair by a guanine-cytosine pair.

One of the most thoroughly studied substitution mutations in humans results in alteration of the amino acid sequence in hemoglobin, thereby leading to sickle-cell anemia. We have previously described the symptoms of this lethal condition, which is determined by an autosomal recessive mutation.

Electrophoretic studies of the hemoglobin from normal red blood cells and from sickle red blood cells showed a distinct difference in the hemoglobins (Fig. 15-4)(Pauling, *et al.* 1949). V. Ingram (1957) extended these studies with a technique that has come to be known as *fingerprinting*. The globin portion of hemoglobin is first broken down into small peptide fragments. The fragments are then subjected to

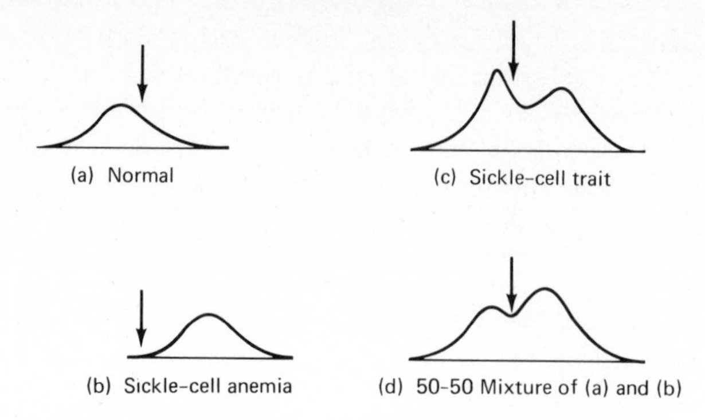

Figure 15-4. Studies on migration patterns of hemoglobins from normal cells, and cells from sickle-cell anemia homozygotes and heterozygotes (sickle-cell trait). Differences in migration indicate molecular differences in the proteins. (From L. Pauling *et al.*, 1949. *Science* 110 : 545; copyright 1949 by the American Association for the Advancement of Science.)

both electrophoresis and chromatography, which allows a finer separation of the fragments that are visualized as distinct spots on the chromatogram (Fig. 15-5).

The results of the studies indicated only a single difference between normal and sickle hemoglobin. Further analysis showed that in hemoglobin S there is valine at position six of the beta chain, instead of the normal glutamic acid. If you look at the genetic code, glutamic acid is coded for by the triplets GAA and GAG; valine by GUA and GUG. Thus a substitution in the second base from adenine to uracil is sufficient to cause significant alteration in the hemoglobin.

With the improvement in techniques in both amino acid sequencing and base

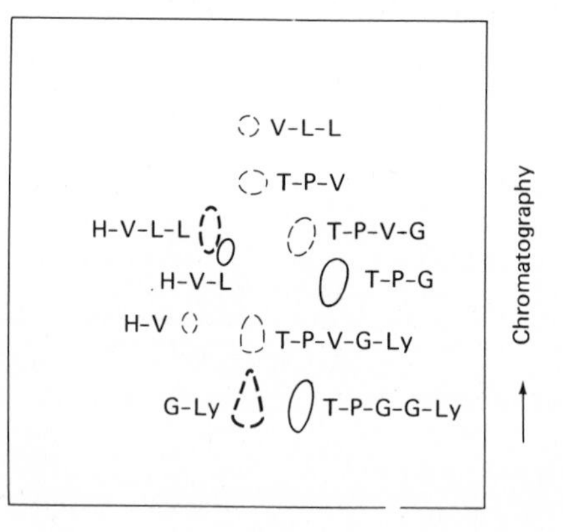

Figure 15-5. Acid degradation and structure of No. 4 peptides from hemoglobins A and S. Hemoglobin A (full lines) : His-Val-Leu-Leu-Thr-Pro-Glu-Lys. Hemoglobin S (broken lines) : His-Val-Leu-Leu-Thr-Pro-Val-Lys. "Fingerprint" of peptide No. 4 from normal hemoglobin and sickle-cell hemoglobin, showing a difference in chromatographic properties. Other peptides from normal and sickle cells were the same. The difference was found to be substitution of glutamic acid by valine in sickle cells. (From V. M. Ingram, 1957. *Nature* 180 : 326.)

sequencing, alterations by mutations in prokaryotes can in many cases be analyzed on the molecular level. Mutations in eukaryotes are still difficult to trace to their molecular basis, however. Hemoglobin alterations are still our best source of material for such analysis because of the accessibility of hemoglobin, and the fact that it is a primary gene product.

Randomness of Mutations

An important question which arose long ago concerning the nature of mutations was whether mutations can be directed. In other words, can the cell, or some outside force, in any way determine changes in its genetic material? When bacteria became a prime source of experimental material, observers noted that a population could seemingly adapt to endless changes in its environment. In the presence of antibiotics, to which wild-type bacteria are susceptible, a resistant mutant could eventually be isolated.

This seemingly endless flexibility of bacteria raised questions as to whether the mutations were *preadaptive* or *postadaptive.* A *preadaptive* mutation is one which exists in an organism *before* the occurrence of an environmental change, but which nevertheless allows adaptation to that change. For example, penicillin resistance would already be present in the population referred to above, but would remain unexpressed until penicillin was encountered in the medium. *Postadaptive* mutations are those occurring *after* an environmental change, and imply an ability to respond to the environment with genetic changes.

Two approaches to this question in the 1940s and 1950s settled the controversy quite clearly in favor of preadaptive mutations. One was the replica plating experiments of the Lederbergs described on page 233. To study mutation to phage resistance, the Lederbergs grew a lawn of bacteria which were then replica-plated to a petri dish in which the agar contained phage particles. Any phage-resistant mutation among the bacterial cells would be represented by growth of colonies of bacteria.

If resistance to phage is preadaptive, one would expect that the mutations found in replica plates would be in bacteria found at identical spots on the original plate. On the other hand, postadaptive mutations would occur spontaneously on the replica plates where the bacteria encounter phages. These mutations therefore should not correlate with the positions of pre-existing mutations on the original plate where there was not phage. The observations all confirmed preadaptive mutation (Lederberg and Lederberg 1952).

The second study was a series of elegant experiments by S. Luria and R. Delbruck reported some ten years earlier than the Lederberg's experiment. They reasoned that if preadaptive mutations exist, since they are spontaneous and occur at different times in different bacterial populations, the number of cells mutant for a particular gene should differ in independent populations. On the other hand, several samples from the same culture should show the same frequency of mutations. Thus, if one were to look at 12 independent populations for phage resistance, each should have variable numbers of phage-resistant cells to begin with. If postadaptive mutations occur, following simultaneous exposure to phages all samples from the same or independent cultures should contain approximately equal numbers of resistant cells.

Table 15-1 shows the data they obtained, which clearly indicate variability in the independent cultures, whereas 12 samples from the same cultures did contain similar numbers of mutations, as one would expect.

Evidence gathered since these experiments has supported preadaptive mutation in all organisms. NO CONVINCING SCIENTIFIC EVIDENCE HAS YET BEEN REPORTED

Table 15-1 Numbers of Bacteriophage-resistant Colonies of *E. coli* Produced from Samples from Single Cultures and Samples from Different Cultures

CONTROL (10 SAMPLES EACH FROM THE SAME CULTURE)				EXPERIMENTAL (SAMPLES FROM DIFFERENT CULTURES)			
	Resistant colonies				Resistant colonies		
Sample	*Expt. 3*	*Expt. 10a*	*Expt. 11a*	*Culture*	*Expt. 1*	*Expt. 11*	*Expt. 15*
1	4	14	46	1	10	30	6
2	2	15	56	2	18	10	5
3	2	13	52	3	125	40	10
4	1	21	48	4	10	45	8
5	5	15	65	5	14	183	24
6	2	14	44	6	27	12	13
7	4	26	49	7	3	173	165
8	2	16	51	8	17	23	15
9	4	20	56	9	17	57	6
10	7	13	47	10		51	10
Mean	3.3	16.7	51.4		26.8	62	26.2
Variance	3.8	15	27		1217	3498	2178

From S. H. Luria and M. Delbruck, 1943. *Genetics* 28: 491.

INDICATING THAT LIVING CELLS CAN RESPOND TO ENVIRONMENTAL CHANGE BY CHANGING THEIR GENOME IN A SPECIFIC WAY.

FACTORS THAT CAUSE MUTATIONS

Since all genetic variability is due ultimately to mutation, agents capable of causing mutations are of keen interest to geneticists. They are of special concern in our society, because we are introducing large numbers of chemicals to, and increasing the amount of radiation in, our environment, and radiation and chemicals represent the two main classes of mutagens.

Radiation as Mutagen

Although artificial mutagens are of great concern to scientists, mutations have occurred spontaneously and naturally throughout the history of the natural world. It is generally agreed that one cause of spontaneous, or natural, mutations is background radiation. Sunlight and radioactive minerals in the crust of the earth are the two main natural sources of background radiation.

Although radiation physics is beyond the scope of our discussion, a few basic facts relevant to mutagenesis must be pointed out here. There are different kinds of radiation, depending on the length of the rays emitted. Generally, all radiation can be mutagenic. For some organisms, such as eukaryotes, short-wavelength radiation (such as X-rays and gamma rays) is of more concern than longer wavelengths (such as UV), since the shorter rays are capable of penetrating our tissues, whereas UV cannot. However, UV radiation can indeed affect our surface tissues.

The effect of radiation on living cells is believed to be due to the collision of high-energy rays with atoms, which releases electrons and forms unstable ions. These ions in turn can lead to formation of abnormal molecules which interfere with normal structure and function of the components of a cell. For example, a common result of irradiation of cells is the appearance of gross chromosomal damage in the form of translocations, fragmentation, and so forth. On the molecular level, the principal abnormalities to DNA from ionizing radiation include chemical damage to the bases and sugars, and single- or double-strand breaks in the DNA.

H. J. Muller and radiation mutagenesis. The first scientific studies on radiation as mutagen came from the laboratory of H. J. Muller. Muller devised a test now known as the *Muller* 5 or *Basc test*. A strain of *Drosophila* known as the Muller-5 strain carries the Bar mutation and a recessive eye mutation at the White Eye locus (known as A*pricot*) on the X chromosome. Wild-type *Drosophila* males are irradiated and mated to a strain of Muller-5 females whose X chromosomes also carry an inversion to suppress crossing-over (Fig. 15-6). F_1 females are then mated to Muller-

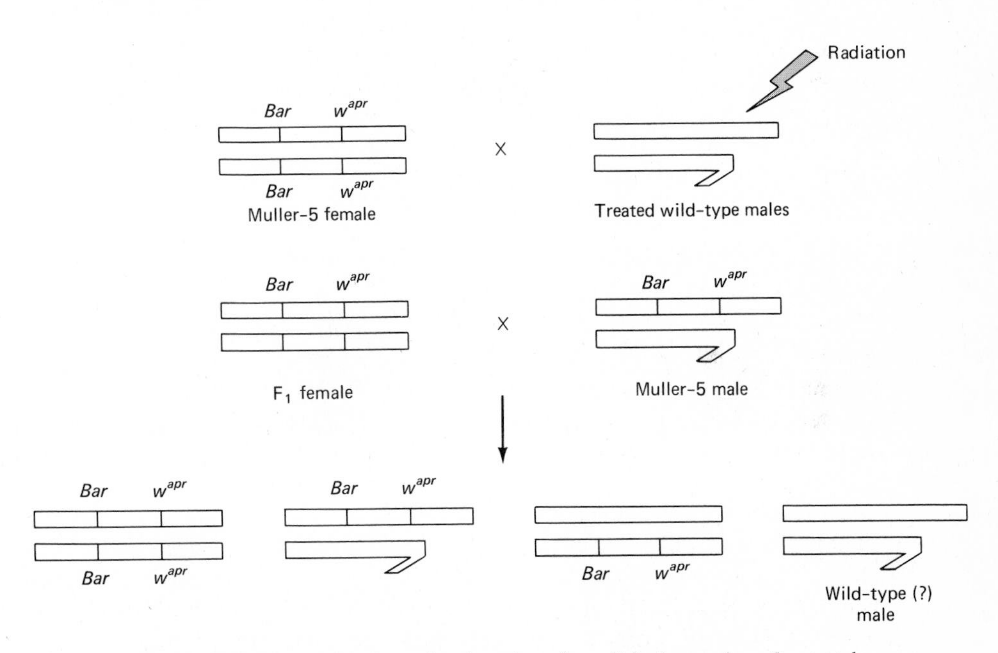

Figure 15-6. The Muller-5 test for detection of sex-linked recessives. See text for explanation.

5 males. If recessive lethal mutations had been induced in the parental wild-type males, one would expect to find a deficiency in F_2 males, which would inherit the X chromosome containing the lethal mutation. Muller found a hundredfold increase in the number of sex-linked recessive mutations in progeny of irradiated males as compared with those of nonirradiated control males. His studies proved conclusively that radiation is mutagenic, and Muller received the Nobel Prize for his contributions to genetics and the study of mutagens.

The specific locus test. The mutagenic effect of radiation on mammals was demonstrated conclusively by studies carried out on mice by W. L. Russell (1958). Russell and his coworkers irradiated male mice from a strain homozygous dominant for seven coat-color genes. These males were then mated to virgin females from a strain homozygous recessive for the same loci. Mutations sustained in the germ cells of the treated males altering the coat-color genes to a recessive form would appear in the offspring. Both X-rays and gamma rays were used on experimental animals. Figure 15-7 shows some of the data obtained, which indicated a distinct increase in frequency of mutations among the offspring of irradiated males. This test is known as the *specific locus test*.

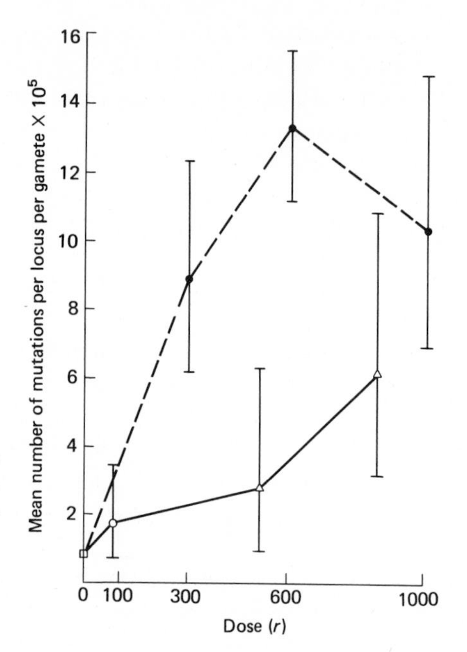

Figure 15-7. Results of radiation experiments on mice. Solid circles represent results from acute X-ray treatment. Open points represent chronic gamma ray results. [After W. L. Russell, 1958. *Science* 128: 1547. (From A. C. Pai, 1974. *Foundations of Genetics*. McGraw-Hill, New York.)]

Characteristics of radiation mutagenicity. These and numerous other studies of the effects of radiation on genetic material have established a number of general characteristics of radiation as a mutagen. ONE SUCH CHARACTERISTIC IS THAT THERE IS A LINEAR INCREASE IN FREQUENCY OF BOTH CHROMOSOMAL AND POINT MUTATIONS WITH INCREASE IN RADIATION DOSAGE. FURTHERMORE, THE LOWEST DOSES STILL INCREASE THE NUMBER OF MUTATIONS SLIGHTLY OVER THE FREQUENCY IN CONTROL ANIMALS.

Another characteristic that has been established is that the effect of a single large dose (acute radiation) is more mutagenic than that of the same dosage broken into several smaller exposures (chronic radiation). Natural, or background, radiation certainly constitutes chronic radiation. Most of the sources of acute radiation exposure are artificial, such as X-rays or nuclear explosions.

Some understanding of the effect of acute radiation on humans has come from studies of survivors of the two atomic bombs which were dropped on heavily populated cities in Japan during the Second World War (Miller 1971). One-third of the survivors of the bombs who were 30 years old or younger at the time of exposure now show somatic mutations in the form of chromosomal abnormalities of somatic cells. This contrasts with the occurrence of somatic mutations in only one percent of a control (unexposed) group. Two-thirds of the people 30 years or older at the time of exposure carry somatic mutations. Even those being carried in utero during exposure show a much higher frequency of chromosomal abnormalities (29 percent) than do those who were not exposed.

In addition, there is an increase among the survivors in the frequency of some forms of cancer, such as leukemia and thyroid cancer (both of which are associated with increased exposure to radiation). Interestingly, children conceived after their parents had been exposed to radiation from the atomic bomb have not shown increased mutation when compared with control groups. However, note that recessive point mutations which may have been induced would not be expressed.

Radiation mutagenesis in prokaryotes. Many studies have demonstrated the effects of radiation on prokaryotes. (See again Beadle and Tatum's experiments on fungi, p. 305). In addition to their vulnerability to the short-wavelength radiation, prokaryotes also are affected by longer-wavelength radiation such as UV light. Because of the single-celled nature of most prokaryotes, they are easily penetrated by UV rays. Prolonged irradiation kills the cells. In fact, laboratories which conduct experiments requiring aseptic conditions, such as cell culture work or bacterial and viral research, use UV light to sterilize their working areas. (We might also point out here that whereas UV does not penetrate the tissues of humans very deeply, overexposure to UV—including in the form of sunlight—can cause skin cancer.)

It has been established that the lethal effects of UV irradiation on microorganisms derive from their damaging the DNA in such a way that replication and transcription cannot take place normally. One of the most common abnormalities resulting from UV radiation of bacteria is the presence of dimers formed by bonding between adjacent bases on the same strand of DNA. Pyrimidine dimers are formed ten times as often as purine dimers, and among pyrimidine dimers, thymine dimers have been found most frequently (Auerbach 1976). The presence of dimers distorts the structure of DNA, and both replication and transcription are obstructed (Fig. 15-8).

Developments in radiation technology have been typical of many aspects of advances in science in the twentieth century. There is no question about the benefits that have been derived from use of X-rays in medical diagnosis and treatment. With the drastic shrinkage of our fossil fuel sources, some experts feel that the most practical source of energy for our world in the immediate future lies in nuclear power plants. Much engineering expertise has been devoted to designing nuclear power plants so as to minimize the risk of increasing radiation in our environment. Yet in each case we must be cognizant of the potential for harm by the misuse of these technologies. We will have to learn to balance the potential benefits with the potential for harm.

Chemicals as Mutagens

A SOURCE OF MUTATION WHICH NOW CONCERNS SOME GENETICISTS MORE THAN RADIATION IS THE INCREASINGLY LARGE AMOUNT OF CHEMICALS POURING FORTH INTO OUR

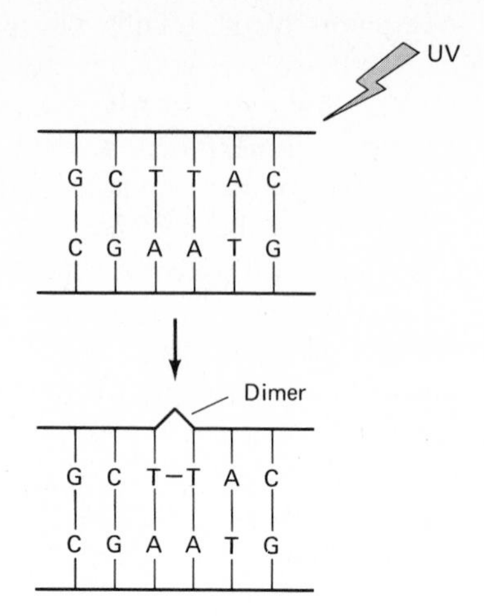

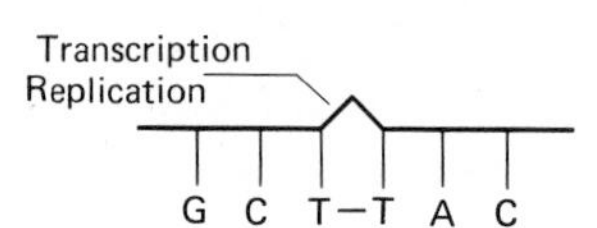

Figure 15-8. Schematic illustration of thymine dimer formed after UV irradiation. The deformity in DNA structure obstructs normal transcription and replication.

ENVIRONMENT. Synthetic chemicals can be found in our food, water, and air, to be ingested, breathed in, and absorbed. The first definitive evidence that chemicals can also be mutagenic came from research on chemical weapons during the Second World War. Charlotte Auerbach and other workers in Scotland obtained data showing that mustard gas increases sex-linked mutations in *Drosophila* by 24 percent (Auerbach and Robson 1947). The damage to chromosomes by mustard gas was very similar to that caused by X-rays.

Since then, studies on the effects of chemicals on both plant and animal cells have confirmed that various chemicals cause mutation. We shall briefly discuss some chemical mutagens, and their effects on cells.

Alkylating agents. Mustard gas belongs to a group of chemical mutagens called *alkylating agents*, which introduce alkyl groups (CH_3, CH_3CH_2, etc.) into the DNA chains. When such a group is attached to a base, mispairing of bases can occur at replication. For example, if guanine is alkylated, it can pair with thymine instead of cytosine (a transition substitution). The next replication event will then result in one of the DNA chains containing a T–A base pair in place of the original G-C base pair (Fig. 15-9). Figure 15-10 shows the comparative effects of mustard gas and X-rays on *Drosophila.* Figure 15-11 shows the frequencies of anaphase aberrations detected in cells of *Vicia faba* after treatment of seed with the alkylating agent ethylmethane sulfonate (EMS), in the presence of divalent metal cations.

Nitrous acid. Chemicals which remove groups from bases can also lead to mutations. For example, nitrous acid has been found to be a mutagen owing to its deamination (removal of the amino group) of cytosine and adenine. Deamination of cytosine, for example, leads to the formation of uracil. At the next replication, uracil would attract adenine, resulting eventually in a substitution mutation (Fig. 15-12).

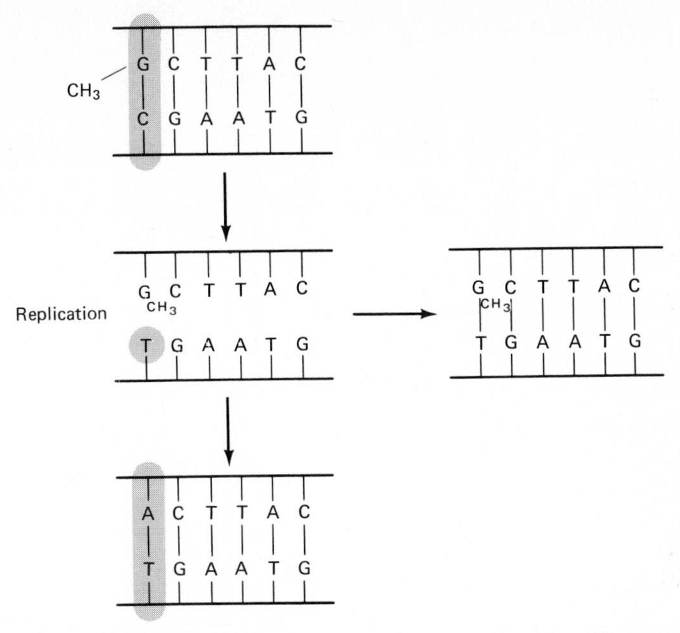

Figure 15-9. The effect of alkylating agents is to cause mispairing of bases during replication, leading to substitutions.

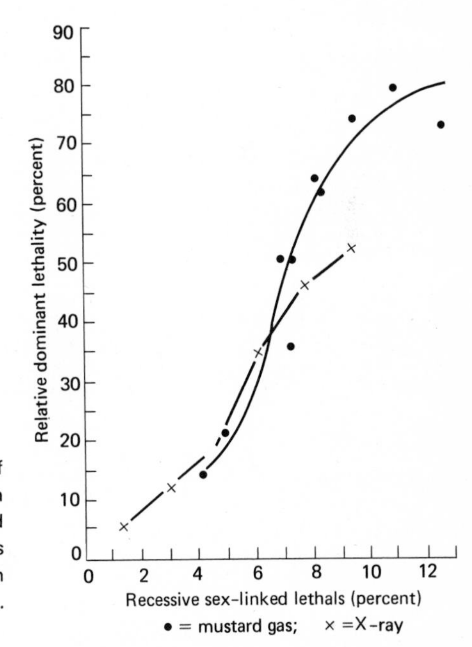

Figure 15-10. Graph showing similarity of effects of mustard gas and X-rays at low doses in inducing dominant lethal mutations and X-linked lethals in *Drosophila*. At higher doses, the effects of mustard gas exceed those of X-rays. (From C. Auerbach, 1976. *Mutation Research*, p. 301. Halsted Press, New York.)

Base analogs. There are some chemicals which are structurally similar to the bases normally found in DNA, which can be incorporated in place of a normal base. For example, 5-bromouracil (BU) or bromodeoxyuridine (BUdR) are analogs of

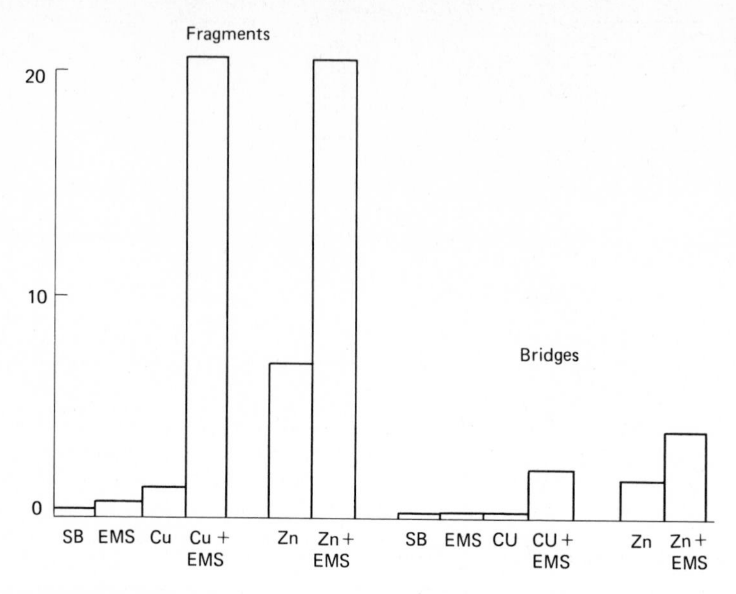

Figure 15-11. Frequencies of anaphase aberrations found in *Vicia faba* after treatment with ethylmethane sulfonate (EMS), an alkylating agent and metal ions SB = Sørensen buffer. (From C. Auerbach, 1976. *Mutation Research*, p. 267. Halsted Press, New York.)

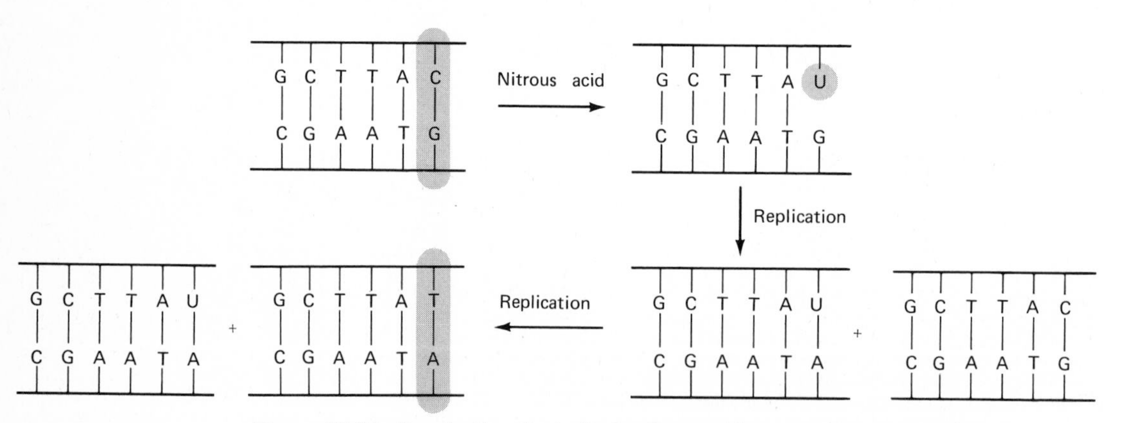

Figure 15-12. Deamination of cytosine by nitrous acid converts it to uracil. Replication causes insertion of adenine, and eventual substitution of an A–T base pair for a C–T base pair.

thymine. If incorporated in place of a thymine, either BU or BUdR can base-pair with guanine during replication. This process, as in the above cases, would lead to a substitution mutation as illustrated in Fig. 15-13.

Other factors. In addition to chemicals and radiation, certain other factors are known to contribute to an increase in mutation. Age, for one, has been discussed

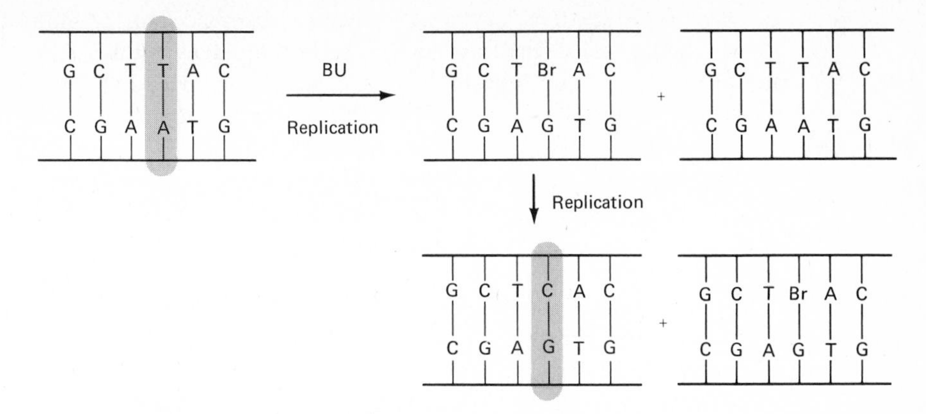

Figure 15-13. Treatment of DNA with base analog 5-bromouracil (BU) causes insertion of BU instead of normal thymine. Since BU can pair with guanine, at the next replication event, a C–G base pair is substituted for an A–T base pair.

previously as related to a higher frequency of aneuploidy. Mutator genes and even heat have been cited as mutagenic. It is certain that the number of mutagens in our world is increasing. As we continue to demand the development of chemicals to improve some aspects of our lifestyle (for surely products such as food additives exist because of the consumer market), we must continuously be wary of their potential as mutagens or carcinogens. Indeed, almost all carcinogens have been found to be mutagenic, and most mutagens have been found to be carcinogenic. (See Table 15-2, p. 489.)

Tests to Identify Mutagens

In an attempt to guard against the unwitting spread of mutagens, a number of different tests have been devised to discern mutagenic from nonmutagenic materials. None of these, however, is considered sufficiently rigorous to allow the results to be accepted without equivocation. We have already described two such tests, the Muller 5 test on *Drosophila* and the specific locus test. Although originally used to determine mutagenicity of radiation, the same tests can also be used for treating males with chemicals. Both these tests are limited in their usefulness, because they allow observations of the effects of a substance on only a limited portion of the genome, namely, X-linked genes and coat-color loci. Mutations occurring at other loci would not be observed. More recent tests (and their limitations) will be discussed in the following sections.

In vitro and cytogenetic tests. A standard test of a substance for human consumption is to apply the material to human cells in vitro and observe possible effects. Since improvements in staining techniques have allowed scientists to identify finer chromosomal rearrangements than formerly possible, it is the goal of these tests to observe changes in the genetic material.

One of the problems with such in vitro monitoring systems is that cells in culture do not represent a whole organism. Chemicals which we ingest, for example, must be exposed to digestive enzymes and then be absorbed into the bloodstream to finally reach interior cells. The in vitro system certainly does not duplicate this complex procedure. In addition, one must always bear in mind that there may be cells which are lethally affected by the substance being tested, but which are not observed because dead cells degenerate quickly in vitro.

Host-mediated assay. A test designed to address some of these problems is known as the *host-mediated assay*. In this test, microorganisms are introduced into a host organism such as a rat. The host is then treated with the substance being tested. After a period of time, the microorganisms are recovered, checked for mutations, and compared to control populations of the same microorganism which were administered the substance directly. Presumably the experimental microorganisms will have encountered the substance after it has been metabolized by the host.

This technique involves various difficulties. If an increase in mutation occurs, the researcher does not really know what has caused the increase, because chemicals are constantly broken down and reused in the synthesis of other substances during metabolism. Also, since we are primarily interested in the mutagenic effects of a substance on higher organisms, we must extrapolate the data from microorganisms to forms such as humans, an obviously difficult task.

Dominant lethal assay. Another method of testing the mutagenicity of a substance is the *dominant lethal assay*. Male mice or rats are treated and mated with untreated virgin females. The females are sacrificed at different times in gestation. Presence of more resorbed embryos and abnormal fetuses than occurs in matings of control animals is taken to indicate that the sperms of treated males carried dominant lethal mutations. However, absence of dominant lethal mutations does not prove that there is no mutagenicity, since recessive mutations would not be identified with this approach.

Ames test in Salmonella. A recent mutagenicity testing method devised by Bruce Ames and his coworkers (1974) uses special strains of the bacterium *Salmonella typhimurium*, and a fraction of liver enzymes from rat liver cells known as the *microsomal* fraction. Tester strains of *Salmonella* are UV-sensitive, carry an R factor which causes ampicillin resistance and seems to increase sensitivity of the cells to mutagens, and are auxotrophic for histidine (his^-). The microsome fraction supplies enzymes that can metabolize the substances being tested, to approximate the in vivo situation.

The mixture of *Salmonella* cells and microsome enzymes is poured into agar plates, and a prescribed amount of the chemical to be tested is added. The numbers of his^+ colonies which arise are taken as an indication of mutagenesis. (Mutations from the mutant state to wild-type are known as *reversion*, or *back*, *mutations*.) It is considered a further indication of mutagenesis if the frequency of mutations to his^+ shows a linear relation to the amount of chemical added to the plates.

The value of the Ames test is of course limited by the fact that it reveals only the effects of an agent on back mutation at one locus. However, because of its relative ease in handling, its rapidity in producing data, and the incorporation of microsome enzymes which indicate the effects of metabolism, the Ames test is currently one of the most frequently used screening devices for mutagenicity and carcinogenicity. Included in Table 15-2 are data obtained upon testing more than 300 chemicals for mutagenicity and carcinogenicity.

Table 15-2 Results of the Ames Test for Carcinogenicity and Mutagenicity of Different Chemicals

Substance	*Car.*	*Mut.*	*Revertants per* *nmol*	*plate*	*Strain*	*S9*
Acridine orange	+	+	66	4370/20	097	A
Captan	+	+	25	820/10	095	—
Carbon tetrachloride	+	0	0.001	$70/10^4$	05	A
Cigarette smoke condensate	+	+	18200/cigarette		8	A
Dieldrin	+	0	0.003	$70/10^4$	0957	A
Dimethylnitrosamine	+	w^+	0.02	1100/4440	03	+
Diethylstilbesterol	+	?	0.38	70/50	0957	A
Ethylmethanesulfonate	+	+	0.16	$12934/10^4$	05	A
Tris	+	+			50	P
Vinyl chloride	+	w^+		G	05	P

From McCann *et al.* PNAS 72: 136–37, 1975. Car. = carcinogenic; Mut. = mutagenic; Revertants = no. his^+/no. micrograms tested; Strain = strain of tester bacteria used; S9 = technique of obtaining microsome fraction; G = volatile gas so that quantitation is difficult; w^+ = weakly mutagenic or carcinogenic.

Potentiation

One major problem in the analysis of mutagenesis we have not yet mentioned is the problem of *potentiation*. This term refers to the phenomenon of two or more factors, chemical and/or physical, which together cause an increase in mutation frequency beyond that which each factor causes individually. For example, there is evidence that caffeine is nonmutagenic when tested by itself. However, in bacteria tested with both caffeine and UV radiation, the mutation frequency increased over that of bacteria exposed to UV alone (Lieb 1961).

Potentiation is a major problem our society must face, since clearly we are ingesting combinations of chemicals, while being exposed to increased levels of radiation. It is necessary to design tests to determine the safety or mutagenicity of not only single factors, but also of the myriad combinations of factors we encounter. The cost of developing and testing a product for commercial sale is already astronomical. Yet there is increasing concern that the tests being used are not adequate. The concerns are real, because in fact the only truly adequate testing would have to be on humans, something that cannot be done. This problem is one of the most difficult of our time.

REPAIR MECHANISMS

Although cells are constantly exposed to mutagens, the frequency of mutations in most species is still very low. Table 15-3 includes some data on the frequencies of various mutations in different species of animals and plants.

Actually, cells sustain far more alteration of their DNA than is reflected in the frequencies listed in Table 15-3. Experiments on mutation in bacteria revealed that several repair mechanisms exist in cells, which detect and repair damage to DNA.

Table 15-3 Spontaneous Mutation Rates per Haploid Genome per Generation for Certain Genes in Various Organisms

Species	*Trait*	*Rate*
Bacteriophage (T2)	Lysis Inhibition $r \longrightarrow r^+$	1×10^{-8}
	Host Range $h^+ \longrightarrow h^-$	3×10^{-9}
E. coli	$lac^- \longrightarrow lac^+$	2×10^{-7}
	$his^- \longrightarrow his^+$	4×10^{-8}
	$his^+ \longrightarrow his^-$	2×10^{-6}
Corn	Shrunken Seeds, Sh $\longrightarrow$ sh	1×10^{-5}
	Purple, P $\longrightarrow$ p	1×10^{-6}
Drosophila	White Eye W $\longrightarrow$ w	4×10^{-5}
	Brown Eye BW $\longrightarrow$ bw	3×10^{-5}
Humans	Normal $\longrightarrow$ Hemophilia	3×10^{-3}
	Normal $\longrightarrow$ Albinism	3×10^{-3}
	Normal $\longrightarrow$ Childhood Progressive Muscular Dystrophy	3.2×10^{-5}
	Retinoblastoma	1.4×10^{-4}
	Waardenburg's Syndrome	3.7×10^{-6}

From R. Sager and F. J. Ryan, 1961, *Cell Heredity*, Wiley, New York.

Photoreactivation

Even before the 1950s, studies had shown that if irradiated colonies of *Streptomyces griseus* were exposed to visible light following irradiation with lethal levels of UV, recovery of colonies occurred. That this phenomenon was directly related to some effect of light on the bacteria could be shown by a direct correlation between time of exposure of *Streptomyces* to visible light and the rate of survival of the bacteria. Table 15-4 displays some of these data, which clearly indicate a beneficial effect of visible light on the irradiated cultures. This phenomenon was termed *photoreactivation*.

Later studies showed that not all bacteria would respond to visible light. For example, *Hemophilus influenzae* showed no photoreactivation, but *E. coli* did. When an extract of *E. coli* was mixed with *Hemophilus* and the mixture exposed to visible light, there was recovery of *Hemophilus* colonies. This indicated that *E. coli* produced an enzyme involved in photoreactivation which was not found in *Hemophilus*. It has since been established that the enzyme splits dimers into monomers (Fig. 15-14), restoring the normal structure of the DNA molecule. Photoreactivating enzymes have been found in many organisms, including *Neurospora*, *Chlamydomonas*, yeast, and humans.

Excision Repair

Studies on UV-resistant and UV-sensitive strains of bacteria revealed that they also possess a mechanism for repair which is not dependent on visible light. In 1964, reports of the presence of dimers in short pieces of nucleotides, found in the cytoplasm of cells irradiated with UV, were published by two different groups (Setlow and Carrier 1964, and Boyce and Howard-Flanders 1964). Other studies have since

Table 15-4 Effect of Duration of Visible Light Illumination on Recovery of *Streptomyces* from Ultraviolet Irradiation Injury

Illumination time (min)	*Viable cells per ml of suspension*	*Relative increase in survival rate*
0	2.5[a]	000 fold
10	2.5×10^3	1000 fold
20	9.2×10^3	3000 fold
30	1.3×10^5	52,000 fold
40	1.6×10^5	64,000 fold
50	2.0×10^5	80,000 fold
60	5.3×10^5	210,000 fold
145	5.5×10^5	220,000 fold
173	7.7×10^5	310,000 fold
240	8.0×10^5	320,000 fold

[a]The count of the non-ultraviolet-irradiated suspension was 4.2×10^6, so that the survival rate at time zero was 6.0×10^{-7}.

From C. Auerbach, 1976, *Mutation Research*, p. 207, Halsted Press.

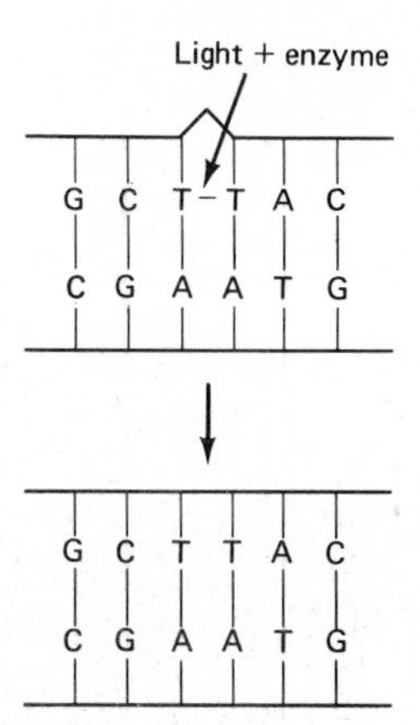

Figure 15-14. Photoreactivation repair, which involves enzymatic action (stimulated by light) breaking dimers into monomers.

confirmed their interpretation of this phenomenon, namely, that enzymes recognize dimers, excise them by nicking the DNA, and then fill in the gaps with the correct bases. Ligase joins the ends of the new piece of DNA with the broken ends of the original DNA molecule (Fig. 15-15). This process is known as *excision repair*, or *dark repair*, since it does not require visible light.

One of the enzymes which plays a major role in excision repair is DNA polymerase I. Recall that when first isolated, it was thought to be the major enzyme in replication, but later this role was assigned to DNA polymerase III. Studies of mutant strains of bacteria which are incapable of dark repair have shown that DNA polymerase I is involved in the gap-filling process. Other loci also taking part in the process are summarized in Table 15-5.

One might expect that factors interfering with repair would increase the frequency of mutations observed in treated cells, and this appears to be the case. For example, we mentioned earlier in discussing potentiation that caffeine increases the

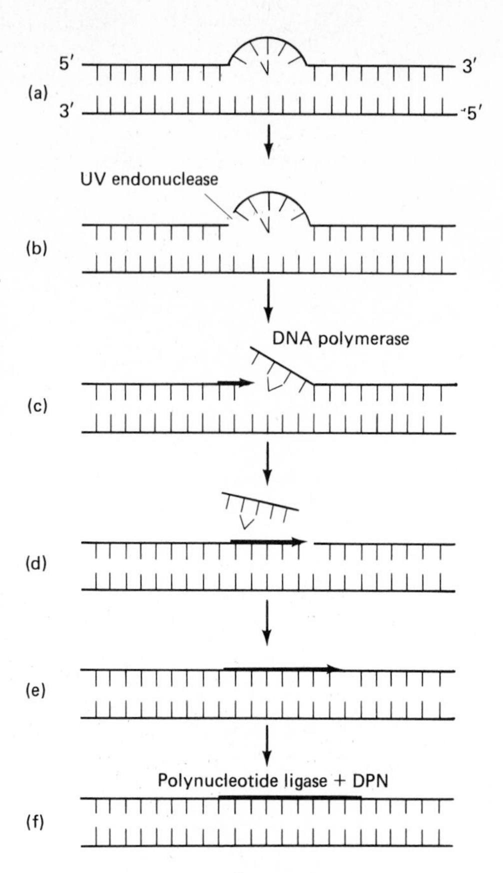

Figure 15-15. A model for excision-repair (after Kelly *et al.*). (a) A UV-irradiated double helix of DNA is represented containing one pyrimidine dimer. (b) The region adjacent and on the 5′ side of the dimer is attacked by an endonuclease specific for UV-irradiated DNA. (c) DNA polymerase binds to DNA at the nicked site, and by addition of 5′ nucleotide triphosphates starts repair polymerization of the damaged strand using the template of the complementary strand. The dimer region becomes displaced. (d) When the 5′ to 3′ exonuclease site of the polymerase reaches the next hydrogen bonded base pair, the phosphodiester bond is hydrolyzed releasing the dimer in a short oligonucleotide. (e) The polymerizing and hydrolytic activities may continue in the 5′ to 3′ direction until (f) the polymerase is displaced by polynucleotide ligase which together with DPN restores the integrity of the phosphodiester backbone of DNA. (From J. M. Boyle, 1972. *Molecular and Cellular Repair Processes*, R. F. Beers, Jr., R. M. Herriott, and R. C. Tilghman, eds., p. 15. Johns Hopkins U. Press, Baltimore.)

mutagenicity of UV irradiation, an effect reported to be due to caffeine's interference with repair processes (Sideropoulos and Shankel 1968, and Lehmann and Kirk-Bell, 1974).

Postreplication Recombination Repair

Observations on bacteria have also indicated that yet another type of repair mechanism exists. For one thing, recombination-deficient (rec^-) mutants have been found to be more sensitive to UV irradiation than rec^+ bacteria. The *rec* loci are apparently involved in repair. Furthermore, some bacteria are capable of tolerating the presence of several dimers in their DNA, while others succumb to the presence of even one dimer.

A model has been proposed to explain these phenomena (Howard-Flanders and W. Rupp 1972). The model invokes repair after replication. It is, accordingly, called *postreplication recombination repair*. During replication, gaps occur in the new strand opposite sites of dimers in the template strand. Following replication, there is recombination and a filling in of the gaps still remaining (Fig. 15-16). Since UV irradiation can create thousands of dimers in the treated DNA, it is thought that excision repair may not excise them all before replication takes place; therefore some

Table 15-5 Some Important Genes and Gene Products Involved in DNA Repair in *E. coli*

Mutant designation	*Gene product*	*Sensitivity*[a] *to UV*	*to ionizing radiation*	*Other important properties*
uvrA, uvrB	UV-specific endonuclease	s	r	
recA	?	s	s	Recombination-deficient
recB, recC	Exonuclease V	s	s	Recombination-deficient
lexA(exrA)	?	s	s	Nonmutable by many agents
polA(resA)	DNA polymerase I	s	s	
polB	DNA polymerase II	r	r	No definite function established
polC(dnaE)	DNA polymerase III	Lethal under restrictive conditions—enzyme involved in DNA replication.		
lig	DNA ligase	Lethal under restrictive conditions—enzyme involved in DNA replication.		

[a] s = sensitive; r = resistant (same as wild-type).

From A. R. Lehmann, 1978. *Mol. Biol. Biochem. Biophys.* 27: 316.

mechanism must exist to repair damage after replication. Presumably, those cells which cannot survive even one dimer are mutant in postreplication recombination enzymes.

Repair in Human Diseases

A study of the effects of chemical carcinogens and mutagens, including ionizing and UV irradiation, on normal human fibroblasts showed that there are two forms of repair (Regan and Setlow 1974). One form involves the insertion of a very short sequence of nucleotides, three or four base pairs, within an hour of exposure. Another form results in excision and replacement of a considerable number of bases, about 100 nucleotides long. The short form was typical of repair following ionizing radiation, and the long form was typical of the effects following UV irradiation. Some chemicals also produced these results; others caused both forms of repair to occur.

It is not clear whether these forms of repair are the same as light and dark repair, both of which have also been detected in human cells. The uncertainty stems from the fact that we have not yet been able to isolate enzymes involved in repair in eukaryotic cells. Nonetheless a study of cells following irradiation has given evidence that dimers form, are excised and replaced.

Further evidence that eukaryotic cells are capable of repair comes from the phenomenon known as *host cell reactivation*. This is the ability of host cells to repair

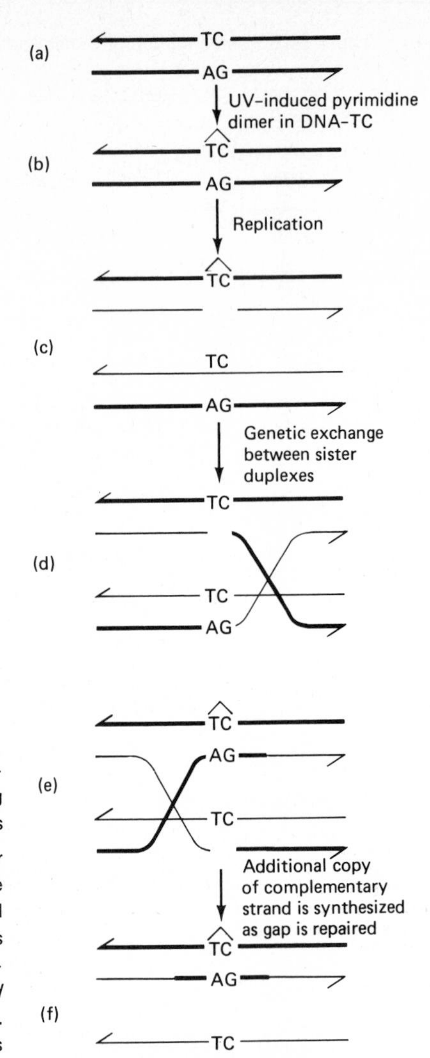

Figure 15-16. Hypothetical scheme for recombinational repair by sister exchanges following DNA replication. Duplex DNA seen at (a) is exposed to UV light which produces a $\widehat{TC}$ dimer seen at (b). Replication results in a gap in the daughter strand opposite the dimer. This is filled in by a sister exchanges and repair synthesis as seen at (d), (e), and (f). (From P. Howard-Flanders and W. D. Rupp, 1972. *Molecular and Cellular Repair Processes*, R. F. Beers, Jr., R. M. Herriott, and R. G. Tilghman, eds., p. 224. Johns Hopkins U. Press, Baltimore.)

damage in viruses treated with physical or chemical agents prior to infection of the cells by the viruses. A high level of reactivation of the viruses indicates repair by the host cell. A low level of reactivation thus implies a deficiency in repair in host cells.

A hereditary skin disease known as *xeroderma pigmentosum* produces acute sensitivity to sunlight in some humans, and leads to the development of skin cancer. The effects of this disease are believed to result from a deficiency in repair processes (Cleaver 1972). Whereas normal fibroblast cells in culture can excise thymine dimers, cells from xeroderma pigmentosum patients manifest a deficiency in this process (Setlow *et al.* 1969, Regan and Setlow 1974). This disease underscores the active role which repair processes normally play in our cells to reduce the effects of radiation damage.

Other human diseases have also been reported to be linked to repair defects,

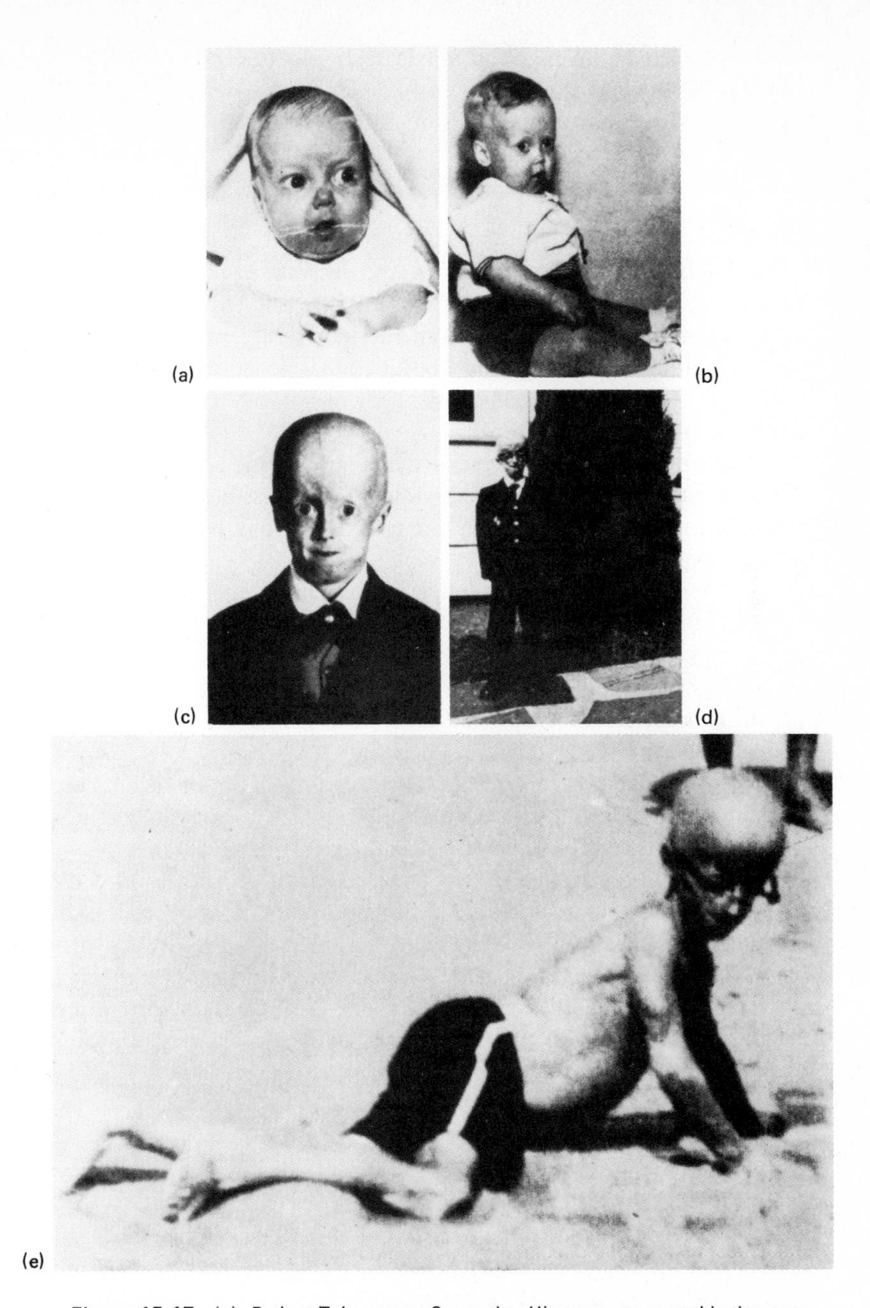

Figure 15-17. (a) Patient T. L., at age 3 months. His appearance at this time was normal. (b) Patient T. L. at age $1\frac{1}{2}$ years. (c) Patient T. L. at age 8 years. His appearance is clearly progeric. Notice the absence of subcutaneous fat, the alopecia, and the prominence of the superficial veins. (d) Patient T. L. at age 10 years. (e) Patient T. L. at age 11 years, when he suffered from severe congestive heart failure shortly before his death. (From W. Reichel, R. Garcia-Bunuel, and J. Dilallo, 1971. "Progeria and Werner's Syndrome as Models for the Study of Normal Human Aging." *J. Am. Ger. Soc.* 19: 371–72, Figs. 1-5.)

such as Fanconi's anemia and ataxia telangiectasia. Interestingly, these diseases are not obviously linked to repair, as in xeroderma pigmentosum. Fanconi's anemia causes anemia, and ataxia leads to abnormalities of the cerebellum and immune system (Setlow 1978).

Repair in Aging

A further fascinating phenomenon is the decrease in repair capabilities in aging cells. It is of course not possible to determine cause and effect in this situation. One could hypothesize that aging results from the increasing inability of cells to repair mutations, resulting in defective and nonfunctional genetic material in older cells. Or one could hypothesize that because of aging, genes determining the enzymes for repair are being progressively shut off.

Along the same line, recall that there is also a dramatic increase in the incidence of cancer among older individuals. The possible association between this fact and increased somatic mutation due to a decrease in repair capabilities of aging cells is being explored.

There has been considerable interest in a genetically determined human disease which presents some symptoms of premature aging. Known as *progeria*, this is a very rare autosomal recessive condition in which children show evidence of aging (Fig. 15-17) such as wrinkling of the skin, and death in the preteen or teenage years of atherosclerosis. In vitro studies indicate some deficiency in repair may be involved, although the results are not clear.

Studies have been done on cocultivating (without fusion) progeria cells with normal young cells and with normal old cells. The results show that whereas the combination of old cell + progeria cell is deficient in repair, that of young cell + progeria cell causes progeria cells to undergo repair (Brown *et al.* 1976). Interestingly, fusion of cells from two different progeria patients also produced increased repair capabilities, indicating some form of complementation. When progeria cells were grown in medium conditioned by normal cells, no effects on repair capabilities were observed, suggesting that cell contact may be necessary for normal cells to cause repair. The research on progeria, aging cells, and their relationship to repair continues today.

PROBLEMS

1. If the *rII* mutants of T4 phage do not lyse strain K of *E. coli*, how would a new spontaneous mutation to an *rII* mutant condition be identified?

2. Assume that there are 6 new mutations, numbered 1–6, at the *rII* locus in T4 phage. You have a strain, #10, known to be an *rII* mutation at the A cistron. Strain #11 is known to be a mutation at the B cistron. Design the experiments necessary to map the new mutations in either the A or the B cistron, and to discern whether any of these mutations are identical to each other.

3. From our discussions in the past few chapters on the molecular aspects of genes, how would you define a gene? (Try to answer in fewer than 300 words!)

4. Explain in detail why Luria and Delbruck found the frequency of phage-resistant colonies in some cultures to be as much as 10 times or more than that found in other (independent) cultures?

5. In *Neurospora*, complementation tests can be carried out by mating two genetically different haploid strains so that a cell at a particular stage contains two nuclei, both functional. Such a cell is called a *heterokaryon*. At a locus we designate as p^+, four mutants have been isolated, called p^1, p^2, p^3, and p^4. Complementation tests yielded the following results:

$$p^1/p^2 \longrightarrow p^+$$
$$p^1/p^3 \longrightarrow p$$
$$p^2/p^3 \longrightarrow p^+$$
$$p^2/p^4 \longrightarrow p$$
$$p^3/p^4 \longrightarrow p^+$$
$$p^4/p^1 \longrightarrow p$$

How many cistrons or complementation groups are there? Which of the mutations are in the same cistron?

6. A cell contains the following sequence of bases in its DNA: C A T.
G T A
What changes in the sequence might be effected by treating the cell with (a) alkylating agents, (b) nitrous acid, or (c) 5-bromouracil? Diagram the steps of the conversions.

7. What effect would there be on the offspring of irradiated males who have sustained a recessive lethal X-linked mutation in their gametes? If they have sustained a dominant lethal X-linked mutation? What would the effect be if females sustained a recessive X-linked mutation in their ova? If they have sustained dominant X-linked mutation?

REFERENCES

AMES, B. N., 1974. "A Combined Bacterial and Liver Test System for Detection and Classification of Carcinogens as Mutagens." *Genetics* 78(1): 91–5.

AUERBACH, C., AND J. M. ROBSON, 1947. "The Production of Mutations by Chemical Substances." *Proc. Roy. Soc. Edinburgh B.* 12: 271–283.

AUERBACH, C., 1976. *Mutation Research Problems, Results and Perspectives.* Chapman & Hall, London.

BEYER, S., 1955. "Fine Structure of a Genetic Region in Bacteriophage." *Proc. Nat. Acad. Sci. U.S.* 41: 344–354. Reprinted in J.A. Peters, ed., 1959. *Class Papers in Genetics.* Prentice-Hall, Inc., Englewood Cliffs, N.J.

BOYCE, R. P., AND P. HOWARD-FLANDERS, 1964. "Release of Ultraviolet Light-Induced Thymine Dimers from DNA in *E. coli* K-12." *Proc. Nat. Acad. Sci. U.S.* 51: 293–300.

BROWN, W. T., J. EPSTEIN, AND J. B. LITTLE, 1976. "Progeria Cells are Stimulated to Repair DNA by Co-Cultivation With Normal Cells." *Exp. Cell Res.* 97: 291–296.

CLEAVER, J. E., 1972. "Excision Repair: Our Current Knowledge Based on Human (Xeroderma Pigmentosum) and Cattle Cells." In R. F. Beers, R. M. Herriott, and R. C. Tilghman, eds., *Molecular and Cellular Repair Processes*. Johns Hopkins U. Press, Baltimore.

HOWARD-FLANDERS, P., AND W. D. RUPP, 1972. "Recombinational Repair in UV-Radiated *Escherichia coli*." In R. F. Beers, R. M. Herriott, and R. C. Tilghman, eds., *Molecular and Cellular Repair Processes*. Johns Hopkins U. Press, Baltimore.

INGRAM, V. M., 1957. "Gene Mutations in Human Hemoglobin: The Chemical Difference Between Normal and Sickle Cell Hemoglobin."*Nature* 180: 326–328.

KARPECHENKO, A., 1928. *Z. Indukt. Abst. Vererb.* 48:27.

LEDERBERG, J. AND E. M. LEDERBERG, 1952. "Replica Plating and Indirect Selection of Bacterial Mutants." *J. Bacteriol.* 63: 399.

LEHMANN, A. R., AND S. KIRK-BELL, 1974. "Effects of Caffeine and Theophylline on DNA Synthesis in Unirradiated and UV Irradiated Mammalian Cells." *Mut. Res.* 26: 73–82.

LIEB, M., 1961. "Enhancement of Ultraviolet-Induced Mutations in Bacteria by Caffeine." *Zeitsche Vererbungslehre* 92: 416–429.

LURIA, S. E., AND M. DELBRUCK, 1943. "Mutations of Bacteria from Virus Sensitivity to Virus Resistance." *Genetics* 28: 491.

MILLER, R. W., 1971. "Delayed Radiation Effects in Atomic Bomb Survivors." *Science* 166: 569–574.

MULLER, H. J., 1927. "Artificial Transmutation of the Gene." *Science* 66: 84–87. Reprinted in J.A. Peters, ed., *Classic Papers in Genetics*. Prentice-Hall, Inc., Englewood Cliffs, N.J.

PAULING, L., H. A. ITANO, S. J. SINGER, AND I. C. WELLS, 1949. "Sickle Cell Anemia, a Molecular Disease." *Science* 110: 543–548.

REGAN, J. D., AND R. B. SETLOW, 1974. "Two Forms of Repair in the DNA of Human Cells Damaged by Chemical Carcinogens and Mutagens." *Canc. Res.* 34: 3318–3325.

RUSSELL, W. L., L. B. RUSSELL, AND E. M. KELLY, 1958. "Radiation Dose Rate and Mutation Frequency." *Science* 128:1546.

SETLOW, R. B., ET AL., 1969. "Evidence that Xeroderma Pigmentosum Cells Do Not Perform the First Step in the Repair of Ultraviolet Damage to Their DNA." *Proc. Nat. Acad. Sci. U.S.* 64: 1035–1041.

SETLOW, R. B., 1978. "Repair Deficient Human Disorders and Cancer." *Nature* 271: 713–717.

SIDEROPOULOS, A. D., AND D. M. SHANKEL, 1968. "Mechanism of Caffeine Enhancement of Mutations Induced by Sublethal Ultraviolet Dosages." *J. Bacteriol.* 96: 198–204.

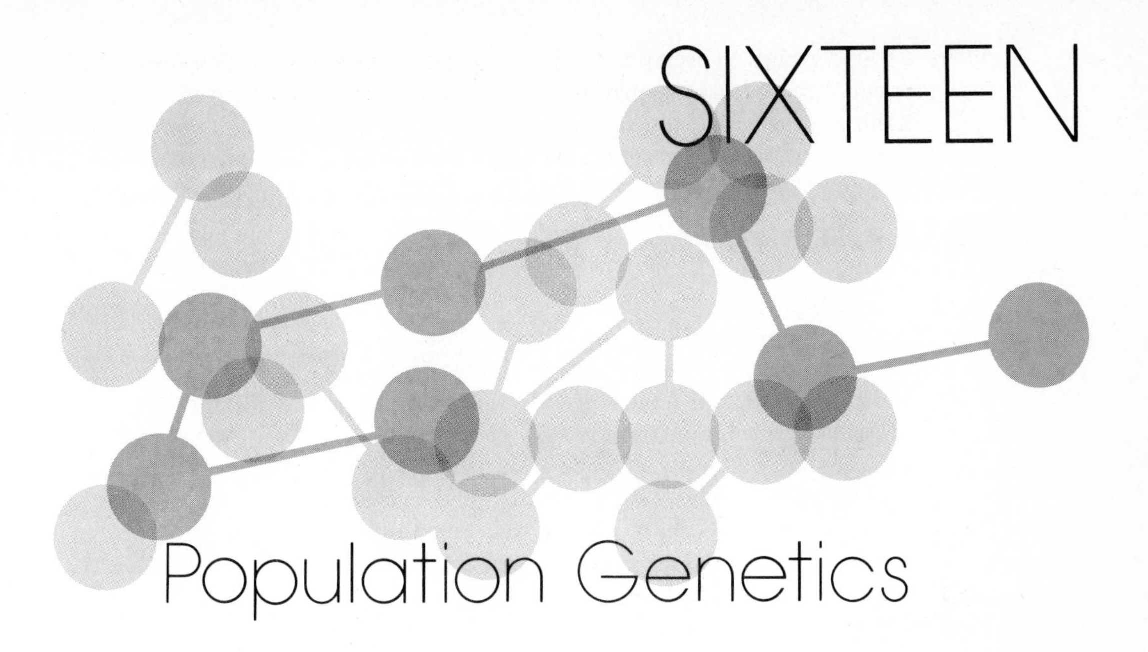

SIXTEEN

Population Genetics

INTRODUCTION

In our previous discussion of the inheritance of traits we have considered only matings among individual pairs of organisms—that is, given the genotypes of both mates, we determined the possible genotypic outcomes and their respective probabilities. In humans for example, we showed that if parents are known to be carriers for Tay Sachs disease ($Tt \times Tt$; t is recessive for Tay Sachs), $\frac{1}{4}$ of their offspring would be expected to develop Tay Sachs (and hence die before the age of two). A similar proportion of children of parents who have the sickle-cell trait ($Ss \times Ss$) would be expected to have sickle-cell anemia.

In Chapter 3 we showed that if parents are carriers for albinism, $\frac{1}{4}$ of their offspring can be expected to be albinos. But we also know that the number of albinos in the world does not equal $\frac{1}{4}$ of the world's population. (Why would you expect it to?) In fact, it is known that one in 20,000 people is an albino, and that in the United States the rate of Tay Sachs births is about 1 in 1600 births in the Jewish population, and 1 in 500,000 births in the rest of the population. The rate of sickle-cell anemia among American Blacks is 1 in 500, whereas that among American Caucasians almost zero.

Looking around us we see that certain traits (eye color or hair color for example, or Tay Sachs disease or sickle-cell anemia) appear in different proportions among different ethnic groups. It seems clear that in the population at large not every allele for a particular trait occurs in equal proportions within every ethnic group.

In this chapter we will discuss the genetics of populations, and try to understand what happens to traits in the general population rather than in individual matings.

GENETIC EQUILIBRIUM AND RANDOM MATING

Anthropological studies suggest that some 400,000 years ago the people of the earth constituted a single population group living in east-central Africa. The descendants of the first group of human beings have spread throughout the world and developed into the various populations (ethnic groups) that exist today. We will discuss several theories that may explain how this diversity developed.

We will start our study of population genetics by considering an ideal case (for example, one group in east-central Africa) and then consider various more realistic cases. An ideal case would be to assume that among the population in our group there is a gene pool in males for each trait and a gene pool in females for each trait, and that mating is random. The *gene pool* is the sum total of the genes in a given breeding population at a particular time. *Random mating* is the assumption that any individual of one sex has an equal chance of mating with any individual of the opposite sex. Random mating is also referred to as *panmixis*, the equivalent of saying that genes are selected randomly and independently from the male gene pool and from the female gene pool. (This is clearly not a realistic assumption, and we will have to consider how modifications of it affect the theory we will develop.) For a particular trait we will denote the dominant trait by H and the recessive trait by h. We will denote the proportion of H genes among males by p and the proportion of h genes among males by q (which equals $1 - p$). Among women the proportions of H and h are denoted by p' and q', respectively. If we assume random mating and use the techniques described in Chapter 3 (tree diagrams), but consider the contributions of gametes to be from the population at large rather than from individuals, we derive the tree diagram in Fig. 16-1:

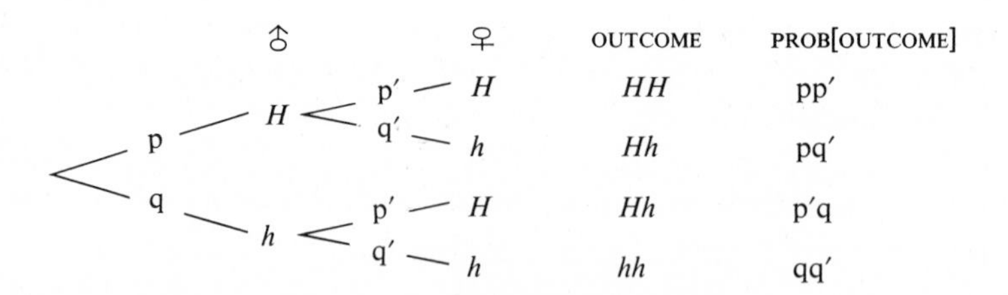

Figure 16-1.

The offspring of the random matings within this population will manifest the following genotypic proportions: $\text{Prob}[HH] = pp'$, $\text{Prob}[Hh] = pq' + p'q$, $\text{Prob}[hh] = qq'$. If we further assume that the sex of these offspring is independent of the genotype (there is no reason to doubt this assumption), the proportion of $HH : Hh : hh$ $(pp' : pq' + p'q : qq')$ in males is the same as the proportion in females.

Suppose that we have a population of plants in which there is incomplete dominance (see Chapter 4). The genotypes are clearly identifiable since the phenotypes are all different. For example, if we let *R* signify the trait Red Flowers, and *r*, the trait White Flowers, *RR* plants appear red, *Rr* plants appear pink, and *rr* plants appear white. Let us assume that 50% of the plants have red flowers, 20% have pink flowers, and 30% have white flowers. What happens to the genotypic (equivalently, phenotypic) proportions if these plants are mated randomly among themselves? Again we use the tree diagram to understand what the genotypic outcomes will be. We know that there are two genes, *R* and *r*, and that there is a certain proportion of each genotype in the population. How do we determine the proportion p of *R* genes in the gene pool and the proportion q of *r* genes in the gene pool? We will assume $p = p'$ and $q = q'$. That is, the proportion of *R* in the male and the female gene pools is the same, as is the proportion of *r*. We represent the mating in question in Fig. 16-2:

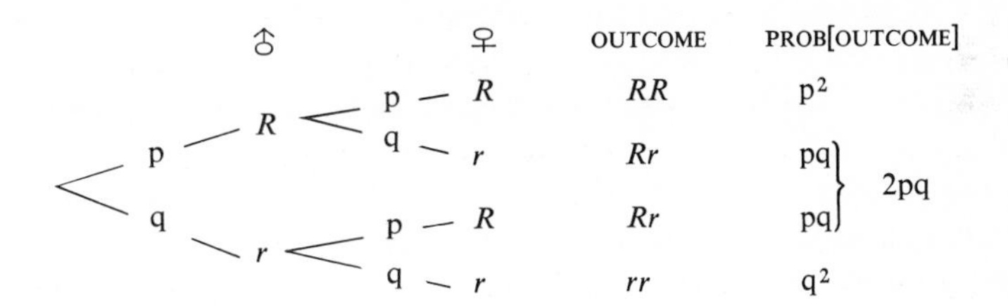

Figure 16-2.

To determine p and q we note that p is equivalent to the probability of choosing an *R* gene if the gene is chosen at random from the gene pool. We will make use of similar equivalences between proportion and probability in the discussion to follow.

In order to estimate p and q we observe that the red flowers consist of all *R* genes, and the pink flowers of both *R* and *r* genes. In fact, in the pink flowers $\frac{1}{2}$ the genes are *R* and $\frac{1}{2}$ the genes are *r*. We can therefore conclude that the proportion of *R* genes in the population is the proportion of red flowers plus half of the proportion of pink flowers. That is, the red flowers (*RR*) contribute 50% to the gene pool of *R* and the pink flowers (*Rr*) contribute $\frac{1}{2}$ of 20%. Therefore we find that

$$\begin{aligned}\text{Prob[R genes]} = p &= \text{Prob}[RR] + \tfrac{1}{2}\text{Prob}[Rr] \\ &= \underbrace{.50}_{\substack{\text{all the} \\ RR \text{ flowers}}} + \underbrace{\tfrac{1}{2}(.20)}_{\substack{\frac{1}{2}\text{ of the} \\ Rr \text{ flowers}}} \\ &= .50 + .10 \\ &= .60\end{aligned}$$

It follows that q, the proportion of *r*, must be .40, since $p + q = 1$. With this information we can now complete Fig. 16-2 by substituting the values for p and q, resulting in Fig. 16-3.

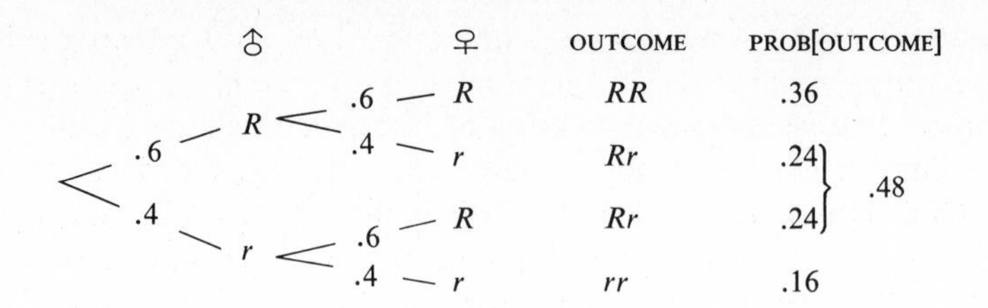

Figure 16-3.

We see that among the offspring we can expect

$$\begin{aligned}
\text{Prob[Red Flowers]} &= \text{Prob}[RR] = p^2 = (.6)^2 = .36 \\
\text{Prob[Pink Flowers]} &= \text{Prob}[Rr] = 2pq = 2(.6)(.4) = .48 \\
\text{Prob[White Flowers]} &= \text{Prob}[rr] = q^2 = (.4)^2 = .16
\end{aligned}$$

What can we expect the genotypic proportions to be if we once again mate the plants randomly? That is, what happens to the genotypic proportions in the F_2 generation? If we think about it we will realize that the tree diagram used in Fig. 16-2 will once again be appropriate. The only thing we will have to do is determine the new value of p (the proportion of *R* genes), and the new value of q (the proportion of *r* genes). Using the same approach we again note that the red flowers contribute only *R* genes, and that only $\frac{1}{2}$ the genes of the pink flowers are *R* genes. We therefore find that:

$$\begin{aligned}
\text{Prob}[R \text{ genes}] = p &= \underbrace{.36}_{\substack{\text{all the} \\ RR \text{ flowers}}} + \underbrace{\tfrac{1}{2}(.48)}_{\substack{\frac{1}{2} \text{ of the} \\ Rr \text{ flowers}}} \\
&= .36 + .24 \\
&= .6
\end{aligned}$$

and Prob[r genes] $= q = 1 - p = .4$. These values of p and q are the same as the old ones. Hence, if we insert them into Fig. 16-2, we will get Fig. 16-3. Therefore the genotypic proportions in the F_2 will be identical to those in the F_1. That is:

$$\begin{aligned}
\text{Prob[Red Flowers]} &= \text{Prob}[RR] = p^2 = (.6)^2 = .36 \\
\text{Prob[Pink Flowers]} &= \text{Prob}[Rr] = 2pq = 2(.6)(.4) = .48 \\
\text{Prob[White Flowers]} &= \text{Prob}[rr] = q^2 = (.4)^2 = .16
\end{aligned}$$

It should by now be obvious that if the conditions of random mating are maintained, the flowers in the F_3 population will appear in the same proportions as in the F_1 and F_2 populations. In fact, if our assumptions are satisfied, these proportions should persist indefinitely—that is, the same proportions of red flowers (.36), pink flowers (.48), and white flowers (.16) should be found after 10, 20, or even 100 random matings! Is this surprising conclusion the result of the particular numbers in our example? The answer is no. In fact, we will be able to verify that, under reasonable assumptions, by at most the F_2 generation, the genotypic proportions of the population will be stabilized and not subject to alteration, the population will be in equilibrium. This phenomenon is called the Hardy-Weinberg law, or Hardy-Weinberg equilibrium. This law was formulated in 1908, when it was independently published by Godfrey Harold Hardy (1908), Professor of Mathematics at Cambridge University in England, and Wilhelm Weinberg (1908), a physician in Stuttgart, Germany. (The law is sometimes referred to as the Castle-Hardy-Weinberg law, since in 1903 William Ernest Castle (1903) of the United States published experimental data of which a part was essentially the same as the Hardy-Weinberg law.)

Underlying Assumptions of the Hardy-Weinberg Law

The following basic assumptions are used to derive the Hardy-Weinberg law:

1. Random Mating—genes are selected randomly and independently from the male and female gene pools.
2. Sex determination and survival are independent of genotype.
3. The proportion of males and females is equal.
4. Among the offspring, the proportion of each trait (dominant or recessive) is the same among males and females.
5. There are no mutations.
6. There is no selection (we will define and discuss this term in more detail later).
7. There is no migration.
8. The population is very large.

If these assumptions are satisfied, equilibrium in the gene populations is established by at most the F_2 generation.

The Hardy-Weinberg Law implies that starting with the F_2, genotypic proportions will be maintained indefinitely—generation after generation—in the ratio $p^2 : 2pq : q^2$ (homozygous dominant : heterozygous : homozygous recessive), if p is the proportion of the dominant allele in the gene pool starting with the second generation. If any of the assumptions are violated (and they usually are, since these assumptions are not totally realistic), deviations from the equilibrium values will occur. In subsequent sections we will discuss the effects on the equilibrium values if various assumptions are violated.

Verification of the Hardy-Weinberg Law[1]

We can easily verify that the Hardy-Weinberg law is valid by considering a population that satisfies the required conditions. We will assume that we are given the following information about the initial population of males and of females (we use H to denote the dominant gene and h to denote the recessive gene): In the males it has been determined that $\text{Prob}[H♂] = p$ and therefore $\text{Prob}[h♂] = 1 - p = q$; and for females it has been determined that $\text{Prob}[H♀] = p'$ and therefore $\text{Prob}[h♀] = q'$. If these

[1]This subsection can be omitted without loss of continuity, though it is recommended that it be included.

populations mate randomly we get the same tree diagram as given in Fig. 16-1. The F_1 offspring will occur in the following genotypic proportions: Prob[*HH*] = pp′, Prob[*Hh*] = pq′ + qp′, and Prob[*hh*] = qq′. Since the expected proportions of males and females are equal, we expect the ratio of *HH* : *Hh* : *hh* (pp′ : pq′ + qp′ : qq′) to be the same for males and females.

To determine the genotypic proportions for the second generation of offspring we must determine p″, the proportion of *H* genes in the F_1 offspring, and q″ = 1 − p″, the proportion of *h* genes in the F_1 offspring. As in our examples of flowers, the proportion p″ of the dominant genes is given by Prob[*HH*] + $\frac{1}{2}$Prob[*Hh*] and q″ = 1 − p″ = Prob[*hh*] + $\frac{1}{2}$Prob[*Hh*]. Applying these formulas we find that

$$p'' = pp' + \tfrac{1}{2}(pq' + qp')$$
$$q'' = qq' + \tfrac{1}{2}(pq' + qp')$$

We use a tree diagram similar to that in Fig. 16-1 and note that by our assumption (4), p″ is the proportion of *H* for both males and females, so we get: Prob[*HH*] = $(p'')^2$, Prob[*Hh*] = $2p''q''$, and Prob[*hh*] = $(q'')^2$.

If our statement of the Hardy-Weinberg law is true, in the third generation we should find that Prob[*HH*] = $(p'')^2$, Prob[*Hh*] = $2p''q''$, and Prob[*hh*] = $(q'')^2$. In fact, these proportions should hold for all succeeding generations. For the third generation we determine p‴ and q‴ = 1 − p‴. If we can show that p‴ = p″ and q‴ = q″, we will have established that the Hardy-Weinberg law is in fact true. To find p‴ and q‴, consider:

$$\begin{aligned} p''' &= \text{Prob[} HH \text{ 2nd generation]} + \tfrac{1}{2}\text{Prob[} Hh \text{ 2nd generation]} \\ &= (p'')^2 + \tfrac{1}{2}(2p''q'') \\ &= p''p'' + p''q'' \\ &= p''(p'' + q'') \\ &= p'' \qquad \text{since} \quad p'' + q'' = 1 \end{aligned}$$

and

$$\begin{aligned} q''' &= \text{Prob[} hh \text{ 2nd generation]} + \tfrac{1}{2}\text{Prob[} Hh \text{ 2nd generation]} \\ &= (q'')^2 + \tfrac{1}{2}(2p''q'') \\ &= q''q'' + p''q'' \\ &= q''(q'' + p'') \\ &= q'' \qquad \text{since} \quad p'' + q'' = 1 \end{aligned}$$

Therefore we see that the genotypic proportions of the F_3 offspring would be exactly those obtained for the F_2 generation. It should now be easy to see that the genotypic proportions for the fourth generation would be the same. In fact, these proportions would persist indefinitely if the assumptions are maintained and nothing disturbs the relative proportions of each gene.

Applications of the Hardy-Weinberg Law

In order to gain further insight into the Hardy-Weinberg law, we will consider one additional example. Consider the M and N blood type factors. These M and N factors are controlled by a single gene. By analyzing the blood we can determine a person's genotype.

An investigator tested 3100 Belgians of the city of Liège (data modified from Boyd 1950 and from Wiener 1943), and found 50.3 percent of type MN, 28.9 percent of type MM, and 20.8 percent of type NN. If these 3100 people (and their ancestors) can be assumed to have mated independently and randomly with respect to M, N alleles, this sample should have attained the Hardy-Weinberg equilibrium values, and the numbers of MM, MN, and NN should be distributed in the ratio $p^2 : 2pq : q^2$.

If we can estimate p and q (the equilibrium values of p and q) from the data, we will be able to determine the expected Hardy-Weinberg equilibrium proportion of each blood type. As previously, we estimate that

$$\begin{aligned}\text{Prob}[M \text{ gene}] = p &= \text{Prob}[MM] + \tfrac{1}{2}\text{Prob}[MN] \\ &= .289 + \tfrac{1}{2}(.503) \\ &= .5405\end{aligned}$$

and so

$$\begin{aligned}\text{Prob}[N \text{ gene}] = q &= 1 - p \\ &= 1 - .5405 \\ &= .4595\end{aligned}$$

Substituting these values for p and q, we find that

$$\begin{aligned}\text{Prob}[MM] = p^2 &= (.5405)^2 \\ &= .2921 \\ \text{Prob}[MN] = 2pq &= 2(.5405)(.4595) \\ &= .4967 \\ \text{Prob}[NN] = q^2 &= (.4595)^2 \\ &= .2111\end{aligned}$$

These numbers differ from the observed proportions .289, .503, and .208, but are they close enough so that we would accept the Hardy-Weinberg law as a reasonably well-fitting model? The chi square goodness of fit test that we presented in Chapter 3 can be used to test whether this sample is from a population that has attained the Hardy-Weinberg equilibrium. In particular, we assume that the proportions 29.21 : 49.67 : 21.11 give the Hardy-Weinberg equilibrium values of MM : MN : NN. The observed data, stated in terms of number observed,[2] are given in Table 16-1:

[2]The χ^2 goodness of fit test cannot be applied to proportions.

Table 16-1 M and N Blood Types in a Sample of 3100 People in Liège, Belgium

Blood type	*Observed*
MM	896
MN	1559
NN	645
	3100

If the Hardy-Weinberg equilibrium has been attained, the expected number of people with each blood type is obtained by multiplying the appropriate proportion from the Hardy-Weinberg equilibrium by 3100. The expected values are given in Table 16-2.

Table 16-2 Expected Number of People with MM, MN, and NN Blood Types

Blood type	*Expected number of people* E_i
MM	905.63
MN	1539.83
NN	654.53

Using $X^2 = \sum \frac{(O_i - E_i)^2}{E_i}$, we can then perform our test. To evaluate X^2, we get

$$X^2 = \frac{(896 - 905.63)^2}{905.63} + \frac{(1559 - 1539.83)^2}{1539.83} + \frac{(645 - 654.53)^2}{654.53}$$

$$= .48$$

To determine if there is significant agreement between the predictions from the Hardy-Weinberg law and the observations, we must compare the value of X^2 to a standard value in the chi square table found on page 82. From our discussion in Chapter 3, recall that to determine the standard value from the χ^2 table we must first determine the degrees of freedom, which in general would be one less than the number of categories. In this particular example we have 3 categories, so our rule would say that we have 2 degrees of freedom. But in this case we only have 1 degree of freedom. We actually imposed an additional constraint on our data, so we lost an additional degree of freedom. The additional constraint is imposed by estimating p = Prob[*M* gene] from the data in order to determine the Hardy-Weinberg equilibrium values. (We get q automatically once we know p, so this imposes no additional constraints.) The general rule for determining the degrees of freedom is given by

number of categories − 1 − number of quantities estimated from the data

In this example we have 3(number of categories) − 1 − 1(estimation of p). The examples discussed in Chapter 3 did not require us to estimate any quantities from the data. All information needed to determine the expected values could be determined from theoretical considerations and did not require the data, and therefore the number of degrees of freedom was just one less than the number of categories. Returning to our example, we find that we have $X^2 = .48$, and 1 degree of freedom. The standard obtained from the χ^2 table with 1 d.f. and .05 level of significance (risk) is 3.84. Therefore we can conclude that at the .05 level of significance we do not have sufficient evidence to conclude that the Hardy-Weinberg equilibrium values are not valid. That is, the random sample of people from Liège, Belgium, appears to be conforming with what would be expected if the Hardy-Weinberg law were in fact a valid model for the inheritance of MM, MN, and NN blood types.

Let us now consider a slightly different set of circumstances. We consider a hypothetical situation in which 1000 friends and relatives decide to leave Liège, and go to an uninhabited island and settle. When the people arrive on the island their blood types are tested with the following results: 100 people are MM, 600 people are MN, and 300 people are NN. For the sake of discussion, let us ask if these observations are consistent with the hypothesis of the island population being in Hardy-Weinberg equilibrium with the proportions 29.21 : 49.67 : 21.11 hypothesized for the population of Liège. We again use the χ^2 goodness of fit test. The expected values are obtained from the proportions determined in the previous example, that is $1000 \times .2921$, $1000 \times .4967$, and $1000 \times .2111$; and the calculated value of χ^2 is 185.25. In this case we will be able to conclude that this sample is not in Hardy-Weinberg equilibrium. We have two degrees of freedom (nothing is estimated from this set of data), and so if we select a 5% level of significance, the χ^2 value from the table is 5.99. Since $X^2 = 185.25$, we can conclude that this group of 1000 friends is not in Hardy-Weinberg equilibrium with the Liège proportions. But if they remain on this isolated island and continue to mate randomly, we can predict that they will attain their own Hardy-Weinberg equilibrium values. In Problem 7, you will be asked to verify that the Hardy-Weinberg equilibrium values for the offspring of this island population will be 16% MM, 48% MN, and 36% NN. We should comment that the reason for having an example in which a group of friends and relatives go to an island was to point out that since this group does not necessarily constitute a random sample, random mating did not occur among their ancestors, so the Hardy-Weinberg equilibrium does not apply.

If we can assume that a population is in Hardy-Weinberg equilibrium, we can expect homozygous dominant : heterozygous : homozygous recessive to occur in a ratio of $p^2 : 2pq : q^2$. In Fig. 16-4 we summarize the Hardy-Weinberg equilibrium values for all values of p. Figure 16-4 holds for all traits that are in Hardy-Weinberg equilibrium.

One additional comment should be made before we consider various deviations from the simplest ideal case. If we can assume that a population is in Hardy-Weinberg equilibrium, and that the genotypes may not be phenotypically distinguishable (that is, homozygous dominant is indistinguishable from the heterozygote), we can still estimate p (the proportion of the dominant gene) and q (the proportion of the

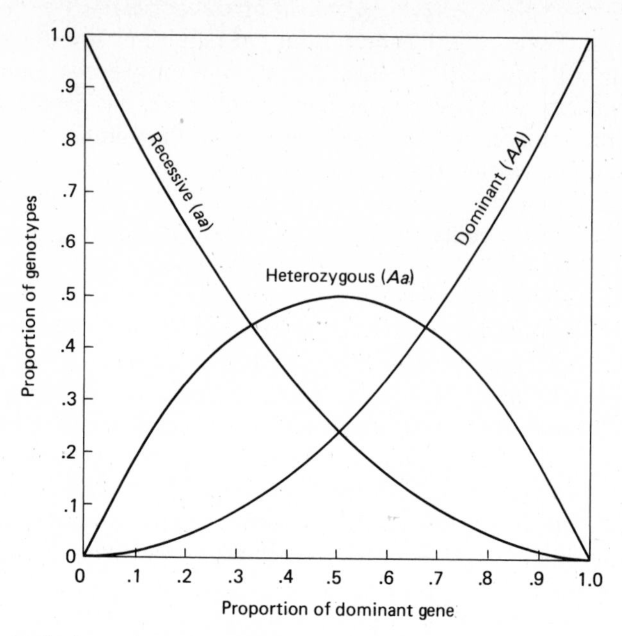

Figure 16-4. Proportions of homozygous dominants, heterozygotes, and homozygous recessives for various proportions of the dominant gene.

recessive gene). (In the examples considered up to this point, all genotypes are phenotypically distinguishable.)

If the homozygous dominants and the heterozygotes are phenotypically alike, and the population is in Hardy-Weinberg equilibrium, we will know that

$$\text{Prob[homozygous dominant and heterozygote]} = p^2 + 2pq$$

and

$$\text{Prob[homozygous recessive]} = q^2$$

How can we estimate p and q? We will use the following example to make our point.

Example: Vestigial Wings in *Drosophila* is a recessive trait. In a certain population, the proportion of adults with vestigial wings is 10^{-4}, and this proportion is fixed from generation to generation. (This would suggest that the population is in Hardy-Weinberg equilibrium.) Predict the proportion of heterozygotes (carriers) that can be expected in the next generation. Suppose that we let v be the recessive allele that controls Vestigial Wings and V be the dominant allele No Vestigial Wings. Since the population is in Hardy-Weinberg equilibrium, we know that

$$\text{Prob}[VV] = p^2, \quad \text{Prob}[Vv] = 2pq, \quad \text{Prob}[vv] = q^2$$

We also know that the proportion of fruit flies with vestigial wings is $10^{-4} = .0001$, therefore the proportion of fruit flies with no vestigial wings is .9999. Symbolically we can write

$$\text{Prob[Vestigial Wings]} = \text{Prob}[vv] = .0001 = q^2$$
$$\text{Prob[No Vestigial Wings]} = \text{Prob}[VV \text{ or } Vv] = .9999 = p^2 + 2pq$$

It should be evident that the equation

$$p^2 + 2pq = .9999$$

may be messy to work with. But if we use the equation $q^2 = .0001$, we can easily solve for q by taking the square root. That is,

$$q = \sqrt{q^2} = \sqrt{.0001}$$

and so

$$q = .01$$

Then $p = 1 - q = .99$. Thus, the proportion of heterozygote (carriers of the trait Vestigial Wings) fruit flies is 2pq, or $2(.99)(.01) = .0198$.

It is interesting to note that a trait such as Vestigial Wings that could be considered relatively rare (1 in 10,000) persists in the population and does not vanish. Intuitively one might be inclined to believe that relatively rare traits such as Vestigial Wings, or Albinism, would disappear from the population. The Hardy-Weinberg law however, suggests that if all its assumptions are satisfied, not only does the trait persist, but it persists at essentially the same rate.

The Hardy-Weinberg law has given us some very interesting insights into how traits are transmitted in populations. In the rest of this chapter we will consider the effects on the Hardy-Weinberg equilibrium of the relaxation or modification of various of its underlying assumptions. The modifications in the following two sections are extensions of simple autosomal (one pair of alleles) inheritance. The modifications in later sections will be formulated to account for realistic exceptions to the assumptions underlying the Hardy-Weinberg equilibrium.

Sex-linked Alleles

What is the long-run behavior of traits that are sex-linked? Is there some sort of Hardy-Weinberg equilibrium for traits such as Color-Blindness and Hemophilia? Such an equilibrium is in fact attained, but it can take much longer for the population to reach the equilibrium values. The difference is that the proportions will vary both in males and in females until they start to stabilize at about the same value.

To explain this phenomenon we will consider an X-linked trait which we will denote by A and a. To make the distinctions between X and Y chromosomes clear, we will use the notation X^A and X^a and Y. Let p_f^n denote the proportion of A genes in the female population in the nth generation and let q_f^n be $1 - p_f^n$, and similarly p_m^n and q_m^n for the male population. If we consider these populations to mate randomly, we can try to determine the Hardy-Weinberg equilibrium. Using tree diagrams and treating male and female offspring separately, we find that

$$p_m^n = p_f^{n-1}$$

and

$$p_f^n = \frac{p_m^{n-1} + p_f^{n-1}}{2}$$

Let us consider some numbers to see what the implications are of the preceding formula. We will start with a hypothetical population in which $p_m^0 = .9$ and $p_f^0 = .3$. The results are given in Table 16-3. We see that the proportion of X^A *alleles* in both

Table 16-3 Simulation of Inheritance of Sex-linked Traits. ($\bar{p}$ is the average proportion of *A* alleles among the men and women, that is, the overall proportion of *A* alleles in both males and females is $\bar{p} = \frac{2}{3}p_f + \frac{1}{3}p_m$.)

			PROPORTION ♀			PROPORTION ♂		
Generation	X^A	X^a	X^AX^A	X^AX^a	X^aX^a	X^AY	X^aY	$X^A_{\bar{p}}$
0	.3	.7	.09	.42	.49	.9	.1	.5
1	.6	.4	.36	.48	.16	.3	.7	.5
2	.45	.55	.2025	.495	.3025	.6	.4	.5
3	.525	.475	.275625	.49875	.225625	.45	.55	.5
4	.4875	.5125	.237656	.499687	.262656	.525	.475	.5
5	.50625	.49375	.256288	.499922	.243790	.4875	.5125	.5
6	.496875	.503125	.246884	.499980	.253135	.50625	.49375	.5
7	.501562	.498438	.251564	.499995	.248441	.496875	.503125	.5
8	.499218	.500782	.249219	.499999	.250782	.501562	.498438	.5
9	.500390	.499610	.250390	.499999	.249610	.499218	.500782	.5
10	.499800	.500200	.249804	.499999	.250196	.500390	.499610	.5
11	.500096	.499904	.250096	.499999	.249904	.499800	.500200	.5
12	.499950	.500050	.249950	.499999	.250050	.500096	.499904	.5
13	.500023	.499977	.250023	.499999	.249977	.499950	.500050	.5
14	.499986	.500014	.249986	.499999	.250014	.500023	.499977	.5
15	.500004	.499996	.250004	.499999	.249996	.499986	.500014	.5
16	.499995	.500005	.249995	.499999	.250005	.500004	.499996	.5
⋮	⋮	⋮	⋮	⋮	⋮	⋮	⋮	
∞	.5	.5	.25	.50	.25	.5	.5	.5

males and females is approaching .5000; that is, an equilibrium value is achieved in the long run. In our example, it is achieved for all practical purposes after 10 generations. The results are graphed in Fig. 16-5.

One more interesting fact can be observed about $\bar{p}$, the overall proportion of X^A alleles in the population. If we think about it, we note that females will contribute twice as many X^A alleles as men. We find that there are really three sets of alleles, two from the women and one from the men. Since there is the same number of both men and women, the overall proportion of X^A alleles in the population can be calculated

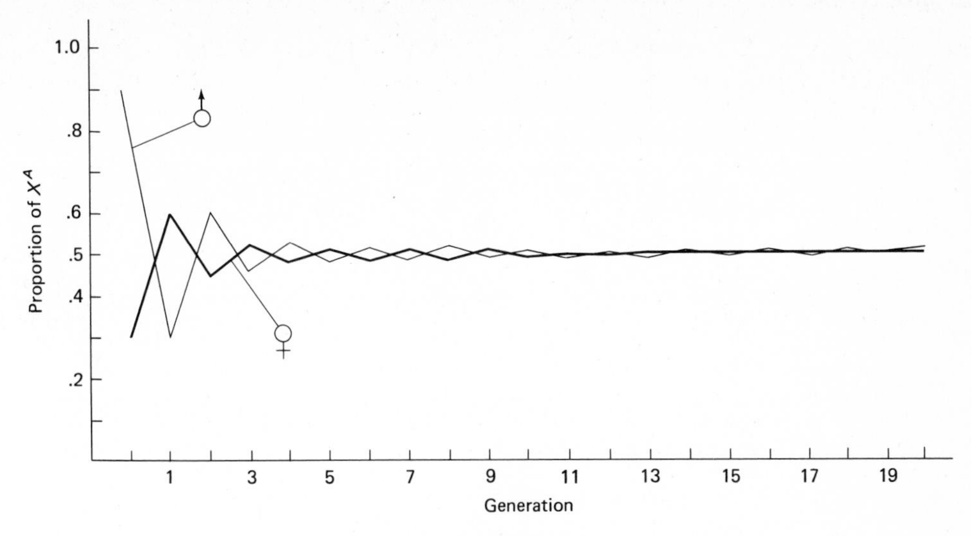

Figure 16-5. Equilibrium values for sex-linked traits.

by weighting the p_f twice and p_m only once. Therefore we find that

$$\bar{p} = \tfrac{2}{3}p_f + \tfrac{1}{3}p_m$$

How about $\bar{p}'$, the overall proportion of X^A alleles in the next generation? It follows that

$$\bar{p}' = \tfrac{2}{3}p'_f + \tfrac{1}{3}p'_m$$

But since $p'_m = p_f$, and $p'_f = \tfrac{1}{2}(p_f + p_m)$, we find that

$$\begin{aligned} \bar{p}' &= \tfrac{2}{3}[\tfrac{1}{2}(p_f + p_m)] + \tfrac{1}{3}p_f \\ &= \tfrac{1}{3}p_f + \tfrac{1}{3}p_m + \tfrac{1}{3}p_f \\ &= \tfrac{2}{3}p_f + \tfrac{1}{3}p_m \\ &= \bar{p} \end{aligned}$$

By similar reasoning we can show that

$$\bar{p}^n = \bar{p}$$

for all n (this will be verified in Problem 9); that is, the overall proportion of X^A alleles remains constant throughout. This fact was verified in Table 16-3 by actually calculating the values of $\bar{p}$ for each generation.

It follows that at equilibrium the ratio of the recessive traits in males and females is $\bar{q} : \bar{q}^2$ (this follows from assumption 2). Since $\bar{q}$ is a number less than 1, it is clear that the proportion of women with the X-linked trait will be much smaller than that of men.

Multiple Alleles

What can we say about the long-run behavior of traits for which there are multiple alleles at one gene location (locus) in only one pair of chromosomes? Is there an equilibrium state that is maintained if we satisfy the conditions of the Hardy–Weinberg law?

One of the better-known examples of such a trait is the ABO blood groups, for which there are three genes at the same locus on the same chromosome. The genes are: I^A for Antigen A, I^B for Antigen B, and i recessive to both I^A and I^B. As we discussed in Chapter 4 and summarized in Table 4-1, genotypes I^AI^A and I^Ai are phenotypically A; genotypes I^BI^B and I^Bi are phenotypically B; genotype I^AI^B is phenotypically AB; and genotype ii is phenotypically O. What is the Hardy-Weinberg equilibrium if we satisfy the assumptions of random mating? We will for simplicity assume the proportions of each allele to be the same in males and females. Let p denote the proportion of I^A genes, q the proportion of I^B genes, and r the proportion of i genes, such that $p + q + r = 1$. (This is a legitimate assumption since we assume that these are all possible alleles. In fact there is more than one type of I^A allele, but we will pool all the I^A alleles into one group.) Using these proportions we can once again use a tree diagram such as that in Fig. 16-6 to see what happens to the proportions of each blood type. Summarizing Fig. 16-6, we find that:

$$\text{Prob[A Blood Type]} = p^2 + 2pr$$
$$\text{Prob[B Blood Type]} = q^2 + 2qr$$
$$\text{Prob[AB Blood Type]} = 2pq$$
$$\text{Prob[O Blood Type]} = r^2$$

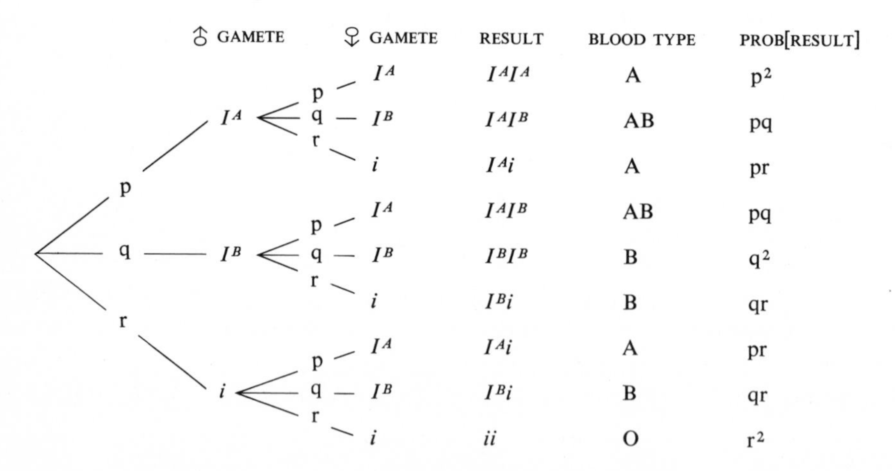

Figure 16-6. Tree diagram for blood types.

What are the expected proportions of each blood type in the next generation? We must once again estimate the proportion of each allele in the population. That is, we want to estimate p′, q′, and r′, the proportion in the F_1 offspring. We use an approach that is quite similar to the technique we used for simple autosomal traits.

$$\text{Prob}\begin{bmatrix} I^A \text{ alleles in} \\ F_1 \text{ offspring} \end{bmatrix} = p' = \underbrace{\text{Prob}[I^AI^A]}_{\text{all the alleles}} + \underbrace{\tfrac{1}{2}\text{Prob}[I^Ai]}_{\frac{1}{2} \text{ the alleles are } I^A} + \underbrace{\tfrac{1}{2}\text{Prob}[I^AI^B]}_{\frac{1}{2} \text{ the alleles are } I^A}$$

$$= p^2 + \tfrac{1}{2}(pr + pr) + \tfrac{1}{2}(pq + pq)$$

$$= p^2 + pr + pq$$

$$= p(p + r + q)$$

$$= p$$

That is, the proportion of I^A alleles in the F_1 offspring is the same as the proportion in the parental generation. A similar derivation shows that $q' = \text{Prob}[I^B \text{ alleles}] = q$ and that $r' = \text{Prob}[i \text{ alleles}] = r$. (This will be verified in Problem 8.) Hence, $p' = p$, $q' = q$, and $r' = r$, implying that an equilibrium has been attained. If the conditions for Hardy-Weinberg are satisfied, p, q, and r will persist and therefore the ratio A : B : AB : O, as given by $p^2 + 2pr : q^2 + 2qr : 2pq : r^2$, will hold for future generations.[3]

By similar arguments we can show that the Hardy-Weinberg equilibrium can be established for more than three alleles at one locus on a pair of chromosomes. (See Problem 9.)

So far we have discussed Hardy-Weinberg equilibrium under ideal conditions, in which, a crucial assumption of the law, that of random mating, has been maintained. In the following sections we will consider what happens in populations if mating is not random. We will first discuss the effects of inbreeding and assortative mating on the inheritance of traits in populations.

NONRANDOM MATING

The basic assumption of random mating suggests that mates are selected in a manner analogous to putting all names of every possible individual of the opposite sex into a giant drum, mixing it, and then picking one name out. This implies that mates are selected without any thought given to phenotypic resemblance, personality, ethnic origins, relatedness, or physical separation. It is obvious that in reality sexually reproducing organisms form mating pairs by violating in many ways the basic Hardy-Weinberg assumption of random mating. There appears to be evidence that even among houseflies, marine fishes, and planktonic crustaceans, where mates might realistically be expected to be unrelated and unassociated, there may be an association between mates in terms of common origin, ecological preference, and morphological resemblance. In fact it is likely that proximity between mates increases the chance

[3]To estimate p, q, and r from actual data is more complicated. The interested reader should consult Spiess (1977).

that mates are related by descent, causes some preference over more distant individuals, or results in preference for either morphological or behavioral tendencies. It is abundantly clear that human beings very rarely if ever select their mates in a completely random manner. (Even mail-order brides had to meet certain specifications.) How, then, does this lack of randomness affect the gene and genotypic pool of the population? In the rest of this chapter we discuss various effects of nonrandom mating on the genetics of populations.

Inbreeding

Inbreeding is the mating of related individuals; it clearly cannot be random mating. The degree of inbreeding and the effects are controlled by how closely related the mates are. Many plants and a few animal species actually practice the most extreme form of inbreeding—self-fertilization. Genetically less extreme cases of inbreeding can occur between brother and sister, father and daughter, mother and son, cousins, uncles, nieces, and so forth. Matings between relatives (not self-fertilization) are often referred to as *consanguineous matings*. We will now discuss the effects of *selfing* and consanguineous matings.

Self-fertilization

Many crop plants such as wheat, rice, barley, oats, peanuts, soybeans, tobacco, tomatoes, peas, beans, cotton, and citrus fruits are naturally self-pollinating. This type of inbreeding is clearly the most extreme form. It is interesting to follow what happens to a population of plants which is self-fertilizing. Let us assume that we start with 4096 hybrid *Bb* plants, and at each generation a plant will fertilize itself and thereby produce only one offspring plant. The offspring of the heterozygous *Bb* plants will always be *BB* : *Bb* : *bb* in the ratio of 1 : 2 : 1, while the homozygous *BB* and *bb* plants will produce all *BB* and *bb* offspring, respectively. The results of reproduction in such a population are given in Table 16-4, where we see very quickly that the genotypic ratios from self-fertilization are dramatically different than those predicted under random mating and the Hardy-Weinberg equilibrium. With repeated self-fertilizations the number of heterozygotes decreases very quickly and the number of homozygous dominants and recessives increases. In fact the number of heterozygotes approaches zero very quickly, since the amount of heterozygosity decreases by one-half each generation. In general, by the nth generation only $(\frac{1}{2})^n$ of the heterozygotes will remain. Table 16-5 summarizes the effects of self-fertilization in terms of percentages.

Self-fertilization can essentially change a heterozygous population into one containing almost all homozygotes (dominant and recessive); that is, self-fertilization leads to pure lines. Such pure lines can be very valuable in agricultural genetics and practical farming. The effects of homozygosity can be either good or bad. It is possible that deleterious recessive genes will be brought to expression (such genes are normally concealed in the heterozygous state), and hence have a bad effect on the population. On the other hand, if deleterious genes are not present, it is possible to develop inbred

Table 16-4 Self-fertilizing Population

Generation	*BB*	*Bb*	*bb*
0		4096	
1	1024	2048	1024
2	1024 + 512		512 + 1024
	1536	1024	1536
3	1536 + 256		256 + 1536
	1792	512	1792
4	1792 + 128		128 + 1792
	1920	256	1920
5	1920 + 64		64 + 1920
	1984	128	1984
6	1984 + 32		32 + 1984
	2016	64	2016
7	2016 + 16		16 + 2016
	2032	32	2032
8	2032 + 8		8 + 2032
	2040	16	2040
9	2040 + 4		4 + 2040
	2044	8	2044
10	2044 + 2		2 + 2044
	2046	4	2046

Table 16-5 Percentages of Each Genotype in a Self-fertilizing Population

Generation	*BB* (%)	*Bb* (%)	*bb* (%)	*Total percent of homozygotes*
0	–	100	–	0
1	25	50	25	50
2	37.5	25	37.5	75
3	43.75	12.5	43.75	87.5
4	46.875	6.25	46.875	93.75
5	48.4375	3.125	48.4375	96.875
6	49.21875	1.5625	49.21875	98.4375

strains with good qualities. The genotypic ratios presented in Tables 16-4 and 16-5 are for the ideal cases of self-fertilization with no interference. Various modifications (selection, mutation, and so forth) can be introduced into the self-fertilization process which will change the rates at which homozygous strains are achieved, but we will not discuss them. Later in this chapter we will discuss these types of modifications for a population that is not inbred.

Sibs and More Distant Relatives

What are the effects of matings between relatives? In fact, some very simple arguments using the multiplication rule (see Chapter 3) can be used to demonstrate that we ourselves are the result of inbreeding. If we assume that each of our ancestors enters our pedigree only once, then 20 to 30 generations ago, each pair of us must have a common relative and so we are all products of a certain amount of inbreeding. However, inbreeding involving such distant relatives can hardly cause an appreciable effect.

We really want to discuss the effects of matings between close relatives: brother-sister, father-daughter, mother-son, cousin-cousin, uncle-niece, and so forth. Is inbreeding of this type harmful? In the United States all states have laws prohibiting marriage of relatives closer than first cousins, and some states (such as Minnesota) forbid marriage if the individuals are closer than second cousins. Why the great concern over inbreeding? The reason is probably a mixture of cultural and biological factors. In the latter case the fear is the potential for bringing out deleterious recessive genes. Before we discuss various consanguineous matings and their increased risks of producing offspring with undesirable homozygous traits, we should mention that one famous consanguineous marriage produced a woman who was considered to represent the ideal of feminine beauty and intelligence. This woman was the legendary Cleopatra. Cleopatra was the result of a long line of inbreeding (since royalty couldn't possibly consider marrying commoners), which in this case seemed not to produce deleterious effects.

Brother-sister mating. As with self-fertilization the effect of inbreeding is to increase the percentage of homozygotes. To demonstrate this, let us consider various matings and determine the probability that the particular union will produce a homozygous offspring. We start with the mating of a brother and sister. A pedigree for such a mating is given in Fig. 16-7. We will assume that in the grandparents' generation the female has a genotype composed of alleles A_1 and A_2 and the male has a genotype composed of alleles A_3 and A_4. (We use different subscripts for genes in different grandparents for ease of distinction.) We want to know the probability that the offspring of the brother-sister mating will be homozygous for a certain trait given the following assumption: Neither of the parents and none of the grandparents is homozygous for that trait and exactly one of the grandparents is a carrier of that gene. (We will call this Assumption A.) In particular we want to know the probability that the offspring will be A_1A_1, A_2A_2, A_3A_3, or A_4A_4. (Some geneticists ask a slightly different variation of the same question. Given that one partner in a mating is a carrier for a particular trait, what is the probability that a related mate is also a carrier? This probability is known as the *coefficient of consanguinity*. Since this question is

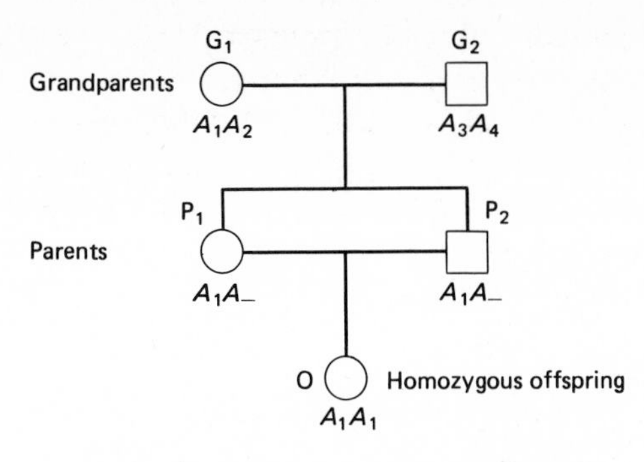

Figure 16-7. Pedigree for brother-sister mating.

similar to ours, we will discuss only our version.[4]) We will first determine the probability of the offspring being A_1A_1. Our assumptions about the pedigree say that A_1 will differ from A_2, from A_3, and from A_4. From Fig. 16-7 we can see that the probability that parent P_1 is A_1A_- and that parent P_2 is A_1A_- is $\frac{1}{2}$ in both cases, where A_- is either A_3 or A_4. If the trait of interest is albinism, our assumptions say that G_1 is a carrier, that P_1 and P_2 are carriers, and that G_2 is neither a carrier nor an albino. Since the matings for each child are independent (see Chapter 3) it follows that

$$\begin{aligned}&\text{Prob[P}_1 \text{ is } A_1A_- \text{ and P}_2 \text{ is } A_1A_-]\\&\quad = \text{Prob[P}_1 \text{ is } A_1A_-]\text{Prob[P}_2 \text{ is } A_1A_-]\\&\quad = (\tfrac{1}{2})(\tfrac{1}{2})\\&\quad = \tfrac{1}{4}\end{aligned}$$

The next step is to determine the probability that the offspring O is actually A_1A_1. If the parents are both A_1A_-, we know that the probability of an A_1A_1 offspring in an $A_1A_- \times A_1A_-$ is $\frac{1}{4}$. So if we combine these events and note that they are all independent, we find that we can multiply our probabilities and get

$$\begin{aligned}&\text{Prob}[A_1A_1 \text{ offspring given our assumptions in Fig. 16-7}]\\&\quad = \text{Prob}[A_1A_1 \text{ offspring given P}_1 \text{ is } A_1A_- \text{ and P}_2 \text{ is } A_1A_-]\\&\quad = \text{Prob}[A_1A_1 \text{ offspring and P}_1 \text{ is } A_1A_- \text{ and P}_2 \text{ is } A_1A_-]\\&\qquad \times \text{Prob[P}_1 \text{ is } A_1A_- \text{ and P}_2 \text{ is } A_1A_-]\\&\quad = (\tfrac{1}{4})(\tfrac{1}{4})\\&\quad = (\tfrac{1}{16})\end{aligned}$$

[4]However, the reader should note that the literature does not always correctly distinguish between these two questions.

Similarly we find that given our pedigree

$$\text{Prob}[A_2A_2 \text{ offspring}] = \tfrac{1}{16}$$
$$\text{Prob}[A_3A_3 \text{ offspring}] = \tfrac{1}{16}$$

and

$$\text{Prob}[A_4A_4 \text{ offspring}] = \tfrac{1}{16}$$

Next suppose that A_1, A_2, A_3, and A_4 are all distinct. The probability of producing a homozygous offspring is simply the probability of getting A_1A_1, or A_2A_2, or A_3A_3, or A_4A_4. This probability is called the *coefficient of inbreeding*. Since A_1, A_2, A_3, and A_4 are distinct, the 4 events A_1A_1, A_2A_2, A_3A_3, and A_4A_4 are mutually exclusive (an offspring can only be one of the four outcomes, either A_1A_1 or A_2A_2 but never both—see Chapter 3). Thus, to compute the coefficient of inbreeding we add the probabilities of each individual event. Therefore we get

$$\begin{aligned}&\text{Prob}[A_1A_1 \text{ or } A_2A_2 \text{ or } A_3A_3 \text{ or } A_4A_4]\\ &\quad = \text{Prob}[A_1A_1] + \text{Prob}[A_2A_2] + \text{Prob}[A_3A_3] + \text{Prob}[A_4A_4]\end{aligned}$$

Since A_1, A_2, A_3, and A_4 are distinct, and Assumption A is satisfied, by our earlier computation this sum is given by $\frac{1}{16} + \frac{1}{16} + \frac{1}{16} + \frac{1}{16} = 4(\frac{1}{16}) = \frac{1}{4}$.

As an example, let us apply our computations to albinism, and determine the probability of producing an albino child from a random mating and from a mating of brother and sister. For comparison's sake we will first discuss a population that is mating randomly and that has achieved the Hardy-Weinberg equilibrium and ascertain the probability that a randomly mating couple who are not albinos will have an albino child. To do this we must determine the probability that the parents are carriers for albinism and then the probability that the child of such a mating will be an albino. We will use the fact that the proportion of albinos is known to be approximately $\frac{1}{20,000}$ to determine the probability that a person is a carrier for albinism. Since albinism is a recessive trait, we utilize the Hardy-Weinberg equilibrium and find that q^2, the proportion of recessives, is equal to $\frac{1}{20,000}$. Therefore,

$$q^2 = \frac{1}{20,000}$$
$$q = \sqrt{\frac{1}{20,000}}$$
$$= .00707$$

and

$$p = .9929289$$

From the Hardy-Weinberg equilibrium we know that the proportion of heterozygotes (carriers) is 2pq, and therefore equal to $.014042 \cong \frac{1}{70}$. If we assume random mating, the probability that a couple will both be carriers is given by

$$\begin{aligned}
&\text{Prob[} P_1 \text{ is a carrier for albinism and } P_2 \text{ is a carrier for albinism]}\\
&\quad = \text{Prob[} P_1 \text{ is a carrier]Prob[} P_2 \text{ is a carrier]}\\
&\quad\quad \text{(since we have independent events—see Chapter 3)}\\
&\quad = (\tfrac{1}{70})(\tfrac{1}{70})\\
&\quad = .000204
\end{aligned}$$

The probability that this couple will have an albino offspring is $\frac{1}{4}$. Putting this all together we find that the probability that two nonalbino partners selected at random will have an albino child is

$$\begin{aligned}
&\text{Prob[} P_1 \text{ carrier and } P_2 \text{ carrier and have an albino child]}\\
&\quad = \text{Prob[} P_1 \text{ and } P_2 \text{ carriers]Prob[albino child given } P_1 \text{ and } P_2 \text{ carriers]}\\
&\quad = (\tfrac{1}{70})(\tfrac{1}{70})(\tfrac{1}{4})\\
&\quad = .000051
\end{aligned}$$

Next let us determine the probability that the mating of a brother and sister will result in an albino offspring. We will consider first the case where G_1 is a carrier for albinism and G_2 is neither a carrier nor an albino. This implies that P_1 and P_2 cannot be albinos. Hence if $A_1 = a$, $A_2 = A$, $A_3 = A$, $A_4 = A$, our original pedigree still holds. We conclude that in a brother-sister mating, the probability is $\frac{1}{16}$ that they will produce an offspring that is aa or A_1A_1. If we can determine the probability that a particular grandparent, G_1, is a carrier and that a second grandparent is neither a carrier nor an albino, we can determine the desired probability. We have already shown that the probability that a person is a carrier for albinism is approximately $\frac{1}{70}$. The probability that G_2 is not a carrier and not an albino is approximately $\frac{69}{70}$. Using the Hardy-Weinberg equilibrium, we know that the probability of homozygous dominant (nonalbino and noncarrier) is $p^2 = .9859078$, which is approximately $\frac{69}{70}$. Therefore the probability that a couple will have an albino child where G_1 is a carrier and G_2 is not a carrier is

$$\begin{aligned}
&\text{Prob}\begin{bmatrix} G_1 \text{ carrier and } G_2 \text{ not a carrier and not an albino and an} \\ \text{offspring of a brother-sister mating is homozygous (recessive) } A_1A_1 \end{bmatrix}\\
&\quad = \text{Prob[} G_1 \text{ carrier and } G_2 \text{ not a carrier and not albino]}\\
&\quad\quad \times \text{Prob}\begin{bmatrix} \text{offspring of a brother-sister mating is } A_1A_1 \text{ given } G_1 \\ \text{is a carrier and } G_2 \text{ is not a carrier and not an albino} \end{bmatrix}\\
&\quad = \text{Prob[}G_1 \text{ carrier]Prob[}G_2 \text{ not a carrier and not albino]}\\
&\quad\quad \times \text{Prob}\begin{bmatrix} \text{offspring of a brother-sister mating is } A_1A_1 \text{ given } G_1 \\ \text{is a carrier and } G_2 \text{ is not a carrier and not an albino} \end{bmatrix}\\
&\quad = (\tfrac{1}{70})(\tfrac{69}{70})(\tfrac{1}{16})\\
&\quad = .00088
\end{aligned}$$

This is not the complete answer since it is possible for G_2 rather than G_1 to be the carrier. It is clear that the probability is also .00088 if G_2 is the carrier and G_1 is neither a carrier nor an albino. Therefore the total probability for an albino offspring if two sibs mate and if one grandparent is a carrier and the other neither a carrier nor an albino is $.00088 + .00088 = .00176$. (The total probability of an albino offspring if two nonalbino parents mate, .00185, is higher than .00176, since one can obtain albino offspring with a different pedigree. This possibility is not included when finding the coefficient of inbreeding.)

If we compare the two results, we find that the probability of producing an albino child is 34.5 times greater in the mating of a brother and sister. Inbreeding dramatically increases the number of homozygotes. (If the trait is deleterious, the harmful effects of such close inbreeding are obvious.)

First-cousin matings. What is the effect on offspring if first cousins mate? Consider Fig. 16-8, which gives the pedigree of a first-cousin mating. For simplicity we will assume that P_1 and P_4 are not related and are not carriers of the trait of interest. Let us assume that the pedigree is such that G_1 is a carrier and G_2 is not, and that of G_2, P_1, P_2, P_3, P_4, C_1, and C_2, none is homozygous for A_1. Let us determine the probability that the offspring in the first-cousin mating will be homozygous A_1A_1. (The coefficient of inbreeding will be 4 times the probability, since it arises by considering the 4 cases of A_1A_1, A_2A_2, A_3A_3, or A_4A_4.) The probability that P_2 will be A_1A_- is $\frac{1}{2}$ (since it results from the mating $A_1A_2 \times A_3A_4$). Note that A^- is A_2, A_3, or A_4, and A_- is A_3 or A_4. Similarly, since by our assumption P_1 is A^-A^-, the probability that C_1 is A_1A^- if P_2 is A_1A_- is $\frac{1}{2}$. Therefore the probability that C_1 is A_1A^- is the probability that P_2 is A_1A^- times the probability that the offspring of $P_1 \times P_2$ is A_1A^-. So

$$\text{Prob[} C_1 \text{ is } A_1A^- \text{ given the pedigree]} = (\tfrac{1}{2})(\tfrac{1}{2}) = \tfrac{1}{4}$$

Similarly

$$\text{Prob[} C_2 \text{ is } A_1A^- \text{ given the pedigree]} = (\tfrac{1}{2})(\tfrac{1}{2}) = \tfrac{1}{4}$$

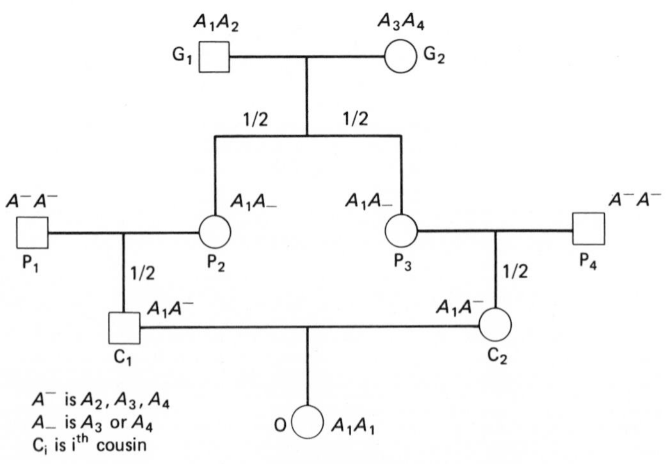

Figure 16-8. Pedigree of first-cousin mating.

Since events C_1 is A_1A^- and C_2 is A_1A^- (given the pedigree) are independent,

$$\begin{aligned}&\text{Prob[} C_1 \text{ is } A_1A^- \text{ and } C_2 \text{ is } A_1A^- \text{ given the pedigree]}\\ &\quad = \text{Prob[} C_1 \text{ is } A_1A^- \text{ given the pedigree]}\\ &\qquad \times \text{Prob[} C_2 \text{ is } A_1A^- \text{ given the pedigree]}\\ &\quad = (\tfrac{1}{4})(\tfrac{1}{4})\\ &\quad = \tfrac{1}{16}\end{aligned}$$

One more step. We know that the probability that the first cousins are both carriers is $\frac{1}{16}$. Now we must find the probability that their offspring will be A_1A_1. This is quite easy since we are looking at the mating $A_1A^- \times A_1A^-$. The probability that the offspring of this mating is A_1A_1 is $\frac{1}{4}$; therefore the probability that an offspring of a first-cousin mating will be A_1A_1 is just the product of the probability that the cousins are both carriers for A_1 given the pedigree and the probability of an A_1A_1 child if the parents are carriers. So we get $(\frac{1}{4})(\frac{1}{4})(\frac{1}{4}) = \frac{1}{64}$. (The coefficient of inbreeding, the probability of being homozygous, is $\frac{1}{64} + \frac{1}{64} + \frac{1}{64} + \frac{1}{64} = \frac{4}{64} = \frac{1}{16}$.)

What is the probability that a first-cousin mating will result in an albino child? This question can be answered by using an approach similar to that for the mating between sibs. Actually we really want to determine the probability that either grandparent, G_1 or G_2, will be a carrier for albinism, that the other is neither a carrier nor an albino,[5] and that the mating of first cousins will produce a homozygous recessive offspring. We already know the probability of each part. The probability that G_1 is a carrier for albinism is $\frac{1}{70}$. The probability that G_2 is neither a carrier nor an albino is $\frac{69}{70}$. Therefore

$$\begin{aligned}&\text{Prob}\begin{bmatrix}G_1 \text{ is a carrier and } G_2 \text{ is neither a carrier nor}\\ \text{an albino and first-cousin mating results in albino}\end{bmatrix}\\ &\quad = \text{Prob[} G_1 \text{ is a carrier and } G_2 \text{ is neither a carrier nor an albino]}\\ &\qquad \times \text{Prob[first-cousin mating results in albino given the pedigree in Fig. 16-8]}\\ &\quad = \text{Prob[} G_1 \text{ is a carrier]} \times \text{Prob[} G_2 \text{ is neither a carrier nor an albino]}\\ &\qquad \times \text{Prob[first-cousin mating results in albino given the pedigree in Fig. 16-8]}\\ &\quad = (\tfrac{1}{70})(\tfrac{69}{70})(\tfrac{1}{64})\\ &\quad = .00022\end{aligned}$$

Therefore the probability of an albino offspring if first cousins mate is

$$\left(\frac{1}{70}\right)\left(\frac{69}{70}\right)\left(\frac{1}{64}\right) + \left(\frac{1}{70}\right)\left(\frac{69}{70}\right)\left(\frac{1}{64}\right) = \frac{2(69)}{(70)(70)(64)} = \frac{138}{313{,}600}$$
$$.00022 + .00022 = .00044$$

We summarize the results in Table 16-6:

[5]The rest of the pedigree follows from this.

Table 16-6 Probability of Albino Offspring for Different Pedigrees

Mating	*Prob[albino]*	*Risk compared to random mating*
Random mating	.000051	1
Sibs	.00176	34.5
First cousins	.00044	8.6

Once again we see that inbreeding tends to increase the chances of homozygosity. The degree to which it increases them depends on the relationship between the mates.

Coefficient of Inbreeding: Summary

The coefficient of inbreeding for various relationships is given in Table 16-7, and gives a feeling for the effect of inbreeding. That is, it shows relatively how much the chances of having a homozygous offspring are increased by various degrees of relationship between mates.

Table 16-7 Coefficient of Inbreeding

Mating	*Coefficient of inbreeding*
Homozygous population	1
Brother-sister	$\frac{1}{4}$
Father-daughter	$\frac{1}{4}$
Uncle-niece	$\frac{1}{8}$
Double first cousins	$\frac{1}{8}$
First cousins	$\frac{1}{16}$
Half first cousins	$\frac{1}{32}$
First cousins once removed	$\frac{1}{32}$
Second cousins	$\frac{1}{64}$
Second cousins once removed	$\frac{1}{128}$
Third cousins	$\frac{1}{256}$

There are some studies that tend to support the hypothesis that consanguineous matings do result in more deleterious effects. The data for two such studies are presented in Tables 16-8 and 16-9.

We see that there is some evidence to suggest caution in consanguineous matings. The effects of consanguineous matings can be felt for more than one generation. But we know that primarily the effect is to increase homozygotes. The rate at which this occurs depends on the degree of relationship between the mates. The long-term effect on the Hardy-Weinberg equilibrium has been studied and is discussed in great detail in Spiess (1977).

Assortative Mating

The last type of nonrandom mating we will discuss is *assortative mating*, which refers to matings based on phenotypic similarities or differences. *Positive assortative*

Table 16-8 Deaths Between 1 and 8 Years Among Offspring of Consanguineous Marriages in Hiroshima, Japan

Relationship of parents	*Number of births*	*Number of deaths*	*Percent of deaths*
First cousins	326	15	.0460
Half first cousins	101	3	.0297
Second cousins	139	3	.0218
Unrelated	544	8	.0147

Source: Anna C. Pai, 1974. *Foundations of Genetics*, McGraw-Hill, New York. Schull, W. J., 1958. "Empirical Risks in Consanguineous Marriages: Sex Ratio, Malformation and Viability." *Am. J.* Human Gen. 10: 294–343.

Table 16-9 Increased Risk of Genetic Defect with Cousin Marriages (Data from Hiroshima and Nagasaki)

	Frequency for unrelated parents	*Frequency for cousin marriages*	*Percent increase*
Congenital malformation	.011	.016	45
Stillbirths	.025	.031	24
Infant deaths	.023	.031	35

Adapted from I. J. Herskowitz, 1965. *Genetics*, 2nd ed., p. 208. Little, Brown, Boston.

mating is the tendency for individuals similar in some phenotypic trait to mate. *Negative assortative mating* is the tendency for dissimilar individuals to mate. The most common example of positive assortative mating is that men and women tend to choose as mates individuals of about their own intelligence level. There is also some evidence that people tend to select mates of similar height. But on the other hand there is the common expression that "opposites attract."

Positive assortative mating tends to be somewhat like the process of self-fertilization, except that in assortative mating selection of mate is based on phenotype and in self-fertilization the selection is based on genotype. The tendency of positive assortative mating would be to increase the proportion of homozygotes if the trait was under genetic control. As in self-fertilization the proportion of each gene in the population remains constant but the Hardy-Weinberg equilibrium does not apply.

Negative assortative mating tends to increase the proportion of heterozygotes in the population.

Realistically speaking, the selection of mates is probably a combination of assortative mating and various other factors. One such obvious factor is geographical limitations which interfere with random selection of mates. Among animals, mate selection almost always requires physical proximity and encounters with mates, and therefore is largely limited by physical location. All these nonrandom aspects of mate selection limit the applicability of the Hardy-Weinberg law. (Although in smaller subgroups of a population the conditions for Hardy-Weinberg equilibrium might more nearly be satisfied.) Nevertheless the Hardy-Weinberg equilibrium does give us various insights into the long-range tendencies of populations.

So far we have discussed how nonrandom mating affects the Hardy-Weinberg equilibrium for populations. We have omitted a discussion of factors which will change gene frequencies, and thereby modify the Hardy-Weinberg equilibrium. In the remainder of this chapter we will discuss such factors, including the effects of mutation, migration, selection, and genetic drift on the genetics of populations.

FACTORS THAT CHANGE GENE FREQUENCIES

Four basic factors act to change gene pools and thereby modify the Hardy-Weinberg equilibrium, and also are the principal agents of biological evolution. These are

1. Mutation
2. Selection
3. Random genetic drift
4. Differential migration

When we stated the Hardy-Weinberg law, though the words were different, we stated or implied that there was no mutation and no selection, and that the population was large so there would be no genetic drift and no migration. That is, in the statement of the Hardy-Weinberg law we established the conditions that would perpetuate the status quo. There would be no external influences or choices that could in any way modify the gene pool.

In the following sections we will consider each of the four factors and their impact on the gene pool and the genetics of populations.

Mutation

In Chapter 15 we discussed how mutations occur and how they modify the gene pool. Now we want to consider the quantitative aspects of mutations. We will consider three cases:

1. Single occurrence of the mutation
2. Recurrent mutation
3. Reverse mutations

Single occurrence. For simplicity let us consider a population consisting of all *LL* individuals, and assume that one of the *L* genes mutates to ℓ. If this occurs the population will consist of all *LL* people except one person who will be $L\ell$. What can happen? If the $L\ell$ individual doesn't mate, the mutant ℓ gene is lost. If the $L\ell$ individual mates and has only one offspring, there are two possibilities: either they have a child who is *LL* (probability .5) or a child who is $L\ell$ (probability .5). Hence there is a 50% chance of the mutant ℓ gene being lost from the population and a 50% chance of the gene remaining in the population. If the couple has two children, the possible outcomes are:

0 $L\ell$ children	and	2 LL children
1 $L\ell$ child	and	1 LL child
2 $L\ell$ children	and	0 LL children

To determine the probabilities of each of the outcomes we have to recall the binomial distribution that we discussed in Chapter 3. We can verify that we have a binomial "experiment" by noting that

1. We have dichotomous outcomes.
2. In a single mating Prob[LL] $= \frac{1}{2}$ and Prob[$L\ell$] $= \frac{1}{2}$.
3. Prob[LL] and Prob[$L\ell$] are constant and don't change for each mating.
4. The genotype of each child is independent of the genotype of any other child.

Therefore we have a binomial "experiment" with two trials, that is, $n = 2$, and p (the probability of an $L\ell$ child) is $\frac{1}{2}$. So we find that

$$\text{Prob}[\ 2L\ell \text{ and } 0LL \text{ children}] = \binom{2}{2}\left(\frac{1}{2}\right)^2\left(\frac{1}{2}\right)^0 = \frac{1}{4}$$

$$\text{Prob}[\ 1L\ell \text{ and } 1LL \text{ child}] = \binom{2}{1}\left(\frac{1}{2}\right)^1\left(\frac{1}{2}\right)^1 = \frac{1}{2}$$

$$\text{Prob}[\ 0L\ell \text{ and } 2LL \text{ children}] = \binom{2}{0}\left(\frac{1}{2}\right)^0\left(\frac{1}{2}\right)^2 = \frac{1}{4}$$

More specifically, we find that the probability is $\frac{1}{4}$ that the mutant gene ℓ will die out and $\frac{3}{4}$ that it will not die out. But there is a 25% chance that there will be two mutant genes. Similar calculations can be made if the couple has more children. Using the binomial we get the values in Table 16-10 that give the probabilities of the mutant gene dying out or remaining in the population. We now see that a single mutant gene may increase or decrease in frequency or be lost from the population. Since the chances at each birth are reasonably high ($\frac{1}{2}$) that the mutant gene will be lost, the fate of a

Table 16-10 Probabilities for a Single Mutant Gene

Number of children	*Prob* [*Mutant gene dies out*]	*Prob* [*Mutant gene survives*]	*Number of possible mutant genes*
1	$\frac{1}{2}$	$\frac{1}{2}$	1
2	$\frac{1}{4}$	$\frac{3}{4}$	1 or 2
3	$\frac{1}{8}$	$\frac{7}{8}$	1, 2, or 3
4	$\frac{1}{16}$	$\frac{15}{16}$	1, 2, 3, or 4

single mutation is often to be eliminated from the population; that is, many mutant genes are lost by chance soon after arising.

Recurrent mutations. The next stage of complexity is to assume that we have a recurring mutation. The rate of mutation per generation varies with different genes (see Table 15-3), but for a particular gene the rate is reasonably constant from generation to generation if conditions remain fairly stable. Therefore we can consider what happens to the proportions of L and ℓ, the mutant gene, if the mutation rate remains constant. We let p_n denote the proportion of L genes in the population in the nth generation and q_n the proportion of ℓ mutant genes in the nth generation, with of course $p_n + q_n = 1$. Also let μ denote the rate of mutation from L to ℓ, and we will assume that μ is constant each generation. Using this notation we find that

$$q_{n+1} = q_n + \mu p_n \tag{1}$$

That is, the proportion of mutant genes in the $(n + 1)^{st}$ generation will equal the proportion in the nth generation plus the proportion that mutated during the nth generation.

Equation (1) provides some insight into the mutation process. We note that q_{n+1} will get larger if μ is greater than zero. Since μ is constant, it follows that the proportion of mutant ℓ genes will increase relatively rapidly when L genes are abundant (p_n will be large), but at a much slower rate when the L genes are rare.

Equation (1) can be rewritten in terms of the proportion of L genes in the population. That is,

$$\begin{aligned} p_{n+1} &= p_n - \mu p_n \\ &= p_n(1 - \mu) \end{aligned} \tag{2}$$

Equation (2) says also that

$$p_n = p_{n-1}(1 - \mu)$$

Substituting this back into (2), we get

$$\begin{aligned} p_{n+1} &= p_{n-1}(1 - \mu)(1 - \mu) \\ &= p_{n-1}(1 - \mu)^2 \end{aligned}$$

If we use the same relationship repeatedly, we will find that

$$p_{n+1} = p_0(1 - \mu)^{n+1} \tag{3}$$

We have related the proportion of L genes in the $(n + 1)$ generation to the original proportion and to the mutation rate. In fact we are assuming that $p_0 = 1$, since all the genes at the zero generation are L, and so

$$p_{n+1} = (1 - \mu)^{n+1}, \text{ and } q_{n+1} = 1 - p_{n+1} \tag{4}$$

If $\mu = 3 \times 10^{-5}$ is the mutation rate of normal genes to hemophilic genes, we can use equation (4) to determine the proportion of hemophilic genes that arise as a result of mutation in each generation (see Table 16-11). The results presented in

Table 16-11 Proportion of Hemophilic Genes Arising in a Population as a Result of Mutations. ($\mu = 3 \times 10^{-5}$. Assume 4 generations equal approximately 100 years.)

Generation	*Number of years*	*Prob* $\left[\begin{matrix}\textit{Normal}\\ \textit{gene}\end{matrix}\right]$	*Prob* $\left[\begin{matrix}\textit{Hemophilic}\\ \textit{gene}\end{matrix}\right]$
0	0	1	0
1	25	.99997	.00003
2	50	.9999400009	.0000599991
3	75	.9999100026	.0000899974
4	100	.9998800053	.0001199947
8	200	.9997600251	.0002399749
16	400	.9995201079	.0004798921
32	800	.9990404462	.0009595538
64	1600	.9980818132	.0019181868
128	3200	.9961673059	.0038326941
256	6400	.9923493015	.0076506985
512	12,800	.9847571362	.0152428638
1024	25,600	.9697466173	.0302533827
2048	51,200	.9404085018	.0595914982
4096	102,400	.8843681503	.1156318497
8192	204,800	.7821070254	.2178929746

this table indicate that for the given value of μ it would take an unbelievably long period of time to reach an appreciable proportion of hemophilic genes. In fact it is possible to determine how long it will take to decrease the proportion of nonmutant genes by a factor of $\frac{1}{4}$ or $\frac{1}{2}$. This calculation is somewhat similar to the calculation of the half-life of a radioactive chemical. In the most general case we use equation (3) and solve for n, the generation at which a certain proportion of nonmutant genes would be attained. That is,

$$p_{n+1} = p_0(1 - \mu)^{n+1}$$

It follows that

$$p_n = p_0(1 - \mu)^n$$

Rewriting, we get

$$\frac{p_n}{p_0} = (1 - \mu)^n$$

If we take the logarithms of both sides of the equation we get

$$\log\left(\frac{p_n}{p_0}\right) = \log(1 - \mu)^n$$
$$= n \log(1 - \mu)$$

Therefore

$$n = \frac{\log\left(\frac{p_n}{p_0}\right)}{\log(1 - \mu)} \tag{5}$$

Equation (5) lets us determine n, the generation at which a certain proportion of non-mutant genes would be achieved. In our hemophilia example, $p_0 = 1$, $\mu = 3 \times 10^{-5}$, and so the number of generations necessary to reach only 99% of the normal genes (or 1% hemophilia genes) is given by

$$n = \frac{\log\left(\frac{.99}{1}\right)}{\log(1 - .00003)}$$
$$= 335$$

Since we are assuming that 4 generations equal about 100 years, it would take about 8375 years to reach 99% normal and 1% hemophilic genes. It would take about 23,105 generations, or about 577,625 years, to attain 50% normal and 50% hemophilic genes in the population. These numbers are almost meaningless in our comprehension of time.

It is clear that if μ is small, as is usually the case, other factors must be operating in order for a mutant gene to reach a significant proportion in the population. We will discuss such factors (selection) in subsequent sections. It should be clear that the proportion of mutant genes in the population depends completely on μ, the mutation rate. Examples of mutation rates were given in Table 15-3, where you can see that the μ's are very small.

Reverse mutations. Lastly we consider reverse mutations. Not only do we study mutations from the normal (wild-type) gene (L) to the mutant gene (ℓ), but mutations from the mutant gene (ℓ) to the normal gene (L) (known as *back mutations*). What happens to the gene frequencies when both types of mutations are occurring?

As soon as mutant genes start to appear in the population there is a certain probability that they will mutate back to the original, or wild-type, gene. The rate of back mutation is usually lower than the original rate from wild-type to mutant. We denote the rate of back (ℓ to L) mutations by λ, and of mutations from wild to mutant (L to ℓ) by μ. As previously we use p to denote the proportion of L genes and q the proportion of ℓ genes. We let Δp denote the change in the proportion of wild-type (L)

genes. We know that Δp should be equal to the difference in the proportion that become L from ℓ, λq (the back mutation rate times the proportion of ℓ), and the proportion that become ℓ from L, μp (as in the previous section). Symbolically, Δp can be expressed as

$$\begin{aligned}\Delta p &= \{\text{change from } \ell \text{ to } L\} - \{\text{change from } L \text{ to } \ell\} \\ &= \lambda q - \mu p\end{aligned}$$

If we consider the mutation rates for *E. coli* given in Table 15-3, we find that μ for histidine requirements $his^+ \longrightarrow his^-$ is 2×10^{-6} and λ for $his^- \longrightarrow his^+$ is 4×10^{-8}. Therefore if we start with the proportion p_0 of his^+ equal to 1 and the proportion q_0 of his^- equal to 0, we find that

$$\begin{aligned}\Delta p_0 &= (4 \times 10^{-8})(0) - (2 \times 10^{-6})(1) \\ &= -2 \times 10^{-6}\end{aligned}$$

The proportion of his^+ for the first generation, p_1, is given by

$$\begin{aligned}p_1 &= (\text{proportion at generation 0}) + (\text{the change during generation 0}) \\ &= p_0 + \Delta p_0 \\ &= 1 - 2 \times 10^{-6} \\ &= .999998\end{aligned}$$

and $q_1 = .000002$. This certainly isn't a very dramatic change. If we consider the next generation, we find that even then the effect on the proportion is negligible

$$\begin{aligned}\Delta p_1 &= \lambda q_1 - \mu p_1 \\ \Delta p_1 &= (4 \times 10^{-8})(.000002) - (2 \times 10^{-6})(.999998) \\ &= -1.99999592 \times 10^{-6}\end{aligned}$$

and

$$\begin{aligned}p_2 &= p_1 + \Delta p_1 \\ &= .999998 - .00000199 \\ &= .999996\end{aligned}$$

and

$$q_2 = .000004$$

The interest in reverse mutations (that is, when both forward and back mutations are occurring) is to determine whether a dynamic equilibrium is reached. Is there a point at which gene frequencies are held stable by opposing forces? Do the proportions of wild-type and mutant genes reach a point such that with the given mutation rates the forward and back mutations balance each other out? At this point, no further changes in proportion will occur in subsequent generations. Mathematically, finding the equilibrium point is equivalent to finding the value for p (and q since $q = 1 - p$) for which $\Delta p = 0$. The equilibrium exists when

$$\Delta p = \lambda q - \mu p = 0$$

or when

$$\lambda q = \mu p$$

At this point the actual number of mutations in each direction is equal, so the net result is that no changes occur in gene frequency. We can solve for p at equilibrium by substituting for $q = 1 - p$. We denote the equilibrium value by $\hat{p}$, and we can show that

$$\lambda(1 - \hat{p}) = \mu\hat{p},$$
$$\lambda - \lambda\hat{p} = \mu\hat{p}$$
$$\hat{p} = \frac{\lambda}{\mu + \lambda} \qquad (6)$$

And since $\hat{q} = 1 - \hat{p}$,

$$\hat{q} = 1 - \frac{\lambda}{\mu + \lambda}$$
$$= \frac{\mu}{\mu + \lambda}$$

It is interesting to note that the equilibrium values do not depend on any of the values of p and q but only on the mutation rates. In fact the same equilibrium values will be attained regardless of the original proportions of each gene. For the previous example of *E. coli* we would find that

$$\hat{p} = \frac{4 \times 10^{-8}}{(2 \times 10^{-6}) + (4 \times 10^{-8})}$$
$$= .0196$$

and $\hat{q} = .9804$. That is, once these values are attained, the proportions $\hat{p}$ and $\hat{q}$ remain fixed.

For simplicity let us consider an example for which the numbers are easier to work with. We will let the back mutation rate, λ, equal $\frac{1}{4}$ and the forward mutation rate, μ, equal $\frac{1}{2}$. The equilibrium value as given by equation (6) is

$$\hat{p} = \frac{\lambda}{\lambda + \mu}$$
$$= \frac{\frac{1}{4}}{\frac{1}{4} + \frac{1}{2}}$$
$$= \frac{1}{3}$$

and

$$\hat{q} = \frac{2}{3}$$

From Table 16-12, which gives the values of p and q and shows how they approach the equilibrium values, we note that at equilibrium the number of mutant genes is twice as great as the number of wild-type genes; but since the back mutation rate (mutant to wild) is $\frac{1}{2}$ that of the forward mutation rate, the total number of mutations in each direction is the same: $(\frac{1}{2})(\frac{1}{3}) = (\frac{1}{4})(\frac{2}{3})$. For any value of p less than $\frac{1}{3}$, the proportion of wild-type genes will increase to reach the equilibrium value of $\frac{1}{3}$. These equilibrium values are different than the Hardy-Weinberg equilibrium.

Table 16-12 Forward and Backward Mutations and Their Effect on Wild-type and Mutant Genes

Generation	*Prob[Wild-type]*	*Prob[Mutant]*
0	1.0000000000	0.0
1	0.50000000000	0.50000000000
2	0.37500000000	0.62500000000
3	0.34375000000	0.65625000000
4	0.33593750000	0.66406250000
5	0.33398437500	0.66601562500
6	0.33349609375	0.66650390625
7	0.33337402344	0.66662597656
8	0.33334350586	0.66665649414
9	0.33333587646	0.66666412354
10	0.33333396912	0.66666603088
11	0.33333349228	0.66666650772
12	0.33333337307	0.66666662693
13	0.33333331347	0.66666668653
14	0.33333331347	0.66666668653
15	0.33333331347	0.66666668653
16	0.33333331347	0.66666668653
17	0.33333331347	0.66666668653
18	0.33333331347	0.66666668653
19	0.33333331347	0.66666668653
20	0.33333331347	0.66666668653
⋮	⋮	⋮
∞	$0.3333333333\bar{3}$	$0.6666666666\bar{6}$

Note: Back Mutation Rate $\lambda = \frac{1}{4}$
Forward Mutation Rate $\mu = \frac{1}{2}$.

As we have noted previously, spontaneous mutation rates are so small that their effects are virtually inconsequential. To reach appreciable proportions in the population the mutation would have to recur for centuries. But there is strong evidence that mutations exist and flourish. How, then, do mutations reach appreciable proportions in populations? One of the answers to this question is the process of selection. If we study selection, and mutation in conjunction with selection, we will have a much better understanding of what actually takes place. First we will consider selection and then the combination of mutation and selection.

Selection

Selection is the sum total of environmental forces acting to determine the fate of an allele. It is sometimes referred to as *natural selection*. Literally this means selection by nature, and is the determination of which individuals or groups shall reproduce and which shall not. Such selection is the result of the cumulative action of all forces that tend to cause individuals of one genotype to leave larger numbers of offspring than do individuals of another genotype. It is important to note that the forces of selection act on the *organism* possessing a particular gene, *not* directly on the gene. Since selection is on the organism, it is possible that certain genes that greatly benefit an organism may actually ensure the survival into the next generation of other genes that are actually deleterious to the organism. It is even possible for a gene that is lethal when homozygous to persist in the population and produce a heterozygous individual that is superior for some other trait.

In haploid organisms, since each gene is expressed, it is subject to the forces of selection. In diploid organisms, the forces of selection are modified by whether a gene is completely dominant, incompletely dominant, or codominant. We will here devote ourselves to a discussion of the effects of selection on populations of diploid organisms.

The study of selection essentially boils down to determining the genotypic proportions of future generations. We recall that if there is no preference for one genotype over the other, and if there is random mating, the proportions of all genes remain constant generation after generation. The Hardy-Weinberg equilibrium state would be a good model of what to expect in the population. But this model assumes that all genotypes are equally able to survive. If there is a differential survival or differential fertility (reproductive rate), what will happen to the Hardy–Weinberg equilibrium?

Fitness and the selection coefficient. The difference in the survival and reproduction of one organism (in particular, a genotype) in a given environment over another organism is measured by a relative quantity known as the *fitness*. The fitness of a particular genotype is denoted by the letter W, and takes on values from 0 to 1. That is, $0 \leq W \leq 1$. A particular genotype that is either lethal or renders the organism sterile has a fitness value of 0 (that is $W = 0$). Those genotypes which have no deleterious effects on the fertility or survival of the organism have a fitness value of 1 ($W = 1$). To better understand the fitness value let us suppose that we have an equal number of two homozygous *Drosophila*, *DD* and *dd*. If we also assume that for every 1000 *DD Drosophila* that survive and reproduce, only 800 *dd Drosophila* survive and reproduce, the fitness of *dd Drosophila* is 80% of the fitness of *DD Drosophila*. The relative fitnesses for these *Drosophila* are given by the following:

$$\begin{aligned} &DD \quad Drosophila \quad W = 1 \\ &dd \quad Drosophila \quad W = \tfrac{800}{1000} = .8 \end{aligned}$$

Fitness is a relative measure, so that one of the genotypes has a fitness of 1, and the less fit genotype has a fitness less than 1.

More commonly one studies the complement of the fitness. That is, population geneticists estimate how a genotype will persist in a population by considering the magnitude of the forces acting to prevent a particular genotype from surviving and reproducing. The measure of such forces is called the *selection coefficient* and is denoted s. The coefficient is defined to be 1 − W. It follows that

$$W = 1 - s.$$

From our definition we see that for a lethal or sterile genotype,

$$s = 1$$

and for a genotype with the highest reproductive efficiency,

$$s = 0$$

As we mentioned previously, the effects of selection depend on the interaction of the genes. We will now discuss how selection affects the genotypic proportions if selection is

1. against homozygous recessives
2. against semidominant genes
3. against dominant genes
4. favoring heterozygotes

Selection against homozygous recessive genes. Let us consider what happens to the proportion of each genotype if the homozygous recessive individuals or organisms (*aa*) of a large population are prevented to some extent from reproducing. We will consider the general case by using $1 - s$ for the fitness of the homozygous recessive. Then s can be any value such that $0 \leq s \leq 1$. We will further assume that there is complete dominance, and that the homozygous dominant and the heterozygous organism produce, on the average, equal numbers of viable and fertile offspring. With these assumptions it follows that the fitness of *AA* and *Aa* organisms is 1. What proportion of each genotype can we expect if the offspring of this population mate randomly and if selection is operating?

We can best answer this question by using a tree diagram which is extended by incorporating a branch to account for fitness, in particular the fitness coefficient. If we assume that the proportion of *A* genes in the gene pool is p, and that the proportion is the same for ♂ and ♀, we can derive the appropriate tree diagram as given in Fig. 16-9. (We shall assume throughout that p is not 0 and not 1, and therefore q is not 1 and not 0; otherwise the processes we are studying are not interesting.)

From Fig. 16-9 we can conclude that the relative proportion of each genotype in the offspring *AA* : *Aa* : *aa* is given by $p^2 : 2pq : q^2(1 - s)$. We see that the homozygous dominant and the heterozygotes appear in the same proportion, but that there will be relatively fewer homozygous recessives, since q^2 is multiplied by a number less than 1. How much smaller the proportion of homozygous recessives will be depends on the magnitude of the selection coefficient, s.

We return to our example to determine what the genotypic proportions will

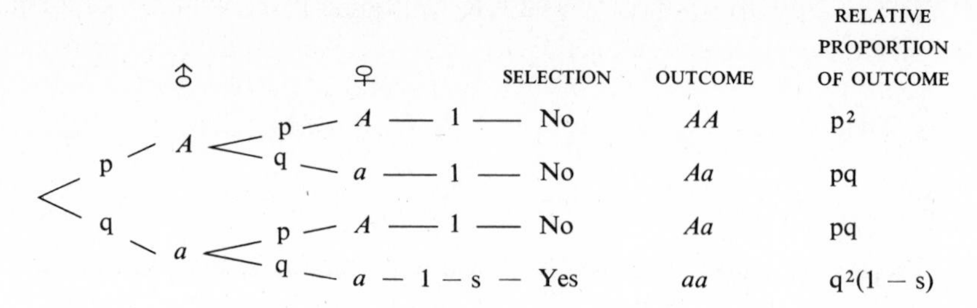

Figure 16-9. Random mating with selection against homozygous recessive genes.

be in the second generation if selection is operating. It should be clear that the same tree diagram is applicable, but that we must determine the values of p and q in the first generation. The same method we have used so far for obtaining p and q can still be used, but we need to make one slight modification. Previously when we found genotypic proportions the sum of all the terms in the proportions was always 1, and $p + q = 1$. That is, we could show that $p^2 + 2pq + q^2 = 1$. For:

$$\begin{aligned} p^2 + pq + pq + q^2 &= p(p + q) + q(p + q) \\ &= p(1) + q(1) \qquad\qquad (\text{since } p + q = 1) \\ &= 1 \end{aligned}$$

But now if we add up the relative genotypic proportions we get

$$\begin{aligned} p^2 + 2pq + q^2(1 - s) &= \underbrace{p^2 + 2pq + q^2}_{1} - q^2 s \\ &= 1 - sq^2 \end{aligned}$$

That is, the sum of these relative genotypic proportions is less than 1. To translate these into true probabilities (that is, numbers which add up to 1), but still maintain the relative magnitudes between each genotype, we divide each proportion by the sum of the relative proportions. That is, the ratio of *AA* : *Aa* : *aa* is given by

$$\frac{p^2}{1 - sq^2} : \frac{2pq}{1 - sq^2} : \frac{q^2(1 - s)}{1 - sq^2} \qquad (7)$$

Therefore the proportion q_2 of *a* genes in the second generation is given by

$$\begin{aligned} q_2 &= \text{Prob}[aa] + \tfrac{1}{2}\text{Prob}[Aa] \\ &= \frac{q^2(1 - s)}{1 - sq^2} + \frac{1}{2}\left[\frac{2pq}{1 - sq^2}\right] \end{aligned}$$

By computation,

$$q_2 = \frac{q(1 - qs)}{1 - sq^2}$$

The proportion p_2 of dominant gene A can be shown by using the same method to be $\frac{p}{1 - sq^2}$. If we use a tree diagram, we can show that the ratio in the F_2 of $AA : Aa : aa$ is given by $p_2^2 : 2p_2q_2 : q_2^2(1 - s)$ and actually equals $\frac{p^2}{(1 - sq^2)^2} : \frac{2pq(1 - sq)}{(1 - sq^2)^2} : \frac{q^2(1 - sq)^2(1 - s)}{(1 - sq^2)^2}$. Using the same approach we can show that q_3, the proportion of a genes among the F_3 offspring, is $\frac{q_2(1 - sq_2)}{1 - sq_2^2}$ (to express q_3 in terms of q would be very messy), and p_3, the proportion of A genes, is $\frac{p_2}{1 - sq_2^2}$. In general,

$$q_{n+1} = \frac{q_n(1 - sq_n)}{1 - sq_n^2} \tag{8}$$

If $s \neq 0$ and $q_n \neq 1$, q_n^2 is smaller than q_n. Hence $\frac{1 - sq_n}{1 - sq_n^2}$ is smaller than 1, and so in each generation the proportion of homozygous recessives will be smaller than that in the previous generation. Thus we do not have an equilibrium in the sense of the Hardy-Weinberg equilibrium. In fact if we look at Fig. 16-10, we see the effects of

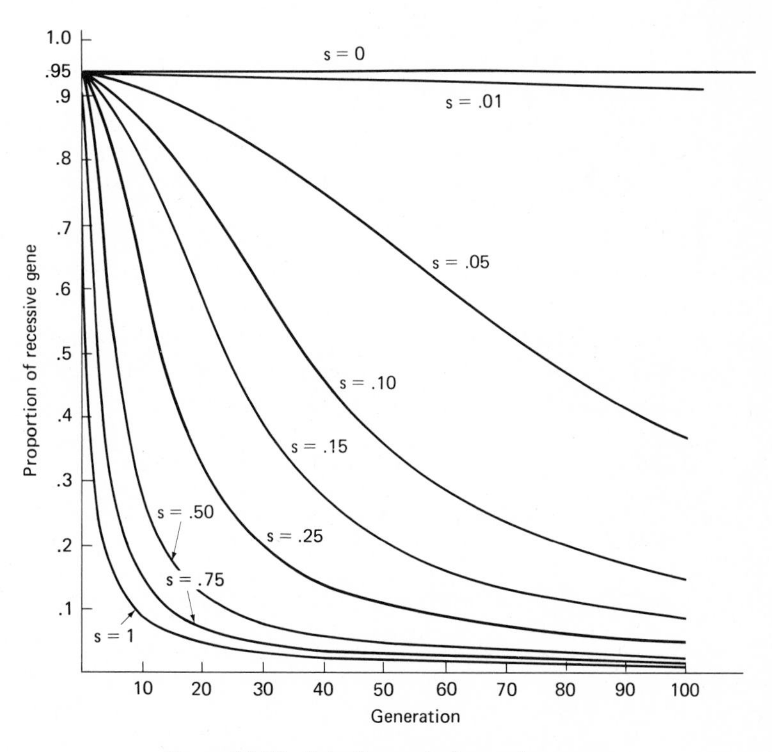

Figure 16-10. Selection against recessive genes.

various selection coefficients on the proportion of the recessive gene. We considered for each case that the proportion at the starting point was .95, but we can use Fig. 16-10 to get any starting value of q. For instance if we want to assume that the starting proportion of the recessive is .5 for a particular value of s, we look up .5 on the vertical axis. The generation corresponding to .5 on the curve for the s value on the horizontal axis is read as 0 rather than the given value, and the number of generations is counted from the new zero value. For example if $s = .75$, the zero generation for a q of .5 corresponds to the third generation. The value of q will be .15 at what would be its seventh generation (actually in Fig. 16-10 it would be the tenth generation). From Fig. 16-10 we note that at first the effect of selection is greater and that later it starts to decrease. This is true for values of $.5 \leq s \leq 1$. For $s = .25$ the effect is much slower. For $s = .01$ the effect on q is negligible but then again an s of .01 (or a fitness $W = 1 - .01 = .99$) is really very small. When studying these curves it is important to keep in mind that in human beings one generation is approximately 25 years, and so 10 generations correspond to 250 years.

One special case will be considered in more detail. Let us consider the implications of a selection coefficient $s = 1$ (fitness $W = 0$). If the selection coefficient is 1 we are saying that no homozygous recessives can reproduce. This might in fact be the case if a eugenics[6] program is instituted to eliminate all homozygous recessives for a particular gene. Ignoring for the moment any moral or ethical considerations, what will be the effect of a eugenics program that forbids all homozygous recessives for a particular gene from reproducing?

Would such a program eliminate the recessive gene? If so, would this occur within a reasonable period of time?

What happens in successive generations if we have a population for which $s = 1$? If we look at formula (7), and the tree diagram in Fig. 16-9, we find that if the starting proportion of *A* genes is p and the proportion of *a* genes is $q = 1 - p$, then

$$\text{Prob}[AA] = \frac{p^2}{1 - q^2} = \frac{p^2}{(1 - q)(1 + q)} = \frac{p^2}{p(1 + q)} = \frac{p}{1 + q}$$

$$\text{Prob}[Aa] = \frac{2pq}{1 - q^2} = \frac{2pq}{(1 - q)(1 + q)} = \frac{2pq}{p(1 + q)} = \frac{2q}{1 + q}$$

$$\text{Prob}[aa] = 0$$

In fact if we look at equation (8) and let $s = 1$, we find that

$$\begin{aligned} q_{n+1} &= \frac{q_n(1 - q_n)}{1 - q_n^2} \\ &= \frac{q_n(1 - q_n)}{(1 + q_n)(1 - q_n)} \qquad \left(\begin{array}{l}\text{since } p_n + q_n = 1 \text{ and} \\ 1 - q_n^2 = (1 - q_n)(1 + q_n)\end{array}\right) \\ &= \frac{q_n}{1 + q_n} \end{aligned} \tag{9}$$

[6] *Eugenics* is a program of selective reproduction or of prohibition of reproduction by individuals with specific genotypes. (See Chapter 19.)

We can simplify equation (9) even further, since $q_1 = \frac{q_0}{1 + q_0}$, and $q_2 = \frac{q_1}{1 + q_1}$.

Substituting for q_1, we get that $q_2 = \frac{\frac{q_0}{1 + q_0}}{1 + \frac{q_0}{1 + q_0}}$ which can be shown to equal $\frac{q_0}{1 + 2q_0}$. Similarly we can show that $q_3 = \frac{q_0}{1 + 3q_0}$. In general it can be shown that

$$q_n = \frac{q_0}{1 + nq_0} \tag{10}$$

We can use equation (10) to determine the percentage of homozygous recessives that will remain in the population if $s = 1$ (see Table 16-13). Note that total negative selection can't immediately eliminate all homozygous recessives, since there will still be heterozygote carriers of the recessive gene. From Table 16-13 we see that even total negative selection will not eliminate a particular gene. Even though the percentage of homozygous recessives decreases, by the 99th generation (about 2475 years in human

Table 16-13 Effects of Total Negative Selection

Generation	*Proportion of recessives in the population*	*Percent of heterozygotes (carriers)*	(*a*) *Percent of homozygous recessives*		(*b*) *Percent*[1] *reduction of homozygous recessives*	*Percent heterozygotes: percent homozygous recessives*
0	.99999[2]	.001	99.999	>	75	~ 0
1	1/2	50	25	>	55.56	2 : 1
2	1/3	44.44	11.11	>	43.74	4 : 1
3	1/4	37.50	6.25	>	36.00	6 : 1
4	1/5	32.00	4.00	>	30.50	8 : 1
5	1/6	27.78	2.78	>	26.62	10 : 1
6	1/7	24.49	2.04	>	23.53	12 : 1
7	1/8	21.88	1.56	>	21.15	14 : 1
8	1/9	19.75	1.23	>	18.70	16 : 1
9	1/10	18.00	1.00	>	17.00	18 : 1
10	1/11	16.53	.83			20 : 1
20	1/21	9.07	.23			40 : 1
50	1/51	3.84	.0384			100 : 1
99	1/100	1.98	.01			198 : 1
999	1/1000	.1998	.0001			1998 : 1
9999	1/10000	.019998	.00000001			19998 : 1

[1]The entry between generations i and $i + 1$ in column (b) represents the difference between the entries for generations i and $i + 1$ in column (a) as a percentage of the entry for generation i in column (a). That is, $\frac{25 - 11.11}{25} = .5556$, or 55.56%.

[2]For ease of calculation we assume that q is 1, but in fact, as we pointed out in the text, q cannot be equal to 1.

beings), .01% of the population will still be homozygous recessive. As the gene becomes rare the effect of negative selection becomes smaller and smaller. We see from Table 16-13 that when the proportion of recessive genes is relatively large, negative selection has a much greater impact than when the proportion of recessive genes is small. In fact most recessive genes for which a eugenics program of this type might be considered are basically very rare.

Let us consider an example for which the proportion of recessive genes is known. Assume that it has been decided that albinos will not be allowed to reproduce. (We are by no means suggesting such a program should be considered or implemented.) Therefore the only way an albino child can be born is if both parents are carriers of the gene for albinism. We know that in the general population of the United States the proportion of recessive genes for albinism is $\frac{1}{141}$. Therefore if $s = 1$, we can use equation (9) to show that the proportion of recessive genes in the next generation is given by

$$q_1 = \frac{q}{1+q} = \frac{\frac{1}{141}}{1+\frac{1}{141}} = \frac{1}{142}$$

Thus the proportion of albinos born if no albinos are allowed to reproduce is given by

$$(q_1)^2 = \left(\frac{1}{142}\right)^2 = \frac{1}{20{,}164}$$

If albinos were allowed to reproduce, Hardy-Weinberg would be in effect and q would equal $\frac{1}{141}$, so that the proportion of albinos born would be

$$q^2 = \left(\frac{1}{141}\right)^2 = \frac{1}{19{,}881}$$

How can we interpret the difference in these two values? We can best see this by considering the United States population, which is about 2×10^8. If there were no eugenics program, we would expect about 10,060 albinos, whereas if a eugenics program were in effect, we would expect about 9919 albinos. Therefore if albinos were not allowed to reproduce, 141 fewer albinos would be born, an essentially insignificant difference. The effects of such a eugenics program over a long period of time are given in Table 16-14. We see that if the proportion of recessive genes is small the effects of total negative selection will take an *extremely* long time to decrease significantly the number of people who are homozygous recessive. Even if we could ignore the moral and ethical considerations of a eugenics program, its very limited effects would not even make it reasonable to consider. The reason why selection against homozygous recessives accomplishes so little is that under a system of random mating the number of heterozygotes far outnumbers the homozygous recessives, especially when the recessive gene is rare. Of course, if mating is not random, and, for example, if homo-

Table 16-14 Number of Albinos in the U.S. Population (Total Negative Selection vs. No Selection)

Generation	*Years*	*Number of albinos born no selection*	*Number of albinos born s = 1*	*Decrease in number of albinos*
0	0	10,060	10,060	> 141 in 25 years
1	25	10,060	9,919	> 138 in 25 years
2	50	10,060	9,781	> 136 in 25 years
3	75	10,060	9,645	> 132 in 25 years
4	100	10,060	9,513	> 130 in 25 years
5	125	10,060	9,383	> 127 in 25 years
6	150	10,060	9,256	> 125 in 25 years
7	175	10,060	9,131	> 122 in 25 years
8	200	10,060	9,009	> 120 in 25 years
9	225	10,060	8,889	> 117 in 25 years
10	250	10,060	8,772	> 1056 in 250 years
20	500	10,060	7,716	> 876 in 250 years
30	750	10,060	6,840	> 1357 in 500 years
50	1250	10,060	5,483	> 2039 in 1250 years
100	2500	10,060	3,444	> 1724 in 2500 years
200	5000	10,060	1,720	> 1233 in 7500 years
500	12,500	10,060	487	> 333 in 12,500 years
1000	25,000	10,060	154	> 146 in 100,000 years
5000	125,000	10,060	8	> 6 in 125,000 years
10,000	250,000	10,060	2	

zygous recessives tend to mate only with homozygous recessives, those conclusions would no longer be valid. If we return to Table 16-13, we see that the ratio of heterozygotes to homozygote recessives increases when total negative selection takes place, and the recessive gene will persist in the population even if there is total negative selection.

Selection against semidominant genes. We will consider now the situation where the recessive gene, *a*, has a deleterious effect on the heterozygote, *Aa*. For simplicity we assume that the selection against the heterozygote is s, and against the homozygous recessive, 2s. (It would be more general to use s_1 and s_2, but we can make a more interesting analysis if we use s and 2s.) If we once again incorporate the fitness into our tree diagram, we get Fig. 16-11. In order to obtain probabilities

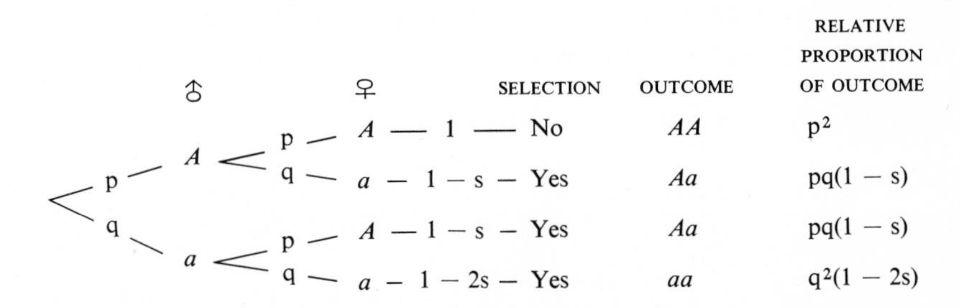

Figure 16-11. Random mating with selection against semidominant genes.

we must compensate for the selection by dividing by the sum of the relative proportions, that is, $p^2 + 2pq(1 - s) + q^2(1 - 2s)$. But we can simplify:

$$\begin{aligned} p^2 + 2pq(1 - s) + q^2(1 - 2s) &= p^2 + 2pq + q^2 - 2pqs - 2q^2s \\ &= (p + q)^2 - 2sq(p + q) \\ &= 1 - 2sq \end{aligned}$$

Therefore we find the ratio of $AA : Aa : aa$ is given by

$$\frac{p^2}{1 - 2sq} : \frac{2pq(1 - s)}{1 - 2sq} : \frac{q^2(1 - 2s)}{1 - 2sq}$$

We are really interested in the long-range effects: Will an equilibrium be reached or will the values of p and q continue to change from generation to generation? Therefore we must find the value of q for the next generation. We get

$$\begin{aligned} q_1 &= \text{Prob}[a \text{ in } F_1] = \text{Prob}[aa] + \tfrac{1}{2}\text{Prob}[Aa] \\ &= \frac{q^2(1 - 2s)}{1 - 2sq} + \frac{1}{2}\left[\frac{2pq(1 - s)}{1 - 2sq}\right] \end{aligned}$$

By computation, $q_1 = \frac{q(1 - s - sq)}{1 - 2sq}$; similarly we can show that $p_1 = \frac{p(1 - sq)}{1 - 2sq}$. And in general $q_{n+1} = \frac{q_n(1 - s - sq_n)}{1 - 2sq_n}$, and $p_n = \frac{p_n(1 - sq_n)}{1 - 2sq_n}$. Hence if $s \neq 0$ and $q_0 \neq 1$, for each generation the values of p and q will change. The values of q are given in Fig. 16-12 for various values of s. (Note that the largest value of $s = \frac{1}{2}$, since any $s > \frac{1}{2}$ would yield a fitness $(1 - 2s)$ which is negative and hence values of q which are negative.) If we compare Fig. 16-12 and Fig. 16-10 for comparable values of s, we find that selection is much more effective in the semidominant case than in the homozygous recessive. Practically speaking, then, a deleterious recessive gene can be controlled if reproduction in both the heterozygotes and the homozygous recessives is limited. (This is more effective than if there is total negative selection limiting homozygous recessives.) Even if moral and ethical considerations were not a problem, however, it would be virtually impossible in most situations to identify the heterozygotes, so although it would be theoretically efficient to select against the heterozygotes and the homozygous recessives, it would not be practical. It should also be clear from studying Fig. 16-12 that the Hardy-Weinberg equilibrium would not be maintained in such a situation since the values of q and hence p are constantly changing.

Selection against dominant genes. If there is selection against dominant genes, the selection will be against the homozygous dominant and the heterozygote, with the same value of s. The selection value for AA is $1 - s$; Aa is $1 - s$; and aa is 1.

If we use an appropriate tree diagram we can easily show that the ratio of $AA : Aa : aa$ is given by $\frac{p^2(1 - s)}{1 - sp(1 + q)} : \frac{2pq(1 - s)}{1 - sp(1 + q)} : \frac{q^2}{1 - sp(1 + q)}$. (Note that the denominator is sometimes written as $1 - s(1 - q^2)$ or as $1 - 2sp + sp^2$, which

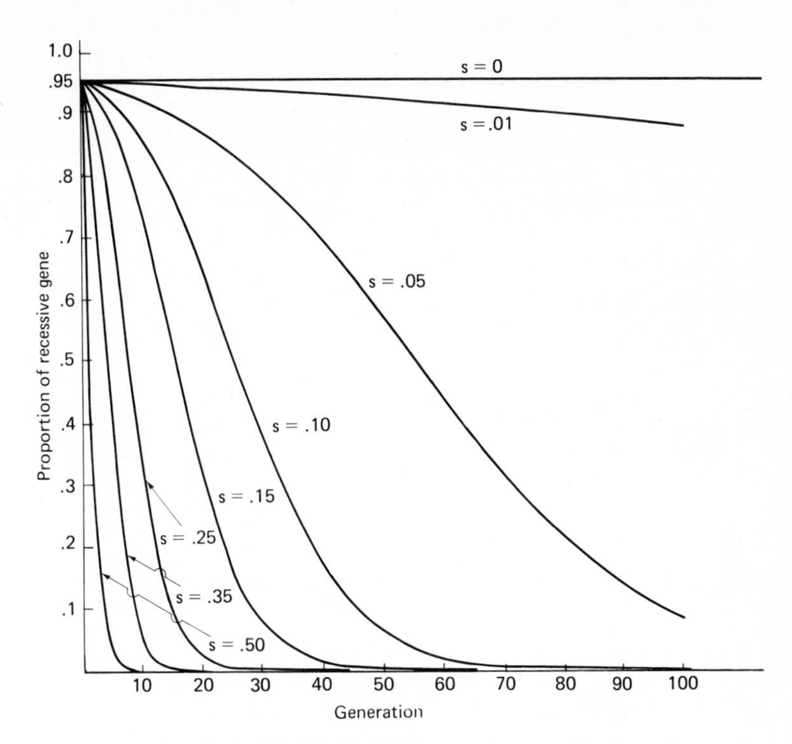

Figure 16-12. Selection against a semidominant gene.

are equivalent forms.) By using the same approach as previously we can show that

$$p_{n+1} = \frac{p_n(1 - s)}{1 - sp_n(1 + q_n)} = \frac{p_n(1 - s)}{1 - s(1 - q_n^2)}$$

(The second term will be easier for calculation purposes.) Once again we see that if $p_0 \neq 0$ and $s \neq 0$, p will vary from generation to generation, so that there will be no Hardy-Weinberg equilibrium. Some examples of how p varies are given in Fig. 16-13. An interesting case is when $s = 1$, that is, the selection is complete against all dominant genes. Intuitively it suggests that only recessive genes will remain after one generation, which is exactly what the formulas and Fig. 16-13 suggest.

If total negative selection ($s = 1$) for dominant genes is so successful and quick, why can't it be employed to eliminate undesirable traits? The problem is that most human abnormalities are not determined by dominant genes.

There are, however, some diseases that are controlled by dominant genes. In some cases deleterious dominant genes are not expressed until relatively late in life. For example, Huntington's chorea is a hereditary disease determined by an autosomal dominant gene. In this disease, mental deterioration is accompanied by uncontrollable, involuntary muscular movements, and the disease is fatal. The average age of onset is 40, though it can develop at any age. If every person who has the gene for Huntington's chorea did not produce children, Huntington's chorea would

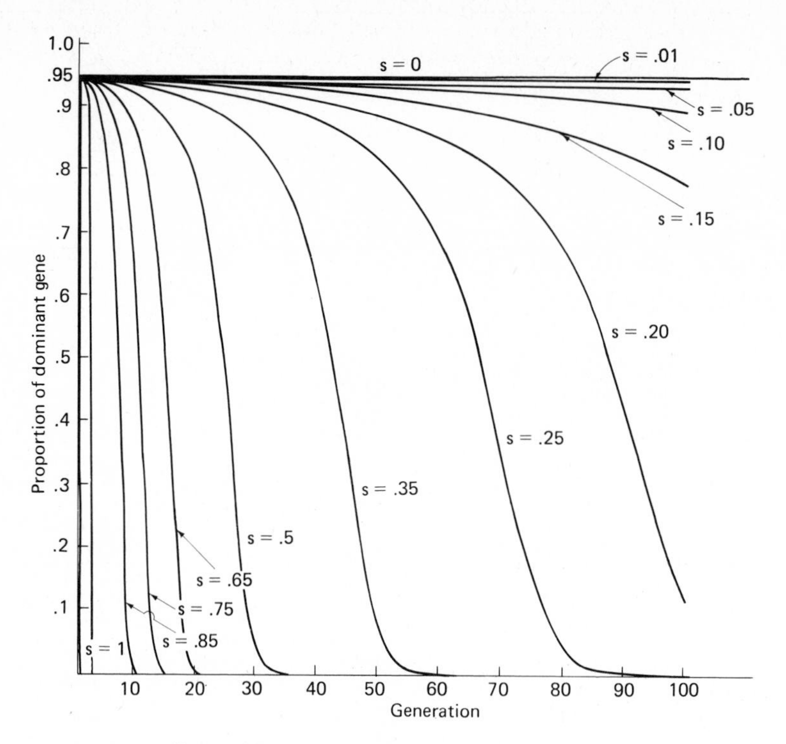

Figure 16-13. Selection against dominant genes.

essentially disappear in one generation. The one major problem with this solution is that many people who will ultimately develop Huntington's chorea do not develop the disease until after they have produced children. Ethically this type of solution presents other problems since it is expected that only $\frac{1}{2}$ the offspring in which one parent has the gene for Huntington's chorea will develop the disease. So even though theoretically we may have a solution there are practical and ethical considerations that must be taken into account.

If we examine selection against homozygous recessive genes (Fig. 16-10) and against dominant genes (Fig. 16-13) for comparable s values, we see that initially selection is much more effective in the homozygous recessive case, but that in the end selection against the dominant is much more complete and eliminates the gene much faster.

Favoring heterozygotes. In this subsection we turn to the process in which selection of the heterozygote is favored and both homozygotes are selected against. This process tends to counter the effect of selection acting against a deleterious gene. In the cases we have discussed thus far, after a long enough period of time the proportion of the gene being selected against would approach zero; now there will be a totally different process occurring. To make our discussion general, we will denote the selection against the homozygous dominant by s_1, and that against the homozygous recessive by s_2.

Once again using a tree diagram, incorporating the fitness values and dividing by the sum of the relative proportions, we can show that the ratio of *AA* : *Aa* : *aa* is given by

$$\frac{p^2(1 - s_1)}{1 - s_1p^2 - s_2q^2} : \frac{2pq}{1 - s_1p^2 - s_2q^2} : \frac{q^2(1 - s_2)}{1 - s_1p^2 - s_2q^2}$$

What can we say about the value of q, the proportion of recessives, over time? We can express q_1 as the sum of the previous proportion and an additional term (which will be convenient). We can show that

$$q_1 = q + \frac{pq(s_1p - s_2q)}{1 - (s_1p^2 + s_2q^2)}$$

Or in general,

$$q_{n+1} = q_n + \frac{p_nq_n(s_1p_n - s_2q_n)}{1 - (s_1p_n^2 + s_2q_n^2)} \qquad (11)$$

Note that the denominator of (11) is positive, since $s_1p_n^2 + s_2q_n^2$ is less than or equal to $s(p_n^2 + q_n^2)$, where s is the larger of s_1 and s_2. This in turn is less than or equal to $s(p_n^2 + q_n^2 + 2p_nq_n)$ which is less than $(p_n + q_n)^2 = 1$, if $s \neq 1$. Hence if $s \neq 1$, the value of q_{n+1} increases or decreases depending on whether s_1p_n is greater than or less than s_2q_n. By inspecting equation (11) we find that some sort of equilibrium is possible. Equilibrium implies that $q_{n+1} = q_n$. The two proportions will be equal only if the term

$$\frac{p_nq_n(s_1p_n - s_2q_n)}{1 - (s_1p_n^2 + s_2q_n^2)}$$

is equal to zero. This term can be equal to zero only if $p_n = 0$, $q_n = 0$, or $s_1p_n = s_2q_n$. If we disregard the former two possibilities, and if we let $\hat{p}$ and $\hat{q}$ denote the equilibrium values, we can show that:

$$\hat{p}_n = \frac{s_2}{s_1 + s_2}, \quad \text{and} \quad \hat{q}_n = \frac{s_1}{s_1 + s_2} \qquad (12)$$

That is, the proportion of genes will remain constant when the proportion of dominant genes is equal to $\frac{s_2}{s_1 + s_2}$. Interestingly enough these equilibrium values are independent of the initial gene proportions and completely determined by the selection values against the homozygotes. (Note that if $s_1 = s_2$, $\hat{p} = \frac{1}{2}$.)

There are a few interesting examples indicating that the heterozygote is more fit than the homozygote. In the first, both homozygotes are lethal ($s_1 = s_2 = 1$, so $W_1 = W_2 = 0$), so that only the heterozygote survives; such a system is known as a *balanced lethal system*. A balanced lethal system occurs at the *T* locus in mice. The *TT* homozygotes and homozygotes for lethal *t* alleles die, while *Tt* produces a viable, tailless individual. For *Tt* × *Tt*, the resultant offspring and the selection coefficients are given in Table 16-15. It is clear that $q_{n+1} = q_n$. If we look at equation (11) we see

Table 16-15 Balanced Lethal System at the T Locus in Mice

Genotype	*TT (die)*	*Tt (viable)*	*tt (die)*
Selection	$1(s_1)$	0	$1(s_2)$
Fitness	0	1	0
Prob[Offspring]	$p^2(0)$	$2pq(1)$	$q^2(0)$
$\frac{\text{Prob[Offspring]}}{\text{Sum of Prob}}$	0	$\frac{2pq}{2pq} = 1$	0

that this means that either $p_n = 0$, $q_n = 0$, or $s_1 p_n - s_2 q_n = 0$, so that the population is in equilibrium. In fact from equation (12) we find that $\hat{q}_n = \frac{s_1}{s_1 + s_2} = \frac{1}{1+1} = \frac{1}{2}$, and so $\hat{p}_n = \frac{1}{2}$. For any balanced lethal system, equilibrium is established after one generation, since only heterozygotes remain. This gives $p_1 = q_1 = \frac{1}{2}$, and from then on, $p_n = \hat{p}_n = q_n = \hat{q}_n = \frac{1}{2}$.

The second interesting example occurs in humans that are carriers of the sickle-cell trait. From our discussion of sickle-cell anemia in Chapter 4, recall that for the homozygous recessive the disease is fatal. Various studies (see Spiess [1977] for references) have indicated that the heterozygote (carrier of the Sickle-cell gene) seems to manifest increased resistance to falciparum malaria compared to persons who are not carriers of the Sickle-cell gene. (Falciparum malaria, caused by *Plasmodium falciparum*, can be fatal.) We can summarize this discussion in Table 16-16, where for simplicity we assume that heterozygote survivorship (selection) is

Table 16-16 Sickle-cell and Falciparum Malaria Selection Favoring Heterozygote

	Hb^A/Hb^A (*Normal*)	Hb^A/Hb^s (*Carrier*)	Hb^s/Hb^s (*Sickle-cell*)
Selection coefficient (against falciparum malaria)	$1 - s_1$	1 relatively	0

1. As a result of this sort of selection, the heterozygotes, in areas where the incidence of malaria was high (the incidence of malaria is much lower today than it was 20 years ago) were much better able to survive and reproduce than were heterozygotes in nonmalarial areas. Therefore the Sickle-cell gene was maintained in relatively high proportions in malarial areas as a result of the selection superiority conferred on the heterozygotes. In environments free of malaria the gene is relatively much rarer, and among American Blacks it seems to be decreasing rapidly. The data in Table 16-17 seem to confirm the observations that the proportion of heterozygotes is greater in malarial areas. If we assume that the populations described in Table 16-17 have achieved Hardy-Weinberg equilibrium (even though it is not appropriate since there

Table 16-17 Percent Distribution of Hb^A/Hb^A, Hb^A/Hb^s, and Hb^s/Hb^s in Dallas, Philadelphia, and Kampala, Uganda

Locality	Hb^A/Hb^A	%	Hb^A/Hb^s	%	Hb^s/Hb^s	%	*Total*
Dallas	1080	92.7	78	6.7	7	.6	1165
Philadelphia	923	92.3	74	7.4	3	.3	1000
Kampala, Uganda	2817	83.8	542	16.1	3	.09	3362

Data from F. B. Livingstone, 1967. *Abnormal Hemoglobins in Human Populations.* Aldine, Chicago.

is selection and some of the Hb^s/Hb^s may have already died and therefore they are not represented appropriately), we find that the relative proportion of Sickle-cell genes in certain cities is as follows:

Dallas	$.006 + \frac{1}{2}(.067) = .0395$
Philadelphia	$.003 + \frac{1}{2}(.074) = .0400$
Kampala, Uganda	$.0009 + \frac{1}{2}(.161) = .0814.$

This is one of the more interesting examples of selection for the heterozygote.

Interaction of Mutation and Selection

We have already discussed the fact that mutation and selection, each acting separately, change gene proportions over the generations. The speed at which the changes occur depends on the selection coefficients and mutation rates. The same will obviously be true when these two factors are considered simultaneously. Previously we assumed that either mutation or selection occurred, but not both. What happens to gene proportions if both mutation and selection occur?

If mutation and selection exert their pressures in the same way (selection favors homozygous recessives and mutation increases the proportion of recessive genes), it follows that the change in gene proportions will be faster than if only one of the two were operational.

If mutation and selection forces oppose each other, their independent effects can cancel each other resulting in a stable equilibrium. We shall briefly consider three possibilities:

1. Mutation and selection involving rare recessives
2. Mutation and selection involving dominants
3. Mutation and selection including reverse mutation

Mutation and selection involving rare recessives. What are the effects on rare recessives if both mutation and selection are operating? We assume that the selection coefficient against the rare recessive, aa, is s, and that selection against homozygous dominants and heterozygotes is 0. We also assume that the mutations from dominant A to recessive a occur at the rate of μ per generation, and lastly that the mutations from recessive a to dominant A occur at the rate of λ per generation. With these three forces operational, what is the net change in recessive genes?

$$\begin{bmatrix}\text{Change in}\\ a \text{ genes}\end{bmatrix} = \begin{bmatrix}\text{Change due to}\\ \text{selection only}\end{bmatrix} + \begin{bmatrix}\text{Change due to}\\ \text{mutation}\\ \text{A} \longrightarrow \text{a}\end{bmatrix} + \begin{bmatrix}\text{Change due to}\\ \text{mutation}\\ \text{a} \longrightarrow \text{A}\end{bmatrix}$$

$$q_{n+1} - q_n = \frac{-sq_n^2p_n}{1 - sq_n^2} + \frac{\mu p_n}{1 - sq_n^2} - \frac{\lambda q_n}{1 - sq_n^2} \qquad \left(\begin{matrix}\text{1st term rewritten}\\ \text{from equation 8}\end{matrix}\right)^7$$

For simplicity we will assume that the reverse mutation rate, λ, is negligible in comparison to μ, and that q_n is small, so we can assume that $\frac{\lambda q_n}{1 - sq_n^2} \sim 0$. Therefore we can write

$$q_{n+1} = q_n + \frac{\mu p_n}{1 - sq_n^2} - \frac{sq_n^2 p_n}{1 - sq_n^2}$$

We see that the proportion of recessive genes can change from generation to generation. But it is possible for the mutation forces and the selection forces to cancel each other, so that $q_{n+1} - q_n = 0$, and an equilibrium will be established. If $q_{n+1} - q_n = 0$, then it follows that

$$\frac{\mu\hat{p}}{1 - s\hat{q}^2} - \frac{s\hat{q}^2\hat{p}}{1 - s\hat{q}^2} = 0$$

where $\hat{p}$ and $\hat{q}$ are the equilibrium values. Solving for $\hat{q}$, we get that $\hat{q} = \sqrt{\frac{\mu}{s}}$. If the homozygous recessive is lethal ($s = 1$), the equilibrium value of q becomes $\hat{q} = \sqrt{\mu}$. The proportion of homozygous recessives in the population becomes

$$\hat{q}^2 = \mu$$

Intuitively this makes sense if the recessives are lethal, since then the only recessives in the population at a given point in time are those that arise by mutations. If the homozygous recessive is semilethal, for example, if $s = \frac{1}{2}$, then

$$\hat{q} = \sqrt{\frac{\mu}{\frac{1}{2}}} = \sqrt{2\mu}$$

which boils down to doubling the mutation rate. Whereas if $s = \frac{1}{4}$, $\hat{q} = \sqrt{4\mu}$. This makes sense since more recessives are surviving.

Mutation and selection involving dominants. What are the consequences if we consider mutation and selection involving dominants? If selection is against only the homozygous dominant and we assume that the dominant gene is rare, by using the

[7] It can be shown that $1 - sq_n^2$ is the appropriate denominator, but the details will be omitted.

same argument we used for the recessive case we can show that if λ is the mutation rate from dominant to recessive, $\hat{p} = \sqrt{\frac{\lambda}{s}}$. If the homozygous dominant is lethal ($s = 1$), $\hat{p} = \sqrt{\lambda}$, and if the heterozygote is also lethal, $\hat{p} = \lambda$. That is, the only dominant genes in the population are those that arise by mutation and they are subsequently eliminated by selection.

Mutation and selection including reverse mutation. Lastly we consider the effects of mutation (forward and back) and selection. We will also assume that the proportions of p and q are not small so that none of the mutations will be negligible. For the case of selection against the dominant gene, we find that

$$\begin{bmatrix}\text{Change in}\\ A \text{ gene}\end{bmatrix} = \begin{bmatrix}\text{Change in } A \text{ due to}\\ \text{selection only}\end{bmatrix} + \begin{bmatrix}\text{Change due to}\\ \text{forward mutation}\\ A \longrightarrow a\end{bmatrix} + \begin{bmatrix}\text{Change due to}\\ \text{back mutation}\\ a \longrightarrow A\end{bmatrix}$$

$$p_{n+1} - p_n = \frac{sp_nq_n^2}{1 - 2sp_n + sp_n^2} - \frac{\mu p_n}{1 - 2sp_n + sp_n^2} + \frac{\lambda q_n}{1 - 2sp_n + sq_n^2}$$

$$= \frac{\lambda q_n + \mu p_n + sp_nq_n^2}{1 - 2sp_n + sq_n^2} \tag{13}$$

At equilibrium $p_{n+1} - p_n = 0$, so that the numerator of equation (13) must equal zero at equilibrium. That is, the equilibrium values $\hat{p}$ and $\hat{q}$ are solved from

$$\lambda\hat{q} - \mu\hat{p} + s\hat{p}\hat{q}^2 = 0 \tag{14}$$

Equation (14) cannot be solved explicitly until we have values for λ, μ, and s. It turns out that there may be one or two equilibrium values for $\hat{q}$. (Note that if $s = 0$—that is, there is no selection—we can again verify that $\hat{q} = \frac{\mu}{\mu + \lambda}$. See page 530).

The reality is that selection removes genes from a population while mutations can replenish them. If these two opposing forces balance each other, an equilibrium is maintained.

Genetic Drift

In our derivation of the Hardy-Weinberg equilibrium we assumed that we were dealing with large populations, and that the gene pool consisted of all possible genes. Realistically we know that mates are not selected from among all possible members of the opposite sex. In real populations of sexually reproducing organisms there is evidence of a correlation between parents, in terms of morphological resemblance, common origin, ecological preference, and various other biases. Proximity must play a very important role in the selection of many mates. Families, tribes, cohorts, and various religious, social, ethnic, and economic groupings subdivide the human populations into subunits. Very possibly selection of mates may often be limited to one or more of these subunits rather than to the whole population. What are the effects on population proportions if the gene pool is small? What happens in small populations?

Genetic drift refers to chance occurrences in small populations which lead to changes in gene frequencies from generation to generation. In small populations the Hardy–Weinberg equilibrium can very easily be upset. Let us consider the smallest possible reproducing population, one male and one female. As a concrete example let one mate be *Aa* and the other *aa*. Therefore Prob[A] $= p_0 = \frac{1}{4} = .25$, and Prob[$a$] $= q_0 = \frac{3}{4} = .75$. In an $Aa \times aa$ mating we expect $Aa : aa$ in a $1 : 1$ ratio (this is actually a long-run average result; see Chapter 3). Now if this mating produces only one offspring something very interesting could happen. The child could be *aa*, and the dominant gene would be lost from the population, unless there were a consanguineous mating. In this case, in the first generation

$$p_1 = \text{Prob}[A \text{ in } F_1] = 0, \text{ and } q_1 = \text{Prob}[a \text{ in } F_1] = 1$$

(An analogous situation would be the following: You take a coin and flip it once, and use only that outcome to describe the Prob[Head]. Would you get a reasonable estimate?) Let us be a bit more realistic (in a totally unrealistic setting) and assume that this couple has two children; that is, they replace themselves. The possible outcomes are:

Child 1	*Child 2*	*Ratio* $Aa : aa$	*Prob*[A] $= p_1$
Aa	*Aa*	2 : 0	$\frac{1}{2}$
Aa	*aa*	1 : 1	$\frac{1}{4}$
aa	*Aa*	1 : 1	$\frac{1}{4}$
aa	*aa*	0 : 2	0

We note that only two of the possible outcomes satisfy the predicted ratio of 1 : 1, and maintain the same proportion of A and a in the population. In the other possibilities there are some very interesting deviations. In one case where both offspring are *Aa*, the proportion of dominant genes has doubled, from $\frac{1}{4}$ to $\frac{1}{2}$. In the last case, where both offspring are homozygous recessives, the dominant gene has been lost from the population. Therefore, if we merely consider chance effects in a small population, we have noted that the proportion of A genes in the gene pool may increase, stay the same, or decrease even to the point of being lost from the population. Chance fluctuations such as these are an illustration of how genetic drift operates.

For further illustration let us consider what might happen if we have a population of 10 men and 10 women. We will assume that each couple has two children, that the proportion of the dominant A gene is $\frac{1}{4}$ and of the recessive a gene is $\frac{3}{4}$, and that the genotypes of the population are in the ratio $3AA : 4Aa : 13aa$. For concreteness we assume that the 10 couples mate in the following way: (1) $AA \times aa$; (2) $aa \times aa$; (3) $aa \times aa$; (4) $aa \times AA$; (5) $Aa \times aa$; (6) $aa \times aa$; (7) $Aa \times AA$; (8) $aa \times Aa$; (9) $aa \times aa$; (10) $aa \times Aa$. These 10 couples will produce 20 offspring who will live to reproduce. For the purposes of our example only (since there are other chance results) we will assume the following outcomes for each mating:

	Mating	*Outcome*	
(1)	$AA \times aa$	Aa, Aa	
(2)	$aa \times aa$	aa, aa	
(3)	$aa \times aa$	aa, aa	
(4)	$aa \times AA$	Aa, Aa	
(5)	$Aa \times aa$	aa, aa	(possible to have *Aa* offspring)
(6)	$aa \times aa$	aa, aa	
(7)	$Aa \times AA$	Aa, Aa	(possible to have *AA* offspring)
(8)	$aa \times Aa$	aa, aa	(possible to have *Aa* offspring)
(9)	$aa \times aa$	aa, aa	
(10)	$aa \times Aa$	aa, aa	(possible to have *Aa* offspring)

If we count up the offspring in this hypothetical example we see that the ratio among the offspring is $0AA : 6Aa : 14aa$. Among these offspring we find that

$$p_1 = \text{Prob}[A \text{ in } F_1] = \tfrac{6}{40} = .15 \text{ or } 15\%$$

and

$$q_1 = \text{Prob}[a \text{ in } F_1] = \tfrac{34}{40} = .85 = 85\%$$

Clearly there has been a change in the gene proportions. In this particular case these changes are due entirely to random fluctuations in a small population. In future generations it would be entirely feasible for the proportion of dominant genes to decrease further and ultimately to be lost from the population; and so the population would consist of only recessive genes. If this should occur we would say that the recessive gene *a* had become *fixed* in the population and that the dominant gene *A* had become *lost*. Of course, some similar deviation from Hardy-Weinberg equilibrium could also occur under random mating in large populations, but it is very unlikely to do so.

To comprehend the essence of genetic drift we can also use a table of random numbers to simulate the behavior of a population of 10 couples (20 people) who each produce two offspring. Using a table of random numbers (a table with many digits comprised of approximately equal numbers of each digit in a random order) we predict the proportion of dominant genes in each generation. The use of the table of random numbers assures (as much as possible) that no biases are creeping in and the results are just chance fluctuations, that is, genetic drift. The results are given in Fig. 16-14. (Note that if this simulation were to be done once more the graph would almost certainly look quite different and it is possible that the dominant gene could be lost and the recessive gene could become fixed.)

It is also very interesting to compare Fig. 16-14, the graph of genetic drift, with Figs. 16-10, 16-12, and 16-13, the graphs for selection. We now can see the results of chance effects. Figures 16-10, 16-12, and 16-13 are essentially smooth curves without any up-and-down jumps, and the general tendencies are predictable, whereas the values in Fig. 16-14 are totally unpredictable and knowledge of one generation gives relatively little information about the next generation.

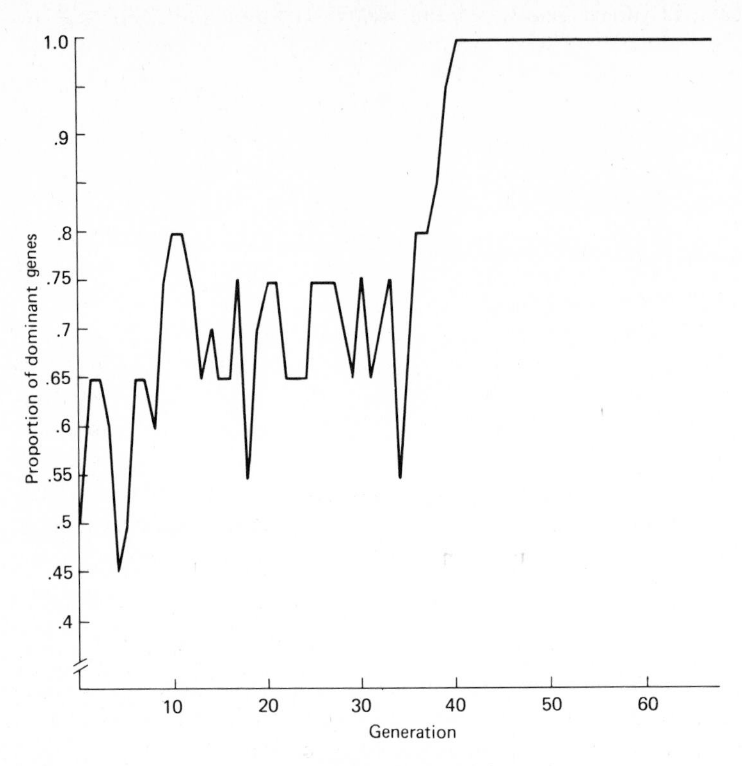

Figure 16-14. Genetic drift.

So far we have discussed hypothetical examples to indicate the effect of genetic drift. We will next discuss some real examples.

Dr. Bentley Glass (1952, 1953, and 1956) from Johns Hopkins University studied the Old Order Dunker (Old German Baptist Brethren) community in Pennsylvania. The Old Order Dunkers are one of a few communities in various regions of the United States that tend to be very closed. People who marry outside the community are excluded from the community. The Dunker population has existed at an average size of 250 to 300 persons for more than 200 years, ever since they migrated from Germany. Glass studied various genetic traits and compared some of them to those in the overall West German and United States populations. For the MN blood groups the results are given in Table 16-18.

Interestingly enough the proportions in West Germany and the United States are essentially the same. Yet among the Dunkers the proportion of people with type M blood is considerably higher than in the two other groups. For the Dunkers there is no apparent reason to explain the selective advantage of being type M, so that this difference is probably explained by chance fluctuation, or genetic drift.

Dr. Glass made some other interesting observations when he divided the Dunkers into three groups that might correspond roughly to generations: 56 years old or older, 28 to 55 years, and 2 to 27 years. He found that among the oldest group the proportion of *M* genes was .54; that among the middle group the proportion

Table 16-18 Comparisons of M, MN, and N Blood Types Among West Germans, Dunkers, and Americans

		Blood Type Percent			Proportion of Gene	
	Number	M	MN	N	M	N
West Germany (Main, Koln, Bonn, Frankfurt)	6,800	29.85	49.90	20.20	54.80	45.20
Dunkers	229	44.50	41.90	13.50	65.45	34.55
United States (Brooklyn, N.Y.; Columbus, Ohio)	6,129	29.16	49.58	21.26	53.95	46.05

From H. B. Glass, M. S. Sacks, E. F. Jahn, and C. Hess, 1952. *Am. Nat.* 86: 145–159.

was about .66; and that among the youngest group the proportion was about .74. This type of fluctuation would strongly suggest genetic drift.

Dr. Glass also showed that the proportion of genes for blood type A was .6 among the Dunkers, and .4 in the United States and West Germany. He also found that among the Dunkers the proportion of the *Rh* gene was .11, whereas it was .15 in the United States and Germany.

As a second example of genetic drift, about 80% of Blackfoot and Blood Indians in Montana have blood type A. Among other North American Indians this percentage ranges from 2 to 22. Since there is no apparent reason to think that blood type A should be advantageous to Blackfoot and Blood Indians, it seems likely that their high percentage is probably due to genetic drift rather than to selection.

Founder Principle

Similar to genetic drift though somewhat different is the *founder principle*, which is also based on randomness, or chance events. It is possible that if an unpopulated area is settled by a nonrandom group of people who differ genetically from the parent population, their offspring and future generations will differ significantly from the parent population. A striking example of the founder principle occurs among the Old Order Amish of Lancaster County, Pennsylvania, who manifest an uncharacteristically high proportion of the recessive gene which in the homozygous state causes a particular form of dwarfism accompanied by extra fingers (polydactyly). Almost all 8000 people in this closed community include among their ancestors three couples who came to the United States before 1770. This form of dwarfism–polydactyly is so rare that outside of Lancaster County only about 50 cases have so far been reported in the medical literature. But among the Amish of Lancaster County 61 definite cases have been recorded; and it is estimated that 13% of the Amish are carriers of the gene. Such evidence suggests that purely by chance one of the "founders" of this settlement in Lancaster must have been a carrier. That is, one person out of 200 was probably a heterozygote. Amish living in Ohio or Indiana do not show this form of dwarfism–polydactyly. The combination of the founder principle and random genetic drift probably account for what has happened to the Amish in Lancaster County.

Genetic Drift and Selection

In many cases various effects can operate simultaneously, for example, both selection and genetic drift could be operating on the same population. Dobzhansky and Pavlovsky (1957) used experimental populations of *Drosophila pseudoobscura* to show the interaction of selection and the "founder" type of genetic drift. In their experiments Dobzhansky and Pavlovsky showed that the effects of selection between two chromosomal arrangements of *Drosophila pseudoobscura* (*AR* and *PP*) were more predictable for a population with a large number of founders than for populations with just a few founders.

Twenty identical populations were kept in uniform environmental conditions. The founders for each of the populations consisted of second-generation hybrids (F_2) from pure parental strains of a Texas population with *PP* arrangements of chromosome #3 and a California (Mather origin) population with the *AR* arrangement in the same chromosome. Ten populations were each founded by 4000 flies, while the other ten populations were each started by only 20 founders. Each of the 20 populations was allowed to breed at random and grow to the maximum population which the cage could hold. All 20 populations started with 50% of *AR* types and 50% of *PP* types. The percentage of *PP* types at the end of five months and of eighteen months are given for each of the 20 populations in Fig. 16-15; means and variances are given in Table 16-19. The results are very interesting. At the end of 18 months the variance of the smaller population is statistically significantly greater than that

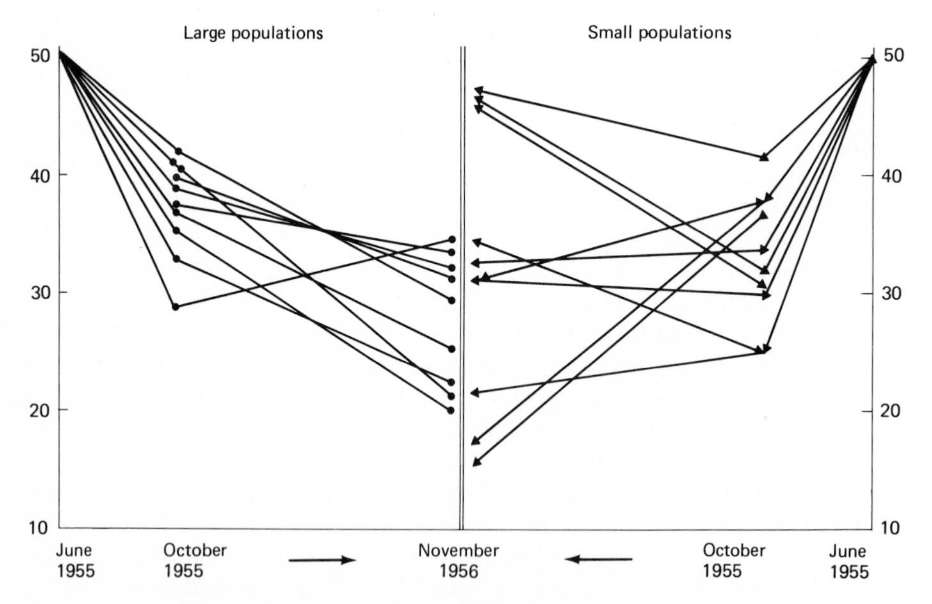

Figure 16-15. The frequencies in percentage of *PP* chromosomes in 20 replicate experimental populations of mixed geographic origin (Texas by California); populations founded by large initial size samples (N = 4000) on the left and by small initial size samples (N = 20) on the right. (From Dobzhansky and Pavlovsky, 1957. *Evolution* 11: 311–319.)

Table 16-19 Percentage of *PP* Chromosomes in Large and Small Populations

Large Populations (Founded by 4000)			Small Populations (Founded by 20)		
Number	*5 months*	*18 months*	*Number*	*5 months*	*18 months*
145	39.3	31.7	155	37.7	18.0
146	42.3	29.0	156	30.7	32.0
147	29.3	34.7	157	31.0	46.0
148	38.0	34.0	158	32.3	46.7
149	33.3	22.7	159	34.3	32.7
150	36.0	20.3	160	41.7	47.3
151	40.3	32.0	161	37.3	16.3
152	41.0	22.3	162	25.3	34.3
153	37.0	25.7	163	37.7	32.0
154	42.0	22.0	164	25.3	22.0
Mean	37.85	27.44	Mean	33.33	32.73
Variance	15.30	26.96	Variance	26.73	118.91

From T. Dobzhansky and O. Pavlovsky 1957. *Evolution* 11: 311–319.

for the large population. (We conclude that the variance is greater at the 5% level of significance.) If we look at Fig. 16-15 again, we see that in the small populations three of the groups seem to be returning to 50% *PP* types and two of the groups may have approached zero, if the experiment had continued long enough. The large population uniformly showed a steady decrease in the percentage of *PP* types.

It should be clear that the effects of genetic drift in small populations can be dramatic, while in large populations the effects average out and tend to be closer to the Hardy-Weinberg predictions.

Migration

Lastly we discuss one additional way in which new genes can be introduced into the gene pool. When we still assumed that the gene pool included all possible organisms, the main source of new genes was mutations. When discussing genetic drift we implied that the gene pool may actually consist of subunits: geographical isolates; religious and ethnic groups; species; and so forth. Since the various subunits are not necessarily totally closed systems, immigration from one unit to the other is possible. By immigration, different genes can be introduced into a population. A single such introduction into a new population is called *gene exchange*; if the migration is constant and recurrent it is known as *gene flow*. The closer the populations (subunits) are related spatially and genetically, the more likely become the chances of gene flow, which is probably at its highest among adjacent subunits of one species. Migration among the subunits would be equivalent to interbreeding. If interbreeding among contiguous subunits is high, it seems very likely that the subunits have similar gene pools. In such cases gene flow will not significantly modify the gene pool.

Geographic isolates (populations separated by geographical barriers) or strict religious groups (Dunkers or Amish) are more independent, and gene flow into and out of them is limited, but nevertheless it may exist. The tendency for divergent isolated breeding populations to develop may be counteracted by migration. Migration pressure results from the invasion of a population by individuals that breed with members of the population and thereby add their genes to the gene pool going to form subsequent generations.

The effectiveness of gene flow depends on the amount of migration and the dissimilarity of the population subunits. We can best see this by deriving a simple model to explain gene flow. Assume that we are studying the effects of gene flow on a semi-isolated island population. The immigrants will come from a large mainland population, and the emigration from the island population to the mainland is so small that it will be assumed to be negligible. All assumptions of the Hardy-Weinberg model apply except that we are now permitting gene flow. We assume that migration is occurring at a constant rate, M, from the mainland, in the sense that in each generation a fixed proportion, M, of the island's population is from the mainland. M is known as the *migration coefficient*,[8] or the *coefficient of replacement*. The proportion of recessive genes, *a*, on the mainland is Q_m. We will assume that the population of the mainland is so large and the migration rate so small that the effect on Q_m is negligible. The proportion of recessive genes, *a*, on the island during the zero generation is denoted by q_0, and this proportion will vary from generation to generation since the migrants and their ancestors will represent a greater proportion of the island's population. To derive our model we note that the total island population at each generation consists of two segments: the new immigrants and the natives. In fact, the proportion will be M immigrants to (1 − M) natives. To determine the proportion of recessive genes in the island we will really consider a weighted average of the two groups. That is,

$$q_1 = MQ_m + (1 - M)q_0 \qquad (15)$$

q_1	M	Q_m	$(1 - M)$	q_0
Proportion of q in first generation	Proportion of immigrants	Proportion of *a* among immigrants	Proportion of natives	Proportion of *a* among natives

We rewrite equation (15) as

$$q_1 = MQ_m + q_0 - Mq_0 \qquad (16)$$

If we look at equation (16), we see that the difference in the proportion of the recessive gene is given by $q_1 - q_0 = -M(q_0 - Q_m)$. So we see that the effect of migration is measured by the migration coefficient and the genetic differences in the populations. Following a similar analysis, we can show that in general

$$q_{n+1} - q_n = -M(q_n - Q_m) \qquad (17)$$

To show the effect of migration we note that

$$q_n = (1 - M)^n(q_0 - Q_m) + Q_m \qquad (18)$$

[8]It is probably more accurate to call it the *immigration* coefficient.

and consider the following hypothetical values. Let $M = 10\%$, $q_0 = .25$, and $Q_m = .75$; then $q_1 = (1 - .10)^1(.25 - .75) + .75 = .30$. Using equation (18) we derive Table 16-20.

Table 16-20 Effects of Migration

Generation	q_n	*Generation*	q_n
0	.25	7	.51085155
1	.30	8	.534766395
2	.345	16	.65734899
3	.3855	32	.73283158
4	.42195	64	.74941049
5	.454755	128	.749999304
6	.4842795	256	.749999999

Under these hypothetical conditions the effect of migration is quite noticeable. We also see from equation (17) that in such a population there would not be an equilibrium unless $q_n = Q_m$. Also note that if M is large and q_n and Q_m are very different, q_{n+1} will vary rapidly. If M is large and q_n and Q_m differ, breeders often introduce new sires or some other source of new genetic material into their population, in order to modify the proportions of particular genes.

It should be noted that migration can occur in conjunction with selection. Formulas can be derived, but they are beyond our level of presentation.

We have attempted in this chapter to discuss, at least briefly, a number of models to explain the fluctuations of genes in populations. Our discussion has centered on the various ways in which the simple Hardy-Weinberg equilibrium can be modified to make the model more closely approximate reality. Even then we have oversimplified the problems. In the next chapter, we shall continue to study fluctuations in populations, but from the point of view of evolution.

PROBLEMS

1. Assume that we have a population of flowers mating randomly. There are two genes, *R* for Red Flowers and *r* for White Flowers. The proportion of *R* genes in males is .4, and the proportion of *R* genes in females is .6. Determine the proportion of *RR*, *Rr*, *rr* when these flowers mate randomly.

2. (a) It is known that about one in 19,881 people is an albino. If the proportion of recessive genes among men and women is equal, determine q, the proportion of recessive (*a*) genes in the population.

(b) Determine p, the proportion of dominant (*A*) genes in the population.

(c) Determine Prob[*AA*], the probability of being homozygous dominant.

(d) Determine Prob[*Aa*], the probability of being a heterozygote.

3. Suppose that we have a population of plants in which there is not complete dominance. We let *R* be the trait Red Flowers, and let *r* be the trait White Flowers, so genotypically *RR* plants appear red, genotypically *Rr* plants appear pink, and genotypically *rr* plants appear white. We assume that 35% of the plants have red flowers, 30% have pink flowers, and 35% have white flowers. Determine the values of the Hardy-Weinberg equilibrium, if the plants mate randomly and all the conditions for the Hardy-Weinberg equilibrium are satisfied.

4. Assume that a population has the following genotypic makeup:

	Number of people		
	NN	*Nn*	*nn*
Male	1200	800	2000
Female	900	600	500

(a) Assume independent random mating and determine:

$$\text{Prob}[NN], \quad \text{Prob}[Nn], \quad \text{and} \quad \text{Prob}[nn]$$

in the F_1 offspring.

(b) If the offspring of each generation mate independently and randomly, determine:

$$\text{Prob}[NN], \quad \text{Prob}[Nn], \quad \text{and} \quad \text{Prob}[nn]$$

in the F_{14} offspring.

5. A group of 500 people migrate to an isolated island. Fifty years later a geneticist determines that 250 people are *SS*, 200 are *Ss*, and 50 people are *ss*. Has the population achieved the Hardy-Weinberg equilibrium values? Explain.

6. One hundred years ago a group of 200 people migrated to an isolated area in Alaska. They had the following distribution of M, MN, and N blood types: 95 of type M, 70 of type MN, and 35 of type N. Today their descendants number 4000. If the conditions of the Hardy-Weinberg law have been satisfied, how many of the 4000 will be expected to have each blood type?

7. For the hypothetical population that immigrated from Liège to the isolated island (page 507) verify that the Hardy-Weinberg equilibrium values are:

16% blood type M
48% blood type MN
36% blood type N

8. On page 510 verify $p_m^n = p_f^{n-1}$ and $p_f^n = (p_m^{n-1} + p_f^{n-1})/2$. Hint: use tree diagrams and treat ♂ and ♀ offspring separately.

9. Verify that $\bar{p}^n = \bar{p}$ on page 511, where $\bar{p}$ is the overall proportion of X^A alleles in the population among ♂ and ♀ for a sex-linked trait.

10. Verify for blood type A, B, AB, and O, that $q' = \text{Prob}[I^B \text{ alleles}] = q$ and $r' = \text{Prob}[i \text{ alleles}] = r$. (See page 512.)

11. Verify the Hardy-Weinberg equilibrium for four alleles at one locus on a pair of chromosomes.

12. Verify the entries in Table 16-5.

13. Verify the Coefficients of Inbreeding given in Table 16-7.

14. (a) Determine the probability of an albino offspring if an uncle and niece were to mate. Compare this value to the value for simple random mating.
(b) Repeat (a) for second cousins.

15. Verify the values given in Table 16-10 and determine the values for five and six children.

16. Redo Problem 1, but assume that selection against the homozygous recessive (*rr*) is (a) 1, (b) .9, (c) .5, (d) .4, (e) .2, and (f) .1. Compare the results.

17. Using the information in Problem 2, determine Prob[*AA*], Prob[*Aa*], and Prob[*aa*] assuming that selection against *aa* is (a) 1, (b) .85, (c) .75, (d) .55, (e) .25, and (f) .05. Compare the results.

18. Redo Problem 3, but assume that
(a) selection against the white flowers is .2;
(b) selection against the *R* gene is .2; then
(c) compare the results.

19. (a) Assume that the population in Problem 1 is in Hardy-Weinberg equilibrium and $p = q = \frac{1}{2}$, and that selection against the homozygous recessive is 1. Determine the proportions of homozygous recessives (*rr*) for generations 1, 2, 3, 4, 5, 6, 7, 8, 9, 10, 20, 50, 100.
(b) Assume selection against the homozygous recessive is .5. Determine the proportion of homozygous recessives (*rr*) for generations 1, 2, 3, 4, 5, 6, 7, 8, 9, 10.
(c) Repeat part (b) assuming that $s = .3$.
(d) Compare the answers to parts (a), (b), and (c).

20. What would be the effect on the number of albinos born if only 50% of the albinos were allowed to reproduce? (Fill in a table like Table 16-14 if $s = .5$)

21. Consider the effects if a eugenics program were instituted for carriers of Sickle-cell anemia or Tay Sachs disease.

22. Consider Figs. 16-10, 12, and 13 and discuss the differences and similarities for the comparable values of s in each case.

23. Discuss the effects of selection favoring the heterozygotes and the Sickle-cell trait in malarial areas. What are the effects in nonmalarial areas?

24. Angus cattle can be black or red. The Red Angus cow is a homozygous recessive. Black Angus cattle is preferred over Red Angus cattle. For many generations the Red Angus cattle have been selected against, but Red Angus cattle have not been eliminated. Assume that originally the dominant and recessive genes exist equally in the population.
(a) Sketch a graph to show the proportion of Red Angus cattle over time.

(b) How might the Red Angus cattle be eliminated more quickly? Justify. Can your solution be implemented? Explain.

25. Assume that sheep breeders want to eliminate yellow fat in sheep. Assume that yellow fat is a homozygous recessive trait, and that 10% of the sheep have yellow fat. Consider various selection policies and pick the most satisfactory policy and explain.

26. Consider the effects of migration if $M = 15\%$, $q_0 = .3$, and $Q_m = .8$. (Determine entries for Table 16-20.)

27. The ability to taste PTC is controlled by a dominant gene *T*. Nontasters are homozygous recessives (*tt*). If 64% of the population are tasters, estimate the proportions of *T* and *t* in the gene pool.

28. Suppose that 20% of the population are Rh-negative (see Chapter 4) (the dominant gene results in Rh-positive blood). If the population has attained Hardy-Weinberg equilibrium, estimate the proportions of dominant and recessive alleles.

29. If the rate of children born with PKU is 1/10,000, determine the probability of a child with PKU, if

(a) 2 unrelated people mate;
(b) brother and sister mate;
(c) first cousins mate;
(d) second cousins mate;
(e) uncle and niece mate; then
(f) compare the results.

REFERENCES

BOYD, W. C., 1950. *Genetics and the Races of Man.* Little Brown, Boston.

CASTLE, W. E., 1903. "The Laws of Galton and Mendel and Some Laws Governing Race Improvement by Selection." *Proc. Amer. Acad. Arts and Sci.* 35: 233–242.

DOBZHANSKY, T., AND O. PAVLOVSKY, 1957. "An Experimental Study of Interaction Between Genetic Drift and Natural Selection." *Evolution* 11: 311–319.

GLASS, H. B., 1953. "The Genetics of the Dunkers." *Sci. Am.* 189: 76–81.

———, 1956. "On the Evidence of Random Genetic Drift in Human Populations." *Am. J. of Phys. Anthrop.* 14: 541–555.

———, M. S. SACKS, E. F. JAHN, AND C. HESS, 1952. "Genetic Drift in a Religious Isolate: An Analysis of the Causes of Variation in Blood Groups and Other Gene Frequencies in a Small Population." *Am. Nat.* 86: 145–159.

HARDY, G. H., 1908. "Mendelian Proportions in a Mixed Population." *Science* 28: 49–50.

HERSKOWITZ, I. J., 1965. *Genetics*, 2nd ed. Little Brown, Boston.

LIVINGSTONE, F. B., 1967. *Abnormal Hemoglobins in Human Populations.* Aldine, Chicago.

PAI, A. C., 1974. *Foundations of Genetics.* McGraw-Hill, New York.

SCHULL. W. J., 1958. "Empirical Risks in Consanguineous Marriages: Sex Ratio, Malformation and Viability." *Am. J. Human Gen.* 10: 294–343.

SPIESS, E. B., 1977. *Genes in Populations.* Wiley, New York.

WEINBERG, W., 1908. "Über den Nachweis der Vererbung beim Menschen." ("On the Demonstration of Heredity in Man.") *Jahreshefte des Vereins für Vaterlandische Naturkunde in Württemberg, Stuttgart* 64: 368–382. Translated in S. H. Boyer, ed., 1963. *Papers on Human Genetics.* Prentice-Hall, Englewood Cliffs, N.J.

WIENER, A. S., 1943–1962. *Blood Groups and Transfusions*, 3rd ed. Charles C. Thomas, Springfield, Ill. Reprinted by Hafner, New York.

WRIGHT, S., 1921. "Systems of mating I–V." *Genetics* 6: 111–178.

SEVENTEEN

The Genetics of Evolution

INTRODUCTION

Variability in populations of plants and animals, and how it comes about, generates much interest in the natural world. Underlying this variability are genetic differences which arise from many of the factors we have discussed in preceding chapters, such as mutations. These differences progressively become so great that we characterize the resulting populations as distinct species. The process of progressive change of phenotypes of populations is known as *evolution*.

In this chapter we shall discuss classical and modern approaches to the analysis of evolutionary changes and relationships. In this analysis, we can hope to gain some insight not only into the manner in which other species have developed, but also into the position of our own species in the evolving natural world.

A HISTORICAL PERSPECTIVE

The analysis of evolutionary changes has been a preoccupation of humans since long before Mendel's studies and the establishment of genetics as a science. Curiosity about different organisms and how they arise inevitably extended to thoughts on the origin of life itself, and such thoughts were first formulated in primitive societies and later developed into philosophies and religions. Most of these dogmas maintain that life as we know it has been placed on our planet by a supernatural force, and that changes which have occurred are determined by this force.

Therefore, implicit in such "creationist" dogmas is the idea that forms of life can change abruptly, and totally. Furthermore, the pattern of such changes is not random, but follows some form of master plan. The superiority of human mental capacity thus would represent some form of epitome in evolution, consigning to humans a special position in the natural world. Most biologists would agree there is no real evidence that allows us to determine how life originated; however, with few exceptions, most biologists today support a concept of change in life forms very different from that of the creationists mentioned above.

The foundations of modern biological concepts of evolution were laid more than a century ago. By the turn of the nineteenth century, geological evidence had accumulated which argued against the idea that catastrophic events led to changes in forms of life on earth. Geologists such as Charles Lyell concluded from their studies of rocks and fossils found in different layers of the earth that change in both rock formation and life forms is gradual, progressive, and continuous.

Lamarckism

In 1809 a French zoologist, Jean Baptiste Lamarck, proposed that constantly changing environments place stresses on organisms. In order to survive, organisms change physically in an attempt to adapt to the stresses. New structures would develop in response to the changing conditions, and unused structures would disappear. This idea, known both as *Lamarckism* and as the *inheritance of acquired characteristics*, contained a serious flaw: the implication that such changes could then be transmitted to the next generation. For example, giraffes needed to stretch to reach leaves of trees during droughts; their offspring were then born with longer necks. In his time, Lamarck's theory seemed quite logical, for it was proposed well before Mendel, and well before genes and their transmission were understood.

Other biologists, however, recognized this flaw, and though they could not propose a countertheory, discounted the development of new forms of life through response to environmental change. The evidence for Lamarckism just did not exist. Surely organisms can adapt to an extent to environmental change, but their capability to do so exists *before* the change occurs. For example, we build shelters as protection from the environment. Our ability to do so is not simply a response to changing environmental conditions, however; it is inherited. Those that cannot adapt simply become extinct. Furthermore, if Lamarck's arguments were true, a person whose arm had been amputated for example could conceivably transmit such a change to his or her offspring; such phenomena are of course not observed.

Darwin and Natural Selection

As a result of observations made on perhaps the most well-known voyage ever taken by a biologist, a book titled *The Origin of Species* was published in 1859, authored by a naturalist, Charles Darwin (Fig. 17-1). The book contained a revolutionary theory on evolution which in many ways directly contradicted religious dogma. All

Figure 17-1. Charles Darwin. (American Museum of Natural History.)

biology textbooks abound with detailed descriptions of this famous trip to the Galapagos Islands, and Darwin's recognition of differences among the flora and fauna of the islands. We shall therefore only summarize here the most important observations that were made, and the conclusions Darwin drew from them on the process of *speciation*, or how different species evolve.

The Galapagos Islands are a group of 15 islands derived from volcanoes. Darwin noted that in the long span of time that had elapsed, populations of birds and other organisms which had at one time been of the same species, had immigrated to the different islands and evolved distinct physical differences which were well adapted to their own island environments (Fig. 17-2).

After years of deliberation (and prompted by a manuscript from Alfred Russel Wallace (Fig. 17-3), who had independently derived essentially the same theory from his observations in the Malaysian islands), Darwin published his analysis of the significance of these differences. He interpreted evolutionary change in the following way: all populations contain individuals with a variety of physical differences. These differences render some individuals better adapted to their environment than others who do not share the same characteristics. Since there is in all natural populations a struggle for survival, only those more fit will survive. These survivors, as they tend to be the ones that reproduce, will then transmit the favorable characteristics to their offspring. The less fit will die, and thus the less fit characteristics will also be removed from the population. Darwin called this process *natural selection*. Beneficial traits thus become fixed in a population over the years as part of the normal phenotype.

Darwin later published another work, called *The Descent of Man*, in which he proposed that humans are also subject to the same forces of natural selection as any other species. His work raised a furor all over the world, and was especially

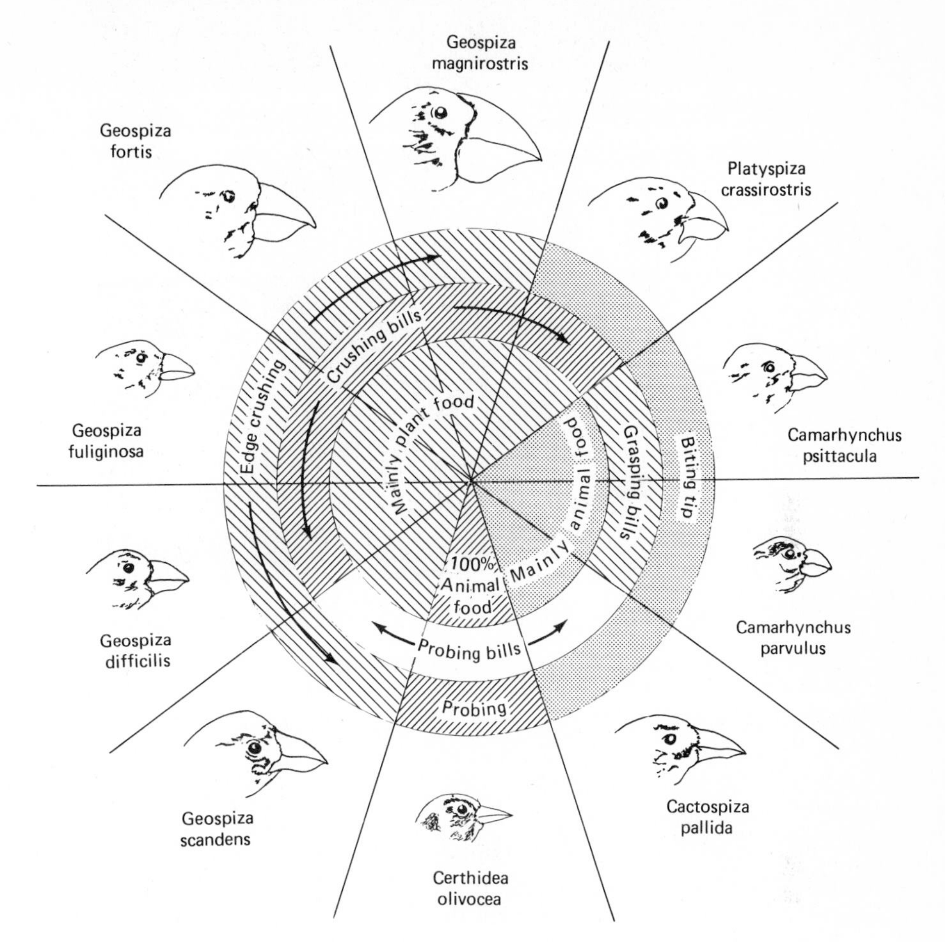

Figure 17-2. Morphological and functional differentiation of the bills of the Galápagos finches. [Adapted from Bowman. (From T. H. Eaton, Jr., 1970. *Evolution*, p. 81. W. W. Norton & Co., Inc., New York.)]

denounced by those who espoused fundamentalist religions. FOR IMPLICIT IN DARWIN'S THEORY IS THAT EVOLUTIONARY CHANGE IS A GRADUAL PROGRESSIVE PROCESS WHICH IS RANDOM, DEPENDENT ON CHANCE VARIABILITY IN POPULATIONS AND INFLUENCED BY FLUCTUATIONS IN THE ENVIRONMENT. THUS THE EXISTENCE OF ALL SPECIES, INCLUDING OURS, HAS COME ABOUT NOT BY DESIGN, BUT BY CHANCE. Furthermore, humans must then be directly related to some of the "lesser" living creatures, such as the apes, to whom we bear more obvious resemblance than to any other species.

Resistance to Darwinism continued with great force into the twentieth century, even after irrefutable evidence supporting the theory of natural selection accumulated. One famous example of this resistance came in the 1925 "Monkey Trial," in which a biology teacher named John Scopes was tried and convicted for teaching Darwinism in his classes, thus breaking a Tennessee law forbidding the inclusion of the theory

Figure 17-3. Alfred Russel Wallace. (American Museum of Natural History.)

of natural selection in biology classes. The confrontation and debate between the opposing lawyers—Clarence Darrow for the defense and William Jennings Bryan for the state—has been depicted in literature, on stage, and in films as a classic representation of opposing camps in the controversy over Darwinism.

Even today, there are groups who promote the creationist theory of evolution as an educational policy. For example, a controversy over the inclusion of creationism in textbooks occurred in California in the early 1970s (Nelkin 1976). Scientists on the whole, however, now feel that the process of change after life appeared on earth (however that might have happened) is one that operates essentially on Darwinian concepts of natural selection. In the following sections, we shall discuss modern modifications of Darwin's theory of evolutionary change.

NEO-DARWINISM

Our modern concept of evolution can be appropriately termed *neo-Darwinism*, because it supports most of the basic concepts originally proposed in *The Origin of Species*. We call it *neo-* ("*new*") Darwinism because its basic tenet is natural selection; however, we have been able to expand and refine Darwin's theories, since we now have the advantage of knowing how the mechanisms of heredity operate.

Sources of Variability in Populations

The major change from Darwinism to neo-Darwinism lies in the fact that where Darwin simply recognized the existence of variability in populations, we now can account for the sources of such variability. We will discuss these sources of variability in the following sections.

Mutation. Ultimately, variability is the result of mutation. We must stress here that mutations, being by definition only genetic changes, are not in themselves beneficial or detrimental to the organisms in whose cells they occur. What causes a given mutation to be "good" or "bad" is its effects on an individual in a particular environment. For example, a mutation to streptomycin resistance in bacteria is beneficial only if there is streptomycin in the environment.

Thus, depending on an organism's environment, the same mutation may be beneficial in one area but detrimental in another, a phenomenon which accounts for the existence of different genes in the gene pools of different populations. First there occurs the *random* genetic changes we call mutation, and secondly, mutations benefiting organisms, depending on their environment, are retained in the gene pool and transmitted to future generations.

Sexual reproduction. Another source of variability for most complex organisms is sexual reproduction. During gametogenesis there is random assortment of genes, and also crossing-over and genetic recombination. These two phenomena, as you know, result in diversity of germ cells. If the randomness of fertilization is also considered, diversity of offspring can be enormous.

Factors of population genetics. In addition to mutation and sexual reproduction, all the factors discussed in the preceding chapter on population genetics serve to change the frequency of genes in populations. Such change in the frequency of genes also leads to variability in populations. For example, we have seen that migration and genetic drift can change the gene pool of a population in dramatic ways.

Evolutionary Advantages of Variability in Populations

We have on occasion alluded to the advantages of variability in populations. We can return to it here by pointing out that if a population is genetically homogeneous, a change in the environment could devastate the entire population. Staying with our simple example of bacteria, if all bacteria in a population were susceptible to streptomycin, the presence of that antibiotic would eliminate that population. But if there is genetic variability in the population in the form of resistant individuals, at least these cells can survive in the presence of streptomycin.

The advantage of genetic variability holds true for populations of all organisms. The ability of an organism to survive in a changing environment is referred to as *adaptation*. As we stressed in our discussion of the fluctuation tests of Delbruck and Luria (p. 479), ADAPTATION RESULTING FROM GENETIC CHANGES IS NOT A RESPONSE ON THE PART OF ORGANISMS TO THE ENVIRONMENT. IT RESULTS FROM THE PRESENCE IN VARIOUS INDIVIDUALS IN THE POPULATION OF PREEXISTING RANDOM MUTATIONS WHICH BY CHANCE RENDER THEM ABLE TO ADAPT TO ENVIRONMENTAL CHANGES.

Thus the greater the variability in a population, the greater the probability that it can survive. Along this line, since sexual reproduction leads to genetic variability, sexually reproducing organisms have a great advantage over asexually reproducing organisms such as bacteria. In the latter, binary fission gives rise to offspring which are genetically identical to the parent cell.

On the other hand, the very persistence of asexually reproducing species testifies to their adaptive capabilities. What sources of genetic variability would be the basis for this adaptiveness? A moment of reflection should lead you to recognize that one source is the very short life cycle and huge numbers of organisms that can be built up in a population of prokaryotes. Random mutations can then be expected to produce enough genetically different organisms in a generation for the population to survive. In addition, remember that genetic recombination takes place between bacteria, mediated by viruses and plasmids.

Interspecies infection by viruses can also contribute to greater genetic variability in higher organisms. For example, studies of endogenous feline leukemia viruses indicate that these sequences have been in existence for a long period of time. The viruses actually are genes derived from a rodent virus which infected a host ancestor of modern cats. These genes were incorporated at some point into the genomes of cats and transmitted over the centuries. (See reference to Todaro 1974 in Chapter 12.)

Along this line, we must recognize that the long life spans and reproductive cycles of mammals and other complex forms of life are a distinct *disadvantage*. If the environment changes at a very rapid rate, there may not be enough time for such species to have produced enough offspring which have sustained or received the mutations necessary to convey fitness for their changing environment. We need only to point out the disturbing numbers of species of birds and mammals which have become extinct, or are threatened with extinction, in our own lifetime to illustrate this point. Indeed, we humans are not exempt from the forces of natural selection, as we shall discuss in more detail later.

Speciation

As we mentioned earlier the process of evolutionary change through which different species develop is known as *speciation*. We can define a *species* in a genetic sense as a group of individuals among whom breeding can take place. Members of a species can breed only with members of the same species and not with members of different species. By *breeding* we mean the production of fertile offspring. Mating can occur between members of some different species, as between the horse and donkey to produce the mule as offspring. However, genetically we do not consider this to be successful breeding, as the mule is sterile. How do populations of apparently related organisms become different species so that they can no longer breed? GENERALLY, AS DARWIN RECOGNIZED, IT IS BELIEVED THAT THE FIRST STEP TOWARD SPECIATION IS GEOGRAPHICAL ISOLATION OF POPULATIONS.

Geographical isolation. As we mentioned earlier, geologic processes are believed to have contributed to the separation of populations. If the environments eventually become sufficiently different so that different traits become the more adaptive in each, it is assumed that these traits, genetically determined, will be retained in the separate gene pools. As centuries pass, and the environments diverge, the genomes of individuals in the different areas will also continue to diverge. Eventually, the divergence leads to the second step in speciation, namely, reproductive isolation.

Reproductive isolation. The genetic divergence which results from natural selection produces noticeable morphological differences between species of related organisms. Differences in bill structure among the finches of the Galapagos Islands provide a classic example (see again Fig. 17-2). BUT MORE IMPORTANT THAN MINOR MORPHOLOGICAL DIFFERENCES, FROM THE GENETIC POINT OF VIEW, ARE DIFFERENCES IN GENOMES WHICH CAN LEAD TO THE INABILITY OF MEMBERS OF DIFFERENT SPECIES TO INTERBREED. THIS INABILITY IS KNOWN AS REPRODUCTIVE ISOLATION.

Isolating mechanisms. Reproductive isolation can be brought about by different *isolating mechanisms* (Savage 1969). One such isolating mechanism is *ecologic* isolation. As an illustration, the pig frog, *Rana grylic*, and gopher frog, *Rana areolata*, live and breed in deep ponds and shallow waters, respectively. These habitat preferences eliminate the possibility of reproductive interaction between these two species, simply because they do not encounter one another.

Other characteristics that may result in reproductive isolation are *differences in behavior*, such as mating calls recognized only by members of the same species. In addition, genetically determined differences can lead to *size differences* that prohibit mating between members of different populations, for example, in different species of domestic dogs.

Isolating mechanisms are known in plants as well. Different species of plants have life cycles that result in the maturation of germ cells at different times of the year, so that pollination between them cannot take place. Obviously, there is also ecologic isolation between aquatic and terrestial plants.

As genetic divergence increases, eventually genomic differences become so great that even if mating occurs, either the offspring are not viable, or they are rendered sterile by genetic imbalances or defective gametogenesis. It is then that we consider two populations to be different species.

Note that because of reproductive isolation, two populations which will not interbreed can exist within the same geographical range. Such populations are known as *sympatric* populations, or better, sympatric species, since they represent the end result of the speciation process. The term *allopatric* refers to species that are actually separated by physical or ecological barriers, as in the case of the Galapagos Islands.

EVIDENCE FOR NEO-DARWINIAN EVOLUTION

Since genetic changes form the basis for evolutionary changes, evidence supporting the neo-Darwinian theory of evolution must come from genetic studies. However, our attempts to derive such evidence cannot follow traditional lines of genetic research. The primary problem is that we cannot make experimental crosses between species, since by definition species are unable to interbreed. Consequently, evidence for evolution based on random genetic changes and natural selection has had to be developed using other approaches, which we shall explore in the following sections.

Chromosomal Evidence for Genetic Divergence

If the basic premise holds true that genetic differences between species arise through the accumulation of different genes, we should expect to observe a spectrum of genetic divergence among populations, ranging from a few differences between closely related species to more and greater differences between less closely related species. There is much evidence that this is indeed the case.

Studies on Drosophila. Because of the advantages offered by studies on *Drosophila*, much work has been done on the genes and chromosomes of different *Drosophila* populations to illustrate genetic divergence both in nature and under laboratory conditions. Figure 17-4 compares haploid chromosome complements of *Drosophila virilis* sub-species. There exist in these chromosomes several inverted sections representing genetic divergence between species. The inversions can be detected by observation of changes of the banding patterns of chromosomes.

Theodosius Dobzhansky also found inversions in wild populations of *Drosophila pseudoobscura* in the southwestern United States. These inversions represent genetic divergence in the different populations of *Drosophila pseudoobscura*, which would presumably increase through time, under the influence of environmental changes, to the point of effecting reproductive isolation.

In one study, two populations of *D. pseudoobscura* which carried different inversions on the #3 chromosome were subjected to different environmental conditions in the laboratory (Dobzhansky and Spassky 1954). The two different gene arrangements on the #3 chromosome (we need not go into details here) were called Standard (*St*) and Chiricahua (*Ch*). The crossing of flies homozygous for *St* and *Ch* produced a generation of flies heterozygous for *St* and *Ch*, which were then crossed to each other at 25°C and 21°C.

At 25°C, heterozygotes showed a distinct survival advantage over both homozygotes for the two gene arrangements. Such a selective advantage would lead to establishment of heterozygotes in the population. If the flies were given a species of yeast known as *Kloeckera*, the selective advantage of heterozygosity (referred to as *heterosis*) was maintained only at 21°C. If the yeast *Zygosaccaromyces* was provided, *St*/*St* homozygotes were found to be as fit as heterozygotes, while *Ch*/*Ch* homozygotes decreased in fitness.

Added to these observations were data, collected in wild populations of *Drosophila pseudoobscura*, which revealed a seasonal influence on the frequency of *St* or *Ch* chromosomal arrangements (Dobzhansky and Levene 1948). For example, in some areas the frequency of *St* waned while *Ch* waxed from March to June; with the reverse occurring from June to September. On the other hand, in populations only 15 miles away, an eight-year study showed that there were no seasonal fluctuations, but rather a steady increase of the *St* gene arrangement and decrease of *Ch*. Differences in the environments of these populations must have resulted in the different patterns of selection.

Cytogenetic studies of *Drosophila* populations have also documented the progressive genetic divergence which one might expect in the process of speciation. Some populations of *Drosophila virilis* show the presence of different chromosomal rearrangements such as inversions shown in Fig. 17-4. The inversions do not prevent successful mating between individuals of such populations, however, and so we do not consider them separate species.

Figure 17-5, on the other hand, shows chromosomes of males of 15 different species of *Drosophila*. Gross differences that have accumulated over the course of

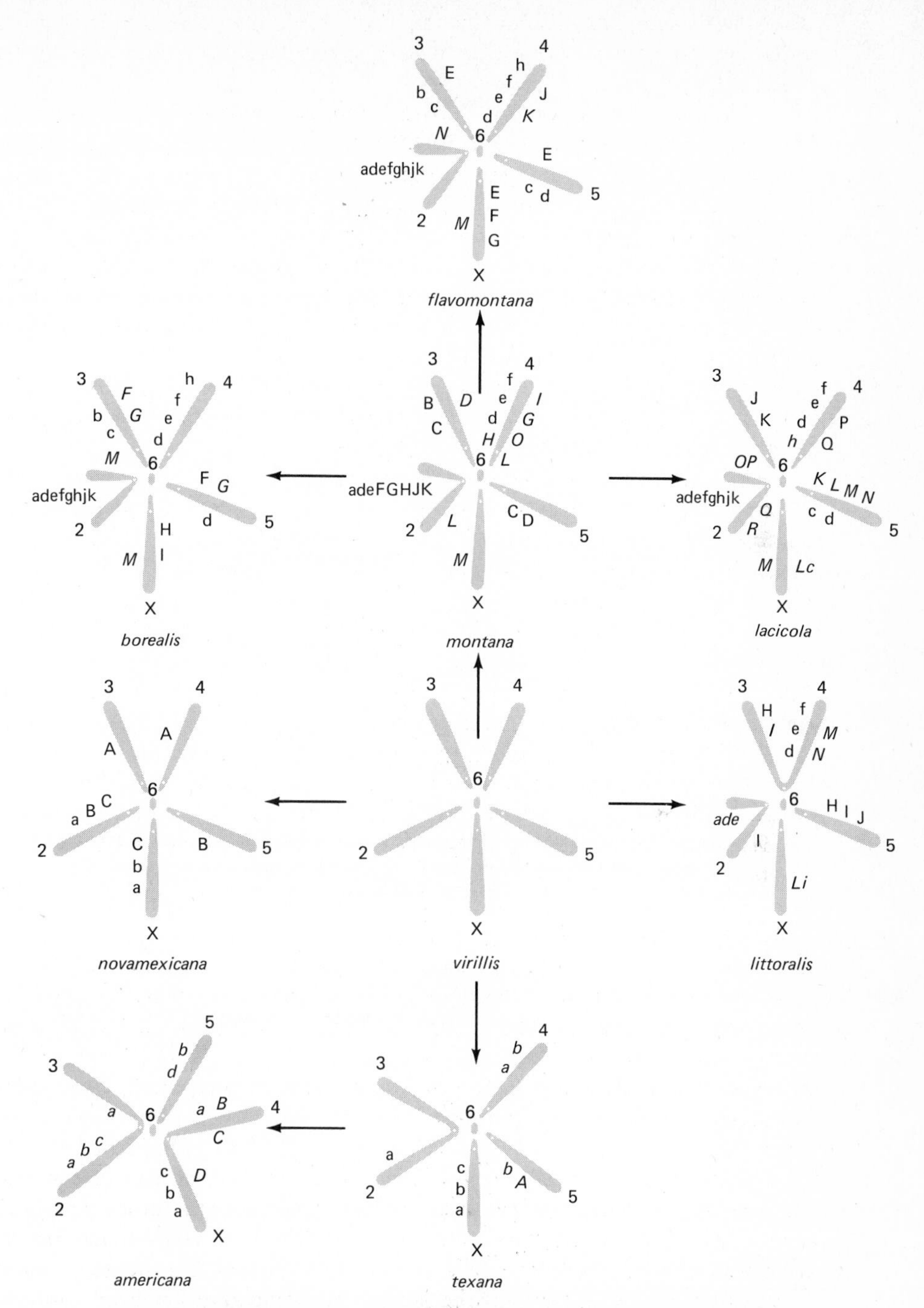

Figure 17-4. Comparison and probable relationships of haploid chromosome complements of the *virilis* group of *Drosophila* species. The letters indicate inverted sections of chromosomes. [After T. C. Hsu. (From E. W. Sinnott, L. C. Dunn, and T. Dobzhansky, 1958. *Principles of Genetics*, p. 297 McGraw-Hill, New York.)]

Figure 17-5. Chromosome complements in males of certain species of *Drosophila*. The drawings are so oriented that the X and Y chromosomes, which are visibly different in some species but not in others, are in the lower portion of each drawing. (1) *Drosophila virilis.* (2) *D. funebris.* (3) *D. repleta.* (4) *D. montana.* (5) *D. pseudoobscura.* (6) *D. miranda.* (7) *D. azteca.* (8) *D. affinis.* (9) *D. putrida.* (10) *D. melanogaster.* (11) *D. willistoni.* (12) *D. prosaltans.* (13) *D. ananassae.* (14) *D. colorata.* (15) *D. spinofemora.* (16) *D. americana.* [After Wharton. (From E. W. Sinnott, L. C. Dunn, and T. Dobzhansky, 1958. *Principles of Genetics*, p. 293. McGraw-Hill, New York.)]

evolution are quite obvious in the structures of the chromosomes, and these differences result in reproductive isolation.

Studies on higher organisms. Recent developments in improved staining and banding techniques for karyotyping have allowed the comparison of chromosomes between different species of higher organisms. Studies on the gross morphology of chromosomes of different species involve a comparison of banding patterns in chromosomes. Another approach is to hybridize DNA from chromosomes of one species to the chromosomes of different species (in situ) to observe the degree of homology of nucleotide sequences between the chromosomes.

Distinct similarities in structure and numbers of chromosomes among closely related primate species, such as humans and the great apes, are apparent in a comparison of their respective karyotypes (Fig. 17-6). For example, human chromosome #6 has a pattern of banding identical to that of a corresponding chromosome in the chimpanzee, gorilla, and organgutan. Interestingly, the gene for the enzyme superoxide dismutase-2, known to be mapped on chromosome #6 in humans, is not on the corresponding chromosome in the chimpanzee. Apparently, during evolution, the gene was translocated or in some way moved in either our predecessor or that of the chimp. Such divergences, correlated to fossil records, provide clues to the closeness of relationship between different species, and can give some indication of when, in the evolutionary past, humans began to differ from other primates.

Improvements in banding techniques have also enabled geneticists to identify gene rearrangements and chromosomal abnormalities which can shed light on the evolutionary relationship between "higher" species. The presence of inversions, for example, and their implication in the evolutionary process of higher organisms, has been reported in primates, cats, rats, and other taxa (Miller 1977). The chromosome banding patterns of the rhesus monkey compared with those of the African green monkey, on the other hand, indicate centric fusion and reciprocal translocations rather than inversions.

Human satellite DNA has been labeled and hybridized to chromosomes of other primates to study the degree of homology that might exist (Figure 17-7 shows the results of some of these experiments, performed by K. W. Jones and his coworkers). Furthermore, human satellite DNA has been found to consist of four groups: I, II, III, and IV. Jones has estimated that satellite III evolved 25–30 million years ago; satellite I, 20–25 million years ago; and satellite II, 9–12 million years ago. Classes I, II, and III are found in chromosomes of the orangutan, but class III is missing in the gibbon. One can therefore speculate that orangutans are more closely related to humans than are gibbons, and that the divergence between the orangutan and gibbon must have occurred after evolutionary establishment of these three satellites.

These are just some of the chromosomal studies now being conducted on innumerable species. Thus far, most estimates based on these studies have correlated well with more traditional sources of evidence for evolutionary change, such as fossil records. As our techniques improve in the analysis of chromosomal structures, so will our understanding of the relationships among species.

Evidence for Natural Selection

The validity of neo-Darwinian evolution theory depends not only on evidence for genetic divergence, but also on evidence that the resultant variability does confer adaptive advantage on some members of a population. In other words, there should be evidence that natural selection does favor the fixation of inherited traits that render the individual more fit in some way to survive and produce young.

Adaptations. Illustrations of adaptation and natural selection are abundant in our natural world. Indeed, ecology is an area of biology where the central interest is the analysis of how organisms are adapted to their environment. We shall give here only a few examples of adaptations.

Adaptations can involve behavioral, morphological, or physiological traits. One of the best-known examples of adaptation on the morphological level is that of

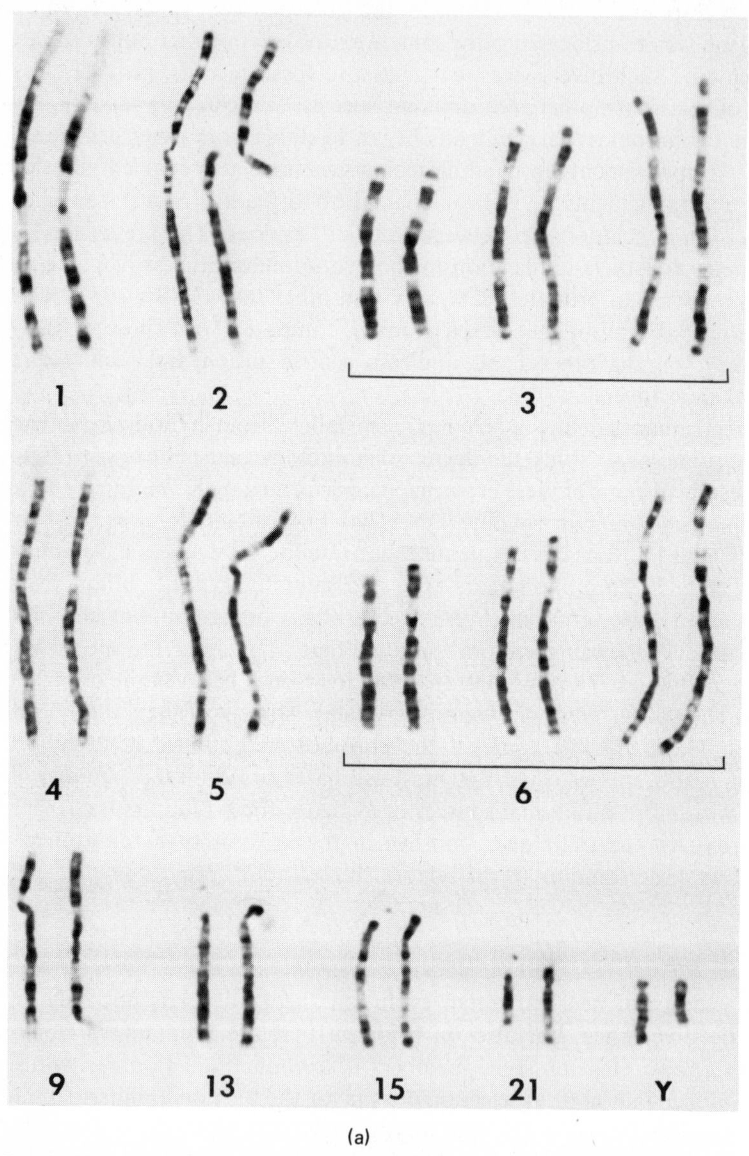

(a)

Figure 17-6. (a) Selected examples of Giemsa banded chromosomes of man and chimpanzee arranged to show maximum homology with the human karyotype. The chromosomes of man are on the left. (b) Schematic representation of late prophase chromosomes of man and chimpanzee arranged to show maximum homology with the human chromosome complement. Human chromosomes are on the left. (Photos courtesy Jorge J. Yunis, M. D. Dept. Laboratory Medicine and Pathology, Medical School, University of Minnesota.)

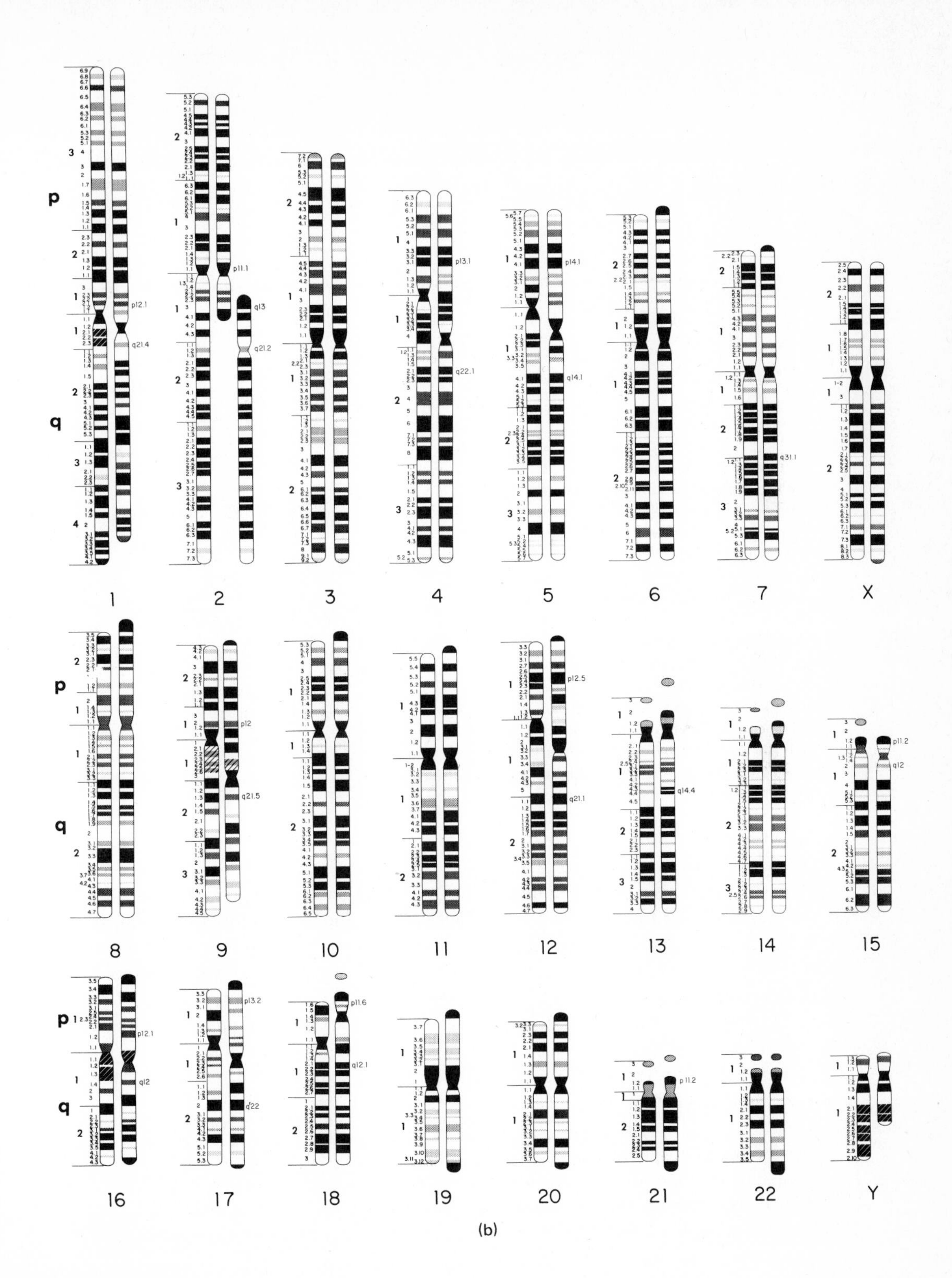

(b)

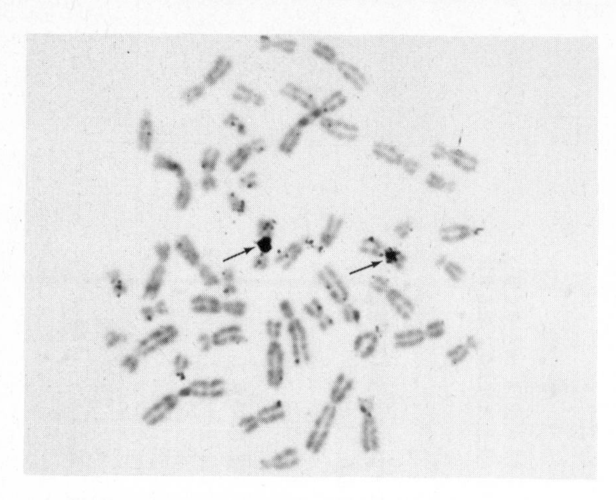

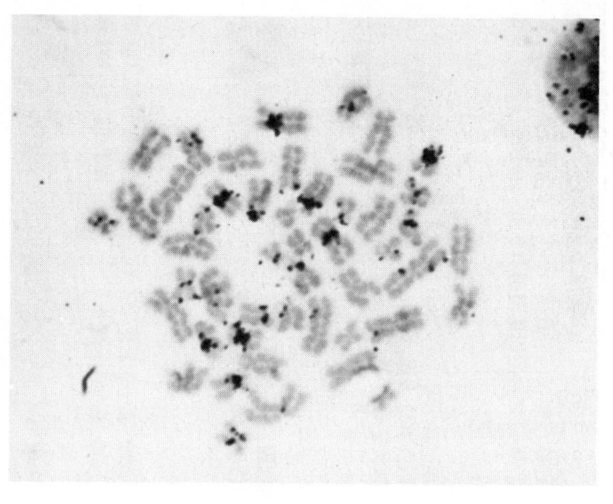

Figure 17-7. The conservation of a satellite DNA in evolution is illustrated by these two chromosome spreads. The upper one shows the hybridization pattern of human satellite DNA III with human chromosomes where it localizes strongly on the #9 pair (arrows). The lower shows a chimpanzee chromosome spread hybridized under identical conditions. It is evident that there are related sequences present which are also in centromeric heterochromatin, but which are more prevalent and widespread. Human satellite III-related sequences are also found in the gorilla and the orangutan. [Adapted from Jones *et al.*, 1972. (From K. W. Jones, 1977. *Molecular Structure of Human Chromosomes*, J. J. Yunis, ed. Academic Press, New York.)]

"industrial melanism" reported in 1961 by B. D. Kettlewell in England. In his research, Kettlewell released equal numbers of light-colored and dark-colored moths in both industrial and rural environments. He then trapped samples of the moth population after a few days and noted a sharp decrease in the proportion of light-colored moths in industrial areas, and of dark-colored moths in rural areas.

Figure 17-8 illustrates the explanation for this shift in frequency of the two phenotypes. It is quite obviously more difficult to see dark moths against the sooty background of an industrial region, and conversely, light-colored moths are camouflaged in the lichen of rural areas. Predators therefore pick off the more visible types in each case. Thus, a phenotype which is selected against in one environment is selected for in another.

An example of adaptation on a physiological level occurs in scorpaenid fish (Fig. 17-9a). One species, *Sebastolobus altivus*, is known to inhabit the north Pacific Ocean at a depth of 550–1300 m; another species, *S. alascanus* is found at 180–440 m

(a) (b)

Figure 17-8. Contrasting backgrounds leading to selection against dark-colored moths on lichen in rural areas (a), and against light-colored moths on sooty background in industrial regions (b). (From American Museum of Natural History.)

(Fig. 17-9b). The skeletal muscle lactate dehydrogenases (LDHs) of the two species have been reported to differ in their effectiveness depending on hydrostatic pressure. Higher hydrostatic pressure is better for LDH activity of *S. altivus*, as might be expected. It has been conjectured that such physiological differences place constraints on the ability of organisms to utilize certain habitats, and serve as isolating mechanisms (Siebenaller and Somero 1978).

Behavioral adaptations are also known, and are among the most interesting examples of natural selection. For example, there are various parasitic species of birds which lay their eggs in the nests of host species. The host then incubates the egg and raises the young as its own. Studies of these relationships have revealed a number of unique adaptations which have evolved that increase the likelihood of the host accepting the egg and young of the parasite (Nicolai 1974). One form of adaptation is in the song of parasitic species which mimics exactly the song of their host. Thus the young are more likely to be accepted by foster parents.

There are additional adaptations in the parasitic species. The mouth markings of the African widow bird young mimic closely the mouth markings of the host's young, and although the adults of the two species are quite different (Figure 17-10),

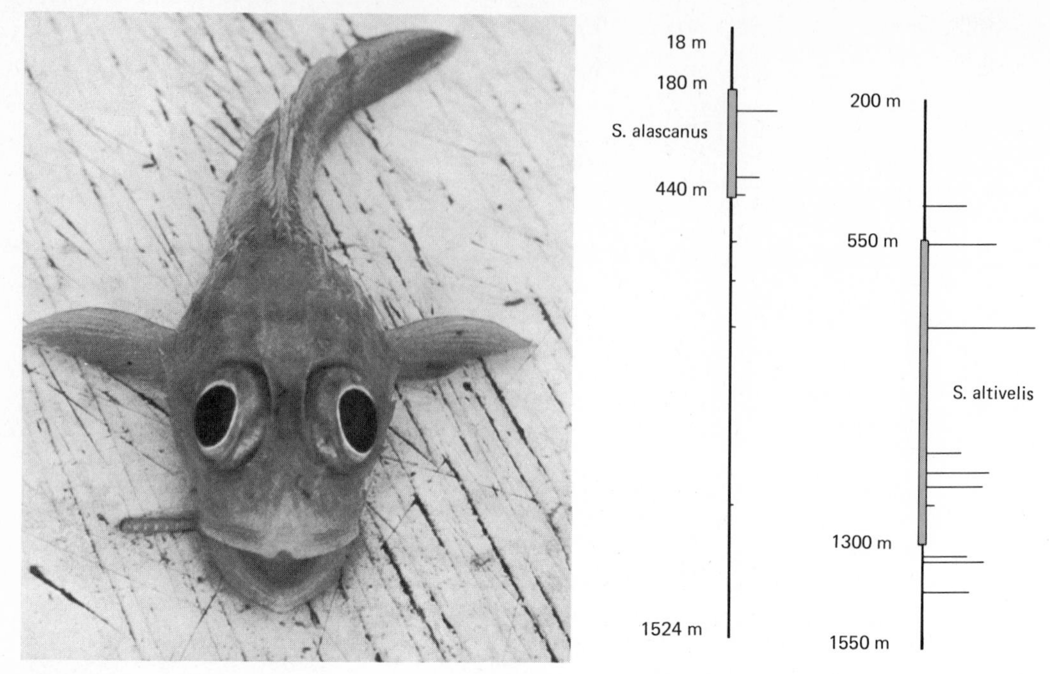

Figure 17-9. (a) The scorpaenid fish *Sebastolobus*, of which there are two species found to live at different depths in the North Pacific. (b) Depth distributions of the *Sebastolobus* species. The widened region on the vertical bars indicates depths at which the species is most common, and the extending bars represent the depths at which any specimens have been taken. The horizontal bars indicate the numbers of specimens taken for an electrophoretic survey of populations. The longest horizontal bar represents 73 specimens; the shortest bar, one specimen. A depth change of 10 m represents a hydrostatic pressure change of approximately 1 atm. The different ecological niches are apparently due to different LDH isozymes in the two species which are sensitive to hydrostatic pressure. (From J. Siebenaller and G. N. Somero, 1978. *Science* 201 : 255; copyright 1978 by the American Association for the Advancement of Science.)

the young are almost identical (Figure 17-11). We stress again here that such mimicry is believed to have arisen randomly. There is no process by which parasitic species could have *intentionally* brought about changes in morphology or behavior to mimic their host. Only those which by chance inherited such traits were able to reproduce young which would be better accepted by host species, and thus survive to reproduce in turn.

Fossils and ontogeny. Fossils and geological formations have been a traditional source of material for the analysis of evolutionary relationships. From such

---

Figure 17-10. Parasitic adults, the female (a) and male (b) paradise widow bird, do not in any way mimic the appearance of the foster parents of their young, the female (c) and male (d) melba finch. The male paradise widow bird is seen in its bright breeding plumage. (From J. Nicolai, 1974. *Sci. Am.* 231 : 94; copyright 1974 by the Scientific American. All rights reserved.)

A
B
C
D

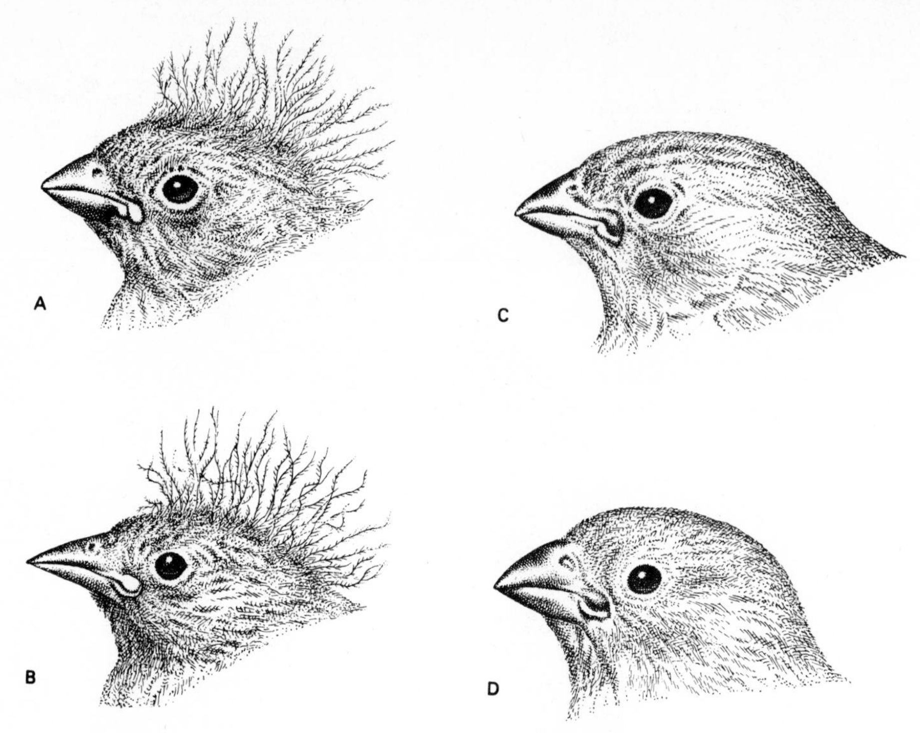

Figure 17-11. Young parasites, unlike adult parasites, closely resemble their host siblings in appearance. Profiles at left show a paradise widow-bird nestling (a) and a melba-finch nestling (b) 13 days after hatching; profiles at right show a straw-tailed widow-bird nestling (c) and a purple-grenadier nestling (d) 15 days old. The young are independent after five weeks. (From J. Nicolai, 1974. *Sci. Am.* 231 : 95; copyright 1974 by the Scientific American. All rights reserved.)

studies concepts of evolutionary trees have been developed, whose branches are formed from the different phyla, classes, and other taxonomic classifications (Fig. 17-12).

As further support for some of the relationships that have been postulated, biologists point to similarities evident in vertebrate embryos at different stages of development (Fig. 17-13), which reflect the evolutionary history of their species. For example, the temporary presence of nonfunctional gill-like formations in embryos of terrestrial species is an indication of their aquatic origin.

This type of analysis inspired the coining of the once-popular phrase, "Ontogeny recapitulates phylogeny." *Ontogeny* refers to the sequence of changes in development, and *phylogeny* to the sequence of changes in evolution between ancestral groups. However, the phrase is not entirely accurate in that embryonic gill formation notwithstanding, fishes are not believed to be ancestral to mammals. It is more likely that our respective ancestors trace back to some common predecessor.

Evolution and the human races. The preceding sections have summarized various lines of evidence supporting neo-Darwinism as modern biology's basic theory

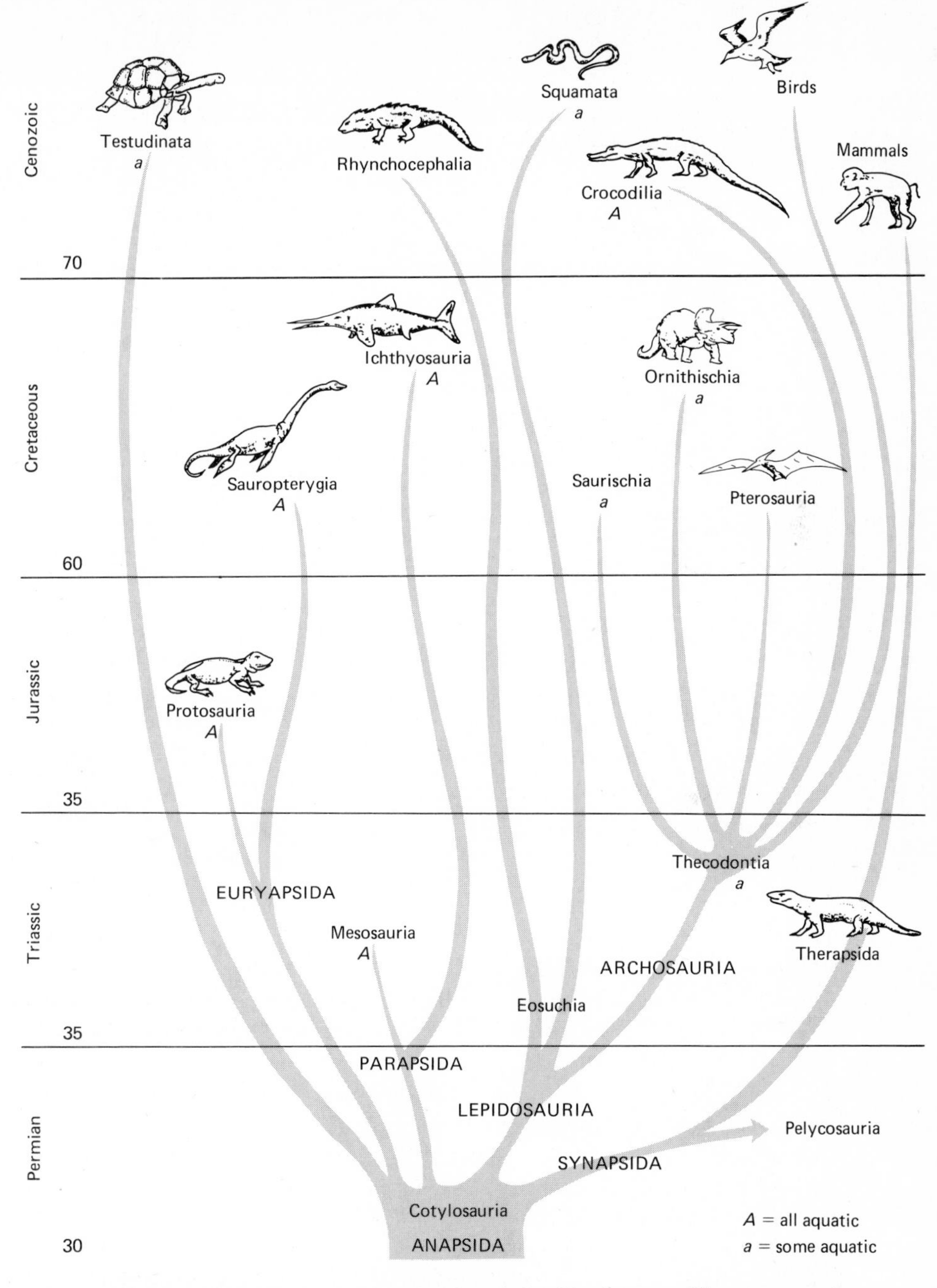

Figure 17-12. Evolutionary tree showing relationships between different groups of reptiles and descendants. (From *Evolution*, 2nd edition by Jay M. Savage. Copyright © 1963, 1969 by Holt, Rinehart & Winston, Inc. Reprinted by permission of Holt, Rinehart & Winston, Inc.)

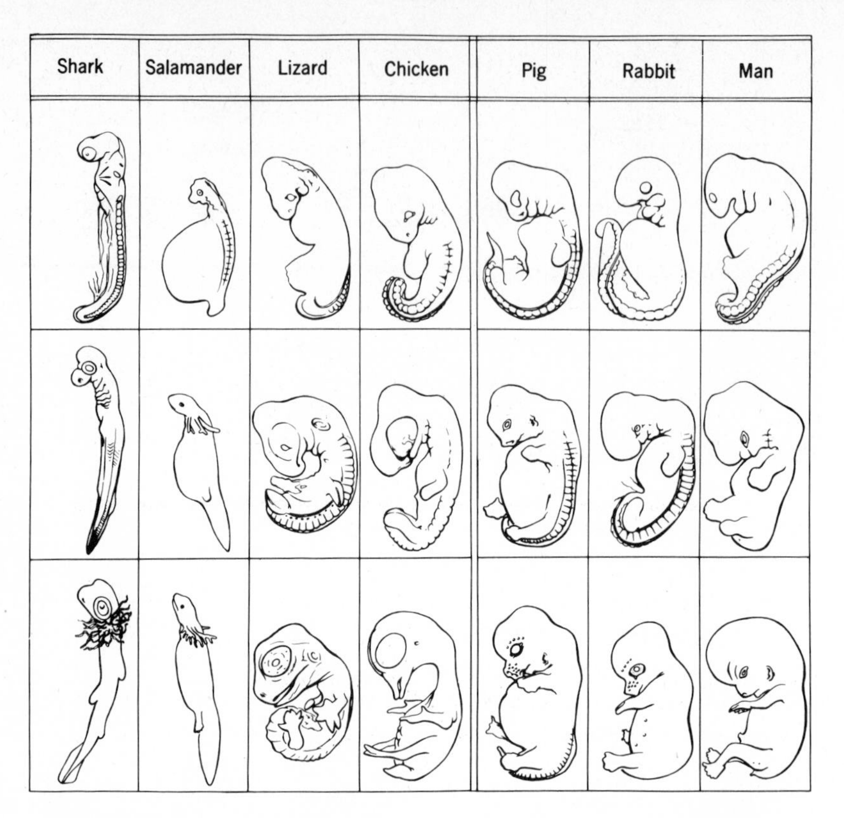

Figure 17-13. Different stages of development in various vertebrates. Similarities of early embryos to each other and to aquatic animals (presence of gill-like structures) have been interpreted in the past as being suggestive of evolutionary relationships. (From Torrey, 1962. *Morphogenesis of Vertebrates*, p. 7. John Wiley, New York.)

for evolution. It is commonly accepted that the forces believed to have brought about evolutionary change in other forms of life are also those that shaped the evolution of our own species. We know through the studies of fossils that there are distinct morphological differences between *Homo sapiens* and our predecessors (Fig. 17-14) which reflect the genetic differences that have accumulated over millions of years.

We see morphological differences between modern human populations, in fact, which reflect genetic divergence brought about by different environments. There is a correlation between skin pigmentation and the intensity of sunlight in different parts of the world. Adaptation to various climates also occurs, such as to severe cold in the protective fat-padded eyelids of Eskimos. Such differences have led to the classification of different populations of humans into *races*. The genetic definition of race is a general one: a *race* is a group of individuals in which some genes are expressed in higher frequencies than in other groups. Note that this definition includes grouping humans into races on the basis of skin pigmentation, which is also the sociological criterion for races. However, it also would allow the grouping of humans into races on the basis of blood types, eye color, and whatever other traits might be under con-

Figure 17-14. Model busts of *Homo sapiens* and ancestral forms of humans. (American Museum of Natural History.)

sideration. After all, the genes determining skin pigmentation consist of DNA, as do genes controlling the expression of any trait.

From the genetic point of view, then, there is little reason to classify people into different races solely on the basis of skin color. And from this viewpoint it would seem absurd that so much human misery has been caused by strife between races. Humans are indeed unique in that they are the only species which has seen strife between its subgroups simply because these groups differ in one trait.

Among the recent controversies generated by those who insist on treating races as being very different populations only on the basis of skin color, have been reports which profess to show differences in IQ between Blacks and Whites. In the following Chapter we shall address ourselves to these reports, considering the validity of both their genetic claims and their statistics. We will also discuss a related topic, that is, the genetic basis of behavior.

NEW DIRECTIONS: MOLECULAR EVOLUTION

Traditionally, evolutionary relationships have been postulated on the basis of similarities in the gross phenotype of various organisms. Homologies of limb structure, for example, reveal relationships between classes of vertebrates. Whether organisms have hair or scales also aids in our classifications of animals.

In recent years, however, because of new techniques developed to allow scientists to determine the exact sequence of amino acids in proteins, and the exact sequence of nucleotides of the DNA of a cell, we have been able to draw inferences on the evolutionary relationships between organisms on the basis of similarities on the molecular level. For example, by studying some of the proteins found in the cells of different primates, we have been able to estimate closeness of relationships, and even approximate times of divergence.

As we shall discuss in the following sections, these new approaches to the study of evolution have revealed some interesting phenomena, involving both evolutionary relationships, and the nature of genes and their function.

Protein Sequences and Evolution

So much evidence has been accumulated on the sequences of proteins and their similarities and differences, that proteins have now been classified into groups such as superfamilies, families, and subfamilies in a manner similar to the classification of organisms into various taxa (Dayhoff 1976). The basic unit of this protein classification is the family: Proteins are grouped within a family if more than half their amino acid sequence is the same. Such proteins generally have similar functions, as might be expected. A good example of a protein family would be the different kinds of cytochrome C proteins (respiratory enzymes) found in the mitochondria of different species. They are believed to be determined by genes that arose through gene duplication.

Similarities in sequences detected in several families cause them to be classified in the same superfamily. Sequences within a family can be placed into subfamilies, if they differ at fewer than 20% of their amino acid positions. Since proteins contain hundreds of amino acids, the analysis of protein sequence is often done mathematically, and with the aid of computers. Figure 17-15 shows an evolutionary tree of the cytochrome C proteins found in different organisms based on studies of their amino acid sequences.

Some proteins, such as histones, have been found to be remarkably stable over the course of evolution, having the same sequence in many different species,

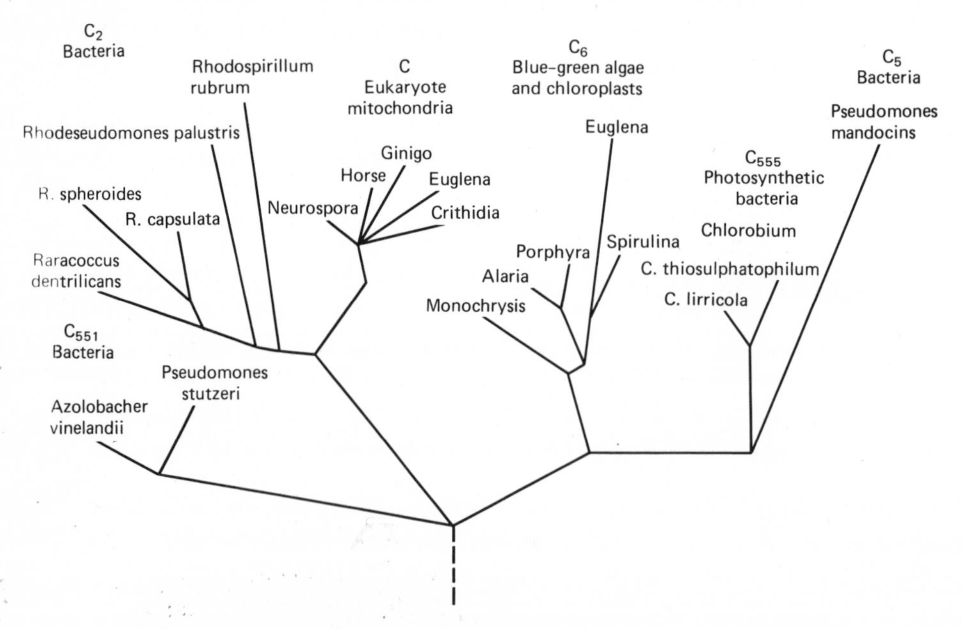

Figure 17-15. A proposed evolutionary relationship of cytochrome C enzymes found in different organisms based on amino-acid sequence studies. (From M. O. Dayhoff, 1976. *Fed. Proc.* 35: 2137.)

while others, such as fibrinopeptides and hemoglobins, have sustained many amino acid changes (Dickerson 1972) (Fig. 17-16). It is not surprising that different species can produce some of the same proteins in their cells. These generally are essential proteins which if rendered defective by mutation would probably result in a lethal condition. Thus such mutations would not be transmitted to future generations.

Multigene Families in Eukaryotes

Fine-structure analysis of genes in eukaryotes has in recent years revealed the existence of a number of systems containing variable numbers of genes with remarkable

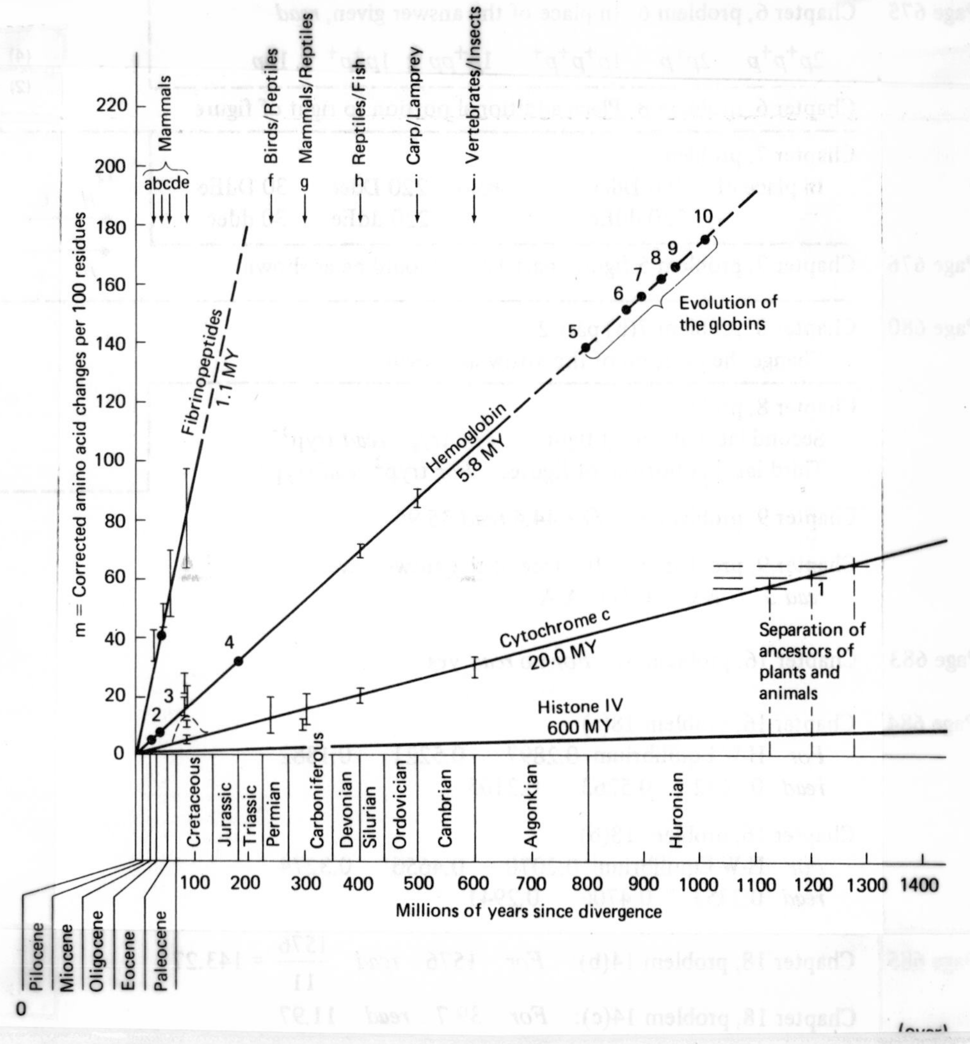

Figure 17-16. Comparison of rates of evolutionary change of different proteins. Histones appear to be least changed of the proteins studied. (From R. D. Dickerson, *Journal of Molecular Evolution,* 1 : 26 : 45, 1971.)

similarities in structure and function. These genes may have arisen during evolution, possibly by duplication of an ancestral gene or through other genetic events. These genes have been referred to as *multigene families* (Hood 1976).

Members of a multigene family share four characteristics: multiple copies of genes, similar fine structure of genes, overlapping functions, and close linkage. A bit of reflection will bring to mind several multigene families which we have discussed in past chapters; one of the most fitting examples would be antibody genes. Recall that variable and constant heavy-chain genes are on one linkage group, and that lambda and kappa light chains are linked on two different chromosomes. Each group contains an unknown number of closely linked genes whose sequences are similar, and which of course determine proteins involved in similar reactions. Figure 17-17 illustrates a hypothetical scheme proposed by L. Hood and his coworkers on how antibody gene families may have arisen during evolution.

Other multigene families that we have discussed would be the genes determining β-, γ-, ϵ- and δ-globin chains of hemoglobin. As discussed earlier (p. 390), these fit the four criteria of a multigene family. Ribosomal genes also can be categorized as a multigene family. The T locus has been mentioned in the context of mutigene families, although the system is not even as clearcut as in antibody genes or

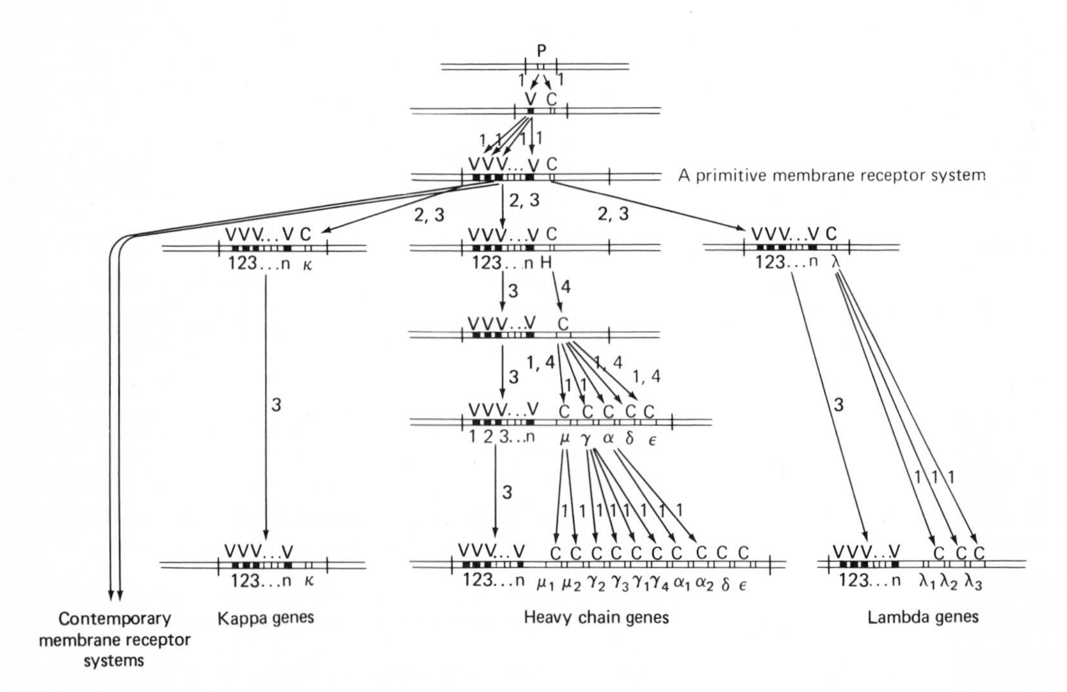

Figure 17-17. A hypothetical scheme for the evolution of the antibody families. The order of gene duplication events is unknown. A number of genetic mechanisms seem to be employed in the evolution of these families as indicated by numbers adjacent to arrows. These are: (1) discrete gene duplication; (2) gene duplication by polyploidization or chromosomal translocation; (3) contiguous gene duplication; and (4) coincidental evolution of multiple genes. (From L. Hood, 1976. *Fed. Proc.* 35: 2161.)

globin genes, since gene products determined by alleles of the T locus have not been well characterized. Similarly, satellite DNAs may belong to a multigene family, although their function is not well understood.

Hood and his coworkers have classified multigene families into three general categories on the basis of structure: "The *simple-sequence* family encompasses segments of DNA derived from 10^3–10^7 repetitions of a short fundamental sequence, generally 6–15 nucleotides in length. The degree of homology among the repeat units is often 80 to 100%. The *multiplication* family consists of 10–10,000 copies of a gene 80–1,000 nucleotides long. The repeat units are essentially identical. An *informational* family has individual members that can differ markedly in sequence from one another, although all are homologous and obviously share an ancient ancestry" (Hood 1976). Satellite DNA would be a simple-sequence family of genes; ribosomal genes would fit into the multiplication category; and antibody genes would constitute an informational family.

How do such multigene families arise during evolution? Two possible mechanisms would be those we have previously mentioned, namely, duplication and/or unequal crossing-over. Existence of antibody genes on different linkage groups also implies that translocation may have been involved, since their structure indicates that they may have arisen from a common ancestral gene even though they are not linked.

The existence of multiple copies of genes with the same or similar structure and function raises a question of the effect of natural selection on these systems. Obviously, a mutation in one of the genes would not be cause for selection against the individual since its effect would be masked by the other normally functioning genes. For selection to have any role in the fixation or loss of a multigene family from a gene pool would require either the accumulation of detrimental mutations in all the genes of the family, or some mechanism to amplify the mutant gene such that it becomes representative of most of the genes in the family. Figure 17-18 illustrates two possible mechanisms for this process, which has been called *coevolution*,

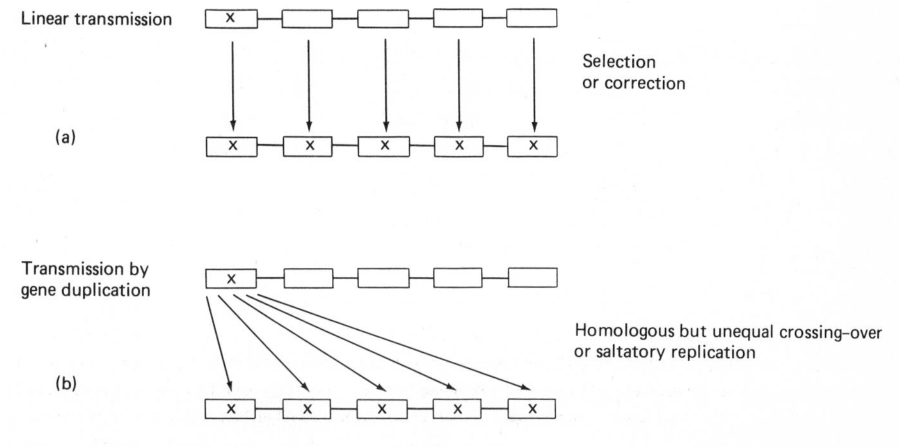

Figure 17-18. Two models for coincidental evolution (see text). (From L. Hood, 1976. *Fed. Proc.* 35: 2165.)

horizontal evolution, *species-specific* evolution (since it presumably would occur only in a few members of a particular species), and *coincidental* evolution (see again Hood 1976).

Among the multigene families that have recently been discovered are those involving genes determining a variety of hormones. We shall dwell briefly on only one example, but the information that has accumulated indicates an additional level of complexity in gene regulation in eukaryotes. We refer here to sequences of DNA which have been found to code for different products in an overlapping manner.

In 1964, a lipotropic hormone, β-lipotropin (β-LPH) was isolated from sheep pituitary glands. Since then, parts of its amino acid sequence have been found to be the same as, or similar to, at least three other peptide hormones: β-melanocyte stimulating hormone (β-MSH); β-endorphins (*endogenous morphine*-like substances); and adrenocorticotropic hormone (ACTH). Recent studies indicate that a precursor molecule is produced which is degraded into β-LPH and ACTH. β-LPH then can be degraded into β-MSH and β-endorphin (Roberts and Herbert 1977; Crine *et al.* 1977). All these peptides have similar properties, such as producing analgesic effects similar to morphine. They also cause behavioral changes in experimental animals similar to those of morphine, such as inducing drowsiness in rats and prolonging a conditioned avoidance response.

Figure 17-19 illustrates the relative positions of the DNA sequences determining these peptides. From the point of view of regulation of gene action, we must

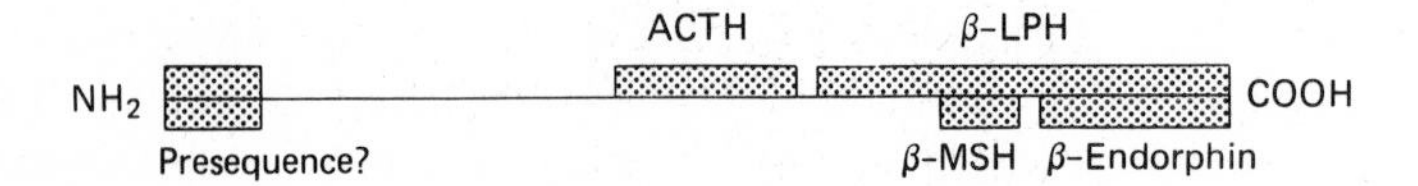

Figure 17-19. Map relationships of genes determining the ACTH–β–LPH family of peptide hormones. (From Roberts and Herbert, 1977. *Proc. Nat. Acad. Sci. U.S.* 74: 5304.)

here consider that post-translational mechanisms must exist which recognize junctions in the amino acid sequence of precursor molecules, and can break the precursor molecules into various component peptides as needed by the organism. This system continues under study today, and promises to provide information on the physiology of various neural processes as well as new regulatory mechanisms for gene-determined cell products. Another such group of peptide hormones, also pituitary in origin, is that which includes placental lactogen, growth hormone, and prolactin, all of which have areas of homology in amino acid sequence as well as overlapping functions.

Evolution and Regulatory Genes

We have previously mentioned the similarity of metabolic processes in all living cells. The vast majority of enzymes (perhaps more than 90%) involved in basic metabolism are the same from species to species, in prokaryotes and eukaryotes alike. In this chapter we have mentioned that some proteins, such as hemoglobin, are exactly the same in humans and chimpanzees. This fact has led many scientists to hypothesize that what makes chimpanzees so different from humans or any other species must not be due to structural-gene evolution and divergence, but to evolutionary differences in regulatory genes.

There is, of course, no hard evidence to support this idea, since we know so little about regulatory systems of eukaryotes in the first place. Nonetheless, the idea is logical, and biologists such as C. Markert of Yale have turned to various systems such as isozymes for supporting evidence. Markert and others had discovered that the different isozymes of LDH were formed from tetrameric combinations of two different polypeptides, and A and B chains (see again p. 383). Since the two proteins exhibit certain homologous amino acid sequences, it is felt that they evolved from an ancestral gene, and diverged through an accumulation of mutations (Markert *et al.* 1975).

A study of LDH isozymes in many species of fish seems to support this idea. For example, the primitive lamprey was found to have only one "A-like" LDH gene, likened to the ancestral gene. The next oldest evolutionary group of fishes, the cartilaginous fishes, have both A and B subunits. Finally, the bony fishes have even a third subunit, C, believed to have arisen by duplication of the B gene.

In primitive bony fishes, the pattern of expression of the B and C subunits in different tissues was more similar than the pattern of expression of the A subunit. As the studies progressed to more advanced forms of bony fish, the C subunit was found to be more restricted in expression, in some species limited to the liver, in others to the eye. This finding has been interpreted to be consistent with the general scheme of gene evolution shown in Fig. 17-20.

In this scheme there is first a duplication of structural genes, with divergence of the duplicated genes from the accumulation of mutations. Owing to different struc-

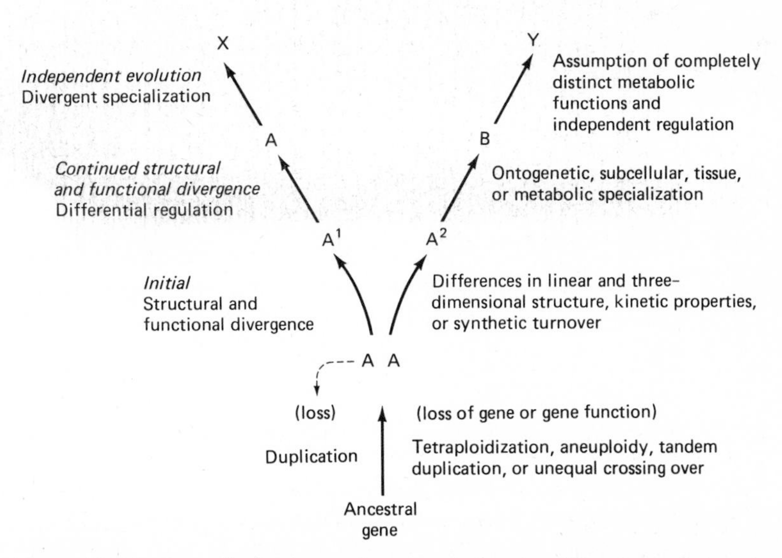

Figure 17-20. Generalized scheme of gene evolution describing the evolution of structure, function, and regulation. (From C. L. Markert *et al.*, 1975. *Science* 189: 112; copyright 1975 by the American Association for the Advancement of Science.)

ture, different functions of the gene products also develop, which lead to expression of the two genes at times when they are most beneficial to various tissues. Essentially this means the evolution of regulatory mechanisms controlling the expression of the structural genes. It is also possible that changes in gene products may cause these products to assume regulatory functions. We must await further developments in research on regulatory mechanisms for verification of this interesting hypothesis.

WHAT LIES AHEAD

There is, of course, no way that we can predict the future of humans or of any other living species on earth, but it might be worthwhile to summarize some of the salient points in the chapter as a reminder of our position in the natural world. The existence and evolution of all living organisms, humans included, are subject to the forces of natural selection. Any species can continue to exist and evolve if its gene pool enables members to adapt to their environment. Any species becomes extinct if its members cannot adapt to their environment.

We cannot change our genomes, but we can, of course, change our environment. The simple truth, however, is that whatever changes we render unto our environment had better be changes to which we can adapt. Otherwise, we can cause our own extinction. We have already caused the extinction of numerous species by changing the environment for our own purposes, and have only recently begun to recognize that some of the changes are detrimental to our species as well. It is not too late to reverse some of our destructive habits, but the reversal cannot be delayed much longer.

In addition, we can look forward to even more improvement in our methodology, which will allow finer analysis of gene structure and function. This in turn should provide greater insight into our evolutionary relationships to all living organisms, and thus to our ultimate origin.

PROBLEMS

1. Theoretically, would it be possible for sympatric populations to evolve into one common species at some point in time? Explain.

2. Assume that there is a small island in the middle of the Atlantic Ocean. It is extremely windy. A study of the insect population showed predominantly nonwinged species. Explain this observation in neo-Darwinian terms.

3. Which cellular proteins would you expect to be the most stable through evolution? Why?

4. Summarize the genetic evolutionary advantages of eukaryotes and of prokaryotes.

5. In view of your answer to the previous question, which groups of organisms would you expect to survive over the long haul of evolution? Why?

6. If there are humanlike organisms elsewhere in our universe, what would you expect their planet to be like? Why?

7. Why do we not see the reappearance of extinct species in our natural world?

REFERENCES

CRINE, P., S. BENJANNET, N. G. SEIDAH, M. LIS, AND M. CHRETIEN, 1977. "In Vitro Biosynthesis of β-Endorphin, γ-Lipotropin, and β-Lipotropin by the Pars Intermedia of Beef Pituitary Glands." *Proc. Nat. Acad. Sci. U. S.* 74: 4276–80.

DARWIN, CHARLES, 1859. *The Origin of Species.*

______, 1871. *The Descent of Man.* Both of Darwin's works have been reprinted by The Modern Library, New York.

DAYHOFF, M. O., 1976. "The Origin and Evolution of Protein Superfamilies." *Fed Proc.* 35: 2132–38.

DICKERSON, R. D., 1972. "The Structure and History of an Ancient Protein." *Sci. Am.* 226: 68.

DOBZHANSKY, T., AND H. LEVENE, 1948. "Genetics of Natural Populations. XVII, Proof of Operation of Natural Selection in Wild Populations of *Drosophila Pseudoobscura.*" *Genetics* 33: 537–547.

DOBZHANSKY, T., AND N. SPASSKY, 1954. "Environmental Modification of Heterosis in *Drosophila Pseudoobscura.*" *Proc. Nat. Acad. Sci. U. S.* 40: 407–415.

GISH, D. T., 1973. "Creation, Evolution, and the Historical Evidence." *Am. Biol. Teacher* 35: 132–140.

HOOD, L., 1976. "Antibody Genes and Other Multigene Families." *Fed. Proc.* 35: 2158–67.

KETTLEWELL, H. B. D., 1961. "The Phenomenon of Industrial Melanism in *Lepidoptera.*" *Ann. Rev. Entomol.* 6: 245–262.

MARKERT, C. L., J. B. SHAKLEE, AND G. S. WHITT, 1975. "Evolution of a Gene." *Science* 189: 102–114.

MILLER, D. A., 1977. "Evolution of Primate Chromosomes." *Science.* 198: 1116–24.

NELKIN, D., 1976. "The Science Textbook Controversies." *Sci. Am.* 234: 33–39.

NICOLAI, J., 1974. "Mimicry in Parasitic Birds." *Sci. Am.* 231: 92–98.

ROBERTS, J. L., AND E. HERBERT, 1977. "Characterization of a Common Precursor to Corticotropin and β-Lipotropin: Identification of β-Lipotropin Peptides and Their Arrangement Relative to Corticotropin in the Precursor Synthesized in a Cell-Free System." *Proc. Nat. Acad. Sci. U. S.* 74: 5300–5304.

SAVAGE, J. M., 1969. *Evolution.* Holt, Rinehart & Winston, New York.

SIEBENALLER, J., AND G. N. SOMERO, 1978. "Pressure-Adaptive Differences in Lactate Dehydrogenases of Congeneric Fishes Living at Different Depths." *Science.* 201: 255–257.

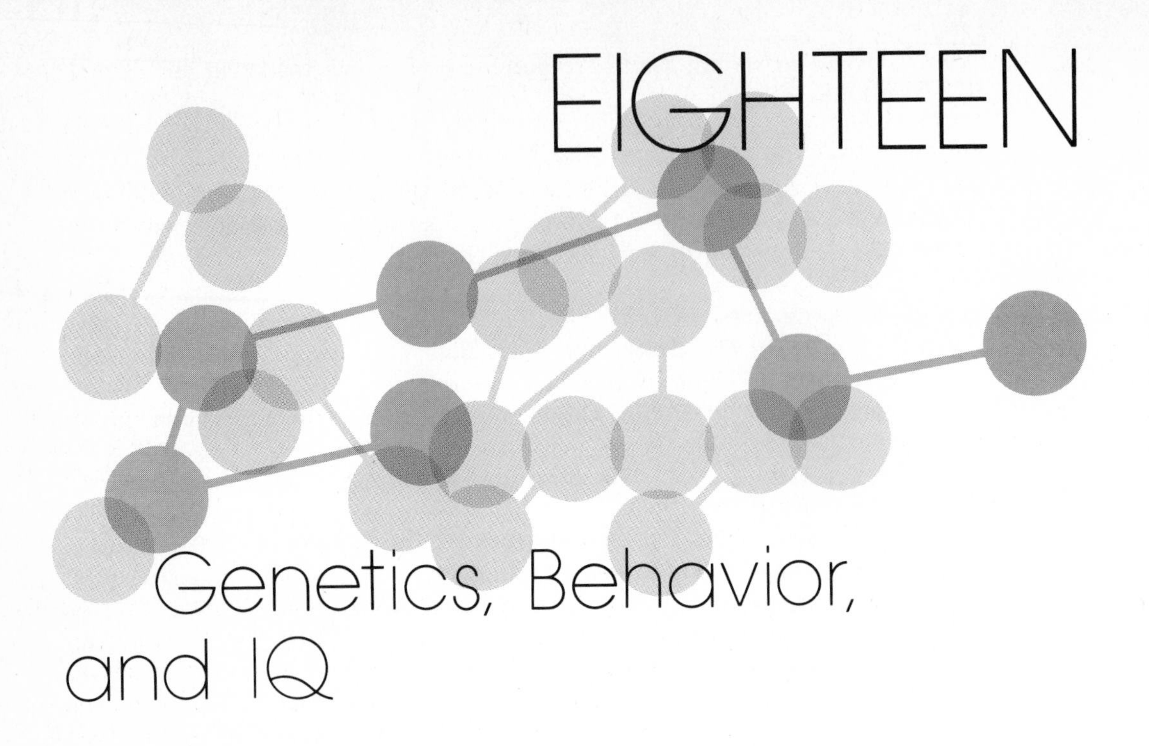

EIGHTEEN

Genetics, Behavior, and IQ

INTRODUCTION

Kennedy, van de Riet, and White (1963), in a frequently quoted study, reported that tests show Blacks have lower IQ's than Whites. If true, is this result due to heredity or environment? The heredity-environment question—sometimes known as the nature-nurture controversy—has flourished for years in the fields of psychology, sociology, and biology. The basic question is: Can behavior be explained purely either by inheritance or by environmental factors? Recently this controversy has been most heated over the relation of IQ and race, and the dispute will probably continue for years to come. Some scientists are committed exclusively to the genetic explanation; others, exclusively to the environmental viewpoint. In fact, it appears that such a stark dichotomy is not realistic. No trait develops in a vacuum void of its environment and no organism develops without genetic factors. A given trait may manifest itself differently under various combinations of genotype and environment and differently at different times. It appears that for certain behaviors either heredity or environment may be relatively more important.

In this chapter we shall study genetic effects on behavior in general. We shall emphasize genetic effects on IQ and the resulting effects on behavior. We will consider the effect of single genes on simple behavioral traits; then we will consider the effect of many genes on more complex behavioral traits. We will focus a large part of our discussion on IQ (intelligence quotient). Why the emphasis on IQ? As we shall see, in order to evaluate the relative contributions of heredity and environment. quantitative information is necessary. IQ measurements provide quantitative data about intelligence (we will discuss the significance of the IQ measure later) and hence are amenable to attempts to quantitatively discuss hereditary and environmental contributions. IQ will provide a good example of the general problems involved in assessing the relative roles of heredity and environment as well as constitute an interesting subject in its own right.

SINGLE GENES AND BEHAVIOR

As a result of studying certain behavior in *Drosophila*, rodents, and man, behavioral geneticists have been able to show that genes affect behavior. Various point mutations have demonstrated genetic control of behavior. On the other hand it is also possible to show that behavior is modified by the environment or by forces other than the gene. Such effects are more easily studied by using simple behavioral traits which are Mendelian in their pattern of inheritance.

Drosophila. Studies have shown that mating success varies among mutant forms of *Drosophila melanogaster*. Bösiger (1957, 1967) showed that *D. melanogaster* individuals with the Cinnabar mutant (autosomal recessive) were more successful in attracting females for copulation than were those with the Vermillion mutant (sex-linked recessive, see pp. 305-307). Bastock (1956), working with purebred strains for yellow body color, showed that males with the Yellow mutant (sex-linked recessive) are less successful in matings with females of the normal gray body color than are normal males. He found that the courtship pattern of the males had been altered by the mutation from wild-type to yellow body color.

Mating preferences are demonstrated by various experiments. Male animals display preferentially to females of their own species, and females normally respond only to the displays of males of their own species. A female *Drosophila*, if kept in isolation until ready to mate, and then offered the simultaneous choice of several species including her own, will almost always accept the courtship of the male of her own species. Experiments with *D. pseudoobsucura* (Pruzan and Ehrman 1974) suggest that age, previous experience, or the environment, may also influence mating choice.

In 1975, Hay performed some maze-learning experiments that suggested that in *D. melanogaster* (1) learning does occur and (2) it is affected by genes. There are a great number of *Drosophila* experiments confirming genetic effects on behavior (see Benzer 1971, 1973; Konopka and Benzer 1971; Pittendrigh 1958).

Honeybees. There is evidence for genetic control of nest-cleaning behavior of honeybees. Rothenbuhler (1964) showed that some honeybees maintain a hygienic environment within the hive. Cells in the hive were inoculated with *Bacillus larvae* causing a disease called American foul brood. Honeybees opened the combs housing afflicted young and evacuated them, to prevent continuous contamination.

Genes at two independently segregating loci account for hygienic or nonhygienic behavior in bees. One gene (*u*, recessive) accounts for the behavior of opening the comb and the second gene (*r*, recessive) accounts for the removal of the infected young. The hygienic individuals are *uurr*, and they are important to maintaining the hive. No other physical or physiological differences have been recorded between totally hygienic, partially hygienic, or totally nonhygienic honeybees. We see that two single-gene loci can apparently affect behavior.

Rodents. In mice there is evidence that a single gene may influence behavior (Thiessen, Owen, and Whitsett 1970). Changes in fur color and/or eye color are accompanied by modified behavior. For example, albino mice show decreased nibbling, delayed audiogenic seizure, decreased water-escape performance, decreased active avoidance, increased passive avoidance, decreased activity, competitive advantage with sex partners, and poor black-white discrimination. Mice with brown fur instead of black show increased grooming, mice with blue-gray fur give evidence of decreased activity, and mice with dilute coat color and tail and belly spots show decreased nibbling. White mice with pink eyes and reduced or brown pigment show decreased staring at examiner, less paw lifting, and more grooming and shaking. Thus, evidence for genetic control of behavior in mice is also being collected.

Humans. There are single genes that affect in a broad sense the behavior of human beings. Various metabolic diseases which are probably controlled by a single gene lead to very distinctive behavior. The best-known of such diseases is phenylketonuria, an autosomal recessive trait (PKU, see p. 305). PKU is caused by a genetic defect of a liver enzyme, phenylalanine hydroxylase, which transforms phenylalanine into tryosine. PKU-afflicted individuals retain a great excess of untransformed phenylalanine in their bloodstreams, and are severely mentally retarded. It was determined by Bickel and his colleagues (1953) that a special diet very low in phenylalanine resulted in normal blood chemistry and prevented retardation when administered to very young PKU patients. Interestingly, Hsia *et al.* (1958) found no improvement in intelligence when the diet was administered to older phenylketonuric children. Today a simple blood test administered to newborn babies (required by law in many states), as we will discuss in Chapter 19, has been very effective in identifying PKU-afflicted babies for immediate treatment. PKU is an interesting example of genetic effects on intelligence and also on behavior that can be mediated by the environment.

Lesch-Nyhan syndrome, which appears to be a sex-linked recessive, is a metabolic abnormality which is characterized by cerebral palsy, mental retardation, choreoathetosis (irregular, spasmodic, involuntary movements of the facial muscles, limbs, and extremities), aggressive behavior, and compulsive biting that results in self-mutilation of the lips and fingers. This condition is characterized by absence of the enzyme HGPRT (hypoxanthine guanine phosphoribosyl-transferase) and an overproduction of uric acid. This disease is not well understood but it is also an example of genetic effects on behavior.

Galactosemia, when not fatal, leads to mental retardation. Tay Sachs and Spielmeyer-Vogt are lethal diseases manifesting profound retardation (among other traits) in children before the children die. These diseases are among a very comprehensive list of mutations collected and tabulated by McKusick (1971).

Some of the traits we have discussed so far have been controlled by single genes, and some have given evidence of environmental influences. Many behavioral traits of interest are controlled not by single genes, but by many genes; we have called such traits quantitative in Chapter 4.

COMPLEX TRAITS

Much behavior exhibited by man, birds, flies, and other animals is very complex. Shortly we will discuss how we measure the relative contribution of genes and the environment in complex behavior, but first we will give an example of complex behavior.

Birds. Nest-building is an example of the genetic effect on complex behavior. Young birds grow up in nests and when they mature they build nests for their own offspring without any instruction from older birds. Baby birds have no opportunity to observe the actual nest construction because it is completed before the eggs are ever laid. Interestingly the sites and the materials chosen for nests are characteristic for each species, and differ widely among species. Though it is possible that some learning is involved, especially about selecting the site, it would be hard to say that this complex behavior is entirely learned; somehow such behavior must be built right into the genotype.

A study involving two species of lovebirds supports the hypothesis of genetic effects and to a much lesser extent environmental effects. Dilger (1962a,b) took two species of lovebirds and the F_1 hybrids between them, and studied their nest-building behavior. *Agapornis personata fischeri* carries strips of nesting material to the nest in its bill, whereas *Agapornis roseicollis* tucks the nesting material into its rump feathers. Dilger studied how the F_1 hybrids from a mating of the two species of lovebirds carried their building material to their nests. The F_1 hybrids at first almost invariably tried to tuck the nesting material into their feathers, but they were never successful. The way in which these birds tried to put the nesting material into their feathers varied. In some, the tucking movements were incomplete, in others the tucking movements were normal but the birds failed to let go of the strip after tucking it in, or they seized the strip in the middle rather than at the end, or they tucked it into the wrong place. The F_1 lovebirds acted as if they were totally confused and unable to decide between the typical *A. personata fischeri* bill-carrying behavior and the *A. roseicollis* behavior of tucking the strips in their rump feathers. Initially the F_1 lovebirds wasted a lot of effort in trying to get the strips of nesting materials to the nest sites. Ninetyfour percent of the time the birds dropped the nesting strips before they reached the sites. Only 6 percent of the time did they successfully carry the materials to the nesting site in their bills. With time, the F_1 apparently learned more efficient behavior, slowly increasing the proportion of strips carried in their bills. As much as three years later the birds still unsuccessfully tried to carry nest strips by tucking them in their wings. Further genetic effects on reproductive behavior could not be studied since the F_1 lovebirds were sterile.

These experiments show that the complex behavior of carrying nesting material to nest sites is largely genetically controlled. But it also suggests that there is an environmental influence, as indicated by the birds' ability to learn.

In Chapter 16 we developed various models to explain deviations from the basic assumptions of the Hardy-Weinberg law. In this section we will try to develop models that might be used to gain insight into determining the relative contributions of genetic and environmental factors to behavior. We will focus much of our attention on intelligence so as to have something concrete to deal with, and because it can be "measured." Models have been proposed to explain certain observed behavioral phenomena. Basically the models attempt to explain the variations or differences observed in phenotypes in terms of genetic and environmental factors. We will briefly discuss a few such models. We consider an idealized model to account for the relative effects of genes and environment.

We let P' denote the *phenotypic value* (that is, the measured value of some characteristic of behavior). For concreteness we will here consider IQ. P' therefore measures how the characteristic is manifested, that is, the phenotype (for example, an individual's IQ score). Also, we let μ denote the average (mean) effect for the population under study (for IQ scores, $\mu = 100$). Deviations from the average IQ of 100 are explained by G, the *genotypic value*[1] (the measured value of the contribution of the genotype; that is, the part of the character or behavior that is attributable to the genotype); E, the *environmental effects*[1] (the measured value of the contribution of the environment; that is, the part of the character or behavior that is attributable to the environment); and ϵ, the *random error* (unexplained, or undefined, error). Using this notation we can write a model for a measurable observed behavior as

$$P' = \mu + G + E + \epsilon \tag{1}$$

By convention, we really consider

$$P' - \mu = G + E + \epsilon$$

that is, we consider $P = P' - \mu$ as the deviation from the population mean, in phenotypic terms. Therefore we can express equation (1) as

$$P = G + E + \epsilon \tag{2}$$

It may be true that there is an interaction, $I_{E \times G}$ (meaning that the responses are modified differently depending on the particular genotype and environment). Therefore the appropriate model might be

$$P = G + E + I_{G \times E} + \epsilon \tag{3}$$

There are other possible models, but we want to mention only a few to explain the general approach. Note that the two models presented might be called *additive*; that is, the various factors act independently and their effects can be added together. It is entirely possible, however, that an additive model may not be valid, and that a multiplicative model might be more appropriate. Then we might say, for instance, that

[1]We are not asserting that G or E can be measured directly.

$P = GE + \epsilon$, or $P = GE^{1/2}$. There are more complex models which we will not discuss here, for such models can be difficult to evaluate. We will assume that the additive model is appropriate.

Geneticists, in studying genetic and environmental effects on behavior, define a very important quantity, H^2 (*heritability*, which we will discuss below), in terms of a statistical quantity known as *variance*. Models (2) and (3) can also be analyzed in terms of variance.

Variance

In statistical terms, variance is a descriptive measure that reflects the variability in a set of observations. It is extremely important in many cases to know much more than an average value. Suppose you had to teach a sixth-grade class or even a high-school class. Which class would you rather teach: (1) a class with an average IQ of 110, with IQ's varying from 90 to 125; or (2) a class with an average IQ of 109.5, with IQ's varying from 75 to 140?

Often it is important to understand how "spread out", or variable a given set of observations may be. For discussion purposes only let us assume that we know the IQ scores of 10 people: 105, 90, 120, 115, 115, 125, 100, 110, 110, 110. To determine the average (mean) we simply add the 10 IQ scores and divide by 10. Recalling our discussion in Chapter 3, we note that if we let X_i denote the ith IQ score, the *mean*, or average, IQ score is given by $\frac{\sum_{i=1}^{10} X_i}{10}$. In general we say that the *mean* (average) $\bar{X}$, is

$$\bar{X} = \frac{\sum_{i=1}^{n} X_i}{n}$$

For our example we find that

$$\begin{aligned}\sum X_i &= 105 + 90 + 120 + 115 + 115 + 125 + 100 + 110 + 110 + 110 \\ &= 1100,\end{aligned}$$

and so

$$\begin{aligned}\bar{X} &= \frac{1100}{10} \\ &= 110\end{aligned}$$

In order to determine the variability, spread, or dispersion of the observations (IQ scores), we calculate the variance, a quantity that uses the mean as a reference point and relates the observations to the mean. That is, we will determine the distance of each observation from the mean. We then square the deviations from the mean (otherwise the sum of the deviations from the mean would be zero), add them up, and then divide by $n - 1$ (the number of observations minus 1). Symbolically we write

$$\text{Variance} = (\text{S. D.})^2 = \frac{\sum (X_i - \bar{X})^2}{n - 1}$$

For this specific example we find that the variance is equal to

$$
\begin{aligned}
&\tfrac{1}{9}\{(105-110)^2+(90-110)^2+(120-110)^2+(115-110)^2+(115-110)^2\\
&\quad+(125-110)^2+(100-110)^2+(110-110)^2+(110-110)^2+(110-110)^2\}\\
&\quad=\tfrac{1}{9}\{(-5)^2+(-20)^2+(10)^2+(5)^2+(5)^2+(15)^2+(-10)^2+0^2+0^2+0^2\}\\
&\quad=\tfrac{1}{9}\{25+400+100+25+25+225+100+0+0+0\}\\
&\quad=\tfrac{900}{9}\\
&\quad=100.
\end{aligned}
$$

Commonly the quantity that is used as a measure of variation is the square root of the variance, known as the *standard deviation.* That is,

$$\text{Standard Deviation} = \text{S. D.} = \sqrt{\text{Variance}}$$

The standard deviation of the IQ scores is $\sqrt{100} = 10$. To get some feeling for the standard deviation let us consider a second group of people with the following IQ scores: 70, 140, 95, 125, 115, 120, 100, 120, 80, 130. Simply by looking at the two groups we find that the IQ scores of the second group are much more variable than those of the first group. Since the standard deviation is a measure of the variability we would expect the standard deviation of the second group to be larger. Calculating the mean of the second group we get 109.5 and for the variance, 508.06. Therefore, the standard deviation for the second group is 22.54. The standard deviation clearly reflects the apparent difference in the variability of the two groups.

In terms of our models we will analyze the variance observed in the phenotypic outcome and try to break it down into its component parts. The phenotypic variance is broken down into the genotypic variance, the environmental variance, and the variance of the random error. So an analogous formulation of model (2) might be

$$V_P = V_G + V_E + V_\epsilon$$

where V_P is the phenotypic variance, V_G the genotypic variance, V_E the environmental variance, and V_ϵ the variance of the random error. This formulation is important in trying to measure the genetic contribution to a given trait or behavior.

HERITABILITY

In large part the controversy over race and intelligence (as measured by IQ) revolves around trying to place a numerical value on the extent to which genetic factors contribute to intelligence. Researchers attempt to measure such a contribution when they *try* to measure heritability. In 1940, J. L. Lush first defined *heritability* as "the fraction of the observed variance (variation) which was caused by differences in heredity." Heritability is not a measure for an individual (we can't calculate the variance for one observation) but rather for the population. Heritability is specific for a given trait at a given time for the population in which it is measured. Lush in

1949 amplified his definition of heritability and divided it into two parts:

1. "heritability in the narrow sense" =

$$\frac{\text{the variance due to the average effects of genes}}{\text{the variance due to individual differences in phenotype}}$$

2. "heritability in the broad sense" =

$$\frac{\text{the variance due to all sources of genetic variance}}{\text{the variance due to individual differences in phenotype}}$$

From these two definitions one argues that genetic variance should be broken down into two component parts. That is, the genetic component should be expressed as

$$G = A + D$$

Here A is called the *additive genetic value* for the character being studied; it represents the sum of the average effects due to genes at different loci. D is called the deviation due to a *dominance effect*. For a single locus, any departure of the additive genetic value from the genotypic value is ascribed to dominance. The dominance deviation D is caused by nonlinear interaction between alleles at a given locus. With this breakdown, heritability in the narrow sense, H_n^2, can be expressed as

$$H_n^2 = \frac{V_A}{V_P}$$

where V_A is the variance due to the additive genetic value and V_P is the phenotypic variance. Primarily, we will discuss heritability in the broad sense, which is commonly referred to simply as heritability, and can be expressed as

$$H^2 = \frac{V_G}{V_P} \tag{4}$$

where V_G is the genotypic variance.

From equation (4), we find that if the heritability is 1, $V_G = V_P$, and all the variation is genetic. If the heritability is 0, $V_G = 0$, and the variation observed must be due to factors other than genetic ones. Heritabilities between 0 and 1 give the proportion of the total variation that is due to genetic factors.

Data for Estimation of Heritability

The crucial problem is how to measure heritability. How do we separate the effects of genetic factors on phenotype from those of all other factors? The presence of a genetic-environmental interaction, and lack of independence, can create problems. If we assume that the genetic and environmental components act independently and that there is no interaction, can we actually determine V_G, the genetic variance? This is a rather difficult problem. Various mathematical approaches have been proposed for estimating V_G and V_P and therefore H^2. To estimate these variances a descriptive statistical measure known as the correlation coefficient is used.

Correlation coefficient. The *correlation coefficient* is a measure of the association between two variables. For purposes of our discussion let us compare the blood pressures of identical twins. Intuitively we might expect the blood pressures of identical twins to be very similar, and hence show a high degree of association. A correlation coefficient, r, measures the agreement or lack of agreement between two sets of observations in the form of a number between -1 and $+1$. Fig. 18-1 presents

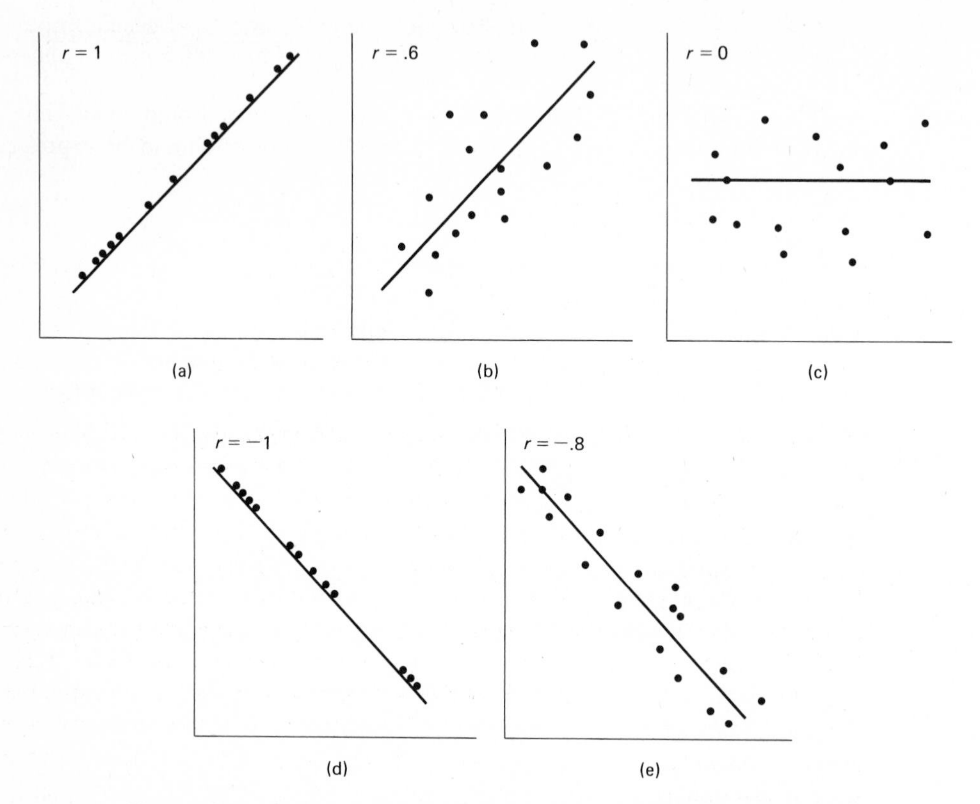

Figure 18-1. Correlation coefficients and straight line fits for various sets of data.

some examples of correlation coefficients for various sets of data. Essentially a correlation coefficient measures the tendency or lack of tendency of points to aggregate around a straight line. Straight lines have been added to Fig. 18-1 to illustrate this point. We see in Fig. 18-1a and 18-1d that the knowledge of one of the values in a pair can tell us the second value exactly. In these cases $r = 1$ or -1. In Fig. 18-1c, knowledge of one of the values of the pair gives us no information about the second value, and $r = 0$. Fig. 18-1b and 18-1e show that we can understand something about a second value in a pair though not as precisely as in case (a) or (d).

In Fig. 18-2 we plot the systolic blood pressures of each of 25 pairs of identical twins. Fig. 18-2 appears to show a reasonable degree of similarity in the systolic blood pressures of the identical twins. It would seem that knowing the systolic

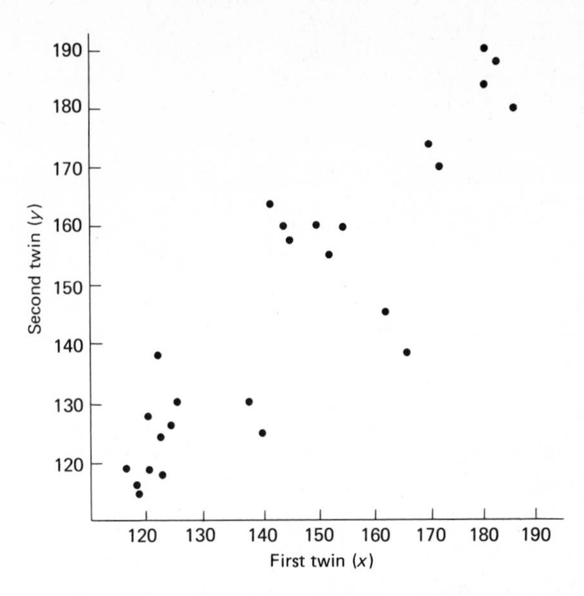

Figure 18-2. Systolic blood pressure of identical twins.

blood pressure of one of a pair of the twins would be a reasonably good predictor of the systolic blood pressure of the other twin. We measure the relationships by calculating r, the correlation coefficient.[2] For the blood pressures of identical twins, given in Fig. 18-2, the calculated correlation coefficient is .93. If we take a random sample of 25 pairs of people (the pair is simply matched as to sex, age, and ethnic origin) determine systolic blood pressure for each pair, and then plot these values (as in Fig. 18-3), we will get a totally different relationship. A priori and by looking at Fig. 18-3, it should be clear that knowledge of the systolic blood pressure of one person in a pair tells us virtually nothing about the other person's systolic blood pressure. The correlation coefficient for this set of observations is $-.0009$ which is approximately 0. As expected there is no association between the blood pressures of people selected at random.

Certain correlation coefficients can be used to estimate values of the several variances needed to estimate heritability. It should be clear that for different relationships a different genetic correlation would be expected. See, for example, Fig. 18-2 (blood pressures of identical twins). Fig. 18-3 (blood pressures of a random sample of matched pairs), and Fig. 18-4, which gives approximate correlation coefficients for intelligence test scores of relatives reared together and reared apart. (Test scores will be discussed later.) As the relationship gets closer the correlation gets stronger, hinting at a genetic component to intelligence. Fig. 18-4 shows the correlation coefficients of intelligence derived from many studies of different genetic relationships. Each dot is derived from one study and the vertical lines indicate the

[2] $$r = \frac{\sum (X_i - \bar{X})(Y_i - \bar{Y})}{\sqrt{\sum (X_i - \bar{X})^2 \sum (Y_i - \bar{Y})^2}}$$

170
160
150
140
130
120
110
100
Second person of the pair
110 120 130 140 150 160 170
First person of the pair

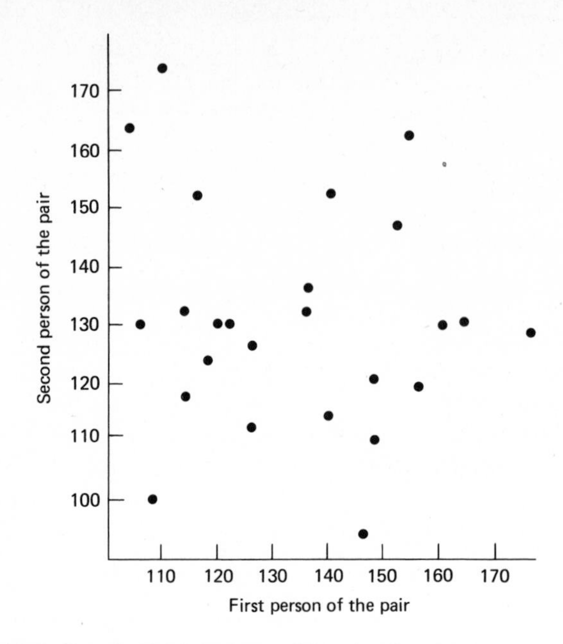

Figure 18-3. Systolic blood pressures of 25 pairs of randomly selected people.

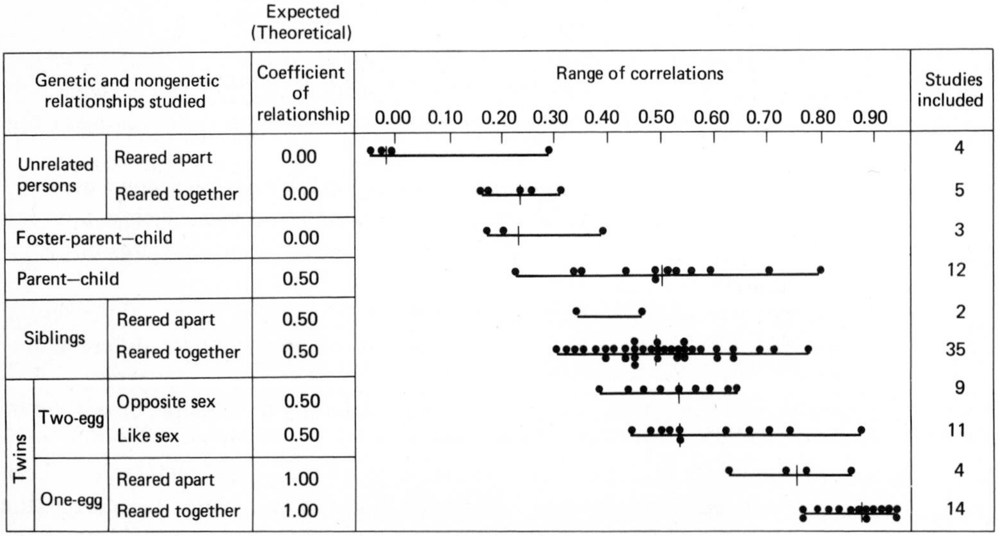

Expected (Theoretical)
Genetic and nongenetic relationships studied
Coefficient of relationship
Range of correlations
0.00 0.10 0.20 0.30 0.40 0.50 0.60 0.70 0.80 0.90
Studies included
Unrelated persons
Reared apart
0.00
4
Reared together
0.00
5
Foster-parent–child
0.00
3
Parent–child
0.50
12
Siblings
Reared apart
0.50
2
Reared together
0.50
35
Twins
Two-egg
Opposite sex
0.50
9
Like sex
0.50
11
One-egg
Reared apart
1.00
4
Reared together
1.00
14

Figure 18-4. A summary of correlation coefficients compiled by L. Erlenmeyer-Kimling and L. F. Jarvik from various sources. The horizontal lines show the range of correlation coefficients in "intelligence" between individuals of various degrees of genetic and environmental relationship. The vertical lines show the medians. (After Erlenmeyer-Kimling and Jarvik, 1963. *Science* 142: 1477–1479. Copyright © 1963 by the American Association for the Advancement of Science.)

median[3] values. Note that Fig. 18-4 gives the expected, or theoretical, genetic correlation for a given relationship as well as the phenotypic (observed, or how intelligence is actually manifested) correlation. Why do the genetic correlations and the phenotypic correlations differ? Phenotypic correlations are functions of genetic, environmental, and unexplained factors (see equations (1) and (2)); the exact equations can be mathematically derived. Some of these correlations can be utilized to estimate heritability. In fact the ratio of an expected correlation to an observed correlation can be used to estimate heritability. We will not go into details in this book, but we briefly want to consider how we can best get the data required to estimate heritability. (The interested reader can refer to McClearn and DeFries 1973, and Bodmer and Cavalli-Sforza 1976.)

Twin studies. What types of data (relationships) will give us the most direct estimate of heritability? Table 18-1 lists various relationships and the quantities the correlation coefficients can estimate. We see that one way to estimate H^2, heritability,[4] is to study monozygotic twins (MZ), even though this may give an overestimate of heritability. The overestimate results because common environment can be confused with genotype.

Table 18-1 Correlation Coefficient Estimates for Various Relationships

Relatives	*Correlation coefficient*	*Correlation coefficient estimate of*	*Type of estimate*
Offspring, 1 parent	r_{OP}	$\frac{\frac{1}{2}V_A}{V_P}$	$\frac{1}{2}H_n^2$ or overestimate
Half Sibs (1 common parent)	r_{HS}	$\frac{\frac{1}{4}V_A}{V_P}$	$\frac{1}{4}H_n^2$ or overestimate
Full Sibs	r_{FS}	$\frac{\frac{1}{2}V_A + \frac{1}{4}V_D}{V_P}$	$\frac{1}{2}H_n^2$ or overestimate
Fraternal Twins (DZ) Dizygotic Twins (same as full sibs except for shared environment in womb)	r_{DZ}	$\frac{\frac{1}{2}V_A + \frac{1}{4}V_D}{V_P}$	$\frac{1}{2}H_n^2$ or overestimate
Identical Twins (MZ) Monozygotic Twins	r_{MZ}	$\frac{V_A + V_D}{V_P}$	H^2 or overestimate

[3]The *median* of a collection of values is the middle value when the values are arranged from lowest to highest (including repetitions). If there is an even number of values the median is the average of the two middle values.

[4]Some other relationships might be used to estimate H^2. For example it can be shown that

$$4r_{FS} - 2r_{OP} \sim \frac{V_G}{V_P} = H^2$$

where the r_{FS} and r_{OP} are correlation coefficients for certain relationships.

Recall that identical, or monozygotic (MZ), twins develop from a single zygote (a single egg fertilized by a single sperm) which divides into two cells each of which is capable of developing into a complete embryo. Fraternal, or dizygotic (DZ), twins arise from two independent zygotes (two eggs fertilized by two independent sperm cells). Dizygotic twins have different genotypes, and are no more similar than any two ordinary siblings. The main difference is that the dizygotic twins are growing up at the same time, (they are even in the uterus at the same time), living much of their childhoods very close together so that their environments may be more alike than the environments of ordinary siblings. Since monozygotic twins have *identical* genotypes, the environment alone might explain possible differences.

For a trait that is almost completely determined by heredity (genotypic) we would expect a very high correlation between MZ twins, a smaller correlation between DZ twins or between nontwin siblings, and the smallest correlation between unrelated people. For a trait with low heritability we would expect to find comparably low correlations for MZs, DZs, siblings, and unrelated people. Even though studying monozygotic twins would seem to be the ideal way to determine heritability, there is a problem. The environmental influences on MZ twins can not be considered to represent adequately the environmental influences on two people chosen at random. Twins (DZ and MZ), siblings, parents, and children usually share a common environment in terms of food, family style, philosophy, customs, income, and so forth. Therefore, studies using MZ twins and studies comparing MZ and DZ twins are probably biased. The best possible situation would be to study MZ twins which have been brought up in separate families.[5] Intuitively we see that this approach makes a lot of sense. Since MZ twins are genetically identical it would seem that to determine the influence of environment on behavioral traits, especially IQ, we would need to study a reasonably large number of identical twins who are raised in separate families (environments). This would be the ideal method (in the sense of what is possible) for determining the heritability of IQ.

Adoption Studies. Another potential method of separating genotypic from environmental factors would be to study adopted children. Correlations between biological relations would give a measure of the total behavior while studies of adoptees might be used to discriminate between genetic and environmental factors.

Ideally it would make sense to study correlations between the adoptees and their biological relations (parents, siblings) and between the adoptees and their foster relations, since the former would be a measure of genotypic inheritance and the latter a measure of other factors. Various complications arise when trying to use this approach: (1) The biological parents aren't always known. (2) The adoptive parents may not have their own children. (3) Adoptees often come from poorer environments and therefore various factors may be confounded and hard to discern. (4) The criteria used for "matching" the child and the adopting family may create problems (matching may be done on the basis of certain skin, eye, or hair color). (5) Adoptions don't always take place immediately after birth. (6) Adoptions are

[5]In fact if we could produce a large number of clones (see Chapter 14) and raise them in a wide spectrum of environments, we might really begin to understand the contributions of the environment and of heredity. Although this would be the ideal situation, it is not possible.

becoming rarer. Even though there are problems with adoption studies, we will briefly discuss them below.

Animal Studies. While heritability is hard to estimate for humans, it is sometimes easier to estimate for animals. For example, there is interest, for breeding purposes, in estimating heritability of birth weight, body weight, litter size, thickness of backfat, thickness of belly, egg production, etc., for such animals as beef and dairy cattle, chickens and swine. For some data, see Spector (1956).

IQ (INTELLIGENCE)

IQ Scores

IQ (intelligence quotient) is a score or number which attempts to predict certain types of performance whose exact relationships with intelligence are unknown. There are many different procedures for testing IQ, but once an IQ test is scored, it is standardized in such a way that a particular score relates to a certain mental age which is divided by the real age and then multiplied by 100. The mental ages are derived by a standardization procedure which is determined by testing a large representative sample from the population being tested. The tests are standardized so that the IQ values have a mean of 100 and a standard deviation of 15 (see page 595), independent of age and sex. The tests are further scaled in such a way that ideally one would expect the frequency distribution of IQ scores to be normally distributed (see Chapter 4); the theoretical breakdown by percentages for various scores is given in Table 18-2. The actual values observed in the population, however, do not follow this distribution exactly. Most of the discrepancies between actual and theoretical distributions of IQ are at the lower range. About .25% of the population have an IQ between 20 and 50 as opposed to the theoretical value of .04%, and about .06% of the population are found to have an IQ below 20 whereas the theory predicts .00002%. Some of these lower IQs can be explained by some of the deleterious genes we dis-

Table 18-2 Percentage Breakdown of IQs

IQ score	*Percent of population with a greater IQ*	*Range IQ score*	*Percent of population with IQ in range*
156	.01	85–115	68.00
146	.10	71–129	95.00
135	1.00	61–139	99.00
125	5.00	20– 50	.04
119	10.00	less than 20	.00002
110	25.00	greater than 160	.00316
100	50.00		
90	75.00		
81	90.00		
75	95.00		
65	99.00		
44	99.99		

cussed on page 592 and some by environmental traumas. Otherwise the distribution of IQ is consistent with quantitative inheritance as discussed in Chapter 4.

What IQ measures and what information we can obtain from an IQ score are subjects of controversy. There is a range of opinions as to the value and meaning of IQ. Some scientists prefer to measure other factors as indicators of a person's intelligence, or to consider their scholastic performance. The first IQ test was developed in 1905 by French psychologist Alfred Binet to measure or predict potential achievement in schools.

If one considers the way questions for IQ tests are developed, a source of difficulty becomes obvious. Some of the statements that an IQ score makes can be made only because the test questions have been selected so as to permit such statements to be made. It seems that during the development and standardization of the test various attempts were made to select questions that actually turned out to suggest certain observed behavior. A teacher's estimates of a subject's abilities and intelligence, and the subject's actual performance later on, were considered and questions that were in reasonable agreement with teachers' estimates and subjects' performance were kept in the test, while other questions were dropped. Thus potential was related very highly to performance, but since potential to perform is not the same as actual performance, a source of bias was apparently introduced.

The American IQ test developed at Stanford University in 1916 by L. M. Terman and others, and revised in 1937 and 1960, also appears to have been developed with some circular reasoning. When this test, known as the Stanford-Binet test, was administered, boys tended to have higher IQs than girls. It was suggested that the IQs of boys and girls are equal; therefore the test must have contained questions that were sex related. In later revisions these types of questions were eliminated and now the IQ distributions for males and for females resulting from the test are essentially identical. However, this does not necessarily mean that the *actual* IQ distributions are identical. It is clear that attempts were made to *design the test* so that there would be no differences in IQ scores between the sexes—that is, the tests are "sex free." We really know nothing about any actual, innate differences between the sexes in terms of intelligence or potential.

There is one additional source for a possible bias. Tests are now considered to be "sex free," but are the tests "culture free" or "culture fair?" The question is: Have the test developers been able to eliminate those questions that could yield significantly different scores for children from different socioeconomic, cultural, racial, or ethnic backgrounds? Studies have pointed out that the Stanford-Binet test includes questions that will elicit different answers from subjects who are not from a White, middle-class, American background. The goal therefore would be to develop questions that can elicit the same response regardless of the subject's social and economic status (SES), or cultural, ethnic, or social background, and still correlate highly with scholastic performance, career achievement, teachers' evaluations, and so forth. On the other hand, if we develop a test which allows all ethnic, cultural, and SES groups to have the same normal distribution of IQ's, will we have obliterated a difference that may in fact exist? We should make it clear that we are *not* suggesting the use of a test which puts certain groups at a disadvantage because of the types of questions that are used. Using IQ tests as strong evidence of cultural, racial, ethnic, or SES differences in intelligence in the various groups might be fallacious, or it may be valid. With the current tests it is hard to distinguish real differences from weaknesses in the tests.

Heritability of IQ (Intelligence)

As we have seen, the best practical way to estimate heritability of intelligence seems to be the study of identical twins who have been raised apart. One of the more famous studies, which recently has been the target of many questions, was reported by Sir Cyril Burt (1958), who studied heritability from two points of view: (1) uniform environment, and (2) uniform heredity. Burt studied children and records of children in London who had lived in orphanages and other residential institutions, and had been placed in these institutions during their first weeks of life. He assumed that such an environment would be to a large degree uniform for all the children. Burt apparently expected the IQs of the 600 children he studied to be very similar, but instead the IQs varied over an unusually wide range. He claimed that in the majority of cases the intelligence of the children correlated with differences in the intelligence of one or both parents. There were quite a few illegitimate children with high ability (often the father had been a casual acquaintance, of a social and intellectual status well above that of the mother, and had taken no further interest in the child). Burt concluded that "in such cases it is out of the question to attribute the high intelligence of the child to the special cultural opportunities furnished by the home environment, since his only home has been the institution."

In order to study uniform heredity Burt studied monozygotic twins (as well as others). He claimed that the mother of twins is not infrequently unable or unwilling to bring up two children at the same time, so that one twin is consequently sent to a relative or to a foster home. (In one study it was reported that in as many as two-thirds of the cases where MZ twins were reared apart, the second twin was reared in the home of a relative with environmental conditions fairly similar to those of the first twin, thereby not really being reared in a different environment.) It is not an easy task to find MZ twins who are reared apart, for even if a family would separate twins, it is not clear that they would be eager to let people know about it. Nevertheless, Burt claimed that he and a colleague collected information on 53 sets of MZ twins reared apart. The data on MZ twins and various other relations are presented in Table 18-3. From his data on MZs reared apart, Burt concluded that the heritability of IQ is about .88—that is, 88 percent of the phenotypic variation found for intelligence as defined by IQ test performance in a population is due to genotypic factors. Recently many very severe critiques have been made of Burt's work (Kamin 1974; Wade 1976; Dorfman 1978). It seems that Burt's reporting may not have been very accurate, his coresearcher doesn't seem to exist, and some of his other data so perfectly fit the theoretical model that they are very suspect. Even Jensen, who relied heavily on Burt's data in his well-known and highly controversial work on IQ and race, admits that he has some questions. (We will discuss Jensen below.) Some other studies involving MZs reared apart are given in Table 18-4. These studies are used as the basis for the figure of .80 which is often quoted as the heritability of IQ.

Table 18-3 Correlation Between Mental and Scholastic Assessments

		MZs together	*MZs apart*	*DZs together*	*Sibs together*	*Sibs apart*	*Unrelateds reared together*
Numbers of pairs		95	53	27	—	—	136
Mental "Intelligence"	Group test	.944	.771	.542	.515	.441	.281
	Individual test	.921	.843	.526	.419	.463	.252
	Final assessment	.925	.876	.551	.538	.517	.269
Scholastic	General attainment	.898	.681	.831	.814	.526	.534
	Reading and spelling	.944	.647	.915	.853	.490	.548
	Arithmetic	.862	.723	.748	.769	.563	.476

From C. Burt, 1958. *American Psychologist* 13: 1–15.

Table 18-4 Summary of Various Studies Estimating H^2 from MZs Reared Apart

	N (pairs)	*Mean IQ*	*S.D.*	*Difference in IQ of MZ twins* $\lvert d \rvert$	*Jensen estimate of* H^2 r_j	*Bodmer Cavalli-Sforza estimate of* H^2
Burt (G.B.)	53	97.7	14.8	5.96	.88	.86
Shields (G.B.)[a]	38	93.0	13.4	6.72	.78	.77
Newman, Freeman & Holzinger (U.S.A.)	19	95.7	13.0	8.21	.67	.69
Juel-Nielson (Denmark)	12	106.8	9.0	6.46	.68	.73
Combined (122 as one sample)	122	96.8	14.2	6.60	.82	.81

[a]Shields study is a well-recorded study but more than two-thirds of the apart twins were raised by close relatives.

From A. R. Jensen, 1970. *Behavior Genetics* 1: 133–148.

It is interesting to look at some additional data collected by Newman *et al.*, Shields, and Burt, which are presented in Table 18-5. One could probably use these data to argue either side of the "nature-nurture controversy." If one looks at MZs together and MZs apart in the Newman *et al.* (1937) study, it appears that being reared apart reduced the similarity of IQ among twins, thereby demonstrating the effect of environmental factors. On the other hand, in the studies by Burt and by Shields, the differences between MZs apart and MZs together are virtually negligible. However, since MZ twins are more alike than DZ twins, we can argue for the influence of heredity factors.

Table 18-5 Correlations of Intelligence for MZ and DZ Twins Reared Together and MZ Twins Reared Separately

Group	NEWMAN ET AL. (1937)		SHIELDS (1962)		BURT (1966)	
	Correlation	*Number of twin pairs*	*Correlation*	*Number of twin pairs*	*Correlation*	*Number of twin pairs*
MZs together	.88	50	.76	44	.92	95
MZs apart	.68	19	.77[a]	44	.87	53
DZs together	.63	50	.51	28	.55	127

[a]Reared apart if reared in separate homes for at least five years, and in many cases the homes of relatives.

From G. E. McClearn and J. C. DeFries, 1973. *Introduction to Behavioral Genetics*, W. H. Freeman and Company, San Francisco. Copyright © 1973.

It is also possible to estimate heritability by using data obtained from MZ and DZ twins (McClearn and DeFries, 1973). The estimates of H^2 (summarized in Vandenberg 1971) using MZ and DZ twins are much more variable (ranging from .34 to .84) than those derived from studies of MZs apart and MZs together. The average of the estimates of H^2 is .52. Since mating isn't random (there is positive assortative mating, see Chapter 16), .52 is probably an overestimate.

One other approach to the determination of heritability is through studies of adopted children. The expected correlation between adopted children and their adopted parents is zero; therefore any positive correlation might be a measure of environmental influences. The best-known study of correlations between IQ of adopted children and mental age of adoptive parent and between control children and their biological parents was done by Burks in 1928. Similar results were also obtained by Leahy (1935). Both results are given in Table 18-6. The evidence for an important environmental contribution is limited.

A study by Honzik (1957) showed approximately the same correlation be-

Table 18-6 Correlations Between Adopted Children, Control Children, and Parents

		Correlation between child's IQ and parental mental age[a]	*Correlation between child's IQ and parental IQ*[b]
Adopted	Father	.07	.15
	Mother	.19	.20
	Midparent*	.20	.18
Control	Father	.45	.51
	Mother	.46	.51
	Midparent*	.52	.60

*Average value of parents.

[a]From B. S. Burks, 1928. *Nature and Nurture: Their Influence upon Intelligence*, pp. 219–316.
[b]From A. M. Leahy, 1935. *Genetic Psychology Monographs* 17: 236–308.

tween foster children and biological parents with whom they had no contact since birth, as between children and biological parents who raised them. The adopted children did not correlate significantly with their adopting parents. A study by Skodak and Skeels (1949) considered children of rather low-IQ mothers (mean 85.75)[6] who were adopted into superior foster homes. These adopted children showed a correlation of .38 with their biological mothers, with whom they had no contact beyond infancy. The mean IQ for these children was 11 points higher than would be predicted if they represented a random sample of offspring from mothers with a mean IQ of 85, and were placed in randomly selected environments. The adoptive homes were selected because of good environmental features. However, there is evidence that these children were selected by the adoption agency as "suitable" for adoption, that is, they may have had higher IQ's than the norm would suggest. Thus the environmental effects are probably not as strong as the 11 points would suggest. Leahy in 1935 showed that illegitimate children who are adopted have an average IQ higher than that of both illegitimate children who are not adopted and legitimate children placed for adoption. It is difficult to interpret these studies.

In summary, we can see that the question of heredity and environment is very complex. There are various methods and techniques for trying to estimate heritability, but the collection of data is very difficult. The ideal conditions (experiments) that would enable us to estimate heritability are impossible to implement, so the best we can do is try to get indirect estimates. Such indirect estimates are limited by the conditions under which the data are collected. There are no completely satisfactory answers.

In the next section we will briefly discuss various studies relating to environmental factors and intelligence. In the last section we will discuss race and IQ.

Environmental Factors and IQ

Some environmental factors seem to have an effect on intelligence as measured by IQ. Examples of environmental effects are the relation between IQ and birth order and the IQs of twins (MZ or DZ) or triplets.

Belmont and Marolla (1973) explored the relationship between birth order and family size relative to intellectual competence. They used a population of about 400,000 19-year-old men in the Netherlands. These men represented almost all male survivors of the children born in the Netherlands in 1944–47 who were still living in the Netherlands at age 19. All 19-year-olds were required to be examined to determine their fitness for military service. The tests included measures of language, arithmetic, mechanical comprehension, perceptual speed, and nonverbal intelligence. The IQ test used was the Raven Progressive Matrices (Dutch modification with 40 items). The raw scores are divided into six classes from one (high score) to six (low score). There was information on Raven score, family size, and birth order for 386,114 individuals; some of the results which Belmont and Marolla derived from this information are given in Fig. 18-5.

If we study Fig. 18-5 we notice three things: (1) The average Raven score of

[6]The biological fathers' contributions to IQ were unknown since they were not available for study. There is some speculation that the IQ of biological fathers might be higher.

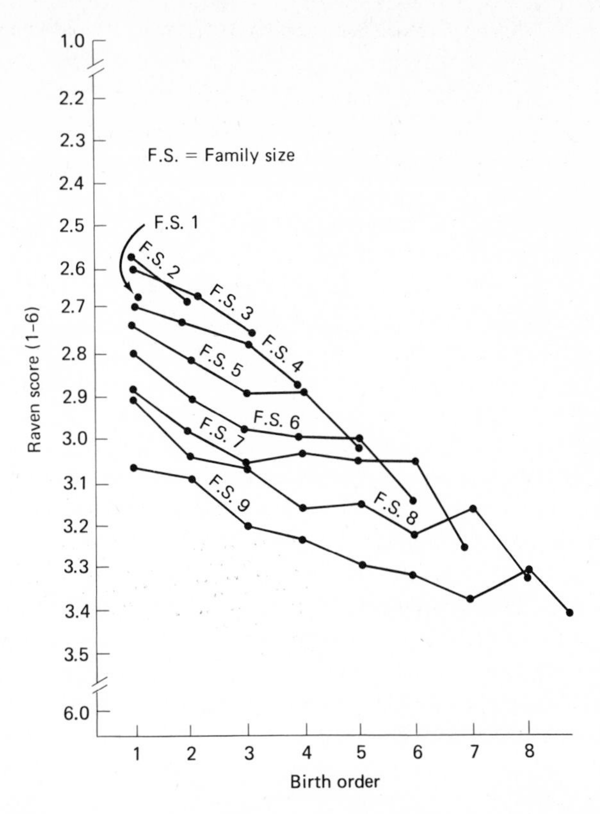

Figure 18-5. Relation between birth order and IQ (Raven score) based on data from Dutch conscripts. Curves are given for different family sizes varying from a total of one to nine children. (From L. Belmont and F. A. Marolla, 1973. *Science* 182 : 1096–1101. Copyright © 1973 by the American Association for the Advancement of Science.)

first-borns exceeded the average of their siblings. (2) In most cases, the Raven score declined with rising birth order. That is, first-borns scored better than second-borns, who in turn scored better than third-borns, and so forth. (3) In general, as family size increased, the Raven score decreased within a particular birth position. Since the hereditary factors would not change, these observations might be explained by the influences of the environment. It could be that larger families' parents have less time to spend with each additional child after the first one.

Another study, by Breland (1974), based on the National Merit Scholarship Qualification Test (NMSQT), gave results (see Table 18-7) similar to those of Belmont and Marolla.

Zajonc and Markus (1975) have proposed models to explain these observations which suggest strong influences of environmental factors such as age gap between children and time spent with children. The larger the age gap between siblings the less depressing the effect.

Twins score consistently lower on IQ tests and other intellectual tests than nontwins. Breland's study of the NMSQT scores noted that twins had an average

Table 18-7 Mean Scores on NMSQT (1965). (N = 800,000.)

	Birth Order				
Family size	*1*	*2*	*3*	*4*	*5*
1	103.76				
2	106.21	104.44			
3	106.14	103.89	102.71		
4	105.59	103.05	101.30	101.18	
5	104.39	101.71	99.37	97.69	96.87
Twins	98.4				

Adapted from H. M. Breland, 1974 (*Child Devel.* 45: 1011); from R. B. Zajonc., 1976. *Science* 192: 227–236.

score of 98.4 and singly born children an average of 103.76. Tabah and Sutter (1954) reported similar results. Record, McKeown, and Edwards (1970) found an average verbal reasoning score of 95.7 for twins and 91.6 for triplets. The Zajonc and Markus model would make sense since the age gap between twins and triplets is certainly for practical purposes nonexistent. According to Zajonc and Markus their model also suggests that the performance of twins who were separated early in life should be higher than that of twins reared together. The study by Record, McKeown, and Edwards indicates that twins whose cotwins were stillborn or died within four weeks of birth seem to have IQs comparable to those of nontwins. The results are given in Table 18-8.

Table 18-8 Mean Verbal Reasoning Scores and Mean Birth Weights of Twins

	Twins Whose Cotwins Were Stillborn or Died in First 4 Weeks			Twins Whose Cotwins Survived		
Sex	*n*	*Verbal reasoning score*	*Birth weight (kg)*	*n*	*Verbal reasoning score*	*Birth weight (kg)*
Males	85	98.2	2.34	976	93.9	2.58
Females	63	99.3	2.22	948	96.5	2.45
Both Sexes	148	98.7	2.29	1924	95.2	2.52

Adapted from R. G. Record, T. McKeown, and J. H. Edwards, 1970. *Annals of Human Genetics* 34: 11–20.

There are also studies showing that one-parent homes constitute an intellectually inferior environment and that the early loss of a parent has a more serious effect than a later loss. Fatherless students scored in the 55th percentile on the American College Entrance Examination, while a comparable group of children from intact homes scored in the 65th percentile (Sutton-Smith *et al.* 1968). Another study (Gerard and Miller 1975) on a combined mathematical and verbal achievement test showed that children from intact homes scored 100.64 (S.D. 15.05) while children from single-parent homes scored 95.37 (S.D. 13.95). Some studies have shown that the longer the father is absent and the younger the child when loss of the parent occurs, the greater the effect on intellectual performance.

Zajonc (1976) suggests that a possible explanation for a single child scoring relatively lower on the NMSQT and on the Netherlands Raven Test might be that a single child has fewer opportunities to be a teacher. Children with younger siblings often have to perform as teachers in many areas. There are some hints that the last-born may be in a situation similar to that of an only child, but at the same time there may be an advantage since some of the siblings may be intellectually mature.

To summarize, we have discussed here some studies which hint at possible environmental influences, on the apparently genetically determined trait of intelligence. Next we want to address ourselves to the questions of race, social class and IQ.

Socioeconomic Class and IQ

Various studies seem to suggest the presence of some relationship between socioeconomic class and IQ. For example, in their study of birth order, family size, and intelligence, Belmont and Marolla (1973) also subdivided the population into three occupational classes, which they related to Raven Scores. The three occupational classes were (1) nonmanual, consisting of professional and white collar workers; (2) manual (skilled, semiskilled, and unskilled workers); and (3) farm (farmers and farm laborers). The results are given in Figure 18-6.

Another similar study reported by Bodmer and Cavalli-Sforza (1970) and based on Burt (1961) showed IQ differences related to social (occupational) class.

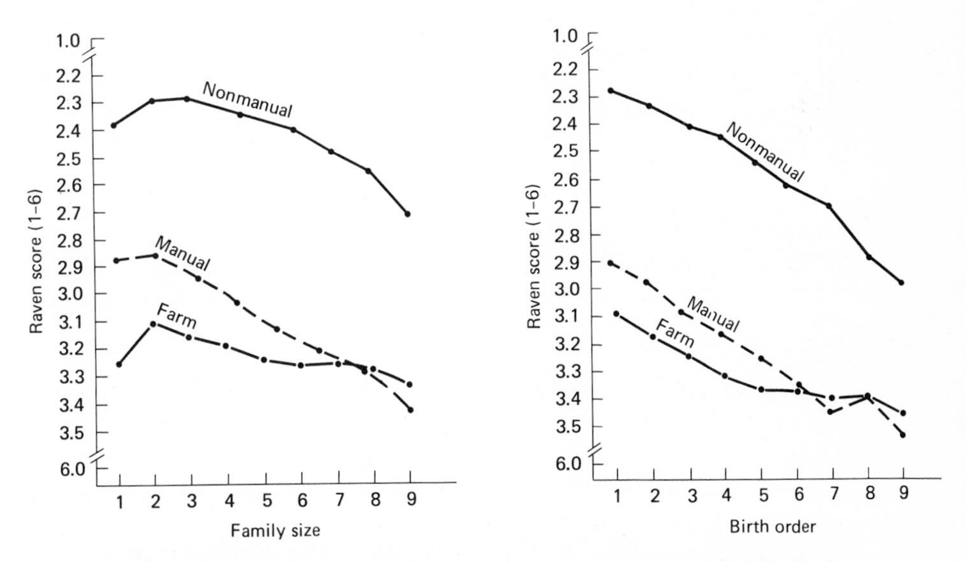

Figure 18-6. Mean Raven score for the three social class groups by (a) family size and (b) birth order (nonmanual, $N = 137{,}823$; manual, $N = 184{,}334$; farm, $N = 45{,}196$). (From L. Belmont and F. A. Marolla, 1973. *Science* 182: 1096–1101; copyright 1973 by the American Association for the Advancement of Science.)

Social class, or socioeconomic status (SES), is largely determined by educational levels, income, and occupation. Various studies estimate the correlation between various measures of SES and IQ at between .40 and .60. If a population is divided into at least four SES groups, the difference in IQ of the extreme groups may range between 15 and 30 points. The range of IQ within each SES class is large and the entire range of abilities is found within each SES group.

There has been a considerable attempt to chart the passage of IQ from parents to children, and to study how this relates to SES. The data given in Fig. 18-7 show that there is a difference in IQ between parents and children, with the IQs of the children tending toward the mean of the whole population. That is, children of parents with high IQs tend to have lower IQs than their parents (regressing to the mean), and children of parents with low IQs tend to have higher IQs than their parents (regressing to the mean). (The phenomenon of regressing to the mean is one that is often observed; it also occurs with height, for example.) The Burt data must be taken with a certain degree of uncertainty, for as mentioned earlier, some critics cast doubt on the credibility of Burt's data.

The great variation that exists between parents and children within a White SES group is very interesting. For example, Stern (1973), in studying a group of adopted children in Minnesota homes, observed that the mean IQ declines for adopted and biological children as the occupational status of parents declines. In going from professional parents to unskilled parents, the decline in IQ of adopted children is approximately five points, but for the biological children the decline is about eighteen points, again showing a very wide range among the IQs.

It is difficult to design an experiment that would show whether the IQ differences between social classes are environmental or genetic. That is, there are no estimates of H^2 for the different SES groups.

As we have mentioned previously, any estimate of heritability of a behavioral trait or intelligence is applicable only to the population in which it is measured. A priori there is no reason to believe that contributing factors to behavior in general, and IQ in particular, measured in one population at a given time will manifest the same proportion of genetic and environmental variance when measured in a second population which differs in genetic or environmental characteristics. Racial and SES groups differ enough so that the validity of applying generalizations drawn from one group to other groups is questionable.

Race and IQ

Racial discrimination and ethnic discrimination in various forms has always existed and continues today throughout the world. The excuse sometimes given for such discrimination is that the group being discriminated against is innately inferior. In the United States over the years various ethnic groups that were once considered undesirable have, to a greater or lesser degree, been assimilated into the population. The assimilation of Blacks has been much slower (social mobility has been much more difficult) and the issue of equal opportunity for Blacks has been and is extremely important. An important facet of equal opportunity is related to whether there are differences in IQ between Blacks and Whites. Various empirical studies suggesting

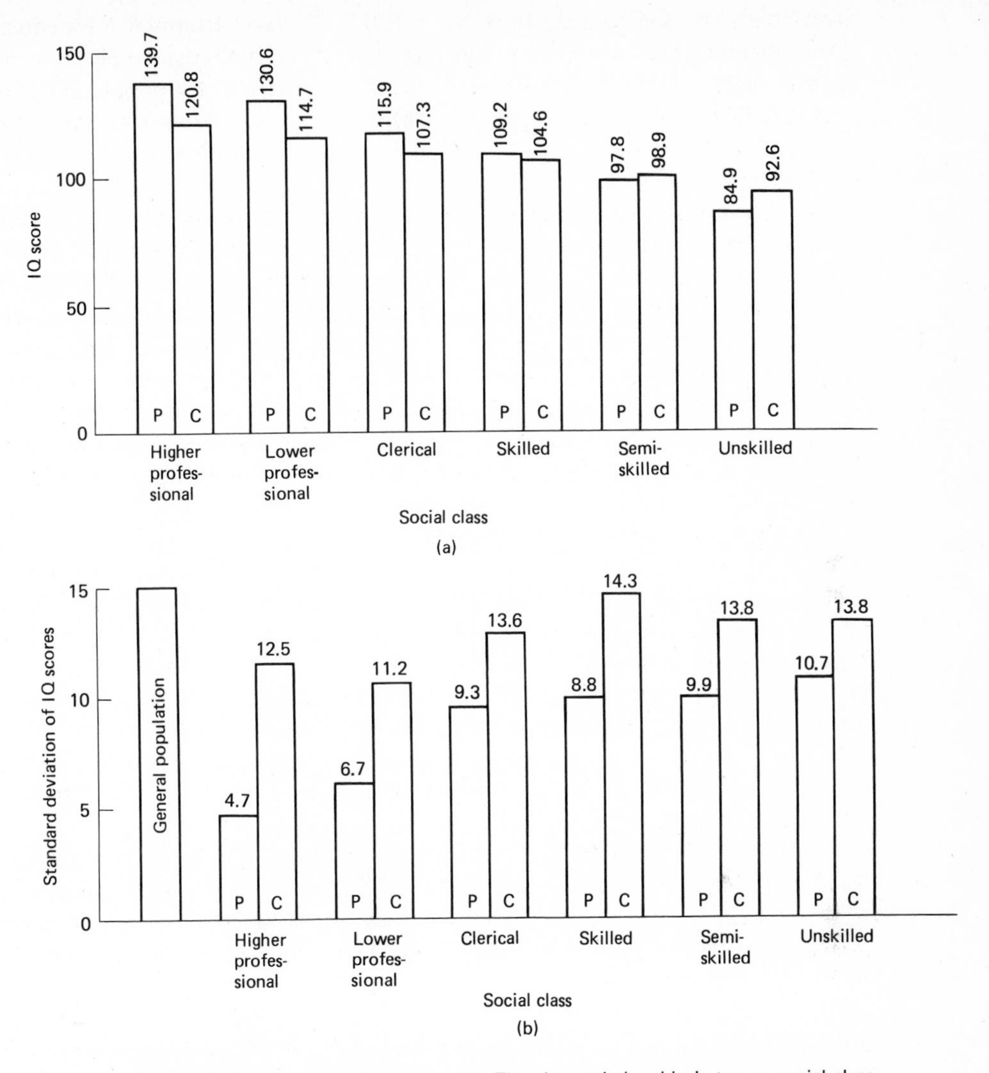

Figure 18-7. IQ and social class. (a) The close relationship between social class and intelligence (as measured by IQ tests) is indicated in these data on London schoolchildren, collected by Cyril Burt (P = parents, C = children). In all classes, the IQ of the children shows a regression toward the mean. Such regression occurs for both genetically transmitted and culturally transmitted traits. (b) The standard deviation of the children's IQ scores is closer to that of the general population than is that of the parental scores. (After W. F. Bodmer, and L. I. Cavalli-Sforza, 1970. *Sci. Am.* 223: 19–29. Copyright © 1970 by Scientific American, Inc. All rights reserved.)

that the distribution of IQs of Blacks is lower than that of Whites will be discussed below.

Attempts to determine whether the differences are genetic or environmental have been hotly debated. We will develop briefly the background needed to understand this important debate. Various researchers have proposed to answer these questions and have aroused considerable furor. The interested reader should read Schockly (1967, 1972, 1973), Jensen (1969, 1973a, 1973b), Eysenck (1971), Herrenstein (1971, 1973), Lewontin (1970, 1976), Coleman (1966) and Loehlin, Lindzey, and Spuhler (1975).

Many studies on IQs of Blacks show that in North America the average IQ of Blacks is about 85, while among Caucasians the average IQ is 100. The results of the study most often quoted, based on IQ tests given to 1800 Black elementary-school children (Kennedy, van de Riet, and White, 1963) in the southern United States, are presented in Fig. 18-8. When the IQs of the 1800 Black children were

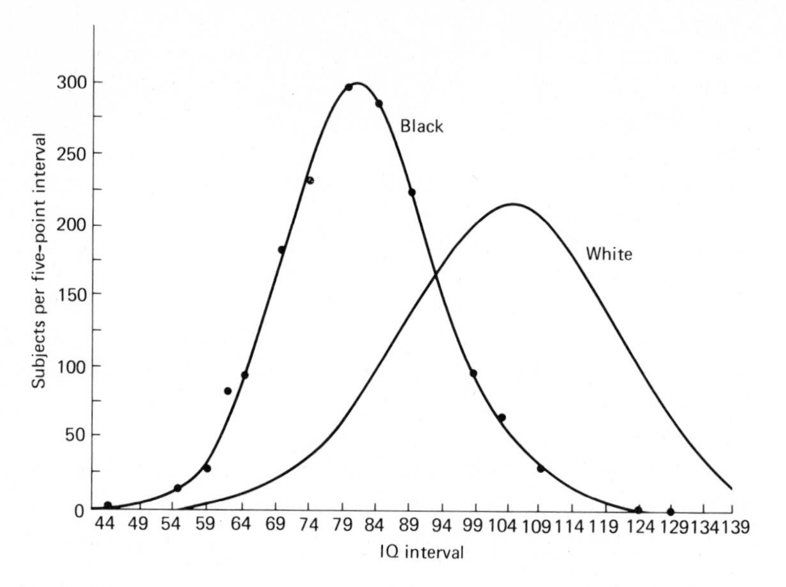

Figure 18-8. IQ in United States Blacks and Whites. Values for Blacks were obtained from 1800 southern school children. Values for Whites reflect "normative" sample of White population. (From W. A. Kennedy, V. van de Riet, and J. C. White, Jr., 1963.)

compared to a sample of White children's scores taken in 1960, the difference in the mean IQs was 21.1 points (the Blacks' being lower). (Most studies show differences of between 10 and 20 points.) In this study there is considerable overlap in the two distributions, but still 95.5 percent of the Blacks have IQs lower than the mean White IQ of 101.8 and 18 percent of the Blacks and only 2 percent of the Whites have an IQ lower than 70. The differences within a race are greater than the differences between races (a bit like SES). The IQ difference between Blacks and Whites in northern states is smaller than that in southern states. During World War I, IQ tests were administered to a large number of randomly selected young United States males inducted into the Army. In every state, Black IQs averaged lower than those of Whites from the same state. A rather striking result was that Blacks from some northern states had average IQs higher than the IQs of Whites from some southern states. An obvious question therefore is whether the tests being employed are really culture free (see p. 604). It is known that the IQ test used on the WWI inductees was especially culture loaded. The evidence that Blacks have lower average IQ scores is fairly strong and cannot be argued. The debatable question is to what this difference is attributable; once that is determined, various options for future action can be considered. (Some believe that the question should *not* be answered or even asked.) We will briefly review some aspects of the problem.

Estimation of Heritability in Blacks

A few studies have been done to estimate the heritability (H^2) of IQ for the Black population. From our previous discussion (see page 602) we recall that the best method for estimating H^2 is to study MZ twins who are reared apart. No such data for Blacks have yet been reported. Some researchers made approximations of H^2 by studying MZ twins and DZ twins, and some used siblings. We will report on a few such studies.

Vandenberg (1969) performed a series of studies on twins in Louisville, Kentucky, in the mid-1960s. He administered a battery of cognitive tests to 137 MZ twins and 99 same-sex DZ twins. Jointly and simultaneously, Osborne and coworkers (1968) presented the battery of tests to 33 MZ twins and 12 same-sex DZ twins in Atlanta, Georgia. The combined samples contained 32 Black MZ twins and 11 Black same-sex DZ twins. These sets of twins (especially the 11 DZ twins) constitute too small a sample for statistical inference, nevertheless the data were used to estimate heritability, H^2, in Blacks. Later, both Vandenberg and the Osborne group published papers based on the combined results. They independently compared the heritabilities of the Blacks and Whites based on the various cognitive tests. They used different methods for estimating H^2 and reached diametrically opposed conclusions. Loehlin, Lindzey, and Spuhler (1975) abstracted the following from the published results:

Vandenberg (1970): "It is clear from this tabulation that there is good evidence for the thesis that the ratio between hereditary potential and realized ability was generally lower for Negroes than Whites."

Osborne: "The H^2 differences are not remarkable, but on seven of the eight spatial tests H^2 was higher for Negroes than for Whites suggesting more rather than less genetic or biological contributions for Negro children than for White children" (Osborne and Gregor, 1968). "The findings cannot support the hypothesis of differential heritability ratios for White and Negro children on tests of numerical ability" (Osborne and Miele, 1969).

Probably the most reasonable conclusion is that the sample sizes were much too small (especially the 11 DZ twins) to give the estimates much credence. (There are even some questions about the validity of some of the results from the 11 DZ twins.)

P. L. Nichols (1970) presented data from a larger Black twin study and a large sample of sibling pairs. Nichols derived data from the Collaborative Study of Cerebral Palsy, Mental Retardation, and Other Neurological and Sensory Disorders of Infancy and Childhood coordinated by the National Institutes of Health. In this study various participating institutions (mostly public clinics) selected every n^{th} pregnancy and any offspring were followed up with various physical, psychological, and neurological examinations until age eight. Nichols selected his sample from among 18,000 White and 20,000 Black mothers in the study at 12 hospitals throughout the United States. Nichols studied 1251 White and 1264 Black pairs of siblings (including twins and half-siblings) who were tested with the short form of the Stanford-Binet Test at age four. He also studied 2030 Black and 1761 White pairs tested with the Bayley Mental and Motor Scales at age eight months.

Using the results of the scores of the four-year-olds on the Stanford-Binet Test, Nichols concluded that there was a lower heritability of intelligence among the Blacks in his sample than among the Whites. For 1100 White sibling pairs (non-twins), he estimated the heritability to be .52, and for 970 Black sibling pairs he estimated it at .37. Such large sample sizes would suggest that these differences are real. But on the other hand other data of Nichols show no such significant differences; in particular, the twins in the study did not give strong evidence of different heritabilities between Blacks and Whites. Some of the data are given in Table 18-9. Also note that the DZ twins include opposite and same sex twins.

Table 18-9 Summary of Nichols Study

	Correlations		*Heritability**	*Number of pairs*	
Sample and test	*MZ*	*DZ*	H^2	*MZ*	*DZ*
Whites, 8-month Bayley Mental Scale	.83	.53	.64	60	97
Blacks, 8-month Bayley Mental Scale	.83	.51	.68	85	117
Whites, 4-year Stanford-Binet	.62	.51	.23	36	65
Blacks, 4-year Stanford-Binet	.77	.52	.53	60	84

*Using Jensen's formula.

From J. C. Loehlin, G. Lindzey, and J. N. Spuhler, 1975. *Race Differences in Intelligence*, W. H. Freeman and Company, San Francisco. Copyright © 1975.

A study by Barker, Scarr-Salapatek and Katz (1974) seems to indicate no significant differences between Blacks and Whites in the heritability of intelligence.

A widely discussed study by Jensen (1973a) on virtually all the children of a school district in California suggested that the heritability of intelligence is roughly equal for Black and White school-age children. Jensen studied 8000 children and obtained the correlations among siblings on a battery of achievement and ability tests. Some of the correlations are given in Table 18-10.

The interpretation of these studies is difficult. There is some evidence for differences between Blacks and Whites in heritability of intelligence, while on the other hand there is evidence implying no differences in such heritability. How does one really measure heritability and will enough subjects ever be available to assure sound studies?

There are hints in some studies that heritability of intelligence and SES may

Table 18-10 Age-Corrected Correlations for Black and White Siblings on Various Tests

Measure	Correlation		Number of families	
	Black	*White*	*Black*	*White*
Ability Test				
Digit Memory	.22	.26	260	429
Figure Copying	.36	.26	277	435
Verbal IQ	.36	.38	346	707
Nonverbal IQ	.34	.39	359	709
Achievement Tests				
Word Meaning	.35	.33	278	206
Spelling	.21	.37	64*	84*
Arithmetic	.45	.31	76*	88*
Physical Measures				
Height	.45	.42	414	744
Weight	.37	.38	414	743

*Relatively small sample.

From A. R. Jensen, 1973a. Table 4-1 (p. 111) from *Educability and Group Differences* by Arthur R. Jensen. Reprinted by permission of Harper & Row, Publishers, Inc.

be related. Evidence of lower IQs in the lower SES groups is found for both Blacks and Whites (see page 612), but it is hard to distinguish between SES and race, since these two factors are so closely interrelated that determining individual effects is difficult. It may also be true that the interaction of genetic and environmental factors plays an important role.

What can we say about the environments of Blacks and Whites? If in the United States schools with predominantly Black students are inferior to schools with predominantly White students, the argument that children who have attended school for six (or ten) years have been exposed to comparable education must seriously be questioned. If Blacks are in lower SES groups, and SES groups and IQ seem to be related, is it the racial group or the SES group that plays a significant role in determining intelligence? If children from one-parent homes perform more poorly on IQ tests, what can we say about Black children from one-parent homes? If there are many environmental factors that may be involved with IQ, how do we distinguish the effects of race and environment? It may be that the interaction of race and environment may be stronger in Blacks than in Whites. There are many unanswered questions here. Furthermore, studying the influence of environments is difficult since humans cannot be put into laboratory situations with carefully designed experimental conditions. In view of such limitations it is difficult to try to settle the great debate of the genetic versus the environmental factors in determining IQ in races.

The controversy surrounding the work of Burt, Jensen, and others on IQ and race is but one societal problem related to genetics. In the next, and last, chapter, we will examine a number of other such problems and how they may yet be solved by genetic research.

PROBLEMS

1. Examine the nature-nurture controversy by taking each side and developing a strong case for each view. What is your point of view?

2. Define heritability and explain it.

3. Take a trait or characteristic and discuss it in terms of heritability. Why is heritability important? Explain the social consequences, if any.

4. Explain carefully why DZ twins are so important when studying the heritability of certain traits. Design the ideal method for studying the heritability of a particular trait of interest to you. Is your ideal experiment feasible in the real world? Explain.

5. How have the data collected on twins been used to separate the effects of genotype and environment on quantitative traits?

6. Why are adoptions important to the study of heritability? Describe the ideal experiments. Explain the limitations of the real world.

7. Explain how IQ is measured. What is IQ supposed to measure?

8. Give an assessment of the value and validity of the estimates of heritability in Whites and in Blacks.

9. Can the estimates of heritabilities found for Whites be used for Black populations? Explain.

10. Discuss the effects of environmental factors on intelligence.

11. Why is the IQ distribution among social classes relatively stable?

12. Propose some models that might be plausible for explaining the observed phenotype. (See for example equations (1), (2), and (3).) Justify your model.

13. If $V_G = 15$ and $V_P = 25$, estimate heritability and interpret.

14. The following are the IQs of 12 eight-year-old children: 105, 94, 112, 85, 120, 102, 114, 95, 128, 100, 99, 106. Calculate (a) the mean, $\bar{X}$; (b) the variance; and (c) the standard deviation.

15. (a) If $H^2 = .9$ and $V_G = 45$, what does V_P equal?
(b) If $H^2 = .3$ and $V_P = 70$, what does V_G equal?

16. A farmer records the number of eggs laid by his prize hen. For 20 consecutive times he records the following number of eggs laid: 3, 5, 2, 4, 0, 5, 1, 2, 4, 3, 3, 3, 0, 1, 2, 5, 6, 2, 5, 2. Calculate (a) the mean, $\bar{X}$; (b) the variance, and (c) the standard deviation.

REFERENCES

BARKER, W., S. SCARR-SALAPATEK, AND S. KATZ, 1974. "On Sampling Factor-analysis and Heritability." Paper presented at Behavior Genetics Association annual meeting, Minneapolis, June, 1974.

BASTOCK, M., 1956. "A Gene Mutation Which Changes a Behavior Pattern." *Evolution*, 10: 421–439.

BELMONT, L., AND F. A. MAROLLA, 1973. "Birth Order, Family Size, and Intelligence." *Science* 182: 1096–1101.

BENZER, S., 1971. "From the Gene to Behavior." *J. Am. Med. Assoc.* 218: 1015–1026.

_____, 1973. "Genetic Dissection of Behavior." *Sci. Am.* 229: 24–37.

BICKEL, H., J. GERRARD, AND E. M. HICKMANS, 1953. "Influence of Phenylalanine Intake on Phenylketonuria. *Lancet* ii: 812–813.

BODMER, W. F., AND L. L. CAVALLI-SFORZA, 1970. "Intelligence and Race." *Sci. Am.* 223: 19–29.

BODMER, W. F., AND L. L. CAVALLI-SFORZA, 1976. *Genetics, Evolution, and Man.* W. H. Freeman & Company Publishers, New York.

BÖSIGER, E., 1957. "Sur L'activité sexuelle des males de plusieurs souches de *Drosophila melangaster.*" *C.R. Acad. Sci.* (*D*), (*Paris*) 244: 1419–1422.

_____, 1967. "La Signification evolutive de la selection sexuelle chez les animaux." *Scientia* 102: 207–223.

BRELAND, H. M., 1974. "Birth Order, Family Configuration, and Verbal Achievement." *Child Devel.* 45: 1011.

BURKS, B. S., 1928. "The Relative Influence of Nature and Nurture upon Mental Development: A Comparative Study of Foster Parent–Foster Child Resemblance and True Parent–True Child Resemblance." *Nature and Nurture: Their Influence upon Intelligence, Twenty-Seventh Yearbook of the National Society for the Study of Education*, Part 1. pp. 219–316.

BURT, C., 1958. "The Inheritance of Mental Ability." *Am. Psychol.* 13: 1–15.

_____, 1961. "Intelligence and Social Mobility." *British J. Stat. Psychol.* 14: 3–24.

_____, 1966. "The Genetic Determination of Differences in Intelligence: A Study of Monozygotic Twins Reared Together and Apart." *British J. Psychol.* 57: 137–153.

COLEMAN, J. S., E. Q. CAMPBELL, C. J. HOBSON, J. MCPARTLAND, A. M. MOOD, F. D. WEINFIELD, AND R. L. YORK, 1966. *Equality of Educational Opportunity.* U.S. Office of Education, Washington, D.C.

DILGER, W., 1962a. "Behavior and Genetics." In E. Bliss, ed., *Roots of Behavior.* Harper & Row, Pub., New York.

_____, 1962b. "The Behavior of Lovebirds." *Sci. Am.* 206: 88–98.

DORFMAN, D. D., 1978. "The Cyril Burt Question: New Findings." *Science* 201: 1177–1186.

ERLENMEYER-KIMLING, L., AND L. F. JARVIK, 1963. "Genetics and Intelligence: A Review." *Science*, 142: 1477–1479.

EYSENCK, H. J., 1971. *The IQ Argument: Race, Intelligence and Education.* Library Press, New York.

GERARD, H. B., AND N. MILLER, 1975. *School Desegregation.* Plenum, New York.

HAY, D. A., 1975. "Strain Differences in the Maze-Learning Abilities in *D. melanogaster.*" *Nature* 257: 44–46.

HERRENSTEIN, R., 1971. "IQ." *Atlantic* 228: 43–64.

_____, 1973. *IQ in the Meritocracy.* Little Brown, Boston.

HONZIK., M. P., 1957. "Developmental Studies of Parent-Child Resemblance in Intelligence." *Child Devel.* 28: 215–228.

HSIA, D. Y.-Y., W. E. KNOX, K. V. QUINN, AND R. S. PAINE, 1958. "A One-Year Controlled Study of the Effect of Low-Phenylalanine Diet on Phenylketonuria." *Pediatrics* 21: 178–202.

JENSEN, A. R., 1969. "How Much Can We Boost IQ and Scholastic Achievement?" *Harvard Educ. Rev.* 39: 1–123.

______, 1970. "IQ's of Identical Twins Reared Apart." *Behavior Genet.* 1: 133–148.

______, 1973a. *Educability and Group Differences.* Harper & Row, Publishers, New York.

______, 1973b. "The Differences are Real." *Psychology Today* 7: 80–86.

JUEL-NIELSEN, N., 1965. "Individual and Environment: A Psychiatric-Psychological Investigation of Monozygous Twins Reared Apart." *Acta Psychiatrica et Neurologica Scandinavica* (Monoyr Suppl. 183).

KAMIN, L., 1974. *The Science and Politics of IQ.* Erlbaum, Potomac, Ma.

KENNEDY, W. A., V. VAN DE RIET, AND J. C. WHITE, JR., 1963. "A Normative Sample of Intelligence and Achievement of Negro Elementary School Children in the Southeastern United States." *Monog. Soc. Res. in Child Devel.* 28, No. 90.

KONOPKA, R. J., AND S. BENZER, 1971. "Clock Mutants of *Drosophila melanogaster*." *Proc Nat. Acad. Sci. U.S.* 68: 2112–2116.

LEAHY, A. M., 1935. "Nature-Nurture and Intelligence." *Genet. Psychol. Monog.* 17: 236–308.

LEWONTIN, R. C., 1970. "Race and Intelligence." *Bull. Atomic Sci.* 20: 2–8.

______, 1976. "The Fallacy of Biological Determinatism." *The Sciences* 16: 6–10.

LOEHLIN, J. C., G. LINDZEY, AND J. N. SPUHLER, 1975. *Race Differences in Intelligence.* W. H. Freeman & Company Publishers, San Francisco.

LUSH, J. L., 1940. "Intra-Sire Correlations or Regressions of Offspring on Dam as a Method of Estimating Heritability of Characteristics." 33rd *Ann. Proc. Am. Soc. Animal Prod.* pp. 293–301.

MCCLEARN, G. E., AND J. C. DEFRIES, 1973. *Introduction to Behavioral Genetics.* W. H. Freeman & Company Publishers, San Francisco.

MCKUSICK, V., 1971. *Mendelian Inheritance in Man.* Johns Hopkins Univ. Press, Baltimore.

NEWMAN, H. H., F. N. FREEMAN, AND K. J. HOLZINGER, 1937. *Twins: A Study of Heredity and Environment.* Univ. of Chicago Press, Chicago.

NICHOLS, P. L., 1970. "The Effects of Heredity and Environment on Intelligence Test Performance in Four- and Seven-Year-Old White and Negro Sibling Pairs." (Doctoral Dissertation, University of Minnesota) Ann Arbor, Mich.: University Microfilms No. 17–18, 874.

OSBORNE, R. T., AND A. J. GREGOR, 1968. "Racial Differences in Heritability Estimates for Tests of Spatial Ability." *Percept. and Mot. Skills* 27: 735–739.

OSBORNE, R. T., AND F. MIELE, 1969. "Racial Differences in Environmental Influences on Numerical Ability as Determined by Heritability Estimates." *Percept. and Mot. Skills* 28: 535–538.

OSBORNE, R. T., A. J. GREGOR, AND F. MIELE, 1968. "Heritability of Factor V: Verbal Comprehension." *Percept. and Mot. Skills* 26: 191–202.

PITTENDRIGH, C. S., 1958. "Adaptation, Natural Selection, and Behavior." In A. Roe, and G. Simpson, eds., *Behavior and Evolution*, pp. 390–416. Yale Univ. Press, New Haven, Conn.

PRUZAN, A., AND L. EHRMAN, 1974. Age, Experience, and Rare-Male Mating Advantages in *Drosophila pseudoobscura. Behav. Genet.* 4: 159–164.

RECORD, R. G., T. MCKEOWN, AND J. H. EDWARDS, 1970. "An Investigation of the Difference in Measured Intelligence Between Twins and Single Births." *Ann. Human Gen.* 34: 11–20.

ROTHENBUHLER, N., 1964. "Behavior Genetics of Nest Cleaning in Honeybees. 4. Responses of F_1 and Backcross Generations to Disease-Killed Brood." *Am. Zool.* 4: 111–123.

SHIELDS, J., 1962. *Monozygotic Twins*. Oxford Univ. Press, London.

SHOCKLY, W., 1967. "A 'Try Simplest Cases' Approach to the Heredity-Poverty-Crime Problem." *Proc. Nat. Acad. Sci. U.S.* 57: 1767–1774.

______, 1972. "Dysgenics, Geneticity, Raceology: A Challenge to the Intellectual Responsibility of Educators." *Phi Delta Kappan* 53: 1297–307.

______, 1973. "Deviations from Hardy-Weinberg Frequencies Caused by Assortative Mating in Hybrid Populations." *Proc. Nat. Acad. Sci. U.S.* 70: 732–736.

SKODAK, M., AND H. M. SKEELS, 1949. "A Final Follow-up Study of One Hundred Adopted Children." *J. Genet. Psychol.* 75: 85–125.

SPECTOR, W. S. (ed.), 1956. *Handbook of Biological Data*. W. B. Saunders, Philadelphia.

STERN, C., 1973. *Principles of Human Genetics*, 3rd ed. W. H. Freeman & Company Publishers, San Francisco.

SUTTON-SMITH, B., B. G. ROSENBERG, AND F. LANDY, 1968. "Father-Absence Effects in Families of Different Sibling Compositions." *Child Dev.* 39: 1213.

TABAH, L., AND J. SUTTER, 1954. "Le Niveau Intellectual des enfants d'une même famille." *Ann. Human Genet.* 19: 120–150.

THIESSEN, D. D., K. OWEN, AND M. WHITSETT, 1970. "Chromosome Mapping of Behavioral Activities." Reprinted in G. Lindzey, and D. D. Thiessen, eds. *Contributions to Behavior Genetic Analysis: The Mouse as a Prototype*, pp. 161–204. Appleton, Century, Crofts, New York.

VANDENBERG, S. G., 1969. "A Twin Study of Spatial Ability." *Multivariable Behav. Res.* 4: 273–294.

______, 1970. "A Comparison of Heritability Estimates of U.S. Negro and White High School Students." *Acta Geneticae Medicae et Gemellologiae* 19: 280–284.

______, 1971. "What Do We Know Today About the Inheritance of Intelligence and How Do We Know It?" Reprinted in R. Cancro, ed. *Intelligence: Genetic and Environmental Influences*, pp. 182–218. Grune & Stratton, New York.

WADE, N., 1976. "IQ and Heredity: Suspicion of Fraud Beclouds Classic Experiment." *Science* 194: 916–919.

ZAJONC, R. B., 1976. "Family Configuration and Intelligence." *Science* 192: 227–236.

______, AND G. B. MARKUS, 1975. "Birth Order and Intellectual Development." *Psychol. Rev.* 82: 74.

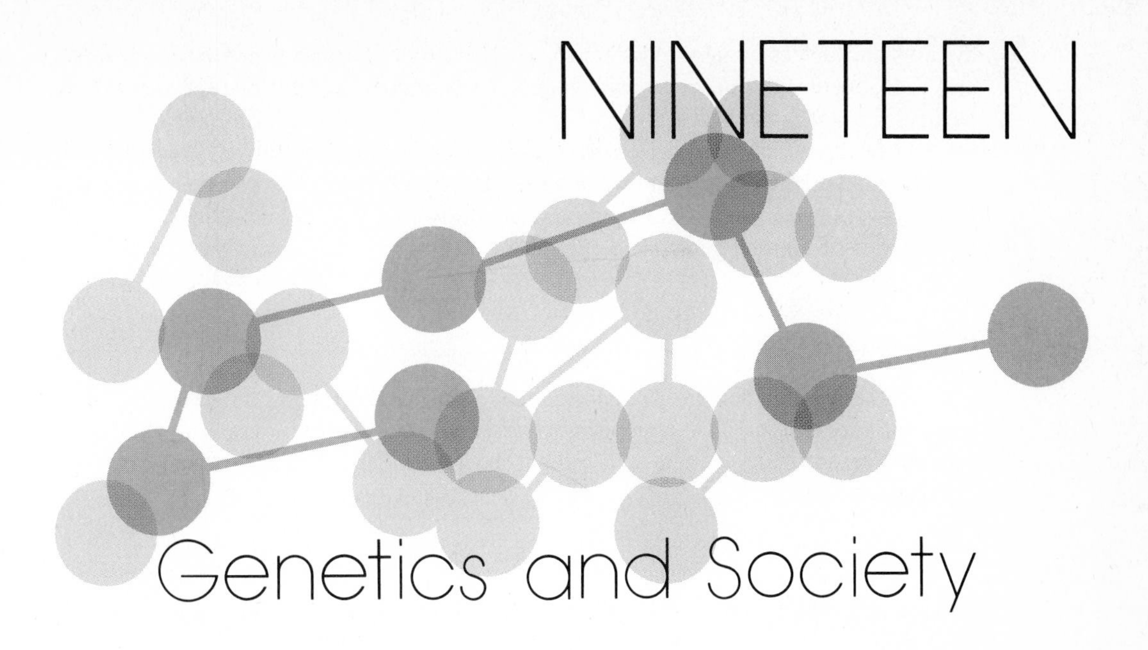

NINETEEN

Genetics and Society

INTRODUCTION

As our understanding of how genes determine life processes has increased in ever greater detail, our perspective of the role and place of the human species has also progressed. We pointed out in Chapter 17 that most biologists have now accepted the concept that human beings are very much an intrinsic part of our natural world, as exposed to the forces of nature as any other species on earth.

Yet we *are* different and unique because of our great intellectual capacity (which we have indeed used to create many of our own problems). We are the only species which has discovered and analyzed the genetic basis to all life. Having attained this knowledge, can we foresee its application to better human lives, to affect the conduct of human society, to improve our world for the benefit of all organisms? These are the possibilities of *applied genetics*.

We must consider not only the applications of genetic phenomena, but also their implications. Our ability to manipulate living cells has raised many debates regarding the ethics of what can be done, and what might be possible in the future. The problems involved are complex, and there are no clearcut answers. But they are our problems, and we need to bring them into focus so that they may be intelligently explored.

Applied genetics consists essentially of three general categories: euphenics, eugenics, and genetic engineering. *Euphenics* is the symptomatic treatment of genetic diseases and defects. *Eugenics* refers to the selection of certain genotypes for mating and reproduction. *Positive eugenics* involves the encouragement of reproduction by particular individuals, whereas *negative eugenics* involves prevention of reproduction by certain individuals. *Genetic engineering* refers to the actual alteration of the genomes of individuals. In this chapter, we shall discuss how these areas of applied genetics are being practiced today and how new techniques already described may influence the practice of applied genetics in the future.

EUPHENICS

The most immediate benefits to mankind may lie in the treatment of medical disorders using information derived from genetic research. Euphenics is certainly not new. In a real sense, many of our well-established treatments for diseases are based on euphenics.

Euphenics Today

The use of antibiotics is an example of a euphenic measure which has been a medical practice for decades. The molecular nature of the activity of some antibiotics is now known; streptomycin, for example, causes malfunctioning of the protein synthesis mechanisms of bacteria. Discovery of resistance transfer factors has alerted clinicians to the possibility that pathogenic organisms might be unaffected by certain antibiotics. In such cases, laboratory tests of the organisms may be used to indicate the most effective program of treatment.

Another example of euphenic measures is the treatment of some genetic diseases by special diets, such as removing milk from the diet of galactosemic patients to prevent the development of symptoms. In other conditions we are able to replace a missing substance, such as insulin in diabetes and clotting factors in hemophilia.

Although much remains to be learned about the function and genetic basis of the immune system, medical science has for years manipulated our immune responses for medical (euphenic) purposes. We are immunized against pathogens with vaccinations either of antibodies, or of the organisms themselves. On the other hand, our immune response can be suppressed with radiation and chemicals, at least temporarily, to facilitate organ transplant procedures.

Euphenics in the Future

Euphenic measures of the future will no doubt involve recombinant DNA methodology. For example, we can foresee the use of cloned or amplified DNA determining the synthesis of needed proteins for some diseases. Already the first steps toward this goal have been achieved in the production of insulin.

Insulin is composed of two polypeptides, chains A and B, but recent studies of the molecular structure of the gene and hormone have revealed the interesting fact that the two chains are actually determined by a single gene. The mRNA transcribed from the insulin gene contains the message for chains A and B, and also for a third

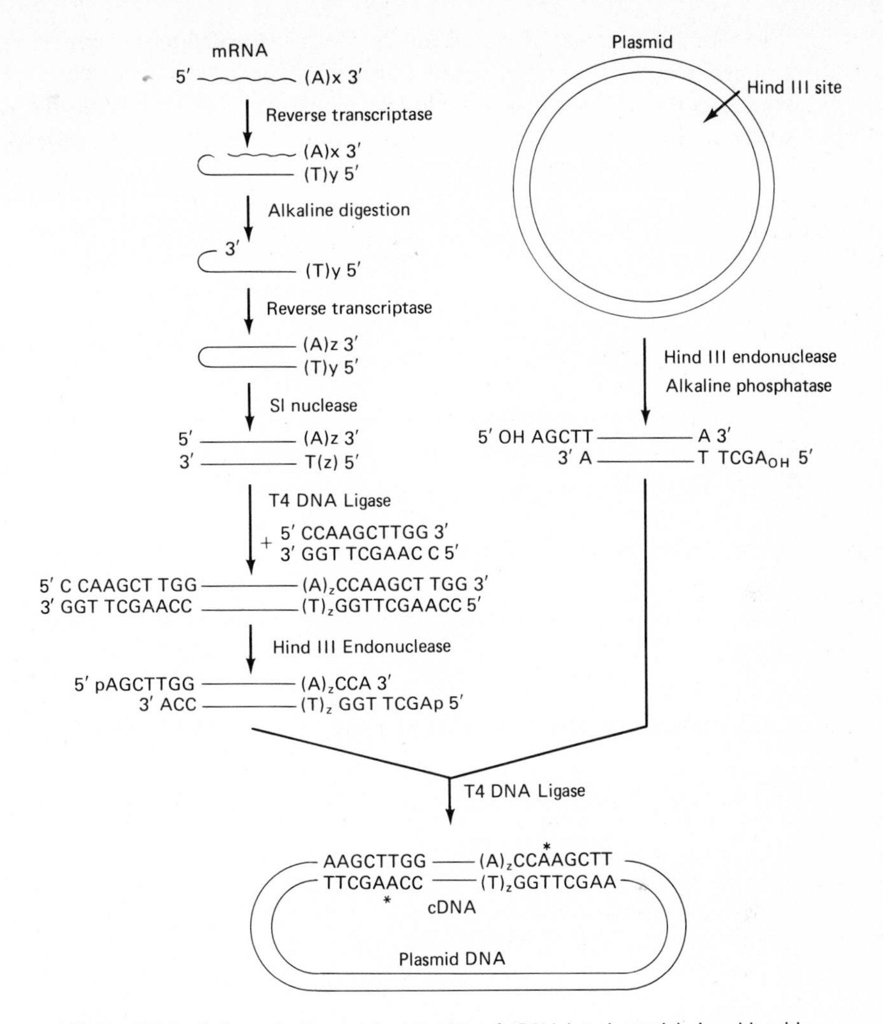

Figure 19-1. Schematic diagram for insertion of cDNA into bacterial plasmids with the use of chemically synthesized restriction site linkers. The asterisks in the recombinant plasmid indicate the position where a phosphodiester bond has not formed because of the absence of a 5′ terminal phosphate. The example is for the Hind III decanucleotide. In certain cases the Eco RI octanucleotide (5′)TGAATTCA(3′) was used; in these cases insertion was made into the Eco RI site of the plasmid. (From Ullrich *et al.*, 1977. *Science* 196:1313; copyright 1977 by the American Association for the Advancement of Science.)

polypeptide, chain C, which connects A and B. There also exists a sequence of 20 amino acids, called a *prepeptide*, at the amino terminus leading into chain B. The sequence of the molecule translated from the complete message thus is NH_2–prepeptide–chain B–chain C–chain A–COOH. This molecule is known as *preproinsulin*. By a mechanism which remains obscure, the prepeptide is removed, and the resultant molecule, which contains chains B, C, and A, is known as *proinsulin*. Chain C is then removed, to form the functional insulin molecule.

In 1977, plasmids containing the sequences coding for rat insulin were constructed by isolating preproinsulin mRNA. The mRNA molecules then served as templates for the synthesis of complementary DNA (cDNA) molecules using reverse transcriptase. The double-stranded DNA molecule generated from cDNA was inserted into plasmids and cloned upon infection of *E. coli* cells (Ullrich *et al.* 1977). Figure 19-1 illustrates significant steps in the research leading to cloning of rat insulin DNA in bacteria.

The potential benefit to diabetic patients from this technology is obvious. Large-scale production of insulin for therapeutic purposes in the future will reduce the cost of such treatment. And insulin is only one such substance. Other proteins produced in large quantities by certain cells could conceivably be synthesized in this manner.

Another protein whose production on a large scale is desirable is interferon, a protein produced by cells in response to viral infection (p. 404). Studies on the effects of interferon have been limited because cells produce so little of the protein that it was difficult to obtain sufficient quantities for valid studies. As we stated in Chapter 14, fibroblast interferon genes, one on the long arm of chromosome #2 and the other on the short arm of chromosome #5 (Slate and Ruddle 1979), have now been cloned. Scientists anticipate having considerable quantities of interferon to test in the very near future.

Genetic Consequences of Advances in Euphenics

There is a genetic consequence of successfully treating individuals with genetic diseases which is not normally considered by most people. This is the effect of increasing the incidence of detrimental genes in future generations. Since many of the conditions we hope to treat are so severe that patients cannot reproduce, natural selection against these genotypes normally prevents their transmission. The incidence of such genes in a population is then maintained only by spontaneous mutation rates which, as we have seen earlier (p. 490), are generally very low.

If we counter the forces of natural selection, the result will undoubtedly be an increase in frequency of the mutant gene in the population. We then must consider the extent to which we will be increasing the incidence of affected individuals in future generations. For example, how much will the treatment of PKU homozygotes with the proper diets (so that they can mature normally and reproduce) increase the incidence of the disease in our society? (Homozygotes, of course, transmit the gene to every child.) What of dominant mutations? For example, suppose we were to discover the basis to the growth stunting characteristic of achondroplastic dwarfism?

The gene frequency of PKU is on the order of $\frac{1}{100}$. A leading human geneticist, James Crow, has estimated that if all PKU homozygotes were successfully treated and later reproduced, the frequency of the gene would increase by one percent in the next generation (Crow 1973). This increase in gene frequency would in turn cause about a 2% increase in the frequency of PKU homozygotes in the next generation. Crow further estimates that with a 2% increase in the frequency of homozygotes per generation (a generation is considered to be 30 years), it would require about 40 generations, or 12 centuries, to double the incidence of PKU genes in the

Table 19-1 Rate of Increase of a Harmful Gene Under Relaxation of Selection (frequencies per Million)

	Dominant Phenotypes				Recessive Phenotype	
Old fitness:	0.00				0.00	
New fitness:	0.50	0.75	0.90	1.00	1.00	
Generation	*Case frequency*				*Case frequency*	*Gene frequency*
0	20	20	20	20	9.97	3157
5	39	66	94	120	10.29	3207
10	40	77	137	220	10.61	3257
15	40	79	163	320	10.93	3307
20	40	80	178	420	11.27	3357

Initial frequencies are based on a forward mutation rate of 1 in 100,000 and a back mutation rate of 1 in 10,000,000. Only fitness is changed.

From E. S. Murphy, 1973. In I. A. Porter, and R. H. Salko, eds., *Heredity and Society*, p. 142. Academic Press, New York.

population. The increase would of course be even slower if some homozygotes escape detection until they are irreversibly affected, or if treatment is not totally successful. The incidence of rare recessive mutations thus would not be expected to increase dramatically with successful euphenic control.

On the other hand, the frequency of conditions determined by dominant mutations are likely to increase faster than those determined by recessive mutations. In the case of mutations that are lethal or prevent reproduction in heterozygotes, the incidence of the gene, as we discussed in Chapter 16, is entirely dependent on the spontaneous mutation rate. If such individuals are successfully restored to normal health and function, in the next generation incidence of the gene would increase 50% by transmission of the mutation from heterozygotes, or 100% by transmission from homozygotes. Table 19-1 estimates the rate of increase of harmful genes, if the phenotypes they determine can be successfully treated; hypothetical back and forward mutation are taken into consideration.

Societal Consequences

Any increase in the incidence of genetic diseases would of course require an increase in our ability to treat such conditions. This necessity poses a major societal problem, which is of great concern to geneticists and clinicians alike, as well as to those directly affected: the economic burden of support for genetically deficient people. A survey published in 1975 by the U.S. Department of Health, Education, and Welfare (HEW) presented the problem in dramatic figures.

Some 12 million Americans suffer from diseases related to defective genes or chromosomes. More than 100,000 spontaneous abortions per year result from gross chromosomal defects. At least 40% of all infant mortality results from genetic factors. Each married couple stands a 3% risk of having a genetically defective child. In addition, diseases such as arthritis, schizophrenia, and heart disease are now believed to have a partly genetic basis. These figures are likely to increase owing to the increase in mutagens in our environment.

The financial burden involved in treatment of these diseases or in the maintenance and support of the untreatable is enormous and growing. The HEW survey pointed out, for example, that all the types of muscular dystrophy have a combined incidence of 2 or 3 per 10,000 live births. There are 200,000 muscular dystrophy patients; the annual cost of their medical care totaled about 400 million dollars in 1975. In the case of patients requiring institutionalization, such as some of those with Down's syndrome, the lifetime cost has been placed at $250,000 per patient. Some 5000 new Down's patients are born each year (on the basis of a rate of 1 per 700 live births).

What of a disease that can be detected and treated? In 1975, the cost of screening newborns for PKU was $1.25 per test; this projects into approximately $17,000 spent to detect each case (since most newborns tested are normal). Once a PKU homozygote has been detected, $8000 to $16,000 must be spent over a period of 10 years to provide a special diet that will prevent mental retardation. Those untreated would presumably require institutionalization, at a cost equal to the amount spent on a Down's syndrome patient.

To discuss human suffering in terms of dollars and cents seems contrary to humanitarian instincts. Nonetheless, the extent of the financial burden resulting from biological and medical advances which sustain life in persons normally selected against cannot be ignored. It is indeed a problem that all of society must bear, since even those not directly affected must shoulder some of the burden through taxes and costs of medical programs.

Still, our advances in understanding the genetic bases to diseases and new technology such as recombinant DNA give us hope of developing programs and approaches that will allow us to overcome these problems. There is in addition another approach that may also help to control the incidence of genetic diseases. This method, which raises serious moral and ethical questions, is to prevent the transmission of detrimental genes to future generations by controlling reproduction.

EUGENICS

The term *eugenics* was first coined by Francis Galton (1812–1911) to designate a process of improving human "stock" by encouraging matings between the "more suitable races or strains of blood" (in Gordon 1972). As mentioned earlier, this approach, involving the selection of persons to reproduce, is known as *positive* eugenics. In the early decades of the twentieth century, eugenics institutes were established in both England and the United States.

We might point out here that "eugenics" as practiced by plant and animal breeders has led to the development of new crops and species which have been of great benefit to society. Recall that there were plant breeders, such as Mendel's father, long before there was a science called Genetics. However, where we consider eugenics desirable in plants and domestic animals, its application to humans is of course a different matter.

The concept of eugenics in humans shortly became associated with *negative* eugenics, advocated by persons associated with various eugenics movements, and with the aim of discouraging or preventing persons considered to possess less desirable traits from mating and reproducing. The danger in such schemes of course is in establishing what is a desirable trait, and what is not. Who would have the right to make this choice? Clearly, the risk of serving the purpose of those with biases against certain populations is inherent in a negative eugenics program. Both positive and negative eugenics, for example, reached their height of absurdity in Nazi Germany.

Because of the atrocities committed in Nazi Germany in the name of eugenics and the control over the freedom to mate and reproduce that eugenics advocates, most people find eugenics objectionable, Yet eugenics has been practiced over the years. In the next few sections we shall explore the various ways in which our society has controlled reproduction in the past, and the form which such control may take in the future.

Genetics and the Law

Although the thought of government control over reproduction is anathema to most of us, laws do exist in many states which prohibit marriage and procreation in specified situations. Laws prohibiting consanguineous marriages, for example, were enacted by many states in the late nineteenth and early twentieth centuries (Farrow and Juberg 1969).

Laws on consanguineous marriages. In law as well as in genetics, consanguinity is recognized as being either lineal or collateral. *Lineal consanguinity* refers to a relationship in which one person is descended from the other, such as father–daugher, mother–son. *Collateral consanguinity* refers to persons descended from a common ancestor, but not from each other, such as cousins. As we discussed in Chapter 16, there are degrees of relatedness as expressed by the coefficient of inbreeding. As summarized in Table 19-2, many states have laws which prohibit marriage to any ascendant or descendant of any degree. Other laws prohibit marriage depending on the degree of relatedness as shown in Table 19-3.

As pointed out by Farrow and Juberg, a number of genetically strange prohibitions are included in the marriage laws which they surveyed. For example, in Georgia, the nephew–aunt marriage is prohibited, while the uncle–niece marriage is not! Furthermore, many states have laws forbidding marriage to affinous relatives (relatives by marriage), as shown in Table 19-4.

In Rhode Island, yet another factor enters into the laws prohibiting certain marriages, namely, religion. The law states that marriages between Jews permitted by their religion can be exceptions to Rhode Island laws applying to all other people.

Laws on interracial marriages. In the past, a number of states passed laws to prohibit marriage of two persons of different races. For example, Whites and Blacks were not allowed to marry each other in Virginia until this statute was challenged in 1966. Richard Loving, a white man who had married a black woman, Mildred Jeter, moved from Washington, D.C., to Virginia, and was convicted of violating a law against interracial marriage. The case was taken to the Supreme Court of the United States, which overturned this obviously discriminatory statute (Lederberg 1976).

Table 19-2 Inclusive Laws Regarding Prohibition of Consanguineous Marriages in the United States and Two Territories

State or territory	A person may not marry: Any ascendant or descendant	A person may not marry: Any relative within and including the following degree
Alaska	—	3rd
Arizona	X	—
California	X	—
Colorado	X	—
Connecticut	—	3rd
Delaware	X	—
Florida	X	—
Hawaii	X	—
Idaho	X	—
Indiana	—	5th
Kentucky	—	5th
Louisiana	X	—
Maryland	X[a]	—
Minnesota	—	5th
Missouri	X	—
Montana	X	—
Nevada	—	5th
New Jersey	X	—
New Mexico	X	—
New York	X	—
North Carolina	—	3rd
North Dakota	X	—
Ohio	—	5th
Oklahoma	X	—
Oregon	—	4th
South Dakota	X	—
Tennessee	X[b]	—
Utah	X	4th
Washington	—	5th
Wisconsin	—	5th
Puerto Rico	X[c]	3rd
Virgin Islands	—	3rd
Total States and Territories	20	14

[a]Or any ascendant or descendant of either parent.
[b]Within and including three degrees.
[c]Or any ascendant or descendant by affinity.

From M. G. Farrow and R. C. Juberg, 1969. *J. Am. Med. Assoc.* 209: 536.

Laws on marriages involving the mentally retarded. Laws also exist in some states prohibiting marriage of the "mentally infirm." As Lederberg points out in his review, restrictions may be based on:

1. Their incompetence to appreciate their responsibilities under the marriage contract, or
2. The protection of the infirm from others who might abuse a marital relationship as servitude, or
3. As an indirect means of inhibiting the procreation of individuals whose offspring may need protection from potentially inadequate parental guidance, or
4. As an indirect means of inhibiting the procreation of individuals whose mental defectiveness may be inherited.

In two states, Nebraska and North Carolina, eugenics takes the form of laws that require sterilization of mentally deficient persons who wish to marry. The question of involuntary sterilization is a highly controversial and emotional one. Also,

Table 19-3 Prohibitions of Marriage to Lineal Relatives in the United States and Two Territories

	A man may not marry his			
	Mother	*Daughter*	*Grandmother*	*Granddaughter*
Coefficient of inbreeding	$\frac{1}{4}$	$\frac{1}{4}$	$\frac{1}{8}$	$\frac{1}{8}$
All states, D.C., and two territories, except GA	X	X	X	X
Georgia	X	—	—	X
TOTAL	53	52	52	53

	A woman may not marry her			
	Father	*Son*	*Grandfather*	*Grandson*
Coefficient of inbreeding	$\frac{1}{4}$	$\frac{1}{4}$	$\frac{1}{8}$	$\frac{1}{8}$
All states, D.C., and two territories, except GA	X	X	X	X
Georgia	—	X	X	—
TOTAL	52	53	53	52

From M. G. Farrow and R. C. Juberg, 1969. *J. Am. Med. Assoc.* 209: 536.

there is a question of where the line will be drawn between those deemed competent and incompetent. Recall that intelligence and behavior are quantitative traits, and that there is a continuum of variation in genotypes. Given our less-than-precise measurements of intelligence, the line between competent and incompetent becomes vague indeed.

Furthermore, we must consider the environmental effect on intelligence development. We cannot determine if mental incompetence is due only to the genotype, or if it is determined partly by cultural deprivation. In addition, it is known that the average child of mentally retarded parents tends to be brighter than the parents.

The question of control over mentally deficient persons currently takes on more singificance since there has been an increasing tendency to accept the idea of "normalization" of the retarded, allowing them to live in the community if they have the capacity to do so (Baron 1976). As an illustration, Baron points to the case of *In Re Cavitt*. A 35-year-old woman, who had an IQ of 71, was given the option of leaving the Nebraska state home in which she had been institutionalized for a short period of time, if she would submit to sterilization. This requirement was not overturned in the court. Although the option given Mrs. Cavitt may be deemed harsh, it must be pointed out that she had borne eight children by her common-law husband before she was institutionalized.

For the great majority of our population, however, such laws restricting marriage and reproduction do not apply and would be hard to enforce. We do have choices to make, and recent developments in genetics have allowed these choices to be more knowledgeable. These developments are in prenatal diagnosis and genetic counseling, to which we will now turn our attention.

Table 19-4 Prohibitions of Marriage to Affinous Relatives in the United States and Two Territories

	State or territory							
	Conn, Okla, SD	*Ala, Miss, Pa, Tex*	*Tenn*	*Iowa NH*	*Ga*	*Va, WVa*	*DC, Me, Md, Mass, Mich, PR, RI, SC, Vt, VI*[a]	*Total*
A man (woman) may not marry his (her)								
Father's wife (mother's husband)	X	X[b]	X	X[b]	X	X	X	23
Grandfather's wife (grandmother's husband)	—	—	—	—	—	—	X	10
Wife's mother (husband's father)	—	—	—	X	X	—	X	13
Wife's grandmother (husband's grandfather)	—	—	—	—	—	—	X	10
Wife's daughter (husband's son)	X	X	X	X	X	X	X	23
Wife's granddaughter (husband's grandson)	—	X	X	—	X	X	X	18
Wife's stepdaughter (husband's stepson)	—	—	—	—	—	X	—	2
Son's wife (daughter's husband)	—	X[c]	X	X[c]	X	X[c]	X	20
Grandson's wife (granddaughter's husband)	—	—	X	X[d]	—	—	X	13
Nephew's wife (niece's husband)	—	—	—	—	—	X[e]	—	2

[a]VI indicates Virgin Islands.
[b]Father's widow specified (Iowa, NH, Tex).
[c]Son's widow specified (Ala, Iowa, NH, Tex, Va).
[d]Grandson's widow specified.
[e]Niece's husband only (Va).

From M. C. Farrow, and R. C. Juberg, 1969. *J. Am. Med. Assoc.* 209: 537.

Prenatal Diagnosis

In the 1960s, a technique was developed which has proven to be very useful in the prenatal diagnosis of certain genetic diseases. The technique is known as *amniocentesis*, and involves the removal of amniotic fluid from the womb of a pregnant woman. The fluid contains fetal cells which are sloughed off as the fetus is suspended in the fluid in early pregnancy. The cells are cultured and then can be studied biochemically and by karyotype for defects. Figure 19-2 illustrates the technique, which now involves only a very small risk to the mother or fetus.

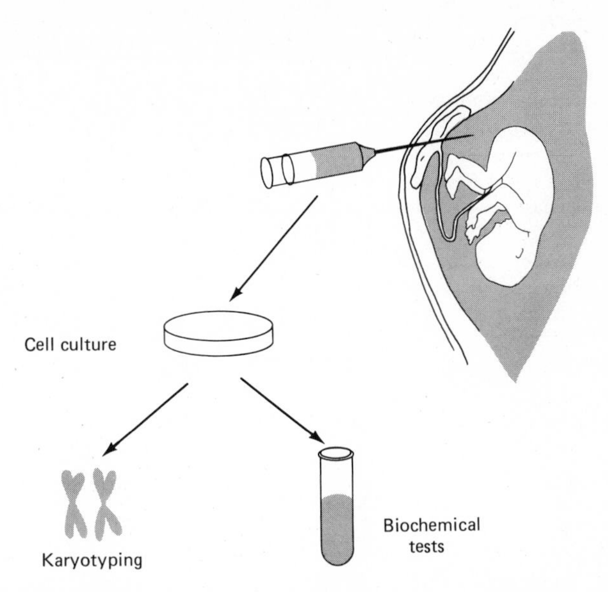

Figure 19-2. Schematic illustration of amniocentesis. Amniotic fluid containing fetal cells is withdrawn and the cells cultured. Biochemical tests and karyo typing can then be performed to test for deficiencies.

Recent surveys, however, reveal a 5% incidence of failure to obtain amniotic fluid at the first attempt, and 1 percent to 2 percent failure to grow the amniotic fluid cells in culture; such failures would require repeating the procedure. The number of spontaneous abortions sustained by women who have undergone amniocentesis is no greater, however, than the number sustained by women who have not undergone the procedure (Epstein and Golbus 1977). Recent developments, such as the use of ultrasound rather than X-rays for more precise location of the fetus, further reduce the risk of injury to the fetus (Devey and Wells 1978).

The technique of amniocentesis has limitations in its usefulness, however, as it can be used in the diagnosis only of conditions with detectable biochemical or chromosomal changes. For example, Tay Sachs disease can be diagnosed on the basis of the absence of an enzyme, hexosaminidase A. (Tay Sachs disease is associated with the Jewish ethnic group and causes progressive neural degeneration of the child victim, eventually leading to death.) All chromosomal or gross mutations can be detected by karyotyping cells obtained from amniotic fluid. Table 19-5 lists biochemical disorders that have been diagnosed prenatally. It must be pointed out, however, that unless there is cause to believe that the child may be defective, amniocentesis is not routinely performed, as there still remains a very small chance of injury to the fetus.

Table 19-5 Biochemical Disorders That Have Been Diagnosed Prenatally

Disorder	*Metabolic defect*
Acid phosphatase deficiency	Lysosomal acid phosphatase deficiency
Adenosine deaminase deficiency (combined immunodeficiency)	Adenosine deaminase deficiency
Adrenogenital syndrome	C-11 or C-21 steriod hydroxylase deficiency
Argininosuccinic aciduria	Argininosuccinase deficiency
Citrullinemia	Argininosuccinic acid synthetase deficiency
Cystinosis	Cystine accumulation
Fabry disease	Ceramidetrihexoside α-galactosidase deficiency
Fucosidosis	α-Fucosidase deficiency
Galactokinase deficiency	Galactokinase deficiency
Galactosemia	Galactose-1-phosphate uridyltransferase deficiency
Generalized gangliosidosis (GM1 gangliosidosis, Type I)	β-Galactosidase deficiency
Juvenile gangliosidosis (GM1 gangliosidosis, Type II)	β-Galactosidase deficiency
Gaucher disease	Glucocerebrosidase deficiency
Glycogen storage disease, Type II (Pompe disease)	α-1, 4-Glucosidase deficiency
Hemoglobinopathy (sickle cell)	Synthesis of hemoglobin S
Hunter syndrome	α-L-Iduronic acid-2 sulfatase deficiency
Hurler syndrome	α-L-Iduronidase deficiency
Hypophosphatasia (some types)	Alkaline phosphatase deficiency
I-cell disease	Multiple lysosomal enzyme deficiencies
Isovaleric acidemia	Isovaleryl CoA dehydrogenase deficiency
Ketotic hyperglycinemia	Propionyl CoA carboxylase deficiency
Krabbe disease	Galactocerebroside β-galactosidase deficiency
Lesch-Nyhan syndrome	Hypoxanthine guanine phosphoribosyltransferase deficiency
Maple syrup urine disease	Branched chain ketoacid decarboxylase deficiency
Maroteaux-Lamy syndrome	Arylsulfatase B deficiency
Menkes disease	Copper accumulation
Metachromatic leukodystrophy	Arylsulfatase A deficiency
Methylmatonic aciduria	Methylmalonic CoA mutase deficiency
Niemann-Pick disease	Sphingomyelinase deficiency
Placental sulfatase deficiency	Placental sulfatase deficiency
Prophyria—acute intermittent type	Uroporphyrinogen I synthetase deficiency
Pyruvate decarboxylase deficiency	Pyruvate decarboxylase deficiency
Sandhoff disease	Hexosaminidase A and B deficiency
Sanfilippo syndrome, Type A	Heparin sulfatase deficiency
Sanfilippo syndrome, Type B	N-acetyl-α-glucosaminidase deficiency
Tay Sachs disease	Hexosaminidase A deficiency
α-Thalassemia	Decreased synthesis of α chain of hemoglobin
β-Thalassemia	Decreased synthesis of β chain of hemoglobin
Wolman disease	Acid lipase deficiency
Xeroderma pigmentosum	UV endonuclease deficiency

From C. J. Epstein and M. S. Golbus, 1977. *Am. Sci.* 65: 708.

Genetic Counseling

The decision to perform amniocentesis is usually a result of discussions between the parents and their physician and/or a genetic counselor. The field of genetic counseling has developed in recent years as a result of increased awareness of the extent of genetic disease, and improved diagnostic methods.

The genetic counselor can be either a medical practitioner who has specialized in inherited diseases, or a person trained specifically in genetics working in conjunction with a doctor. The role of the counselor is to communicate to the prospective parents the risk of their producing an affected child. Such people generally may be brought to the attention of genetic counselors for a number of reasons. For example, they may already have produced an affected child, there may be a history of a condition in one or both families, they may be members of a population which is associated with one of the ethnic diseases, or they may belong to an advanced age group. Such people are considered to be *at risk* with regard to an inherited disease.

The counselor should communicate three important ideas to parents who are at risk (Murphy 1972). One is the intensity of the burden represented by the disease. For example, color-blindness is relatively trivial; Tay Sachs is severe. The second is the duration of the burden. Some diseases are lethal even at developmental stages; others, such as Down's syndrome, could last a normal lifetime. Finally, the counselor must explain the probability which the parents face in producing an affected child. These are the same probabilities which we discussed in the early chapters on transmission genetics, and also the probabilities as determined by statistical studies on a particular population. The results of a study undertaken to determine the estimated frequency of genetic disorders in newborns are summarized in Table 19-6.

It is important to stress that the role of genetic counseling is simply to inform; it is not to determine the right decision for the parents. Only the parents at risk have the right to decide whether to conceive children, or whether to carry a child to term or abort. Prenatal diagnosis and counseling have helped many prospective parents in making this decision more intelligently.

Genetic Screening

Because of the relatively high incidence of genetic diseases, mass programs to detect and counsel carriers of detrimental genes in a population have been advocated. The mass testing of a population for either carriers or affected individuals is known as *genetic screening*. We shall first discuss programs that are already in existence for the detection of affected newborns.

Screening of newborns. The first disease for which a screening program was instituted was PKU, a disease that satisfies several requirements for a screening program to be effective. First, the defect is a simple one that can be detected easily. In PKU, elevated levels of phenylpyruvic acid in urine and serum of newborn babies is used as the criterion for diagnosis. Second, the disease has a treatment, which calls for the elimination of the amino acid phenylalanine from the diet. By 1966, 43 states had passed laws requiring the testing of newborn babies for PKU.

Initially, the development of screening programs received support from lay groups and legislators as well as geneticists, as they were seen as a means of reducing

Table 19-6 Estimated Frequency of Genetic Disorders in the Newborn Population

Disorder	*Percentage*
Chromosomal disorders[a]	
Autosomal	
Trisomy 21	0.13
Trisomy 18	0.03
Trisomy 13	0.02
Other	0.02
	0.20
Sex chromosome	
X0 and X deletions	0.02
Other "severe" defects	0.01
XXY	0.1
XXX	0.1
XYY	0.1
Others	0.015
	0.35
Total chromosomal disorders	0.55
Monogenic (Mendelian) disorders	
Autosomal recessive	
Mental retardation, severe	0.08
Cystic fibrosis	0.05
Deafness, severe (several forms)	0.05
Blindness, severe (several forms)	0.02
Adrenogenital syndrome[b]	0.01
Albinism	0.01
Phenylketonuria	0.01
Other amino acidurias[b]	0.01
Mucopolysaccharidoses (all forms)[a]	0.005
Tay Sachs disease[a]	0.001
Galactosemia[a]	0.0005
	0.25
X-linked	
Duchenne muscular dystrophy[c]	0.02
Hemophilias A and B[c]	0.01
Others[b,c]	0.02
	0.05
Autosomal dominant	
Blindness (several forms)	0.01
Deafness (several forms)	0.01
Marfan syndrome	0.005
Achondroplasia	0.005
Neurofibromatosis	0.005
Myotonic dystrophy	0.005
Tuberous sclerosis	0.005
All others	0.015
	0.06
Total genic disorders	0.36

Table 19-6 *(continued)*

Disorder	*Percentage*
Multifactorial disorders	
Congenital malformations	
Spina bifida and anencephaly[a]	0.45[d]
Congenital heart defects	0.4
Pyloric stenosis	0.3
Club feet	0.3
Cleft lip and palate	0.1
Dislocated hips	0.1
	1.65
"Common disease" (e.g., diabetes, malignancy, allergies)	?
Total multifactorial disorders	1.65
Total frequency of genetic disorders	2.56

[a]Diagnosable in utero. [b]Some diagnosable in utero.
[c]Fetal sex can be determined. [d]Figure for U.K.; in U.S., 0.1%–.0.2%.

From C. J. Epstein, and M. S. Golbus, 1977. *Am. Sci.* 65: 701.

the cost of treatment of affected persons. As mentioned previously, the cost of identifying a PKU child is still much less costly than the expenses of institutionalizing a mentally retarded PKU child. In addition to PKU, several states also include a number of different genetic disorders in their screening programs, as shown in Table 19-7.

Screening a population. Another approach to the identification of affected individuals or carriers is to test all the adult members of a population. Since there exist such a multitude of genetic diseases, it is not possible to screen a general population for all diseases. Most screening programs, therefore, have been aimed at members of a certain group within the population which are known to have a higher incidence of a particular disease than other groups. We have previously described conditions associated with certain groups as "ethnic diseases." For example, Blacks would be screened for sickle-cell anemia, and Northern European Jews would be screened for Tay Sachs, both of which are caused by mutations which can be detected in the heterozygote.

The basic goal of these programs is to alert individuals to the possibility of transmitting mutations to their children. Barring new mutations those found not to carry the detrimental gene can be assured that their children will be free from these conditions. Mass screening would appear to be an approach without reproach; however, in recent times, many voices have been raised in caution or objection to mass screening both of newborns and adults.

Problems in genetic screening. There are a number of reasons behind the opposition to genetic screening, some of which are economic, some scientific, and some emotional. As for the economic issue, we can look at the case of a program in Rhode Island for the screening of newborns for maple sugar urine disease, an inherited condition that leads to mental retardation. The incidence of this disorder is only 1

Table 19-7 Current State Programs in Mass Neonatal Screening (Other Than PKU)

	Adenosine deaminase deficiency (?)	*Galacto-semia* (1: 75,000)[b]	*Homo-cystinuria* (1: 220,000)[b]	*Sickle-cell anemia* (1: 500)[a, b]	*Maple-syrup-urine disease* (1: 200,000)[b]	*Tyrosi-nemia* (1: 1,000,000)[b] (?)	*Histidi-nemia* (1: 24,000)[b]
Alabama				X			
Alaska					X		
Connecticut		X					
Georgia				X			
Kentucky				X			
Louisiana				X			
Maryland					X		
Massachusetts[d]		X	X		X	X	X
Mississippi				X			
Montana		X	X		X	X	
New York	X	X	X	X	X		X
Ohio		X	X				
Oregon		X	X		X	X	
Rhode Island[e]			X		X		

[a]Only American Blacks screened.
[b]Estimated frequency.
[c]Plans to expand screening for aminoacidurias in 1975.
[d]Screens for many disorders not listed in this table.
[e]Screens for "some aminoacidurias."

From P. Reilly, 1976. In *Genetics and the Law*, p. 1641. Plenum Press, New York.

in 300,000 births. In a small state such as Rhode Island, the disease can be expected to be detected only once in 10 years. One might question the cost-benefit ratio in this case (unless one happened to be a parent of the one case).

Opposition to screening based on scientific reasons cites the fact that a number of the diseases detectable in fetuses or newborns by screening procedures such as combined immune deficiency disease or sickle-cell anemia, cannot be treated. Unless parents have already determined to abort fetuses manifesting such incurable diseases, screening serves only to detect the existence of the problem earlier.

Perhaps equally important is that false positive or false negative results can lead to disastrous consequences. For example, of the approximately 3 million babies born each year in the United States that are screened for PKU, 3000 will give positive results; of these only 200 will actually have classic PKU (Culliton 1976). Most of the cases of elevated phenylpyruvic acid do not appear to be harmful to the child, and are traceable to factors other than the PKU mutation. In these cases, removal of phenylalanine from the diet would be unnecessary, expensive, and possibly harmful. False negatives obviously mean that the condition goes undetected and the disease therefore develops.

Much of the opposition to mass screening has been associated with screening for ethnic diseases, especially for carriers of the diseases. For example, many Blacks have voiced objections to screening for carriers of sickle-cell anemia. They point out that such information can be extremely disruptive of their lives. To be known as a sickle-cell carrier may cause them to lose job opportunities, and to sustain higher insurance rates. They feel stigmatized by the label of carrier, and fear detrimental social consequences.

Dr. R. F. Murray, a critic of screening, has pointed out the case of an unusual town in Greece, where the incidence of sickle-cell anemia was found to be as high as in Black populations (Murray 1975). Some 25% of the people were found to be carriers; 1% of all live births were sickle-cell homozygotes. Following a screening program each person was counseled as to his or her genotype. Seven years later, the scientists returned and found no change in the frequency of births of sickle-cell anemia children. The people had disregarded the counseling and even lied about their genotypes in order to avoid ostracism.

This story emphasizes several difficulties with screening for ethnic diseases. First, the high incidence of sickle-cell anemia in the Greek town shows that the disease is not found solely in Black populations. Since the disease is still predominantly one of Black populations, however, should we ignore all other populations, and would this not be a form of reverse discrimination? This would be true of testing for any ethnic disease.

Second, there is the reluctance of some of those screened to accept their genotypes. In many cases, one or two sessions of counseling is simply not adequate to communicate the risks which people face, and the necessity of indeed facing these risks, rather than ignoring them.

There are many hard questions which society must consider in regard to genetic screening. Do we have a right to make screening of certain persons mandatory? Arguments against mandatory screening would follow the same reasoning as we mentioned in our discussion of screening for sickle-cell anemia.

Arguments in support of mandatory screening would emphasize that increasing the incidence of disease affects society as a whole, not only those directly affected. The burden of the cost sooner or later falls on everyone. If all are to contribute to treatment, then ought not all have a voice in the control of the problem? Legislation has been proposed to authorize federal support for treatment of or research on various ethnic diseases. If this should come to pass, what then of nonethnic diseases such as hemophilia, or muscular dystrophy? Should not these and the myriad other genetic diseases receive equal support?

Most scientists agree that genetic screening can be helpful if properly conducted. Such efforts, however, must include well-thought-out programs for the education of the populations to be screened. The test itself should be one that can be easily administered, and should be as accurate as possible. Whether certain ethnic populations should be screened for diseases more commonly associated with them than with other groups remains debatable.

And what of those who as a result of screening are found to be carriers of detrimental genes? Do we have the right to demand, for the sake of society as a whole, that such people be prohibited from reproducing and transmitting the defective gene to future generations? What restrictions should be imposed, and by whom? Should carriers of defective diseases be restricted from marrying other carriers? What then of carriers of dominant diseases which are expressed in adult years, such as Huntington's chorea (an autosomal disease resulting in degeneration of the nervous system)? The probability that such people would transmit the gene to their offspring is $\frac{1}{2}$. Some have advocated that such people should adopt genetically normal children, and at the same time alleviate overpopulation. Counterarguments, however, point out again the infringements upon individual freedom, and the fact that very common diseases such as heart disease, cancer, and diabetes are believed to have a genetic basis. If we were to include all these diseases, the effect on society would be drastic. The question is whether such an approach should be legislated into law.

It would seem that the ideal solution would be a voluntary decision by each individual, based on a thorough knowledge of the problem. Education and communication would be absolutely necessary, either by individuals such as practicing physicians and genetic counselors, or on a larger scale, in courses to be made available in educational institutions or local organizations.

Germinal Choice

Genetics in the future may provide couples with some choice as to the type of progeny they decide to have, a situation referred to as *germinal choice*. Although the decision to abort genetically defective children or of carriers not to have children is a form of germinal choice, most of the time germinal choice is mentioned in the context of positive eugenics. It is postulated that someday we may be able to distinguish X-carrying sperm from Y-bearing sperm. Through artificial insemination with the proper sperm, a couple could be assured of producing the sex of child which they desire.

The late geneticist H. J. Muller even went so far as to publicly advocate establishing sperm banks. Muller was convinced that nuclear war is unavoidable. In the later years of his life, he proposed that mankind should provide for the future by obtaining sperm and storing them. It is now possible to freeze sperm at very low temperatures without reducing the effectiveness of the germ cells.

The aspect of Muller's proposals that was considered with astonishment by many geneticists was that he proposed storing sperm from men in different walks of life such as musicians, engineers, and so forth, so that children conceived by artificial insemination of women with these sperm would have the genetic potential to enter such careers. One of the problems here, of course, as T. Dobzhansky pointed out, is who would be president of such a bank? In other words, who is to decide which men should be the donors? And where would the eggs come from, assuming women to be equally vulnerable to nuclear war?

If the idea of sperm banks and germinal choice seem more in the realm of science fiction, consider that scientists have now developed methods of freezing and storing whole mouse embryos, thawing them out, and implanting them successfully into a foster mother. Although the thought of applying this technique to human embryos is too bizarre to consider (although genetically there would be less chance for chromosomal defects if couples could conceive all their children while young), it is clear that the procedure offers advantages in the breeding and transportation of domestic animal stock. Embryos conceived from a particularly fine specimen of sheep or horse, for example, could be stored and transplanted to mothers for development. Bovine embryos are now temporarily implanted in rabbits for shipment, then recovered and implanted into cows when they reach their destination.

Along the same line, new techniques have been developed which allow scientists to create embryo chimeras by fusing blastomeres from different embryos (Mintz 1967). By using embryos of mice from strains with different coat colors, the results of the fusion experiments can be dramatically demonstrated (Fig. 19-3). Again, embryos from strains possessing different desirable characteristics could be fused to produce a chimera which hopefully would manifest all the desirable traits. This approach would be useful only for species in which inbred strains can be developed whose phenotypes are well documented.

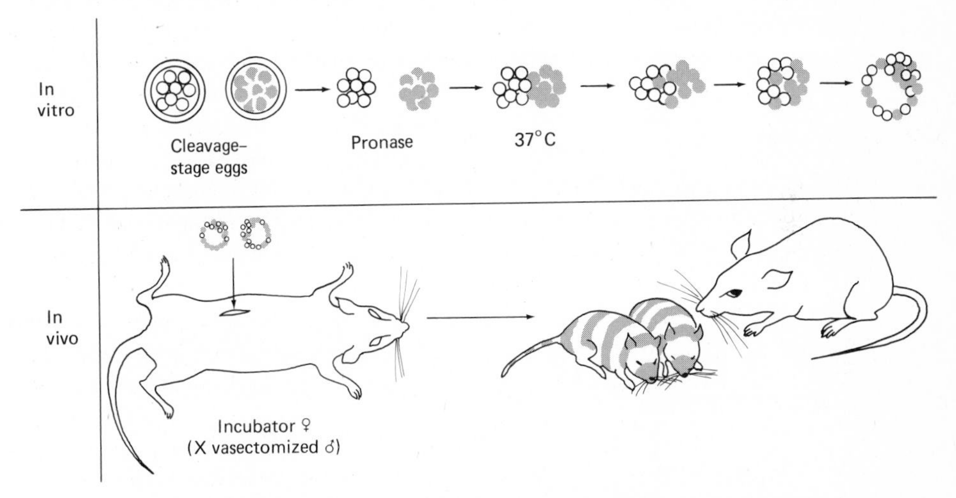

Figure 19-3. Diagram of the experimental procedures for producing allophenic mice. (From B. Mintz, 1967. *Proc. Nat. Acad. Sci. U.S.* 58:345.)

A slightly different approach is one pursued by a number of different laboratories which are attempting to create an in vitro environment that simulates the uterine environment of viviparous species (Sherman 1978). By so doing, it is hoped that events in early development can be better studied and understood. This knowledge can then eventually be applied toward reducing developmental trauma that can lead to birth defects of a serious nature. One might also project such a technique to the mass production of domestic mammals, in the same manner that chickens and other domestic fowl are now mass-produced by incubation.

GENETIC ENGINEERING

Yet another area of applied genetics, which is today further from practical realization than either euphenics or eugenics, is genetic engineering, the actual alteration of the genome of an individual. We have discussed a number of new techniques which can achieve this remarkable change in prokaryotes, or on the cellular level in eukaryotes.

Among these techniques are those of recombinant DNA technology and somatic cell fusion. In each case the genetic constitution of cells is changed. Where these approaches to changing genomes have been and continue to be extremely helpful in many areas of genetic research, so far their application, in terms of being able to affect the genotype or phenotype of a whole eukaryotic organism, remains limited. We must first determine the specific location of genes to be modified, then we must develop the methodology to alter the DNA in germ cells or embryos to effect the change in all cells of an individual.

Recently there have been reports of successful injection of exogenous DNA into amphibian oocytes, a technique known as *microinjection* which may lead to interesting developments in the future. J. B. Gurdon, D. D. Brown, and their coworkers have been able to inject both SV40 DNA and 5S RNA genes into *Xenopus laevis* oocytes. The DNA was inserted into the nucleus manually by micropipettes, and tests showed that it was not only conserved, but in the case of the DNA for 5S RNA was faithfully transcribed (Mertz and Gurdon 1977; Wyllie *et al.* 1978; Gurdon and Brown 1978). The amphibian egg, as we pointed out earlier, is very large, and whether the technique can be adapted to the much smaller eggs of mammalian species remains to be seen.

Fears expressed by many laymen that continued development of genetic engineering methodology will lead to the creation of monsters are on the whole unfounded. Our understanding of, and therefore our ability to control, life processes is, at best, incomplete. Furthermore, the urge to exploit new techniques will hopefully not supercede the concern of scientists that their work is to benefit mankind, not destroy it.

SOCIETY AND SCIENCE

Here we can turn our attention to a final problem which confronts science and society. Many of the conflicts pointed out in this and other chapters condense into a fundamental question: Which has the higher priority, the rights of an individual, or the rights of society? In many instances, such as screening and restrictions on reproduction, the interests of society may be in direct conflict with the interests of the individual. In danger of regulation are freedom of choice of such dearly held individual rights as choice of mate, of reproduction, of a particular way of life. Regulation of such activities would be in the name of the good of society as a whole.

This question also applies to science today. In many ways the freedom of scientists has been encroached upon because of the far-reaching potential of their work, especially in genetics. For the first time, decisions are being made by non-scientists on the type of research scientists may pursue. The effect of this scrutiny (which some scientists feel is interference) on the rate of progress in scientific research remains to be seen.

Scientists on the whole are fearful that such encroachment upon intellectual activities sets a precedent with unclear and perhaps unlimited bounds. Besides difficult questions such as who shall make the decisions on the type of research allowed, there is fear that problems in enforcing guidelines may deprive even the most qualified laboratories of support. If research should come to a standstill, one thing is certain: the problems at which the research would be aimed would continue to exist, and probably increase.

The coming years will be looked upon with great interest. The potential for crucial breakthroughs is a source of excitement. The development and application of newly gained knowledge is a source of concern. Of more importance than ever before is the communication of information between the scientific community and the society it serves. There is little doubt that we humans, who have created so many problems, with better understanding of what we are and where we stand in our natural world, can solve many of these problems. Indeed, it appears that we have no alternative.

PROBLEMS

1. What scientific reasons caused geneticists to be dismayed at Muller's proposals for the establishment of sperm banks containing sperm from men trained in different fields?

2. If genetic screening were to be proposed in your community, would you be in favor or not in favor? Why?

3. Would you agree that legislation banning reproduction by carriers of detrimental genes should be enacted? Why?

4. What is your reaction to involuntary sterilization of mentally retarded persons? Do you think that the mentally retarded mother of eight should have been sterilized? Give your reasons.

5. Would you be willing to pay taxes to support treatment of patients with genetic diseases? Discuss.

6. If you knew you were at risk for carrying a detrimental gene, would you want to have this information? Why or why not? If you were found to be a carrier, would you still plan on having children? Why?

7. Do you agree that scientific efforts should come under public scrutiny and control? Why?

REFERENCES

BARON, C. H., 1976. "Voluntary Sterilization of the Mentally Retarded." In A. Milunsky, and J. G. Arnas, eds., *Genetics and the Law*. Plenum Press, New York.

CROW, J., 1973. "Genetics and Medicine." In A. S. Baer, ed., *Heredity and Society*. Macmillan, New York.

CULLITON, B. J., 1976. "Genetic Screening: States May Be Writing the Wrong Kind of Laws." *Science* 191: 926–929.

G. B. DEVEY AND P. N. T. WELLS, 1978. "Ultrasound in Medical Diagnosis." *Sci. Am.* 238 (5): 98–112.

EPSTEIN, C. J., AND M. S. GOLBUS, 1977. "Prenatal Diagnosis of Genetic Diseases." *Am. Sci.* 65: 703–711.

FARROW, M. C., AND R. C. JUBERG, 1969. "Genetics and Laws Prohibiting Marriage in the United States." *J. Am. Med. Assoc.* 209: 534–538.

GORDON, H., 1972. "Genetics and Civilization in Historical Perspective." In I. H. Porter, and R. G. Skalko, eds., *Heredity and Society*, pp. 3–44. Academic Press, New York.

GURDON, J. B., AND D. D. BROWN, 1978. "The Transcription of 5S DNA Injected into *Xenopus* Oocytes." *Dev. Biol.* 67: 346-356.

LASKEY, R. A., AND J. B. GURDON, 1970. "Genetic Content of Adult Somatic Cells Tested by Nuclear Transplantation from Cultured Cells." *Nature* 228: 1332–4.

LEDERBERG, S., 1976. "State Channeling of Gene Flow by Regulation of Marriage and Procreation." In A. Milunsky, and G. J. Arnes, eds., *Genetics and the Law*. Plenum Press, New York.

MERTZ, G. E., AND J. B. GURDON, 1977. "Purified DNAs are Transcribed After Microinjection into *Xenopus* Oocytes." *Proc. Nat. Acad. Sci. U. S.* 74: 1502–1506.

MINTZ, B., 1967. "Gene Control of Mammalian Pigmentary Differentiation. I. Clonal Origin of Melanocytes." *Proc. Nat. Acad. Sci. U. S.* 58: 344–351.

MURPHY, E. A., 1972. "The Effects of Genetic Counseling." In Porter, I. H., and R. G. Skalko, eds., *Heredity and Society*, pp. 119–142. Academic Press, New York.

MURRAY, R. F., JR., 1975. "Genetic Screening and Counseling in Industrial Societies." *Symposium on Ethics of Current Genetic Research*. Rockland Comm. College Suffern, N. Y.

SHERMAN, M. I., 1978. "Implantation of Mouse Blastocysts in Vitro." In Daniel, J. C., Jr., ed., *Methods in Mammalian Reproduction*. Academic Press, New York.

SLATE, D. L., AND F. H. RUDDLE, 1979. "Fibroblast Interferon in Man Is Coded by Two Loci on Separate Chromosomes." *Cell* 16: 171–180.

ULLRICH, A., ET AL., 1977. "Rat Insulin Genes: Construction of Plasmids Containing the Coding Sequences." *Science* 196: 1313–1319.

"What Are the Facts About Genetic Disease?" U. S. Dept. HEW Publication No. (NIH) 75–370, 1975.

WYLLIE, A. H., A. LASKEY, J. FINCH, AND J. B. GURDON, 1978. "Selective DNA Conservation and Chromatin Assembly After Injection of SV 40 DNA into *Xenopus* Oocytes." *Dev. Biol.* 64: 178–188.

Glossary

abortive infection: An animal cell virus infection that is spontaneously lost.

acrocentric chromosome: Chromosome with a centromere near one end.

adaptation: Ability of organisms to survive in changing environments.

allele: Alleles are different forms of the same gene.

allotype: Subclasses of immunoglublins which exhibit different amino acid sequences within their constant regions.

antibody: Protein effector molecule of the immune response which is able to interact specifically with antigen.

anticodon: A triplet of bases on one loop of tRNAs which interacts with a codon on mRNA.

antigen: An immunologically foreign agent.

assortative mating: Mating based on phenotypic similarities or differences.

attached X (chromosome): Permanent nondisjunction of the X chromosome.

attenuator region: A region of DNA which controls transcription of the structural genes, in operons involved in the biosynthesis of amino acids.

autogenous regulation: Regulation of the production of a substance by the substance itself.

autosome (adj. autosomal): Chromosome occurring in homologous pairs (i.e., not a sex chromosome).

backcross: A cross between a parent and its offspring.

back mutation: Mutation from the mutant gene to the normal (wild-type) gene.

balbiani ring (= chromosome puff): Expanded region of the dipteran giant chromosome related to gene activity.

barr body: A darkly staining nuclear chromatin body found in the somatic cells of female mammals during interphase and thought to be the inactive X chromosome.

binomial experiment: An experiment with the following properties:

1. Experiment repeated under identical conditions (trials);
2. Dichotomous outcomes (Success and Failure);
3. Prob [Success] = p and is constant from trial to trial, and Prob [Failure] = $1 - p = q$.
4. Trials are independent.

bivalents: Closely associated homologous pairs of chromosomes formed during the zygotene substage of prophase I in meiosis.

breakage-reunion: A model of recombination in which the homologous DNA molecules break at identical sites and reanneal with each other.

caps: Methyl groups on bases at the 5′ end of eukaryotic mRNA.

carcinogens: Factors which can cause cancer.

cell cycle: The cell's cyclical pattern of synthesis and division consisting of G1 (gap), S (synthesis), G2, and M (mitosis) phases.

cell-mediated immune reaction: An immune reaction characterized by the production of lymphocytes which interact specifically with the antigen.

chiasma (plural, chiasmata): Points of crossing over between nonsister chromatids of homologous chromosomes during the diplotene substage of meiotic prophase I.

chromocenter: The point of attachment of the component chromosomes via their centromeres in dipteran giant chromosomes.

chromomere: A region of heavy coiling within the chromosome related to the banding pattern.

cistron: Unit of gene function, synonymous with structural gene.

clone (vb. cloning): A group of genetically identical cells or organisms.

codominance: Expression of the phenotype of both alleles in an undiluted manner in heterozygotes.

codon: A triplet of bases on mRNA which determines the specific amino acid to be inserted during translation.

coefficient of consanguinity: The probability that a related mate is a carrier for a particular trait if the other partner is known to be a carrier for the trait.

coefficient of inbreeding: The probability that offspring will be homozygous for a given trait if neither the parents nor the grandparents are homozygous for that trait and exactly one of the grandparents is a carrier for that gene.

coefficient of replacement: See migration coefficient.

cointegrate plasmid: A plasmid carrying both genes for antibiotic resistance and genes for transfer.

colinearity: A characteristic of the genetic code whereby there is a direct correlation in the position of codons on the DNA and the position of amino acids on the protein which the DNA determines.

collateral consanguinity: Marriage of persons descended from a common ancestor.

competence: In bacterial transformation—state of cells which are able to take up

DNA and consequently be transformed. (In development—state of cells which interact with other cells in an inductive system and are induced for a further differentiated state.)

complementation: The interaction of two mutant genes to produce an apparently normal phenotype.

conjugation: A process of genetic exchange between bacteria through a physical bridge, mediated by the sex factor.

consanguineous mating: Mating among relatives (see inbreeding).

constitutive strain: A strain which continuously produces a gene product.

contact inhibition: The phenomenon whereby contact between cells in culture prevents their further movement.

coordinate regulation: The concomitant control over transcription of several structural genes.

copy choice: A model for recombination in which the replication process switches from one template molecule to another.

correlation coefficient: Measure of the agreement or lack of agreement of two sets of observations:

$$\text{correlation coefficient} = r = \frac{\sum (X_1 - \bar{X})(Y_1 - \bar{Y})}{\sqrt{\sum (X_1 - \bar{X})^2 \sum (Y_1 - \bar{Y})^2}}$$

coupling phase (cis position): Location of the dominant alleles of linked genes on the same homolog.

crossing-over: The phenomenon of chiasmata formation between nonsister chromatids of homologous chromosomes during the diplotene substage of meiotic prophase I.

cytological map: The assignment of genes to particular chromosomes by observation of chromosomal morphology.

cytoplasmic inheritance: Traits determined by non-nuclear genetic elements.

degeneracy: A characteristic of the genetic code whereby more than one codon specifies a particular amino acid.

determination: The restriction of developmental potential.

dichotomous outcomes: an experiment with two outcomes, generally referred to as Success and Failure.

dihybrid: an individual heterozygous for two pairs of alleles.

diploid: The number of chromosomes present in the somatic cells of normal sexually reproducing adults; 2n.

dominant: An allele which is expressed over recessive counterparts in a heterozygote.

dosage compensation: The regulation of X-linked gene activity in females to compensate for their double dose (compared with the male's) of such genes.

dyad: A chromosome consisting of two sister chromatids.

endogenote: The recipient chromosome in a conjugation event.

endomitosis: Mitotic duplication of chromosomes without separation, resulting in polyteny as in dipteran giant chromosomes.

endonuclease: An enzyme which can cause a nick within a double-stranded DNA molecule.

environmental effects: The measured value of the contribution of the environment.

epistasis: The phenomenon whereby the expression of one gene is prevented by the expression of another at a different locus.

equational division: The second cell division cycle of meiosis in which the daughter cells retain the haploid chromosome number of the parent.

eugenics: The selection of certain genotypes in humans for mating and reproduction.

euphenics: The symptomatic treatment of genetically determined diseases.

event: One or a combination of outcomes in an experiment.

excision: The process whereby a prophage is released from the bacterial chromosome.

exogenote: That portion of the bacterial donor chromosome which is transferred during conjugation.

exon: Sequences of nucleotides of structural genes which are retained in mature mRNA, and translated into proteins.

exonuclease: An enzyme which degrades nucleic acids from an exposed end.

experiment (statistically speaking): A process which produces some outcome or result which can be observed or imagined.

feedback inhibition: A system in which the presence of a product inhibits its further production, by interacting with an enzyme that is involved in its production.

fitness: A relative quantity measuring the difference in the reproduction and survival of one organism (in particular, a genotype) in a given environment over another organism:

$$\text{Fitness} = 1 - \text{selection coefficient}$$

$$w = 1 - s$$

fixed genes: A situation in which all the genes at a certain locus are the same, the other allele having been lost from the population.

founder principle: If a new colony is founded by emigration from a population, the emigrants (founders) may not possess a typical (random) sample of the genes present in the population; therefore, by chance the colony may differ genetically from the parent population.

gene: A unit of the chromosome which determines specific traits by determining the structure of proteins or by serving regulatory functions.

gene amplification: Disproportionate replication of specific genes.

gene exchange: A single occurrence of the introduction of new genes into a new population by immigration.

gene flow: Introduction of new genes into a population by constant and recurrent migration.

gene pool: Sum total of the genes in a given breeding population at a particular time.

generalized transduction: Random transfer of host genes by the phage vector.

genetic code: The sequences of bases in nucleic acids which code for specific amino acids.

genetic counseling: Estimation on the basis of pedigree and population studies of the likelihood that a particular detrimental gene will be transmitted to the children.

genetic drift: Chance occurrences in small populations which lead to changes in gene frequencies from generation to generation.

genetic engineering: Alteration of the genome of a cell or an organism.

genetic screening: Mass testing of a population to detect carriers of genetic diseases and those afflicted with genetic diseases.

genotype: The genetic constitution of an individual.

genotypic value: The measured value of the contribution of the genotype.

germinal vesicles: The nuclei of amphibian eggs during early oogenesis.

graft-versus-host reaction: A reaction in which foreign lymphocytes (such as those which are passengers in a tissue transplant) mount an immune response against the cells of the host organism.

gynandromorphism: The condition in which an organism consists of a mixture of genetically male and female cells.

haploid: The number of chromosomes present in gametes following meiotic division of the germ cells; n number.

haplotype: The entire set of genes contained within the major histocompatibility complex.

Hardy-Weinberg law: The relationship between gene frequencies and genotype frequencies in a population under prescribed conditions, as expressed in the ratio $p^2 : 2pq : q^2$ (homozygous dominant, heterozygous, homozygous recessive).

hemizygous: Possessing only one allele of a particular gene, as in the case of X-linked genes in the heterogametic sex.

heritability: The fraction of the observed variance (variation) which is caused by differences in heredity. Heritability is a measure for a population.

heterochromatin: Darkly staining regions of chromosomes believed to be more highly condensed and probably transcriptionally inactive.

heteroduplex DNA: A region of mismatched base pairs in parts of the DNA molecule.

heterogametic sex: The sex which possesses dissimilar sex chromosomes.

heterogeneous nuclear RNA (hnRNA, hetRNA): Newly transcribed RNA present in the nucleus as very long molecules containing poly-A.

heterogenote: A partially diploid bacterium, with different alleles at the diploid locus, resulting from transduction.

heterozygote (adj. heterozygous): A diploid individual with different alleles of a particular gene pair.

histocompatibility antigens: Cell-surface antigens which are specific for an individual genotype and which can stimulate a cell-mediated immune response.

HLA (HLA-A, HLA-B, HLA-C): The genes which specify the major histocompatibility antigens of humans.

H-2 (H-2D and H-2K): The genes which specify the major histocompatibility antigens of mice.

histones: Small basic polypeptides which contribute to the structure of chromosomes.

holandric (y-linked) inheritance: The pattern of inheritance of genes on the Y chromosome.

homogametic sex: The sex which possesses a pair of identical sex chromosomes.

homogenote: A partially diploid bacterium with the same alleles at the diploid locus, resulting from transduction.

homozygote (adj. homozygous): A diploid individual with two identical alleles of a particular gene pair.

host-cell reactivation: The repair of damaged DNA of viruses by host-cell repair mechanisms.

humoral (circulating) immune reaction: An immune reaction characterized by the production of circulating antibody.

hybrid: An individual heterozygous for a particular gene pair or gene pairs.

inborn errors of metabolism: Inherited diseases caused by abnormal metabolic processes.

inbreeding: Mating of related individuals (see consanguinuous matings).

incomplete (partial) dominance: Expression of an intermediate phenotype in heterozygotes when neither allele shows dominance over the other.

independent (random) assortment: The independent distribution to gametes of genes which are located on different chromosomes.

inducer (effector): A substance which stimulates the activity of certain genes.

induction: Of viruses—the process of causing proviruses to reproduce. Of gene expression—initiation of gene expression due to the presence of an effector substance.

initiation complex: A complex of mRNA, 30S ribosome particle, and tRNA with amino acid formed at the onset of translation.

insertion element: An extrachromosomal genetic element that can be inserted into sites of a host chromosome, and that causes alteration of expression of neighboring genes.

insertion mutation: Mutation resulting from insertion of an extrachromosomal genetic element (insertion element) into a specific site on bacterial chromosomes.

interferon: A glycoprotein which is produced by animal cells in response to viral infection and which protects against subsequent viral infections.

intervening sequences (introns): Sequences of bases in structural genes which are transcribed but removed from precursor messenger RNA before translation.

isolating mechanisms: Geographical and genetic factors which prevent organisms from mating and breeding.

isozyme (isoenzyme): An enzyme which can exist in more than one molecular form, each of which catalyzes the same reaction.

karyotype: A compilation of the entire set of chromosomes from a cell.

lamarckism: Theory of evolution proposed by Lamarck which espouses the inheritance of acquired characteristics.

lampbrush chromosome: A chromosome during the diplotene of prophase I in oogenesis, which resembles a brush because of the presence of loops of transcribing DNA.

leader region: A region of regulatory DNA involved in amino acid synthesis, within the promoter-operator region, which is transcribed even in the presence of a repressor.

lethal gene: A gene which causes death of individuals of certain genotypes (for example, in individuals homozygous for a lethal recessive allele).

level of significance: Risk of incorrectly rejecting an appropriate model.

ligase: An enzyme which rejoins cut ends of DNA molecules.

lineal consanguinity: Marriage of persons one of whom is descended from the other.

linkage: The concept that each chromosome contains many genes which are thus linked.

linkage groups: Groups of genes which are transmitted together because of linkage.

locus: The defined chromosomal position of a particular gene.

lost gene: A situation in which all the genes at a certain locus are the same and the other allele has been permanently eliminated from the population.

lysis: The breaking open of a host cell owing to reproduction of viruses.

lysogeny (adj. lysogenic): The phenomenon whereby a bacterial cell harbors dormant viruses in the temperate phase of their life cycle.

lytic phase (= vegetative phase): The phase of the viral life cycle during which viral replication and host-cell lysis occur.

major histocompatibility complex: A genetic region which determines a variety of immune phenomena, including the histocompatibility antigens which elicits the strongest reactions.

meiosis: The form of cell division which is involved in the production of gametes and which gives rise to daughter cells containing half the number of chromosomes of the parent cell; one chromosome from each pair of homologs.

merodiploidy: Temporary partial diploidy in bacteria due to the presence of transferred genes.

merozygote: A bacterium merodiploid for certain genes as a result of gene transfer.

messenger RNA: The class of RNA which interacts with ribosomes and tRNA in protein synthesis, and which contains a sequence of bases complementary to the sequence of the DNA from which it is transcribed.

metacentric chromosome: A chromosome with a central centromere.

migration coefficient: The constant proportion of individuals migrating from one population to another.

minimal medium: A base nutritive medium which will support the growth of wild-type cells.

mitosis: The form of cell division which gives rise to two daughter cells each with the same number of chromosomes as the parent cell.

modifier gene: A gene which acts to modify the expression of other genes.

monohybrid: An individual heterozygous for a single pair of alleles.

monozygotic (identical) twins: Genetically identical twins produced by the splitting of a single zygote.

mosaics: Organisms consisting of a mixture of cells with different genetic (chromosomal) constitutions.

multigene family: Groups of genes which are similar in fine structure, have overlapping functions, and are closely linked.

multiple allelism: The existence of more than two different forms (alleles) of a gene.

multiplication rule: If an event occurs in n_1 ways, and a totally independent event can occur in n_2 ways, the number of ways both events can occur at the same time is $n_1 \times n_2$.

mutagens: Agents which cause mutation.

mutation: The alteration of a gene.

muton: Unit of genetic mutation.

natural selection: The process by which better-adapted individuals in a population will survive and produce reproductive offspring, while the less well-adapted individuals will die.

negative assortative mating: The tendency for dissimilar individuals to mate.

negative control: The control of gene expression by the prevention of transcription, for example with a repressor molecule.

Neo-Darwinism: The theory that evolutionary change is due to variability of individuals from mutation and natural selection.

nondisjunction: Failure of homologs (in anaphase I) or sister chromatids (in anaphase II) to separate, resulting in abnormal numbers of chromosomes in gametes.

nucleic acids: Polymers of nucleotides which contribute to the structure of chromosomes (for example, DNA).

nucleoside: Nucleic acid subunit composed of a five-carbon sugar and a nitrogenous base.

nucleosomes (nu bodies): Beadlike subunits of interphase chromatin composed of coiled DNA complexed with histones.

nucleotide: Nucleic acid subunit composed of a five-carbon sugar, a phosphate group, and a nitrogenous base.

operator region: A region of DNA which controls the transcription of a structural gene.

operon: All the genes (regulatory and structural) involved in the production of functional polypeptides.

origin: In replication—that point in the chromosome where DNA replication commences. In conjugation—the region of a donor cell chromosome which is adjacent to the integrated sex factor and which is transferred first during conjugation.

palindrome: Region of DNA duplex in which the complementary strands possess identical base sequences when read in the 3′ to 5′ direction or the 5′ to 3′ direction.

parthenogenesis: The development of an unfertilized egg.

partial diploids: Bacteria containing two alleles for a particular locus owing to gene transfer by transduction or sexduction.

pedigree: A means of analyzing the pattern of genetic transmission in humans by charting the transmission of particular phenotypes in members of a family.

penetrance: The expression or lack of expression of a particular expected trait.

phenocopy: A phenotypic condition induced by nongenetic factors but which closely resembles another which is genetically determined.

phenotype: The appearance of an individual.

phenotypic value: The measured value of some characteristic of behavior.

photoreactivation: A decrease in mutations in bacteria exposed to UV irradiation, owing to the activation of cellular repair processes by visible light.

pilus (pl. pili): A cytoplasmic bridge connecting bacteria during conjugation.

plasmid: An extrachromosomal genetic element which can exist independently of chromosomal genes or can be incorporated into the host chromosome.

pleiotropy: The ability of alleles at a single locus to affect a large number of characteristics.

ploidy: The number of sets of chromosomes.

polar body: A small haploid cell formed by the first and second meiotic divisions of the oogonia (female germ cells).

polar mutation: A mutation which causes changes in expression of genes located to only one side of the mutation.

polycistronic mRNA: A single mRNA molecule produced by the transcription of more than one cistron.

polygenic trait (quantitative inheritance): A trait not readily separable into discrete groups but which may be continuously variable and often fits a normal distribution curve. Such a trait is controlled by more than one pair of genes and may also be modified by environmental factors.

polypeptide: A polymer of amino acids joined by peptide bonds.

polytene chromosome (n. polyteny): Giant dipteran chromosomes which consist of many chromatids attached in parallel.

position effect (cis-trans, Lewis effect): The phenomenon whereby different phenotypes result from the same combination of pseudoalleles depending on whether they are in the cis (same homolog) or trans (different homologs) position; modification of the effect of a gene by its translocation to another chromosome.

positive assortative mating: The tendency for individuals similar in some phenotypic trait to mate.

potentiation: Phenomenonof increased mutagenesis or carcinogenesis is when two or more factors are combined, compared to their individual effects.

primary nondisjunction: Nondisjunction occurring in cells that are normally diploid at the onset of division.

primer DNA: An asymmetric piece of DNA which acts as a template and a primer for the *in vitro* synthesis of DNA by DNA polymerase I.

probability: The probability of an event E is defined as

$$\text{Probability of an event E} = \frac{n(\text{E})}{n}$$

where n is the number of equally likely outcomes, and n(E) is the number of outcomes that make up the event E.

proband (m., propositus; f., proposita): The first individual in a pedigree noted to be expressing a particular trait.

productive infection: An infection of animal cells by a virus which results in the production of new viral particles without lysis of the host cell.

promoter region: A region responsible for the initiation of the transcription of a structural gene binding with RNA polymerase.

prophase: A phage integrated into the host cell chromosome.

pseudoalleles (position pseudoalleles): Alleles of very closely linked genes which can be separated by crossing over and which exhibit position (cis-trans) effect.

qualitative traits: A trait easily classified into discrete groups.

quantitative characteristic: A continuously varying characteristic; a measuring instrument is often used to distinguish quantitative characteristics.

race: A group of individuals who have a higher frequency of certain genes than other groups.

random mating (= panmixis): A situation in which any individual of one sex has an equal chance of mating with any individual of the opposite sex.

recessive: An allele which is not expressed in the presence of a dominant allele of the same gene in a heterozygote.

recombinant DNA: Molecules of DNA composed of fragments from different chromosomes, which are joined by enzymatic reactions.

recon: Unit of genetic recombination.

reductional division: The first division of meiosis, resulting in two daughter cells each containing the haploid number of chromosomes.

regulator gene: The gene which codes for the repressor molecule.

regulatory gene: A gene which controls the activity of a structural gene.

repair mechanism: An enzymatic mechanism for the repair of abnormal DNA.

repetitive sequence (satellite DNA): A base sequence of nucleic acids which may be repeated many times in the genome.

replicating form (RF): The double-stranded form of single-stranded DNA viruses (for example, ϕX174).

repressor: A protein which binds to the operator gene and thus prevents its transcription (negative control).

repulsion phase (trans position): Location of the dominant alleles of linked genes on different homologs.

restriction endonuclease: An enzyme which cleaves the DNA molecule at specific base sequences.

restriction map (= cleavage map): Genetic map determined by use of fragments from restriction endonuclease digestion of chromosomes.

retrovirus (oncornavirus): A group of RNA virus which induces cancer.

reverse transcriptase: An RNA-dependent DNA polymerase, the activity of which is commonly associated with infection by RNA viruses since it is determined by a viral gene.

reversion mutation (= back mutation): Mutation of a mutant gene to original wild-type.

satellite: A portion of the chromosome separated from the main body of the chromosome by secondary constrictions.

secondary nondisjunction: The phemonenon which results in gametes with extra chromosomes owing to the abnormal karyotype of the original germ cell and not to a nondisjunctional event during gamete formation.

segregation: The result of the movement of homologous chromosomes (dyads) to opposite ends of the meiotic spindle during the first cell division cycle of meiosis, which leads to separation of homologs into different gametes.

selection: The sum total of environmental forces acting to determine the fate of an allele.

selection coefficient: A measure of all the forces acting to prevent a particular genotype from surviving and reproducing.

$$\text{Selection coefficient} = 1 - \text{fitness}$$

$$s = 1 - w$$

semiconservative replication: The mode of replication of the DNA molecule, in which one strand from the parent molecule is conserved and acts as a template for the synthesis of a complementary daughter strand.

sex chromosomes: A heteromorphic chromosome pair involved in the determination of the sex of an individual.
sexduction: Gene transfer between bacteria mediated by the sex factor (*F*).
sex factor (*F*) (= sex particle): An episome of bacteria which, when integrated into the chromosome, causes chromosome breakage at conjugation and facilitates genetic exchange; also exists autonomously in F^+ cells.
sex-influenced: Autosomal traits expressed predominantly in one or the other sex owing to sex hormones.
sex-limited: Traits which are governed by autosomal genes but expressed in one or the other sex owing to the influence of sex hormones.
sex-linkage: Genes which are located on the sex chromosomes (specifically on the X chromosome in X-linkage).
shotgun experiment: An experiment in which the entire genome of an organism is exposed to restriction endonuclease fragmentation, then random ligation to vectors for cloning into host cells.
siblings: The children of one generation in a single family.
sibship: A generation of children in a single family.
sister chromatids: The two identical component strands of duplicated chromosomes during cell division.
somatic pairing: The phenomenon of close association of homologs in somatic cells, as seen in dipteran giant chromosomes.
specialized (restricted) transduction: The transduction of specific genes.
species (sing. and pl.): A group of individuals so genetically different from another group that matings between the two cannot produce viable or reproductive offspring.
standard deviation (S. D.)= measure of variability of observations

$$\text{S. D.} = \sqrt{\text{variance}}$$

structural gene: A gene which determines the structure of a polypeptide.
submetacentric chromosome: A chromosome with a centromere somewhat nearer to one end than to the other.
supercoiling (n. supercoil, = superhelix): Twisting of the double helical DNA molecule to form secondary coils (superhelices).
synapsis: The pairing of chromosome homologs during the zygotene substage of meiosis.
synaptinemal complex: An organized structure of nucleic acids and proteins which appears between the closely associated homologous chromosomes at the pachytene substage of meiotic prophase I.
telocentric chromosome: A chromosome with a terminal centromere.
temperate phase: A phase of the viral life cycle in which the host cell is not lysed.
template: The conserved strand onto which a complementary daughter strand is synthesized during DNA replication.
terminalization: The phenomenon by which chiasmata formed during diplotene move to the end of the chromatids during diakinesis.
testcross: A cross between an individual homozygous for the recessive allele of a

particular gene, and an individual expressing the dominant trait whose genotype is unknown.

tetrad: A bivalent during the diplotene substage in meiosis, when the component homologs each consist of two sister chromatids.

totipotency: The ability of embryonic cells to differentiate into a variety of distinct cell types.

transcription: The synthesis of an RNA molecule using DNA as a template.

transduction: Bacteriophage-mediated gene transfer between host cells.

transfer RNA (tRNA): A class of small, long-lived RNA molecules which serve a variety of functions in cells, such as carrying amino acids to mRNA during translation.

transformation: Of bacteria—incorporation into the bacterial chromosome of exogenous genes absorbed from the environment. Of eukaryotic cells—process of changes in structure and function of cells resulting in malignancy.

transition: Mutation resulting from the substitution of one purine for another purine, or of one pyrimidine for another pyrimidine.

translation: The process of protein synthesis involving interaction between mRNA, tRNA, ribosomes, and amino acids.

translocation: Of genes—the attachment of a portion of one chromosome to another, nonhomologous chromosome. Of tRNA—the movement of tRNA from the amino acyl site to the peptidyl site on ribosomes during translation.

transposon (= insertion element): An extrachromosomal genetic element which is incorporated in specific sites on the bacterial chromosome.

transsexual: A person who has adopted the appearance of the opposite sex with the aid of surgical procedures and hormone therapy.

transversion: Mutation resulting from substitution of a purine for a pyrimidine or a pyrimidine for a purine.

triploidy ($3n$): The possession of three haploid sets of chromosomes.

variance: Measure of variability.

$$\text{Variance} = (\text{S. D.})^2 = \frac{\Sigma(X_i - \bar{X})^2}{n - 1}$$

vector: An extrachromosomal genetic element which, when incorporated into a recombinant DNA molecule, can cause transfer of the DNA into a host cell.

virogene: The complete viral genome.

wobble hypothesis: A hypothesis that the first two bases of the codon pair firmly with the anticodon, whereas the third base-pairing is weak, allowing pairing with different anticodons.

zygote: The fertilized egg cell, resulting from the fusion of the ovum with a sperm.

Appendix A

Nearest Neighbor Analysis and DNA Structure

In addition to establishing the direction of DNA synthesis, Josse, Kaiser, and Kornberg (1961) used nearest neighbor analysis to obtain evidence supporting the key concept of the Watson-Crick helix, that of complementary base pairing.

Cultures of the bacterium *Mycobacterium phlei* were grown in media containing the four nitrogenous bases. In each of four media, one of the bases carried a radioactive label in the phosphate. The DNA of bacteria grown in the media then was extracted and degraded enzymatically into the component nucleotides. For the same reasons that we mentioned above, the nucleotide carrying the label following degradation was known to be the nearest neighbor to the base that was originally labeled. For example, if we take the culture in which adenine nucleosides carried the label, we can obtain the frequencies of the following two-base pairs (or dinucleotide sequences): A*–T, A*–C, A*–G, A*–A (the asterisk represents the originally labeled base). Table A-1 shows the frequencies of all two-base sequences from all four cultures.

If complementary base-pairing actually occurs in DNA, we can expect the frequency of A*–A sequences, for example, to be the same as T*–T sequences, because wherever a sequence of A–A exists on one chain, T–T should be its counterpart on the complementary chain. As you can see in Table A-1, this appears to be so. The same would hold true for G*–G and C*–C dinucleotide frequencies as well.

Nearest neighbor analysis and polarity of DNA strands. You may have noticed, however, that in Table A-1, the frequency of G*–A dinucleotides equals the frequency of C–T*, but not C*–T. Analysis of this phenomenon led to confirmation of

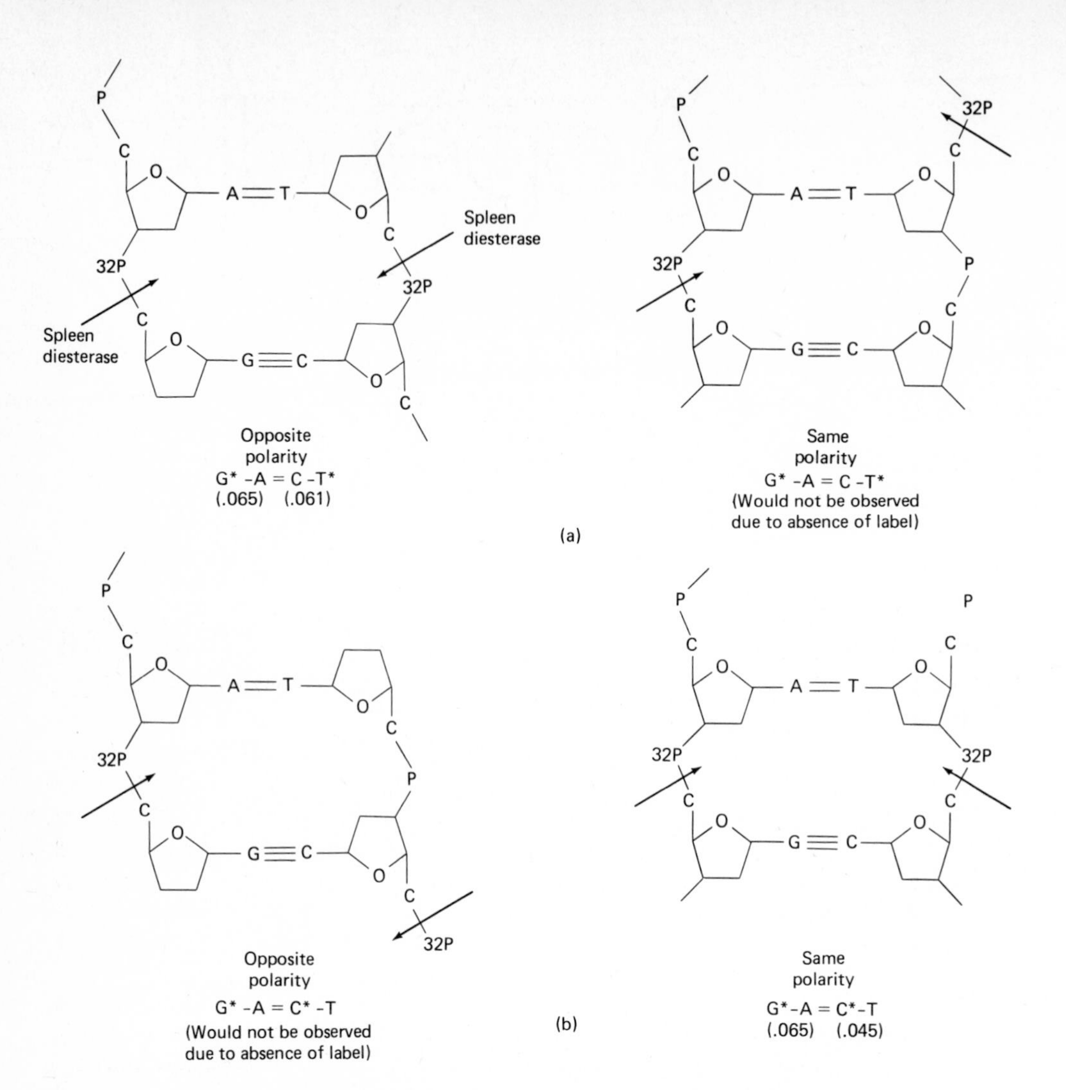

Figure A-1. Dinucleotide frequencies obtained by Josse, Kaiser, and Kornberg showed that the frequency of G*–A dinucleotides was very similar to the frequency of C–T* dinucleotides, and not that of C*–T. These results strongly supported the concept of opposte polarity of the two strands of DNA as conceived by Watson and Crick. Numbers in parentheses are actual frequencies obtained. Arrows point to sites of breakage by the enzyme spleen diesterase.

opposite polarity of complementary strands in DNA, as Crick and Watson had originally hypothesized.

As illustrated in Fig. A-1, the frequency of G–A dinucleotides should be very similar to the frequency of C–T dinucleotides because of specific base pairing. If there is opposite polarity on the two strands of DNA, C–T* dinucleotide frequency should equal G*–A frequency. The label of C–T* dinucleotides could not be picked up if the polarity were the same, since the spleen diesterase would cause the ^{32}P to be attached to the nucleotide on the other side of the thymine from the cytosine.

Table A-1 Dinucleotide Frequencies in *M. phlei* DNA

	A	G	T	C
A*	A*–A .024	A*–G .045	A*–T .013	A*–C .064
G*	G*–A .065	G*–G .090	G*–T .026	G*–C .061
T*	T*–A .012	T*–G .063	T*–T .026	T*–C .061
C*	C*–A .063	C*–G .129	C*–T .045	C*–C .090

From J. Josse, A. D. Kaiser, and A. Kornberg, 1961. *J. Biol. Chem.* 236 : 864.

Using the same reasoning, if C*–T base combinations are used, the dinucleotide would be observed only if the polarities of the two strands of the DNA were the same. If the polarities were opposite, again the label would be attached to the base on the other side of cytosine from thymine and would thus not be observed.

The data obtained by Josse, Kaiser, and Kornberg (Table A-1) clearly shows a similarity in frequency between G*–A dinucleotides and C–T* dinucleotides, and noticeable difference from the frequency of C*–T dinucleotides. THEREFORE, WE CAN CONCLUDE THAT THE TWO CHAINS OF THE DNA HELIX HAVE OPPOSITE POLARITIES. This analysis can be applied to any of the complementary dinucleotide combinations in Table A-1, leading to the same conclusion.

Reference

JOSSE, J., A. D. KAISER, A. KORNBERG, 1961. "Enzymatic Synthesis of Deoxyribonucleic Acid. VIII Frequencies of Nearest Neighbor Base Sequences in Deoxyribonucleic Acid." *J. Biol. Chem.* 236: 864–75.

Appendix B

Immune Reactions Determined by the Major Histocompatibility Complex

Recognition of foreign thymus-dependent antigens in T-B cell interactions which we discussed in Chapter 12 is determined by *IR* or *immune response*, genes of the MHC. How the product of *IR* genes controls recognition is not yet understood. *IR* genes are involved in both positive (helper) and negative (suppressor) effects of T cells on antibody production (Benacerraf and McDevitt 1972).

IR genes are found in a region between the *H-2D* and *H-2K* loci in mice, which is called the *I* region. The most complex of the areas of the MHC, the *I* region is implicated in a number of immune reactions such as graft-vs.-host reaction. The *I* region also appears to contain genes which cause genetically different cells to react in tissue culture (a reaction immunologists refer to as *mixed leucocyte reaction*); these genes are called *LD* genes (for *lymphocyte defined*). In addition, the *I* region contains genes that determine the presence of membrane glycoproteins different from the SD antigens, on the surface of cells of the immune system. The genes and the glycoprotein antigens are termed *Ia* (for *immune-response-region-associated*), and the latter are found predominantly on B cell lymphocytes and macrophages.

There is a region known as the *S*, or *Ss*, region in the MHC which is believed to determine several components of proteins found in serum known as *complement*. Complement is a group of approximately 11 proteins which interact with antibody molecules in immune reactions such as the lysis of bacterial cells. When antibody molecules bind to antigens on the surface of the bacteria, they activate complement proteins which in turn are thought to enzymatically open holes in the cell membrane, causing the cells to burst.

Finally, between strains of mice there are also genetically determined differences associated with the MHC genes involving susceptibility or resistance to certain diseases. For example, L. Old and his coworkers (1964) showed that inbred mice with different haplotypes differed in their resistance to the Gross leukemia virus. (The term *haplotype* refers to the entire set of genes found in the MHC of an individual in the cis position.) They reported that a gene *rgv-1*, mapping close to the *H-2K* locus, determined resistance to leukemia. Since then, susceptibility or resistance to a growing list of human diseases, such as coeliac disease, ankylosing spondylitis, and juvenile diabetes mellitus, has also been found associated with the HLA complex. Table B-1 summarizes traits controlled by MHC genes.

Part of the difficulty encountered in mapping genes of the MHC is that we are actually dealing with a very small region of the chromosome. The *H-2D* and *H-2K* loci in mice are only 0.5 map units apart. Add to this the highly polymorphic nature of immune genes (as many as 100 variant forms of both *H-2D* and *H-2K* antigens have been found), and problems associated with mammalian systems, and it is not surprising that gene maps of the MHC in both humans and mice are somewhat vague.

Table B-1 Assignment of Immunological Phenomena to Regions of the Major Histocompatibility Complex (MHC)

Functions of MHC	*H-2 regions*
Antigens stimulating tissue rejection and specific cytolytic T cells	K and D I
Cytolytic T cells specific for cell membrane antigens: virally coded, chemically modified, tumor-associated, or minor histocompatibility	K and D
Mixed leukocyte reaction; graft-versus-host reactions	I K and D
Specific immune responses to thymus-dependent antigens (IR genes)	I (I–A, I–B, I–C)
Specific immune suppression by T cells, immune suppression genes	I
Antigenic specificity on suppressor cells and suppressor factor	I–J
T and B cell cooperative interactions in secondary IgG responses	I–A
Antigen presentation to T cells by macrophages and specificity contribution in secondary responses and delayed hypersensitivity	I–A
Complement (C) components linked to the MHC in humans, mice, and guinea pigs: serum levels of the components of complement C_4, C_2, and C_3; factor B	S and other loci outside the MHC

From W. E. Paul and B. Benacerraf, 1977. *Science* 195: 1294.

We know very little about regulatory mechanisms controlling the activity of MHC genes. The importance of immune reactions is so central to our well-being and that of other mammals, however, that research can be expected to expand and increase in the coming years.

References

BENACERRAF, B., H. O. MCDEVITT, 1972. "Histocompatibility-Linked Immune Response Genes." *Science* 175: 273–279.

OLD, L., E. A. BOYSE, F. LILLY., 1964. "Formation of Cytoxic Antibody Against Leukemias Induced By Friend Virus." *Cancer Res.* 23: 10638.

Answers to Problems

CHAPTER 1

1. $2^3 = 8$
   All cells would have the same chromosomes.

2. (a) 46 chromosomes, 92 chromatids
   (b) 23 chromosomes
   (c) 23 chromosomes, 46 chromatids
   (d) 23 chromosomes, 46 chromatids
   (e) 46 chromosomes, 92 chromatids
   (f) 46 chromosomes
   (g) 23 chromosomes, 46 chromatids
   (h) 46 chromosomes, 92 chromatids
   (i) 46 chromosomes, 92 chromatids
   (j) 92 chromosomes

3. (a) Prophase II, meiosis, $2n = 6$
   (b) Metaphase I, meiosis, $2n = 2$
   (c) Metaphase II, meiosis, $2n = 8$, or metaphase, mitosis, $2n = 4$
   (d) Metaphase II, meiosis, $2n = 10$

4. 1 normal : 1 abnormal

6. To align homologs in such a way that the partner chromosomes will be transmitted into different daughter cells. This ensures daughter cells of receiving one member of each pair of homologous chromosomes.

7. Lack of gene-to-gene pairing due to genetic differences in chromosomes from different species results in abnormal or absent synapsis. If genetic differences between species are great enough, absence of alleles in interspecific zygote which determine essential metabolic processes would result in lethality.

8. Dispermic zygotes would be 3n. This causes difficulties in synapsis and division. The resulting daughter cells would contain different combinations of chromosomes. Too many or too few chromosomes usually lead to abnormalities and/or death.

9. See Chapter 17.

10. They are genetically identical as their cells are derived from the same zygote. Non-identical twins are the result of fertilization of different eggs by different sperm. There must have been at least 3 separation events of blastomeres during early development.

11. Diploidy would arise by chromosome duplication without cell division. The animals must of necessity be homologous at all loci due to the fact that sister chromatids are genetically identical.

CHAPTER 2

1. $TtWw \times ttww$
   $ttWW \times TTww$
   $ttWw \times TTww$
   $Ttww \times ttWw$

2. Mother = $FfTt$
   Father = $FfTt$

(a)

	FT	Ft	fT	ft	
FT	$FFTT$	$FFTt$	$FfTT$	$FfTt$	9 Freckled Taster
Ft	$FFTt$	$FFtt$	$FfTt$	$Fftt$	3 Freckled Non-Taster
fT	$FfTT$	$FfTt$	$ffTT$	$ffTt$	3 Non-Freckled Taster
ft	$FfTt$	$Fftt$	$ffTt$	$fftt$	1 Non-Freckled Non-Taster

(b)

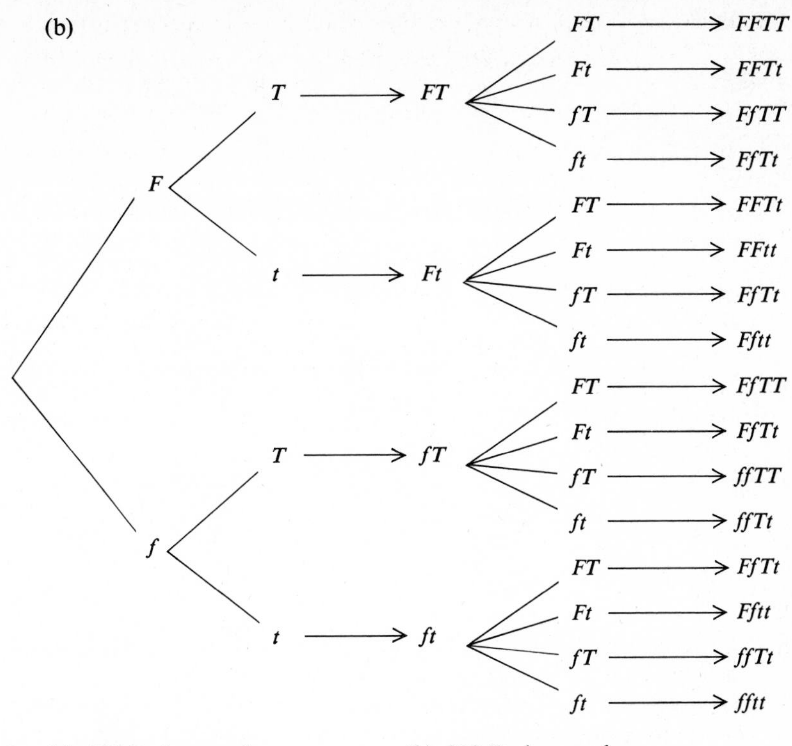

3. (a) 400 Red normal
400 Red vestigial
400 Brown normal
400 Brown vestigial
Same for genotypes

(b) 900 Red normal
300 Red vestigial
300 Brown normal
100 Brown vestigial

100 *BBVV*
200 *BBVv*
100 *BBvv*
200 *BbVv*
400 *BbVv*
200 *Bbvv*
100 *bbVV*
200 *bbVv*
100 *bbvv*

4. Mutations would not be detrimental. These species traits are probably determined by genes homozygous in every member.

5. *AaBb* × *aabb*

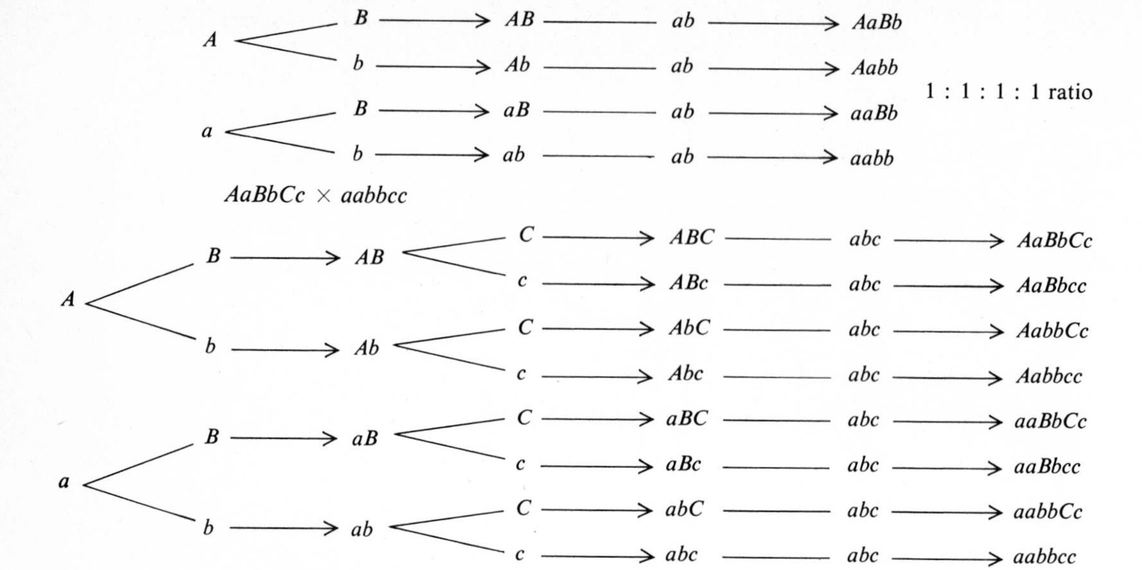

6. A. (a) recessive. (b) both y and the affected son of z had unaffected parents. (c) x = heterozygote; y = homozygous recessive; z = heterozygote.
   B. (a) dominant. (b) every affected person has 1 affected parent; when not transmitted to one generation, is lost; no skipping of generations. (c) x = heterozygote; y = homozygous recessive; z = heterozygote.
   C. (a) recessive. (b) none of the sibship of z is affected; every individual in sibship of z's mother is affected. (c) x and y therefore probably homozygous recessive; z = heterozygote.

7. Remate parents and backcross mutant (if possible) and sibs to appropriate parent. (a) A spontaneous dominant mutation is unlikely to occur again, so only crosses of the abnormal mouse would transmit the abnormal trait. (b) Both parents would have to be heterozygous. They should produce more abnormal young. Also, heterozygous litter mates will produce abnormal young when backcrossed. If the abnormal mouse can be backcrossed, its offspring would be expected to be 50% abnormal. (c) Unlikely reappearance of abnormality.

9. (a) cows: $2^{30} = 1{,}073{,}741{,}824$
   spider monkeys: $2^{17} = 131{,}072$
   oppossums: $2^{11} = 2{,}048$
   (b) cows: $2^{30} \times 2^{30} = (2^2)^{30} = 2^{60} = 1.1529 \times 10^{18}$
   spider monkeys: $2^{17} \times 2^{17} = (2^2)^{17} = 2^{34} = 17{,}179{,}869{,}184$
   oppossums: $2^{11} \times 2^{11} = (2^2)^{11} = 4{,}194{,}304$

10. (a) $3 \times 3 = 3^2 = 9$ (b) $3 \times 3 \times 3 = 3^3 = 27$
    (c) $3 \times 3 \times 3 \times 3 \times 3 = 3^5 = 243$ (d) $\underbrace{3 \times 3 \times 3 \times \ldots \times 3}_{n \text{ times}} = 3^n$

11. (a) $2 \times 2 = 2^2 = 4$ (b) $2 \times 2 \times 2 = 2^3 = 8$

(c) $2 \times 2 \times 2 \times 2 \times 2 = 2^5 = 32$ (d) $\underbrace{2 \times 2 \times 2 \times \ldots \times 2}_{n \text{ times}} = 2^n$

12. *Genotype.*

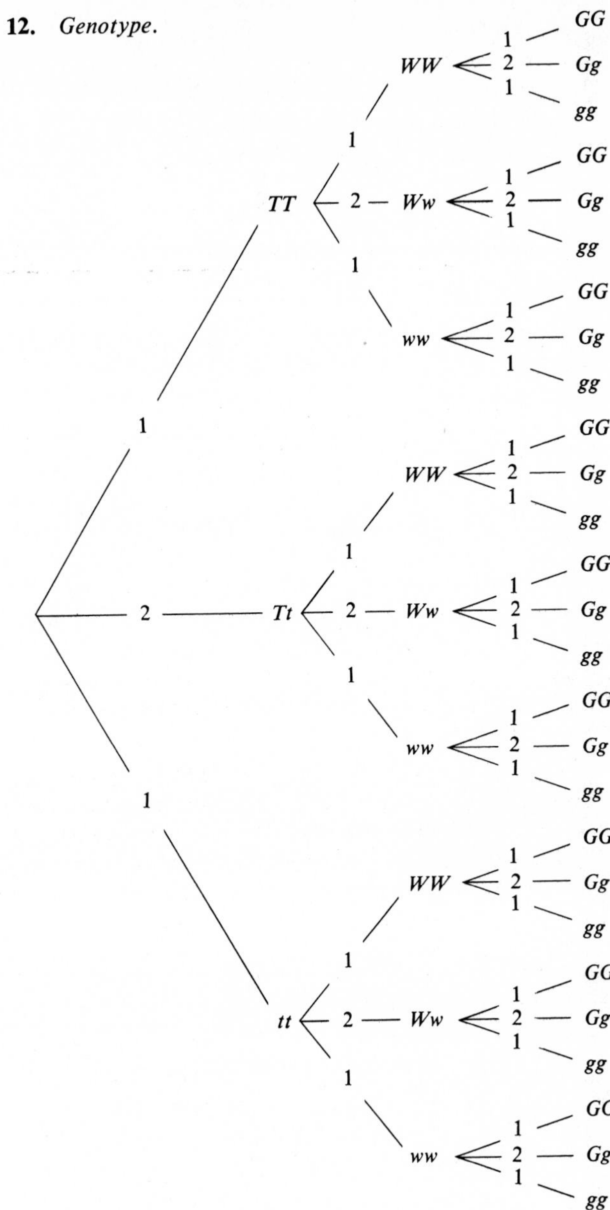

Results:

Genotype

TTWWGG 1	TtWWGG 2	ttWWGG 1
TTWWGg 2	TtWWGg 4	ttWWGg 2
TTWWgg 1	TtWWgg 2	ttWWgg 1
TTWwGG 2	TtWwGG 4	ttWwGG 2
TTWwGg 4	TtWwGg 8	ttWwGg 4
TTWwgg 2	TtWwgg 4	ttWwgg 2
TTwwGG 1	TtwwGG 2	ttwwGG 1
TTwwGg 2	TtwwGg 4	ttwwGg 2
TTwwgg 1	Ttwwgg 2	ttwwgg 1

27 distinct genotypic possibilities $= 3 \times 3 \times 3 = 3^3 = 27$

Phenotype

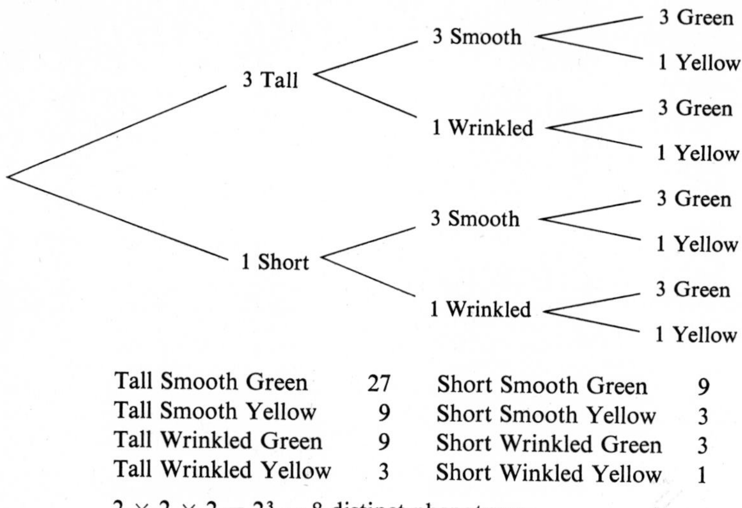

Tall Smooth Green	27	Short Smooth Green	9
Tall Smooth Yellow	9	Short Smooth Yellow	3
Tall Wrinkled Green	9	Short Wrinkled Green	3
Tall Wrinkled Yellow	3	Short Winkled Yellow	1

$2 \times 2 \times 2 = 2^3 = 8$ distinct phenotypes

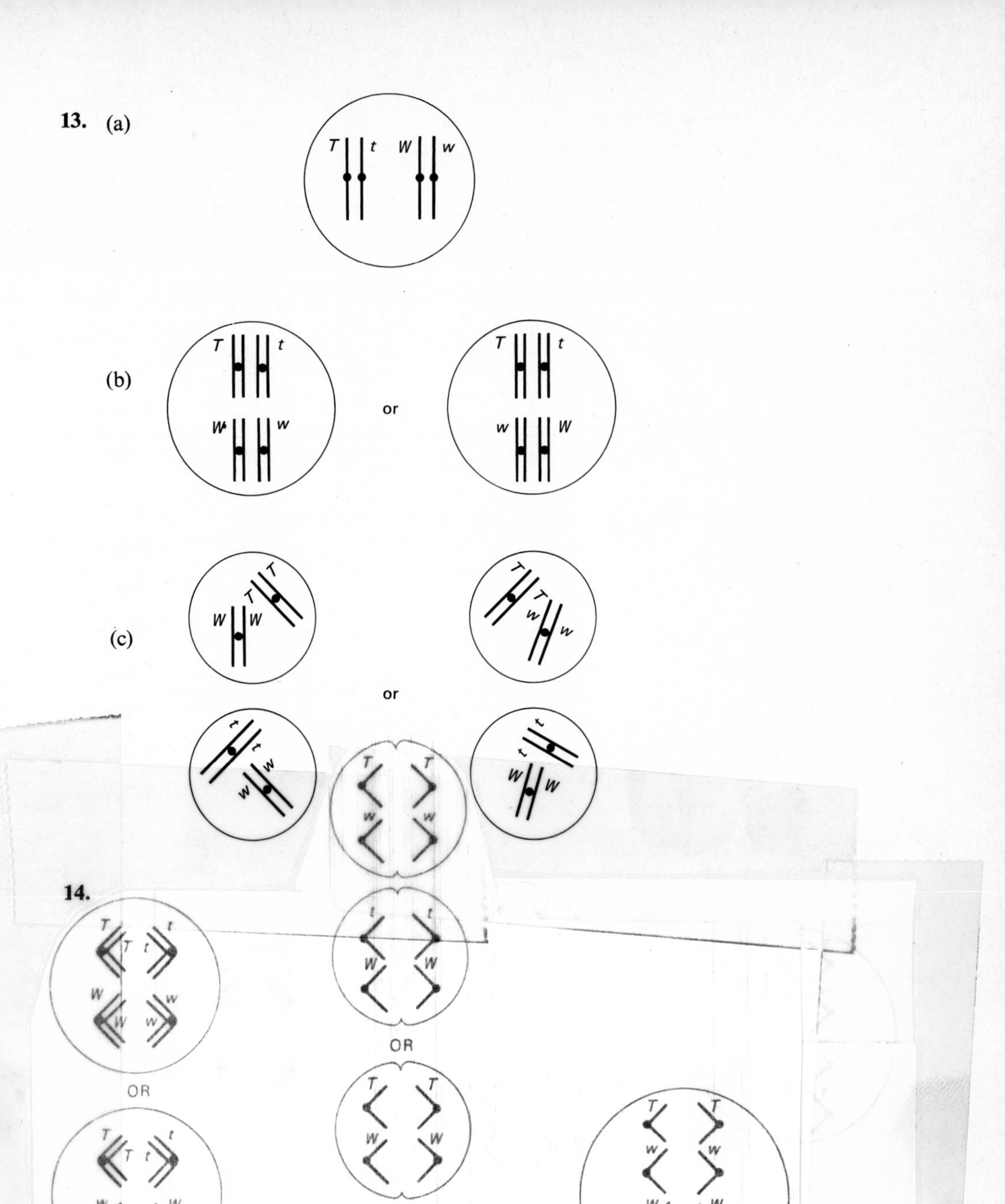

CHAPTER 3

1. (a) Prob [Albino Child] $= \frac{1}{2}$ (b) Prob [Normally Pigmented] $= \frac{1}{2}$

2. Prob [Short fingers] $= \frac{3}{4}$
Prob [Normal fingers] $= \frac{1}{4}$

3. Prob [Male with Brachydactyly] $= \frac{3}{8}$

4. (a)

Genotypes	*Prob*
BbAA	$\frac{1}{8}$
BbAa	$\frac{1}{4}$
Bbaa	$\frac{1}{8}$
bbAA	$\frac{1}{8}$
bbAa	$\frac{1}{4}$
bbaa	$\frac{1}{8}$

(b)

Phenotype	*Prob*
Brach. & Albino	$\frac{1}{8}$
Brach. & Normal Pigmentation	$\frac{3}{8}$
Normal Finger & Albino	$\frac{1}{8}$
Normal Finger & Normal Pigmentation	$\frac{3}{8}$

Each probability multiplied by $\frac{1}{2}$ if we are interested in the sex of each offspring.

5. (a) $\frac{1}{2}$ (b) $\frac{1}{12}$ (c) $\frac{5}{12}$

6. 2 traits (height and color)

Tall-Green	*TTGG*	$\frac{1}{16}$	*TtGG*	$\frac{2}{16}$
	TTGg	$\frac{2}{16}$	*TtGg*	$\frac{4}{16}$
Tall-Yellow	*TTgg*	$\frac{1}{16}$	*Ttgg*	$\frac{2}{16}$
Short-Green	*ttGG*	$\frac{1}{16}$	*ttGg*	$\frac{2}{16}$
Short-Yellow	*ttgg*	$\frac{1}{16}$		

Phenotype:

Tall Green	$\frac{9}{16}$
Tall Yellow	$\frac{3}{16}$
Short Green	$\frac{3}{16}$
Short Yellow	$\frac{1}{16}$

7. $\frac{10}{32} = 0.3125$

8. (a) $\frac{1}{256} = 0.0039$ (b) $\frac{81}{256} = 0.3164$
(c) $\frac{54}{256} = 0.2109$

9. (a) $(\frac{1}{16})^3 = 0.000244$ (b) $(\frac{1}{32})^3 = 0.0000305$

10. (a) $\frac{1}{8} = 0.125$ (b) $\frac{1}{8} = 0.125$
(c) $3(\frac{1}{2})^3 = 0.375$

11.

Poly. Brach.	$\frac{3}{8}$
Poly. Normal	$\frac{1}{8}$
Normal Brach.	$\frac{3}{8}$
Normal Normal	$\frac{1}{8}$

12. $\frac{1}{64} = 0.015625$

13.

Number of children with normal fingers	*Probability*	*Number of children with normal fingers*	*Probability*
0	0.0563	6	0.0162
1	0.1887	7	0.0031
2	0.2816	8	~0.0004
3	0.2503	9	~0.00003
4	0.1460	10	~0.000001
5	0.0584		

14. $X^2 = 0.4545$ Do not reject ratio 1 : 2 : 1 if level of significance 5%

15. $X^2 = 2.453$ Do not reject ratio 9 : 3 : 3 : 1 if level of significance 5%

16. $X^2 = 8$ Reject ratio of 1 : 1 if level of significance 5%

17. $X^2 = 1.444$ Do not reject ratio of 9 : 3 : 4 if level of significance 5%

18. $X^2 = 8.64$ Reject ratio of 15 : 1 if level of significance 5%

19. $X^2 = 3.611$ Do not reject ratio of 9 : 3 : 4 if level of significance 5%

CHAPTER 4

1. 1 pink : 1 white
2. (a) Remate the frizzled fowl; test cross frizzled fowl with normal; backcross frizzled offspring.
   (b) The test cross should produce 1 frizzle : 1 normal
3. 2 *A* : 1 *AB* : 1 *B*

	I^A	i
I^A	I^AI^A	I^Ai
I^B	I^AI^B	I^Bi

4.

	NI^A	NI^B	MI^A	MI^B
MI^A	MNI^AI^A	MNI^AI^B	MMI^AI^A	MMI^AI^B
Mi	MNI^Ai	MNI^Bi	MMI^Ai	MMI^Bi

2 MA : 2 MNA : 1 MNAB : 1 MNB : 1 MAB : 1 MB

5. *BBcc* × *bbCC*
6. *aaBBcc*
7. (a)

	T	t^9
t^9	Tt^9	t^9t^9
t^x	Tt^x	t^xt^9

2 tailless : 1 normal-tailed

(b) All tailless ~~2 tailless : 1 normal-tailed~~

8. ssbrbr × SsBrBr

↓

	SBr	*sBr*
sbr	*SsBrbr*	*ssBrbr*
	wild type	scarlet

9. (a) 3 : 1
   (b) 1 : 1 : 2
   (c) 2 : 1 : 1
   (d) 3 : 1
   (e) 1 : 2 : 1
   (f) 1 : 1 : 1 : 1
   (g) 1 : 1 : 1 : 1
   (h) 1 : 1 : 1 : 1

10. They are homozygous for different recessive albinism genes.

12. (a) 1.625 (b) 0.5

(c)

Frequency Distribution of Lengths of Ears of Corn

Lengths	*Relative number of ears*
8	1
9.125	16
10.25	120
11.375	560
12.5	1,820
13.625	4,368
14.75	8,008
15.875	11,440
17.0	12,870
18.125	11,440
19.25	8,008
20.375	4,368
21.5	1,820
22.625	560
23.75	120
24.875	16
26	1
	65,536

CHAPTER 5

1. Yes, if her father were colorblind and mother at least heterozygous.

2. $XX^c \times XY$ $\frac{1}{4}$ of the grandchildren would be colorblind
$\frac{1}{2}$ of the grandsons would be colorblind

3. Not possible, as the gene was transmitted on the X chromosome of the mother.

4. $I^AI^BHh \times I^AiHY$

	I^AH	I^Ah	I^BH	I^Bh
I^AH	I^AI^AHH	I^AI^AHh	I^AI^BHH	I^AI^BHh
I^AY	I^AI^AHY	I^AI^AhY	I^AI^BHY	I^AI^BhY
iH	I^AiHH	I^AiHh	I^BiHH	I^BiHh
iY	I^AiHY	I^AihY	I^BiHY	I^BihY

Girls: 4 A normal blood : 2 AB normal blood : 2B normal blood
Boys: 2 A normal blood : 2 A hemophiliac : 1 AB normal blood:
1 AB hemophiliac : 1 B normal blood : 1 B hemophiliac

5. The female was $\widehat{XX}Y$; the white eye gene is on both X chromosomes which are permanently non-disjoined.

6. The white eye is determined by an X-linked dominant mutation.

7. The mutation may affect some basic aspect of pigment formation. For example, the mutation may cause abnormal development of cells which produce pigment.

8. In Y-linked transmission, the pattern is from father to son; affected sons would have affected fathers. Sex-limited traits may be expressed in males whose fathers are normal, having inherited the gene from their mothers.

9. The father is at least heterozygous. If heterozygous, his wife is at least heterozygous. They cannot both be homozygous, as some of the children are not bald. If the father is homozygous, the mother is most probably heterozygous.

10.

	X^1	X^2	Y	X^1Y	X^2Y	X^1X^2
X^1	X^1X^1	X^1X^2	X^1Y	X^1X^1Y	X^1X^2Y	$X^1X^1X^2$
X^2	X^1X^2	X^2X^2	X^2Y	X^1X^2Y	X^2X^2Y	$X^1X^2X^2$

11.

	X^cX^c	X	X^cX	X^c
X	XX^cX^c	XX	X^cXX	XX^c
Y	X^cX^cY	XY	X^cXY	X^cY

4 normal-visioned females:
2 normal-visioned males;
2 colorblind males

12. Two sperms, one Y-bearing, one X-bearing, fertilized the ovum and a polar body which became incorporated as a single embryo. The Y or X is lost from one of the early blastomeres to result in an X0 cell. Or, the Y chromosome was lost from a blastomere in an XY embryo; in one of the subsequent X0 daughter cells, there was non-disjunction of X sister chromatids resulting in the XX cells.

13. Yes, because it is unlikely that the maternal and paternal X-chromosomes are genetically identical; therefore only material alleles would be expressed.

14. Females. Less likelihood of homozygosity for detrimental genes.

15.

	Z	W
Z	ZZ	ZW
W	ZW	WW

Two females to one male.

16. A: autosomal dominant
B: autosomal recessive
C: X-linked recessive

17. (A) 1. *Aa* or *AA* 2. *aa* 3. *aa* 4. *Aa* 5. *Aa* 6. *aa* 7. *Aa* 8. *aa* 9. *Aa* 10. *aa* 11. *aa*.
(B) 1. *Aa* or *AA* 2. *Aa* or *AA* 3. *AA* 4. *Aa* 5. *Aa* or *AA* 6. *Aa* 7. *Aa* 8. *Aa* or *AA* 9. *Aa* 10. *Aa* 11. *aa* 12. *Aa* or *AA* 13. *aa* 14. *Aa* or *AA* 15. *Aa* 16. *Aa* 17. *aa*.

(C) 1. aY 2. Aa or AA 3. AY 4. Aa 5. AY 6. Aa or AA 7. Aa 8. AY 9. aY 10. aY 11. AY.

18. $CcHh \times CcHY$

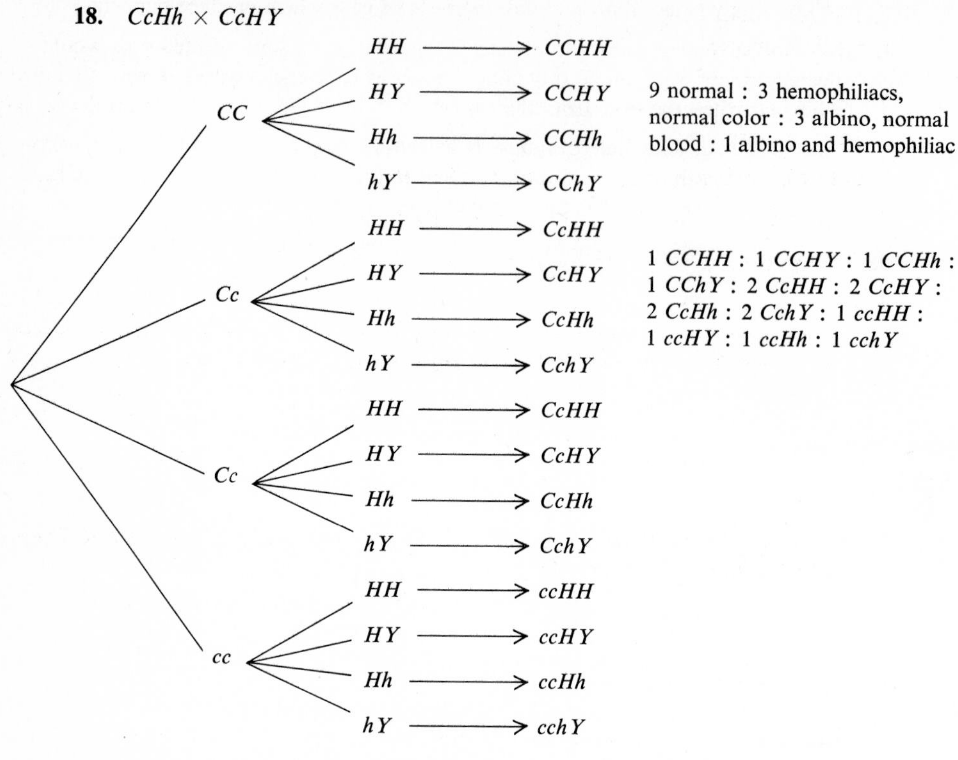

9 normal : 3 hemophiliacs, normal color : 3 albino, normal blood : 1 albino and hemophiliac

1 $CCHH$: 1 $CCHY$: 1 $CCHh$: 1 $CChY$: 2 $CcHH$: 2 $CcHY$: 2 $CcHh$: 2 $CchY$: 1 $ccHH$: 1 $ccHY$: 1 $ccHh$: 1 $cchY$

19. Dispermic fertilization, perhaps of the ovum and a polar body.

CHAPTER 6

2.

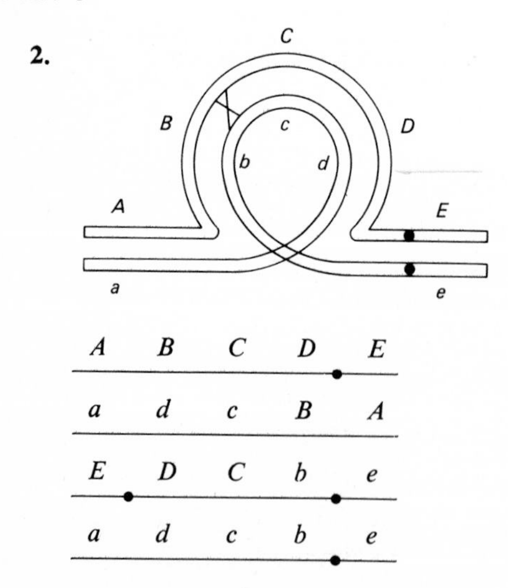

3.

	XX	X
X	XXX	XX
Y	XXY	XY

4. Probably lethal, because the offspring would be trisomic for some chromosome parts, and monosomic for others.

5. 6th chromosome, short arm, region 2, 9th band.

6. $2\ p^+p^+p^+ : 2\ p^+p : 1\ p^+p^+p^+ : 1\ p^+pp : 1\ p^+p^+ : 1\ pp$

7. $2\ p^+pp : 2\ p^+p : 1\ p^+p^+p : 1\ ppp : 1\ p^+p^+ : 1\ pp$

8.

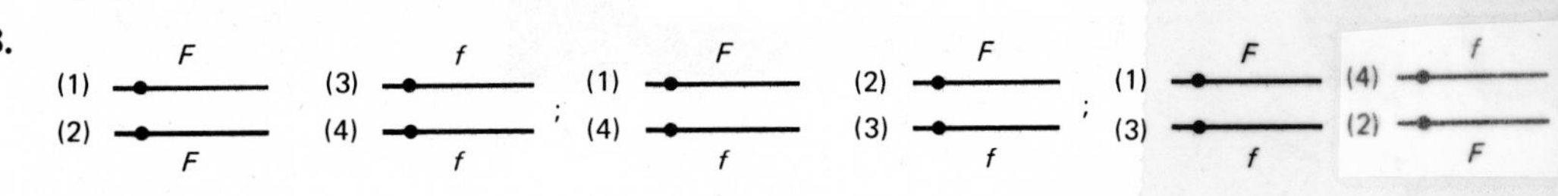

9. 8 *Ff* : 2 *FF* : 2 *ff*

10. $\frac{1}{36}$

11. An equal number of bivalents (from the New World species) and individual chromosomes (from the Old World species).

CHAPTER 7

1. $\frac{D\quad e}{d\quad E}$ × $\frac{d\quad e}{d\quad e}$ ~~250 *Ddee*~~ 220 Ddee 30 DdEe
~~250 *ddEe*~~ 220 ddEe 30 ddee

2. Non-linkage. If the genes are linked, the back-cross would yield only black and albino offspring in equal proportions.

3. *ppLL* × *PPll* F_1 = purple, long
F_2 = 1 red long : 2 purple long : 1 purple round

4.

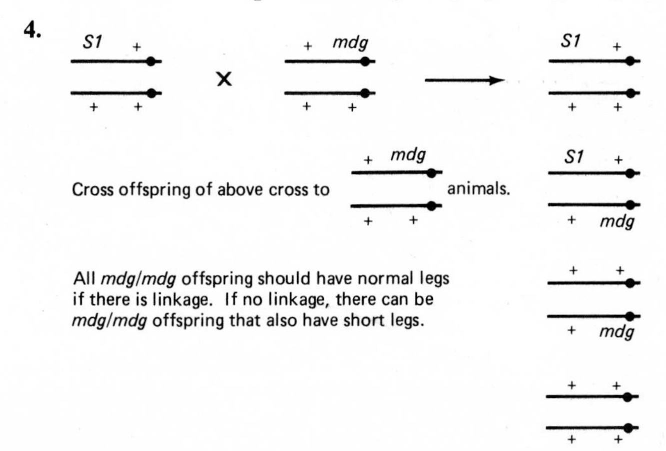

5.

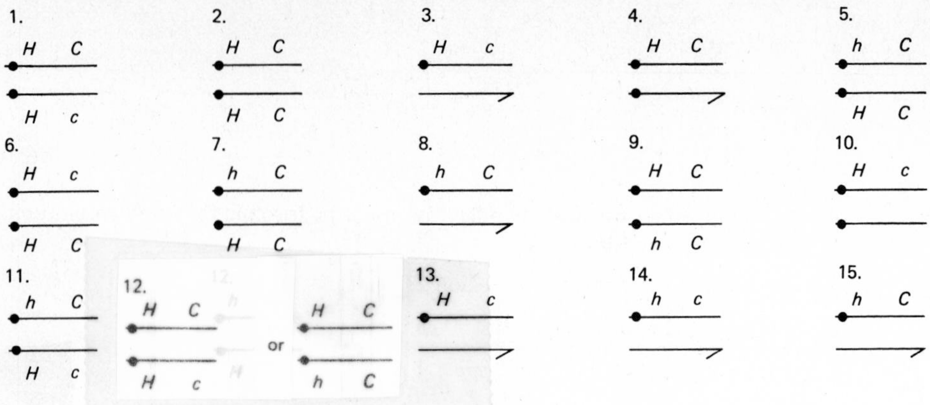

1.	2.	3.	4.	5.
H C / H c	H C / H C	H c / (Y)	H C / (Y)	h C / H C

6.	7.	8.	9.	10.
H c / H C	h C / H C	h C / (Y)	H C / h C	H c / —

11.	12.	13.	14.	15.
h C / H c	H C / H c or H C / h C	H c / (Y)	h c / (Y)	h C / (Y)

6.

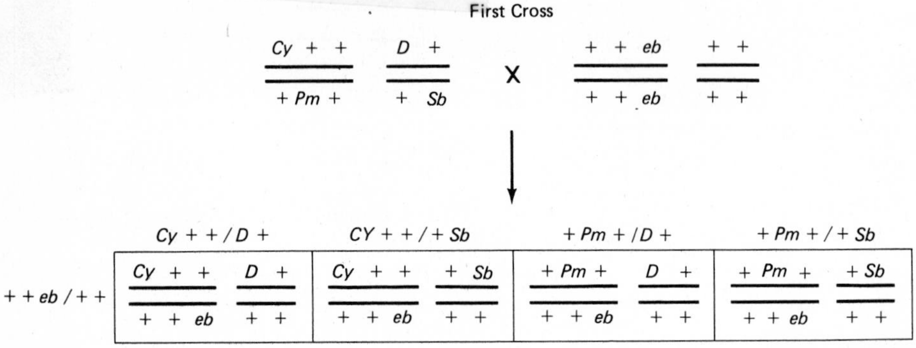

First Cross

$\frac{Cy + +}{+ Pm +}\ \frac{D +}{+ Sb}$ X $\frac{+ + eb}{+ + eb}\ \frac{+ +}{+ +}$

↓

	Cy + + / D +	CY + + / + Sb	+ Pm + / D +	+ Pm + / + Sb
+ + eb / + +	$\frac{Cy + +}{+ + eb}\ \frac{D +}{+ +}$	$\frac{Cy + +}{+ + eb}\ \frac{+ Sb}{+ +}$	$\frac{+ Pm +}{+ + eb}\ \frac{D +}{+ +}$	$\frac{+ Pm +}{+ + eb}\ \frac{+ Sb}{+ +}$

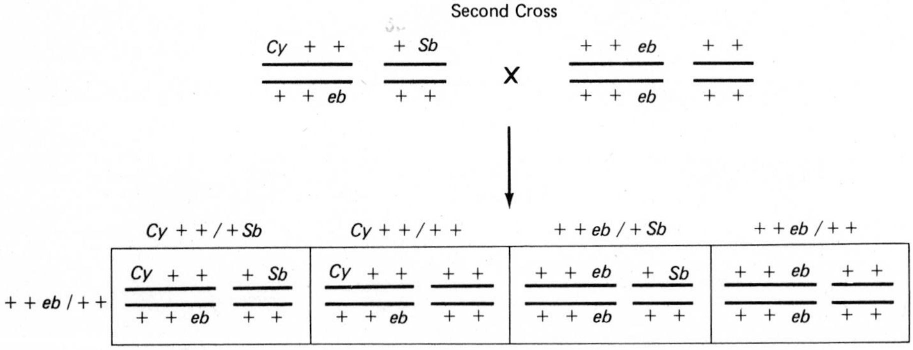

Second Cross

$\frac{Cy + +}{+ + eb}\ \frac{+ Sb}{+ +}$ X $\frac{+ + eb}{+ + eb}\ \frac{+ +}{+ +}$

↓

	Cy + + / + Sb	Cy + + / + +	+ + eb / + Sb	+ + eb / + +
+ + eb / + +	$\frac{Cy + +}{+ + eb}\ \frac{+ Sb}{+ +}$	$\frac{Cy + +}{+ + eb}\ \frac{+ +}{+ +}$	$\frac{+ + eb}{+ + eb}\ \frac{+ Sb}{+ +}$	$\frac{+ + eb}{+ + eb}\ \frac{+ +}{+ +}$

7.

First Cross

$\frac{Cy\ +}{+\ Pm}\ \frac{D\ +\ +}{+\ Sb\ +}$ X $\frac{+\ +}{+\ +}\ \frac{+\ +\ eb}{+\ +\ eb}$

↓

	Cy + / D + +	Cy + / + Sb +	+ Pm / D + +	+ Pm / + Sb +
+ + / + + eb	$\frac{Cy\ +}{+\ +}\ \frac{D\ +\ +}{+\ +\ eb}$	$\frac{Cy\ +}{+\ +}\ \frac{+\ Sb\ +}{+\ +\ eb}$	$\frac{+\ Pm}{+\ +}\ \frac{D\ +\ +}{+\ +\ eb}$	$\frac{+\ Pm}{+\ +}\ \frac{+\ Sb\ +}{+\ +\ eb}$

Second Cross

$\frac{Cy\ +}{+\ +}\ \frac{+\ Sb\ +}{+\ +\ eb}$ X $\frac{+\ +}{+\ +}\ \frac{+\ +\ eb}{+\ +\ eb}$

↓

	Cy +/+ Sb +	Cy +/+ + eb	+ +/+ Sb +	+ +/+ + eb
+ + / + + eb	$\frac{Cy\ +}{+\ +}\ \frac{+\ Sb\ +}{+\ +\ eb}$	$\frac{Cy\ +}{+\ +}\ \frac{+\ +\ eb}{+\ +\ eb}$	$\frac{+\ +}{+\ +}\ \frac{+\ Sb\ +}{+\ +\ eb}$	$\frac{+\ +}{+\ +}\ \frac{+\ +\ eb}{+\ +\ eb}$

8.

Single crossovers	*S M n*	66
	s m N	66
	S m n	91
	s M N	91
Double crossovers	*S M N*	9
	s m n	9
Parentals	*S m N*	334
	s M n	334
		1,000 Total

9. (a) Repulsion (b) $C - A - B$
(c) $C - 10 - A - 30 - B$ (d) No interference

10. (a) 34% (b) 34% (c) 1.5% (d) 6%

11. If unlinked:

		or		
$+^a$	$+^b$		$+^a$	b
$+^a$	$+^b$		$+^a$	b
$+^a$	$+^b$		$+^a$	b
$+^a$	$+^b$	or	$+^a$	b
a	b		a	$+^b$
a	b		a	$+^b$
a	b		a	$+^b$
a	b		a	$+^b$

If linked:

$+^a$	$+^b$
$+^a$	$+^b$
$+^a$	$+^b$
$+^a$	$+^b$
a	b
a	b
a	b
a	b

12. Very close.

13. (a)

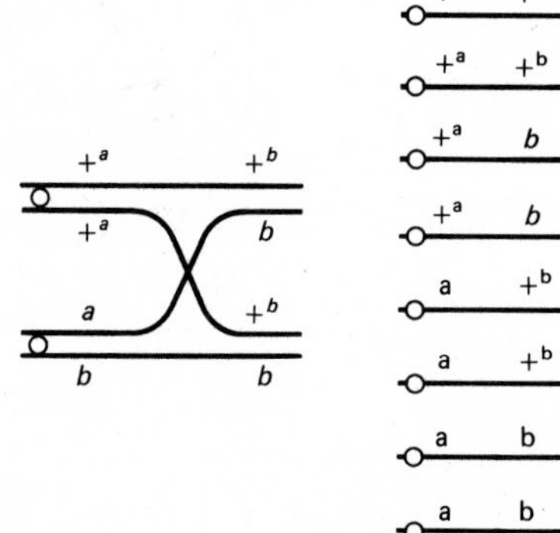

(b)

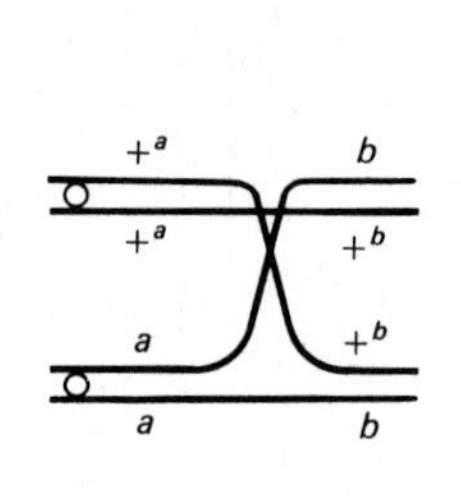

$+^a$	b
$+^a$	b
$+^a$	$+^b$
$+^a$	$+^b$
a	$+^b$
a	$+^b$
a	b
a	b

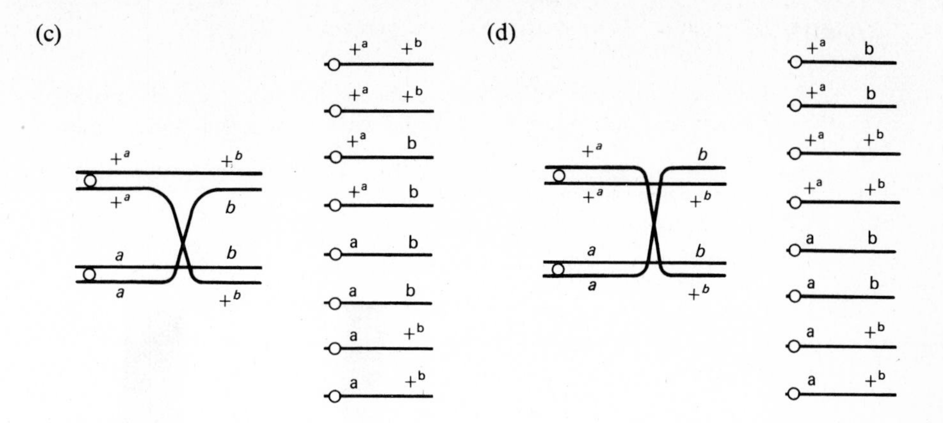

14. 80

15. (a) A, F = PD; D, E = NPD; B, C, G = T

(b) Distance between r and the centromere is 6.4 units.

Distance between s and the centromere is 26.9; therefore, the distance between r and s is 20.5 m.u.

16. Yes, genes for enzymes II and IV are syntenic. II and IV assigned to chromosome 2; I to chromosome 1: Enzyme III does not seem to be linked to any others or to any of the 3 chromosomes.

17. (a) ♀ ♂

H C H c n i n i

H C N i n i or H C n i N i

(b)

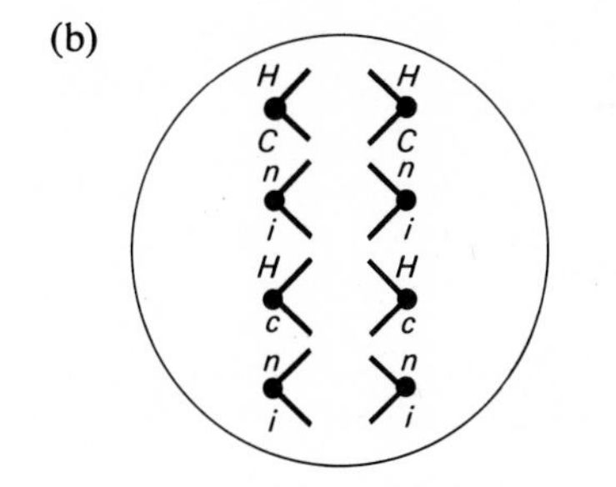

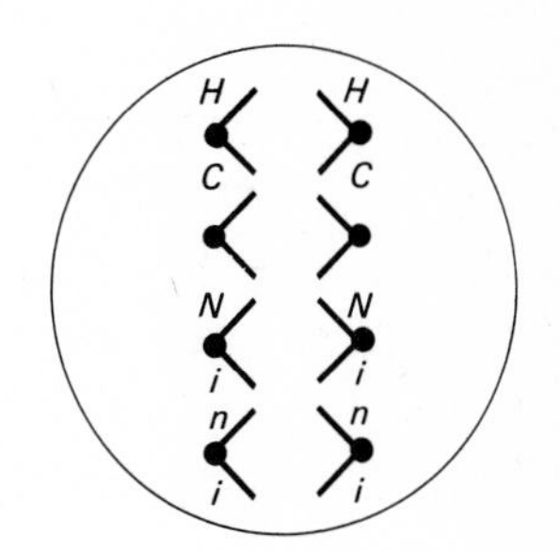

CHAPTER 8

3. Plate cells on media with streptomycin and no mannitol, and on media without streptomycin and no mannitol, and media with streptomycin and mannitol.

4. They are rarely if ever simultaneously seen in transformed cells, if DNA from a genetically different strain is used for transformation.

5. $\frac{10}{280} = \frac{1}{28} = 0.04 = 4$ m.u.

6. Determine pilus formation.

7. (a) (b) 1. 2. 3.

p m n o   p m n o   p m n o   p m n o

8. Either by using an F^-lac^- that is due to a deletion of the lac locus, or by observing phenotype of the daughter cells of the sexduced cell that may have lost the F particle.

9. A double cross-over:

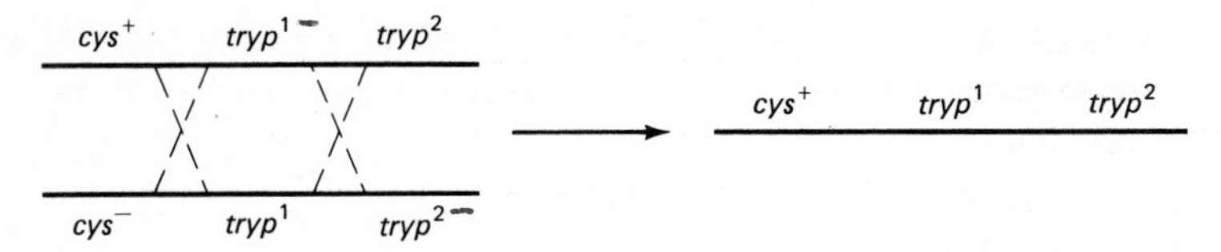

10. To select against donors so that all colonies are known to be exconjugants.

11. (a) $a^- b^+ str^r$ (b) $a^+ b^- str^r$ (c) $a^+ b^+ str^r$

12.
20 10
c a b

13. Limited amount of DNA of transducing phages.

14. By studying the effect of regulatory genes when transduced to different parts of the DNA.

CHAPTER 9

1. Around 29,412 base pairs.

2. 29–30 genes.

3. ~~44.4~~ 35.95

4. 3′U A G C C G G A A 5′

5. Two bands after one replication, 2, 5 replications. The ^{15}N band would increase in quantity in relation to the ^{14}N band.

6. C—G
|
T
|
A

7. Yes. One would expect to find molecules longer than the viral chromosome due to the replication of part of the new + strand, while the original + strand is still partially bonded to the − strand (Fig. 9-24).
8. Use a transducing virus to remove the inserted genes. If it had replaced the original DNA, the result would be a deletion of those loci (no reversions).

CHAPTER 10

1. Yes, if you assume the missing substance produced by normal pigment cells can diffuse into the transplanted cells from the albino.
2. (a) $\longrightarrow B \longrightarrow C \longrightarrow D \longrightarrow A$ (b) $\xrightarrow{\downarrow 1} B \xrightarrow{\downarrow 2} C \xrightarrow{\downarrow 3} D \xrightarrow{\downarrow 4} A$
3. (a) 5′ A U G C U A A C 3′ (b) A U G
   (c) A U G
4. In vitro studies showed DNA polymerase can only add to pre-existing nucleotide chain.
5. Proline = 0.64; arginine = 0.16; alanine = 0.16; glycine = 0.04.
6. Only that CGC and GCG codes for one or the other. With the probability being equal for the two in problem 5, it would still not be possible to assign specific codons for the two amino acids.
7. (a) No effect. (b) Incomplete protein.
   (c) No protein synthesis.
8. Conservation of substrates and energy. Red blood cells and ova.
9. Deficiency of amino acids leads to deficiency of protein synthesis and lack of growth.

CHAPTER 11

2. No induction of lac enzymes in presence of lactose.
3. Both trans and cis dominant because the gene product is abnormal.
4. There may be a mutation leading to abnormal cAMP.
5. Both the lac and ara operons would be shut off due to repressor protein interaction with the operator genes, and the absence of cAMP to interact with the promoter region.
7. For the regulation of the proper length of DNA to be transcribed.
9. An inactive virus, resulting probably in its loss. Loss of lytic ability of virus.
10. *cro*.
11. *cII* and *cIII* needed for initiation of lysogeny; *cI* needed for maintenance of lysogeny.
12. Yes, because there would be complementation.

CHAPTER 12

3. Place milipore filters (with openings too small for cells to pass through) between the components.

4. Uniformity of molecular structure.
7. Embryos are genetically different from maternal tissues due to paternal genes.
8. Similarity of base sequence between portions of cellular and viral DNA's.
9. The DNA is inserted into the cellular genome, and eventually transcribed into RNA.
10. Young cells contain enzymes that can rescue old or progeria cells by repairing damage of the DNA. Progeria^1 and progeria^2 cells lack or are deficient in different substances, and thus complement each other.

CHAPTER 13

1. (a) All sinistral; Dd (b) All dextral; 1 DD : 2 Dd : 1 dd
2. (a) All offspring die. (b) All offspring survive.
3. Random distribution of cytoplasmic components.
4. No, due to equal contribution by mother and father, with the exception of X-linked or Y-linked inheritance.
5. $\frac{3}{4}$ are killers.
6. Decrease in number of particles in some daughter cells.

CHAPTER 15

1. Presence of colonies of strain K after infection by T_4.
2. Co-infection with #10 or #11 with new mutations. Complementation indicates mutations at different genes. Then co-infection of new mutations with known deletion mutations of A and B loci.
5. Two cistrons; p^2 and p^4 are in one cistron, p^1 and p^3 are in a second cistron.
6. (a)

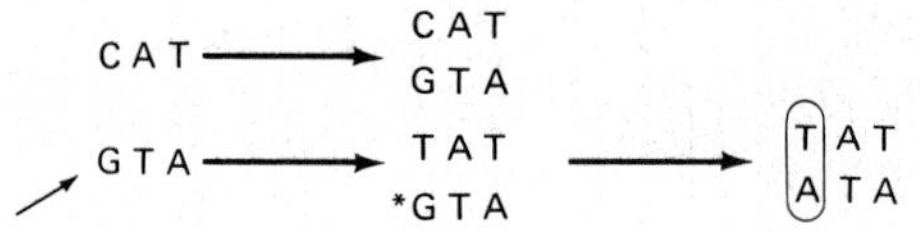

CAT → CAT / GTA

GTA → TAT / *GTA → TAT / ATA

(b)

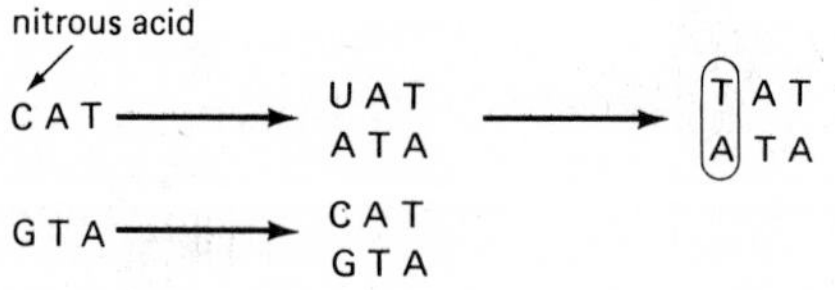

nitrous acid

CAT → UAT / ATA → TAT / ATA

GTA → CAT / GTA

(c)

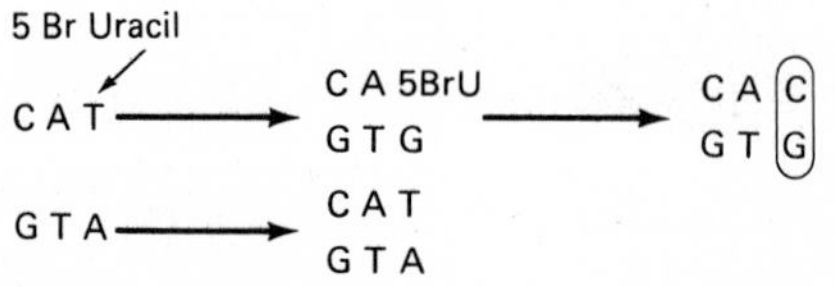

5 Br Uracil

CAT → CA 5BrU / GTG → CAC / GTG

GTA → CAT / GTA

7. (a) No apparent effect.
(b) A deficiency of female offspring.

(c) A deficiency of male offspring.
(d) Equal effect on male and female offspring.

CHAPTER 16

1. $RR : Rr : rr$ is given by .24 : .52 : .24

2. (a) $q = \frac{1}{141}$, (b) $p = \frac{140}{141}$, (c) .9858659 (d) .0140837

3.

Genotype	RR	Rr	rr
H-W Equilibrium	0.25	0.50	0.25

4. (a) F_1 : Prob[NN] = 0.24, Prob[Nn] = 0.52, Prob[nn] = 0.24
(b) F_{14} : Prob[NN] = 0.25, Prob[Nn] = 0.50, Prob[nn] = 0.25

5. ~~No.~~ YES

6.

Blood Type	M	MN	N
Expected Number	1,690	1,820	490

14. (a) Uncle-Niece Prob[albino offspring] = 0.00088
Random Mating Prob[albino offspring] = 0.000051
Uncle-Niece : Random Mating 17.256 to 1
(b) 2nd Cousins Prob[albino offspring] = 0.00011
2nd Cousins : Random Mating 2.157 to 1

15. *5 offspring:*
Prob[5 LL and 0 $L\ell$] = 1/32 Prob[2 LL and 3 $L\ell$] = 10/32
Prob[4 LL and 1 $L\ell$] = 5/32 Prob[1 LL and 4 $L\ell$] = 5/32
Prob[3 LL and 2 $L\ell$] = 10/32 Prob[0 LL and 5 $L\ell$] = 1/32
6 offspring:
Prob[6 LL and 0 $L\ell$] = 1/64 Prob[2 LL and 4 $L\ell$] = 15/64
Prob[5 LL and 1 $L\ell$] = 6/64 Prob[1 LL and 5 $L\ell$] = 6/64
Prob[4 LL and 2 $L\ell$] = 15/64 Prob[0 LL and 6 $L\ell$] = 1/64
Prob[3 LL and 3 $L\ell$] = 20/64

Number of children	Prob[*Mutant Gene Dies Out*]	Prob[*Mutant Gene Survives*]	*Number of possible mutant genes*
5	1/32	31/32	1, 2, 3, 4, or 5
6	1/64	63/64	1, 2, 3, 4, 5, or 6

16. *Relative Proportion*

	RR	Rr	rr	Prob[RR]	Prob[Rr]	Prob[rr]
(a)	.24	.52	0	.32	.68	0
(b)	.24	.52	.024	.306	.663	.031
(c)	.24	.52	.12	.273	.591	.136
(d)	.24	.52	.144	.265	.575	.159
(e)	.24	.52	.192	.252	.546	.202
(f)	.24	.52	.216	.246	.533	.221

17. *Relative Proportion*

	RR	*Rr*	*rr*	Prob[*RR*]	Prob[*Rr*]	Prob[*rr*]
(a)	.9858659	.0140838	0	.9859154	.0140845	0
(b)	.9858659	.0140838	.0000075	.9859080	.0140844	.0000075
(c)	.9858659	.0140838	.0000125	.9859030	.0140843	.0000125
(d)	.9858659	.0140838	.0000226	.9858931	.0140841	.0000226
(e)	.9858659	.0140838	.0000377	.9858783	.0140839	.0000377
(f)	.9858659	.0140838	.0000477	.9858683	.0140838	.0000477

18. (a) Genotype *RR* *Rr* *rr*

H-W Equilibrium ~~0.2897~~ 0.2632 ~~0.5221~~ 0.5263 ~~0.1882~~ 0.2105

(b) Genotype *RR* *Rr* *rr*

H-W Equilibrium ~~0.2070~~ ~~0.4656~~ ~~0.3274~~

0.2353 0.4706 0.2941

19.

	(a) s = 1		(b) s = .5		(c) s = .3	
Generation	q_n	Prob[*rr*]	q_n	Prob[*rr*]	q_n	Prob[*rr*]
0	$\frac{1}{2}$	.25	$\frac{1}{2}$	.25	$\frac{1}{2}$	.25
1	$\frac{1}{3}$	.1111	.4286	.1837	.4595	.2111
2	$\frac{1}{4}$	.0625	.3708	.1375	.4229	.1789
3	$\frac{1}{5}$	.04	.3243	.1052	.3902	.1522
4	$\frac{1}{6}$	.0277	.2868	.0823	.3610	.1303
5	$\frac{1}{7}$	.0204	.2562	.0657	.3350	.1122
6	$\frac{1}{8}$	.0156	.2310	.0534	.3118	.0972
7	$\frac{1}{9}$	.0123	.2099	.0441	.2912	.0848
8	$\frac{1}{10}$	.01	.1921	.0369	.2727	.0743
9	$\frac{1}{11}$	.0083	.1769	.0313	.2561	.0656
10	$\frac{1}{12}$	.0069	.1638	.0268	.2411	.0581
20	$\frac{1}{22}$	.0021				
50	$\frac{1}{52}$	.0004				
100	$\frac{1}{102}$	.0001				

20.

Generation	*Year*	*No selection*	*Selection s = 0.5*
0	0	10,060	10,060
1	25	10,060	9,989
2	50	10,060	9,919
3	75	10,060	9,850
4	100	10,060	9,781
5	125	10,060	9,714
6	150	10,060	9,646
7	175	10,060	9,580
8	200	10,060	9,513
9	225	10,060	9,448
10	250	10,060	9,384
11	275	10,060	9,320

26.

Generation	q_n
0	0.3
1	0.375
2	0.43875
3	0.492935
4	0.538996875
5	0.578147343
6	0.611425242
7	0.639711455
8	0.666375474
16	0.762874456
32	0.797243388
64	0.799984802
128	0.799999999

27. Proportion of T is 0.4 } if we assume
Proportion of t is 0.6 } H-W equilibrium

28. Proportion of Dominant is 0.5528
Proportion of Recessive is 0.4472

29. (a) .00009801, (b) .0024257, (c) .0006064, (d) .0001516, (e) .0012128

CHAPTER 17

1. Yes, through accumulation of convergent genetic mutations.

2. Winged species blown away; only chance non-winged survive.

3. Those determining replication without which there can be no continuing life.

6. Similar to ours which would support organisms showing our traits.

7. Because the environment constantly changes.

CHAPTER 18

13. $H^2 = 0.6$

14. (a) $\bar{X} = 105$ (b) Variance = 1576
(c) Standard deviation = 39.7

15. (a) $V_p = 50$ (b) $V_G = 21$

16. (a) 2.9, (b) 3.0421 (c) 1.74416

CHAPTER 19

1. Mainly lack of predictability due to random assortment and taking the environment into account.

Index